Critical Values of t

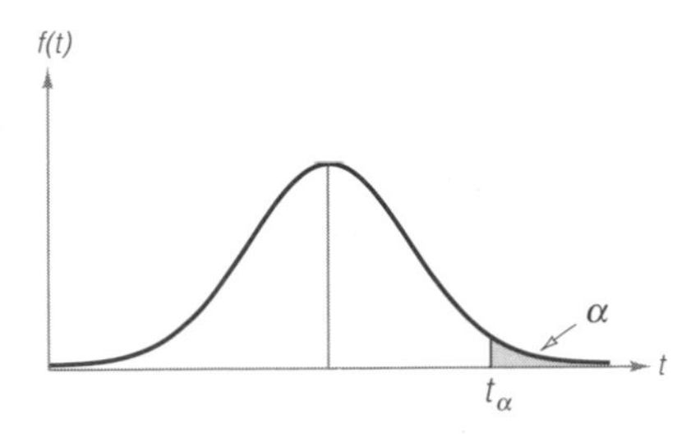

ν	$t_{.100}$	$t_{.050}$	$t_{.025}$	$t_{.010}$	$t_{.005}$	$t_{.001}$	$t_{.0005}$
1	3.078	6.314	12.706	31.821	63.657	318.31	636.62
2	1.886	2.920	4.303	6.965	9.925	22.326	31.598
3	1.638	2.353	3.182	4.541	5.841	10.213	12.924
4	1.533	2.132	2.776	3.747	4.604	7.173	8.610
5	1.476	2.015	2.571	3.365	4.032	5.893	6.869
6	1.440	1.943	2.447	3.143	3.707	5.208	5.959
7	1.415	1.895	2.365	2.998	3.499	4.785	5.408
8	1.397	1.860	2.306	2.896	3.355	4.501	5.041
9	1.383	1.833	2.262	2.821	3.250	4.297	4.781
10	1.372	1.812	2.228	2.764	3.169	4.144	4.587
11	1.363	1.796	2.201	2.718	3.106	4.025	4.437
12	1.356	1.782	2.179	2.681	3.055	3.930	4.318
13	1.350	1.771	2.160	2.650	3.012	3.852	4.221
14	1.345	1.761	2.145	2.624	2.977	3.787	4.140
15	1.341	1.753	2.131	2.602	2.947	3.733	4.073
16	1.337	1.746	2.120	2.583	2.921	3.686	4.015
17	1.333	1.740	2.110	2.567	2.898	3.646	3.965
18	1.330	1.734	2.101	2.552	2.878	3.610	3.922
19	1.328	1.729	2.093	2.539	2.861	3.579	3.883
20	1.325	1.725	2.086	2.528	2.845	3.552	3.850
21	1.323	1.721	2.080	2.518	2.831	3.527	3.819
22	1.321	1.717	2.074	2.508	2.819	3.505	3.792
23	1.319	1.714	2.069	2.500	2.807	3.485	3.767
24	1.318	1.711	2.064	2.492	2.797	3.467	3.745
25	1.316	1.708	2.060	2.485	2.787	3.450	3.725
26	1.315	1.706	2.056	2.479	2.779	3.435	3.707
27	1.314	1.703	2.052	2.473	2.771	3.421	3.690
28	1.313	1.701	2.048	2.467	2.763	3.408	3.674
29	1.311	1.699	2.045	2.462	2.756	3.396	3.659
30	1.310	1.697	2.042	2.457	2.750	3.385	3.646
40	1.303	1.684	2.021	2.423	2.704	3.307	3.551
60	1.296	1.671	2.000	2.390	2.660	3.232	3.460
120	1.289	1.658	1.980	2.358	2.617	3.160	3.373
∞	1.282	1.645	1.960	2.326	2.576	3.090	3.291

Source: This table is reproduced with the kind permission of the Trustees of Biometrika from E. S. Pearson and H. O. Hartley (eds.), *The Biometrika Tables for Statisticians,* Vol. 1, 3d ed., *Biometrika,* 1966.

A First Course in
STATISTICS

FIFTH EDITION

James T. McClave
Info Tech, Inc.
University of Florida

Terry Sincich
University of South Florida

Englewood Cliffs, New Jersey 07632

Library of Congress Cataloging-in-Publication Data
McClave, James T.
A first course in statistics / James T. McClave/Terry Sincich.— 5th ed.
Includes index.
ISBN 0-02-379195-0
1. Statistics. I. Sincich, Terry. II. Title.
QA276.12.M3997 1994
519.5—dc20 94-41307
CIP

On the cover is *Notes 2/85 I* by Brookes Byrd. The piece is mixed media on paper and measures 40″ × 26″. The paintings in the *Notes* series are an exploration of the layering nature of painting. Byrd uses grids, transparencies, and a variety of media, including collage of her own paintings, to expose the layers. Along one edge she keeps a record of each layer with small marks of paint. The series is titled *Notes* because the aim is to make visual notes of the art and ideas that interest her. Byrd has shown in museums and galleries in Connecticut, Texas, San Francisco, Los Angeles, Chicago, New York, and Colorado. She has studied at Smith College, NYU, and the University of California, Davis. She is represented by the John Natsoulas Gallery in Davis, California.

A Simon & Schuster Company
Englewood Cliffs, New Jersey 07632

Printed in the United States of America

10 9 8 7 6 5 4 3 2 1

ISBN 0-02-379195-0

Prentice-Hall International (UK) Limited, *London*
Prentice-Hall of Australia Pty. Limited, *Sydney*
Prentice-Hall Canada Inc., *Toronto*
Prentice-Hall Hispanoamericana, S.A., *Mexico*
Prentice-Hall of India Private Limited, *New Delhi*
Prentice-Hall of Japan, Inc., *Tokyo*
Simon & Schuster Asia Pte. Ltd., *Singapore*
Editora Prentice-Hall do Brasil, Ltda., *Rio de Janeiro*

Contents

CHAPTER FOUR

CHAPTER FIVE

CHAPTER SIX

CHAPTER SEVEN

Inferences Based on a Single Sample: Tests of Hypotheses 311

CHAPTER EIGHT

Comparing Two Population Means 369

CHAPTER NINE

Comparing Population Proportions 443

CHAPTER TEN

Simple Linear Regression 487

APPENDIX A

Tables 561

APPENDIX B

Demographic Data Set 581

APPENDIX C

ASP Tutorial 585

Case Studies

List of Exercise Data Sets

Most of the exercise data sets that contain 30 or more measurements are listed below and are available on a computer disk (ASCII format). Instructors who adopt this text may obtain a copy of the disk from the publisher.

Preface

Instructors who teach a single one-semester (one-quarter) introductory statistics course may find that many of the available texts have been written for a two-semester sequence. *A First Course in Statistics* is designed to cope with this problem.

Our fifth edition is the second published in the 1990s, and we are proud and humbled by the fact that editions have been published in each of the last three decades. This causes us to wonder what R. A. Fisher and other early "founders" of the science of statistics would think of statistics as we find it today. We hope they would agree with us, that statistical inference continues to be the theme and primary objective of statistical inquiry, but that the power of modern-day computers and sophisticated statistical software have combined to make computationally intensive descriptive and analytical techniques an important road on the journey to that objective. Some texts, it seems to us, spend so much time looking at the data from every angle and with every computer technique that limited time and space are devoted to statistical inference. With each edition we attempt to incorporate more and more of these techniques without losing sight of the final goal: statistical inference.

Revisions

The fifth edition of *A First Course in Statistics* contains the following major changes:

1. **Mutually exclusive and independent events separated (Chapter 3).** The concepts of mutually exclusive and independent events are put into separate sections to make the probability chapter more readable.
2. **Poisson distribution added (Chapter 4).** As an optional section (Section 4.4), we now include material on the discrete Poisson probability distribution.
3. **More emphasis on *p*-values generated by computer.** We have included a number of exercises and examples utilizing computer software and the interpretation of observed significance levels (*p*-values) that are typically produced by statistical software.
4. **Many new exercises with applications.** We have added a number of new exercises from the social sciences: psychology, criminal justice, sociology, and political science applications from the research literature are featured in the fifth edition.
5. **Answers to all exercises provided.** At the suggestion of several users, we now include short answers to both the odd- and even-numbered exercises.
6. **ASP statistical software disk included with text.** (See details on ASP provided later in this preface.)

Our reviewers have made many other suggestions for improved pedagogy, readability, and clarification; we gratefully acknowledge our users' insights, and we have incorporated many of their recommendations.

The flexibility of past editions is maintained in this edition. Sections that are not prerequisite to succeeding sections and chapters are marked "(Optional)." For example, an instructor who wishes to devote significant time to exploratory data analysis might cover all topics in Chapter 2. In contrast, an instructor who wishes to move rapidly into inferential procedures might omit the optional section on box plots and devote only several lectures to this chapter.

We have maintained the features of this text that we believe make it unique among introductory statistics texts. These features, which assist the student in achieving an overview of statistics and an understanding of its relevance in the sciences, business, and everyday life, are as follows:

Features

1. **Case Studies.** (See the list of case studies on page vii.) Many important concepts are emphasized by the inclusion of case studies, which consist of brief summaries of actual applications of statistical concepts and are often drawn directly from the research literature. These case studies allow the student to see applications of important statistical concepts immediately after their introduction. The case studies also help to answer by example the often asked questions, "Why should I study statistics? Of what relevance is statistics to my program?" Finally, the case studies constantly remind the student that each concept is related to the dominant theme—statistical inference.
2. **The Use of Examples as a Teaching Device.** We have introduced and illustrated almost all new ideas by examples. Our belief is that most students will better understand definitions, generalizations, and abstractions *after* seeing an application. In most sections, an introductory example is followed by a general discussion of the procedures and techniques, and then a second example is presented to solidify the understanding of the concepts.
3. **A Simple, Clear Style.** We have tried to achieve a simple and clear writing style. Subjects that are tangential to our objective have been avoided, even though some may be of academic interest to those well versed in statistics. We have not taken an encyclopedic approach in the presentation of material.
4. **Many Exercises—Labeled by Type.** The text has a large number of exercises illustrating applications in almost all areas of research. However, we believe that many students have trouble learning the mechanics of statistical techniques when problems are all couched in terms of realistic applications—the concept becomes lost in the words. Thus, the exercises at the ends of all sections are divided into two parts:
   a. **Learning the Mechanics.** These exercises are intended to be straightforward applications of the new concepts. They are introduced in a few words and are unhampered by a barrage of background information designed to make them

"practical," but which often detracts from instructional objectives. Thus, with a minimum of labor, the student can recheck his or her ability to comprehend a concept or a definition.

b. **Applying the Concepts.** The mechanical exercises described above are followed by realistic exercises that allow the student to see applications of statistics across a broad spectrum. Once the mechanics are mastered, these exercises develop the student's skills at comprehending realistic problems that describe situations to which the techniques may be applied.

5. **On Your Own . . .** Each chapter ends with an exercise entitled **On Your Own . . . .** The intent of this exercise is to give the student some hands-on experience with an application of the statistical concepts introduced in the chapter. In most cases, the student is required to collect, analyze, and interpret data relating to some real application.

6. **Using the Computer.** Another feature at the end of most chapters encourages the use of computers in the analysis of real data. A demographic database, consisting of 1,000 observations on 15 variables, has been described in Appendix B and is available on diskette from the publisher. Most chapters include a suggested computer application, **Using the Computer . . .** , which provides one or more computer exercises that utilize the data in Appendix B and enhance the new material covered in the chapter.

7. **Where We've Been . . . Where We're Going . . .** The first page of each chapter is a "unification" page. Our purpose is to allow the student to see how the chapter fits into the scheme of statistical inference. First, we briefly show how the material presented in previous chapters helps us to achieve our goal (Where We've Been). Then, we indicate what the next chapter (or chapters) contributes to the overall objective (Where We're Going). This feature allows us to point out that we are constructing the foundation block by block, with each chapter an important component in the structure of statistical inference. Furthermore, this feature provides a series of brief résumés of the material covered as well as glimpses of future topics.

8. **Footnotes.** Although the text is designed for students with a noncalculus background, footnotes explain the role of calculus in various derivations. Footnotes are also used to inform the student about some of the theory underlying certain results. The footnotes allow additional flexibility in the mathematical and theoretical level at which the material is presented.

9. **Supplementary Material.** Solutions manuals, a Minitab supplement, an integrated software system, a computer-generated test system, a test bank, and a 1,000-observation demographic database are available.

   a. **Student Solutions Manual** (by Nancy S. Boudreau). The student solutions manual presents detailed solutions to most odd-numbered exercises in the text. Many points are clarified and expanded to provide maximum insight into and benefits from each exercise. ISBN 0-02-312717-1.

   b. **Instructor's Solutions Manual** (by Nancy S. Boudreau). The instructor's solutions manual presents the full solutions to the even-numbered exercises contained in the text. For adopters, the manual is complimentary from the publisher. ISBN 0-02-379196-9.

c. **Minitab Supplement** (by Ruth K. Meyer and David D. Krueger). The Minitab computer supplement was developed to be used with Minitab Release 8.0, a general-purpose statistical computing system. The supplement, which was written especially for the student with no previous experience with computers, provides step-by-step descriptions of how to use Minitab effectively as an aid in data analysis. Each chapter begins with a list of new commands introduced in the chapter. Brief examples are then given to explain new commands, followed by examples from the text illustrating the new and previously learned commands. Where appropriate, simulation examples are included. Exercises, many of which are drawn from the text, conclude each chapter.

   A special feature of the supplement is a chapter describing a survey sampling project. The objectives of the project are to illustrate the evaluation of a questionnaire, provide a review of statistical techniques, and illustrate the use of Minitab for questionnaire evaluation. ISBN 0-02-381001-7.

d. **ASP statistical software diskette.** New to this edition, the text includes a 3½″ diskette containing the ASP program, *A Statistical Package for Business, Economics, and the Social Sciences*. ASP, from DMC Software, Inc., is a user-friendly, totally menu-driven program that contains all of the major statistical applications covered in the text, plus many more. ASP runs on any IBM-compatible PC with at least 512K of memory. With ASP, students with no knowledge of computer programming can create and analyze data sets easily and quickly. Appendix C contains start-up procedures and a short tutorial on the use of ASP. Full documentation is provided complimentary to adopters of the text.

e. **ASP Tutorial and Student Guide** (by George Blackford). Most students have little trouble learning to use ASP without documentation. Some, however, may want to purchase the *ASP Tutorial and Student Guide*. Bookstores can order the tutorial from DMC Software, Inc., 6169 Pebbleshire Drive, Grand Blanc, MI 48439.

f. **Test Bank** (by Mark Dummeldinger). This manual provides a large number of test items utilizing real data. ISBN 0-02-379197-7.

g. **PH Test Manager IBM 3.5″.** This program allows users to generate tests and quizzes by chapter or section number, choosing from numerous questions. Full editing and graphing capabilities are included; instructors may modify existing questions or create their own questions as well as original graphics. ISBN 0-13-359704-0.

h. **Database.** A demographic data set was assembled based on a systematic random sample of 1,000 U.S. zip codes. Demographic data for each zip code area selected were supplied by CACI, an international demographic and market information firm. Fifteen demographic measurements (including population, number of households, median age, median household income, variables related to the cost of housing, educational levels, the work force, and purchasing potential indexes based on the Bureau of the Census Consumer Expenditure Surveys) are presented for each zip code area.

Some of the data are referenced in the **Using the Computer** sections. The objectives are to enable the student to analyze real data in a relatively large sample using the computer, and to gain experience using the statistical techniques and concepts on real data.

i. **Data Sets on Disk.** All of the large exercise data sets, as well as the demographic data set, are available on a computer disk. A list of the exercise data sets, with the exercise number, the page number, and the file name for each data set, is given on page viii. ISBN 0-02-379198-5.

Full Version Available

A full-sized one- or two-term general statistics text is also available: *Statistics, 6th Edition* by James T. McClave (University of Florida) and Frank H. Dietrich (Northern Kentucky University); © 1994, 925 pp. cloth, ISBN 0-02-379211-6 (U4609-6).

Contents. 1: What Is Statistics? 2: Methods for Describing Sets of Data. 3: Probability. 4: Discrete Random Variables. 5: Continuous Random Variables. 6: Sampling Distributions. 7: Inferences Based on a Single Sample: Estimation. 8: Inferences Based on a Single Sample: Tests of Hypotheses. 9: Inferences Based on Two Samples: Estimation and Tests of Hypotheses. 10: Analysis of Variance: Comparing More Than Two Means. 11: Nonparametric Statistics. 12: The Chi-Square Test and the Analysis of Contingency Tables. 13: Simple Linear Regression. 14: Multiple Regression. 15: Model Building. Appendix A: Tables. Appendix B: Calculation Formulas for Analysis of Variance. Appendix C: Demographic Data Set. Appendix D: ASP Tutorial. Answers to Selected Exercises. Index. Call Prentice Hall Faculty Services at 1-800-526-0485, or your local representative.

Acknowledgments

Thanks are due to many individuals who helped in the preparation of this text. Among them are the reviewers, whose names are listed below. Special thanks to Mary Jay McClave and Mark Dummeldinger for their assistance in finding new applications from the social sciences. Susan Reiland has our appreciation and admiration for managing the production of this book. Her work defies explanation; you have to see to believe the care and professionalism with which she works. We thank Brenda Dobson for converting our scribbling to a polished manuscript. Finally, we thank the thousands of students who have helped us to form our ideas about teaching statistics. Their most common complaint seems to be that texts are written for the instructors rather than the student. We hope that this book is an exception.

William Applebaugh
University of Wisconsin,
Eau Claire

David Atkinson
Olivet Nazarene University

Beverly Barker
Moorpark College

William H. Beyer
University of Akron

P. K. Bhattacharya
University of California, Davis

Patricia M. Buchanan
Pennsylvania State University

Nancy Carter
California State University, Chico

Kathryn Chaloner
University of Minnesota

Rosalee Clark
College of San Mateo, California

Larry Dion
California State University, Chico

John Dirkse
California State University at Bakersfield

Michael J. Doviak
Old Dominion University

N. B. Ebrahimi
Northern Illinois University

Dale Everson
University of Idaho

Rudy Gideon
University of Montana

Larry Griffey
Florida Community College

David Groggel
Miami University at Oxford

John E. Groves
California Polytechnic State University at San Luis Obispo

Shu-ping Hodgson
Central Michigan University

Jean L. Holton
Virginia Commonwealth University

John H. Kellermeier
State University College at Plattsburgh

Timothy J. Killeen
University of Connecticut

William G. Koellner
Montclair State University

James R. Lackritz
San Diego State University

Diane Lambert
AT&T/Bell Laboratories

James Lang
Valencia Junior College

Glenn Larson
University of Regina

Maryke Lee
Valencia Community College

John J. Lefante, Jr.
University of South Alabama

Pi-Erh Lin
Florida State University

R. Bruce Lind
University of Puget Sound

Rhonda Magel
North Dakota State University

Linda C. Malone
University of Central Florida

Allen E. Martin
California State University at Los Angeles

Leslie Matekaitis
Cal Genetics

E. Donice McCune
Stephen F. Austin State University

Satya Narayan Mishra
University of South Alabama

A. Mukherjea
University of South Florida

Bernard Ostle
University of Central Florida

William B. Owen
Central Washington University

Won J. Park
Wright State University

John J. Peterson
Smith Kline & French Laboratories

Chandler Pike
Mercer University

Andrew Rosalsky
University of Florida

C. Bradley Russell
Clemson University

Rita Schillaber
University of Alberta

James R. Schott
University of Central Florida

Susan C. Schott
University of Central Florida

George Schultz
St. Petersburg Junior College

Carl James Schwarz
University of Manitoba

Mike Seyfried
Shippensburg University

Lewis Shoemaker
Millersville University

Charles W. Sinclair
Portland State University

Robert K. Smidt
California Polytechnic State
University at San Luis Obispo

Vasanth B. Solomon
Drake University

W. Robert Stephenson
Iowa State University

Barbara Treadwell
Western Michigan University

Dan Voss
Wright State University

Augustin Vukov
University of Toronto

Dennis D. Wackerly
University of Florida

Theophil J. Worosz
Metropolitan State College
of Denver

CHAPTER ONE

What Is Statistics?

Contents

Case Studies

Where We're Going

Statistics? Is it a field of study, a group of numbers that summarizes the state of our national economy, the performance of a football team, the social conditions in a particular locale, or, as the title of a popular book (Tanur et al., 1989) suggests, "a guide to the unknown"? We will see in Chapter 1 that each of these descriptions has some applicability in understanding statistics. We will see that *descriptive statistics* focuses on developing numerical summaries that describe some phenomenon, whereas *inferential statistics* uses these numerical summaries to assist in making decisions. The primary theme of this text is inferential statistics. Thus, we concentrate on showing you how statistics can be used to interpret and use data to make decisions. Since many jobs in industry, government, medicine, and other fields require this facility, we show you how statistics can be beneficial to you.

1.1 Statistics: What Is It?

What does statistics mean to you? Does it bring to mind batting averages, Gallup polls, unemployment figures, numerical distortions of facts (lying with statistics!), or simply a college requirement you have to complete? We hope to convince you that statistics is a meaningful, useful science with a broad, almost limitless scope of application to business, government, and the physical and social sciences. We also want to show that statistics lie only when they are misapplied. Finally, our objective is to paint a unified picture of statistics to leave you with the impression that your time was well spent studying a subject that will prove useful to you in many ways.

Statistics means "numerical descriptions" to most people. Monthly unemployment figures, the failure rate of a particular type of steel-belted automobile tire, and the proportion of women who favor the Equal Rights Amendment all represent statistical descriptions of large sets of data collected on some phenomenon. Often the purpose of calculating these numbers goes beyond the description of the set of data. Frequently, the data are regarded as a sample selected from some larger set of data whose characteristics we may wish to estimate. For example, the ages of a sampling of customers at a video store would allow you to estimate the average age of *all* customers of the store. This estimate could then be used to target the store's advertisements to the appropriate age group. So, the applications of statistics can be divided into two broad areas: (1) describing large masses of data (**descriptive statistics**) and (2) drawing conclusions (making estimates, decisions, predictions, etc.) about some set of data based on sampling (**inferential statistics**).

Descriptive statistics utilizes numerical and graphical methods to look for patterns, to summarize, and to present the information in a set of data.

Inferential statistics utilizes sample data to make estimates, decisions, predictions, or other generalizations about a larger set of data.

Although both descriptive and inferential statistics are discussed in the following chapters, the primary theme of the text is **inference.** Let us examine some case studies that illustrate applications of statistics.

CASE STUDY 1.1 / A Survey: Where "Women's Work" Is Done by Men

The 1980 February/March issue of *Public Opinion* describes the results of a survey of several hundred married men from each of nine countries who responded to the following question:

In the following list, which household jobs would you say it would be reasonable that the man would often take over from his wife: washing up (doing dishes), changing baby's napkin (diaper), cleaning house, ironing, organizing meal, staying at home with sick child, shopping, none of these?

The graphs in Figure 1.1 provide an effective summary of the thousands of opinions obtained and allow for an easy comparison of attitudes across countries. The graphs, which summarize and describe the data, are examples of *descriptive statistics*. The area of statistics concerned with the summarization and description of data is called **descriptive statistics.**

FIGURE 1.1 ▶
"Women's Work" Is Rarely Done by Men in Italy, Germany . . .
Note: The sample size for each country exceeded 900, except in Luxembourg, where the sample size was 334.
Source: Survey by the European Economic Community Commission. "Women and Men of Europe in 1978," October–November 1977, as shown in *Public Opinion*, February–March 1980, p. 37. Reprinted with the permission of the American Enterprise Institute for Public Policy Research, Washington, D.C.

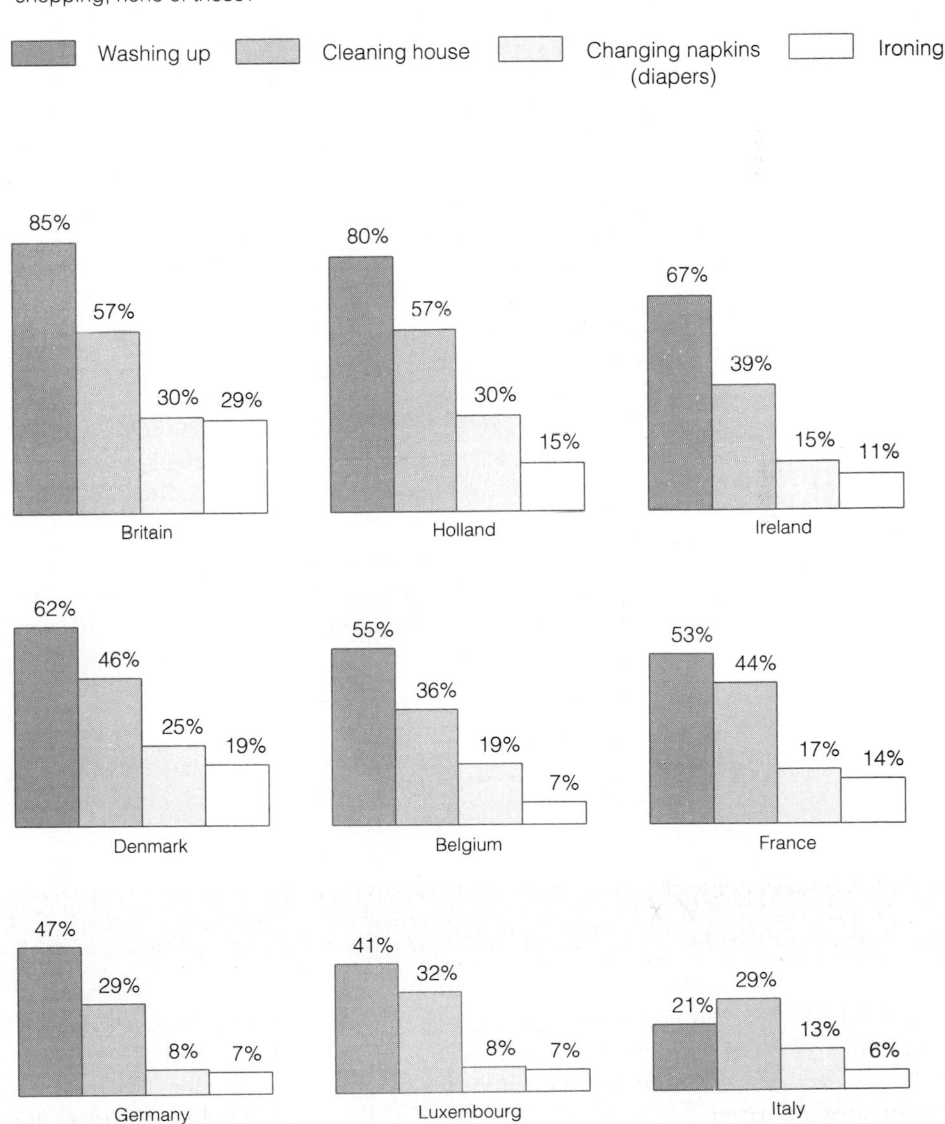

CASE STUDY 1.2 / An Experiment: Investigating an Effect of Smoking During Pregnancy

In an article in the *Journal of the American Medical Association*, M. Sexton and J. R. Herbel (1984) reported on their research into the effects of maternal smoking on the birth weight of babies. Their experiment randomly assigned 935 pregnant women smokers into two groups; one continued smoking throughout pregnancy and the other, the **control group**, received smoking intervention (i.e., assistance to reduce or eliminate smoking). From the measured baby weights of these two groups of women, Sexton and Hebel inferred that "some fetal growth retardation can be overcome by the provision of antismoking assistance to pregnant women."

In making this statement, Sexton and Hebel made an *inference* about the impact of a smoking intervention program on all pregnant women who smoke, an inference based on the comparison of the samples of baby weights for the smoking and the control groups of women. Thus, these authors applied *inferential statistics* in the analysis of their data.

CASE STUDY 1.3 / Does Judicial Action Affect the Probability of Conviction?

Defense attorneys often remind juries that accused criminals are innocent until proven guilty. But if the judge waits to deliver his or her own version of the reminder until after the testimony is in, jurors may be more likely to render a guilty verdict than if they had heard the reminder before the trial.

This quote is from C. Rubenstein's article in the June 1980 issue of *Psychology Today*, which discusses a study conducted by two social psychologists at the University of Kansas.

Involved in the study were 107 student jurors, some of whom heard the judge's reminder at the start of a taped trial, some at the end, and some not at all. An analysis of the 107 student verdicts prompted the quote just given.

Note that a **sample** of 107 student jurors was observed and that their verdicts were used to *infer* that the jurors in any trial may be more likely to render a guilty verdict if they hear a reminder from the judge at the end of the trial. This case study, like Case Study 1.2, is an example of the use of *inferential statistics*.

CASE STUDY 1.4 / Taste Preference for Beer: Brand Image or Physical Characteristics of the Beer?

Two sets of data of interest to the marketing department of a food products firm are (1) the set of taste-preference scores given by consumers to their product and their competitors' products when all brands are clearly labeled and (2) the taste-preference scores given by the same set of consumers when all brand labels have been removed and the consumer's only means of product identification is taste. With such information the marketing department should be able to determine whether taste preference arose because of perceived physical differences in the products or as a result of the consumer's image of the brand. (Brand image is, of course, largely a result of a firm's marketing efforts.) Such a determination should help the firm develop marketing strategies for their product.

A study using these two sets of data was conducted by Ralph Allison and Kenneth Uhl (1965) in an effort to determine whether beer drinkers could distinguish among major brands of unlabeled beer. A sample of 326 beer drinkers was randomly selected from the set of beer drinkers identified as males who drank beer at least three times a week. During the first week of the study each of the 326 participants was given a six-pack of unlabeled beer containing three major brands and was asked to taste-rate each beer on a scale from 1 (poor) to 10 (excellent). During the second week the same set of drinkers was given a six-pack containing six major brands. This time, however, each bottle carried its usual label. Again, the drinkers were asked to taste-rate each beer from 1 to 10. From a statistical analysis of the two sets of data yielded by the study, Allison and Uhl concluded that the 326 beer drinkers studied could not distinguish among brands by taste on an overall basis. This result enabled them to infer statistically that such was also the case for beer drinkers in general. Their results also indicated that brand labels and their associations did significantly influence the tasters' evaluations. These findings suggest that physical differences in the products have less to do with their success or failure in the marketplace than the image of the brand in the consumers' mind. As to the benefits of such a study, Allison and Uhl note, "to the extent that product images, and their changes, are believed to be a result of advertising . . . the ability of firms' advertising programs to influence product images can be more thoroughly examined."

This case study, like Case Studies 1.2 and 1.3, is an example of the use of *inferential statistics*. Using data collected from a sample of 326 beer drinkers, Allison and Uhl make inferences about the ability of all beer drinkers to distinguish among major brands of unlabeled beer.

These case studies provide four examples of the uses of statistics. Note that each involves an analysis of data, either for the purpose of describing the data set (Case Study 1.1) or for making inferences about a much larger data set based on sampling (Case Studies 1.2, 1.3, and 1.4). They thus provide realistic examples of the two broad areas of statistical applications.

1.2 The Elements of Statistics

Statistical methods are particularly useful for studying, analyzing, and learning about **populations**.

Definition 1.1

A **population** is a set of units (usually, people, objects, transactions, or events).

Examples of populations include (1) all employed workers in the United States, (2) all registered voters in California, (3) everyone who is afflicted with AIDS, (4) all the cars produced last year by a particular assembly line, (5) the current stock of spare parts at United Airlines' maintenance facility, (6) all sales made at the drive-in window of a fast-food restaurant during a given year, and (7) the set of all accidents occurring on a particular stretch of interstate highway during a holiday period. Notice that the first three population examples (1–3) are sets (groups) of people; the next two (4–5) are sets of objects; the next (6) is a set of transactions; and the last (7) is a set of events.

In studying a population, we focus on one or more characteristics or properties of the units in the population. For example, we may be interested in the age, gender, and/or the number of years of education of the people currently unemployed in the United States. We call such characteristics **variables.**

Definition 1.2

A **variable** is a characteristic or property of an individual population unit. The name *variable* is derived from the fact that any particular characteristic may "vary" among the units in a population.

In studying a particular variable it is helpful—as we will see in forthcoming chapters—to be able to obtain a numerical representation for the variable. Thus, when numerical representations are not readily available, the process of measurement plays an important supporting role in statistical studies. **Measurement** is the process by which numbers are assigned to variables of individual population units. Measurement may entail asking a registered voter to rate the performance of the president on a scale from 1 to 10 or simply asking a worker how old she is. Frequently, however, it involves the use of instruments such as stop watches, scales, and calipers. We discuss measurement in more detail in the next chapter.

If the population you wish to study is small in size, then it is feasible to measure a variable for every unit in the population. For example, if you are measuring the GPA for all incoming freshmen at your university, it is at least feasible to obtain every GPA. When we measure a variable for every unit of a population, it is called a **census** of the population. However, the populations of interest in most applications are typically much larger, involving perhaps many thousands or even an infinite number of units. Examples of large populations are those given under Definition 1.1, as well as all graduates of your university or college, all potential buyers of a new facsimile machine, and all pieces of first-class mail handled by the U.S. Post Office. In studying such populations it would typically be too time-consuming and/or too costly to conduct a census. A more reasonable alternative would be to select and study a subset (a portion) of the units in the population.

Definition 1.3

A **sample** is a subset of the units of a population.

Thus, for example, instead of polling all 120,000,000 registered voters in the United States during a presidential election year, a pollster may select and examine a sample of only 1,500 voters. If he is interested in the variable "presidential preference," then he would record (measure) the preference of each sampled vote.

The method of selecting the sample is called the sampling procedure, or sampling plan. One very important sampling procedure is **random sampling**, one that assures every subset of fixed size in the population has the same chance of being included in the sample. Thus, if the pollster samples 1,500 of the 120,000,000 voters in the population so that every voter (and subset of 1,500 voters) has an equal chance of being included in the sample, he has devised a random sample. Random sampling is discussed in Chapter 3.

After selecting the sample and measuring the variable(s) of interest for every sampled unit, the information contained in the sample is used to make *inferences* about the population.

Definition 1.4

A **statistical inference** is an estimate, prediction, or other generalization about a population based on information contained in a sample.

That is, *we use the information contained in the smaller sample to learn about the larger population.** Thus, from the sample of 1,500 voters, the pollster may estimate the percentage of all the voters who would vote for each presidential candidate if the election were held on the day the poll was conducted or predict the outcome on election day.

The preceding definitions identify four of the five elements of an inferential statistical problem: a population, one or more variables of interest, a sample, and an inference. The fifth, and perhaps most important, is a measure of *reliability* for the inference. This is the topic of Section 1.3.

EXAMPLE 1.1

A sociologist hypothesizes that the average annual income of households in a particular large city is less than $25,000 per year. To test her hypothesis, she samples 500 households in the city and determines the income of each.

a. Describe the population.
b. Describe the variable of interest.
c. Describe the sample.
d. Describe the inference.

Solution

a. The population is the set of units of interest to the sociologist, which is the set of all households in the city.
b. The total annual income of each household is the variable of interest to the sociologist.

*The terms *population* and *sample* are often used to refer to the sets of measurements themselves, in addition to the units on which the measurements are made. For applications in which a single variable of interest is being measured, this will cause little confusion. When the terminology is potentially ambiguous, the measurements are referred to as *population data sets* and *sample data sets*, respectively.

c. The sample must be a subset of the population. In this case, it is the 500 households selected by the sociologist. If the 500 households represent a random sample, then the sampling procedure used must be such that each of the households in the city had an equal chance of being included in the sample.

d. The inference of interest involves the *generalization* of the information contained in the sample of 500 households to the population of all households in the city. In particular, the sociologist wants to estimate the average income of the households in the city in order to determine whether it is less than $25,000. This might be accomplished by calculating the average income in the sample and using the sample average to estimate the population average. Of course, this sample estimate is not likely to be exact, and therefore some measure of its reliability is needed. Measuring the reliability of an inference is the subject of Section 1.3.

EXAMPLE 1.2

Cola wars is the popular media term for the intense competition between the marketing campaigns of Coca-Cola and Pepsi. The campaigns have featured movie and television stars, rock videos, athletic endorsements, and claims of consumer preference based on taste tests. Suppose, as part of a Pepsi marketing campaign, 1,000 cola consumers are given a "blind" taste test (i.e., a taste test in which the two brand names are disguised). Each consumer is asked to state a preference for brand A or brand B.

a. Describe the population.

b. Describe the variable of interest.

c. Describe the sample.

d. Describe the inference.

Solution

a. The population of interest is the collection or set of all consumers.

b. The characteristic of each cola consumer that Pepsi wants to measure is the consumer's cola preference as revealed under the conditions of a blind taste test. Thus, cola preference is the variable of interest.

c. The sample is the 1,000 cola consumers selected from the population of all cola consumers.

d. The inference of interest is the *generalization* of the cola preferences of the 1,000 sampled consumers to the population of all cola consumers. In particular, the preferences of the consumers in the sample can be used to *estimate* the percentage of all cola consumers who prefer each brand.

1.3 Statistics: Witchcraft or Science?

The *primary objective of statistics is inference*. In the previous section we described inference as making generalizations about populations based on information contained in a sample. But making the inference is only part of the story. We also need to know how good the inference is. The only way we can be reasonably certain that an inference about a population is correct is to include the entire population in our sample. But, due to resource constraints (i.e., insufficient time and/or money), this is generally not an option. In basing inferences on only a portion of the population (a sample), we introduce an element of uncertainty into our inferences. In general, the smaller the sample size, the less certain we are about the inference. Thus, an inference based on a sample of size 5 is (usually) less reliable than an inference based on a sample of size 100. Consequently, whenever possible, it is important to determine and report the **reliability** of each inference made; this is the fifth element of inferential statistical problems.

The measure of reliability that accompanies an inference separates the science of statistics from the art of fortune-telling. A palm reader, like a statistician, may examine a sample (your hand) and make inferences about the population (your life). However, unlike statistical inferences, no measure of reliability can be attached to the palm-reader's inferences.

Suppose as in Example 1.1 we are interested in estimating the average income of a population of households from the average income of a sample of households. Using statistical methods, we can determine a *bound on the estimation error*. This bound is simply a number that our estimation error (the difference between the average income of the sample and the average income of the population of households) is not likely to exceed. We will see in later chapters that this bound is a measure of the uncertainty of our inference. The reliability of statistical inferences is discussed throughout this text. For now, we simply want you to realize that an inference is incomplete without a measure of its reliability.

We conclude this section with a summary of the elements of inferential statistical problems and an example to illustrate a measure of reliability.

Five Elements of Inferential Statistical Problems

1. The population of interest
2. One or more variables (characteristics of the population units) that are to be investigated
3. The sample of population units
4. The inference about the population based on information contained in the sample
5. A measure of reliability for the inference

EXAMPLE 1.3

Refer to Example 1.2, in which 1,000 consumers indicated their cola preferences in a taste test. Describe how the reliability of an inference concerning the preferences of all cola consumers in the Pepsi bottler's marketing region could be measured.

Solution

When the preferences of 1,000 consumers are used to estimate the preferences of all consumers in the region, the estimate will not exactly mirror the preferences of the population. For example, if the taste test shows that 56% of the 1,000 consumers preferred Pepsi, it does not follow (nor is it likely) that exactly 56% of all cola drinkers in the region prefer Pepsi. Nevertheless, we may be able to use sound statistical reasoning (which is presented later in the text) to ensure that the sampling procedure used will generate estimates that are almost certainly within a specified limit of the true percentage of all consumers who prefer Pepsi. For example, such reasoning might assure us that the estimate of the preference for Pepsi is almost certainly within 5% of the actual population preference. The implication is that the actual preference for Pepsi is between 51% [i.e., (56 − 5)%] and 61% [i.e., (56 + 5)%]. This interval represents a measure of reliability for the inference.

EXAMPLE 1.4

Refer to Case Study 1.2, in which research on the effects of maternal smoking on birth weight is discussed. Assume that 500 of the pregnant women receive smoking intervention therapy during their pregnancies, whereas 435 receive no therapy or treatment and continue to smoke during their pregnancies.

a. Describe the population for this research.
b. Describe the variable of interest.
c. Describe the sample.
d. Describe the type of inference of interest to the researchers.
e. Discuss how the reliability of the inference could be measured.

Solution

a. Because the objective of this research is to compare these data between two groups of babies, those born to smoking mothers and those born to mothers receiving smoking intervention therapy, two populations are pertinent: The first consists of all babies born to smoking mothers, and the second consists of all babies born to mothers receiving smoking intervention therapy.
b. The characteristic of the babies being measured in this research is their birth weight. Thus, the variable of interest is *birth weight*.
c. Two samples were selected, one from each population. The first sample consisted of 435 babies born to smoking mothers; the second sample consisted of the 500 babies born to mothers receiving intervention therapy.
d. The inference of interest might be to compare the average weight of babies born to smoking mothers to the average weight of babies born to mothers receiving smoking

intervention therapy. Specifically, the research is aimed at determining whether the average weight of babies born to smoking mothers is less than that of those born to mothers receiving therapy.

e. The reliability of the inference must address the issue of the precision with which the difference between the average weights of the two populations is estimated. For example, suppose the two samples indicate that babies born to smoking mothers average .8 pound less at birth than those born to mothers receiving smoking intervention therapy. It is premature to assert that this implies that the average weight of babies born to *all* mothers receiving therapy will exceed that of babies born to all smoking mothers; we must first know the precision associated with the difference in sample averages. For example, suppose that sound statistical reasoning (which we develop in this text) is used to show that the sample estimate of the difference is almost certainly within .5 pound of the true difference between the two population averages. This implies that the true difference is (almost certainly) within the interval from $(.8 - .5) = .3$ to $(.8 + .5) = 1.3$ pounds, which supports the inference that babies born to smoking mothers will, on average, weigh between .3 and 1.3 pounds less than those born to mothers receiving therapy. This interval is therefore a measure of reliability. Note that there remains uncertainty not only about the exact value of the difference but also about the cause of the difference. The cause might be attributed to smoking, but the reliability of that aspect of the inference depends on more than statistical reasoning. Might another factor, such as diet, be different for smoking mothers than for mothers receiving therapy, and might diet rather than smoking be the *cause* of the babies' average weight difference? The issue of causation requires the expertise of medical scientists as well as statistical inference.

1.4 Why Study Statistics?

Why study statistics? The growth in data collection associated with scientific phenomena as well as the operations of business and government (quality control, statistical auditing, forecasting, etc.) has been truly remarkable over the past several decades. Published results of political, economic, and social surveys as well as increasing government emphasis on drug and product testing provide vivid evidence of the need to be able to evaluate data sets intelligently. Consequently, you will want to develop a discerning sense of rational thought that will enable you to evaluate numerical data. You may be called upon to use this ability to make intelligent decisions, inferences, and generalizations. For this reason, the study of statistics is an essential preparation for a role in modern society.

Exercises 1.1 – 1.18

Learning the Mechanics

1.1 Explain the difference between descriptive and inferential statistics.

1.2 List and define the five elements of an inferential statistical analysis.

1.3 Explain how populations and variables differ.

1.4 Explain how populations and samples differ.

1.5 Why would a statistician consider an inference incomplete without an accompanying measure of its reliability?

1.6 Consider the set of all students enrolled in your statistics course this term. Suppose you are interested in learning about the current grade point averages (GPAs) of this group.

a. Define the population and variable of interest.
b. Suppose you determine the GPA of every member of the class. Would this represent a census or a sample?
c. Suppose you determine the GPA of 10 members of the class. Would this represent a census or a sample?
d. If you determine the GPA of every member of the class and then calculate the average, how much reliability does this have as an "estimate" of the class average GPA?
e. If you determine the GPA of 10 members of the class and then calculate the average, will the number you get necessarily be the same as the average GPA for the whole class? On what factors would you expect the reliability of the estimate to depend?

1.7 Refer to Exercise 1.6. What must be true in order for the sample of 10 students you select from your class to be considered a random sample?

Applying the Concepts

1.8 Pollsters regularly conduct opinion polls to determine the popularity rating of the current president. Suppose a poll is to be conducted tomorrow in which 2,000 individuals will be asked whether the president is doing a good or bad job.

a. What is the relevant population?
b. What is the variable of interest? Is it numerical or nonnumerical?
c. What is the sample?
d. What is the inference of interest to the pollster?

1.9 Refer to Exercise 1.8. Suppose the poll is conducted as described therein, except that each of the 2,000 individuals polled is asked to rate the job performance of the president on a scale from 0 to 100. How do your answers to parts **a–d** change, if at all?

1.10 To evaluate the current status of the dental health of schoolchildren, the American Dental Association conducted a survey to estimate the average number of cavities per child in grade school in the United States. One thousand schoolchildren from across the country were selected, and the number of cavities for each was recorded.

a. Describe the population of interest to the American Dental Association.
b. What is the variable of interest?
c. Describe the sample.

1.11 A first-year chemistry student conducts an experiment to determine the amount of hydrochloric acid necessary to neutralize 2 milliliters of a basic solution. The student prepares five 2-milliliter portions of the solution and adds a known concentration of hydrochloric acid to each. The amount of acid necessary to achieve neutrality of the solution is recorded for each of the five portions.

a. Describe the population of interest to the student.
b. What is the variable of interest?
c. Describe the sample.

1.12 A manufacturer of vacuum cleaners has decided that an assembly line is operating satisfactorily if less than 2% of the cleaners produced per day are defective. If 2% or more of the cleaners are defective, the line must be shut down and proper adjustments made. To check every cleaner as it comes off the line would be costly and time-consuming. The manufacturer decides to choose 30 cleaners at random from a specific day's production and test for defects.

a. Describe the population of interest to the manufacturer.
b. Identify the variable of interest.
c. Describe the sample.
d. Give an example of an inference the manufacturer might make.

1.13 An insurance company would like to determine the proportion of all medical doctors who have been involved in one or more malpractice suits. The company selects 500 doctors at random from a professional directory and determines the number in the sample who have ever been involved in a malpractice suit.

a. Describe the population of interest to the insurance company.
b. Identify the variable of interest.
c. Describe the sample.
d. Give an example of an inference the insurance company might make.

1.14 A Gallup Youth Poll was conducted to determine the topics that teenagers most want to discuss with their parents. The findings show that 46% would like more discussion about the family's financial situation, 37% would like to talk about school, and 30% would like to talk about religion. The survey was based on a national sampling of 505 teenagers.

a. Describe the sample.
b. Describe the population from which the sample was selected.
c. What is the variable of interest?
d. How is the inference expressed?
e. Newspaper accounts of most polls usually give a margin of error (a percentage) for the survey result. What is the purpose of the "margin of error" and what is its interpretation?

1.15 Myron Gable and Martin T. Topol (1988) sampled 218 department store executives in order to study the relationship between job satisfaction and the degree of *Machiavellian* orientation. Briefly, the Machiavellian orientation is one in which the executive exerts very strong control, even to the point of deception and cruelty, over the employees he or she supervises. The authors administered a questionnaire to each of the sampled executives and obtained both a job-satisfaction score and a Machiavellian rating. They concluded that those with higher job satisfaction scores are likely to have a lower "Mach" rating.

a. What is the population from which the sample was selected?
b. What variables were measured by the authors?
c. Identify the sample.
d. What inference was made by the authors?

1.16 Dr. A. Lewis Rhodes conducted research on the relationship between religion and environmental concern, and presented a report of the results at the 1985 meeting of the Society for the Study of Social Problems. Surveys were conducted to determine both religion and environmental concern, where environmental concern was measured by the willingness to have government spend more money for environmental protection.

a. Describe the population of interest for this research.
b. Describe the sample for the study.
c. What is the variable of interest?
d. What inference do you think is of interest in this research?

1.17 A *U.S. News & World Report* (June 21, 1982) article describes a new method of treating a major form of blindness in elderly people. The process, using laser beams to seal abnormal blood vessels in the eye, was tried on 224 patients. Of these, only 14% went blind in 1 year. In a control group of similar untreated patients, 42% went blind in 1 year. Therefore, to determine whether the laser beam treatment was effective, research physicians wished to compare the proportions of patients going blind in 1 year for *two* different populations.

a. Describe the two populations that the research physicians want to compare.
b. Identify the samples.
c. What is the variable of interest?
d. Describe the inference that the researchers will make.

1.18 In order to monitor the quality of care provided to Medicare recipients, 5% of all Medicare surgical cases performed in hospital outpatient departments and ambulatory surgical centers are to be sampled each year on an ongoing basis (Cronin, 1988). Peer-review organizations will evaluate the sampled surgeries and will assign a quality rating to each.

a. Describe the population being studied.
b. Describe the variable of interest.
c. Describe the sample in terms of process output.
d. Describe the inference of interest.
e. Before any sound statistical conclusions about the quality of Medicare can be drawn, what should accompany the inference?

On Your Own

Scan a recent issue of a daily newspaper and look for articles that contain numerical data. The data might be a summary of the results of a public opinion poll, the results of a vote by the United States Senate, crime rates, birth or death rates, an election result, etc. For each article containing data that you find, answer the following questions:

a. Do the data constitute a sample or an entire population? If a sample has been taken, clearly identify both the sample and the population; otherwise, identify the population.
b. If a sample has been observed, does the article present an explicit (or implied) inference about the population of interest? If so, state the inference made in the article.
c. If an inference has been made, has a measure of reliability been included? What is it?

References

Allison, R. I., and Uhl, K. P. "Influence of beer brand identification on taste perception." *Journal of Marketing Research*, Aug. 1965, pp. 36–39.

Careers in Statistics. American Statistical Association and the Institute of Mathematical Statistics, 1974.

Cronin, Carol. "PRO's focus on quality sharpened with new federal contracts." *Business and Health*, March 1988, p. 47.

Gable, Myron, and Topol, Martin T. "Machiavellianism and the Department Store Executive." *Journal of Retailing*, Spring 1988, pp. 68–84.

Rubenstein, C. "The presumption of innocence needs prompting." *Psychology Today*, June 1980, p. 30.

Sexton, M., and Hebel, J. R. "A clinical trial of change in maternal smoking and its effect on birth weight." *Journal of the American Medical Association*, Feb. 17, 1984, Vol. 251, No. 7, pp. 911–915.

Tanur, J. M., Mosteller, F., Kruskal, W. H., Link, R. F., Pieters, R. S., and Rising, G. R. *Statistics: A Guide to the Unknown*. (E. L. Lehmann, special editor.) San Francisco: Holden-Day, 1989.

CHAPTER TWO

Methods for Describing Sets of Data

Contents

Case Studies

Where We've Been

In Chapter 1 we examined some typical examples of the use of statistics. We discussed the role that statistics plays in supporting decision-making. We introduced you to descriptive and inferential statistics and to the five elements of inferential statistics: a population, one or more variables, a sample, an inference, and a measure of reliability for the inference. We described the primary goal of inferential statistics as using sample data to make inferences (estimates, predictions, or other generalizations) about the population from which the sample was drawn.

Where We're Going

Before we make an inference, we must be able to describe a data set. Both graphic and numerical methods for describing sets of data are discussed in this chapter. As you will learn in Chapter 6, we will use some sample numerical descriptive measures to estimate the values of corresponding population descriptive measures. Therefore, our efforts in this chapter will ultimately lead to statistical inference.

2

Suppose we wish to evaluate the mathematical capabilities of a class of 1,000 college freshmen based on their quantitative Scholastic Aptitude Test (SAT) scores. How would you describe these 1,000 measurements? You can see that this is not an easy question to answer. The 1,000 scores provide too many bits of information for our minds to comprehend. It is clear that we need some method for summarizing the information in a data set. Methods for describing data sets are also essential for statistical inference. Most populations are large data sets. Consequently, if we are going to make descriptive statements (inferences) about a population based on information contained in a sample, we will once again need methods for describing a data set.

Two methods for describing data are presented in this chapter, one **graphic** and the other **numerical**. As you will subsequently see, both play an important role in statistics.

In Section 2.1 we define four different types of data. Then we present graphical methods for describing data in Section 2.2. Numerical descriptive methods are presented in Sections 2.3–2.7. We conclude this chapter with a section on the *misuse* of descriptive techniques.

2.1 Types of Data

In Chapter 1 you learned that statistics, both descriptive and inferential, is concerned with the measurements of one or more variables of a sample of units drawn from a population. These measurements are referred to as **data**. We generally classify data as one of four types: **nominal**, **ordinal**, **interval**, or **ratio**.

Definition 2.1

Nominal data are measurements that simply classify the units of the sample (or population) into categories.

Nominal data (also referred to as **categorical data**) are labels or names that identify the category to which each unit belongs. The following are examples of nominal data:

1. The political party affiliation of each individual in a sample of 50 registered voters
2. The gender of each individual in a sample of seven applicants for a computer programming job
3. The brand of toothpaste preferred by each individual in a sample of 100 consumers

Note that in each case—political party, gender, and brand—the measurement is no more than a categorization of each sample unit. Nominal data are often reported as nonnumerical labels, such as Democrat, women, and Crest. Even if the labels are converted to numbers, as they often are for ease of computer entry and analysis, the numerical values are simply codes. They cannot be meaningfully added, subtracted, multiplied, or divided. For example, we might code Democrat = 1, Republican = 2, and other = 3. These are simply numerical codes for each of the categories into which units may fall and have no utility beyond that.

Definition 2.2

Ordinal data are measurements that enable the units of the sample (or population) to be ordered with respect to the variable of interest.

Ordinal data are measurements that indicate the *relative* amount of a property possessed by the units. The following are examples of ordinal data:

1. The size of car rented by each individual in a sample of 30 business travelers: compact, subcompact, midsize, or full-size
2. A taste-tester's ranking of four brands of barbecue sauce for a panel of 10 tasters
3. A supervisor's annual performance rating on a scale of 1 (lowest) to 10 (highest) for each of 20 employees

Note that in each case—size, flavor preference, and performance rating—more than a categorization of units is involved. In addition to a categorization, the measurements actually rank the units. For example, we know that a midsize car is larger than a subcompact, and that an employee with a performance ranking of 9 performed better, in the opinion of the supervisor, than one with a ranking of 7. We also know that a taster preferred brand C barbecue sauce to brand A if he gives brand C a higher flavor preference ranking than brand A.

Ordinal data are said to represent a "higher" level of measurement than nominal data because ordinal data contain all the information of nominal data (i.e., category labels that differentiate units) *plus* an ordering of the units. As with nominal measurements, the distance between ordinal measurements is not meaningful. For example, we do not know whether the difference in size between a full-size and midsize car is the same as the difference between a midsize and subcompact. Nor do we know whether the degree of flavor preference between barbecue sauce brands C and A is the same as that between brands A and B if a taster reports his flavor preference as $C > A > B > D$, where "$>$" means "more flavorful than."

As with nominal data, ordinal data can be reported with or without numbers. For example, the automobile sizes are nonnumerically labeled, whereas the supervisor's performance ratings are numerical. Even if numbers are used, we must again be careful; they simply provide an ordering or ranking of the units in the sample or population. The arithmetic operations of addition, subtraction, multiplication, and addition are not meaningful for ordinal data.

Definition 2.3

Interval data are measurements that enable the determination of how much more or less of the characteristic being measured is possessed by one unit of the sample (or population) than another.

Interval data are always numerical, and the numbers assigned to two units can be subtracted to determine the difference between the units with respect to the variable being measured. The following are examples of interval data:

1. The temperature (in degrees Fahrenheit) at which each of a sample of 20 pieces of heat-resistant plastic begins to melt
2. The scores of a sample of 150 law school applicants on the LSAT, a standardized law school entrance exam administered nationwide
3. The time at which the 5 P.M. Washington to New York air shuttle arrives at LaGuardia on each of a sample of 30 weekdays

Note that in each case—temperature, score, and arrival time—more than a ranking is involved. The difference between the numerical values assigned to the units is meaningful. For example, the difference between scores of 600 and 580 on the LSAT is the same as that between scores of 520 and 500. Also, the morning shuttle due at 9 A.M. but arriving at 9:20 A.M. is just as late as the afternoon shuttle due at 5:30 P.M. and arriving at 5:50 P.M. Note in each case that the difference is the key, not the numerical measurement itself.

Interval data represent a higher level of measurement than ordinal data, because in addition to ranking the units, interval data reflect the difference between the units with respect to the variable being measured. Although adding and subtracting interval data are valid, multiplying and dividing them are not. This is because the zero point (the origin, or 0) does not indicate an absence of the characteristic being measured by the data. For example, the origin on the temperature scale differs for the Fahrenheit and Celsius scales and does not indicate an absence of heat on either scale. Temperatures lower than 0° (e.g., −10°C and −10°F) indicate that less heat is present, so 0° cannot mean no heat. The result is that we cannot say that a temperature of 100°F indicates twice the heat of 50°F. Similarly, since LSAT scores range from 200 to 800, a zero score is not even possible, and thus has no meaning. The result is that a score of 600 cannot be interpreted as 50% higher than a score of 400.

Most scientific and business data are measured on scales for which the origin is meaningful. Thus, most numerical measurements encountered in science and business are **ratio data**.

Definition 2.4

Ratio data are measurements that enable the determination of how many times as much of the characteristic being measured is possessed by one unit of the sample (or population) than another.

Ratio data are always numerical, and the ratio between the numbers assigned to two units can be interpreted as the multiple by which the units differ. The following are examples of ratio data:

1. The annual income for each member of a sample of 500 households
2. The unemployment rate (reported as a percentage) in the United States for each of the past 60 months
3. The number of female executives employed in each of a sample of 50 industrial companies

Note that in each case—dollars of income, unemployed percentage, and count of female executives—the scale measures the absolute amount of the characteristic possessed by the unit. The result is that the ratio of measurements between units is meaningful. That is, a household with annual income of $60,000 has twice that of a household with $30,000 annual income. Similarly, an unemployment rate of 8% means twice as many unemployed as a rate of 4%. And a company with 30 female executives has 1.5 times as many as one with 20 female executives.

Ratio data represent the highest level of measurement. The numbers can be used to categorize, rank, differentiate, and measure multiples of one unit with respect to another. All arithmetic operations performed on ratio data are meaningful. *The key to differentiating interval and ratio data is that the zero point, or origin, denotes an absence of the characteristic being measured by the ratio data.* For example, zero income, zero unemployment, and zero female executives mean *absence* of income, unemployment, and female executives, respectively. Most measurement scales yield ratio data: measures of monetary value, distance, weight, height, percentages, and numerical counts all usually generate ratio data.

The four types of data are often combined into two classes that are sufficient for most statistical applications. Nominal and ordinal data are often referred to as **qualitative data**, whereas interval and ratio data are called **quantitative data**.

The properties of the four types of data are summarized in the box on page 22. As you would expect, the methods for describing and reporting data depend on the type of data being analyzed. Describing qualitative data generally consists of calculating the percentage of the sample measurements falling in each category. However, most data are quantitative, and a number of methods exist to summarize and describe these data. We devote the remainder of this chapter to graphical and numerical methods for describing quantitative data.

Types of Data	
Nominal	Classification of sample (or population) units into categories Often labels rather than numbers
Ordinal	Rank orders the sample (or population) units May be verbal labels or numbers
Interval	Enables comparison of sample (or population) units according to differences between values Always numerical, but the zero point on the scale does not indicate an absence of the measured characteristic
Ratio	Enables comparison of sample (or population) units according to multiples of the values Always numerical, and the zero point on the scale denotes an absence of the measured characteristic
Qualitative	Includes nominal and ordinal data types
Quantitative	Includes interval and ratio data types

Exercises 2.1–2.8

Learning the Mechanics

2.1
a. Explain the difference between nominal and ordinal data.
b. Explain the difference between interval and ratio data.
c. Explain the difference between qualitative and quantitative data.

2.2 Each of the descriptions of data defines one of the following types: nominal, ordinal, interval, ratio. Match the correct type to each description.
a. Data that enable the units of the sample to be compared by the differences between their numerical values.
b. Data that enable the units of the sample to be classified into categories
c. Data that enable the units of the sample to be rank-ordered
d. Data that enable the units of the sample to be compared by computing the ratios of the numerical values

2.3 Suppose you are provided a data set that classifies each sample unit into one of four categories: A, B, C, or D. You plan to create a computer database consisting of these data, and you decide to code the data as

A = 1, B = 2, C = 3, and D = 4 for inputting them into the computer. Are the data consisting of the classifications A, B, C, and D qualitative or quantitative? After the data are input as 1, 2, 3, or 4, are they qualitative or quantitative? Explain your answers.

Applying the Concepts

2.4 As most students know, colleges and universities are requiring an ever-increasing amount of information about applicants during the process of making acceptance and financial aid decisions. Classify each of the following types of data required on a college application as nominal, ordinal, interval, or ratio.
a. High school GPA
b. High school class rank
c. Applicant's score on the Scholastic Aptitude Test (SAT)
d. Gender of applicant
e. Parents' income
f. Age of applicant

2.5 Classify the following samples of data as nominal, ordinal, interval, or ratio. Justify your classifications.
a. Ten college freshmen were asked to indicate the brand of jeans they prefer.
b. Fifteen television cable companies were asked how many hours of sports programming they carry in a typical week.
c. Fifty students were asked what percentage of their time they spend studying each week.
d. The number of long-distance phone calls made from each of 100 public telephone booths on a particular day was recorded.

2.6 Classify the following examples of data as either qualitative or quantitative:
a. The bacteria count in the water at each of 30 city swimming pools
b. The occupation of each of 200 shoppers at a supermarket
c. The marital status of each person living on a city block
d. The number of months between auto maintenance for each of 100 sales representatives

2.7 Classify the following examples of data as nominal, ordinal, interval, or ratio.
a. The brand of stereo speaker for which each of 25 college students indicated a preference
b. The loss (in dollars) incurred in each of the last 5 years by a department store as a result of shoplifting
c. The color of interior house paint (other than white) that each of the five largest manufacturers of paint says generates the most sales revenue for the firm
d. The final ranking of the 10 football teams in the Southeastern Conference at the end of the season

2.8 A food-products company is considering marketing a new snack food. To see how consumers react to the product, the company conducted a taste test using a sample of 100 randomly selected shoppers at a suburban shopping mall. The shoppers were asked to taste the snack food and then fill out a short questionnaire that requested the following information:
a. What is your age?
b. Are you the person who typically does the food shopping for your household?
c. How many people are in your family?
d. How would you rate the taste of the snack food on a scale of 1 to 10, where 1 is least tasty?
e. Would you purchase this snack food if it were available on the market?
f. If you answered yes to part **e**, how often would you purchase it?
Each of these questions defines a variable of interest to the company. Classify the data generated for each variable as nominal, ordinal, interval, or ratio. Justify your classifications.

2.2 Graphic Methods for Describing Quantitative Data: Histograms and Stem-and-Leaf Displays

Recall from Section 2.1 that quantitative data sets consist of either interval or ratio data. Most data are quantitative, so that methods for summarizing quantitative data are especially important.

Before we can use the information in a sample to make inferences about a population, we need methods to summarize, or describe, a set of data. For example, the Environmental Protection Agency (EPA) performs extensive tests on all new car models to determine their mileage rating. Suppose that the 100 measurements in Table 2.1 represent the results of such tests on a certain new car model. How can we summarize the information in this rather large sample?

TABLE 2.1 EPA Mileage Ratings on 100 Cars

36.3	41.0	36.9	37.1	44.9	36.8	30.0	37.2	42.1	36.7
32.7	37.3	41.2	36.6	32.9	36.5	33.2	37.4	37.5	33.6
40.5	36.5	37.6	33.9	40.2	36.4	37.7	37.7	40.0	34.2
36.2	37.9	36.0	37.9	35.9	38.2	38.3	35.7	35.6	35.1
38.5	39.0	35.5	34.8	38.6	39.4	35.3	34.4	38.8	39.7
36.3	36.8	32.5	36.4	40.5	36.6	36.1	38.2	38.4	39.3
41.0	31.8	37.3	33.1	37.0	37.6	37.0	38.7	39.0	35.8
37.0	37.2	40.7	37.4	37.1	37.8	35.9	35.6	36.7	34.5
37.1	40.3	36.7	37.0	33.9	40.1	38.0	35.2	34.8	39.5
39.9	36.9	32.9	33.8	39.8	34.0	36.8	35.0	38.1	36.9

A visual inspection of the data indicates some obvious facts. Most of the mileages are in the 30's, for example, with a smaller fraction in the 40's. But it is difficult to provide much additional information on the 100 mileage ratings without resorting to some method of summarizing the data.

A **relative frequency histogram** for these 100 EPA mileage readings is shown in Figure 2.1. The horizontal axis of Figure 2.1, which gives the miles per gallon for a given automobile, is divided into intervals commencing with the interval from 29.95 to 31.45 and proceeding in intervals of equal size to 43.45 to 44.95 miles per gallon. The vertical axis gives the proportion (or **relative frequency**) of the 100 readings that fall in each interval. Thus, you can see that .33, or 33%, of the owners obtained a mileage between 35.95 and 37.45. This interval contains the highest relative frequency, and the intervals tend to contain a smaller fraction of the measurements as the mileages get smaller or larger.

By summing the relative frequencies in the intervals 34.45–35.95, 35.95–37.45,

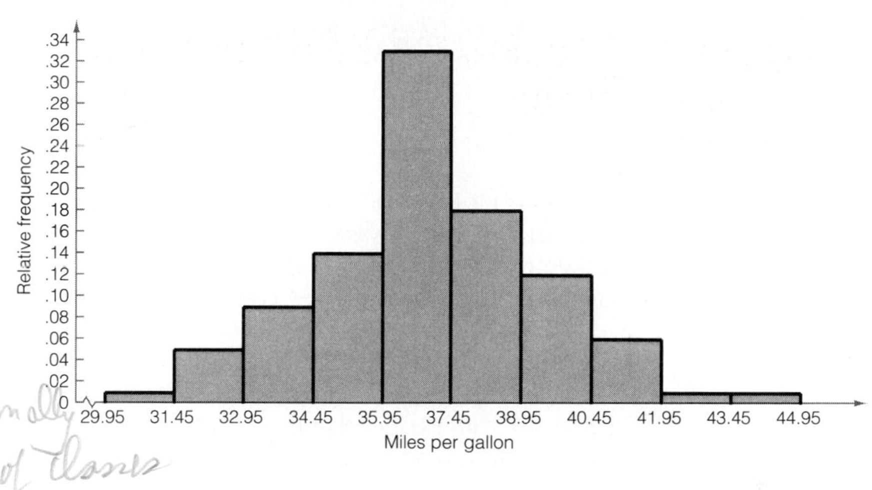

FIGURE 2.1 ▶
Histogram for EPA mileage data

and 37.45–38.95, you can see that 65% of the mileages are between 34.45 and 38.95. Similarly, only 2% of the cars obtained a mileage rating over 41.95. Many other summary statements can be made by further study of the histogram.

Another graphic representation of these same data, a **stem-and-leaf display**, is shown in Figure 2.2. In this display the **stem** is the portion of the observation to the left of the decimal point, while the remaining portion to the right of the decimal point is the **leaf**.

FIGURE 2.2 ▶
A stem-and-leaf display for the EPA mileage ratings on 100 cars

Stem	*Leaf*
30	0
31	8
32	7 5 9 9
33	9 1 8 9 2 6
34	8 0 4 8 2 5
35	5 9 3 9 7 6 2 0 6 1 8
36	3 2 3 5 8 9 9 0 7 6 4 8 5 4 6 1 8 7 7 9
37	0 1 3 9 2 6 3 1 9 4 0 0 1 6 8 7 0 2 4 7 5
38	5 6 2 3 0 2 7 8 4 1
39	9 0 8 4 0 7 3 5
40	5 3 7 2 5 1 0
41	0 0 2
42	1
43	
44	9

The stems and leaves for the mileage readings 36.3, 32.7, and 40.5 are shown here:

Stem	*Leaf*
36	3

Stem	*Leaf*
32	7

Stem	*Leaf*
40	5

The stems for the data set are listed in a column from the smallest (30) to the largest (44). Then the leaf for each observation is recorded in the row of the display corresponding to the observation's stem. For example, the leaf (3) of the first observation in Table 2.1 is written in the row corresponding to the stem (36). Similarly, the leaf (7) for the second observation in Table 2.1 is recorded in the row of Figure 2.2 corresponding to the stem (32).

The stem-and-leaf display (Figure 2.2) presents a compact picture of the data set. You can see at a glance that the 100 mileage readings were distributed between 30.0 and 44.9, with most of them falling in stem rows 35 to 39. The six leaves in stem row 34 indicate that six of the 100 readings were at least 34.0 but less than 35.0. Similarly, the 11 leaves in stem row 35 indicate that 11 of the 100 readings were at least 35.0 but less than 36.0. Only five cars had readings equal to 41 or larger, and only one was as low as 30.

The definitions of the stem and leaf for a data set can be modified to alter the graphic description. For example, suppose we had defined the stem as the tens digit for the gas mileage data, rather than the ones and tens digits. With this definition, the stems and leaves corresponding to the measurements 36.3 and 32.7 would be as follows:

Stem	*Leaf*
3	6

Stem	*Leaf*
3	2

Note that the decimal portion of the numbers has been dropped. Generally, only one digit is displayed in the leaf.

If you look at the data, you will see why we did not define the stem this way. All the mileage measurements fall in the 30's and 40's, so all the leaves would fall into only two stem rows in this display. The resulting picture would not be nearly as informative as Figure 2.2.

Both the histogram and stem-and-leaf displays provide useful graphic descriptions of quantitative data. Since most statistical software packages can be used to construct these displays, we will focus on their interpretation rather than their construction.

Histograms can be used to display either the **frequency** or **relative frequency** of the measurements falling into specified intervals (called **measurement classes**). The frequency is just a count of the number of measurements in a class, while the relative frequency is the proportion, or fraction, of measurements in the class. The measurement classes, frequencies, and relative frequencies for the EPA car mileage data are shown in Table 2.2.

TABLE 2.2 Measurement Classes, Frequencies, and Relative Frequencies for the Car Mileage Data

Measurement Class	Frequency	Relative Frequency
29.95–31.45	1	.01
31.45–32.95	5	.05
32.95–34.45	9	.09
34.45–35.95	14	.14
35.95–37.45	33	.33
37.45–38.95	18	.18
38.95–40.45	12	.12
40.45–41.95	6	.06
41.95–43.45	1	.01
43.45–44.95	1	.01
	100	1.00

By looking at a histogram (say, the relative frequency histogram in Figure 2.1), you can see two important facts. First, note the total area under the histogram and then note the proportion of the total area that falls over a particular interval of the *x*-axis. You will see that the proportion of the total area that falls above an interval is equal to the relative frequency of measurements falling in the interval. For example, the relative frequency for the class interval 35.95–37.45 is .33. Consequently, the rectangle above the interval contains .33 of the total area under the histogram.

Second, you can imagine the appearance of the relative frequency histogram for a very large set of data (say, a population). As the number of measurements in a data set is increased, you can obtain a better description of the data by decreasing the width of the class intervals. When the class intervals become small enough, a relative frequency histogram will (for all practical purposes) appear as a smooth curve (see Figure 2.3).

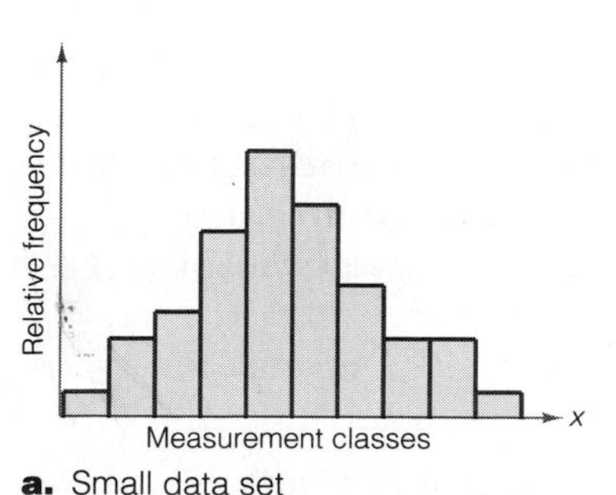

a. Small data set

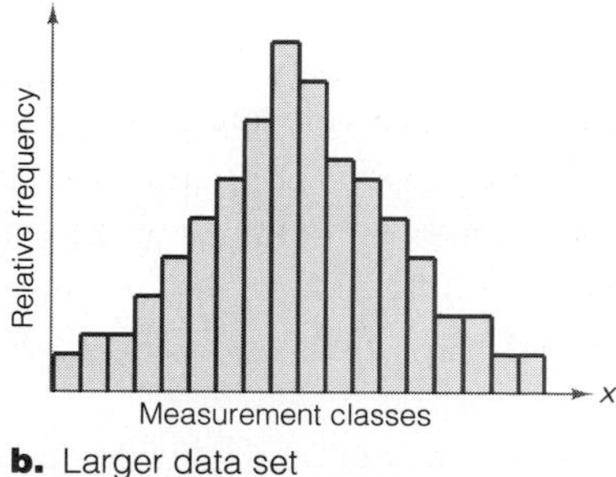

b. Larger data set

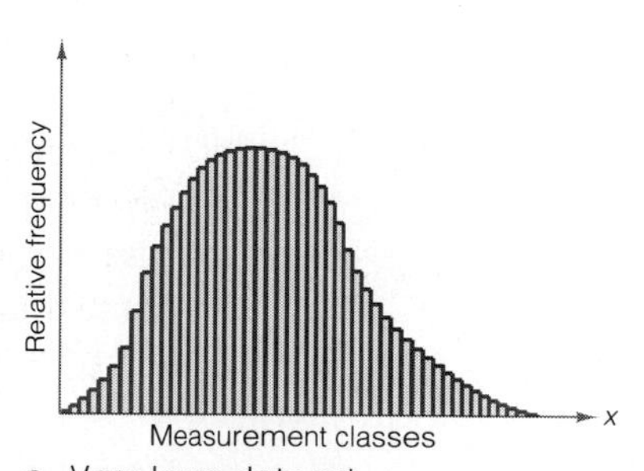

c. Very large data set

FIGURE 2.3 ▲ The effect of the size of a data set on the outline of a histogram

While histograms provide good visual descriptions, particularly of very large data sets, individual measurements cannot be identified by looking at a histogram. In contrast, each of the original measurements is "visible" in a stem-and-leaf display. The stem-and-leaf display arranges the data in ascending order, which enables easy location of the individual measurements. For example, in Figure 2.2 we can easily see that two of the gas mileage measurements are equal to 36.3, while that fact is not evident by inspection of the histogram in Figure 2.1. However, stem-and-leaf displays can become unwieldy for very large data sets. A very large number of stems and leaves causes the vertical and horizontal dimensions of the display to become cumbersome, so that the usefulness of the visual display is diminished.

EXAMPLE 2.1

The data in Table 2.3 give the percentages of the total number of college or university student loans by state that are in default.

TABLE 2.3 Percentage of Student Loans (per State) in Default

State	%	State	%	State	%	State	%
Ala.	12.0	Ill.	9.3	Mont.	6.4	R.I.	8.8
Alaska	19.7	Ind.	6.7	Nebr.	4.9	S.C.	14.1
Ariz.	12.1	Iowa	6.2	Nev.	10.1	S.Dak.	5.5
Ark.	12.9	Kans.	5.7	N.H.	7.9	Tenn.	12.3
Calif.	11.4	Ky.	10.3	N.J.	12.0	Tex.	15.2
Colo.	9.5	La.	13.5	N.Mex.	7.5	Utah	6.0
Conn.	8.8	Maine	9.7	N.Y.	11.3	Vt.	8.3
Del.	10.9	Md.	16.6	N.C.	15.5	Va.	14.4
D.C.	14.7	Mass.	8.3	N.Dak.	4.8	Wash.	8.4
Fla.	11.8	Mich.	11.4	Ohio	10.4	W.Va.	9.5
Ga.	14.8	Minn.	6.6	Okla.	11.2	Wis.	9.0
Hawaii	12.8	Miss.	15.6	Oreg.	7.9	Wyo.	2.7
Idaho	7.1	Mo.	8.8	Pa.	8.7		

Source: National Direct Student Loan Program

a. Use a statistical computer software package to create a relative frequency histogram for these data.

b. Use a statistical computer software package to create a stem-and-leaf display for these data.

c. Compare and interpret the two graphic displays of these data.

Solution

a. We used SAS to generate the relative frequency histogram in Figure 2.4.

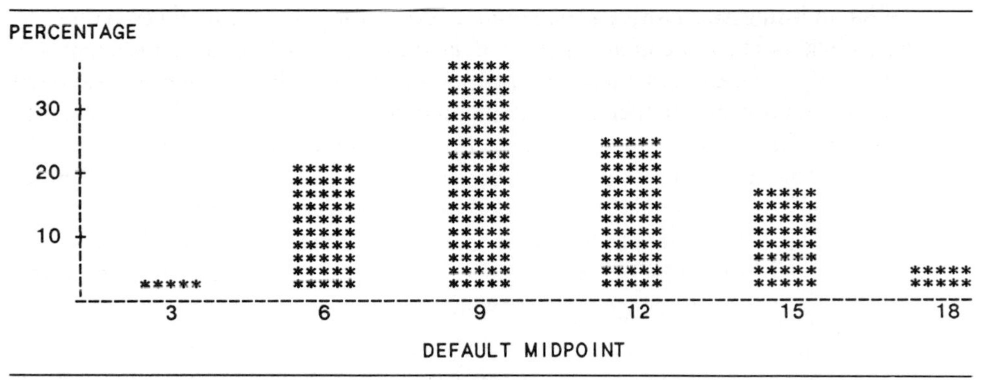

FIGURE 2.4 ▲ **SAS printout of relative frequency histogram for student loan default data**

Note that six classes were formed by the SAS program. The classes are identified by their midpoints rather than their endpoints. Thus, the first interval has a midpoint of 3, the second of 6, etc. The corresponding measurement classes are therefore 1.5 to 4.5, 4.5 to 7.5, etc. Note too that the SAS program labels the vertical axis "Percentage" rather than "Relative frequency." Thus, we see that the measurement class with a midpoint of 9 (ranging from 7.5 to 10.5) contains more than 30% of the measurements.

b. We used Minitab to generate the stem-and-leaf display in Figure 2.5.

FIGURE 2.5 ▶ **Computer-generated stem-and-leaf display for defaulted student loan data**

```
STEM-AND-LEAF DISPLAY OF DEFAULT
LEAF DIGIT UNIT =     .1000
1 2 REPRESENTS 1.2


   1    2  7
   1    3
   3    4  89
   5    5  57
  10    6  02467
  14    7  1599
  21    8  3347888
 (5)    9  03557
  25   10  1349
  21   11  23448
  16   12  001389
  10   13  5
   9   14  1478
   5   15  256
   2   16  6

       HI  197,
```

Note that the stem is the second column in the printout, and has been defined as the number to the left of the decimal. The leaf is the third column in the printout, and is the number to the right of the decimal. Minitab also provides several other useful descriptive aids. The most extreme observation, 19.7, is displayed separately, with the label of "HI" to indicate that the measurement is disconnected from the main body of the data.* Also, Minitab indicates the cumulative number of measurements (first column of the printout) from the nearest tail of the distribution to each stem. For the stem that is in the middle of the data set, the number of leaves is indicated in parentheses. The program also prints a key (at the top), giving the units of the leaf and an example showing where the decimal goes.

c. As is usually the case for data sets that are not too large (say, fewer than 100 measurements), the stem-and-leaf display provides more detail than the histogram without being unwieldy. For the student default rate data, note that the stem-and-leaf display in Figure 2.5 clearly indicates that most states' rates are relatively evenly distributed from 6 to 13 (note that between four and seven measurements are "attached" to each stem from 6 to 12). The stem-and-leaf display also clearly indicates one low default rate (2.7) and one high rate (19.7) that are distinctly separate from the main body of data.

   The histogram in Figure 2.4 also indicates a relatively even distribution of rates over the four measurement classes with midpoints from 6 to 15, but the boundaries are somewhat more difficult to determine since the individual measurements cannot be "seen" in the display. Similarly, the two small bars over the outside classes with midpoints of 3 and 18 indicate the presence of some extreme measurements, but the exact number and location of these measurements is not displayed. Of course, we could obtain more detail by increasing the number of measurement classes. However, histograms are most useful for displaying very large data sets, when the overall shape of the distribution of measurements is more important than the identification of individual measurements.

Most statistical software packages can be used to generate either histograms or stem-and-leaf displays. Both are useful tools for graphically describing data sets. We recommend generating and comparing both displays when feasible. The histogram will generally be more useful for very large data sets, while the stem-and-leaf display provides useful detail for smaller data sets.

*The designation of a "HI" or "LO" measurement by Minitab is related to definitions of "outliers" in box plots of the data. Box plots (and outliers) are discussed in optional Section 2.7. You should exercise caution in the interpretation of these designations until we have covered their definitions.

CASE STUDY 2.1 / Mercury Poisoning and the Dental Profession

The hazards to health traced to environmental pollution constitute an area of major national concern. Mercury has been identified as one source of environmental contamination. Recognition of mercury as a hazard can be traced to Theophrastus, a Greek scientist living about 400 B.C. The physiologic effects of mercury poisoning are a matter of record with death as the ultimate possibility.

This is the introductory paragraph to an article that appeared in the *Journal of the American Dental Association* (Miller et al., 1974). The article discusses mercury vapor contamination levels in the air of dental offices and the resulting threat to the health of dentists and dental assistants. Such contamination might result from spills in handling mercury, the unprotected storage of scrap amalgam, or aerosols created by the use of high-speed rotary cutting instruments in removing old amalgam fillings.

The cumulative absorption of small quantities of mercury can result in serious medical problems. The constant daily exposure of dentists and their auxiliary personnel to possible mercury contamination is therefore an important concern. A level of .05 milligram of mercury per cubic meter of air is the largest amount considered safe for those working a 40-hour week.

A determination of mercury vapor levels was made in 60 dental offices in San Antonio, Texas. A relative frequency histogram summarizing the information provided by these measurements is given in Figure 2.6. The histogram clearly shows that an alarming fraction (6/60, or 1/10) were above the danger level of .05 milligram. You can see that these data have been effectively summarized and clearly indicate a need for strict policing of mercury vapor levels in dental offices.

Case Study 2.1 illustrates a shortcoming of relative frequency histograms and graphic data displays in general. Suppose you wish to use the relative frequency histogram to infer the nature of the population relative frequency distribution—that is, the distribution of the mercury vapor levels in the offices of all dentists practicing in the United States. It is true that the sample and population relative frequency distributions will be similar. But how similar? How can you explain how similar the two figures will be? Or, equivalently, how can you measure the reliability of the inference? In Section 2.3 we will explain how you can use one or

FIGURE 2.6 ▶
Relative frequency histogram

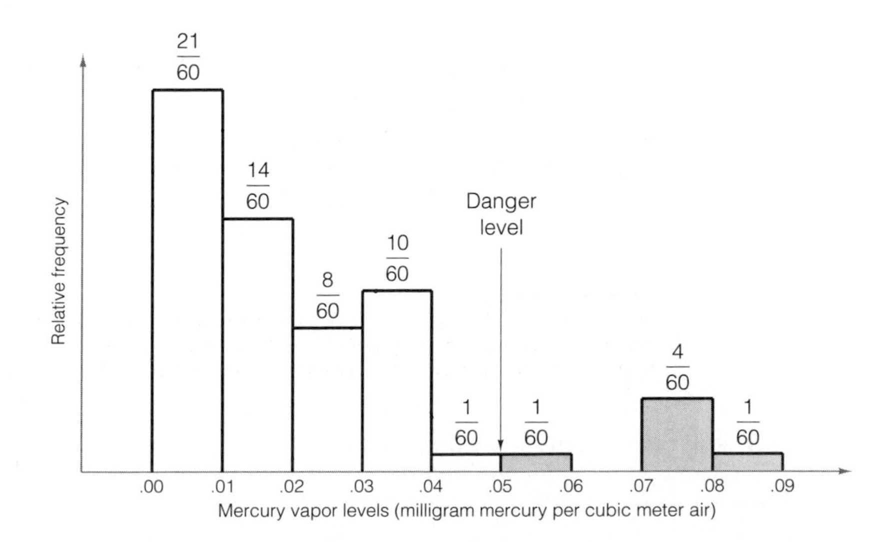

more numbers (numerical descriptive measures) to describe a distribution of measurements. Further, you will see in Chapter 6 that we can use sample numerical descriptive measures to make inferences about their population counterparts and that we can measure the reliability of these inferences. In short, you will see that numerical descriptive measures are superior to graphic descriptive measures when you want to use the sample data to make inferences about the population from which the sample was selected.

Exercises 2.9–2.20

Note: Starred (*) exercises require the use of a computer.

Learning the Mechanics

2.9 Graph the relative frequency histogram for the 500 measurements summarized in the accompanying relative frequency table.

Measurement Class	*Relative Frequency*
.5– 2.5	.10
2.5– 4.5	.15
4.5– 6.5	.25
6.5– 8.5	.20
8.5–10.5	.05
10.5–12.5	.10
12.5–14.5	.10
14.5–16.5	.05

2.10 Refer to Exercise 2.9. Calculate the number of the 500 measurements falling into each of the measurement classes. Then graph a frequency histogram for these data.

2.11 SAS was used to generate the stem-and-leaf display shown here. Note that SAS arranges the stems in descending order.

Stem	*Leaf*
5	1
4	4 5 7
3	0 0 0 3 6
2	1 1 3 4 5 9 9
1	2 2 4 8
0	0 1 2

a. How many observations were in the original data set?

b. In the bottom row of the stem-and-leaf display, identify the stem, the leaves, and the numbers in the original data set represented by this stem and its leaves.

Applying the Concepts

2.12 The graph summarizes the scores obtained by 100 students on a questionnaire designed to measure aggressiveness. (Scores are integer values that range from 0 to 20. A high score indicates a high level of aggression.)

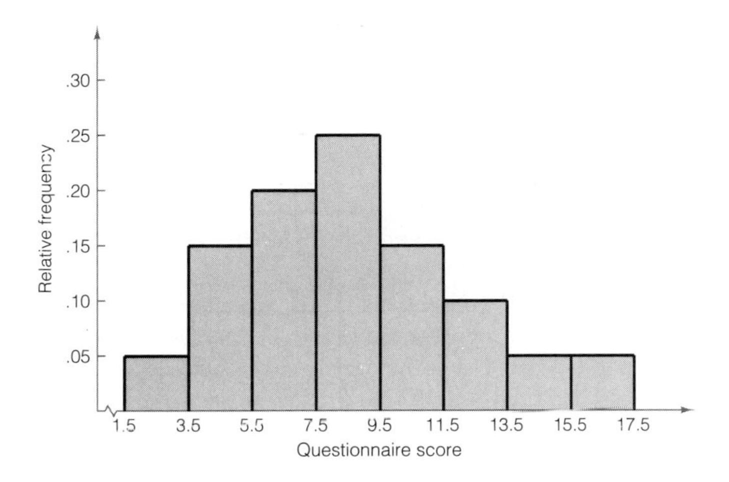

a. Which measurement class contains the highest proportion of test scores?
b. What proportion of the scores lie between 3.5 and 5.5?
c. What proportion of the scores are higher than 11.5?
d. How many students scored less than 5.5?

2.13 While producing many economic benefits to the state of Florida, gypsum and phosphate mines also produce a harmful byproduct: radiation. It has been known for a number of years that the mine tailings (waste) contain radioactive radon 222. In fact, new housing complexes built over the leveled piles of residue have shown disturbing radiation levels within the houses. The radiation levels in waste gypsum and phosphate mounds in Polk County, Florida, are regularly monitored by the Eastern Environmental Radiation Facility (EERF), and by the Polk County Health Department (PCHD), Winter Haven, Florida. Shown in the table are measurements of the exhalation rate (a measure of radiation) of soil samples taken on waste piles in Polk County, Florida. They represent part of the data contained in a report by Thomas R. Horton of EERF.

Exhalation Rate of Soil Samples

1,709.79	4,132.28	2,996.49	2,796.42	3,750.83	961.40	1,096.43	1,774.77
357.17	1,489.86	2,367.40	11,968.23	178.99	5,402.35	2,315.52	2,617.57
1,150.94	3,017.48	599.84	2,758.84	3,764.96	1,888.22	2,055.20	205.84
1,572.69	393.55	538.37	1,830.78	878.56	6,815.69	752.89	1,977.97
558.33	880.84	2,770.23	1,426.57	1,322.76	1,480.04	9,139.21	1,698.39

Source: Horton, T. R. "A preliminary radiological assessment of radon exhalation from phosphate gypsum piles and inactive uranium mill tailings piles." EPA–520/5–79–004. Washington, D.C.: Environmental Protection Agency, 1979.

SPSS/PC+ was used to generate the following stem-and-leaf display for these data:

```
Stem-and-leaf display for variable .. EXHLRATE

   0 . 22445668990123455677889
   2 . 013468880088
   4 . 14
   6 . 8
   8 . 1
  10 .
  12 . 0
```

a. Interpret the display. Which digit(s) was used for the stem and which for the leaf? Find the largest measurement in the data set, and interpret its representation in the display.

b. Note that the presence of a measurement well removed from the main body of data somewhat distorts the display, since most of the data set is compressed into a small portion of it. The largest measurement was removed from the data set, and a new SPSS/PC+ stem-and-leaf display was generated, as shown here. Interpret this display, identifying the stem and leaf used. Using both displays, give a verbal description of the data.

```
Stem-and-leaf display for variable .. EXHLRATE

   0 . 2244566899
   1 . 0123455677889
   2 . 01346888
   3 . 0088
   4 . 1
   5 . 4
   6 . 8
   7 .
   8 .
   9 . 1
```

2.14 Refer to Exercise 2.13, where we constructed a stem-and-leaf display for the 40 exhalation rate measurements on waste gypsum and phosphate mounds in Florida. The accompanying figure is a SAS-generated relative frequency histogram for these data.

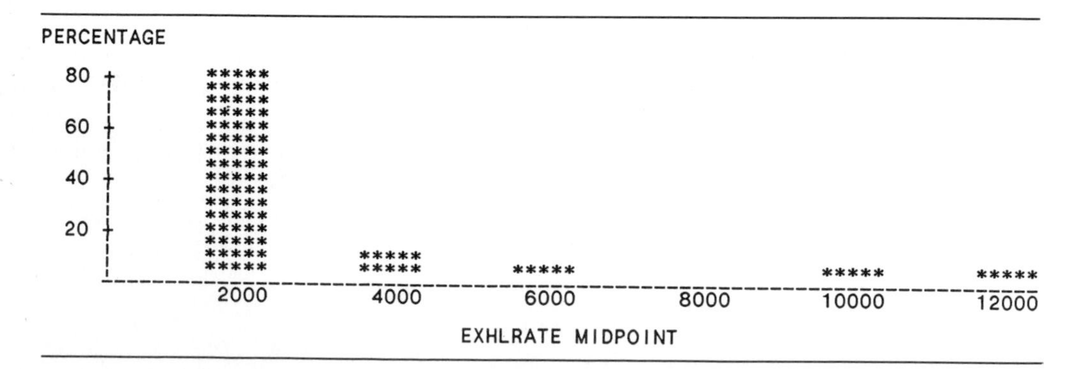

a. Interpret the histogram. What fraction of the measurements fall in the first measurement class? What are the boundaries for that class? [*Note:* Many computer programs, including SAS, do not require that the first and last class have well-defined outer boundaries.]
b. Compare the histogram to the stem-and-leaf display for these data in Exercise 2.13**a**.

2.15 Health care issues are receiving much attention in both academic and political arenas. A sociologist recently conducted a survey of citizens over 60 years of age whose net worth is too high to qualify for Medicaid, but who have no private health insurance. The ages of 25 uninsured senior citizens were as follows:

68 73 66 76 86 74 61 89 65 90 69 92 76
62 81 63 68 81 70 73 60 87 75 64 82

a. Construct a stem-and-leaf display of the ages.
b. Based on your stem-and-leaf display of these data, are the ages relatively evenly distributed, or are they "skewed" toward the younger or older segment of the senior citizen population? [*Note:* We will discuss the concepts of symmetry and skewness of distributions in Section 2.3.]

2.16 Educators are constantly evaluating the efficacy of public schools in the education and training of American students. One quantitative assessment of change over time is the difference in scores on the Scholastic Aptitude Test (SAT), which has been used for decades by colleges and universities as one criterion for admission. The following table shows the average SAT scores for each of the 50 states for 1975 and 1990:

State	*1975*	*1990*	*State*	*1975*	*1990*	*State*	*1975*	*1990*
Ala.	883	984	La.	947	993	Ohio	955	949
Alaska	942	914	Maine	908	886	Okla.	994	1001
Ariz.	1021	942	Md.	907	908	Oreg.	908	923
Ark.	992	981	Mass.	903	900	Pa.	900	883
Calif.	908	903	Mich.	949	968	R.I.	901	883
Colo.	994	969	Minn.	1058	1019	S.C.	794	834
Conn.	913	901	Miss.	980	996	S.Dak.	1084	1061
Del.	915	903	Mo.	965	995	Tenn.	988	1008
Fla.	915	884	Mont.	1047	987	Tex.	898	874
Ga.	824	844	Nebr.	966	1030	Utah	1069	1031
Hawaii	892	885	Nev.	962	921	Vt.	915	897
Idaho	1017	968	N.H.	934	928	Va.	894	895
Ill.	970	994	N.J.	878	891	Wash.	1011	923
Ind.	881	867	N.Mex.	1002	1007	W.Va.	964	933
Iowa	1091	1088	N.Y.	925	882	Wis.	1036	1019
Kans.	1043	1040	N.C.	827	841	Wyo.	1054	977
Ky.	977	994	N.Dak.	1064	1069			

We used Minitab to construct the following stem-and-leaf displays of the two sets of SAT scores:

```
         1975                                1990

                       Leaf Unit = 10

   1      7 9
   1      8
   3      8 22                   1     8 3
   3      8                      3     8 44
   4      8 7                    5     8 67
   9      8 88999               14     8 888888999
  20      9 00000001111         20     9 000001
  22      9 23                  25     9 22223
 (4)      9 4445                25     9 44
  24      9 666677              23     9 6667
  18      9 88999               19     9 88899999
  13     10 011                 11    10 00011
  10     10 23                   6    10 33
   8     10 4455                 4    10 4
   4     10 66                   3    10 66
   2     10 89                   1    10 8
```

Interpret the two displays, focusing on how the distributions of average state scores changed over this 15-year period.

2.17 Refer to Exercise 2.16. As another method of comparing the 1975 and 1990 average SAT scores, **paired differences** are calculated by subtracting the 1975 score from the 1990 score for each state. The stem-and-leaf display for these differences is shown below:

```
Differences (1990 minus 1975)

Leaf Unit = 10

   1   -0 8
   4   -0 776
   7   -0 444
  16   -0 333322222
 (15)  -0 111111110000000
  19    0 00000111111
   8    0 2223
   4    0 44
   2    0 6
   1    0
   1    1 0
```

a. Interpret the display. How do your conclusions compare to those you reached when comparing the two displays in Exercise 2.16?

b. What is the largest improvement in the average score between 1975 and 1990 as indicated on the display? With which state is this improvement associated in the table of data (Exercise 2.16)?

2.18 Refer to Exercises 2.16 and 2.17. SPSS was used to generate frequency histograms for the average SAT scores for 1975 and 1990. The histograms are shown below:

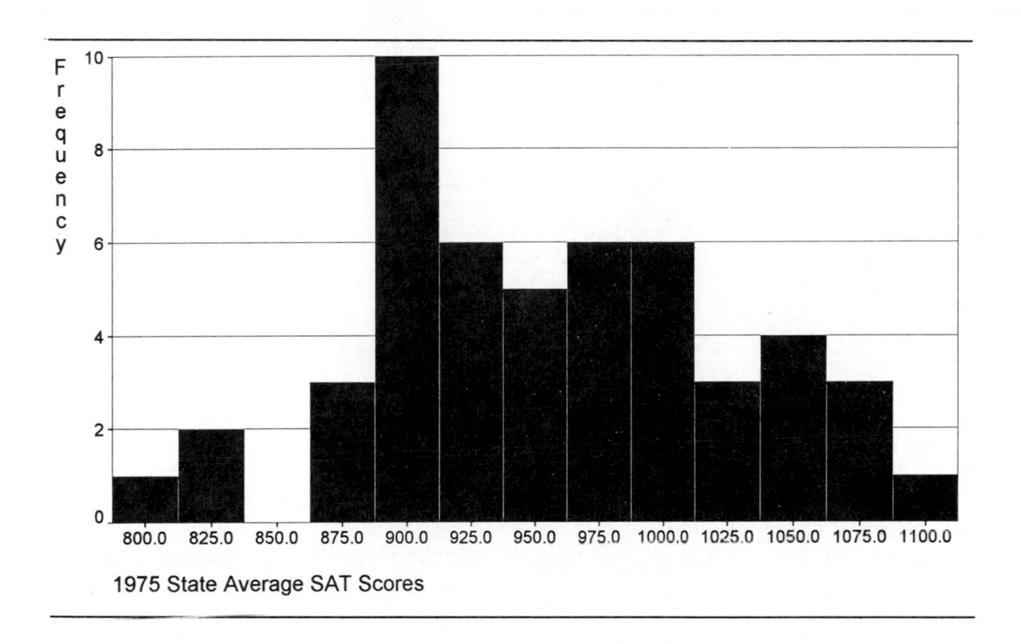

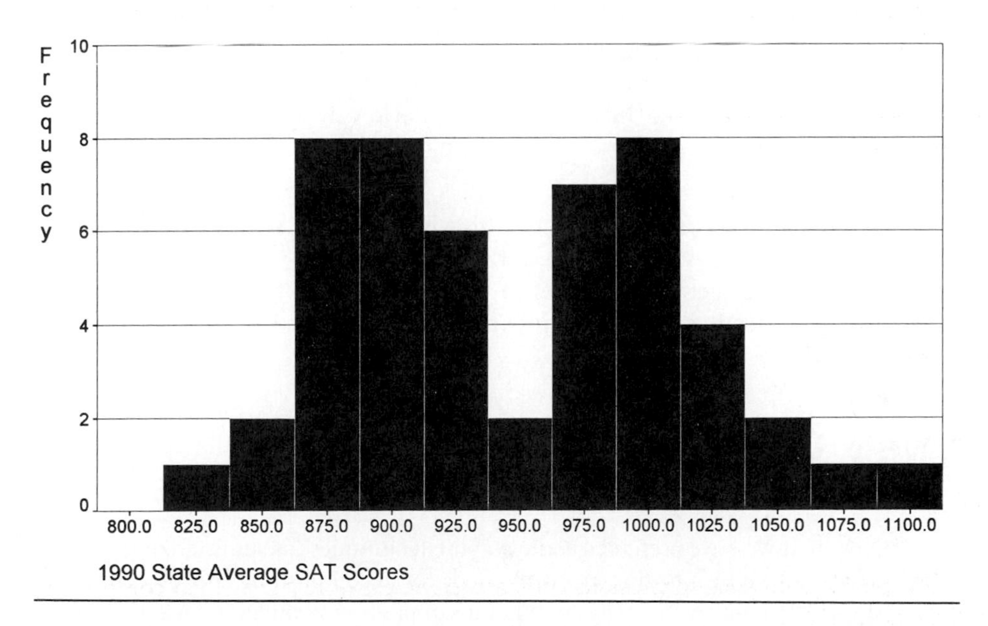

Interpret the histograms, focusing on the difference between the displays for 1975 and 1990. Do you find the stem-and-leaf displays or the histograms more useful for describing these data?

*2.19 According to the U.S. Department of Education, the national dropout rate for high school students fell by more than 1%, from 30.3% to 29.1%, between 1982 and 1984. The accompanying table shows the dropout rate, defined as the percentage of ninth graders that do not graduate, for each state (and the District of Columbia) in 1982 and 1984.

State	*1982*	*1984*	*State*	*1982*	*1984*	*State*	*1982*	*1984*
Ala.	36.6	37.9	Ky.	34.1	31.6	N.Dak.	16.1	13.7
Alaska	35.7	25.3	La.	38.5	43.3	Ohio	22.5	20.0
Ariz.	36.6	35.4	Maine	27.9	22.8	Okla.	29.2	26.9
Ark.	26.6	24.8	Md.	25.2	22.2	Oreg.	27.6	26.1
Calif.	39.9	36.8	Mass.	23.6	25.7	Pa.	24.0	22.8
Colo.	29.1	24.6	Mich.	28.4	27.8	R.I.	27.3	31.3
Conn.	29.4	20.9	Minn.	11.8	10.7	S.C.	37.4	35.5
Del.	25.3	28.9	Miss.	38.7	37.6	S.Dak.	17.3	14.5
D.C.	43.1	44.8	Mo.	25.8	23.8	Tenn.	32.2	29.5
Fla.	39.8	37.8	Mont.	21.3	17.9	Tex.	36.4	35.4
Ga.	35.0	36.9	Nebr.	18.1	13.7	Utah	25.0	21.3
Hawaii	25.1	26.8	Nev.	35.2	33.5	Vt.	20.4	16.9
Idaho	25.6	24.2	N.H.	23.0	24.8	Va.	26.2	25.3
Ill.	23.9	25.5	N.J.	23.5	22.3	Wash.	23.9	24.9
Ind.	28.3	23.0	N.Mex.	30.6	29.0	W.Va.	33.7	26.9
Iowa	15.9	14.0	N.Y.	36.6	37.8	Wis.	16.9	15.5
Kans.	19.3	18.3	N.C.	32.9	30.7	Wyo.	27.6	24.0

a. Use statistical software to generate stem-and-leaf displays for the 1982 and 1984 dropout rates. Compare the displays, focusing on the change in the rates over the 2-year period.

b. Calculate the paired differences between the rates in 1984 and 1982. Use the software to construct the stem-and-leaf display for the differences. Interpret the display. How does your interpretation differ from that in part **a**?

*2.20 Refer to Exercise 2.19.

a. Use statistical software to generate relative frequency histograms for the 1982 and 1984 dropout rates. Compare the histograms, focusing on the change in the rates over the 2-year period.

b. Calculate the paired differences between the rates in 1984 and 1982. Use the software to construct a relative frequency histogram for the differences. Interpret the histogram. How does your interpretation differ from that in part **a**?

2.3 Numerical Measures of Central Tendency

Now that we have presented some graphic techniques for summarizing and describing data sets, we turn to numerical methods for accomplishing this objective. When we speak of a data set, we refer to either a sample or a population. If statistical inference is our goal, we will wish ultimately to use sample numerical descriptive measures to make inferences about the corresponding measures for a population.

As you will see, there are a large number of numerical methods available to describe data sets. Most of these methods measure one of two data characteristics:

1. The **central tendency** of the set of measurements, i.e., the tendency of the data to cluster or to center about certain numerical values.
2. The **variability** of the set of measurements, i.e., the spread of the data.

In this section we concentrate on measures of central tendency. In the next section, we discuss measures of variability.

The most popular and best understood measure of central tendency for a quantitative data set is the **arithmetic mean** (or simply the **mean**) of a data set.

Definition 2.5

The **mean** of a set of quantitative data is equal to the sum of the measurements divided by the number of measurements contained in the data set.

In everyday terms, the mean is the average value of the data set.

Before calculating the mean (or other numerical descriptive measures) of data sets, we present some shorthand notation that will simplify our calculation instructions. Remember that such notation is used for only one reason—to avoid having to repeat the same verbal descriptions over and over. If you mentally substitute the verbal definition of a symbol each time you read it, you will soon become accustomed to its use.

We will denote the measurements of a data set as follows:

$$x_1, x_2, x_3, \ldots, x_n$$

where x_1 is the first measurement in the data set, x_2 is the second measurement in the data set, x_3 is the third measurement in the data set, . . ., and x_n is the nth (and last) measurement in the data set. Thus, if we have five measurements in a set of data, we will write x_1, x_2, x_3, x_4, x_5 to represent the measurements. If the actual numbers are 5, 3, 8, 5, and 4, we have $x_1 = 5$, $x_2 = 3$, $x_3 = 8$, $x_4 = 5$, and $x_5 = 4$.

To calculate the mean of a set of measurements, we must sum them and divide by n, the number of measurements in the set. The sum of measurements $x_1, x_2, \ldots, x_n$ is

$$x_1 + x_2 + \cdots + x_n$$

To shorten the notation, we will write this sum as

$$x_1 + x_2 + \cdots + x_n = \sum_{i=1}^{n} x_i$$

where $\sum$ is the symbol for the summation. Verbally translate $\sum_{i=1}^{n} x_i$ as follows: "The sum of the measurements, whose typical member is x_i, beginning with the member x_1 and ending with the member x_n."

Finally, we will denote the mean of a sample of measurements by $\bar{x}$ (read "x-bar"), and represent the formula for its calculation as follows:

$$\bar{x} = \frac{\sum_{i=1}^{n} x_i}{n}$$

EXAMPLE 2.2

Calculate the mean of the following five sample measurements: 5, 3, 8, 5, 6.

Solution

Using the definition of sample mean and the shorthand notation, we find

$$\bar{x} = \frac{\sum_{i=1}^{5} x_i}{5} = \frac{5 + 3 + 8 + 5 + 6}{5} = \frac{27}{5} = 5.4$$

sum of all sample meas / # of meas in sample

Thus, the mean of this sample is 5.4.*

EXAMPLE 2.3

Calculate the sample mean for the 100 EPA mileages given in Table 2.1.

Solution

The mean gas mileage for the 100 cars is

$$\bar{x} = \frac{\sum_{i=1}^{100} x_i}{100} = \frac{36.3 + 41.0 + \cdots + 38.1 + 36.9}{100} = \frac{3{,}699.4}{100} = 36.99$$

Given this information, you would be able to visualize a distribution of gas mileage readings centered in the vicinity of $\bar{x} = 37.0$. An examination of the relative frequency histogram (Figure 2.1) confirms that $\bar{x}$ does in fact fall near the center of the distribution.

*In the examples given here, $\bar{x}$ is sometimes rounded to the nearest tenth, sometimes the nearest hundredth, sometimes the nearest thousandth. There is no specific rule for rounding when calculating $\bar{x}$ because $\bar{x}$ is specifically defined to be the sum of all measurements divided by n; i.e., it is a specific fraction. When $\bar{x}$ is used for descriptive purposes, it is often convenient to round the calculated value of $\bar{x}$ to the number of significant figures used for the original measurements. When $\bar{x}$ is to be used in other calculations, however, it may be necessary to retain more significant figures.

The sample mean will play an important role in accomplishing our objective of making inferences about populations based on sample information. For this reason, it is important to use a different symbol when we want to discuss the **mean of a population**—i.e., the mean of the set of measurements on every unit in the population. We use the Greek letter μ (mu) for the population mean. We will adopt a general policy of using Greek letters to represent population numerical descriptive measures and Roman letters to represent corresponding descriptive measures for the sample.

$\bar{x}$ = Sample mean μ = Population mean

The sample mean, $\bar{x}$, will often be used to estimate (make an inference about) the population mean, μ. For example, the EPA mileages for the population consisting of *all* cars has a mean equal to some value, μ. Our sample of 100 cars yielded mileages with a mean of $\bar{x} = 36.99$. If, as is usually the case, we did not have access to the measurements for the entire population, we could use $\bar{x}$ as an estimator or approximator for μ. Then we would need to know something about the reliability of our inference. That is, we would need to know how accurately we might expect $\bar{x}$ to estimate μ. In Chapter 6, we will find that this accuracy depends on two factors:

1. *The size of the sample.* The larger the sample, the more accurate the estimate will tend to be.
2. *The variability, or spread, of the data.* All other factors remaining constant, the more variable the data, the less accurate the estimate.

Definition 2.6

The **median** is another important measure of central tendency. In general terms, the median is the middle number when the measurements in a data set are arranged in ascending (or descending) order.

The median is of most value in describing large data sets. If the data set is characterized by a relative frequency histogram (Figure 2.7), the median is the point on the x-axis such that half the area under the histogram lies above the median and half lies below. [*Note:* In Section 2.2 we observed that the relative frequency associated

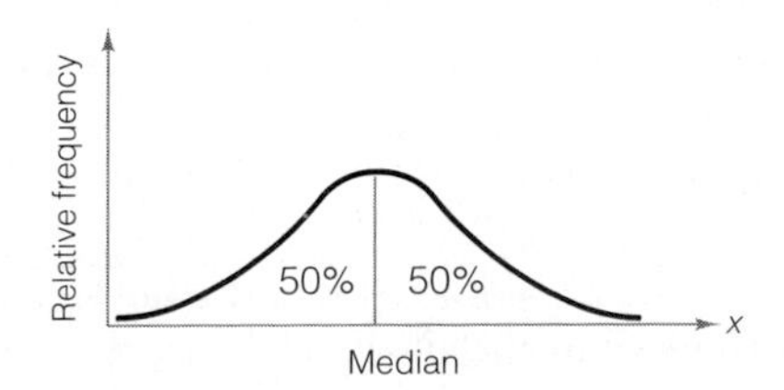

FIGURE 2.7 ► Location of the median

with a particular interval on the x-axis is proportional to the amount of area under the histogram that lies above the interval.]

Calculating a Median

Arrange the n measurements from the smallest to the largest.

1. If n is odd, the median is the middle number.
2. If n is even, the median is the mean of the middle two numbers.

EXAMPLE 2.4

Consider the following sample of $n = 7$ measurements: 5, 7, 4, 5, 20, 6, 2.

a. Calculate the median of this sample.

b. Eliminate the last measurement (the 2) and calculate the median of the remaining $n = 6$ measurements.

Solution

a. The seven measurements in the sample are ranked in ascending order:

2, 4, 5, 5, 6, 7, 20

Because the number of measurements is odd, the median is the middle measurement. Thus, the median of this sample is 5.

b. After removing the 2 from the set of measurements, we rank the sample measurements in ascending order as follows:

4, 5, 5, 6, 7, 20

Now the number of measurements is even, so we average the middle two measurements. The median is $(5 + 6)/2 = 5.5$.

In certain situations, the median may be a better measure of central tendency than the mean. In particular, the median is less sensitive than the mean to extremely large or small measurements. To illustrate, note that all but one of the measurements in part **a** of Example 2.3 center about $x = 5$. The single relatively large measurement, $x = 20$, does not affect the value of the median, 5, but it causes the mean, $\bar{x} = 7$, to lie to the right of most of the measurements.

As another example of data for which the central tendency is better described by the median than the mean, consider the household incomes of a community being studied by a sociologist. The presence of just a few households with very high incomes will affect the mean more than the median. Thus, the median will provide a more accurate picture of the typical income for the community. The mean could exceed the vast majority of the sample measurements (household incomes), making it a misleading measure of central tendency.

If you were to calculate the median for the 100 gas mileage readings in Table 2.1, you would find that the median, 37.0, and the mean, 36.99, are almost equal. This fact indicates that the data form an approximately **symmetric** distribution. (Compare the center figure in the next box with Figure 2.1 on page 25.) As indicated in the box, a comparison of the mean and median gives an indication of the **skewness** (nonsymmetry) of a data set.

Comparing the Mean and the Median

If the median is less than the mean, the data set is skewed to the right:

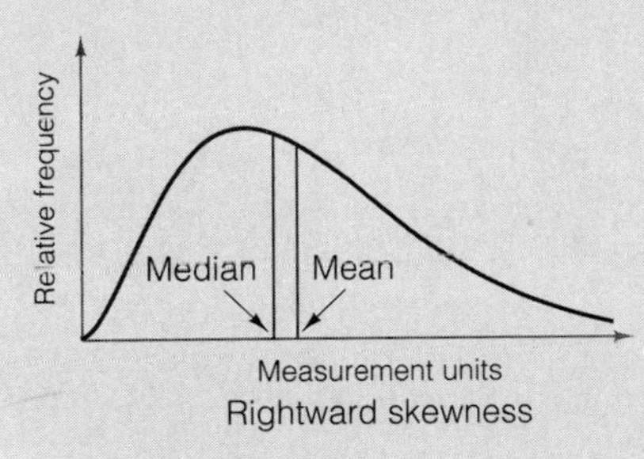

Rightward skewness

For symmetric data sets, the mean equals the median:

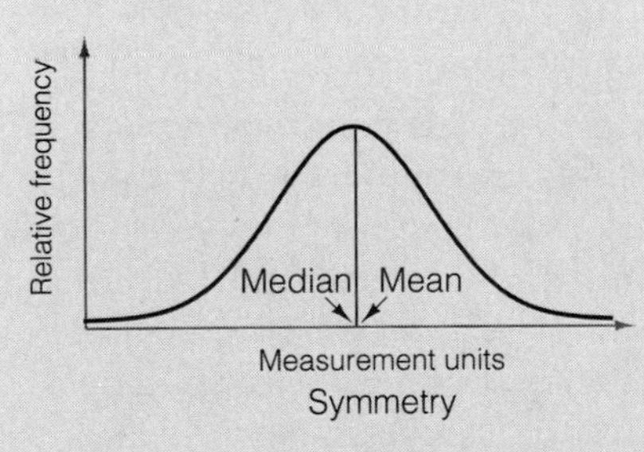

Symmetry

If the median is greater than the mean, the data set is skewed to the left:

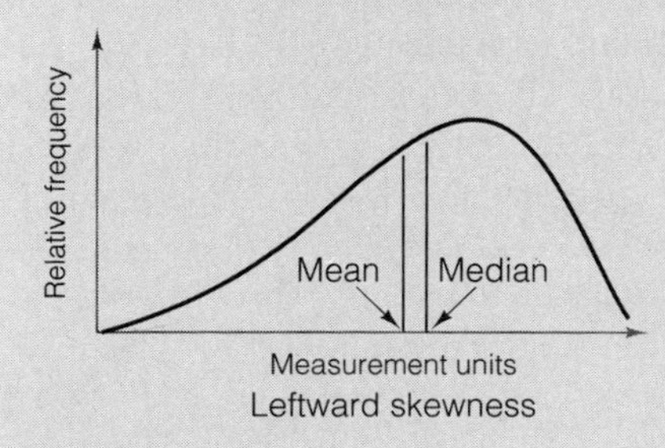

Leftward skewness

A third measure of central tendency is the **mode** of a set of measurements.

Definition 2.7

The **mode** is the measurement that occurs most frequently in the data set.

Therefore, the mode shows where the data tend to concentrate.

EXAMPLE 2.5 Calculate the mode for the following ten quiz grades:

8, 7, 9, 6, 8, 10, 9, 9, 5, 7

Solution Since 9 occurs most often, the mode is 9.

The mode is of primary value in describing large data sets. When a large data set has been described using a relative frequency histogram, the mode will be located in the class containing the largest relative frequency. This is called the **modal class.** Several definitions exist for locating the position of the mode within a modal class, but the simplest is to define the mode as the midpoint of the modal class. For example, examine the relative frequency histogram for the EPA mileage readings (Figure 2.1). You can see that the modal class is the interval 35.95–37.45. The mode (the midpoint) is 36.7. Note that this measure of central tendency is very close to the mean, 36.99, and the median, 37.0.

Because it emphasizes data concentration, the mode is often used with large data sets to locate the region in which much of the data is concentrated. A retailer of men's clothing would be interested in the modal neck size and sleeve length of potential customers. A supermarket manager is interested in the cereal brand with the largest share of the market, i.e., the modal brand. The modal income class of the American worker is of interest to the Labor Department. Thus, the mode provides a useful measure of central tendency for various applications.

Note that the cereal brand example involves a nominal scale measurement. The mode can be used to describe the central tendency of all four types of measurements (nominal, ordinal, interval, and ratio), while the mean and median are primarily useful for interval and ratio data.

Exercises 2.21–2.38

Learning the Mechanics

2.21 Calculate the mode, mean, and median of the following data:

18 10 15 13 17 15 12 15 18 16 11

2.22 Calculate the mean and median of the following grade-point averages:

3.2 2.5 2.1 3.7 2.8 2.0

2.23 Explain the difference between the calculation of the median for an odd and an even number of measurements. Construct one data set consisting of five measurements and another consisting of six measurements for which the medians are equal.

2.24 Explain how the relationship between the mean and median provides information about the symmetry or skewness of the data's distribution.

2.25 Calculate the mean for samples where:
a. $n = 10,\ \Sigma x = 85$ **b.** $n = 16,\ \Sigma x = 400$ **c.** $n = 45,\ \Sigma x = 35$
d. $n = 18,\ \Sigma x = 242$

2.26 Calculate the mean, median, and mode for each of the following samples:
a. 7, −2, 3, 3, 0, 4
b. 2, 3, 5, 3, 2, 3, 4, 3, 5, 1, 2, 3, 4
c. 51, 50, 47, 50, 48, 41, 59, 68, 45, 37

2.27 Describe how the mean compares to the median for a distribution as follows:
a. Skewed to the left **b.** Skewed to the right **c.** Symmetric

Applying the Concepts

2.28 Demographics play a key role in the recreation industry. A recent article indicates that difficult times may lay ahead for the industry (Murdock et al., 1991). The article reports the median age of the population in the United States was 30 in 1980, but will be about 36 by the year 2000.
a. Interpret the value of the median for both 1980 and 2000, and explain the trend.
b. If the recreation industry relies on the 18–30 age group for much of its business, explain what effect this shift in the median age will have on the recreation industry.

2.29 Each year advertisers spend billions of dollars purchasing commercial time on network sports television. In the first 6 months of 1988, advertisers spent $1.1 billion. Who were the largest spenders? In a recent article, Real and Mechikoff (1992) list the top 10 leading spenders (in millions of dollars):

Chrysler	$72.0	AT&T	$26.9
General Motors	63.1	Sears	25.0
Philip Morris	54.7	U.S. Armed Forces	23.9
Anheuser-Busch	54.3	McDonald's	23.0
Ford	29.0	American Express	20.0

a. Calculate the mean and median amounts spent by the top ten.
b. What does the relationship between the mean and median imply about the symmetry or skewness of these 10 measurements?
c. What percentage of the measurements exceed the mean? What percentage exceed the median? Is either of these percentages always the same no matter what the data? Explain.

2.30 A psychologist has developed a new technique intended to improve rote memory. To test the method against other standard methods, 20 high school students are selected at random, and each is taught the new technique. The students are then asked to memorize a list of 100 word phrases using the technique. The following are the number of word phrases memorized correctly by the students:

91 64 98 66 83 87 83 86 80 93
83 75 72 79 90 80 90 71 84 68

a. Define the terms *mean*, *median*, and *mode* in the context of this problem.
b. Construct a relative frequency histogram for the data.
c. Compute the mean, median, and mode for the data set and locate them on the histogram. Do these measures of central tendency appear to locate the center of the distribution of data?

2.31 Would you expect the data sets described below to possess relative frequency distributions that are symmetric, skewed to the right, or skewed to the left? Explain.

a. The salaries of all persons employed by a large university
b. The grades on an easy test
c. The grades on a difficult test
d. The amounts of time students in your class studied last week
e. The ages of automobiles on a used car lot
f. The amounts of time spent by students on a difficult examination (maximum time is 50 minutes)

2.32 The scores for a statistics test are as follows:

87 76 96 77 94 92 88 85 66 89
79 95 50 91 83 88 82 58 18 69

a. Compute the mean, median, and mode for these data.
b. Which of the three measures of central tendency do you think would best represent the achievement of the class?
c. Eliminate the two lowest scores, and again compute the mean, median, and mode. Which measure of central tendency is most affected by extremely low scores?

2.33 Ten presumably trained rats were released in a maze. Their times to escape (in seconds) are recorded below. The N's represent two rats that had still not escaped by the end of the experiment.

100 38 N 122 95 116 56 135 104 N

a. Can you calculate the mean for these data? Explain.
b. Is the median a meaningful measure of central tendency for these data? Explain. Calculate the median.

2.34 The salaries of superstar professional athletes receive much attention in the media. The million-dollar annual contract is becoming more commonplace among this elite group with each passing year. Nevertheless, rarely does a year pass without one or more of the players' associations negotiating with team owners for additional salary and fringe benefit considerations for *all* players in their particular sports.

a. If a players' association wanted to support its argument for higher "average" salaries, which measure of central tendency do you think it should use? Why?
b. To refute the argument, which measure of central tendency should the owners apply to the players' salaries? Why?

2.35 In 1990, U.S. consumers redeemed 6.49 billion manufacturers' coupons and saved themselves $2.24 billion. Find the mean amount saved per coupon.

2.36 The table contains the price per acre of farmland for a sample of states that includes nine eastern states and 11 western states (i.e., west of the Mississippi River).

State	*Price per Acre (1984)*	*State*	*Price per Acre (1984)*
Ariz.	$ 265	Nebr.	$ 444
Calif.	1,726	Nev.	229
Colo.	435	N.H.	1,419
Conn.	3,208	N.J.	3,525
Del.	1,642	N.Mex.	163
Fla.	1,527	N.Dak.	360
Kansas	466	Pa.	1,510
Mass.	2,372	R.I.	3,335
Md.	2,097	S.Dak.	250
Mont.	222	Wyo.	177

Data: U.S. Department of Agriculture.

a. Find the mean and median price per acre for the sample of 20 states.
b. Find and compare the mean price per acre for the eastern states and the western states. Also, compare these means to the mean you found in part a. What do your comparisons reveal about the value of farmland in the United States?
c. As a measure of central tendency, the mean of a data set is frequently used to characterize the data set or to represent a typical measurement in the data set. Examine the data set and determine whether you would use the mean of these data to characterize a typical measurement. Explain.

2.37 Refer to Exercises 2.16–2.18, in which we used stem-and-leaf displays and histograms to compare the average SAT scores in 1975 to those in 1990. We now use SPSS to obtain the mean and median for the two sets of average SAT scores:

```
TOTSAT75

Mean           955.300      Median        952.000

Valid cases         50      Missing cases       0

-----------------------------------------------

TOTSAT90

Mean           947.460      Median        937.50

Valid cases         50      Missing cases       0
```

a. Use the mean and median to compare the central tendencies of the two data sets.
b. What does the relationship of the mean and median imply about the symmetry or skewness of the data sets? Look at the stem-and-leaf displays and histograms in Exercises 2.16 and 2.18 to assist in your evaluation.

2.38 Refer to Exercise 2.19 (page 38). We used Minitab to compute the mean and median for the dropout rates in 1982 and 1984. Part of the Minitab output is shown.

	DROP1982	DROP1984
N	51	51
MEAN	28.12	26.50
MEDIAN	27.60	25.30

R R

a. Interpret these measures of central tendency.

b. What, if anything, do the relationships between the mean and median tell you about the skewness in these two data sets? Is your answer supported by your stem-and-leaf displays (Exercise 2.19) and histograms (Exercise 2.20)?

c. In which direction does the distribution appear to have shifted from 1982 to 1984? Do you think the shift is indicative of some trend due to changes in policies and/or attitudes, or is the shift possibly nothing more than a result of random variation in the rates over time? [*Note:* We learn in Chapter 8 how to distinguish between sample differences caused by true population differences and those caused by random sampling variation.]

2.4 Numerical Measures of Variability

Measures of central tendency provide only a partial description of a quantitative data set. The description is incomplete without a measure of the variability, or spread, of the data set.

If you examine the two histograms in Figure 2.8, you will notice that both hypothetical data sets are symmetric with equal modes, medians, and means.

FIGURE 2.8 ▶ Hypothetical data sets

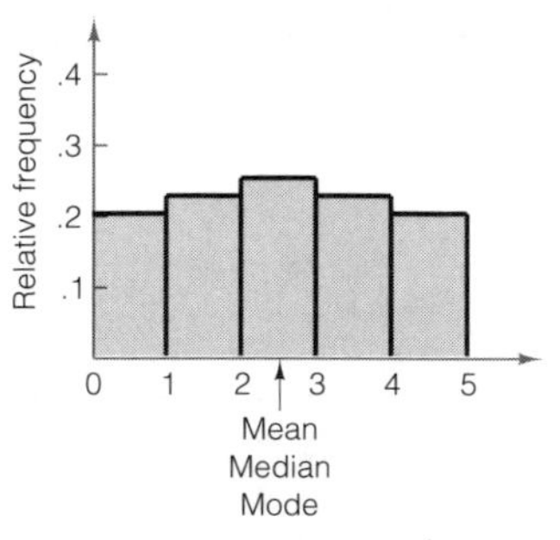

a. Data set 1

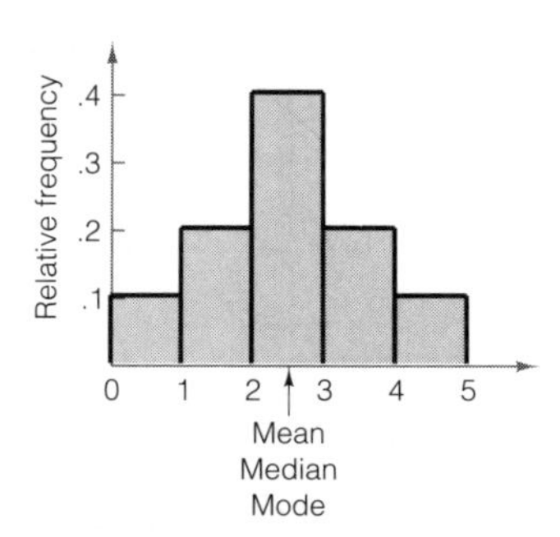

b. Data set 2

However, data set 1 in Figure 2.8(a) has measurements spread with almost equal relative frequency over the measurement classes, while data set 2 in Figure 2.8(b) has most of its measurements clustered about its center. Thus, data set 2 is *less variable* than data set 1. Consequently, you can see that we need a measure of variability as well as a measure of central tendency to describe a data set.

Perhaps the simplest measure of the variability of a quantitative data set is its **range**.

Definition 2.8

The **range** of a data set is equal to the largest measurement minus the smallest measurement.

The range is easy to compute and easy to understand, but it is a rather insensitive measure of data variation when the data sets are large. This is because two data sets can have the same range and be vastly different with respect to data variation. This phenomenon is demonstrated in Figure 2.8. Both distributions of data shown in the figure have the same range, but most of the measurements in data set 2 tend to concentrate near the center of the distribution. Consequently, the data are much less variable than the data in set 1. Thus, you can see that the range does not always detect differences in data variation for large data sets.

Let us see if we can find a measure of data variation that is more sensitive than the range. Consider the two samples in Table 2.4: each has five measurements. (We have ordered the numbers for convenience.) Note that both samples have a mean of 3 and that we have also calculated the distance between each measurement and the mean. What information do these distances contain? If they tend to be large in magnitude, as in sample 1, the data are spread out, or highly variable. If the distances are mostly small, as in sample 2, the data are clustered around the mean, $\bar{x}$, and therefore do not exhibit much variability. You can see that these distances, displayed graphically in Figure 2.9, provide information about the variability of the sample measurements.

TABLE 2.4 Two Hypothetical Data Sets

	Sample 1	Sample 2
Measurements	1, 2, 3, 4, 5	2, 3, 3, 3, 4
Mean	$\bar{x} = \frac{1+2+3+4+5}{5} = \frac{15}{5} = 3$	$\bar{x} = \frac{2+3+3+3+4}{5} = \frac{15}{5} = 3$
Distances of measurement values from $\bar{x}$	(1 − 3), (2 − 3), (3 − 3), (4 − 3), (5 − 3) or −2, −1, 0, 1, 2	(2 − 3), (3 − 3), (3 − 3), (3 − 3), (4 − 3) or −1, 0, 0, 0, 1

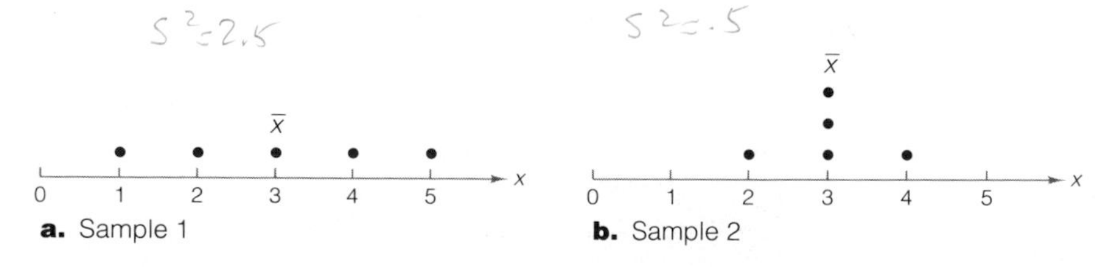

FIGURE 2.9 ▶ Dot diagrams for two data sets

The next step is to condense the information in these distances into a single numerical measure of variability. Averaging the distances from $\bar{x}$ will not help because the negative and positive distances cancel; i.e., the sum of the deviations (and thus the average deviation) is always equal to zero.

Two methods come to mind for dealing with the fact that positive and negative distances from the mean cancel. The first is to treat all the distances as though they were positive, ignoring the sign of the negative distances. We do not pursue this line of thought because the resulting measure of variability (the mean of the absolute values of the distances) presents analytical difficulties beyond the scope of this text. A second method of eliminating the minus signs associated with the distances is to square them. The quantity we can calculate from the squared distances will provide a meaningful description of the variability of a data set and presents fewer analytical difficulties in inference-making.

To use the squared distances calculated from a data set, we first calculate the **sample variance**.

Definition 2.9

The **sample variance** for a sample of n measurements is equal to the sum of the squared distances from the mean divided by $(n - 1)$. In symbols, using s^2 to represent the sample variance,

$$s^2 = \frac{\sum (x - \bar{x})^2}{n - 1}$$

Referring to the two samples in Table 2.4, you can calculate the variance for sample 1 as follows:

$$s^2 = \frac{(1 - 3)^2 + (2 - 3)^2 + (3 - 3)^2 + (4 - 3)^2 + (5 - 3)^2}{5 - 1}$$

$$= \frac{4 + 1 + 0 + 1 + 4}{4} = 2.5$$

The second step in finding a meaningful measure of data variability is to calculate the **standard deviation** of the data set:

Definition 2.10

The **sample standard deviation**, s, is defined as the positive square root of the sample variance, s^2. Thus,

$$s = \sqrt{s^2} = \sqrt{\frac{\sum (x - \bar{x})^2}{n - 1}}$$

The population variance, denoted by the symbol σ^2 (sigma squared), is the average of the squared distances of the measurements on *all* units in the population from the mean, μ, and σ (sigma) is the square root of this quantity. Since we rarely know the numerical values of the population variance or standard deviation, we generally denote these two quantities by their respective symbols, σ^2 and σ.

formulas

s^2 = Sample variance	s = Sample standard deviation
σ^2 = Population variance	σ = Population standard deviation

Notice that in contrast to the variance, the standard deviation is expressed in the original units of measurement. For example, if the original measurements are in dollars, the variance is expressed in the peculiar units "dollars squared," but the standard deviation is expressed in dollars.

You may wonder why we use the divisor $(n - 1)$ instead of n when calculating the sample variance. Although the use of n may seem logical, since then the sample variance would be the average squared distance from the mean, the use of n tends to produce an underestimate of the population variance, σ^2. The use of $(n - 1)$ in the denominator provides the appropriate correction for this tendency.* Since the primary use of sample statistics like s^2 is to estimate population parameters like σ^2, $(n - 1)$ is preferred to n when defining the sample variance.

EXAMPLE 2.6 Calculate the standard deviation of the following sample: 2, 3, 3, 3, 4.

$3 = \bar{x}$ $\frac{(2-3)^2}{4}$

Solution For this data set, $\bar{x} = 3$. Then

$$s = \sqrt{\frac{(2-3)^2 + (3-3)^2 + (3-3)^2 + (3-3)^2 + (4-3)^2}{5-1}}$$

$$= \sqrt{\frac{2}{4}} = \sqrt{.5} = .71$$

As the number of measurements increases, calculating s^2 and s becomes very tedious. Fortunately, as we show in Example 2.9, we can get around the difficulty of calculating s^2 and s by using a statistical software package (or most calculators). However, if you must calculate it by hand, there is a shortcut formula for computing the sample variance.

*Appropriate here means that s^2 with a divisor of $(n - 1)$ is an **unbiased estimator of σ^2**. We define and discuss unbiasedness of estimators in Chapter 5.

Shortcut Formula for Sample Variance

$$s^2 = \frac{\begin{pmatrix}\text{Sum of squares of} \\ \text{sample measurements}\end{pmatrix} - \dfrac{(\text{Sum of sample measurements})^2}{n}}{n-1}$$

$$= \frac{\sum x^2 - \dfrac{\left(\sum x\right)^2}{n}}{n-1}$$

EXAMPLE 2.7

Use the shortcut formula to calculate the standard deviation for the sample used in Example 2.6: 2, 3, 3, 3, 4.

Solution

We need two summations to use the shortcut formula: $\Sigma\, x$ and $\Sigma\, x^2$. These can easily be obtained from the following type of tabulation:

x	x^2
2	4
3	9
3	9
3	9
4	16
$\Sigma\, x = 15$	$\Sigma\, x^2 = 47$

Then we use

$$s^2 = \frac{\sum x^2 - \dfrac{\left(\sum x\right)^2}{n}}{n-1} = \frac{47 - \dfrac{(15)^2}{5}}{5-1} = \frac{2}{4} = .5$$

$$s = \sqrt{.5} = .71$$

Note that the result is identical to that obtained in Example 2.6.

EXAMPLE 2.8

Calculate the sample variance s^2 and the sample standard deviation s for the 100 gas mileage readings given in Table 2.1.

Solution

Recall that $\bar{x} = 36.99$. Clearly it would be a formidable task to calculate $(x - \bar{x})^2$ for all 100 readings. Likewise, a tabulation such as that used in Example 2.7 would be tedious. However, we can use the shortcut formula to follow a step-by-step procedure using a calculator.

Step 1 Calculate the sum of the squared measurements. With the aid of a calculator, we obtain

$$\sum x^2 = 137{,}434.38$$

Step 2 Calculate the square of the sum of the measurements.

$$\left(\sum x\right)^2 = (3{,}699.4)^2 = 13{,}685{,}560.36$$

Step 3 Find $\Sigma (x - \bar{x})^2 = \Sigma x^2 - [(\Sigma x)^2/n]$, the numerator of s^2 in the shortcut formula.

$$\sum (x - \bar{x})^2 = \sum x^2 - \frac{\left(\sum x\right)^2}{n}$$

$$= 137{,}434.38 - \frac{13{,}685{,}560.36}{100} = 578.7764$$

We can now calculate s^2 by dividing this quantity by $(n - 1)$:

$$s^2 = \frac{578.7764}{99} = 5.85$$

$$s = \sqrt{5.85} = 2.42$$

Note that the shortcut formula requires only the sum of the sample measurements, Σx, and the sum of the squares of the sample measurements, Σx^2. Be careful when you calculate these two sums. Rounding the values of x^2 that appear in Σx^2 or rounding the quantity $(\Sigma x)^2/n$ can lead to substantial errors in the calculation of s^2.

EXAMPLE 2.9

Use a statistical software package to compute the mean, median, variance, and standard deviation of the gas mileage data in Table 2.1.

Solution The SPSS/PC+ printout for the median, mean, variance, and standard deviation of the gas mileage data is shown in Figure 2.10.

FIGURE 2.10 ► SPSS/PC+ printout of descriptive statistics

Mean	**36.994**	Std Err	.242	**Median**	**37.000**
Mode	37.000	**Std Dev**	**2.418**	**Variance**	**5.846**
Kurtosis	.770	S E Kurt	.478	Skewness	.051
S E Skew	.241	Range	14.900	Minimum	30.000
Maximum	44.900	Sum	3699.400		
Valid Cases	100	Missing Cases	0		

Note that the answers are identical to those obtained previously, with the exception that different numbers of decimal places are carried in some instances.

One question occurs over and over again about the calculation of s^2. How many decimal places should you carry? The answer has *nothing* to do with the number of decimal places retained in the sample measurements. If the sample measurements are rounded to the nearest hundredth, tenth, or whatever, rounding simply adds to the variability of the measurements. This added variation is reflected in the value of s^2, a number calculated precisely according to Definition 2.9.

Theoretically, s^2 is a number regardless of how many decimal places it takes to specify its value. In practice, we round the calculated value. There are no rules for the rounding procedure but, keeping in mind that we wish to calculate the standard deviation s, it is reasonable to retain twice as many decimal places in s^2 as you wish to have in s. If you wish to calculate s to the nearest hundredth (two decimal places), for example, you should calculate s^2 to the nearest ten-thousandth (four decimal places).

Generally, in our examples and exercises we round final answers so that one or two more significant digits are displayed in the answers than in the original measurements. For example, we displayed the gas mileage answers in Examples 2.3 and 2.8 to two decimals, even though the data were given to the nearest decimal.

You now know that the standard deviation measures the variability of a set of data and how to calculate it. But how can we interpret and use the standard deviation? This is the topic of Section 2.5.

Exercises 2.39–2.53

Learning the Mechanics

2.39 What is the primary disadvantage of using the range to compare the variability of data sets?

2.40 Describe the sample variance using words rather than a formula. Do the same with the population variance.

2.41 Can the variance of a data set ever be negative? Explain. Can the variance ever be smaller than the standard deviation? Explain.

2.42 Calculate the range, variance, and standard deviation for the following samples:

a. 4, 2, 1, 0, 1
b. 1, 6, 2, 2, 3, 0, 3
c. 8, −2, 1, 3, 5, 4, 4, 1, 3, 3
d. 0, 2, 0, 0, −1, 1, −2, 1, 0, −1, 1, −1, 0, −3, −2, −1, 0, 1

2.43 Calculate the variance and standard deviation for samples where:

a. $n = 10, \quad \Sigma x^2 = 84, \quad \Sigma x = 20$ **b.** $n = 40, \quad \Sigma x^2 = 380, \quad \Sigma x = 100$
c. $n = 20, \quad \Sigma x^2 = 18, \quad \Sigma x = 17$

2.44 Calculate the range, variance, and standard deviation for the following samples:

a. 39, 42, 40, 37, 41 **b.** 100, 4, 7, 96, 80, 3, 1, 10, 2 **c.** 100, 4, 7, 30, 80, 30, 42, 2

2.45 Given the following information about two data sets, compute $\bar{x}$, the sample variance, and the standard deviation for each:

a. $n = 25, \quad \sum_{i=1}^{n} x_i^2 = 1{,}000, \quad \sum_{i=1}^{n} x_i = 50$

b. $n = 80, \quad \sum_{i=1}^{n} x_i^2 = 270, \quad \sum_{i=1}^{n} x_i = 100$

2.46 Compute $\bar{x}$, s^2, and s for each of the following data sets. If appropriate, specify the units in which your answer is expressed.

a. 3, 1, 10, 10, 4
b. 8 feet, 10 feet, 32 feet, 5 feet
c. −1, −4, −3, 1, −4, −4
d. ⅕ ounce, ⅕ ounce, ⅕ ounce, ⅖ ounce, ⅕ ounce, ⅘ ounce

2.47 Using only integers between 0 and 10, construct two data sets with at least 10 observations each that have the same mean but different variances. Construct dot diagrams for each of your data sets (see Figure 2.9), and mark the mean of each data set on its dot diagram.

2.48 Using only integers between 0 and 10, construct two data sets with at least 10 observations each that have the same range but different means. Construct a dot diagram for each of your data sets (see Figure 2.9), and mark the mean of each data set on its dot diagram.

2.49 Consider the following sample of five measurements: 2, 1, 1, 0, 3

a. Calculate the range, s^2, and s.
b. Add 3 to each measurement and repeat part **a**.
c. Subtract 4 from each measurement and repeat part **a**.
d. Considering your answers to parts **a**, **b**, and **c**, what seems to be the effect on the variability of a data set by adding the same number to or subtracting the same number from each measurement?

Applying the Concepts

2.50 Improving public education was an important plank in President Clinton's campaign platform of 1992. In an article in *U.S. News & World Report*, Gergen (1990) reported on many factors that have led to the breakdown of public education. One study mentioned in the article found that over 90% of the nation's school districts reported that their students were scoring above the national average on standardized tests. Using your knowledge of measures of central tendency, explain why the schools' reports are incorrect. Does whether the term "average" refers to the mean or the median make any difference to your analysis? Nationally, what effect would this misinformation have on the perception of the nation's schools?

2.51 The final grades given by two professors in introductory statistics courses have been carefully examined. The students in the first professor's class had a grade-point average of 3.0 and a standard deviation of .2. Those in the second professor's class had grade points with an average of 3.0 and a standard deviation of 1.0. If you had a choice, which professor would you take for this course? Explain.

2.52 Consider the following two samples:

Sample 1: 10, 0, 1, 9, 10, 0, 8, 1, 1, 9

Sample 2: 0, 5, 10, 5, 5, 5, 6, 5, 6, 5

a. Examine both samples and identify the one that you believe has the greater variability.
b. Calculate the range for each sample. Does the result agree with your answer to part **a**? Explain.
c. Calculate the standard deviation for each sample. Does the result agree with your answer to part **a**? Explain.
d. Which of the two, the range or the standard deviation, provides a better measure of variability? Why?

2.53 The U.S. Federal Trade Commission has recently begun assessing fines and other penalties against weight-loss clinics that make insupportable or misleading claims about the effectiveness of their programs. Suppose that you have brochures from two weight-loss clinics that both advertise "statistical evidence" about the effectiveness of their programs. Clinic A advertises that the **mean** weight loss during the first month is 15 pounds, while Clinic B advertises a **median** weight loss of 10 pounds.
a. Assuming the statistics are accurately calculated and that you had no other information, which clinic would you recommend? Why?
b. Upon further research, the median and standard deviation for Clinic A are found to be 10 pounds and 20 pounds, respectively, while the mean and standard deviation for Clinic B are found to be 10 and 5 pounds, respectively. Both are based on samples of more than 100 clients. Describe the two clinics' weight-loss distributions as completely as possible given this additional information. What would you recommend to a prospective client now? Why?
c. Note that nothing has been said about how the sample of clients upon which the statistics are based was selected. What additional information would be important regarding the sampling techniques employed by the clinics?

2.5 Interpreting the Standard Deviation

As we have seen, if we are comparing the variability of two samples selected from a population, the sample with the larger standard deviation is the more variable of the two. Thus, we know how to interpret the standard deviation on a relative or comparative basis, but we have not explained how it provides a measure of variability for a single sample.

To understand how the standard deviation provides a measure of variability of a data set, consider a specific data set and answer the following questions: How many measurements are within 1 standard deviation of the mean? How many measurements are within 2 standard deviations? For example, consider the 100 mileage per gallon readings given in Table 2.1.

Recall that $\bar{x} = 36.99$ and $s = 2.42$. Then

$$\bar{x} - s = 34.57 \qquad \bar{x} - 2s = 32.15$$
$$\bar{x} + s = 39.41 \qquad \bar{x} + 2s = 41.83$$

If we examine the data, we find that 68 of the 100 measurements, or 68% of the measurements, are in the interval

$$\bar{x} - s \quad \text{to} \quad \bar{x} + s$$

Similarly, we find that 96, or 96%, of the 100 measurements are in the interval

$$\bar{x} - 2s \quad \text{to} \quad \bar{x} + 2s$$

These intervals are usually written $(\bar{x} - s, \bar{x} + s)$ and $(\bar{x} - 2s, \bar{x} + 2s)$.

These observations identify criteria for interpreting a standard deviation that apply to any set of data, whether a population or a sample. The criteria, expressed as a mathematical theorem and as a rule of thumb, are presented in Tables 2.5 and 2.6 on page 58. In these tables we give two sets of answers to the questions of how many measurements fall within 1, 2, and 3 standard deviations of the mean. The first, which applies to *any* set of data, is derived from a theorem proved by the Russian mathematician Chebyshev. The second, which applies only to mound-shaped distributions of data, is based upon empirical evidence that has accumulated over the years. The frequency histogram of a mound-shaped sample is approximately symmetric, with a clustering of measurements about the midpoint of the distribution (the mean, median, and mode should all be about the same), tailing off rapidly as we move away from the center of the histogram. Thus, the histogram will have the appearance of a mound or bell, as shown in Figure 2.11. The percentages given for the intervals in Table 2.6 provide remarkably good approximations even when the distribution of the data is slightly skewed or asymmetric.

FIGURE 2.11 ▶
Histogram of a mound-shaped sample

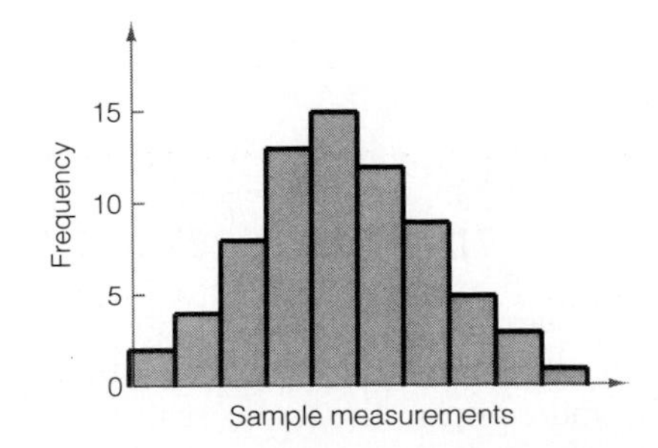

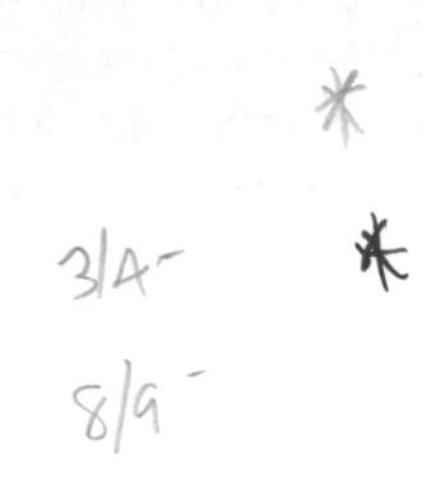

TABLE 2.5 An Aid to Interpretation of a Standard Deviation: Chebyshev's Rule

Chebyshev's Rule

Chebyshev's rule applies to any sample of measurements, regardless of the shape of the frequency distribution:

a. It is possible that very few of the measurements will fall within 1 standard deviation of the mean $(\bar{x} - s, \bar{x} + s)$.
b. At least ¾ of the measurements will fall within 2 standard deviations of the mean $(\bar{x} - 2s, \bar{x} + 2s)$.
c. At least 8⁄9 of the measurements will fall within 3 standard deviations of the mean $(\bar{x} - 3s, \bar{x} + 3s)$.
d. Generally, at least $1 - 1/k^2$ of the measurements will fall within k standard deviations of the mean $(\bar{x} - ks, \bar{x} + ks)$ for any number k greater than 1.

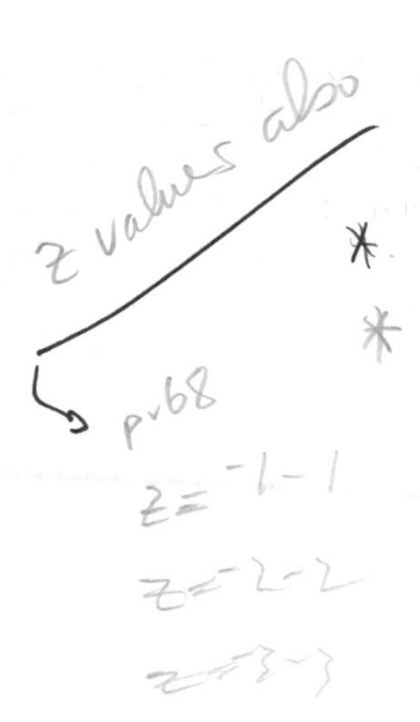

TABLE 2.6 An Aid to Interpretation of a Standard Deviation: The Empirical Rule

The Empirical Rule

The Empirical Rule is a rule of thumb that applies to samples with frequency distributions that are mound-shaped:

a. Approximately 68% of the measurements will fall within 1 standard deviation of the mean $(\bar{x} - s, \bar{x} + s)$.
b. Approximately 95% of the measurements will fall within 2 standard deviations of the mean $(\bar{x} - 2s, \bar{x} + 2s)$.
c. Essentially all the measurements will fall within 3 standard deviations of the mean $(\bar{x} - 3s, \bar{x} + 3s)$.

EXAMPLE 2.10

Thirty students in an experimental psychology class use various techniques to train a rat to move through a maze. At the end of the course, each student's rat is timed through the maze, with the following results (in minutes):

1.97	.60	4.02	3.20	1.15	6.06	4.44	2.02	3.37	3.65
1.74	2.75	3.81	9.70	8.29	5.63	5.21	4.55	7.60	3.16
3.77	5.36	1.06	1.71	2.47	4.25	1.93	5.15	2.06	1.65

The mean and standard deviation of these data are 3.74 minutes and 2.20 minutes, respectively. Calculate the fraction of the 30 measurements in the intervals $\bar{x} \pm s$, $\bar{x} \pm 2s$, and $\bar{x} \pm 3s$, and compare the results with those in Tables 2.5 and 2.6.

Solution

We first form the interval

$$(\bar{x} - s, \bar{x} + s) = (3.74 - 2.20, 3.74 + 2.20) = (1.54, 5.94)$$

A check of the measurements shows that 23 of the times are within this 1 standard deviation interval around the mean. This number represents $23/30 \approx 77\%$ of the sample measurements.

The next interval of interest is

$$(\bar{x} - 2s, \bar{x} + 2s) = (3.74 - 4.40, 3.74 + 4.40) = (-.66, 8.14)$$

All but two of the times are within this interval, so $28/30$, or approximately 93%, are within 2 standard deviations of $\bar{x}$.

Finally, the 3 standard deviation interval around $\bar{x}$ is

$$(\bar{x} - 3s, \bar{x} + 3s) = (3.74 - 6.60, 3.74 + 6.60) = (-2.86, 10.34)$$

All of the times fall within 3 standard deviations of the mean.

These 1, 2, and 3 standard deviation percentages (77, 93, and 100) agree fairly well with the approximations of 68%, 95%, and 100% given by the Empirical Rule (Table 2.6) for mound-shaped distributions. If you look at the frequency histogram for this data set in Figure 2.12, you will note that the distribution is not really mound-shaped, nor is it extremely skewed. Thus, we get reasonably good results from the mound-shaped approximations. Of course, we know from Chebyshev's rule (Table 2.5) that no matter what the shape of the distribution, we would expect at least 75% and 89% of the measurements to lie within 2 and 3 standard deviations of $\bar{x}$, respectively.

FIGURE 2.12 ▶
Histogram for times for rats to move through maze

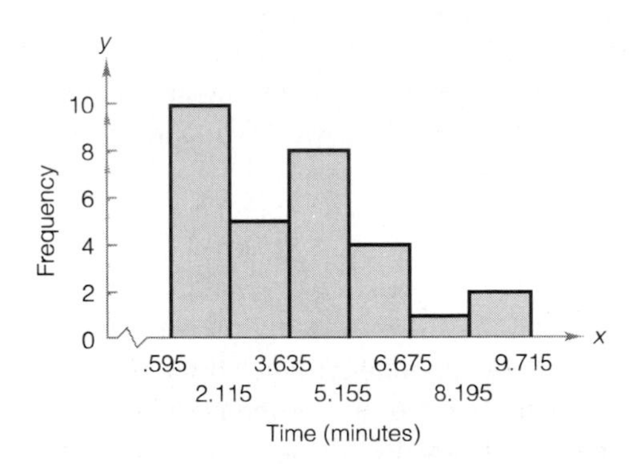

EXAMPLE 2.11

Chebyshev's rule and the Empirical Rule (Tables 2.5 and 2.6) are useful as a check on the calculation of the standard deviation. For example, suppose we calculated the standard deviation for the gas mileage data (Table 2.1) to be 5.85. Are there any "clues" in the data that enable us to judge whether this number is reasonable?

Solution

The range of the mileage data is $44.9 - 30.0 = 14.9$. From Chebyshev's rule and the Empirical Rule we know that most of the measurements (approximately 95% if the distribution is mound-shaped) will be within 2 standard deviations of the mean. And, regardless of the shape of the distribution and the number of measurements, almost all of them will fall within 3 standard deviations of the mean. Consequently, we would expect the range of the measurements to be between 4 (i.e., $\pm 2s$) and 6 (i.e., $\pm 3s$) standard deviations in length (see Figure 2.13). For the car mileage data, this means that s should fall between

$$\frac{\text{Range}}{6} = \frac{14.9}{6} = 2.48 \quad \text{and} \quad \frac{\text{Range}}{4} = \frac{14.9}{4} = 3.73$$

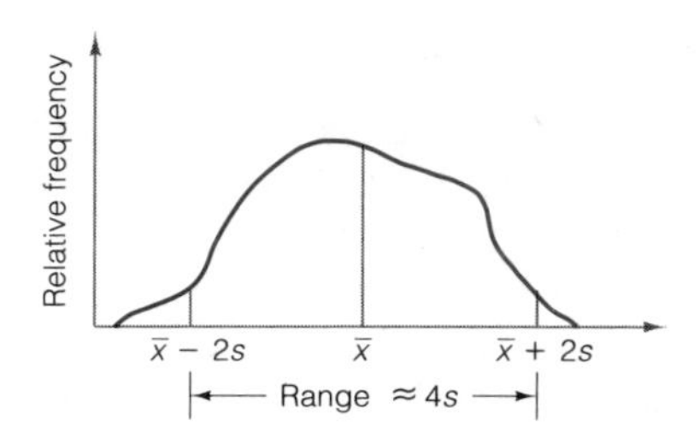

FIGURE 2.13 ▶ The relation between the range and the standard deviation

In particular, the standard deviation should not be much larger than one-fourth of the range, particularly for a data set with 100 measurements. Thus, we have reason to believe that the calculation of 5.85 is too large. A check of our work reveals that 5.85 is the variance s^2, not the standard deviation s (see Example 2.8). We "forgot" to take the square root (a common error); the correct value is $s = 2.42$. Note that this value is slightly smaller than the range divided by 6 (2.48). The larger the data set, the greater the tendency for very large or very small measurements (extreme values) to appear, and when they do, the range may exceed 6 standard deviations.

In examples and exercises we will sometimes use $s \approx$ range/4 to obtain a crude, and usually conservatively large, approximation for s. However, we stress that this is no substitute for calculating the exact value of s when possible.

Finally, and most importantly, we will use the concepts in Chebyshev's rule and the Empirical Rule to build the foundation for statistical inference-making. The method is illustrated in Example 2.12.

EXAMPLE 2.12

A manufacturer of automobile batteries claims that the average length of life for its grade A battery is 60 months. However, the guarantee on this brand is for only 36 months. Suppose the standard deviation of the life length is known to be 10 months, and the frequency distribution of the life-length data is known to be mound-shaped.

a. Approximately what percentage of the manufacturer's grade A batteries will last more than 50 months, assuming the manufacturer's claim is true?
b. Approximately what percentage of the manufacturer's batteries will last less than 40 months, assuming the manufacturer's claim is true?
c. Suppose your battery lasts 37 months. What could you infer about the manufacturer's claim?

Solution

If the distribution of life length is assumed to be mound-shaped with a mean of 60 months and a standard deviation of 10 months, it would appear as shown in Figure 2.14. Note that we can take advantage of the fact that mound-shaped distributions are

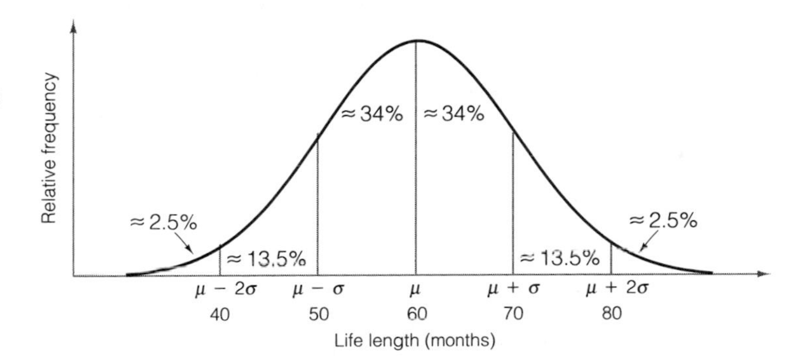

FIGURE 2.14 ▶
Battery life-length distribution: Manufacturer's claim assumed true

(approximately) symmetric about the mean, so that the percentages given by the Empirical Rule can be split equally between the halves of the distribution on each side of the mean. The approximations given in Figure 2.14 are more dependent on the assumption of a mound-shaped distribution than those given by the Empirical Rule (Table 2.6), because the approximations in Figure 2.14 depend on the (approximate) symmetry of the mound-shaped distribution. We saw in Example 2.10 that the Empirical Rule can yield good approximations even for skewed distributions. This will *not* be true of the approximations in Figure 2.14; the distribution *must* be mound-shaped and (approximately) symmetric.

For example, since approximately 68% of the measurements will fall within 1 standard deviation of the mean, the distribution's symmetry implies that approximately ½(68%) = 34% of the measurements will fall between the mean and 1 standard deviation on each side. This concept is illustrated in Figure 2.14. The figure also shows that 2.5% of the measurements lie beyond 2 standard deviations in each direction from the mean. This result follows from the fact that if approximately 95% of the measurements fall within 2 standard deviations of the mean, then about 5% fall outside 2 standard deviations; if the distribution is approximately symmetric, then about 2.5% of the measurements fall beyond 2 standard deviations on each side of the mean.

a. It is easy to see in Figure 2.14 that the percentage of batteries lasting more than 50 months is approximately 34% (between 50 and 60 months) plus 50% (greater than

60 months). Thus, approximately 84% of the batteries should have life length exceeding 50 months.

b. The percentage of batteries that last less than 40 months can also be easily determined from Figure 2.14. Approximately 2.5% of the batteries should fail prior to 40 months, assuming the manufacturer's claim is true.

c. If you are so unfortunate that your grade A battery fails at 37 months, one of two inferences can be made: Either your battery was one of the approximately 2.5% that fail prior to 40 months, or something about the manufacturer's claim is not true. Because the chances are so small that a battery fails before 40 months, you would have good reason to have serious doubts about the manufacturer's claim. A mean smaller than 60 months and/or a standard deviation longer than 10 months would both increase the likelihood of failure prior to 40 months.*

Example 2.12 is our initial demonstration of the statistical inference-making process. At this point you should realize that we will use sample information (in Example 2.12, your battery's failure at 37 months) to make inferences about the population (in Example 2.12, the manufacturer's claims about the life length for the population of all batteries). We will build on this foundation as we proceed.

Exercises 2.54–2.68

Learning the Mechanics

2.54 To what kind of data sets can Chebyshev's rule be applied? The Empirical Rule?

2.55 The output from a statistical computer program indicates that the mean and standard deviation of a data set consisting of 200 measurements are \$1,500 and \$300, respectively.

a. What are the units of measurement of the variable of interest? Based on the units, what type of data is this: nominal, ordinal, interval, or ratio?

b. What can be said about the number of measurements between \$900 and \$2,100? Between \$600 and \$2,400? Between \$1,200 and \$1,800? Between \$1,500 and \$2,100?

2.56 For any set of data, what can be said about the percentage of the measurements contained in each of the following intervals?

a. $\bar{x} - s$ to $\bar{x} + s$ b. $\bar{x} - 2s$ to $\bar{x} + 2s$ c. $\bar{x} - 3s$ to $\bar{x} + 3s$

2.57 For a set of data with a mound-shaped relative frequency distribution, what can be said about the percentage of the measurements contained in each of the intervals specified in Exercise 2.56?

*The assumption that the distribution is mound-shaped and symmetric may also be incorrect. However, if the distribution were skewed to the right, as life-length distributions often tend to be, the percentage of measurements more than 2 standard deviations *below* the mean would be even less than 2.5%.

2.58 The following is a sample of 25 measurements:

7	6	6	11	8	9	11	9	10	8	7	7	5
9	10	7	7	7	7	9	12	10	10	8	6	

a. Compute $\bar{x}$, s^2, and s for this sample.

b. Count the number of measurements in the intervals $\bar{x} \pm s$, $\bar{x} \pm 2s$, $\bar{x} \pm 3s$. Express each count as a percentage of the total number of measurements.

c. Compare the percentages found in part **b** to the percentages given by the Empirical Rule and Chebyshev's rule.

d. Calculate the range, and use it to obtain a rough approximation for s. Does the result compare favorably with the actual value for s found in part **a**?

2.59 Given a data set with a largest value of 760 and a smallest value of 135, what would you estimate the standard deviation to be? Explain the logic behind the procedure you used to estimate the standard deviation. Suppose the standard deviation is reported to be 25. Is this feasible? Explain.

Applying the Concepts

2.60 Refer to Exercise 2.16, in which we compared the 50 states' average SAT scores in 1975 and 1990. The data are repeated here:

State	*1975*	*1990*	*State*	*1975*	*1990*	*State*	*1975*	*1990*
Ala.	883	984	La.	947	993	Ohio	955	949
Alaska	942	914	Maine	908	886	Okla.	994	1001
Ariz.	1021	942	Md.	907	908	Oreg.	908	923
Ark.	992	981	Mass.	903	900	Pa.	900	883
Calif.	908	903	Mich.	949	968	R.I.	901	883
Colo.	994	969	Minn.	1058	1019	S.C.	794	834
Conn.	913	901	Miss.	980	996	S.Dak.	1084	1061
Del.	915	903	Mo.	965	995	Tenn.	988	1008
Fla.	915	884	Mont.	1047	987	Tex.	898	874
Ga.	824	844	Nebr.	966	1030	Utah	1069	1031
Hawaii	892	885	Nev.	962	921	Vt.	915	897
Idaho	1017	968	N.H.	934	928	Va.	894	895
Ill.	970	994	N.J.	878	891	Wash.	1011	923
Ind.	881	867	N.Mex.	1002	1007	W.Va.	964	933
Iowa	1091	1088	N.Y.	925	882	Wis.	1036	1019
Kans.	1043	1040	N.C.	827	841	Wyo.	1054	977
Ky.	977	994	N.Dak.	1064	1069			

The mean and standard deviation of these data are:

1975 SAT Scores	*1990 SAT Scores*
$\bar{x} = 955.3$	$\bar{x} = 947.5$
$s = 69.5$	$s = 64.0$

Calculate the percentages of measurements in the intervals $\bar{x} \pm s$, $\bar{x} \pm 2s$, and $\bar{x} \pm 3s$ for each data set (1975 and 1990). Check the agreement of these percentages with both Chebyshev's rule and the Empirical Rule.

2.61 Professor Jon K. Mills (1991) of the Illinois School of Professional Psychology measured the Locus of Control (LOC), a measure of one's perception of control over factors affecting one's life, for two groups of individuals undergoing weight reduction for treatment of obesity. The mean and standard deviation for a sample of 46 adults were 6.45 and 2.89, respectively, while the mean and standard deviation for a sample of 19 adolescents were 10.89 and 2.48, respectively. A lower score on the LOC scale indicates a perception of internal, or self, control, while a higher score indicates a perception that external factors are more in control of one's life.

a. Calculate the 1- and 2-standard-deviation intervals around the means for each group. Plot these intervals on a line graph using different colors or symbols to represent each group.

b. Assuming that the distributions of LOC scores are approximately mound-shaped, estimate the numbers of individuals within each interval.

c. Based on your answers to parts **a** and **b**, do you think an inference can be made that *all* adults and adolescents undergoing weight reduction treatment differ with respect to LOC? What factors did you consider in making this inference? [We will reconsider this exercise in Chapter 8 to show how to measure the reliability of this inference.]

2.62 A study reported in the *Journal for Research in Mathematics Education* (1984) was designed to investigate the effects of two variables—(1) a student's level of mathematical anxiety and (2) teaching method—on a student's achievement in a course in mathematics. Students who had a low level of mathematical anxiety and were taught using the traditional expository method obtained a mean score of 183.43 and a standard deviation of 52.27. Use the Empirical Rule, along with this information, to give a verbal description of the data set.

2.63 For each day of last year, the number of vehicles passing through a certain intersection was recorded by a city engineer. One objective of this study was to determine the percentage of days that more than 425 vehicles used the intersection. If the mean for the data was 375 vehicles per day and the standard deviation was 25 vehicles:

a. What can be said about the percentage of days that more than 425 vehicles used the intersection? Assume that nothing is known about the shape of the relative frequency distribution for the data.

b. What is your answer to part **a** if you know that the relative frequency distribution for the data is mound-shaped?

2.64 A buyer for a lumber company must decide whether to buy a piece of land containing 5,000 pine trees. If 1,000 of the trees are at least 40 feet tall, the buyer will purchase the land; otherwise, he will not. The owner of the land reports that the height of the trees has a mean of 30 feet and a standard deviation of 3 feet. Based on this information, what is the buyer's decision?

2.65 A chemical company produces a substance composed of 98% cracked corn particles and 2% zinc phosphide for use in controlling rat populations in sugarcane fields. Production must be carefully controlled to maintain the 2% zinc phosphide because too much zinc phosphide will cause damage to the sugarcane and too little will be ineffective in controlling the rat population. Records from past production indicate that the distribution of the actual percentage of zinc phosphide present in the substance is approximately mound-shaped, with a mean of 2.0% and a standard deviation of .08%. If the production line is operating correctly, approximately what proportion of batches from a day's production will contain less than 1.84% zinc phosphide? Suppose one batch chosen randomly actually contains 1.80% zinc phosphide. Does this indicate that there is too little zinc phosphide in today's production? Explain your reasoning.

– **2.66** Solar energy is considered by many to be the energy of the future. A recent survey was taken to compare the cost of solar energy to the cost of gas electric energy. Results of the survey revealed that the distribution of the amount of the monthly utility bill of a 3-bedroom house using gas or electric energy had a mean of \$125 and a standard deviation of \$10.

a. If nothing is known about the distribution of the amounts of monthly utility bills, what can you say about the fraction of all 3-bedroom homes using gas or electric energy having bills between \$95 and \$155?

b. If it is reasonable to assume that the distribution of the amounts of monthly utility bills is mound-shaped, approximately what proportion of 3-bedroom homes would have monthly bills less than \$135?

c. Suppose that three houses with solar energy units had the following monthly utility bills: \$101, \$98, \$104. Does this suggest that solar energy units might result in lower utility bills? Explain. [*Note:* We present a statistical method in Chapter 7 for testing this conjecture.]

2.67 The following data sets have been invented to demonstrate that the lower bounds given by Chebyshev's rule are appropriate. Notice that the data are contrived and would not be encountered in a real-life problem.

a. Consider a data set that contains ten 0's, two 1's, and ten 2's. Calculate $\bar{x}$, s^2, and s. What percentage of the measurements are in the interval $\bar{x} \pm s$? Compare this result to Chebyshev's rule.

b. Consider a data set that contains five 0's, thirty-two 1's, and five 2's. Calculate $\bar{x}$, s^2, and s. What percentage of the measurements are in the interval $\bar{x} \pm 2s$? Compare this result to Chebyshev's rule.

c. Consider a data set that contains three 0's, fifty 1's, and three 2's. Calculate $\bar{x}$, s^2, and s. What percentage of the measurements are in the interval $\bar{x} \pm 3s$? Compare this result to Chebyshev's rule.

d. Draw a histogram for each of the data sets in parts **a**, **b**, and **c**. What do you conclude from these graphs and the answers to parts **a**, **b**, and **c**?

2.68 The results of a study to compare the effects of fructose and glucose on the high endurance of women athletes were presented in *Research Quarterly for Exercise and Sport* (1983). Six women athletes received 300 milliliters each of a certain drink (say, water and glucose) and then ran until exhausted. Various measurements were then taken on each athlete. The mean and standard deviation of the performance times (in minutes) for the six athletes after receiving the glucose drink were 61.9 and 20.3, respectively. What do the mean and standard deviation tell you about this small data set?

2.6 Measures of Relative Standing

As we have seen, numerical measures of central tendency and variability describe the general nature of a data set (either a sample or a population). We may also be interested in describing the relative quantitative location of a particular measurement within a data set. Descriptive measures of the relationship of a measurement to the rest of the data are called **measures of relative standing**.

One measure of the relative standing of a measurement is its **percentile ranking**. For example, suppose you scored an 80 on an examination and you want to know how you fared in comparison with others in your class. If the instructor tells you that you scored in the 90th percentile, it means that 90% of the examination grades were less than yours and 10% were greater. Thus, if the examination scores were described by the relative frequency histogram in Figure 2.15, the 90th percentile would be located at a point such that 90% of the total area under the relative frequency histogram lies below the 90th percentile and 10% lies above. If the instructor tells you that you scored in the 50th percentile (the median of the data set), 50% of the examination grades would be less than yours and 50% would be greater.

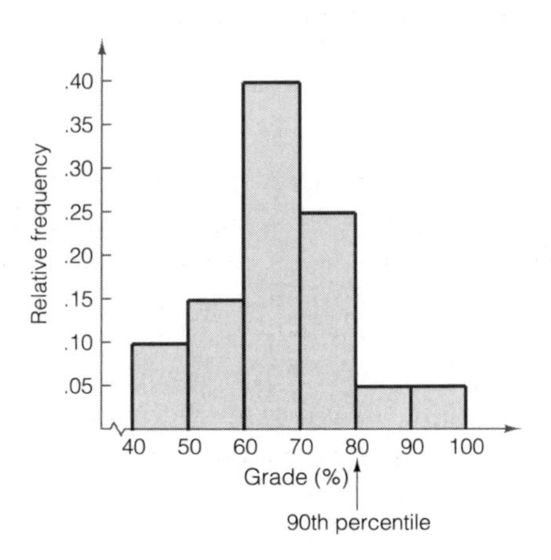

FIGURE 2.15 ▶ Location of 90th percentile for examination grades

Percentile rankings are of practical value only for large data sets. Finding them involves a process similar to the one used in finding a median. The measurements are ranked in order and a rule is selected, similar to that used in locating a median, to define the location of each percentile. Since we are primarily interested in interpreting the percentile rankings of measurements (rather than finding particular percentiles for a data set), we will define the ***p*th percentile** of a data set as shown in Definition 2.11.

Definition 2.11

For any set of n measurements (arranged in ascending or descending order), the ***p*th percentile** is a number such that $p\%$ of the measurements fall below the pth percentile and $(100 - p)\%$ fall above it.

Another measure of relative standing in popular use is the ***z*-score**. As you can see in Definition 2.12, the z-score makes use of the mean and standard deviation of the data set in order to specify the relative location of a measurement:

Definition 2.12

The **sample *z*-score** for a measurement x is

$$z = \frac{x - \bar{x}}{s}$$

The **population *z*-score** for a measurement x is

$$z = \frac{x - \mu}{\sigma}$$

Note that the z-score is calculated by subtracting $\bar{x}$ (or μ) from the measurement x and then dividing the result by s (or σ). The final result, the z-score, represents the distance between a given measurement x and the mean, expressed in standard deviations.

EXAMPLE 2.13

Suppose a sample of 2,000 high school seniors' verbal SAT (Scholastic Aptitude Test) scores is selected. The mean and standard deviation are

$$\bar{x} = 550 \qquad s = 75$$

Suppose Joe Smith's score is 475. What is his sample z-score?

Solution

You can see that Joe Smith's score lies below the mean score of the 2,000 seniors:

325 ($\bar{x} - 3s$) — 475 (Joe Smith's score) — 550 ($\bar{x}$) — 775 ($\bar{x} + 3s$)

We compute

$$z = \frac{x - \bar{x}}{s} = \frac{475 - 550}{75} = -1.0$$

which tells us that Joe Smith's score is 1.0 standard deviation *below* the sample mean, or in short, his sample z-score is -1.0.

The numerical value of the z-score reflects the relative standing of the measurement. A large positive z-score implies that the measurement is larger than almost all other measurements, whereas a large negative z-score indicates that the measurement is smaller than almost every other measurement. If a z-score is 0 or near 0, the measurement is located at or near the mean of the sample or population.

We can be more specific if we know that the frequency distribution of the measurements is mound-shaped. In this case, the following interpretation of the z-score can be given:

Interpretation of z-Scores for Mound-Shaped Distributions of Data

1. Approximately 68% of the measurements will have a z-score between -1 and 1.
2. Approximately 95% of the measurements will have a z-score between -2 and 2.
3. All or almost all the measurements will have a z-score between -3 and 3.

Note that this interpretation of z-scores is identical to that given by the Empirical Rule for samples from mound-shaped distributions. The statement that a measurement falls in the interval $(\mu - \sigma)$ to $(\mu + \sigma)$ is equivalent to the statement that a measurement has a population z-score between -1 and 1, since all measurements between $(\mu - \sigma)$ and $(\mu + \sigma)$ are within 1 standard deviation of μ.

We end this section with an example and a case study that indicate how z-scores may be used to accomplish our primary objective—the use of sample information to make inferences about the population.

EXAMPLE 2.14

Suppose a female bank employee believes that her salary is low as a result of sex discrimination. To substantiate her belief, she collects information on the salaries of her male counterparts in the banking business. She finds that their salaries have a mean of \$34,000 and a standard deviation of \$2,000. Her salary is \$27,000. Does this information support her claim of sex discrimination?

Solution

The analysis might proceed as follows: First, we calculate the z-score for the woman's salary with respect to those of her male counterparts. Thus,

$$z = \frac{\$27,000 - \$34,000}{\$2,000} = -3.5$$

The implication is that the woman's salary is 3.5 standard deviations *below* the mean of the male salary distribution. Furthermore, if a check of the male salary data shows that the frequency distribution is mound-shaped, we can infer that very few salaries in this distribution should have a z-score less than -3, as shown in Figure 2.16. Therefore, a z-score of -3.5 represents either a measurement from a distribution different from the male salary distribution or a very unusual (highly improbable) measurement for the male salary distribution.

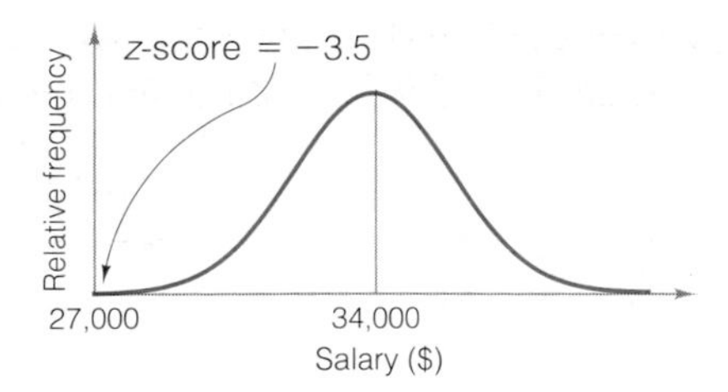

FIGURE 2.16 ▶ **Male salary distribution**

Which of the two situations do you think prevails? Do you think the woman's salary is simply unusually low in the distribution of salaries, or do you think her claim of sex discrimination is justified? Most people would probably conclude that her salary does not come from the male salary distribution. However, the careful investigator should require more information before inferring sex discrimination as the cause. We would want to know more about the data collection technique the woman used and more about her competence at her job. Also, perhaps other factors such as length of employment should be considered in the analysis.

Examples 2.12 and 2.14 exemplify an approach to statistical inference that might be called the **rare event approach**. An experimenter hypothesizes a specific frequency distribution to describe a population of measurements. Then a sample of measurements is drawn from the population. If the experimenter finds it unlikely that the sample came from the hypothesized distribution, the hypothesis is concluded to be false. Thus, in Example 2.14 the woman believes her salary reflects sex discrimination. She hypothesizes that her salary should be just another measurement in the distribution of her male counterparts' salaries if no discrimination exists. However, it is so unlikely that the sample (in this case, her salary) came from the male frequency distribution that she rejects that hypothesis, concluding that the distribution from which her salary was drawn is different from the distribution for the men.

This rare event approach to inference-making is discussed further in later chapters. Proper application of the approach requires a knowledge of probability, the subject of our next chapter.

CASE STUDY 2.2 / Statistics and Air Quality Standards

H. E. Neustadter and S. M. Sidik (1974) discuss calculation procedures for obtaining an air quality standard (AQS) to use in judging whether a company is exceeding air pollution standards. A simplification of a method they discuss is outlined here:

1. Obtain a large set of daily measurements from a company known to be complying with established air pollution standards.
2. If necessary, transform the data so that they form a mound-shaped distribution to which the aids in Table 2.6 may be applied. For example, if the logarithm of each measurement is calculated for typical air quality data, the new set of data often forms a mound-shaped distribution.
3. Calculate the mean $\bar{x}$ and standard deviation s of the mound-shaped set of data and use this information

to obtain the AQS. For example, if it is desired to obtain a standard that would be exceeded on approximately 2.5% of the industry's operating days, calculate

$$AQS = \bar{x} + 2s$$

The number of standard deviations to be added to $\bar{x}$ can be adjusted to reflect the percentage of days a company is to be permitted to exceed the standard. As Neustadter and Sidik point out, "a major objective of air quality monitoring is often to determine compliance with air quality standards which may in part consist of a 24-hour level not to be exceeded more than once a year." They show that one should calculate the mean plus approximately 2.7 standard deviations to meet this standard.

Exercises 2.69–2.78

Learning the Mechanics

2.69 Compute the z-score corresponding to each of the following values of x:

a. $x = 40, \quad s = 5, \quad \bar{x} = 30$ **b.** $x = 90, \quad \mu = 89, \quad \sigma = 2$
c. $\mu = 50, \quad \sigma = 5, \quad x = 50$ **d.** $s = 4, \quad x = 20, \quad \bar{x} = 30$
e. In parts **a–d**, state whether the z-score locates x within a sample or a population.
f. In parts **a–d**, state whether each value of x lies above or below the mean and by how many standard deviations.

2.70 Give the percentage of measurements in a data set that are above and below each of the following percentiles:
a. 75th percentile **b.** 50th percentile **c.** 20th percentile **d.** 84th percentile

2.71 What is the 50th percentile of a quantitative data set called? What is another name for the 50th percentile?

2.72 Compare the z-scores to decide which of the following x values lie the greatest distance above the mean and the greatest distance below the mean.
a. $x = 100, \quad \mu = 50, \quad \sigma = 25$ **b.** $x = 1, \quad \mu = 4, \quad \sigma = 1$
c. $x = 0, \quad \mu = 200, \quad \sigma = 100$ **d.** $x = 10, \quad \mu = 5, \quad \sigma = 3$

2.73 At one university, the students are given z-scores at the end of each semester rather than the traditional GPAs. The mean and standard deviation of all students' cumulative GPAs, on which the z-scores are based, are 2.7 and .5, respectively.
a. Translate each of the following z-scores to a corresponding GPA: $z = 2.0$, $z = -1.0$, $z = .5$, $z = -2.5$.
b. Students with z-scores below -1.6 are put on probation. What is the corresponding probationary GPA?
c. The president of the university wishes to graduate the top 16% of the students with *cum laude* honors and the top 2.5% with *summa cum laude* honors. Where (approximately) should the limits be set in terms of z-scores? In terms of GPAs? What assumption, if any, did you make about the distribution of the GPAs at the university?

2.74 Suppose that 40 and 90 are two elements of a population data set and that their z-scores are -2 and 3, respectively. Using only this information, is it possible to determine the population's mean and standard deviation? If so, find them. If not, explain why it is not possible.

Applying the Concepts

2.75 The distribution of scores on a nationally administered college achievement test has a median of 520 and a mean of 540.

a. Explain why it is possible for the mean to exceed the median for this distribution of measurements.
b. Suppose you are told that the 90th percentile is 660. What does this mean?
c. Suppose you are told that you scored at the 94th percentile. Interpret this statement.

2.76 Many firms use on-the-job training to teach their employees computer programming. Suppose you work in the personnel department of a firm that just finished training a group of its employees to program, and you have been requested to review the performance of one of the trainees on the final test that was given to all trainees. The mean and standard deviation of the test scores are 80 and 5, respectively, and the distribution of scores is mound-shaped.

a. The employee in question scored 65 on the final test. Compute the employee's z-score.
b. Approximately what percentage of the trainees will have z-scores equal to or less than the employee of part **a**?
c. If a trainee were arbitrarily selected from those who had taken the final test, is it more likely that he or she would score 90 or above, or 65 or below?

2.77 A city librarian claims that books have been checked out an average of 7 (or more) times in the last year. You suspect he has exaggerated the checkout rate (book usage) and that the mean number of checkouts per book per year is, in fact, less than 7. Using the card catalog, you randomly select one book and find that it has been checked out 4 times in the last year. Assume that the standard deviation of the number of checkouts per book per year is approximately 1.

a. If the mean number of checkouts per book per year really is 7, what is the z-score corresponding to 4?
b. Considering your answer to part **a**, do you have reason to believe that the librarian's claim is incorrect?
c. If you knew that the distribution of the number of checkouts were mound-shaped, would your answer to part **b** change? Explain.
d. If the standard deviation of the number of checkouts per book per year were 2 (instead of 1), would your answers to parts **b** and **c** change? Explain.

2.78 Polychlorinated biphenyls (PCBs), considered to be extremely hazardous to humans, are often used in the insulation of large electrical transformers. On March 24, 1984, the *Gainesville Sun* reported on the discovery of a particularly high PCB count at a salvage company in Clay County, Florida. The company, which salvaged the copper in electrical transformers, allowed oil contaminated with PCBs to seep into the soil in and around the salvage site. One soil sample in the vicinity registered 200 parts per million (ppm) of PCBs, four times the safe limit established by the Florida Department of Environmental Regulation.

Suppose that the PCB count in samples of soil in the vicinity of the salvage operation has a distribution with mean equal to 25 ppm and standard deviation equal to 5 ppm of PCBs. Would a soil sample showing 200 ppm be classified as an extreme observation? Explain.

2.7 Box Plots: Graphic Descriptions Based on Quartiles (Optional)

The **box plot**, a relatively recent introduction to the methodology of descriptive measures, is based on the **quartiles** of a data set. Quartiles are values that partition the data set into four groups, each containing 25% of the measurements. The lower quartile Q_L is the 25th percentile, the middle quartile is the median M (the 50th percentile), and the upper quartile Q_U is the 75th percentile (see Figure 2.17).

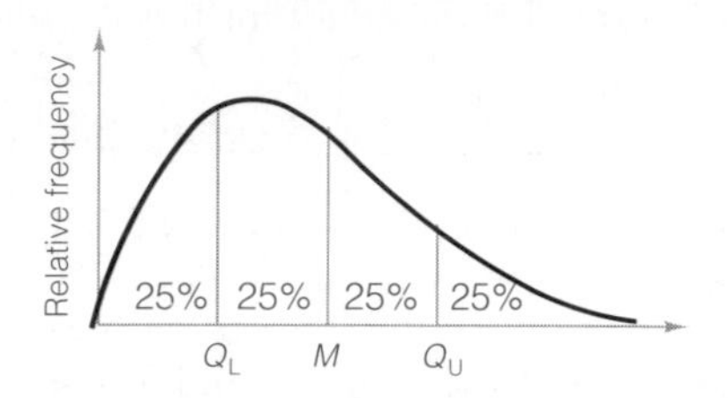

FIGURE 2.17 ► The quartiles for a data set

Definition 2.13

The **lower quartile Q_L** is the 25th percentile of a data set. The **middle quartile M** is the median. The **upper quartile Q_U** is the 75th percentile.

A box plot is based on the **interquartile range (IQR)**, the distance between the lower and upper quartiles:

$$\text{IQR} = Q_U - Q_L$$

Definition 2.14

The **interquartile range (IQR)** is the distance between the lower and upper quartiles:

$$\text{IQR} = Q_U - Q_L$$

The box plot for the gas mileage data (Table 2.1) is given in Figure 2.18. It was generated by the Minitab statistical software package for personal computers.* Note that a rectangle (the **box**) is drawn, with the ends of the rectangle (the **hinges**, represented by the "I's" at the ends of the box drawn by the Minitab program) drawn at the quartiles Q_L and Q_U. By definition, then, the "middle" 50% of the observations—those between Q_L and Q_U—fall inside the box. For the gas mileage data, these quartiles appear to be at (approximately) 35.5 and 38.5. Thus,

$$\text{IQR} = 38.5 - 35.5 = 3.0 \quad \text{(approximately)}$$

Note that the median is shown at about 37 by a + sign within the box.

*Although box plots can be generated by hand, the amount of detail required makes them particularly well suited for computer generation. We use computer software to generate the box plots in this section.

FIGURE 2.18 ▶
Minitab box plot for gas mileage data

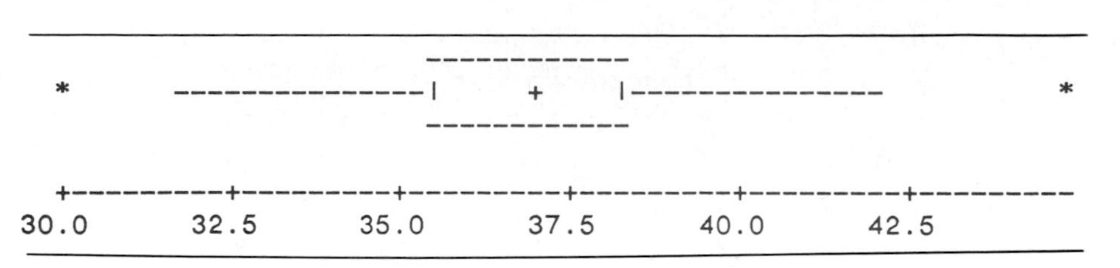

To guide the construction of the "tails" of the box plot, two sets of limits, called **inner fences** and **outer fences**, are used. Neither set of fences actually appears on the box plot. Inner fences are located at a distance of 1.5(IQR) from the hinges. Emanating from the hinges of the box are dashed lines called the **whiskers**. The two whiskers extend to the most extreme observation inside the inner fences. For example, the inner fence on the lower side of the gas mileage box plot is (approximately):

$$\begin{aligned}\text{Lower inner fence} &= \text{Lower hinge} - 1.5(\text{IQR}) \\ &\approx 35.5 - 1.5(3.0) \\ &= 35.5 - 4.5 = 31.0\end{aligned}$$

The smallest measurement *inside* this fence is the second smallest measurement, 31.8. Thus, the lower whisker extends to 31.8. Similarly, the upper whisker extends to 42.1, the largest measurement inside the upper inner fence at about 38.5 + 4.5 = 43.0.

Values that are beyond the inner fences receive special attention because they are extreme values that represent relatively rare occurrences. In fact, for mound-shaped distributions, fewer than 1% of the observations are expected to fall outside the inner fences. Two of the 100 gasoline mileage measurements, 30.0 and 44.9, fall beyond the inner fences, one on each end of the distribution. These measurements are represented by asterisks (*).

The other pair of imaginary fences, the outer fences, are defined at a distance 3(IQR) from each end of the box. Measurements that fall beyond the outer fences are represented by 0's and are very extreme measurements that require special analysis. Less than one-hundredth of 1% (.01%, or .0001) of the measurements from mound-shaped distributions are expected to fall beyond the outer fences. Since no measurement in the gas mileage box plot (Figure 2.18) is represented by a 0, we know that none of the measurements fall outside the outer fences.

Generally, any measurements that fall beyond the inner fences—and certainly any that fall beyond the outer fences—are considered potential **outliers**. Outliers are extreme measurements that stand out from the rest of the sample and may be faulty—incorrectly recorded observations, members of a different population than the rest of the sample or, at the least, very unusual measurements from the same population. For example, the two gasoline mileage measurements beyond the inner fences may be considered outliers. When we analyze these measurements, we find that they are correctly recorded. Perhaps they represent mileages that correspond to exceptional models of the automobile being tested or to unusual gasoline mixtures. Outlier analysis often reveals useful information of this kind and therefore plays an important role in the statistical inference-making process.

The elements (and nomenclature) of box plots are summarized in the next box. Some aids to the interpretation of box plots are also given.

Elements of a Box Plot

1. A rectangle (the **box**) is drawn with the ends (the **hinges**) drawn at the lower and upper quartiles (Q_L and Q_U). The median of the data is shown in the box, usually by a "+".
2. The points at distances 1.5(IQR) from each hinge mark the **inner fences** of the data set. Horizontal lines (the **whiskers**) are drawn from each hinge to the most extreme measurement inside the inner fence.
3. A second pair of fences, the **outer fences**, exist at a distance of 3 interquartile ranges, 3(IQR), from the hinges. One symbol (usually "*") is used to represent measurements falling between the inner and outer fences, and another (usually "0") is used to represent measurements beyond the outer fences. Thus, outer fences are not shown unless one or more measurements lie beyond them.
4. The symbols used to represent the median and the extreme data points (those beyond the fences) will vary depending on the software you use to construct the box plot. (You may use your own symbols if you are constructing a box plot by hand.) You should consult the program's documentation to determine exactly which symbols are used.

Aids to the Interpretation of Box Plots

1. Examine the length of the box. The IQR is a measure of the sample's variability and is especially useful for the comparison of two samples (see Example 2.16).
2. Visually compare the lengths of the whiskers. If one is clearly longer, the distribution of the data is probably skewed in the direction of the longer whisker.
3. Analyze any measurements that lie beyond the fences. Fewer than 5% should fall beyond the inner fences, even for very skewed distributions. Measurements beyond the outer fences are probably **outliers**, with one of the following explanations:
   a. The measurement is incorrect. It may have been observed, recorded, or entered into the computer incorrectly.
   b. The measurement belongs to a population different from that from which the rest of the sample was drawn (see Example 2.16).
   c. The measurement may be correct and from the same population as the rest but represents a rare event. Generally, we accept this explanation only after carefully ruling out all others.

EXAMPLE 2.15

Use a statistical software package to draw a box plot for the student loan default data, Table 2.3.

Solution

The Minitab box plot for the student loan default rates is shown in Figure 2.19. Note that the median appears to be about 9.5, and, with the exception of a single extreme observation, the distribution appears to be symmetrically distributed between approximately 3% and 17%. The single outlier is beyond the inner fence but inside the outer fence. Examination of the data reveals that this observation corresponds to Alaska's default rate of 19.7%.

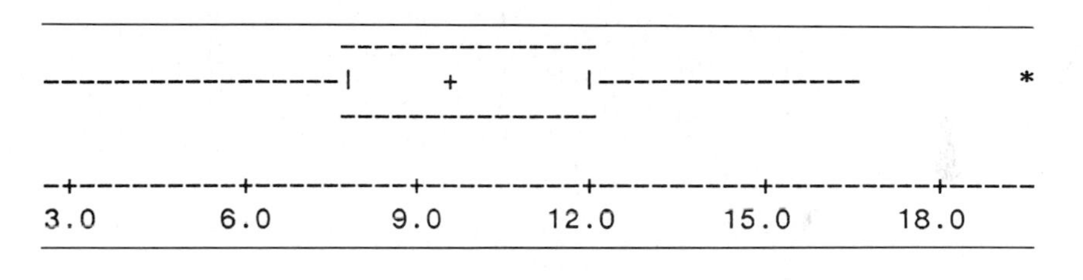

FIGURE 2.19 ▶ Minitab box plot for student loan default rates

EXAMPLE 2.16

A Ph.D. student in psychology conducted a stimulus reaction experiment as a part of her dissertation research. She subjected 50 subjects to a threatening stimulus and 50 to a nonthreatening stimulus. The reaction times of all 100 students were recorded electronically to the nearest tenth of a second. Box plots of the two resulting samples of reaction times are shown in Figure 2.20. Interpret the box plots.

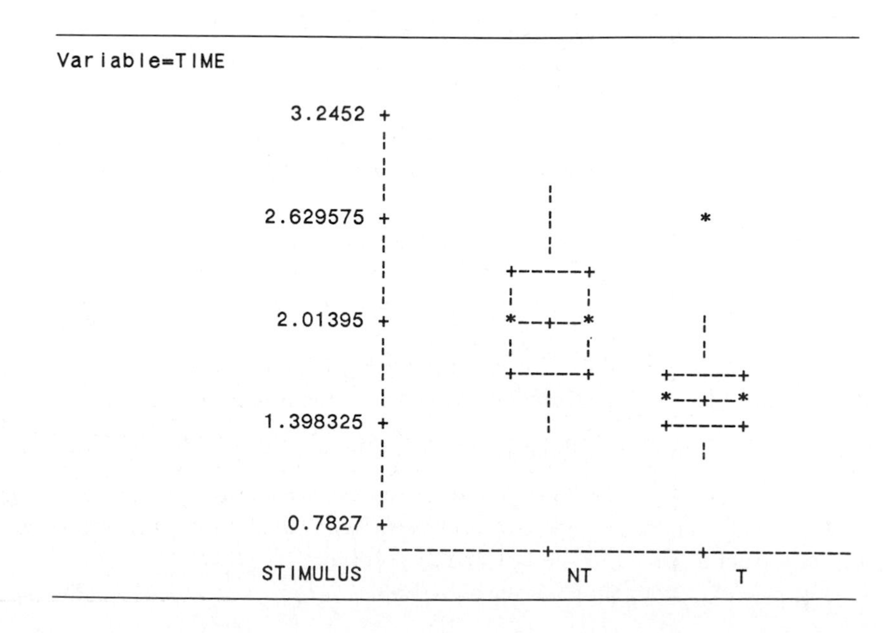

FIGURE 2.20 ▶ SAS box plots for reaction time data

Solution

Perhaps the first thing you notice about the two box plots is that they are arranged vertically rather than horizontally. Some statistical software packages, including the SAS System used here, use this arrangement. Also, note that the median is represented by a dashed line through the box. The plus (+) symbol represents the mean in the SAS box plot. Analysis of the box plots on the same numerical scale reveals that the distribution of times corresponding to the threatening stimulus lies below that of the nonthreatening stimulus. The implication is that the reaction times tend to be faster to the threatening stimulus. Note, too, that the upper whiskers of both samples are longer than the lower whiskers, indicating that the reaction times are skewed to the right. The box length corresponding to the threatening stimulus is smaller than that for the nonthreatening stimulus, indicating less variability in the reaction times to the threatening stimulus.

No observations in the two samples fall between the inner and outer fences (denoted by 0 in SAS). However, note that one of the observations corresponding to the threatening stimulus is beyond the outer fence (denoted by *). When the researcher carefully examined her notes for the experiments, she found that the subject whose time was beyond the outer fence had mistakenly been given the nonthreatening stimulus. You can see in Figure 2.20 that his time would have been within the upper whisker if moved to the box plot corresponding to the nonthreatening stimulus. Of course, the box plots should be reconstructed since they will both change slightly when the misclassified reaction time is moved from one sample to the other.

The researcher concluded that the reactions to the threatening stimulus were faster and more predictable (less variable) than those to the nonthreatening stimulus. However, she was asked by her Ph.D. committee whether the results were *statistically significant*. Their question addresses the issue of whether the observed difference between the samples might be attributable to chance or sampling variation rather than to real differences between the populations. To answer their question, the researcher must use inferential statistics rather than graphic descriptions. We discuss how to compare two samples using inferential statistics in Chapter 8.

Exercises 2.79–2.88

Note: Starred (*) exercises require the use of a computer.

Learning the Mechanics

2.79 Define the 25th, 50th, and 75th percentiles of a data set. Explain how they provide a description of the data.

2.80 Suppose a data set consisting of exam scores has a lower quartile $Q_L = 60$, a median $M = 75$, and an upper quartile $Q_U = 85$. The scores on the exam ranged from 18 to 100. Without having the actual scores available to you, construct as much of the box plot as possible.

2.81 Minitab was used to generate the following box plot:

```
                          -----------
 * *       ---------------I    +   I--------
                          -----------

+----------+----------+----------+----------+
0.0       15.0       30.0       45.0       60.0
```

a. What is the median of the data set (approximately)?
b. What are the upper and lower quartiles of the data set (approximately)?
c. What is the interquartile range of the data set (approximately)?
d. Is the data set skewed to the left, skewed to the right, or symmetric?
e. What percentage of the measurements in the data set lie to the right of the median? To the left of the upper quartile?

2.82 Minitab was used to generate the accompanying box plots. Compare and contrast the frequency distributions of the two data sets. Your answer should include comparisons of the following characteristics: central tendency, variation, skewness, and outliers.

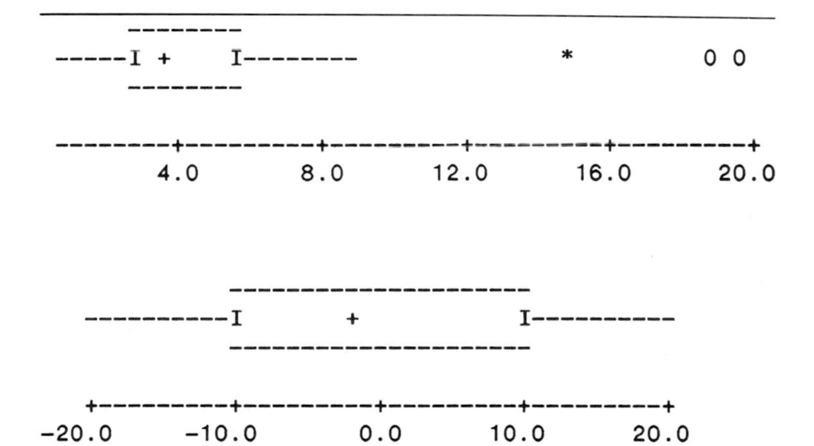

***2.83** Consider the following two sample data sets:

Sample A			*Sample B*		
121	171	158	171	152	170
173	184	163	168	169	171
157	85	145	190	183	185
165	172	196	140	173	206
170	159	172	172	174	169
161	187	100	199	151	180
142	166	171	167	170	188

a. Use a statistical software package to construct a box plot for each data set.
b. Using the information reflected in your box plots, describe the similarities and differences in the two data sets.
c. Identify any outliers that may exist in the two data sets.

Applying the Concepts

2.84 In Exercises 2.13 and 2.14 we constructed a stem-and-leaf display and a relative frequency histogram for 40 radiation (exhalation rate) measurements on waste gypsum and phosphate mounds in Florida. The accompanying figure, constructed using SAS, is a box plot for the same data.

```
Variable=EXHLRATE

      12312.5 +              *
              |
              |
              |
     9234.375 +              *
              |
              |
              |              0
      6156.25 +
              |              |
              |              |
              |              |
     3078.125 +           +-----+
              |           |  +  |
              |           *-----*
              |           +-----+
            0 +              |
               ------------+-----------
```

a. Use the box plot to estimate the lower quartile, median, and upper quartile of these data.
b. Does the distribution appear to be skewed? Explain.
c. Are there any outliers among these data? Explain.
d. Examine the graphic displays for these data in Exercise 2.13 (stem-and-leaf display), Exercise 2.14 (relative frequency histogram), and the box plot here. Which do you prefer as a description of these radiation measurements?

2.85 In Exercises 2.19 and 2.20 we used stem-and-leaf displays and relative frequency histograms to compare the high school dropout rates for the 50 states and the District of Columbia in 1982 and 1984. Box plots, constructed using SAS, for these rates are shown at the top of page 79.
a. How do the median dropout rates compare for the 2 years? [*Hint:* Recall that the SAS program used a dashed line through the box to represent the median.]
b. How do the variabilities of the rates compare for the 2 years?
c. The standard deviations of the rates are 7.29 and 7.95 for 1982 and 1984, respectively. Do the standard deviations agree with the interquartile ranges (part **b**) with regard to the comparison of the variabilities of the rates?
d. Is there evidence of outliers in either of the distributions?

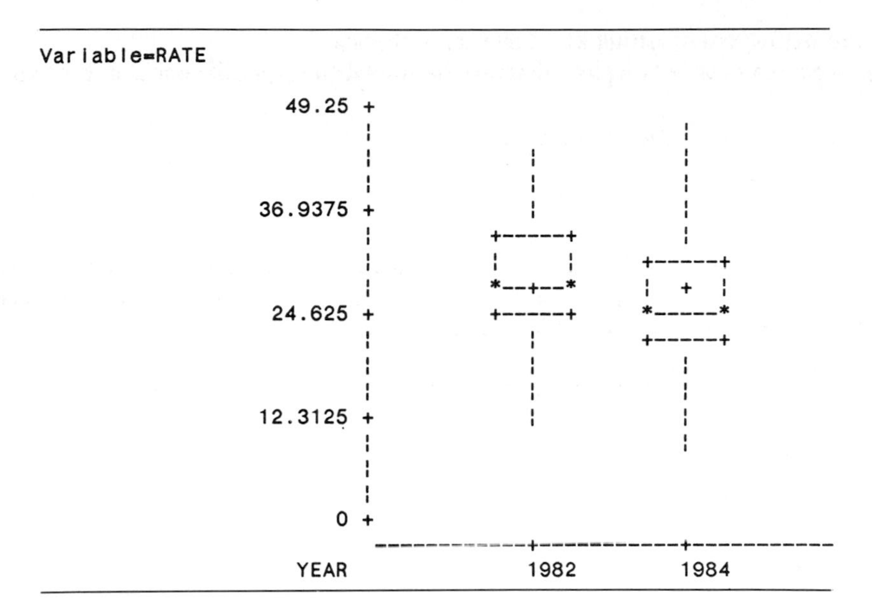

***2.86** A manufacturer of minicomputer systems is interested in improving its customer support services. As a first step, its marketing department has been charged with the responsibility of summarizing the extent of customer problems in terms of system down time. The 40 most recent customers were surveyed to determine the amount of down time (in hours) they had experienced during the previous month. These data are listed in the table.

Customer Number	*Down Time*	*Customer Number*	*Down Time*	*Customer Number*	*Down Time*
230	12	244	2	257	18
231	16	245	11	258	28
232	5	246	22	259	19
233	16	247	17	260	34
234	21	248	31	261	26
235	29	249	10	262	17
236	38	250	4	263	11
237	14	251	10	264	64
238	47	252	15	265	19
239	0	253	7	266	18
240	24	254	20	267	24
241	15	255	9	268	49
242	13	256	22	269	50
243	8				

a. Use a statistical software package to construct a box plot for these data. Use the information reflected in the box plot to describe the frequency distribution of the data set. Your description should address central tendency, variation, and skewness.

b. Use your box plot to determine which customers are having unusually lengthy down times.
c. Find and interpret the z-scores associated with customers you identified in part **b**.

2.87 Refer to Exercises 2.16 and 2.60, in which we compared the 50 states' average SAT scores in 1975 and 1990. The data are repeated here:

State	*1975*	*1990*	*State*	*1975*	*1990*	*State*	*1975*	*1990*
Ala.	883	984	La.	947	993	Ohio	955	949
Alaska	942	914	Maine	908	886	Okla.	994	1001
Ariz.	1021	942	Md.	907	908	Oreg.	908	923
Ark.	992	981	Mass.	903	900	Pa.	900	883
Calif.	908	903	Mich.	949	968	R.I.	901	883
Colo.	994	969	Minn.	1058	1019	S.C.	794	834
Conn.	913	901	Miss.	980	996	S.Dak.	1084	1061
Del.	915	903	Mo.	965	995	Tenn.	988	1008
Fla.	915	884	Mont.	1047	987	Tex.	898	874
Ga.	824	844	Nebr.	966	1030	Utah	1069	1031
Hawaii	892	885	Nev.	962	921	Vt.	915	897
Idaho	1017	968	N.H.	934	928	Va.	894	895
Ill.	970	994	N.J.	878	891	Wash.	1011	923
Ind.	881	867	N.Mex.	1002	1007	W.Va.	964	933
Iowa	1091	1088	N.Y.	925	882	Wis.	1036	1019
Kans.	1043	1040	N.C.	827	841	Wyo.	1054	977
Ky.	977	994	N.Dak.	1064	1069			

SPSS was used to generate the box plots for the 50 states' SAT scores for 1975 and 1990 and are shown here:

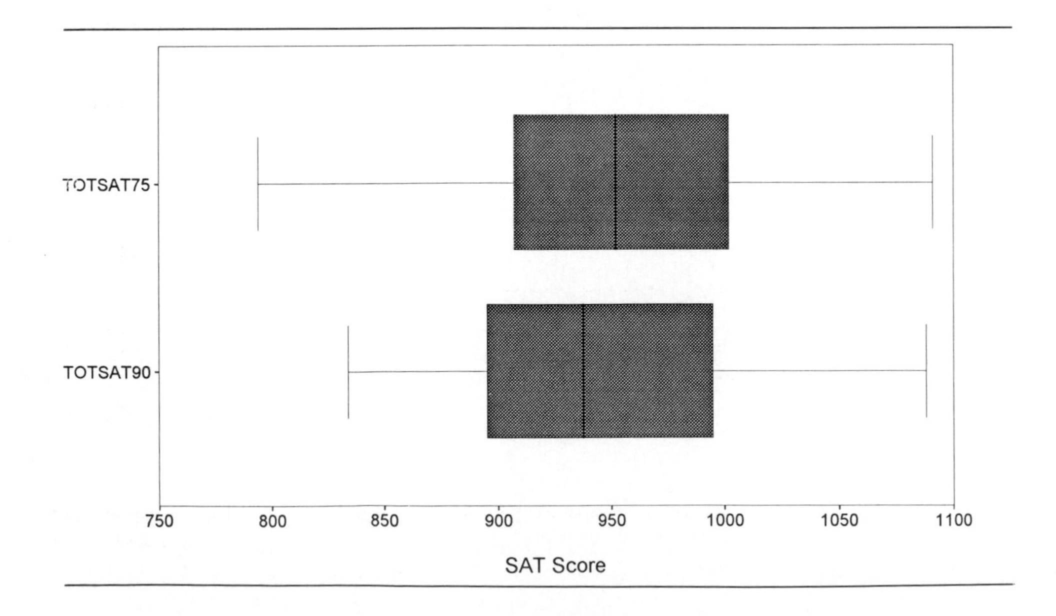

a. Compare the central tendency of the SAT scores for the 2 years. [*Note:* SPSS box plots show only the median, represented by the line within each box.]
b. Compare the variability of the SAT scores for the 2 years.
c. Are any states' SAT scores outliers in either year? If so, identify them.

2.88 The accompanying Minitab-generated box plots describe the U.S. Environmental Protection Agency's 1986 automobile mileage estimates for all models manufactured by Ford and Honda.

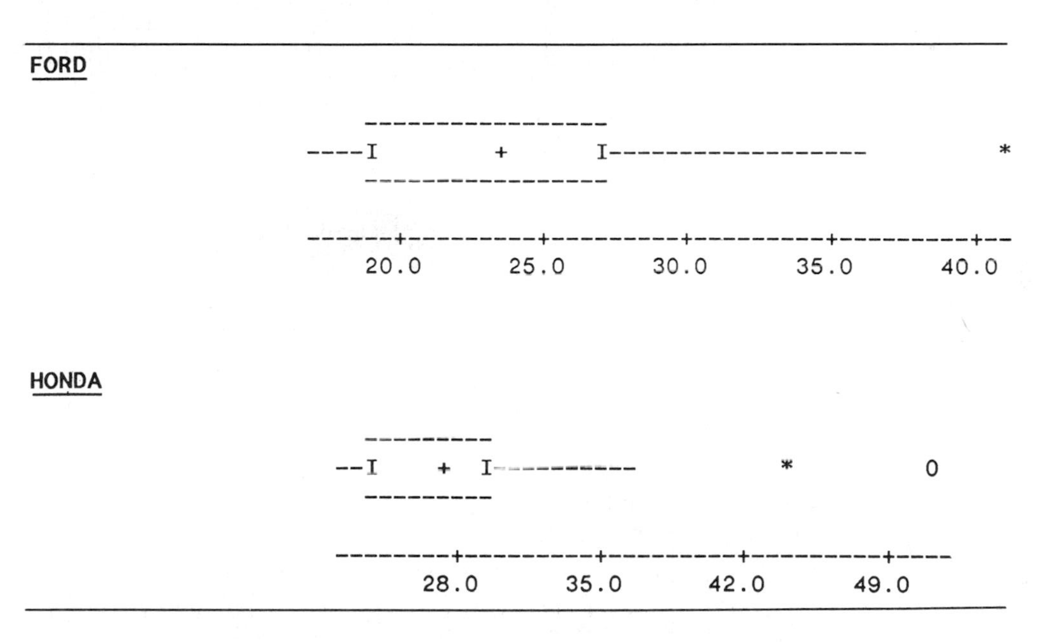

a. Which manufacturer has the higher median mileage estimate?
b. Which manufacturer's mileage estimates have the greater range?
c. Which manufacturer's mileage estimates have the greater interquartile range?
d. Which manufacturer has the model with the highest mileage estimate? Approximately what is that mileage?

2.8 Distorting the Truth with Descriptive Techniques

While it may be true in telling a story that "a picture is worth a thousand words," it is also true that pictures can be used to convey a colored and distorted message to the viewer. So the old adage applies: "Let the buyer (reader) beware." Examine relative frequency histograms and, in general, all graphic descriptions with care.

We will mention a few of the pitfalls to watch for when interpreting a chart or graph. But first we should mention the **time series graph**, which is often the object of distortion. This type of graph records the behavior of some variable of interest recorded over time. Examples of variables commonly graphed as time series abound: economic indices, the U.S. food surplus, defense spending, presidential popularity index, etc. Since the time series graphs often appear in newspapers or magazines, we will use some of them to demonstrate several ways in which pictures are commonly distorted.

One common way to change the impression conveyed by a graph is to change the scale on the vertical axis, the horizontal axis, or both. For example, Figure 2.21 is a

FIGURE 2.21 ▶
Firm A's market share from 1985 to 1990—packed vertical axis

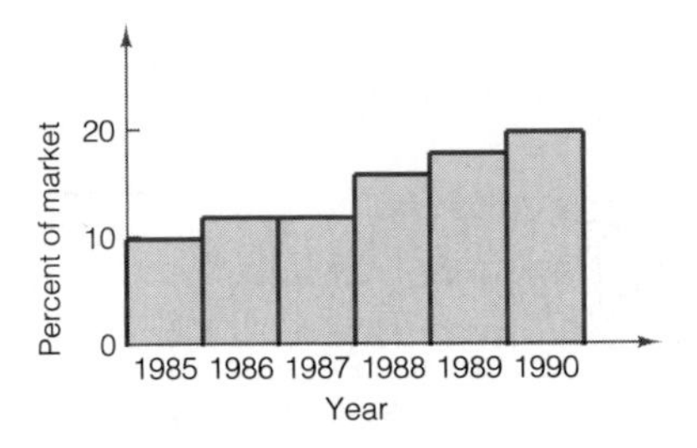

bar graph that shows the market share of sales for a company for each of the years 1985 to 1990. If you want to show that the change in firm A's market share over time is moderate, you should pack in a large number of units per inch on the vertical axis. That is, make the distance between successive units on the vertical scale small, as shown in Figure 2.21. You can see that a change in the firm's market share over time is barely apparent.

If you want to use the same data to make the changes in firm A's market share appear large, you should increase the distance between successive units on the vertical axis. That is, stretch the vertical axis by graphing only a few units per inch as in Figure 2.22. A telltale sign of stretching is a long vertical axis, but this is often hidden by starting the vertical axis at some point above 0, as shown in Figure 2.23(a). The same effect can be achieved by using a broken line—called a *scale break*—for the vertical axis, as shown in Figure 2.23(b).

FIGURE 2.22 ▶
Firm A's market share from 1985 to 1990—stretched vertical axis

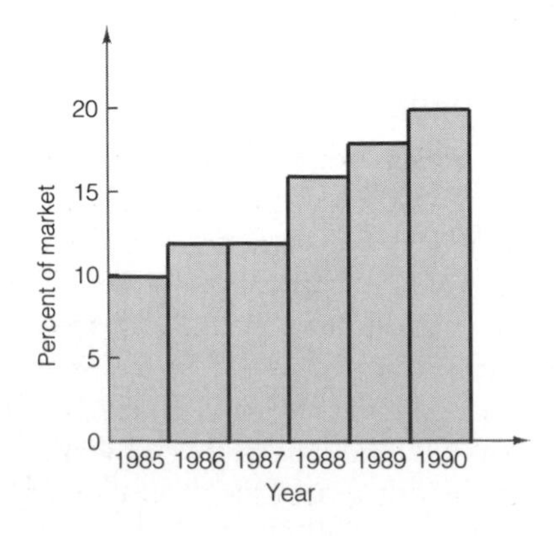

FIGURE 2.23 ▶
Changes in money supply from January to June

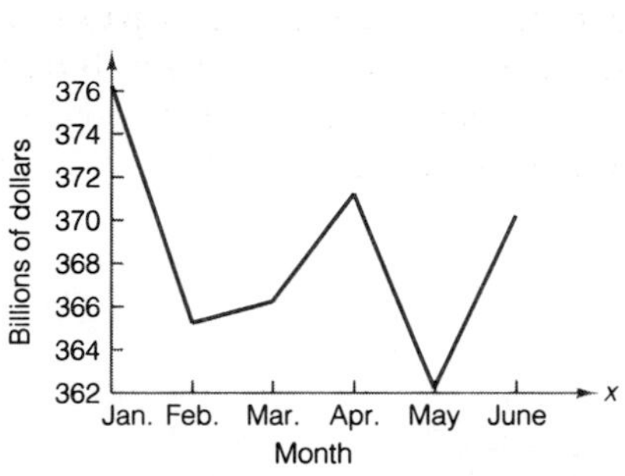

a. Vertical axis started at a point greater than zero

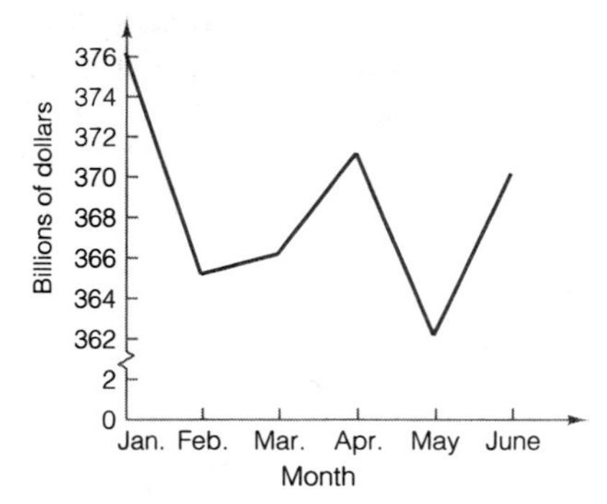

b. Gap in vertical axis

Stretching the horizontal axis (increasing the distance between successive units) may also lead you to incorrect conclusions. For example, Figure 2.24(a) depicts the change in the Gross National Product (GNP) from the first quarter of 1979 to the last quarter of 1980. If you increase the size of the horizontal axis, as in Figure 2.24(b), the change in the GNP over time seems less pronounced.

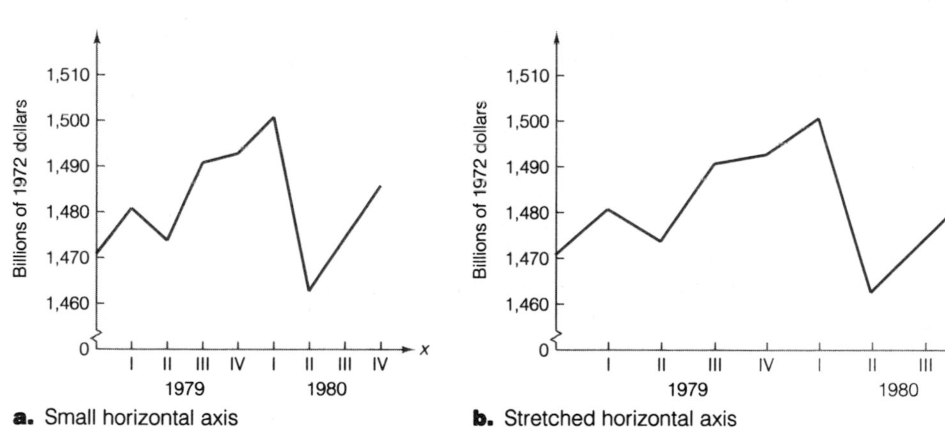

a. Small horizontal axis

b. Stretched horizontal axis

FIGURE 2.24 ▲ Gross national product from 1979 to 1980

The changes in categories indicated by a bar graph can also be emphasized or deemphasized by stretching or shrinking the vertical axis. Another method of achieving visual distortion with bar graphs is by making the width of the bars proportional to their height. For example, look at the bar chart in Figure 2.25(a) on page 84, which depicts the percentage of a year's total automobile sales attributable to each of the four major manufacturers. Now suppose we make the width as well as the height grow as the market share grows. This change is shown in Figure 2.25(b). The reader may tend to equate the *area* of the bars with the relative market share of each manufacturer. In fact, the true relative market share is proportional only to the *height* of the bars.

FIGURE 2.25 ▶ Relative share of the automobile market for each of four major manufacturers

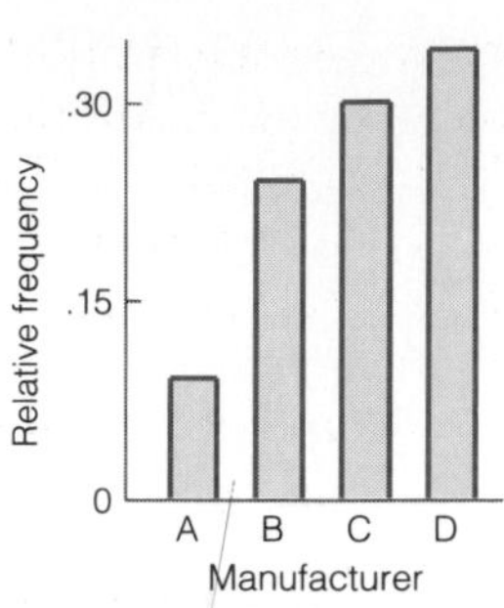

a. Bar chart

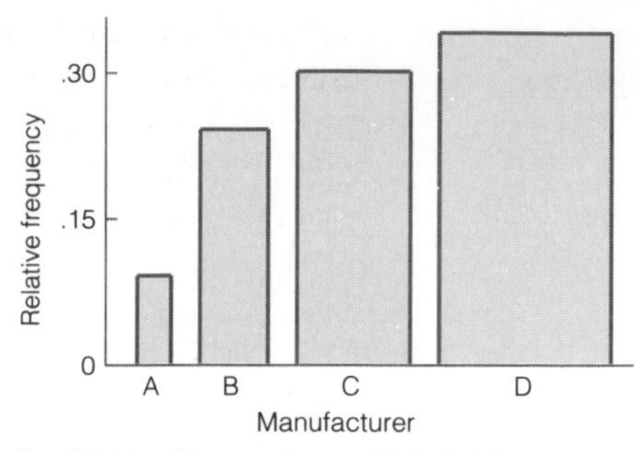

b. Width of bars grows with height

We have presented only a few of the ways that graphs can be used to convey misleading pictures of phenomena. However, the lesson is clear. Examine all graphic descriptions of data with care. Particularly, check the axes and the size of the units on each axis. Ignore the visual changes and concentrate on the actual numerical changes indicated by the graph or chart. The information in a data set can also be distorted by using numerical descriptive measures, as Example 2.17 indicates.

EXAMPLE 2.17

Suppose you are considering working for a small law firm that presently has a senior member and three junior members. You inquire about the salary you could expect to earn if you join the firm. Unfortunately, you receive two answers:

Answer A: The senior member tells you that an "average employee" earns \$57,500.

Answer B: One of the junior members later tells you that an "average employee" earns \$45,000.

Which answer can you believe? The confusion exists because the phrase "average employee" has not been clearly defined. Suppose the four salaries paid are \$45,000 for each of the three junior members and \$95,000 for the senior member. Thus,

$$\bar{x} = \frac{3(\$45{,}000) + \$95{,}000}{4} = \frac{\$230{,}000}{4} = \$57{,}500$$

$$\text{Median} = \$45{,}000$$

You can now see how the two answers were obtained. The senior member reported the mean of the four salaries, and the junior member reported the median. The information you received was distorted because neither person stated which measure of central tendency was being used.

Another distortion of information in a sample occurs when *only* a measure of central tendency is reported. Both a measure of central tendency and a measure of variability are needed to obtain an accurate mental image of a data set.

Suppose you want to buy a new car and are trying to decide which of two models to purchase. Since energy and economy are both important issues, you decide to purchase model A because its EPA mileage rating is 32 miles per gallon in the city, whereas the mileage rating for model B is only 30 miles per gallon in the city.

However, you may have acted too quickly. How much variability is associated with the ratings? As an extreme example, suppose that further investigation reveals that the standard deviation for model A mileages is 5 miles per gallon, whereas that for model B is only 1 mile per gallon. If the mileages form a mound-shaped distribution, they might appear as shown in Figure 2.26. Note that the larger amount of variability associated with model A implies that more risk is involved in purchasing model A. That is, the particular car you purchase is more likely to have a mileage rating that will greatly differ from the EPA rating of 32 miles per gallon if you purchase model A, while a model B car is not likely to vary from the 30 miles per gallon rating by more than 2 miles per gallon.

FIGURE 2.26 ▶
Mileage distributions for two car models

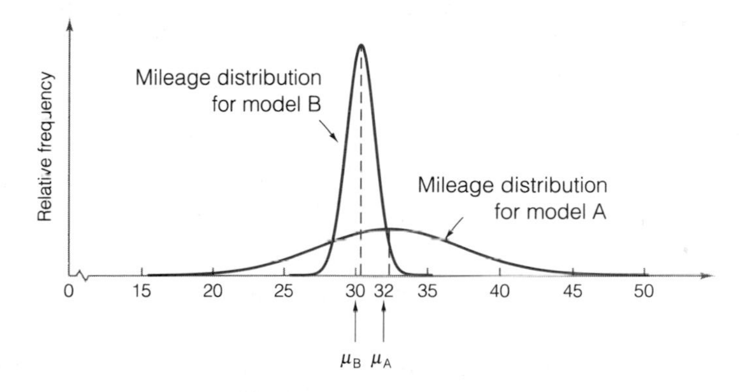

CASE STUDY 2.3 / Children Out of School in America: Making an Ugly Picture Look Worse

David L. Martin (1975) points out another method of distorting the truth with descriptive techniques in his article, "Firsthand report: How flawed statistics can make an ugly picture look even worse." In his critique of the Children Defense Fund's (CDF) 1973 report, *Children Out of School in America*, Martin quotes (boldface comments are Martin's):

25 percent of the 16 and 17 year olds in the Portland, Me., Bayside East Housing Project were out of school. ***Only eight children were surveyed; two were found to be out of school.***

Of all the secondary school students who had been suspended more than once in census tract 22 in Columbia, S.C., 33 percent had been suspended two times and 67 percent had been suspended three or more times. ***CDF found only three children in that entire census tract who had been suspended; one child was suspended twice and the other two children, three or more times.***

In the Portland Bayside East Housing Project, CDF says that 50 percent of all the secondary school children who had been suspended more than once had been suspended three or more times. ***The survey found two secondary school children had been suspended in that area; one of them had been suspended three or more times.***

In each of these examples the reporting of percentages instead of the numbers themselves is misleading. Any inference one might draw from the cited examples would not be reliable. (We will see how to measure the reliability of estimated percentages in Chapter 6.) In short, either the numbers alone should be reported instead of percentages, or, better yet, the report should state that the numbers were too small to report by region. If several regions were combined, the numbers (and percentages) would be more meaningful.

Summary

Data may be classified as one of four types: **nominal**, **ordinal**, **interval**, and **ratio**. Nominal data simply classify the sample or population units, whereas ordinal data enable the units to be ranked. Nominal and ordinal data are often referred to collectively as **qualitative**. Interval data are numbers that enable the comparison of sample or population units by calculation of their numerical differences, whereas ratio data enable the comparison using multiples, or ratios, of the numerical values. Interval and ratio data are often referred to collectively as **quantitative**. Since we want to use sample data to make inferences about the population from which it is drawn, it is important for us to be able to describe the data. **Graphic methods** are important and useful tools for describing data sets. Our ultimate goal, however, is to use the sample to make inferences about the population. We are wary of using graphic techniques to accomplish this goal, since they do not lend themselves to a measure of the reliability for an inference. We therefore developed **numerical measures** to describe a data set.

These numerical methods for describing **quantitative** data sets can be grouped as follows:

1. Measures of central tendency
2. Measures of variability

The measures of central tendency we presented were the **mean**, **median**, and **mode**. The relationship between the mean and median provides information about the **skewness** of the frequency distribution. For making inferences about the population, the sample mean will usually be preferred to the other measures of central tendency. The **range**, **variance**, and **standard deviation** all represent numerical measures of variability. Of these, the variance and standard deviation are in most common use, especially when the ultimate objective is to make inferences about a population.

The mean and standard deviation may be used to make statements about the fraction of measurements in a given interval. For example, we know that at least 75% of the measurements in a data set lie within 2 standard deviations of the mean. If the frequency distribution of the data set is mound-shaped, approximately 95% of the measurements will lie within 2 standard deviations of the mean.

Measures of relative standing provide still another dimension on which to describe a data set. The objective of these measures is to describe the location of a specific measurement relative to the rest of the data set. By doing so, you can construct a mental image of the relative frequency distribution. **Percentiles** and **z-scores** are important examples of measures of relative standing.

The **rare event** concept of statistical inference means that if the chance that a particular sample came from a hypothesized population is very small, we can conclude either that the sample is extremely rare or that the hypothesized population is not the one from which the same was drawn. The more unlikely it is that the sample came from the hypothesized population, the more strongly we favor the conclusion that the hypothesized population is not the true one. We need to be able to assess accurately the rarity of a sample, and this requires knowledge of probability, the subject of our next chapter.

Finally, we gave some examples that demonstrated how descriptive statistics may be used to distort the truth. You should be very critical when interpreting graphic or numerical descriptions of data sets.

Supplementary Exercises 2.89–2.109

Note: Starred (*) exercises require the use of a computer.

Learning the Mechanics

2.89 Classify the following data as one of four types: nominal, ordinal, interval, or ratio.

a. The length of time it takes each of 15 telephone installers to hook up a wall phone
b. The style of music preferred by each of 30 randomly selected radio listeners
c. The arrival time of the 5 P.M. train from New York to Newark
d. A sample of 100 customers in a fast-food restaurant is asked to rate their hamburger on the following scale: poor, fair, good, excellent.
e. Classify each of the data sets in parts **a–d** as qualitative or quantitative.

2.90 Discuss the conditions under which the median is preferred to the mean as a measure of central tendency.

2.91 Construct a relative frequency histogram for the data summarized in the accompanying table.

Measurement Class	*Relative Frequency*	*Measurement Class*	*Relative Frequency*
.00– .75	.02	5.25–6.00	.15
.75–1.50	.01	6.00–6.75	.12
1.50–2.25	.03	6.75–7.50	.09
2.25–3.00	.05	7.50–8.25	.05
3.00–3.75	.10	8.25–9.00	.04
3.75–4.50	.14	9.00–9.75	.01
4.50–5.25	.19		

2.92 If it is not examined carefully, the graphical description of U.S. peanut production shown here can be misleading.

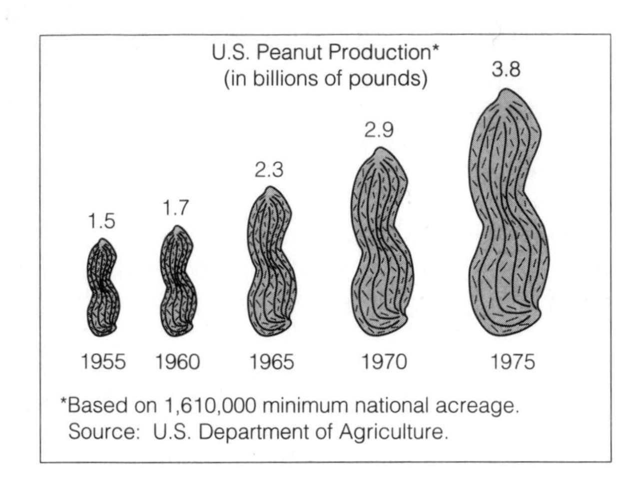

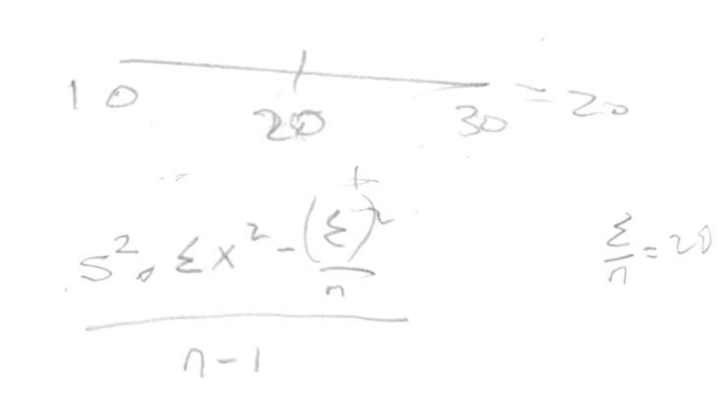

a. Explain why the graph may mislead some readers.
b. Construct an undistorted graph of U.S. peanut production for the given years.

2.93 Consider the following three measurements: 50, 70, 80. Find the z-score for each measurement if they are from a population with a mean and standard deviation equal to:
a. $\mu = 60$, $\sigma = 10$ b. $\mu = 60$, $\sigma = 5$ c. $\mu = 40$, $\sigma = 10$ d. $\mu = 40$, $\sigma = 100$

2.94 If the range of a set of data is 20, find a rough approximation to the standard deviation of the data set.

2.95 Compute s^2 for data sets with the following characteristics:

a. $\sum_{i=1}^{n} x_i^2 = 246, \quad \sum_{i=1}^{n} x_i = 63, \quad n = 22$

b. $\sum_{i=1}^{n} x_i^2 = 666, \quad \sum_{i=1}^{n} x_i = 106, \quad n = 25$

c. $\sum_{i=1}^{n} x_i^2 = 76, \quad \sum_{i=1}^{n} x_i = 11, \quad n = 7$

2.96 For each of the following data sets, compute $\bar{x}$, s^2, and s:
a. 13, 1, 10, 3, 3 b. 13, 6, 6, 0 c. 1, 0, 1, 10, 11, 11, 15 d. 3, 3, 3, 3
e. For each data set in parts **a–d**, form the interval $\bar{x} \pm 2s$, and calculate the percentage of the measurements that fall in the interval.

2.97 For each of the following data sets, compute $\bar{x}$, s^2, and s. If appropriate, specify the units in which your answers are expressed.
a. 4, 6, 6, 5, 6, 7
b. −\$1, \$4, −\$3, \$0, −\$3, −\$6
c. $\frac{3}{5}\%$, $\frac{4}{5}\%$, $\frac{2}{5}\%$, $\frac{1}{5}\%$, $\frac{1}{16}\%$
d. Calculate the range of each data set in parts **a–c**.

2.98 Explain why we generally prefer the standard deviation to the range as a measure of variability for quantitative data.

Applying the Concepts

2.99 In some locations, radiation levels in homes are measured at well above normal background levels in the environment. As a result, many architects and builders are making design changes to assure adequate air exchange so that radiation will not be "trapped" in homes. In one such location, 50 homes' levels were measured, and the mean level was 10 parts per billion (ppb), the median was 8 ppb, and the standard deviation was 3 ppb. Background levels in this location are at about 4 ppb.

a. Based on these results, is the distribution of the 50 homes' radiation levels symmetric, skewed to the left, or skewed to the right? Why?

b. Use both Chebyshev's rule and the Empirical Rule to describe the distribution of radiation levels. Which do you think is most appropriate in this case? Why?

c. Use the results from part **b** to approximate the number of homes in this sample that have radiation levels above the background level.

d. Suppose another home is measured at a location 10 miles from the one sampled, and has a level of 20 ppb. What is the z-score for this measurement relative to the 50 homes sampled in the other location? Is it likely that this new measurement comes from the same distribution of radiation levels as the other 50? Why? How would you go about confirming your conclusion?

2.100 A recent study (Bergin, 1992) reported in the *Journal of Leisure Research* investigated the relationship between academic performance and leisure activities. One hundred fifty-nine high school students were given a list of 43 leisure activities (sports, fishing, music, drama, photography, writing, watching TV, etc.). Each was asked to state how many they participated in each week. From this list, activities that involved reading, writing, or arithmetic were selected to form another variable, "academic leisure activities." Some of the results of the study are presented below:

	$\bar{x}$	s
GPA	2.96	.71
Number of leisure activities	12.38	5.07
Number of academic leisure activities	2.77	1.97

a. For GPA, calculate the intervals $\bar{x} \pm s$, $\bar{x} \pm 2s$, and $\bar{x} \pm 3s$. Based on these intervals, comment on the skewness or symmetry you would expect in these data. (Remember, the range of GPAs is 0 to 4.) Approximately what percentage of the students would you expect to find in each interval?

b. For number of leisure activities, calculate the intervals $\bar{x} \pm s$, $\bar{x} \pm 2s$, and $\bar{x} \pm 3s$. Based on these intervals, comment on the skewness or symmetry you would expect in these data. (Remember, the number of leisure activities cannot be negative.) Approximately what percentage of students would you expect to find in each interval?

c. For number of academic leisure activities, calculate the intervals $\bar{x} \pm s$, $\bar{x} \pm 2s$, and $\bar{x} \pm 3s$. Based on these intervals, comment on the skewness or symmetry you would expect in these data. (Remember, the number of leisure activities cannot be negative.) Approximately what percentage of students would you expect to find in each interval?

d. Based on your answers, which of the variables' distributions would you expect to be most skewed? Why?

2.101 A radio station claims that the amount of advertising per hour of broadcast time has an average of 3 minutes and a standard deviation equal to 2.1 minutes. You listen to the radio station for 1 hour, at a randomly selected time, and carefully observe that the amount of advertising time is equal to 7 minutes. Does this observation appear to disagree with the radio station's claim? Explain.

2.102 Various state and national automobile associations regularly survey gasoline stations to determine the current retail price of gasoline. Suppose one such national association decides to survey 200 stations in the United States and intends to determine the price of regular unleaded gasoline at each station.

a. Identify the population of interest.
b. Identify the sample.
c. Identify the variable of interest.
d. In the context of this problem, define the following numerical descriptive measures: μ, σ, $\bar{x}$, s.
e. Suppose the sample of 200 stations is selected, and the mean and standard deviation of their regular unleaded prices (per gallon) are \$1.39 and \$.12, respectively. Interpret these descriptive statistics and describe the probable distribution of the 200 prices at the time of the survey.
f. One station in the southeast priced unleaded gasoline at \$1.09 per gallon at the time of the survey. Describe the relative standing of this price in the national price distribution as indicated by the sample.

2.103 A severe drought affected several western states for 3 years. A Christmas tree farmer is worried about the drought's effect on the size of his trees. To decide whether the growth of the trees has been retarded, the farmer decides to take a sample of the heights of 25 trees and obtains the following results (recorded in inches):

60	57	62	69	46	54	64	60	59	58	75	51	49
67	65	44	58	55	48	62	63	73	52	55	50	

The following descriptive statistics and stem-and-leaf display were obtained by using SPSS to analyze the 25 tree heights:

```
HEIGHT

Valid cases:    25.0    Missing cases:  .0    Percent missing:  .0

Frequency     Stem &   Leaf

    1.00         4 *   4
    3.00         4 .   689
    4.00         5 *   0124
    6.00         5 .   557889
    6.00         6 *   002234
    3.00         6 .   579
    1.00         7 *   3
    1.00         7 .   5

Stem width:       10.00
Each leaf:         1 case(s)

Mean       58.2400   Std Err      1.6179   Min        44.0000
Median     58.0000   Variance    65.4400   Max        75.0000
5% Trim    58.1000   Std Dev      8.0895   Range      31.0000
```

a. Use the stem-and-leaf display to give a verbal description of the data set. Do any of the measurements appear to be outliers?
b. Examine the computer output to determine the value of $\bar{x}$, the median, s^2, and s for the tree heights. [*Note:* SPSS also produces some descriptive statistics that we have not covered; ignore these for now.]
c. Based on the stem-and-leaf display, and the relationship of the mean and median, is the distribution of these tree height data approximately mound-shaped, or is it skewed to the right or left?
d. Based on your answer to part **c**, what percentage of the measurements do you expect to find in the intervals $\bar{x} \pm s$, $\bar{x} \pm 2s$, and $\bar{x} \pm 3s$?
e. Count the number of measurements that actually fall in each interval of part **d** and express each interval count as a percentage of the total number of measurements. Compare these results to your estimates from part **d**.

2.104 A small computing center has found that the number of jobs submitted per day to its computers has a distribution that is approximately mound-shaped, with a mean of 83 jobs and a standard deviation of 10.
a. On approximately what percentage of days will the number of jobs submitted be between 73 and 93?
b. On approximately what percentage of days will the number of jobs submitted be between 63 and 83?
c. On approximately what percentage of days will the number of jobs submitted be greater than 93?

2.105 The Community Attitude Assessment Scale (CAAS) measures citizens' attitudes toward 15 life areas (e.g., education, employment, and health) on four dimensions—importance, influence, equality of opportunity, and satisfaction. In order to develop the CAAS, a number of households in each of 25 communities were randomly selected and sent questionnaires. Because relatively low response rates suggest that there could be a substantial but unknown opinion bias in the reported data, the percentage of the sample responding to the survey was determined in each community. The results are given here (in percent):

21 14 18 20 14 16 6 22 28 16 26 14 13
15 25 21 14 7 12 8 15 14 21 22 10

*a. Use statistical software to construct a stem-and-leaf display for the data. Use it to give a verbal description of the data set.
*b. Use the software to construct a relative frequency histogram for the data given, locating the mean, median, and mode.
c. Find the range for the data and use it to calculate an approximate value for s. Use this value to check your answer to part **d**.
d. Calculate the variance and standard deviation for the data.
e. Find the proportion of the measurements that fall in the interval $\bar{x} \pm 2s$.

***2.106** Refer to Exercise 2.105. Use a statistical software package to construct a box plot for the percent responses. Use the box plot to describe the distribution of responses.

2.107 A professor believes that if a class is allowed to work on an examination as long as desired, the times spent by the students would be approximately mound-shaped with mean 40 minutes and standard deviation 6 minutes. Approximately how long should be allotted for the examination if the professor wants almost all (say, 97.5%) of the class to finish?

2.108 By law, a box of cereal labeled as containing 16 ounces must contain at least 16 ounces of cereal. It is known that the machine filling the boxes produces a distribution of fill weights that is mound-shaped, with mean equal to the setting on the machine and with a standard deviation equal to .03 ounce. To ensure that most of the boxes contain at least 16 ounces, the machine is set so that the mean fill per box is 16.09 ounces.

a. What percentage of the boxes will contain less than 16 ounces if the machine is set so that $\mu = 16.09$?
b. If the machine is set so that $\mu = 16.09$, is it likely that a randomly selected box would contain less than 16 ounces?
c. If the machine is set so that $\mu = 16.09$, is it likely that a randomly selected box of cereal would contain as little as 16.05 ounces? Explain.

2.109 Most people living in metropolitan areas receive impressions of what is happening in their area primarily through their major newspapers. A study was conducted to determine whether the *Uniform Crime Report*, compiled by the Federal Bureau of Investigation, and the daily newspaper gave consistent information about the trend and distribution of crime in a metropolitan area. An attention score, based on the amount of space devoted to a story, was calculated for each paper's coverage of murders, assaults, robberies, etc. Suppose μ, the average murder attention score of metropolitan newspapers across the country in 1990, was 60, with $\sigma = 4.5$. One metropolitan newspaper in the midwest had a 1990 murder attention score of 69.
a. Approximately what percentage of the newspapers had a murder attention score higher than 69 in 1990? (Make no assumptions about the nature of the distribution of scores.)
b. Repeat part **a**, assuming attention scores were mound-shaped.

On Your Own

We list here several sources of real-life data sets that have been obtained from Wasserman and Bernero's *Statistics Sources*. This index of data sources is very complete and is a useful reference for anyone interested in finding almost any type of data. First we list some almanacs:

CBS News Almanac
Information Please Almanac
World Almanac and Book of Facts

United States Government publications are also rich sources of data:

Agricultural Statistics
Digest of Educational Statistics
Handbook of Labor Statistics
Housing and Urban Development Yearbook
Social Indicators
Uniform Crime Reports for the United States
Vital Statistics of the United States
Business Conditions Digest
Economic Indicators
Monthly Labor Review
Survey of Current Business
Bureau of the Census Catalog

Main data sources are published on an annual basis:

Commodity Yearbook
Facts and Figures on Government Finance

Municipal Yearbook
Standard and Poor's Corporation, Trade and Securities: Statistics

Some sources contain data that are international in scope:

Compendium of Social Statistics
Demographic Yearbook
United Nations Statistical Yearbook
World Handbook of Political and Social Indicators

Utilizing the data sources listed, sources suggested by your instructor, or your own resourcefulness, find one real-life quantitative data set that stems from an area of particular interest to you.

a. Describe the data set by using a relative frequency histogram.
b. Find the mean, median, variance, standard deviation, and range of the data set.
c. Use Tables 2.5 and 2.6 to describe the distribution of this data set. Count the actual number of observations that fall within 1, 2, and 3 standard deviations of the mean of the data set and compare these counts with the description of the data set you developed in part **b**.

Using the Computer

We have described a set of demographic data in Appendix B (available on magnetic tape or diskette from the publisher) that will be used as a source for the **Using the Computer** exercises at the end of most chapters. Briefly, the data consist of 15 demographic variables measured for 1,000 U.S. zip code areas. Included are such variables as population size, number of households, average household size, average income, percentage of college graduates, percentage of women in the work force, and purchasing potential indexes for groceries, sporting goods, and home improvements.

a. Consider the percentage of women in the work force. Use a statistical software package to generate a stem-and-leaf display, a relative frequency histogram, and a box plot for these 1,000 percentages. Compare the graphical descriptions you obtain, and discuss what each reveals about the distribution of the percentage of women in the work force in the sample of 1,000 zip codes.

b. Repeat part **a** for one of the census regions, perhaps the one in which you currently reside. Compare the regional distribution to the national distribution you obtained in part **a**.

c. Finally, repeat part **a** for the zip codes corresponding to one state, perhaps the one in which you currently reside. Compare the state distribution to the national and regional distributions of parts **a** and **b**.

d. Compute the mean and standard deviation of the percentage data over the 1,000 zip codes and over the zip codes in the region you selected in part **b**. Use the computer to count the number of zip codes' percentages in the intervals $\bar{x} \pm s$, $\bar{x} \pm 2s$, and $\bar{x} \pm 3s$. Compare the results with those given by Chebyshev's rule and the Empirical Rule (Tables 2.5 and 2.6).

References

Bergin, D. A. "Leisure activity, motivation, and academic achievement in high school students." *Journal of Leisure Research*, 1992, Vol. 24, No. 3, pp. 225–239.

Clute, P. S. "Mathematics anxiety: Instructional method and achievement in a survey course in college mathematics." *Journal for Research in Mathematics Education*, 1984, 15.

Gergen, D. R. "Lake Wobegon's schools." *U.S. News & World Report*, Mar. 5, 1990, p. 74.

Hoff, D. *How to Lie with Statistics*. New York: Norton, 1954.

Koopmans, L. H. *An Introduction to Contemporary Statistics*. North Scituate, Mass.: Duxbury, 1981, Chapters 1 and 2.

Martin, D. L. "Children out of school: Firsthand report: How flawed statistics can make an ugly picture look even worse." *American School Board Journal*, 1975, 162, pp. 57–59.

McMurray, R. G., Wilson, J. R., and Kitchell, B. S. "The effects of fructose and glucose on high-endurance performance." *Research Quarterly for Exercise and Sport*, 1983, 54.

Mendenhall, W. *Introduction to Probability and Statistics*, 8th ed. North Scituate, Mass.: Duxbury, 1991, Chapter 3.

Miller, S. L. et al. "Mercury vapor levels in the dental office: A survey." *Journal of the American Dental Association*, Nov. 1974, 89, pp. 1084–1091.

Mills, J. K. "Difference in locus of control between obese adults and adolescent females undergoing weight reduction." *Journal of Psychology*, Mar. 1991, 125(2), pp. 195–197.

Minitab Reference Manual, Release 8. State College, Penn.: Minitab, Inc., 1991.

Murdock, S. H. et al. "The implications of change in population size and composition on future participation in outdoor recreational activities." *Journal of Leisure Research*, 1991, Vol. 23, No. 3, pp. 238–259.

Neustadter, H. E. and Sidik, S. M. "On evaluating compliance with air pollution levels 'not to be exceeded more than once a year'." *Journal of the Air Pollution Control Association*, 1974, Vol. 24, No. 6, pp. 559–563.

Nie, N., Hull, C. H., Jenkins, J. G., Steinbrenner, K., and Bent, D. H. *Statistical Package for the Social Sciences*, 2nd ed. New York: McGraw-Hill, 1975.

Real, M. R., and Mechikoff, R. A. "Deep fan: Mythic identification, technology, and advertising in spectator sports." *Sociology of Sport Journal*, 1992, Vol. 9, No. 4.

SAS User's Guide: Statistics, Version 5 ed. Cary, N.C.: SAS Institute, Inc. 1985.

Wasserman, P., and Bernero, J. *Statistics Sources*, 5th ed. Detroit: Gale Research Company, 1978.

CHAPTER THREE

Probability

Contents

Case Studies

Where We've Been

We have identified inference, from a sample to a population, as the goal of statistics. To reach this goal, we must be able to describe a set of measurements. The use of graphic and numerical methods for describing data sets and for phrasing inferences was the topic of Chapter 2.

Where We're Going

Now that we know how to phrase an inference about a population, we turn to the problem of making the inference. What is it that permits us to make the inferential jump from sample to population and then to give a measure of reliability for the inference? As you will subsequently see, the answer is *probability*. This chapter is devoted to a study of probability—what it is and some of the basic concepts of the theory behind it.

3

You will recall that statistics is concerned with decisions about a population based on sample information. Understanding how this is accomplished will be easier if you understand the relationship between population and sample. This understanding is enhanced by reversing the statistical procedure of making inferences from sample to population. In this chapter we assume the population *known* and calculate the chances of obtaining various samples from the population. Thus, probability is the reverse of statistics: In probability, we use the population information to infer the probable nature of the sample.

Probability plays an important role in inference making. To illustrate, suppose you have an opportunity to invest in an oil exploration company. Past records show that out of 10 previous oil drillings (a sample of the company's experiences), all 10 resulted in dry wells. What do you conclude? Do you think the chances are better than 50–50 that the company will hit a producing well? Should you invest in this company? We think your answer to these questions will be an emphatic no. If the company's exploratory prowess is sufficient to hit a producing well 50% of the time, a record of 10 dry wells out of 10 drilled is an event that is just too improbable. Do you agree?

As another illustration, suppose you are playing poker with what your opponents assure you is a well-shuffled deck of cards. In three consecutive five-card hands, the person on your right is dealt four aces. Based on this sample of three deals, do you think the cards are being adequately shuffled? Again, we think your answer will be no and that you will reach this conclusion because dealing three hands of four aces is just too improbable, assuming that the cards were properly shuffled.

Note that the decisions concerning the potential success of the oil drilling company and the decision concerning the card shuffling were both based on probabilities, namely, the probabilities of certain sample results. Both situations were contrived so that you could easily conclude that the probabilities of the sample results were small. Unfortunately, the probabilities of many observed sample results are not so easy to evaluate intuitively. For these cases we will need the assistance of a theory of probability.

3.1 Events, Sample Spaces, and Probability

We begin our treatment of probability with simple examples that are easily described, thus eliminating any discussion that could be distracting. With the aid of simple examples, important definitions are introduced and the notion of probability is more easily developed.

Suppose a coin is tossed once and the up face is recorded. This is an **observation**, or **measurement**. Any process of making an observation is called an **experiment**. Our definition of experiment is broader than that used in the physical sciences, where you would picture test tubes, microscopes, etc. Other practical examples of statistical experiments are recording whether a customer prefers one of two brands of electronic calculators, recording a voter's opinion on an important political issue, measuring the amount of dissolved oxygen in a polluted river, observing the closing price of a stock, counting the number of errors in an inventory, and observing the fraction of insects killed by a new insecticide. This list of statistical experiments could be continued, but the point is that our definition of experiment is very broad.

Definition 3.1

An **experiment** is an act or process that leads to a single outcome that cannot be predicted with certainty.

Consider another simple experiment consisting of tossing a die and observing the number on the up face. The six basic possible outcomes to this experiment are:

1. Observe a 1 2. Observe a 2
3. Observe a 3 4. Observe a 4
5. Observe a 5 6. Observe a 6

Note that if this experiment is conducted once, *you can observe one and only one of these six basic outcomes, and the outcome cannot be predicted with certainty*. Also, these possibilities cannot be decomposed into more basic outcomes. The basic possible outcomes to an experiment are called **simple events**.

Definition 3.2

A **simple event** is the most basic outcome of an experiment.

EXAMPLE 3.1

Two coins are tossed, and their up faces are recorded. List all the simple events for this experiment.

Solution

Even for a seemingly trivial experiment, we must be careful when listing the simple events. At first glance the basic outcomes seem to be: Observe two heads; Observe two tails; or Observe one head and one tail. However, further reflection reveals that the last of these, Observe one head and one tail, can be decomposed into Head on coin 1, Tail on coin 2 and Tail on coin 1, Head on coin 2.* Thus, the simple events are as follows:

1. Observe HH 2. Observe HT
3. Observe TH 4. Observe TT

where H in the first position means "Head on coin 1," H in the second position means "Head on coin 2," etc.

*Even if the coins are identical in appearance, there are, in fact, two distinct coins. Thus, the designation of one coin as coin 1 and the other as coin 2 is legitimate in any case.

We often wish to refer to the collection of all the simple events of an experiment. This collection is called the **sample space** of the experiment. For example, there are six simple events in the sample space associated with the die-toss experiment. The sample spaces for the experiments discussed thus far are shown in Table 3.1.

TABLE 3.1 Experiments and Their Sample Spaces

Experiment: Observe the up face on a coin.

Sample space:
1. Observe a head
2. Observe a tail

This sample space can be represented in set notation as a set containing two simple events:

S: $\{H, T\}$

where H represents the simple event Observe a head and T represents the simple event Observe a tail.

Experiment: Observe the up face on a die.

Sample space:
1. Observe a 1
2. Observe a 2
3. Observe a 3
4. Observe a 4
5. Observe a 5
6. Observe a 6

This sample space can be represented in set notation as a set of six simple events:

S: $\{1, 2, 3, 4, 5, 6\}$

Experiment: Observe the up faces on two coins.

Sample space:
1. Observe HH
2. Observe HT
3. Observe TH
4. Observe TT

This sample space can be represented in set notation as a set of four simple events:

S: $\{HH, HT, TH, TT\}$

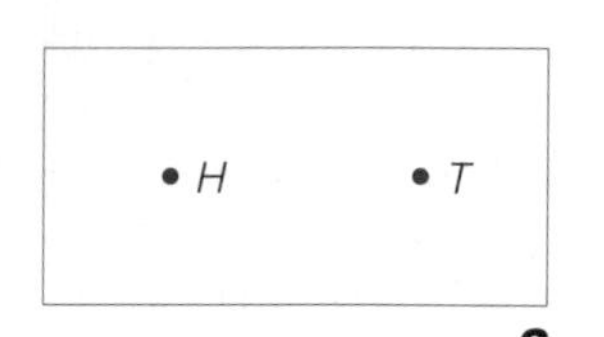

a. Experiment: Observe the up face on a coin

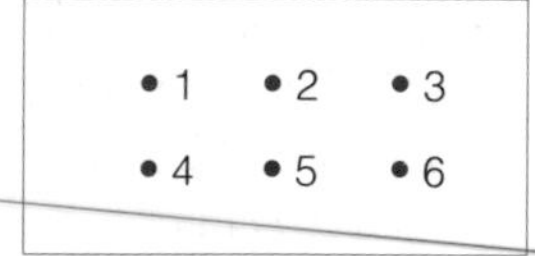

b. Experiment: Observe the up face on a die

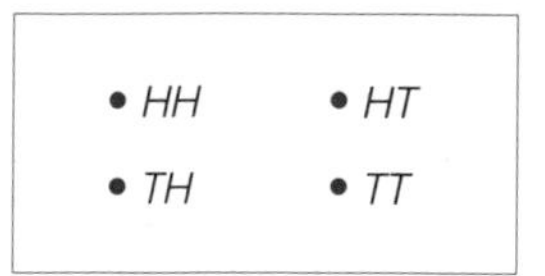

c. Experiment: Observe the up faces on two coins

FIGURE 3.1 ▲
Venn diagrams for the three experiments from Table 3.1

Definition 3.3

The **sample space** of an experiment is the collection of all its simple events.

Just as graphs are useful in describing sets of data, a pictorial method for presenting the sample space and its simple events will often be useful. Figure 3.1 shows such a representation for each of the experiments in Table 3.1. In each case, the sample space is shown as a closed figure, labeled S, containing a set of points, called **sample points,** with each point representing one simple event. Note that the number of sample points in a sample space S is equal to the number of simple events associated with the respective experiment: two for the coin toss, six for the die toss, and four for the two-coin toss. These graphic representations are called **Venn diagrams**.

Now that we have defined simple events as the basic outcomes of the experiment and the sample space as the collection of all the simple events, we are prepared to discuss the probabilities of simple events. You have undoubtedly used the term *probability* and have some intuitive idea about its meaning. Probability is generally used synonymously with "chance," "odds," and similar concepts. We will begin our treatment of probability using these informal concepts and then solidify what we mean later. For example, if a fair coin is tossed, we might reason that both the simple events, Observe a head and Observe a tail, have the same chance of occurring. Thus, we might state that "the probability of observing a head is 50%" or "the odds of seeing a head are 50–50." Both these statements are based on an informal knowledge of probability.

The probability of a simple event is a number between 0 *and* 1 *that measures the likelihood that the event will occur when the experiment is performed.* This number is usually taken to be the relative frequency of the occurrence of a simple event in a very long series of repetitions of an experiment. When this information is not available, we select the number based on experience. For example, if we are assigning probabilities to the two simple events in the coin-toss experiment (Observe a head and Observe a tail), we might reason that if we toss a balanced coin a very large number of times, the simple events Observe a head and Observe a tail will occur with the same relative frequency of .5. Thus, the probability of each simple event is .5.

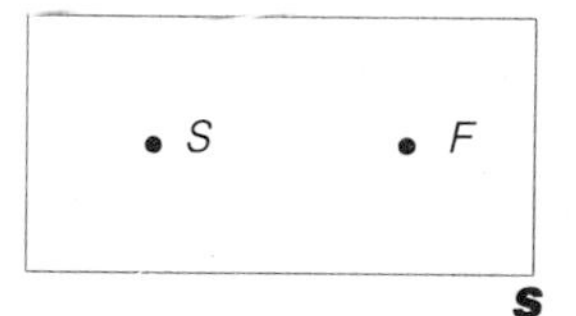

FIGURE 3.2 ▲
Experiment: Invest in a business venture and observe whether it succeeds (*S*) or fails (*F*)

For some experiments, we may assign probabilities to the simple events based on general information about the experiment. For example, if the experiment is to invest in a business venture and to observe whether it succeeds or fails, the sample space would appear as in Figure 3.2. We are unlikely to be able to assign probabilities to the simple events of this experiment based on a long series of repetitions since unique factors govern each performance of this kind of experiment. Instead, we may consider factors such as the personnel managing the venture, the general state of the economy at the time, the rate of success of similar ventures, and any other information deemed pertinent. If we finally decide that the venture has an 80% chance of succeeding, we assign a probability of .8 to the simple event Success. This probability can be interpreted as a measure of our degree of belief in the outcome of the business venture. Such subjective probabilities should be based on expert information and must be carefully assessed. If they are not, we may be misled on any decisions based on these probabilities or based on any calculations in which they appear. [*Note:* For a text that deals in detail with the subjective evaluation of probabilities, see Winkler (1972) or Lindley (1985).]

Bloom County Probabilities

The issue of whether probability should be defined as the relative frequency in a long series of repetitions of an experiment or as a subjective measure of belief is one that has been debated for many years by probabilists, statisticians, and even philosophers. A considerably lighter side of this debate was illustrated in a Bloom County comic strip.

© 1986, Washington Post Writers Group, reprinted with permission.

No matter how you assign the probabilities to simple events, the probabilities assigned must obey two rules:

1. All simple event probabilities *must* lie between 0 and 1.
2. The probabilities of all the simple events within a sample space *must* sum to 1.

Assigning probabilities to simple events is easy for some experiments. For example, if the experiment is to toss a fair coin and observe the up face, we would probably all agree to assign a probability of ½ to the two simple events, Observe a head and Observe a tail. However, many experiments have simple events whose probabilities are more difficult to assign.

EXAMPLE 3.2

A retail computer store owner sells two basic types of microcomputers: IBM personal computers (IBM PCs) and IBM compatibles (PCs that run all or most of the same software as an IBM PC but that are not manufactured by IBM). One problem facing the owner is deciding how many of each type of PC to stock. An important factor affecting the solution is the proportion of customers who purchase each type of PC.

Show how this problem might be formulated in the framework of an experiment with simple events and a sample space. Indicate how probabilities might be assigned to the simple events.

Solution

If we use the term *customer* to refer to a person who purchases one of the two types of PCs, the experiment can be defined as the entrance of a customer and the observation of which type of PC is purchased. There are two simple events in the sample space corresponding to this experiment:

1. *I*: {The customer purchases an IBM PC}
2. *C*: {The customer purchases a compatible}

The difference between this and the coin-toss experiment becomes apparent when we attempt to assign probabilities to the two simple events. What probability should we assign to the simple event *I*? If you answer .5, you are assuming that the events *I* and *C* should occur with equal likelihood, just as the simple events Heads and Tails in the coin-toss experiment. The assignment of simple event probabilities for the PC purchase experiment is not so easy. Suppose a check of the store's records indicates that 80% of its customers purchase IBM PCs. Then it might be reasonable to approximate the probability of the simple event *I* as .8 and that of the simple event *C* as .2. The important points are that simple events are not always equally likely and that the probabilities of simple events are not always easy to assign, particularly for experiments that represent real applications (as opposed to coin- and die-toss experiments).

Although the probabilities of simple events are often of interest in their own right, it is usually probabilities of collections of simple events that are important. Example 3.3 demonstrates this point.

EXAMPLE 3.3

A fair die is tossed, and the up face is observed. If the face is even, you win $1. Otherwise, you lose $1. What is the probability that you win?

Solution

Recall that the sample space for this experiment contains six simple events:

S: {1, 2, 3, 4, 5, 6}

Since the die is balanced, we assign a probability of $\frac{1}{6}$ to each of the simple events in this sample space. An even number will occur if one of the simple events, Observe a 2, Observe a 4, or Observe a 6, occurs. A collection of simple events such as this is called an **event**, and we denote this event by the letter A. Since the event A contains three simple events—all with probability $\frac{1}{6}$—and since no simple events can occur simultaneously, we reason that the probability of A is the sum of the probabilities of the simple events in A. Thus, the probability of A is $\frac{1}{6} + \frac{1}{6} + \frac{1}{6} = \frac{1}{2}$. This implies that *in the long run* you will win $1 half the time and lose $1 half the time.

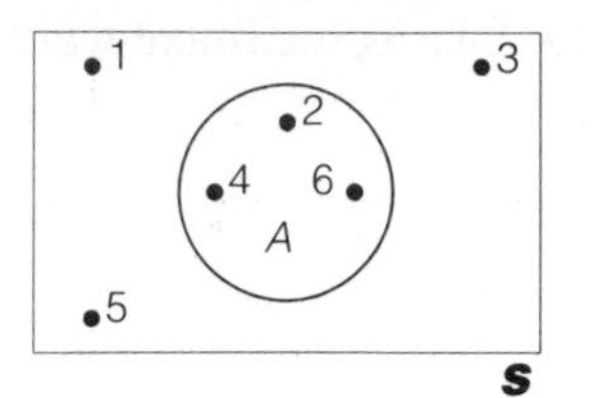

FIGURE 3.3 ▲
Die-toss experiment with event *A*: Observe an even number

Figure 3.3 is a Venn diagram depicting the sample space associated with a die-toss experiment and the event A, Observe an even number. The event A is represented by the closed figure inside the sample space S. This closed figure A contains all the simple events that comprise it.

How do you decide which simple events belong to the set associated with an event A? Test each simple event in the sample space S. If event A occurs when a particular simple event occurs, then that simple event is in the event A. For example, the event A, Observe an even number, in the die-toss experiment will occur if the simple event Observe a 2 occurs. By the same reasoning, the simple events Observe a 4 and Observe a 6 are also in event A.

To summarize, we have demonstrated that an event can be defined in words or it can be defined as a specific set of simple events. This leads us to the following general definition of an *event*:

Definition 3.4

An **event** is a specific collection of simple events.

EXAMPLE 3.4

Simple Event	Probability
HH	4/9
HT	2/9
TH	2/9
TT	1/9

Consider the experiment of tossing two coins. Suppose the coins are *not* balanced and the correct probabilities associated with the simple events are given in the table. [*Note:* The necessary properties for assigning probabilities to simple events are satisfied.]

Consider the events

A: {Observe exactly one head}

B: {Observe at least one head}

Calculate the probability of A and the probability of B.

Solution

Event A contains the simple events *HT* and *TH*. Since two or more simple events cannot occur at the same time, we can easily calculate the probability of event A by summing the probabilities of the two simple events. Thus, the probability of observing exactly one head (event A), denoted by the symbol P(A), is

$$P(A) = P(\text{Observe } HT) + P(\text{Observe } TH)$$
$$= 2/9 + 2/9 = 4/9$$

Similarly, since B contains the simple events *HH*, *HT*, and *TH*,

$$P(B) = 4/9 + 2/9 + 2/9 = 8/9$$

The preceding example leads us to a general procedure for finding the probability of an event A:

> The probability of an event A is calculated by summing the probabilities of the simple events in A.

Thus, we can summarize the steps for calculating the probability of any event, as indicated in the next box.

Steps for Calculating Probabilities of Events

1. Define the experiment, i.e., describe the process used to make an observation and the type of observation that will be recorded.
2. List the simple events.
3. Assign probabilities to the simple events.
4. Determine the collection of simple events contained in the event of interest.
5. Sum the simple event probabilities to get the event probability.

EXAMPLE 3.5

In a poll of "computer-familiar" adults who do not own a home computer, each was asked to identify which of 10 electronic appliances was his or her highest priority purchase, if any. The results are summarized in Table 3.2.

TABLE 3.2 Electronic Appliances of Highest Priority to Purchase by Computer-Familiar Adults

Electronic Appliance	Response[a]
Home computer (HC)	24%
Microwave oven (MO)	13%
Compact disc player (CDP)	4%
Phone-answering machine (PAM)	6%
Car telephone (CT)	5%
Programmable phone (PP)	3%
Video cassette recorder (VCR)	17%
Video camera (VC)	9%
Big screen TV (BSTV)	7%
Movie camera (MC)	3%
None (N)	9%

[a]Response percentages in the *USA Today* article did not add to 100% due to rounding. We have added 1% to the two smallest responses to facilitate the solution to this example.
Source: "Buying a computer is in our budget," *USA Today*, Sept. 25, 1985. © 1985, USA Today. Excerpted with permission.

a. Define the experiment that generated the data in Table 3.2, and list the simple events.

b. Assign probabilities to the simple events.

c. What is the probability that a telephonic appliance is of highest priority?

d. What is the probability that a video appliance is of highest priority?

Solution

a. The experiment is the act of polling a computer-familiar adult. The simple events, the simplest outcomes of the experiment, are the 11 response categories listed in Table 3.2. They are shown in the Venn diagram in Figure 3.4.

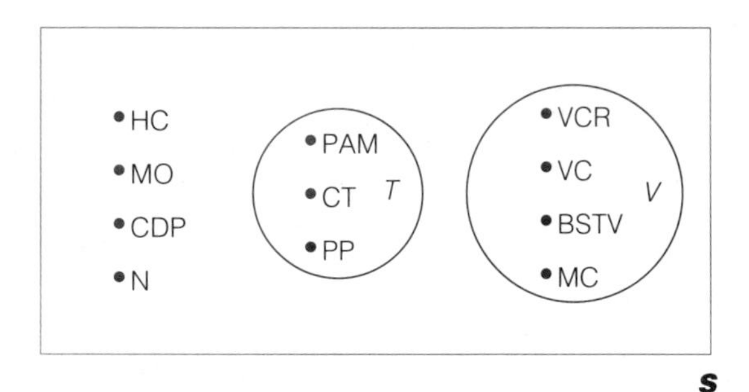

FIGURE 3.4 ► Venn diagram for the electronic appliance poll

b. In Example 3.1 the simple events were assigned equal probabilities. If we were to assign equal probabilities in this case, each of the response categories would be assigned a probability of one-eleventh ($1/11$), or .09. However, you can see by examining Table 3.2 that equal probabilities are not reasonable in this case because the response percentages were not even approximately the same in the 11 classifications. It is more reasonable to assign a probability equal to the response percentage in each class, as shown in Table 3.3.*

TABLE 3.3 Simple Event Probabilities for Electronic Appliance Poll

Simple Event	Probability
HC	.24
MO	.13
CDP	.04
PAM	.06
CT	.05
PP	.03
VCR	.17
VC	.09
BSTV	.07
MC	.03
N	.09

c. The event T that a telephonic appliance is the highest priority is not a simple event because it consists of more than one of the response classifications (the simple events). In fact, as shown in Figure 3.4, T consists of three simple events. The probability of T is defined to be the sum of the probabilities of the simple events in T:

$$P(T) = P(\text{PAM}) + P(\text{CT}) + P(\text{PP})$$
$$= .06 + .05 + .03 = .14$$

*Since the response percentages were based on a sample of households, these assigned probabilities are estimates of the true population response percentages. You will learn how to measure the reliability of probability estimates in Chapter 6.

d. The event V that a video appliance is identified as the highest-priority purchase consists of four simple events, and the probability is the sum of the corresponding simple event probabilities:

$$P(\text{V}) = P(\text{VCR}) + P(\text{VC}) + P(\text{BSTV}) + P(\text{MC})$$
$$= .17 + .09 + .07 + .03 = .36$$

For the experiments discussed thus far, listing the simple events has been easy. For more complex experiments, the number of simple events may be so large that listing them is impractical. In solving probability problems for experiments with many simple events we employ the same principles as for experiments with few simple events. The only difference is that we need **counting rules** for determining the number of simple events without actually enumerating all of them. Counting rules are beyond the scope of this introductory text. For those who are interested, several of the more useful counting rules can be found in the references at the end of this chapter.

CASE STUDY 3.1 / Comparing Subjective Probability Assessments with Relative Frequencies

Preston and Baratta (1948) performed an experiment with the objective of comparing how an individual's subjective assessment of the probability of an event compares with the known probability (relative frequency of occurrence) of the event. The individuals selected for the experiment ranged from undergraduates with no training in probability theory to professors of mathematics and statistics with a "substantial acquaintance with probability theory." Each individual participated in a game in which he or she bet part of an initial stake on one of seven different outcomes of a combination card–dice game. The probabilities of these seven different events, known only to the experimenters, were .01, .05, .25, .50, .75, .95, and .99. From the amount the subject is willing to bet, the individual's subjective probabilities can be determined and then compared with the actual probabilities of the events.

In Figure 3.5 on page 106 we reproduce the author's figure depicting the average subjective probability assessed by the experimental subjects compared to the true probabilities. Some of the conclusions reached were:

1. Events with probabilities less than .25 were subjectively overestimated.
2. Events with probabilities more than .25 were subjectively underestimated.
3. These conclusions were the same for both the probabilistically naive and sophisticated subjects.

For events with probabilities that are not clearly defined (such as the probability of rain tomorrow), it is important to have information on general tendencies in subjectively evaluating the probabilities. Of course, much more evidence has been collected on the subjective evaluation of probabilities since the Preston and Baratta article (not all of which corroborate their conclusions), but it remains an interesting evaluation of a person's ability to evaluate probabilities subjectively.

FIGURE 3.5 ▶
Observed relationship between true and subjective probabilities

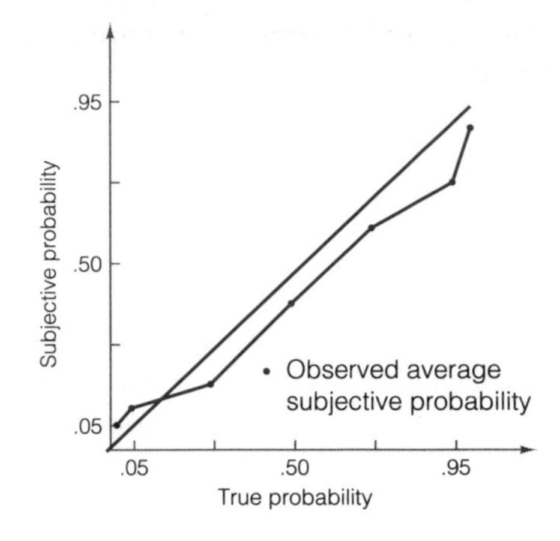

Exercises 3.1–3.17

Learning the Mechanics

3.1 An experiment results in one of the following simple events: E_1, E_2, E_3, E_4, or E_5.

a. Find $P(E_3)$ if $P(E_1) = .1$, $P(E_2) = .2$, $P(E_4) = .1$, and $P(E_5) = .1$. .5

b. Find $P(E_3)$ if $P(E_1) = P(E_3)$, $P(E_2) = .1$, $P(E_4) = .2$, and $P(E_5) = .1$. .3

c. Find $P(E_3)$ if $P(E_1) = P(E_2) = P(E_4) = P(E_5) = .1$. .6

3.2 The diagram describes the sample space of a particular experiment and events A and B.

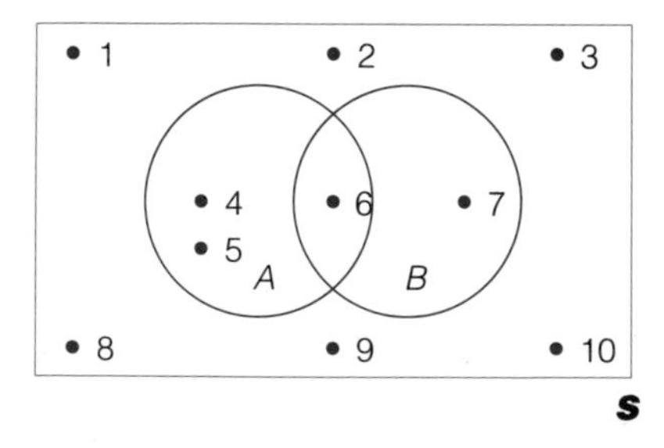

a. What is this type of diagram called?

b. Suppose the simple events are equally likely. Find $P(A)$ and $P(B)$.

c. Suppose $P(1) = P(2) = P(3) = P(4) = P(5) = 1/20$ and $P(6) = P(7) = P(8) = P(9) = P(10) = 3/20$. Find $P(A)$ and $P(B)$.

3.3 The sample space for an experiment contains five simple events with probabilities as shown in the table. Find the probability of each of the following events:

Simple Events	*Probabilities*
1	.05
2	.20
3	.30
4	.30
5	.15

A: {Either 1, 2, or 3 occurs}
B: {Either 1, 3, or 5 occurs}
C: {4 does not occur}

3.4 Consider the experiment of tossing a die and observing the up face.
a. Draw a Venn diagram for the experiment. On your diagram indicate the event Observe a number greater than 4. Call this event A. Also indicate the event Observe an even number. Call this event *B*.
b. We would all agree that the probability of observing a 3 on the toss of a "fair" die is ⅙. Explain what it means for a die to be fair (or balanced) and explain how knowing that a die is fair leads us to $P(3) = 1/6$.
c. For your Venn diagram of part **a**, assume the die is fair and find $P(A)$ and $P(B)$.
d. If you knew that a particular die were unfair (i.e., "loaded"), how would you determine the probability of observing a 3?

3.5 The Venn diagram depicts an experiment with six simple events. The events A and *B* are also shown.

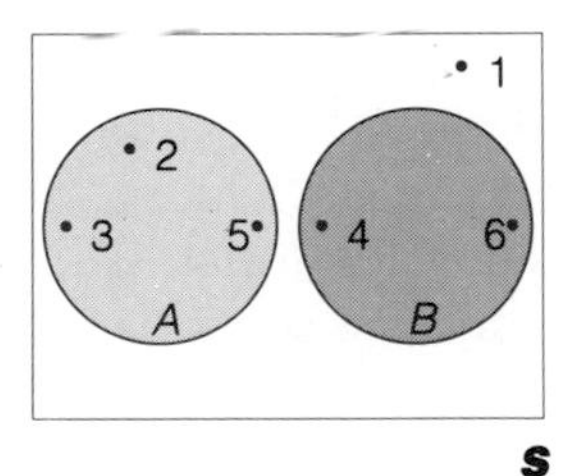

The probabilities of the simple events are as follows:

$P(1) = P(2) = P(4) = 2/9 \qquad P(3) = P(5) = P(6) = 1/9$

a. Find $P(A)$. b. Find $P(B)$.
c. Find the probability that the events A and *B* occur *simultaneously*.

3.6 Two fair dice are tossed, and the up face on each die is recorded.
a. List the 36 simple events contained in the sample space.
b. Find the probability of observing each of the following events:

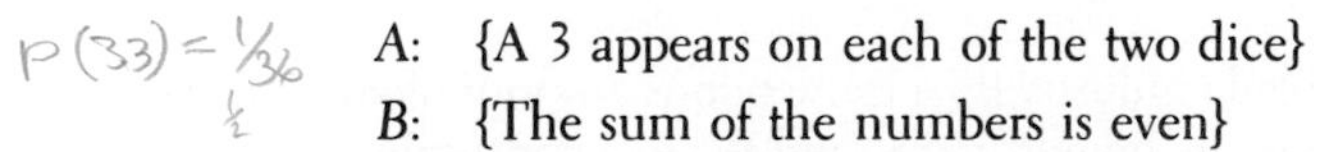

A: {A 3 appears on each of the two dice}
B: {The sum of the numbers is even}
C: {The sum of the numbers is equal to 7}

D: {A 5 appears on at least one of the dice}

E: {The sum of the numbers is 10 or more}

3.7 Consider the experiment composed of one roll of a fair die followed by one toss of a fair coin. List the simple events. Assign a probability to each simple event. Determine the probability of observing each of the following events:

A: {6 on the die; *H* on the coin}

B: {Even number on the die; *T* on the coin}

C: {Even number on the die}

D: {*T* on the coin}

3.8 Two marbles are drawn from a box containing two blue marbles and three red marbles. Determine the probability of observing each of the following events:

A: {Two blue marbles are drawn}

B: {A red and a blue marble are drawn}

C: {Two red marbles are drawn}

3.9 Simulate the experiment described in Exercise 3.8 using any five identically shaped objects, two of which are one color and three are another. Mix the objects, draw two, record the results, and then replace the objects. Repeat the experiment a large number of times (at least 100). Calculate the proportion of time events A, B, and C occur. How do these proportions compare with the probabilities you calculated in Exercise 3.8? Should these proportions equal the probabilities? Explain.

Applying the Concepts

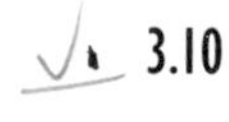

3.10 A hospital reports that two patients have been admitted who have contracted Legionnaire's disease. Suppose our experiment consists of observing whether the patients survive or die as a result of the disease. The simple events and probabilities of their occurrence are shown in the table (where *S* in the first position means that patient 1 survives, *D* in the first position means that patient 1 dies, etc.).

Simple Events	*Probabilities*
SS	.81
SD	.09
DS	.09
DD	.01

Find the probabilities of each of the following events:

A: {Both patients survive the disease}

B: {At least one patient dies}

C: {Exactly one patient survives the disease}

3.11 The corporations in the highly competitive razor blade industry do a tremendous amount of advertising each year. Corporation G gave a supply of the three top name brands, G, S, and W, to a consumer and asked him to use them and rank them in order of preference. The corporation was, of course, hoping the consumer would prefer its brand and rank it first, thereby giving them some material for a consumer interview advertising campaign. If the consumer did not prefer one blade over any other, but was still required to rank the blades, what is the probability that:

a. The consumer ranked brand G first?
b. The consumer ranked brand G last?
c. The consumer ranked brand G last and brand W second?
d. The consumer ranked brand W first, brand G second, and brand S third?

3.12 An individual's genetic makeup is determined by the genes obtained from each parent. For every genetic trait, each parent possesses a gene pair; and each contributes one-half of this gene pair, with equal probability, to their offspring, forming a new gene pair. The offspring's traits (eye color, baldness, etc.) come from this new gene pair, where each gene in this pair possesses some characteristic.

For the gene pair that determines eye color, each gene trait may be one of two types: dominant brown (*B*) or recessive blue (*b*). A person possessing the gene pair *BB* or *Bb* has brown eyes, whereas the gene pair *bb* produces blue eyes.

a. Suppose both parents of an individual are brown-eyed, each with a gene pair of the type *Bb*. What is the probability that a randomly selected child of this couple will have blue eyes? [*Hint:* Construct the sample space for the experiment.]
b. If one parent has brown eyes, type *Bb*, and the other has blue eyes, what is the probability that a randomly selected child of this couple will have blue eyes?
c. Suppose one parent is brown-eyed, type *BB*. What is the probability that a child has blue eyes?

3.13 Three people play a game called "Odd Man Out." In this game, each player flips a fair coin until the outcome (heads or tails) for one of the players is not the same as the other two players'. This player is then "the odd man out" and loses the game. Find the probability that the game ends (i.e., either exactly one of the coins will fall heads or exactly one of the coins will fall tails) after only one toss by each player. Suppose one of the players, hoping to reduce his chances of being the odd man, uses a two-headed coin. Will this ploy be successful? Solve by listing the simple events in the sample space.

3.14 The breakdown of workers in a particular state according to their political affiliation and type of job held is shown here. Suppose a worker is selected at random within the state and the worker's political affiliation and type of job are noted.

		Political Affiliation		
		Republican	Democrat	Independent
Type of Job	White Collar	12%	12%	6%
	Blue Collar	23%	43%	4%

a. List all simple events for this experiment.
b. What is the set of all simple events called?
c. Let A be the event that the worker is a white-collar worker. Find *P*(A).
d. Let *B* be the event that the worker is a Republican. Find *P*(*B*).
e. Let C be the event that the worker is a Democrat. Find *P*(C).
f. Let D be the event that the worker is a white-collar worker and a Democrat. Find *P*(*D*).

3.15 According to David Dreman (*Forbes*, October 27, 1980, pp. 202–203), investment in new issues (the stock of newly formed companies) can be both suicidal and rewarding. Dreman based his comments on a Securities and Exchange Commission (SEC) study of 500 new issues that went public during the 1961–1962 stock boom. The SEC found that of the 500 companies, 43% went bankrupt, 25% were operating at losses, and only 20% showed a profit. Only 12 companies of the 500 appeared to have outstanding prospects. Suppose back in 1961 you had randomly selected two of five new issues for investment and that, unknown to you, only two of the five would eventually show a profit. What is the probability that:

a. Both of the new issues in which you invested will eventually show a profit?
b. Neither of the two issues will eventually show a profit?
c. At least one of the two issues you selected will eventually show a profit?

3.16 Before placing a person in a highly skilled position, a company gives the applicants a series of three examinations. The first is a physical examination, and each applicant is classified as satisfactory or unsatisfactory. The other two are verbal and quantitative examinations, and the scores are used to classify each applicant as high, medium, or low in each area. Thus, each individual will receive a health score, a verbal score, and a quantitative score.

a. List the different sets of classifications that can result from this battery of examinations.
b. If all applicants who take the examinations are equally qualified, and all the variation in test scores is random, what is the probability that an applicant receives the lowest classification on all three examinations?
c. If an applicant scores in the highest category on at least two of the three examinations, the applicant will get a position. What is the probability that a randomly selected applicant will get a position? (Make the same assumption as in part **b**.)

3.17 Often probabilities are expressed in terms of **odds**, especially in gambling settings. For example, handicappers for horse races express their belief about the probabilities of each horse winning a race in terms of odds. If the probability of event E is $P(E)$, then the **odds in favor of *E*** are $P(E)$ to $1 - P(E)$. Thus, if a handicapper assesses a probability of .25 that Snow Chief will win the Belmont Stakes, the odds in favor of Snow Chief are $25/100$ to $75/100$, or 1 to 3. It follows that the **odds against *E*** are $1 - P(E)$ to $P(E)$, or 3 to 1 against a win by Snow Chief. In general, if the odds in favor of event E are a to b, then $P(E) = a/(a + b)$.

a. A second handicapper assesses the probability of a win by Snow Chief to be $1/3$. According to the second handicapper, what are the odds in favor of a Snow Chief win?
b. A third handicapper assesses the odds in favor of Snow Chief to be 1 to 1. According to the third handicapper, what is the probability of a Snow Chief win?
c. A fourth handicapper assesses the odds against Snow Chief winning to be 3 to 2. Find this handicapper's assessment of the probability that Snow Chief will win.

3.2 Unions and Intersections

An event can often be viewed as a composition of two or more other events. Such events are called **compound events**; they can be formed (composed) in two ways, as defined in the boxes and illustrated here.

Definition 3.5

The **union** of two events A and B is the event that occurs if either A or B or both occur on a single performance of the experiment. We denote the union of events A and B by the symbol $A \cup B$.

Definition 3.6

The **intersection** of two events A and B is the event that occurs if both A and B occur on a single performance of the experiment. We write $A \cap B$ for the intersection of events A and B.

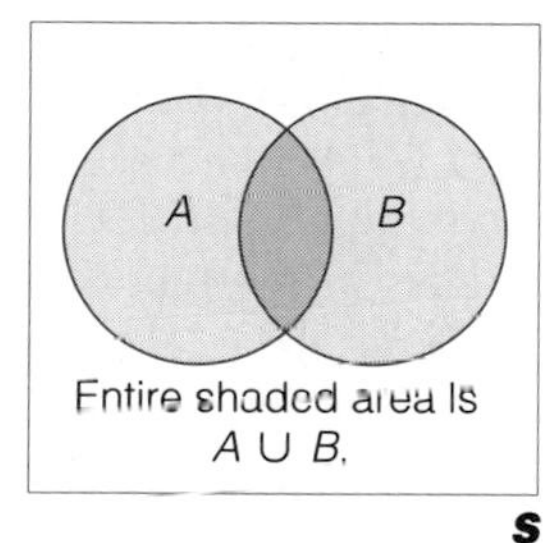

Entire shaded area is $A \cup B$.

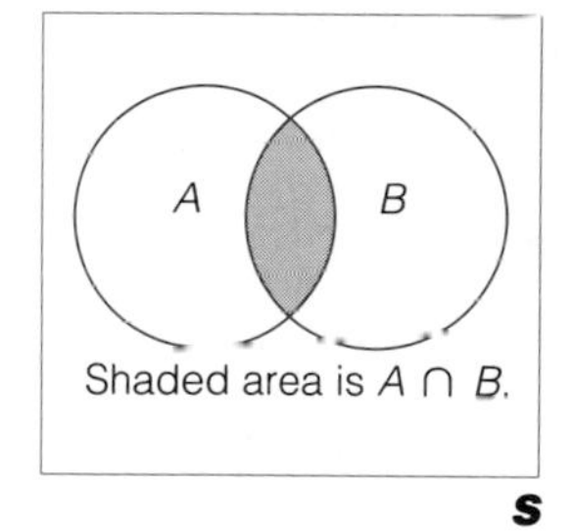

Shaded area is $A \cap B$.

EXAMPLE 3.6

Consider the die-toss experiment. Define the following events:

A: {Toss an even number}

B: {Toss a number less than or equal to 3}

a. Describe $A \cup B$ for this experiment.

b. Describe $A \cap B$ for this experiment.

c. Calculate $P(A \cup B)$ and $P(A \cap B)$ assuming the die is fair.

Solution

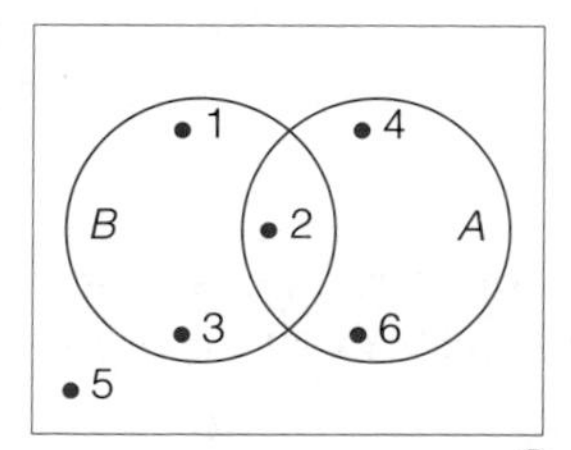

Draw the Venn diagram as shown in the margin.

a. The union of A and B is the event that occurs if we observe either an even number, a number less than or equal to 3, or both on a single throw of the die. Consequently, the simple events in the event $A \cup B$ are those for which A occurs, B occurs, or both A and B occur. Testing the simple events in the entire sample space, we find that the collection of simple events in the union of A and B is

$$A \cup B = \{1, 2, 3, 4, 6\}$$

b. The intersection of A and B is the event that occurs if we observe *both* an even number and a number less than or equal to 3 on a single throw of the die. Testing the simple events to see which imply the occurrence of *both* events A and B, we see that the intersection contains only one simple event:

$$A \cap B = \{2\}$$

In other words, the intersection of A and B is the simple event Observe a 2.

c. Recalling that the probability of an event is the sum of the probabilities of the simple events of which the event is composed, we have

$$P(A \cup B) = P(1) + P(2) + P(3) + P(4) + P(6)$$
$$= 1/6 + 1/6 + 1/6 + 1/6 + 1/6 = 5/6$$

and

$$P(A \cap B) = P(2) = 1/6$$

Unions and intersections can be defined for more than two events. For example, the event $A \cup B \cup C$ represents the union of three events, A, B, and C. This event, which includes the set of simple events in A, B, or C, will occur if any one or more of the events A, B, or C occurs. Similarly, the intersection $A \cap B \cap C$ is the event that all three of the events A, B, and C occur. Therefore, $A \cap B \cap C$ is the set of simple events that are in all three of the events, A, B, and C.

EXAMPLE 3.7

Refer to Example 3.6 and define the event

C: {Toss a number greater than 1}

Find the simple events in:

a. $A \cup B \cup C$ b. $A \cap B \cap C$

where

A: {Toss an even number}
B: {Toss a number less than or equal to 3}

Solution

a. Event C contains the simple events corresponding to tossing a 2, 3, 4, 5, or 6, and event B contains the simple events 1, 2, and 3. Therefore, the event that either A, B, or C occurs contains all six simple events in S—that is, those corresponding to tossing a 1, 2, 3, 4, 5, or 6.

b. You can see that you will observe all of the events A, B, and C only if you observe a 2. Therefore, the intersection $A \cap B \cap C$ contains the single simple event Toss a 2.

EXAMPLE 3.8

Many firms have undertaken direct marketing campaigns to promote their products. The campaigns typically involve mailing information to millions of households. The response rates are carefully monitored to determine the demographic characteristics of respondents. By studying tendencies to respond, the firms can better target future mailings to those segments of the population most likely to purchase the products.

Suppose a distributor of mail-order tools is analyzing the results of a recent mailing. The probability of response is believed to be related to income and age. The percentages of the total number of respondents to the mailing are given by income and age classification in Table 3.4.

TABLE 3.4 Percentages of Respondents in Age–Income Classes

		Income		
		<\$25,000	\$25,000–\$50,000	>\$50,000
Age	<30 yrs	5%	12%	10%
	30–50 yrs	14%	22%	16%
	>50 yrs	8%	10%	3%

Define the following events:

A: {A respondent's income is more than \$50,000}
B: {A respondent's age is 30 or more}

a. Find $P(A)$ and $P(B)$.
b. Find $P(A \cup B)$.
c. Find $P(A \cap B)$.

Solution

Following the steps for calculating probabilities of events, we first note that the objective is to characterize the income and age distribution of respondents to the mailing. To accomplish this, we define the experiment to consist of selecting a respondent from the collection of all respondents and observing which income and age class he or she occupies. The simple events are the nine different age–income classifications:

E_1: {<30 yrs, <\$25,000}
E_2: {30–50 yrs, <\$25,000}
⋮ ⋮
E_9: {>50 yrs, >\$50,000}

Next, we assign probabilities to the simple events. If we blindly select one of the respondents, the probability that he or she will occupy a particular age–income classification is just the proportion, or relative frequency, of respondents in the classification. These proportions are given (as percentages) in Table 3.4. Thus,

$$P(E_1) = \text{Relative frequency of respondents in age–income class } \{<30 \text{ yrs}, <\$25{,}000\} = .05$$

$$P(E_2) = .14$$

and so forth. You may verify that the simple event probabilities add to 1.

a. To find $P(A)$, we first determine the collection of simple events contained in event A. Since A is defined as $\{>\$50{,}000\}$, we see from Table 3.4 that A contains the three simple events represented by the last column of the table. In words, the event A consists of the income classification $\{>\$50{,}000\}$ in all three age classifications. The probability of A is the sum of the probabilities of the simple events in A:

$$P(A) = .10 + .16 + .03 = .29$$

Similarly, event B consists of the six simple events in the second and third rows of Table 3.4:

$$P(B) = .14 + .22 + .16 + .08 + .10 + .03 = .73$$

b. The union of events A and B, $A \cup B$, consists of all simple events in *either* A *or* B *or both* A *and* B. That is, the union of A and B consists of all respondents whose income exceeds $50,000 *or* whose age is 30 or more. In Table 3.4 this is any simple event found in the third column *or* the last two rows. Thus,

$$P(A \cup B) = .10 + .14 + .22 + .16 + .08 + .10 + .03 = .83$$

c. The intersection of events A and B, $A \cap B$, consists of all simple events in *both* A *and* B. That is, the intersection of A and B consists of all respondents whose income exceeds $50,000 *and* whose age is 30 or more. In Table 3.4 this is any simple event found in the third column *and* the last two rows. Thus,

$$P(A \cap B) = .16 + .03 = .19$$

3.3 The Additive Rule and Mutually Exclusive Events

In the previous section we showed how to determine which simple events are contained in a union. Then we showed that the probability of the union can be calculated by adding the probabilities of the simple events in the union. It is also possible to obtain the probability of the union of two events by using the additive rule, as illustrated in the following example.

EXAMPLE 3.9

A loaded (unbalanced) die is tossed and the up face is observed. The following two events are defined:

A: {Observe an even number}
B: {Observe a number less than 3}

Suppose $P(A) = .4$, $P(B) = .2$, and $P(A \cap B) = .1$. Find $P(A \cup B)$. [*Note:* Assuming that we would know these probabilities in a practical situation is not very realistic, but the example will illustrate a point.]

Solution

By studying the Venn diagram in Figure 3.6, we can obtain information that will help us find $P(A \cup B)$. We can see that

$$P(A \cup B) = P(1) + P(2) + P(4) + P(6)$$

Also, we know that

$$P(A) = P(2) + P(4) + P(6) = .4$$
$$P(B) = P(1) + P(2) = .2$$
$$P(A \cap B) = P(2) = .1$$

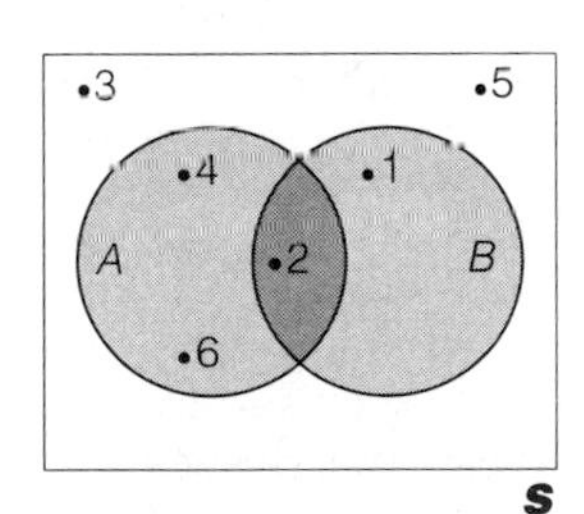

FIGURE 3.6 ▲
Venn diagram for die toss

If we add the probabilities of the simple events that comprise events A and B, we find

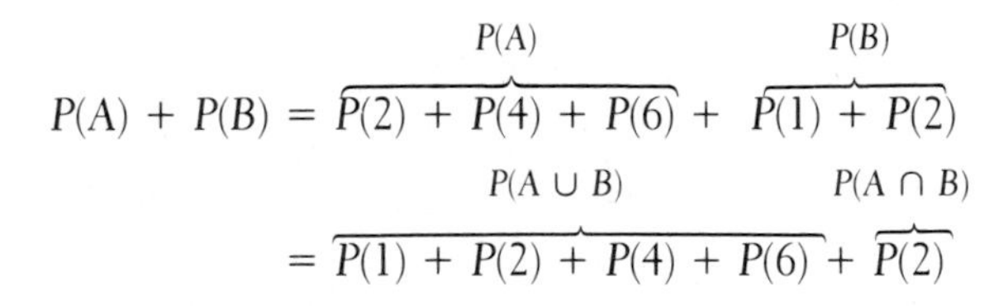

$$P(A) + P(B) = \overbrace{P(2) + P(4) + P(6)}^{P(A)} + \overbrace{P(1) + P(2)}^{P(B)}$$
$$= \overbrace{P(1) + P(2) + P(4) + P(6)}^{P(A \cup B)} + \overbrace{P(2)}^{P(A \cap B)}$$

Thus, by subtraction, we have

$$P(A \cup B) = P(A) + P(B) - P(A \cap B)$$
$$= .4 + .2 - .1 = .5$$

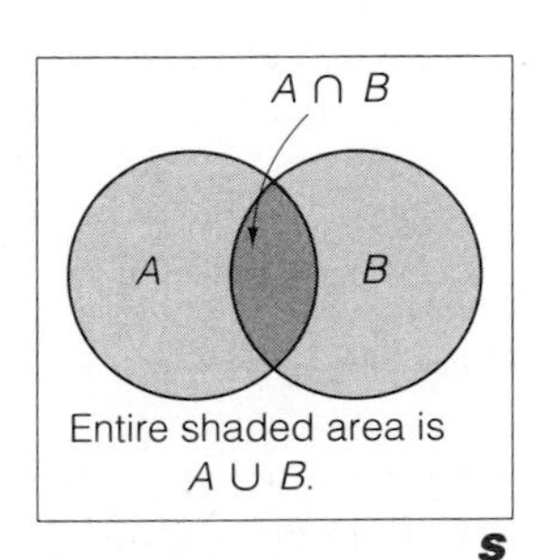

FIGURE 3.7 ▲
Venn diagram of union

By studying the Venn diagram in Figure 3.7, you can see that the method used in Example 3.9 can be generalized to find the union of two events for any experiment. The probability of the union of two events, A and B, can always be obtained by summing $P(A)$ and $P(B)$ and subtracting $P(A \cap B)$. We must subtract $P(A \cap B)$ because the simple event probabilities in $A \cap B$ have been included twice—once in $P(A)$ and once in $P(B)$.

The formula for calculating the probability of the union of two events, often called the **additive rule of probability**, is given in the box.

Additive Rule of Probability

The probability of the union of events A and B is the sum of the probability of events A and B minus the probability of the intersection of events A and B; i.e.,

$$P(A \cup B) = P(A) + P(B) - P(A \cap B)$$

EXAMPLE 3.10

Hospital records show that 12% of all patients are admitted for surgical treatment, 16% are admitted for obstetrics, and 2% receive both obstetrics and surgical treatment. If a new patient is admitted to the hospital, what is the probability that the patient will be admitted either for surgery, obstetrics, or both?

Solution

Consider the following events:

A: {A patient admitted to the hospital receives surgical treatment}
B: {A patient admitted to the hospital receives obstetrics treatment}

Then, from the given information,

$$P(A) = .12 \qquad P(B) = .16$$

and the probability of the event that a patient receives both obstetrics and surgical treatment is

$$P(A \cap B) = .02$$

The event that a patient admitted to the hospital receives either surgical treatment, obstetrics treatment, or both is the union $A \cup B$. The probability of $A \cup B$ is given by the additive rule of probability:

$$P(A \cup B) = P(A) + P(B) - P(A \cap B)$$
$$= .12 + .16 - .02 = .26$$

Thus, 26% of all patients admitted to the hospital receive either surgical treatment, obstetrics treatment, or both.

A very special relationship exists between events A and B when $A \cap B$ contains no simple events. In this case, we call the events A and B **mutually exclusive** events.

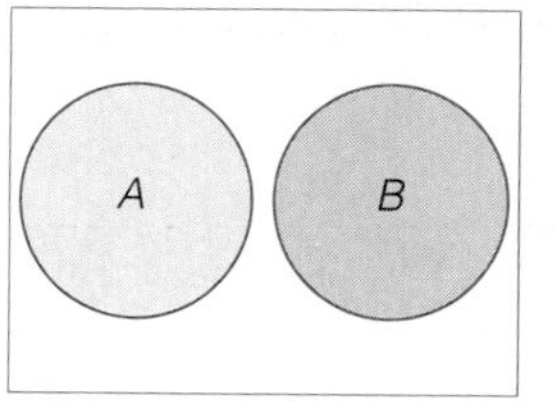

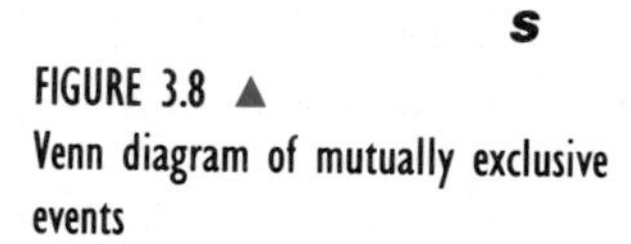

FIGURE 3.8 ▲
Venn diagram of mutually exclusive events

Definition 3.7

Events A and B are **mutually exclusive** if $A \cap B$ contains no simple events.

Figure 3.8 shows a Venn diagram of two mutually exclusive events. The events A and B have no simple events in common, i.e., A and B cannot occur simultaneously, and $P(A \cap B) = 0$. Thus, we have the important relationship given in the box.

If two events A *and* B *are mutually exclusive,* the probability of the union of A and B equals the sum of the probabilities of A and B; that is,

$$P(A \cup B) = P(A) + P(B)$$

EXAMPLE 3.11

Consider the experiment of tossing two balanced coins. Find the probability of observing *at least* one head.

Solution

Define the events

A: {Observe at least one head}
B: {Observe exactly one head}
C: {Observe exactly two heads}

Note that

$$A = B \cup C$$

and that $B \cap C$ contains no simple events (see Figure 3.9). Thus, B and C are mutually exclusive, so that

$$P(A) = P(B \cup C) = P(B) + P(C)$$
$$= \tfrac{1}{2} + \tfrac{1}{4} = \tfrac{3}{4}$$

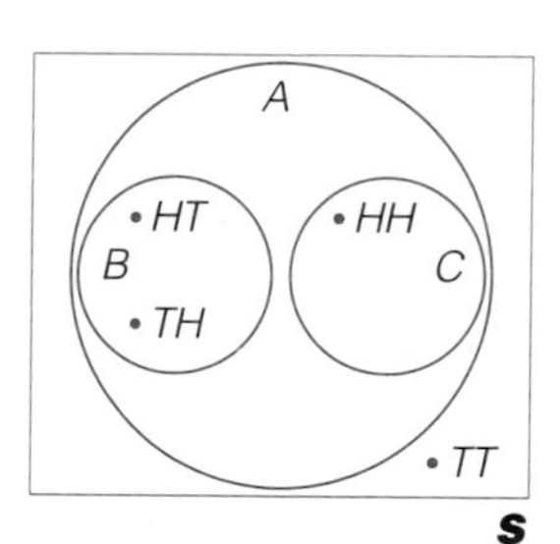

FIGURE 3.9 ▲
Venn diagram for coin toss experiment

Although Example 3.11 is very simple, the concept of writing events with verbal descriptions that include the phrases "at least" or "at most" as unions of mutually exclusive events is a very useful one. This enables us to find the probability of the event by adding the probabilities of the mutually exclusive events.

3.4 Complementary Events

A very useful concept in the calculation of event probabilities is the notion of **complementary events**:

Definition 3.8

The **complement** of an event A is the event that A does not occur—i.e., the event consisting of all simple events that are not in event A. We denote the complement of A by A′.

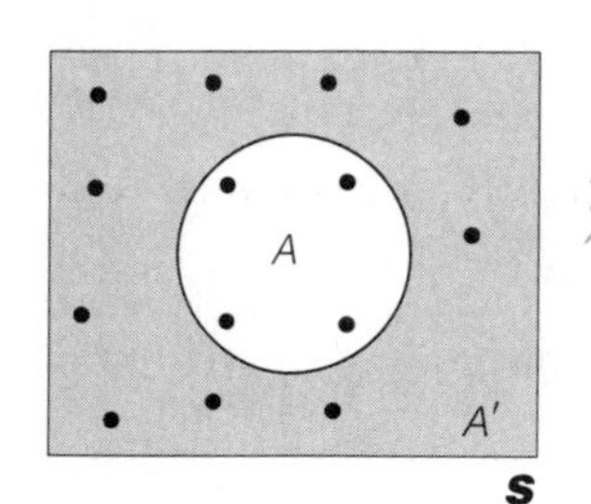

FIGURE 3.10 ▲ Venn diagram of complementary events

An event A is a collection of simple events, and the simple events included in A′ are those that are not in A. Figure 3.10 demonstrates this idea. You will note from the figure that all simple events in S are included in *either* A or A′ and that *no* simple event is in both A and A′. This leads us to conclude that the probabilities of an event and its complement *must sum to* 1:

The sum of the probabilities of complementary events equals 1; i.e.,

$$P(A) + P(A') = 1$$

In many probability problems it is easier to calculate the probability of the complement of the event of interest rather than the event itself. Then, since

$$P(A) + P(A') = 1$$

we can calculate $P(A)$ by using the relationship

$$P(A) = 1 - P(A')$$

EXAMPLE 3.12

Consider the experiment of tossing two fair coins. Calculate the probability of event A: {Observing at least one head} by using the complementary relationship.

Solution

We know that the event A: {Observing at least one head} consists of the simple events

A: {*HH*, *HT*, *TH*}

The complement of A is defined as the event that occurs when A does not occur. Therefore,

A′: {Observe no heads} = {*TT*}

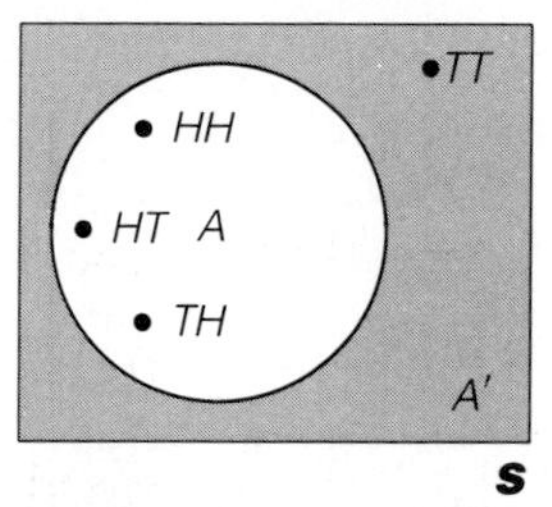

FIGURE 3.11 ▲
Complementary events in the toss of two coins

This complementary relationship is shown in Figure 3.11. Assuming the coins are balanced,

$$P(A') = P(TT) = 1/4$$

and

$$P(A) = 1 - P(A') = 1 - 1/4 = 3/4$$

EXAMPLE 3.13

A fair coin is tossed 10 times, and the up face is recorded after each toss. What is the probability of event A: {Observe at least one head}?

Solution

We solve this problem by following the five steps for calculating probabilities of events (see Section 3.1).

Step 1 Define the experiment. The experiment is to record the results of the 10 tosses of the coin.

Step 2 List the simple events. A simple event consists of a particular sequence of 10 heads and tails. Thus, one simple event is

HHTTTHTHTT

which denotes head on first toss, head on second toss, tail on third toss, etc. Others are *HTHHHTTTTT* and *THHTHTHTTH*. There is obviously a very large number of simple events—too many to list. It can be shown (proof omitted) that there are $2^{10} = 1{,}024$ simple events for this experiment.

Step 3 Assign probabilities. Since the coin is fair, each sequence of heads and tails has the same chance of occurring, and therefore all the simple events are equally likely. Then

$$P(\text{Each simple event}) = \frac{1}{1{,}024}$$

Step 4 Determine the simple events in event A. A simple event is in A if at least one *H* appears in the sequence of 10 tosses. However, if we consider the complement of A, we find that

A′: {No heads are observed in 10 tosses}

Thus, A′ contains only one simple event:

$$A': \{TTTTTTTTTT\} \quad \text{and} \quad P(A') = \frac{1}{1{,}024}$$

Step 5 Since we know the probability of the complement of A, we use the relationship for complementary events:

$$P(A) = 1 - P(A') = 1 - \frac{1}{1{,}024} = \frac{1{,}023}{1{,}024} = .999$$

That is, we are virtually certain of observing at least one head in 10 tosses of the coin.

Exercises 3.18–3.31

Learning the Mechanics

3.18 A fair coin is tossed three times and the events A and B are defined as follows:

A: {At least one head is observed}
B: {The number of heads observed is odd}

a. Identify the simple events in the events A, B, $A \cup B$, A', and $A \cap B$.
b. Find $P(A)$, $P(B)$, $P(A \cup B)$, $P(A')$, and $P(A \cap B)$ by summing the probabilities of the appropriate simple events.
c. Find $P(A \cup B)$ using the additive rule, and compare your answer to the one you obtained in part **b**.
d. Are the events A and B mutually exclusive? Why?

3.19 What are mutually exclusive events? Give a verbal description, and then draw a Venn diagram.

3.20 A pair of fair dice is tossed. Define the following events:

A: {You will roll a 7} (i.e., the sum of the numbers of dots on the upper faces of the two dice is equal to 7)
B: {At least one of the two dice is showing a 4}

a. Identify the simple events in the events A, B, $A \cap B$, $A \cup B$, and A'.
b. Find $P(A)$, $P(B)$, $P(A \cap B)$, $P(A \cup B)$, and $P(A')$ by summing the probabilities of the appropriate simple events.
c. Find $P(A \cup B)$ using the additive rule. Compare your answer to that for the same event in part **b**.
d. Are A and B mutually exclusive events? Why?

3.21 Consider the Venn diagram, where $P(E_1) = P(E_2) = P(E_3) = \frac{1}{5}$, $P(E_4) = P(E_5) = \frac{1}{20}$, $P(E_6) = \frac{1}{10}$, and $P(E_7) = \frac{1}{5}$. Find each of the following probabilities.

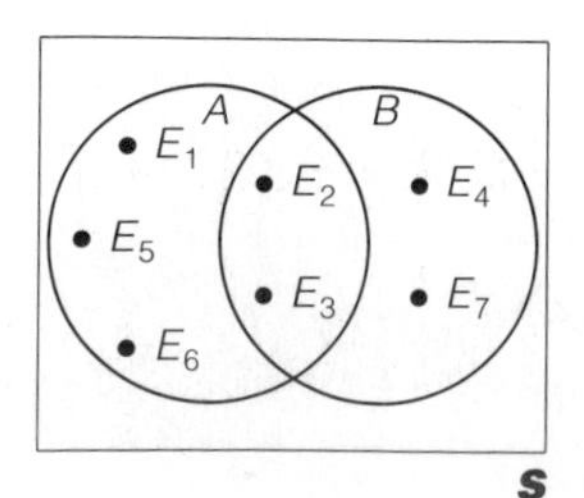

a. $P(A)$ b. $P(B)$ c. $P(A \cup B)$ d. $P(A \cap B)$
e. $P(A')$ f. $P(B')$ g. $P(A \cup A')$ h. $P(A' \cap B)$

3.22 Consider the Venn diagram, where $P(E_1) = .13$, $P(E_2) = .05$, $P(E_3) = P(E_4) = .2$, $P(E_5) = .06$, $P(E_6) = .3$, and $P(E_7) = .06$.

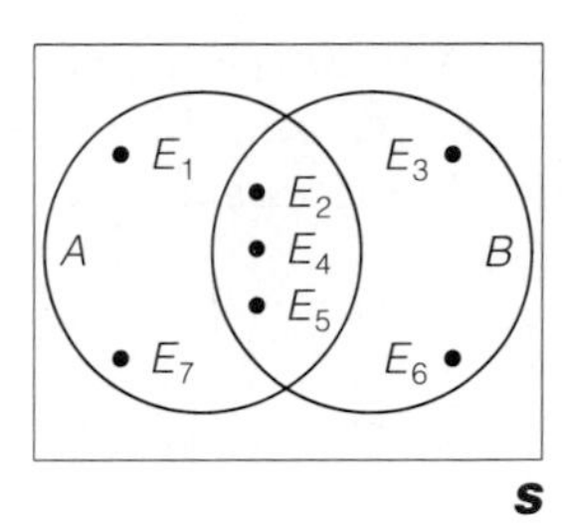

Find each of the following probabilities:

a. $P(A')$ b. $P(B')$ c. $P(A' \cap B)$ d. $P(A \cup B)$
e. $P(A \cap B)$ f. $P(A' \cup B')$ g. Are events A and B mutually exclusive? Why?

3.23 The table describes the adult population of a small suburb of a large southern city.

		Income		
		<$20,000	$20,000–$50,000	>$50,000
Age	<25	950	1,000	50
	25–45	450	2,050	1,500
	>45	50	950	1,000

A marketing research firm plans to randomly select one adult from the suburb to evaluate a new food product. For this experiment the nine age–income categories are the simple events. Consider the following events:

A: {Person is under 25}
B: {Person is between 25 and 45}
C: {Person is over 45}
D: {Person has income under $20,000}
E: {Person has income of $20,000–$50,000}
F: {Person has income over $50,000}

Convert the frequencies in the table to relative frequencies and use them to calculate the following probabilities:

a. $P(B)$ b. $P(F)$ c. $P(C \cap F)$ d. $P(B \cup C)$ e. $P(A')$ f. $P(A' \cap F)$
g. Consider each pair of events (A and B, A and C, etc.) and list the pairs of events that are mutually exclusive. Justify your choices.

3.24 Refer to Exercise 3.23. Use the same event definitions to solve the following.

a. Write the event that the person selected is under 25 with an income over \$50,000 as an intersection of two events.

b. Write the event that the person selected is age 25 or older as the union of two mutually exclusive events. As the complement of an event.

3.25 Three fair coins are tossed. We wish to find the probability of the event A: {Observe at least one head}.

a. Express A as the union of three mutually exclusive events. Find the probability of A using this expression.

b. Express A as the complement of an event. Find the probability of A using this expression.

Applying the Concepts

3.26 A buyer for a large metropolitan department store must choose two firms from the four available to supply the store's fall line of men's slacks. The buyer has not dealt with any of the four firms before and considers their products equally attractive. Unknown to the buyer, two of the four firms are having serious financial problems that may result in their not being able to deliver the fall line of slacks as soon as promised. The four firms are identified as G_1 and G_2 (firms in good financial condition) and P_1 and P_2 (firms in poor financial condition). Simple events identify the pairs of firms selected. If the probability of the buyer selecting a particular firm from among the four is the same for each firm, the simple events and their probabilities for this buying experiment are those listed in the table. We will define the following events:

A: {At least one of the selected firms is in good financial condition}

B: {Firm P_1 is selected}

Simple Events	*Probability*
G_1G_2	1/6
G_1P_1	1/6
G_1P_2	1/6
G_2P_1	1/6
G_2P_2	1/6
P_1P_2	1/6

a. Define the event $A \cap B$ as a specific collection of simple events.

b. Define the event $A \cup B$ as a specific collection of simple events.

c. Define the event A' as a specific collection of simple events.

d. Find $P(A)$, $P(B)$, $P(A \cap B)$, $P(A \cup B)$, and $P(A')$ by summing the probabilities of the appropriate simple events.

e. Find $P(A \cup B)$ using the additive rule. Are events A and B mutually exclusive? Why?

3.27 A state energy agency mailed questionnaires on energy conservation to 1,000 homeowners in the state capital. Five hundred questionnaires were returned. Suppose an experiment consists of randomly selecting one of the returned questionnaires. Consider the events:

A: {The home is constructed of brick}

B: {The home is more than 30 years old}

C: {The home is heated with oil}

Describe each of the following events in terms of unions, intersections, and complements ($A \cup B$, $A \cap B$, A', etc.):

a. The home is more than 30 years old and is heated with oil.
b. The home is not constructed of brick.
c. The home is heated with oil or is more than 30 years old.
d. The home is constructed of brick and is not heated with oil.

3.28 One game that is very popular in many American casinos is *roulette*. Roulette is played by spinning a ball on a circular wheel that has been divided into 38 arcs of equal length, bearing the numbers 00, 0, 1, 2, . . . , 35, 36. The number of the arc on which the ball comes to rest is the outcome of one play of the game. The numbers are also colored in the following manner:

Red: 1, 3, 5, 7, 9, 12, 14, 16, 18, 19, 21, 23, 25, 27, 30, 32, 34, 36
Black: 2, 4, 6, 8, 10, 11, 13, 15, 17, 20, 22, 24, 26, 28, 29, 31, 33, 35
Green: 00, 0

Players may place bets on the table in a variety of ways, including bets on odd, even, red, black, high, low, etc. Define the following events:

A: {Outcome is an odd number} (00 and 0 are not considered odd or even)
B: {Outcome is a black number}
C: {Outcome is a low number (1–18)}

a. Define the event $A \cap B$ as a specific set of simple events.
b. Define the event $A \cup B$ as a specific set of simple events.
c. Find $P(A)$, $P(B)$, $P(A \cap B)$, $P(A \cup B)$, and $P(C)$ by summing the probabilities of the appropriate simple events.
d. Define the event $A \cap B \cap C$ as a specific set of simple events.
e. Find $P(A \cup B)$ using the additive rule. Are events A and B mutually exclusive? Why?
f. Find $P(A \cap B \cap C)$ by summing the probabilities of the simple events given in part **d**.
g. Define the event $(A \cup B \cup C)$ as a specific set of simple events.
h. Find $P(A \cup B \cup C)$ by summing the probabilities of the simple events given in part **g**.

3.29 After completing an inventory of three warehouses, a golf club shaft manufacturer described its stock of 12,246 shafts with the percentages given in the table. Suppose a shaft is selected at random from the 12,246 currently in stock, and the warehouse number and type of shaft are observed.

		Type of Shaft		
		Regular	Stiff	Extra Stiff
Warehouse	1	19%	8%	3%
	2	14%	8%	2%
	3	28%	18%	0%

a. List all the simple events for this experiment.
b. What is the set of all simple events called?

c. Let C be the event that the shaft selected is from warehouse 3. Find $P(C)$ by summing the probabilities of the simple events in C.
d. Let F be the event that the shaft chosen is an extra stiff type. Find $P(F)$.
e. Let A be the event that the shaft selected is from warehouse 1. Find $P(A)$.
f. Let D be the event that the shaft selected is a regular type. Find $P(D)$.
g. Let E be the event that the shaft selected is a stiff type. Find $P(E)$.

3.30 Refer to Exercise 3.29. Define the characteristics of a golf club shaft portrayed by the following events, and then find the probability of each. For each union, use the additive rule to find the probability. Also, determine whether the events are mutually exclusive.
a. $A \cap F$ b. $C \cup E$ c. $C \cap D$ d. $A \cup F$ e. $A \cup D$

3.31 Identifying managerial prospects who are both talented and motivated is difficult. A personnel manager constructed the following two-way table to define nine combinations of talent–motivation levels. The number in a cell is the manager's estimate of the probability that a managerial prospect will fall in that category. Suppose the personnel manager has decided to hire a new manager. Define the following events:

A: {Prospect places in high motivation category}
B: {Prospect places in high talent category}
C: {Prospect is average or better in both categories}
D: {Prospect rates poor in at least one category}
E: {Prospect places highest in both categories}

		Talent		
		High	Medium	Low
Motivation	High	.05	.16	.05
	Medium	.19	.32	.05
	Low	.11	.05	.02

a. Does the sum of the cell probabilities equal 1?
b. List the simple events in each of the events described above and find their probabilities.
c. Find $P(A \cup B)$, $P(A \cap B)$, and $P(A \cup C)$.
d. Find $P(A')$ and explain what this means from a practical point of view.
e. Consider each pair of events (A and B, A and C, etc.). Which of the pairs are mutually exclusive? Why?

3.5 Conditional Probability

The event probabilities we have been discussing give the relative frequencies of the occurrences of the events when the experiment is repeated a very large number of times. They are called **unconditional probabilities** because no special conditions are assumed, other than those that define the experiment.

Sometimes we may wish to alter the probability of an event when we have additional knowledge that might affect its outcome. This probability is called the **conditional probability** of the event. For example, we have shown that the probability of observing an even number (event A) on a toss of a fair die is ½. However, suppose you are given the information that on a particular throw of the die the result was a number less than or equal to 3 (event B). Would you still believe that the probability of observing an even number on that throw of the die is equal to ½? If you reason that making the assumption that B has occurred reduces the sample space from six simple events to three simple events (namely, those contained in event B), the reduced sample space is as shown in Figure 3.12. Because the simple events for the die-toss experiment are equally likely, each of the three simple events in the reduced sample space is assigned an equal *conditional probability* of ⅓. Since the only even number of the three in the reduced sample space B is the number 2 and the die is fair, we conclude that the probability that A occurs *given that B occurs* is ⅓. We use the symbol $P(A \mid B)$ to represent the probability of event A given that event B occurs. For the die-toss example

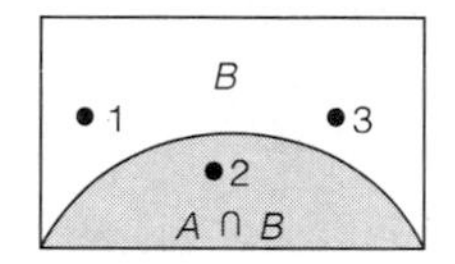

FIGURE 3.12 ▲
Reduced sample space for the die toss experiment—given that event *B* has occurred

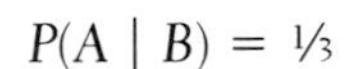

$$P(A \mid B) = 1/3$$

To get the probability of event A given that event B occurs, we proceed as follows. We divide the probability of the part of A that falls within the reduced sample space B, namely, $P(A \cap B)$, by the total probability of the reduced sample space, namely, $P(B)$. Thus, for the die-toss example with event A: {Observe an even number} and event B: {Observe a number less than or equal to 3}, we find

$$P(A \mid B) = \frac{P(A \cap B)}{P(B)} = \frac{P(2)}{P(1) + P(2) + P(3)} = \frac{1/6}{3/6} = \frac{1}{3}$$

The formula for $P(A \mid B)$ is true in general:

To find the *conditional probability that event A occurs given that event B occurs*, divide the probability that *both* A and B occur by the probability that B occurs, that is,

$$P(A \mid B) = \frac{P(A \cap B)}{P(B)} \qquad \text{[We assume that } P(B) \neq 0.]$$

The formula adjusts the probability of $A \cap B$ from its original value in the complete sample space S to a conditional probability in the reduced sample space B. If the simple events in the complete sample space are equally likely, then the formula will assign equal probabilities to the simple events in the reduced sample space, as in the die-toss experiment. If, on the other hand, the simple events have unequal probabilities, the formula will assign conditional probabilities proportional to the probabilities in the complete sample space. This is illustrated by the following examples.

EXAMPLE 3.14

Many medical researchers have conducted experiments to examine the relationship between cigarette smoking and cancer. Let A represent the event that an individual smokes, and let C represent the event that an individual develops cancer. Therefore, $A \cap C$ is the simple event that an individual smokes and develops cancer; $A \cap C'$ is the simple event that an individual smokes and does not develop cancer, etc. Assume that the probabilities associated with the four simple events are as shown in the table for a certain section of the United States. How can these simple event probabilities be used to examine the relationship between smoking and cancer?

Simple Events	*Probabilities*
$A \cap C$	.15
$A \cap C'$	.25
$A' \cap C$	.10
$A' \cap C'$	.50

Solution

One method of determining whether these probabilities indicate that smoking and cancer are related is to compare the conditional probability that an individual acquires cancer given that he or she smokes with the conditional probability that an individual acquires cancer given that he or she does not smoke.

First, we consider the reduced sample space A corresponding to smokers. The two simple events $A \cap C$ and $A \cap C'$ are contained in this reduced sample space, and the adjusted probabilities of these two simple events are the two conditional probabilities:

$$P(C \mid A) = \frac{P(A \cap C)}{P(A)} \quad \text{and} \quad P(C' \mid A) = \frac{P(A \cap C')}{P(A)}$$

The probability of event A is the sum of the probabilities of the simple events in A:

$$P(A) = P(A \cap C) + P(A \cap C') = .15 + .25 = .40$$

Then the values of the two conditional probabilities in the reduced sample space A are

$$P(C \mid A) = \frac{.15}{.40} = .375 \quad \text{and} \quad P(C' \mid A) = \frac{.25}{.40} = .625$$

These two numbers represent the probabilities that a smoker develops cancer and does not develop cancer, respectively. Notice that the conditional probabilities .625 and .375 are in the same 5 to 3 ratio as the original (unconditional) probabilities, .25 and .15. The conditional probability formula simply adjusts the unconditional probabilities so that they add to 1 in the reduced sample space, A, of smokers.

In a like manner, the conditional probabilities of a nonsmoker developing cancer and not developing cancer are:

$$P(C \mid A') = \frac{P(A' \cap C)}{P(A')} = \frac{.10}{.60} = .167$$

$$P(C' \mid A') = \frac{P(A' \cap C')}{P(A')} = \frac{.50}{.60} = .833$$

Notice that the conditional probabilities .833 and .167 are in the same 5 to 1 ratio as the unconditional probabilities .5 and .1.

Two of the conditional probabilities give some insight into the relationship between cancer and smoking: the probability of developing cancer given that the individual is a smoker, and the probability of developing cancer given that the individual is not a smoker. The conditional probability that a smoker develops cancer (.375) is more than twice the probability that a nonsmoker develops cancer (.167). This does not imply that smoking *causes* cancer, but it does suggest a pronounced link between smoking and cancer.

EXAMPLE 3.15

The investigation of consumer product complaints by the Federal Trade Commission (FTC) has generated much interest by manufacturers in the quality of their products. A manufacturer of an electromechanical kitchen aid conducted an analysis of a large number of consumer complaints and found that they fell into the six categories shown in Table 3.5. If a consumer complaint is received, what is the probability that the cause of the complaint was product appearance given that the complaint originated during the guarantee period?

TABLE 3.5 Distribution of Product Complaints

	Reason for Complaint			Totals
	Electrical	Mechanical	Appearance	
During Guarantee Period	18%	13%	32%	63%
After Guarantee Period	12%	22%	3%	37%
Totals	30%	35%	35%	100%

Solution

Let A represent the event that the cause of a particular complaint is product appearance, and let B represent the event that the complaint occurred during the guarantee period. Checking Table 3.5, you can see that $(18 + 13 + 32)\% = 63\%$ of the complaints occur during the guarantee period. Hence, $P(B) = .63$. The percentage of complaints that were caused by appearance and occurred during the guarantee period (the event $A \cap B$) is 32%. Therefore, $P(A \cap B) = .32$.

Using these probability values, we can calculate the conditional probability $P(A \mid B)$ that the cause of a complaint is appearance given that the complaint occurred during the guarantee time:

$$P(A \mid B) = \frac{P(A \cap B)}{P(B)} = \frac{.32}{.63} = .51$$

Consequently, you can see that slightly more than half the complaints that occurred during the guarantee period were due to scratches, dents, or other imperfections in the surface of the kitchen devices.

CASE STUDY 3.2 / Purchase Patterns and the Conditional Probability of Purchasing

In his doctoral dissertation, Alfred A. Kuehn (1958) examined sequential purchase data to gain some insight into consumer brand switching. He analyzed the frozen orange juice purchases of approximately 600 Chicago families during 1950–1952. The data were collected by the *Chicago Tribune* Consumer Panel. Kuehn was interested in determining the influence of a consumer's last four orange juice purchases on the next purchase. Thus, sequences of five purchases were analyzed.

Table 3.6 summarizes the data collected for Snow Crop brand orange juice and part of Kuehn's analysis of the data. In the column labeled "Previous Purchase Pattern" an S stands for the purchase of Snow Crop by a consumer and an O stands for the purchase of a brand other than Snow Crop. Thus, for example, SSSO is used to represent the purchase of Snow Crop three times in a row followed by the purchase of some other brand of frozen orange juice. The column labeled "Sample Size" lists the number of occurrences of the purchase sequences in the first column. The column labeled "Frequency" lists the number of times the associated purchase sequence in the first column led to the next purchase (i.e., the fifth purchase in the sequence) being Snow Crop.

The column labeled "Observed Approximate Conditional Probability of Purchase" contains the relative frequency with which each sequence of the first column led to the next purchase being Snow Crop. These relative frequencies, which give approximate condi-

TABLE 3.6 Observed Approximate Conditional Probability of Purchasing Snow Crop Given the Four Previous Brand Purchases

Previous Purchase Pattern S = Snow Crop O = Other brand	Sample Size	Frequency	Observed Approximate Conditional Probability of Purchase *P*{Purchase | Previous purchase pattern}
SSSS	1,047	844	.806
OSSS	277	191	.690
SOSS	206	137	.665
SSOS	222	132	.595
SSSO	296	144	.486
OOSS	248	137	.552
SOOS	138	78	.565
OSOS	149	74	.497
SOSO	163	66	.405
OSSO	181	75	.414
SSOO	256	78	.305
OOOS	500	165	.330
OOSO	404	77	.191
OSOO	433	56	.129
SOOO	557	86	.154
OOOO	8,442	405	.048

tional probabilities, are computed for each sequence of the first column by dividing the frequency of the sequence by the sample size of the sequence. For example, .806 is the approximate conditional probability that the next purchase will be Snow Crop given that the previous four purchases were also Snow Crop.

An examination of the approximate conditional probabilities in the fourth column indicates that both the most recent brand purchased and the number of times a brand is purchased have an effect on the next brand purchased. It appears that the influence on the next brand of orange juice purchased by the second most recent purchase is not as strong as the most recent purchase, but it is stronger than the third most recent purchase. In general, it appears that the probability of a particular consumer purchasing Snow Crop the next time he or she buys orange juice is inversely related to the number of consecutive purchases of another brand he or she made since last purchasing Snow Crop and is directly proportional to the number of Snow Crop purchases among the four purchases.

Kuehn conducts a more formal statistical analysis of these data, which we do not pursue here. We simply want you to see that probability is a basic tool for making inferences about populations using sample data.

Exercises 3.32–3.42

Learning the Mechanics

3.32 Consider the experiment defined by the accompanying Venn diagram, with the sample space S containing five simple events. The simple events are assigned the following probabilities: $P(E_1) = .1$, $P(E_2) = .1$, $P(E_3) = .2$, $P(E_4) = .5$, $P(E_5) = .1$.

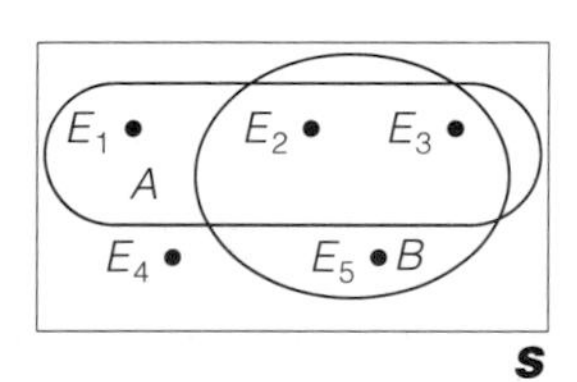

a. Calculate $P(A)$, $P(B)$, and $P(A \cap B)$.

b. Suppose we know event A has occurred, so the reduced sample space consists of the three simple events in A: E_1, E_2, and E_3. Use the formula for conditional probability to determine the probabilities of these three simple events given that A has occurred. Verify that the conditional probabilities are in the same ratio to one another as the original simple event probabilities.

c. Calculate the conditional probability $P(B \mid A)$ in two ways: First, add the adjusted (conditional) probabilities of the simple events in the intersection $A \cap B$, since these represent the event that B occurs given that A has occurred. Second, use the formula for conditional probability:

$$P(B \mid A) = \frac{P(A \cap B)}{P(A)}$$

Verify that the two methods yield the same result.

3.33 Given that $P(A) = .3$, $P(B) = .6$, and $P(A \cap B) = .15$, find $P(A \mid B)$ and $P(B \mid A)$.

3.34 A sample space contains six simple events and events A, B, and C as shown in the Venn diagram. The probabilities of the simple events are $P(1) = .20$, $P(2) = .05$, $P(3) = .25$, $P(4) = .10$, $P(5) = .15$, $P(6) = .25$.

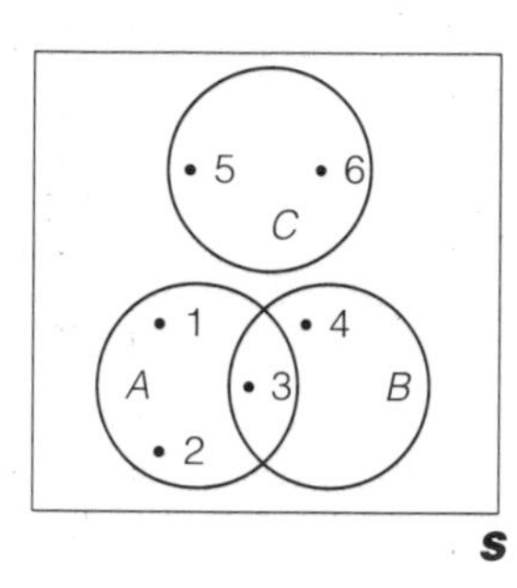

Use the Venn diagram and the probabilities of the simple events to find:

a. $P(A)$, $P(B)$, and $P(C)$

b. $P(A \cap B)$, $P(A \cap C)$, and $P(B \cap C)$

c. Suppose you know that event A has occurred. Assign conditional probabilities to the three simple events contained in A. Verify that they add to 1 and are in the same ratio as the original (unconditional) probabilities.

d. Use the conditional probabilities from part c to calculate $P(B \mid A)$. Use the formula for $P(B \mid A)$ to verify your answer.

e. Use the formula for conditional probability to calculate $P(C \mid A)$ and $P(C \mid A')$. Verify the results by inspection of the Venn diagram, remembering that the "given event" is a reduced sample space for a conditional probability.

3.35 Two fair coins are tossed and the events A and B are defined as follows:

A: {At least one head appears}

B: {Exactly one head appears}

a. Draw a Venn diagram for the experiment, labeling each simple event and showing events A and B. Assign probabilities to the simple events.

b. Find $P(A)$, $P(B)$, and $P(A \cap B)$.

c. Use the formula for conditional probability to find $P(A \mid B)$ and $P(B \mid A)$. Verify your answers by inspecting the Venn diagram and using the concept of reduced sample spaces.

3.36 A box contains two white, two red, and two blue poker chips. Two chips are randomly chosen without replacement and their colors are noted. Define the following events:

A: {Both chips are of the same color}

B: {Both chips are red}

C: {At least one chip is red or white}

Find $P(B \mid A)$, $P(B \mid A')$, $P(B \mid C)$, $P(A \mid C)$, and $P(C \mid A')$.

Applying the Concepts

3.37 In "Crime, Race and Reporting to the Police," R. Shah and K. Pease examined the relationship between the race of the attacker, the race of the victim, and the degree of the injury sustained in reported crimes in a 1975 British crime survey. The following table is cited in the article:

		Attacker/Victim				
		White/ White	White/ Nonwhite	Nonwhite/ Nonwhite	Nonwhite/ White	Totals
Degree of Injury	Fatal	183	18	18	18	237
	Serious	580	49	111	141	881
	Slight	4,336	440	594	1,656	7,026
	None	1,422	136	214	1,801	3,573
	Totals	6,521	643	937	3,616	11,717

Source: Shah, R., and Pease, K., "Crime, Race and Reporting to the Police," *The Howard Journal of Criminal Justice*, Volume 31, No. 3, Aug. 1992.

a. What is the probability that a reported crime involved a white attacker and a white victim?
b. What is the probability that a reported crime involved serious injuries?
c. Given that both the attacker and victim were white, what is the probability that a reported crime involved fatalities?
d. If no injury was reported, what is the probability that a reported crime involved a nonwhite attacker and a white victim?

3.38 A soap manufacturer has decided to market two new brands. An analysis of current market conditions and a review of the firm's past successes and failures with new brands have led the manufacturer to believe that the simple events and the probabilities of their occurrence in this marketing experiment are as shown in the table (where S means the brand succeeds and F means the brand fails in the first year). Define the following events:

A: {Both new brands are successful in the first year}

B: {At least one new brand is successful in the first year}

Simple Events	*Probabilities*
SS	.09
SF	.21
FS	.21
FF	.49

a. Find $P(A)$, $P(B)$, and $P(A \cap B)$. b. Find $P(A \mid B)$ and $P(B \mid A)$.

3.39 Six people apply for two identical positions in a company. Four are minority applicants and the remainder are nonminority. Define the following events:

A: {Both persons selected are nonminority candidates}

B: {Both persons selected are minority candidates}

C: {At least one of the persons selected is a minority candidate}

If all the applicants are equally qualified and the choice is essentially a random selection of two applicants from the six available, find:

a. $P(A)$ **b.** $P(B)$ **c.** $P(C)$ **d.** $P(B \mid C)$

e. Assume that the minority candidates are numbered 1, 2, 3, 4 for purposes of identification. Define the event

D: {Minority candidate 1 is selected}

Find $P(D \mid C)$.

3.40 An article in *Business Week* (September 12, 1983) reports on the problems that evolve from the failure to inform patients adequately of both the proper application of prescription drugs and the precautions to take in order to avoid potential side effects. This failure results in numerous cases of serious illness and, in some cases, even death. One study revealed that 300,000 U.S. hospital admissions each year are caused by adverse reactions to prescription drugs. Another study concluded that 7% of all hospital admissions are related to drug-induced problems resulting from imprudent prescriptions. One method of increasing patients' awareness of the problem is for physicians to provide Patient Medication Instruction (PMI) sheets. The American Medical Association, however, has found that only 20% of the doctors who prescribe drugs frequently distribute PMI sheets to their patients. Assume that 20% of all patients receive the PMI sheet with their prescriptions and that 12% receive the PMI sheet and are hospitalized because of a drug-related problem. What is the probability that a person will be hospitalized for a drug-related problem given that the person has received the PMI sheet?

3.41 A fast-food restaurant chain with 700 outlets in the United States describes the geographic location of its restaurants with the accompanying table of percentages. A restaurant is to be chosen at random from the 700 to test market a new style of chicken.

		Region			
		NE	SE	SW	NW
Population of City	<10,000	5%	6%	3%	0%
	10,000–100,000	15%	15%	12%	5%
	>100,000	20%	4%	10%	5%

a. Given that the restaurant chosen is in a city with population over 100,000, what is the probability that it is located in the northeast?

b. Given that the restaurant chosen is in the southeast, what is the probability that it is located in a city with population under 10,000?

c. If the restaurant selected is located in the southwest, what is the probability that the city it is in has a population of 100,000 or less?
d. If the restaurant selected is located in the northwest, what is the probability that the city it is in has a population of 10,000 or more?

3.42 There are several methods of typing, or classifying, human blood. The most common procedure types blood into the general classifications of A, B, O, or AB. A method that is not as well known examines phosphoglucomutase (PGM) and classifies the blood into one of three main categories, 1-1, 2-1, or 2-2. Suppose a certain geographic region of the United States has the PGM percentages shown in the accompanying table. A person is to be chosen at random from this region.

		1-1	2-1	2-2
Race	White	46.3%	39.2%	4.0%
	Black	6.7%	3.4%	.4%

a. What is the probability that a black person is chosen?
b. Given that a black is chosen, what is the probability he or she is PGM type 1-1?
c. Given that a white is chosen, what is the probability he or she is PGM type 1-1?

3.6 The Multiplicative Rule and Independent Events

The probability of an intersection of two events can be calculated using the **multiplicative rule**, which employs the conditional probabilities we defined in the previous section, as shown in the following example.

EXAMPLE 3.16

An agriculturist, who is interested in planting wheat next year, is concerned with the following events:

B: {The production of wheat will be profitable}
A: {A serious drought will occur}

Based on available information, the agriculturist believes that the probability is .01 that production of wheat will be profitable *assuming* a serious drought will occur in the same year and that the probability is .05 that a serious drought will occur. That is,

$$P(B \mid A) = .01 \qquad \text{and} \qquad P(A) = .05$$

Based on the information provided, what is the probability that a serious drought will occur *and* that a profit will be made? That is, find $P(A \cap B)$, the probability of the intersection of events A and B.

Solution

As you will see, we have already developed a formula for finding the probability of an intersection of two events. Recall that the conditional probability of B given A is

$$P(B \mid A) = \frac{P(A \cap B)}{P(A)}$$

Multiplying both sides of this equation by $P(A)$, we obtain a formula for the probability of the intersection of events A and B. This is often called the **multiplicative rule of probability** and is given by

$$P(A \cap B) = P(A)P(B \mid A)$$

Thus,

$$\begin{aligned} P(A \cap B) &= (.05)(.01) \\ &= .0005 \end{aligned}$$

The probability that a serious drought occurs *and* the production of wheat is profitable is only .0005. As we might expect, this intersection is a very rare event.

Multiplicative Rule of Probability

$$P(A \cap B) = P(A)P(B \mid A) = P(B)P(A \mid B)$$

Intersections often contain only a few simple events. In this case, the probability of an intersection is easy to calculate by summing the appropriate simple event probabilities. However, the formula for calculating intersection probabilities plays a very important role, particularly in an area of statistics known as **Bayesian statistics**. (More detailed discussions of Bayesian statistics are contained in the references at the end of the chapter.)

EXAMPLE 3.17

Consider the experiment of tossing a fair coin twice and recording the up face on each toss. The following events are defined:

A: {First toss is a head}

B: {Second toss is a head}

Does *knowing* that event A has occurred affect the probability that B will occur?

Solution

Intuitively the answer should be no, since what occurs on the first toss should in no way affect what occurs on the second toss. Let us check our intuition. Recall the sample space for this experiment:

1. Observe HH 2. Observe HT
3. Observe TH 4. Observe TT

Each of these simple events has a probability of ¼. Thus,

$$P(B) = P(HH) + P(TH) = \tfrac{1}{4} + \tfrac{1}{4} = \tfrac{1}{2} \quad \text{and} \quad P(A) = P(HH) + P(HT) = \tfrac{1}{4} + \tfrac{1}{4} = \tfrac{1}{2}$$

Now, what is $P(B \mid A)$?

$$P(B \mid A) = \frac{P(A \cap B)}{P(A)} = \frac{P(HH)}{P(A)} = \frac{1/4}{1/2} = \frac{1}{2}$$

We can now see that $P(B) = ½$ and $P(B \mid A) = ½$. Knowing that the first toss resulted in a head does not affect the probability that the second toss will be a head. The probability is ½ whether or not we know the result of the first toss. When this occurs, we say that the two events A and B are **independent**.

Definition 3.9

Events A and B are **independent** if the occurrence of B does not alter the probability that A has occurred; i.e., events A and B are independent if

$$P(A \mid B) = P(A)$$

When events A and B are independent, it is also true that

$$P(B \mid A) = P(B)$$

Events that are not independent are said to be **dependent**.

EXAMPLE 3.18

Consider the experiment of tossing a fair die and let

A: {Observe an even number}

B: {Observe a number less than or equal to 4}

Are events A and B independent?

Solution

The Venn diagram for this experiment is shown in Figure 3.13 on page 136. We first calculate

$$P(A) = P(2) + P(4) + P(6) = \tfrac{1}{2}$$

$$P(B) = P(1) + P(2) + P(3) + P(4) = \tfrac{4}{6} = \tfrac{2}{3}$$

$$P(A \cap B) = P(2) + P(4) = \tfrac{2}{6} = \tfrac{1}{3}$$

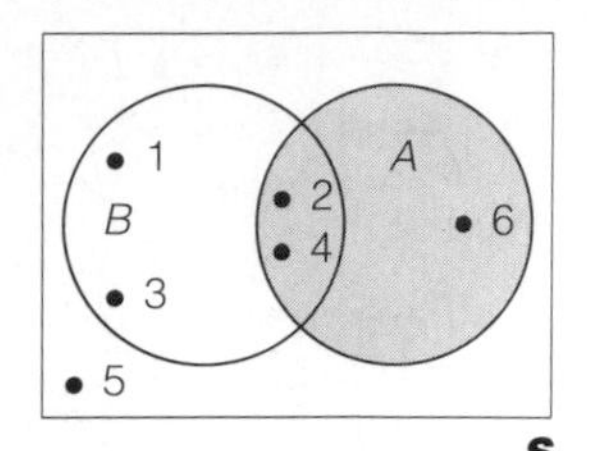

FIGURE 3.13 ▲
Venn diagram for die-toss experiment

Now assuming B has occurred, the conditional probability of A given B is

$$P(A \mid B) = \frac{P(A \cap B)}{P(B)} = \frac{1/3}{2/3} = \frac{1}{2} = P(A)$$

Thus, assuming that event B occurs does not alter the probability of observing an even number—it remains ½. Therefore, the events A and B are independent. Note that if we calculate the conditional probability of B given A, our conclusion is the same:

$$P(B \mid A) = \frac{P(A \cap B)}{P(A)} = \frac{1/3}{1/2} = \frac{2}{3} = P(B)$$

EXAMPLE 3.19

Refer to the consumer product complaint study in Example 3.15. The percentages of complaints of various types during and after the guarantee period are shown in Table 3.5. Define the following events:

A: {Cause of complaint is product appearance}

B: {Complaint occurred during the guarantee term}

Are A and B independent events?

Solution

Events A and B are independent if $P(A \mid B) = P(A)$. We calculated $P(A \mid B)$ in Example 3.15 to be .51, and from Table 3.5 we see that

$$P(A) = .32 + .03 = .35$$

Therefore, $P(A \mid B)$ is not equal to $P(A)$, and A and B are not independent events.

To gain an intuitive understanding of independence, think of situations in which the occurrence of one event does not alter the probability that a second event will occur. For example, new medical procedures are often tested on laboratory animals. The scientists conducting the tests generally try to perform the procedures on the animals so the results for one animal do not affect the results for the others. That is, the event that the procedure is successful on one animal is *independent* of the result for another. In this way, the scientists can get a more accurate idea of the efficacy of the procedure than if the results were dependent, with the success or failure for one animal affecting the results for other animals.

As a second example, consider an election poll in which 1,000 registered voters are asked their preference between two candidates. Pollsters try to use procedures for selecting a sample of voters so that the responses are independent. That is, the objective of the pollster is to select the sample so the event that one polled voter prefers candidate A does not alter the probability that a second polled voter prefers candidate A.

In the world of sports, do you think the results of a batter's successive trips to the plate in baseball or of a basketball player's successive shots at the basket, are independent? If a basketball player makes two successive shots, is the probability of making the next shot altered from its value if the result of the first shot were not known? If a player makes two shots in a row, we are likely to assign a probability to his making the third shot different from what we would assign if we knew nothing about the first two shots. Research has shown that many such results in sports tend to be *dependent* because players (and even teams) tend to get on "hot" and "cold" streaks, during which their probabilities of success may increase or decrease significantly.

We will make three final points about independence. The first is that the property of independence, unlike the mutually exclusive property, cannot be shown on or gleaned from a Venn diagram, and you cannot trust your intuition. In general, the only way to check for independence is by performing the calculations of the probabilities in the definition.

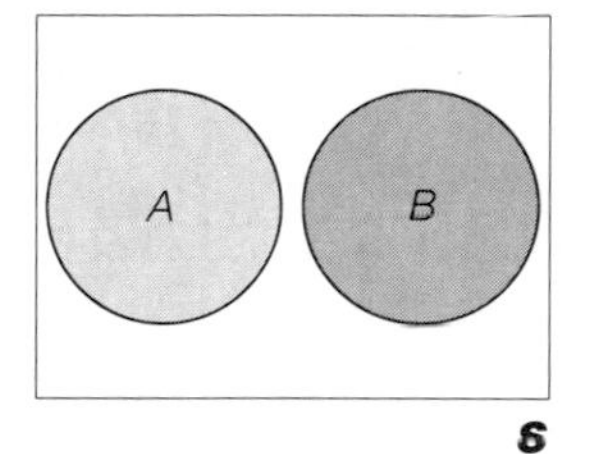

FIGURE 3.14 ▲ Mutually exclusive events are dependent events

The second point concerns the relationship between the mutually exclusive and independence properties. Suppose that events A and B are mutually exclusive, as shown in Figure 3.14. Are these events independent or dependent? That is, does the assumption that B occurs alter the probability of the occurrence of A? It certainly does, because if we assume that B has occurred, it is impossible for A to have occurred simultaneously. *Thus, mutually exclusive events are dependent events.*

The third point is that the probability of the intersection of independent events is very easy to calculate. Referring to the formula for calculating the probability of an intersection, we find

$$P(A \cap B) = P(A)P(B \mid A)$$

Thus, since $P(B \mid A) = P(B)$ when A and B are independent, we have the following useful rule:

> *If events A and B are independent,* the probability of the intersection of A and B equals the product of the probabilities of A and B; that is,
>
> $$P(A \cap B) = P(A)P(B)$$
>
> The converse is also true: If $P(A \cap B) = P(A)P(B)$, then events A and B are independent.

In the die-toss experiment, we showed in Example 3.18 that the events A: {Observe an even number} and B: {Observe a number less than or equal to 4} are independent if the die is fair. Thus,

$$P(A \cap B) = P(A)P(B) = (1/2)(2/3) = 1/3$$

This agrees with the result

$$P(A \cap B) = P(2) + P(4) = 2/6 = 1/3$$

that we obtained in the example.

EXAMPLE 3.20

Almost every retail business has the problem of determining how much inventory to purchase. Insufficient inventory may result in lost business, and excess inventory may have a detrimental effect on profits. Suppose a retail computer store owner is planning to place an order for personal computers (PCs). She is trying to decide how many IBM PCs and how many IBM compatibles (personal computers that run all or most of the same software as the IBM PC but are not manufactured by IBM) to order.

The owner's records indicate that 80% of the previous PC customers purchased IBM PCs and 20% purchased compatibles.

a. What is the probability that the next two customers will purchase compatibles?

b. What is the probability that the next 10 customers will purchase compatibles?

Solution

a. Let C_1 represent the event that customer 1 will purchase a compatible and C_2 represent the event that customer 2 will purchase a compatible. The event that *both* customers purchase compatibles is the intersection of the two events, $C_1 \cap C_2$. From the records the store owner could reasonably conclude that $P(C_1) = .2$ (based on the fact that 20% of past customers have purchased compatibles), and the same reasoning would apply to C_2. However, in order to compute the probability of $C_1 \cap C_2$, we need more information. Either the records must be examined for the occurrence of consecutive purchases of compatibles, or some assumption must be made to enable the calculation of $P(C_1 \cap C_2)$ from the multiplicative rule. It seems reasonable to make the assumption that the two events are independent, since the decision of the first customer is not likely to affect the decision of the second customer. Assuming independence, we have

$$P(C_1 \cap C_2) = P(C_1)P(C_2) = (.2)(.2) = .04$$

b. To see how to compute the probability that 10 consecutive purchases will be compatibles, first consider the event that three consecutive customers purchase compatibles. If C_3 represents the event that the third customer purchases a compatible, then we want to compute the probability of the intersection of $C_1 \cap C_2$ with C_3. Again assuming independence of the purchasing decisions, we have

$$P(C_1 \cap C_2 \cap C_3) = P(C_1 \cap C_2)P(C_3) = (.2)^2(.2) = .008$$

Similar reasoning leads to the conclusion that the intersection of 10 such events can be calculated as follows:

$$P(C_1 \cap C_2 \cap \cdots \cap C_{10}) = P(C_1)P(C_2) \cdot \cdots \cdot P(C_{10})$$
$$= (.2)^{10} = .0000001024$$

Thus, the probability that 10 consecutive customers purchase IBM compatibles is about 1 in 10 million, assuming the probability of each customer's purchase of a compatible is .2 and the purchase decisions are independent.

Exercises 3.43–3.60

Learning the Mechanics

3.43 An experiment results in one of five simple events with the following probabilities: $P(E_1) = .22$, $P(E_2) = .31$, $P(E_3) = .15$, $P(E_4) = .22$, and $P(E_5) = .1$. The following events have been defined:

A: $\{E_1, E_3\}$
B: $\{E_2, E_3, E_4\}$
C: $\{E_1, E_5\}$

Find each of the following probabilities:

a. $P(A)$ b. $P(B)$ c. $P(A \cap B)$ d. $P(A \mid B)$ e. $P(B \cap C)$ f. $P(C \mid B)$

g. Consider each pair of events: A and B, A and C, and B and C. Are any of the pairs of events independent? Why?

3.44 Three fair coins are tossed and the following events are defined:

A: {Observe at least one head}
B: {Observe exactly two heads}
C: {Observe exactly two tails}
D: {Observe at most one head}

a. Sum the probabilities of the appropriate simple events to find: $P(A)$, $P(B)$, $P(C)$, $P(D)$, $P(A \cap B)$, $P(A \cap D)$, $P(B \cap C)$, and $P(B \cap D)$.
b. Use your answers to part **a** to calculate $P(B \mid A)$, $P(A \mid D)$, and $P(C \mid B)$.
c. Which pairs of events, if any, are independent? Why?

3.45 Two hundred shoppers at a large suburban mall were asked two questions: (1) Did you see a television ad for the sale at department store X during the past 2 weeks? (2) Did you shop at department store X during the past 2 weeks? The responses to the questions are summarized in the table.

	Shopped at X	Did Not Shop at X
Saw ad	100	25
Did not see ad	25	50

One of the 200 shoppers questioned is to be chosen at random.

a. What is the probability that the person selected saw the ad?
b. What is the probability that the person selected saw the ad and shopped at store X?
c. Find the conditional probability that the person shopped at store X given that the person saw the ad.
d. What is the probability that the person selected shopped at store X?
e. Use your answers to parts **a**, **b**, and **d** to check the independence of the events Saw ad and Shopped at X.
f. Are the two events Did not see ad and Did not shop at X mutually exclusive? Explain.

3.46 Two fair dice are tossed, and the following events are defined:

A: {Sum of the numbers showing is odd}

B: {Sum of the numbers showing is 9, 11, or 12}

Are events A and B independent? Why?

3.47 A sample space contains six simple events and events A, B, and C as shown in the Venn diagram. The probabilities of the simple events are $P(1) = .20$, $P(2) = .05$, $P(3) = .30$, $P(4) = .10$, $P(5) = .10$, $P(6) = .25$.

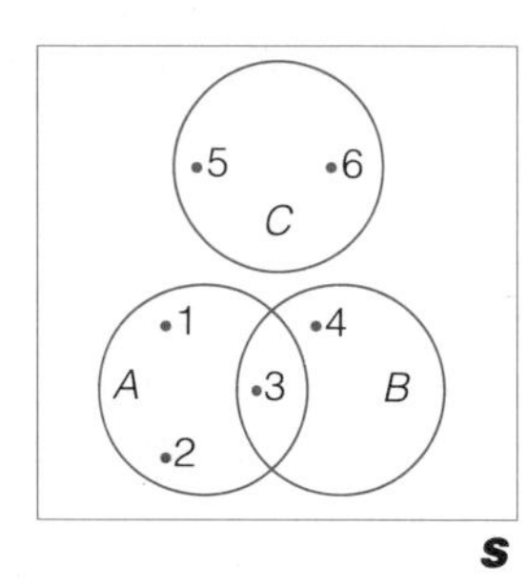

a. Which pairs of events, if any, are mutually exclusive? Why?
b. Which pairs of events, if any, are independent? Why?
c. Find $P(A \cup B)$ by adding the probabilities of the simple events and then by using the additive rule. Verify that the answers agree. Repeat for $P(A \cup C)$.

3.48 Defend or refute each of the following statements:
a. Dependent events are always mutually exclusive.
b. Mutually exclusive events are always dependent.
c. Independent events are always mutually exclusive.

3.49 For two events, A and B, $P(A) = .4$ and $P(B) = .2$.
a. If A and B are independent, find $P(A \cap B)$, $P(A \mid B)$, and $P(A \cup B)$.
b. If A and B are dependent, with $P(A \mid B) = .6$, find $P(A \cap B)$ and $P(B \mid A)$.

Applying the Concepts

3.50 Each of 50 people in a random sample was asked to name his or her favorite soft drink. The responses are shown below:

Pepsi-Cola	18
Coca-Cola	16
Seven-Up	6
Mr. Pibb	4
Sprite	4
Nehi Orange	1
Dr Pepper	1

Suppose a person is selected at random from the survey. Let A be the event that the person preferred a soft drink bottled by the Coca-Cola Company (Coca-Cola, Mr. Pibb, or Sprite). Let B be the event that the person did *not* choose a cola (either Pepsi-Cola or Coca-Cola).

a. Find $P(A)$. b. Find $P(B)$.

c. Describe the event $A \cap B$, and find its probability.

d. Describe the event $A \cup B$ and use the additive rule to find its probability.

e. Use the multiplicative rule to find $P(A \mid B)$.

f. Use the multiplicative rule to find $P(A \mid B')$.

3.51 The probability that a certain microchip for a computer fails when first used is .10. If it does not fail immediately, the probability that it lasts 1 year is .99. Define the experiment as observing whether a microchip fails or not during its first year, and define the events:

A: {Microchip lasts through first use}

B: {Microchip lasts from first use through end of first year}

a. List the simple events of the experiment in terms of the events A and B. [*Hint:* The event that the chip fails during the first year is *not* a simple event because it can be decomposed into two more basic events. There are three simple events.]

b. Assign probabilities to the simple events. [*Hint:* You will need to use the multiplicative rule for some of the simple events.]

c. Use the simple event probabilities to find the probability that the microchip does not fail during the first year.

3.52 Some strings of holiday lights are wired in series; thus, if one bulb fails the entire string goes out. Suppose the probability of an individual bulb failing during a certain period of time is .05. What is the probability that a string of 10 lights goes out during that period of time? What assumption did you make concerning the light bulbs? [*Hint:* The complement of the event, "At least one of the 10 lights fails," is the event, "None of the 10 lights fails."]

3.53 Scoring a hole-in-one is the single greatest shot a golfer can make. In the 1989 U.S. Open, four players each made a hole-in-one on the sixth hole at Oak Hill Country Club. Lois Hains, in an article in *Golf Digest* (1990), reports that the estimated probability of making a hole-in-one is approximately 1/2,780 for male professional golfers and 1/2,920 for female professional golfers. Suppose we randomly select four professional male golfers.

a. What is the probability that these four golfers each make a hole-in-one on the sixth hole during the same round? What assumption did you make to calculate this probability?

b. What is the probability that none of these golfers makes a hole-in-one on the sixth hole?

c. What is the probability that at least one of these golfers makes a hole-in-one on the sixth hole?

d. Repeat parts **a–c** for four female professional golfers.

3.54 On April 25, 1980, the *Bakersfield Californian* reported on a record payout in the Pennsylvania lottery. The winning number in the state lottery that day, with options from 000 to 999, was 666. Harried state officials were denying the possibility of tampering with the lottery, but many bookies in Pittsburgh were refusing bets on any numbers involving 4's or 6's. Each digit of the winning number was determined by allowing a blast of air to propel one of 10 Ping-Pong balls, numbered 0, 1, 2, . . . , 9, into a basket. On this particular day, the TV host (eventually indicted and convicted) and some friends injected liquid into all the balls except those numbered 4 and 6. Thus, only numbers involving the 4 and the 6 digits could be winners.

a. If the balls had not been fixed, what is the probability that a number involving only 4's and 6's would win? What assumption about the determination of the three digits did you make to solve this problem?
b. Given the conditions of part **a**, what is the probability that the number 666 would win?
c. Since the TV announcer and collaborators knew that the winning number would involve only 4's and 6's, what is the probability that the winning number would be 666?

• **3.55** A dice game is played in the following manner: A player starts by rolling two dice. If the result is a 7 or 11, the player wins. For any other sum appearing on the dice, the player continues to roll the dice until that outcome recurs (in which case the player wins) or until a 7 or 11 occurs (in which case the player loses). If on any roll the outcome is 2 ("snake-eyes"), the game is over, and the player loses.
a. What is the probability that a player wins the game on the first roll of the dice? (Assume the dice are balanced.)
b. What is the probability that a player loses the game on the first roll of the dice?
c. If the player throws a total of 3 on the first roll, what is the probability that the game ends on the next roll?

3.56 Despite penicillin and other antibiotics, bacterial pneumonia still kills thousands of Americans every year. The U.S. Food and Drug Administration has approved the use of a new antipneumonia vaccine called Pneumovax. It is designed especially for elderly or debilitated patients who are usually the most vulnerable to bacterial pneumonia. Field trials proved the new vaccine to be 90% effective in stimulating the production of antibodies to pneumonia-producing bacteria (i.e., 90% successful in preventing a person exposed to pneumonia-producing bacteria from acquiring the disease). Suppose the probability of an elderly or debilitated person being exposed to these bacteria is .40 (whether inoculated or not), and after being exposed, the probability of each person contracting bacterial pneumonia if not inoculated with the vaccine is .95. Find the probability that an elderly or debilitated person inoculated with this new vaccine acquires pneumonia. What is the probability if this person has not been inoculated?

3.57 As many as one in every 500 blacks in the United States is reported to have sickle-cell anemia. One in 10 is a carrier of the trait, and carriers who marry have a 1 in 4 chance of passing the disease to their children. Suppose a carrier has three children and the transmission of the disease from the carrier to any one child is independent of whether or not the carrier transmits to another. Find the probability that:
a. None of the children acquire the disease.
b. All three acquire the disease.
c. Exactly one acquires the disease.

3.58 One of the problems encountered in organ transplants is rejection by the body of the transplanted tissue, i.e., the tendency of the white blood cells to attack the foreign tissue. The key to reception or rejection is the nature of the antigens attached to the tissue cells. If the antigens of the donor and receiver match, the body will accept the transplanted tissue. The antigens in twins always match; the probability of a match in siblings is .25, and of a match in two people from the population at large is .001. Suppose you need a kidney, and you have two brothers and a sister.
a. If one of the three siblings offers a kidney, what is the probability that the antigens will match?
b. If all three siblings offer a kidney, what is the probability that all three antigens will match?
c. If all three siblings offer a kidney, what is the probability that none of the antigens will match?

3.59 Refer to Exercise 3.58. Answer parts **b** and **c** but this time assume that the three donors were obtained from the population at large.

3.60 Refer to the kidney transplant exercise (Exercise 3.58). Sandoz, a pharmaceutical firm, has developed a drug, cyclosporine, that appears to retard a body's immune system from rejecting transplanted organs. Sandoz reports that kidney transplant patients who receive the drug have an 80% chance of living through the first year, and it has also improved the survival rates for other types of transplants. Suppose a hospital performs four kidney transplants, and all patients receive the drug cyclosporine. What is the probability that:

a. All four patients are alive at the end of 1 year?
b. None of the four patients is alive at the end of 1 year?
c. At least one of the patients is alive at the end of 1 year?

3.7 Probability and Statistics: An Example

We have introduced a number of new concepts in the preceding sections, and this makes the study of probability a particularly arduous task. It is therefore essential to establish clearly the connection between probability and statistics, which we will do in the remaining chapters. However, we present one brief example in this section so that you can begin to understand why some knowledge of probability is important in the study of statistics.

Suppose a psychologist is researching the hypothesis that rats which have been trained will pass on at least part of the training to their offspring. To test the hypothesis, three offspring (no two with the same parents) of trained rats are randomly selected and subjected to a training test. It is known from many previous experiments that the relative frequency distribution of the scores for untrained rats is mound-shaped with a mean of 60 and a standard deviation of 10. Suppose all three of the trained rats' offspring score more than 70 on the test. What can the research psychologist conclude?

The relative frequency distribution of the scores for untrained rats is shown in Figure 3.15. If the distribution is mound-shaped and approximately symmetric about

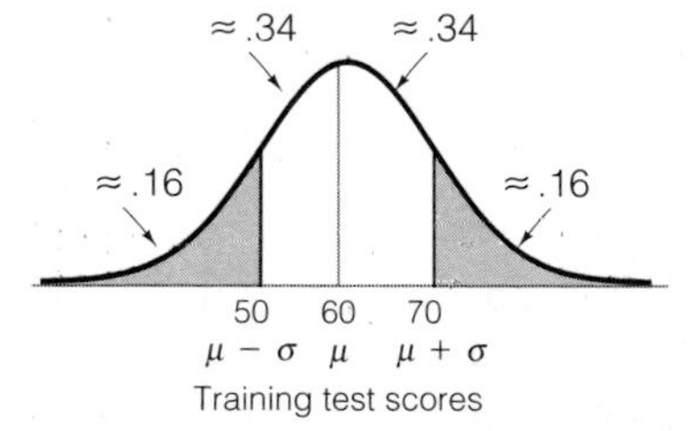

FIGURE 3.15 ▶ Relative frequency distribution of training test scores

the mean, we can conclude that approximately 16% of untrained rats will score more than 70 on the test (see Table 2.6). Now define the events

A_1: {Offspring 1 scores more than 70}
A_2: {Offspring 2 scores more than 70}
A_3: {Offspring 3 scores more than 70}

We want to find $P(A_1 \cap A_2 \cap A_3)$, the probability that all three offspring score more than 70 on the training test.

Since the offspring are selected so that they have different parents, it may be plausible to assume that the events A_1, A_2, and A_3 are independent. That is,

$$P(A_2 \mid A_1) = P(A_2)$$

In words, knowing that the first offspring scores more than 70 on the test does not affect the probability that the second offspring scores more than 70. With the assumption of independence, we can calculate the probability of the intersection by multiplying the individual probabilities:

$$\begin{aligned} P(A_1 \cap A_2 \cap A_3) &= P(A_1)P(A_2)P(A_3) \\ &\approx (.16)(.16)(.16) = .004096 \end{aligned}$$

Thus, the probability that the research psychologist will observe all three offspring scoring more than 70 is only about .004 *if the offspring are untrained*. If this event were to occur, the psychologist might conclude that it lends credence to the theory that the offspring inherit some of the parents' training *since it is so unlikely to occur if they are untrained*. Such a conclusion would be an application of the rare event approach to statistical inference. You can see that the basic principles of probability play an important role.

3.8 Random Sampling

How a sample is selected from a population is of vital importance in statistical inference because the probability of an observed sample will be used to infer the characteristics of the sampled population. To illustrate, suppose you deal yourself four cards from a deck of 52 cards and all four cards are aces. Do you conclude that your deck is an ordinary bridge deck, containing only four aces, or do you conclude that the deck is stacked with more than four aces? It depends on how the cards were drawn. If the four aces were always placed at the top of a standard bridge deck, drawing four aces is not unusual—it is certain. On the other hand, if the cards are thoroughly mixed, drawing four aces in a sample of four cards is highly improbable. The point, of course, is that in order to use the observed sample of four cards to draw inferences about the population (the deck of 52 cards), you need to know how the sample was selected from the deck.

One of the simplest and most frequently employed sampling procedures is implied in the previous examples and exercises. It produces what is known as a **random sample**.

Definition 3.10

If n elements are selected from a population in such a way that every set of n elements in the population has an equal probability of being selected, the n elements are said to be a **random sample.***

EXAMPLE 3.21

Suppose a lottery consists of 10 tickets. (This number is small to simplify our example.) One ticket stub is to be chosen, and the corresponding ticket holder will receive a generous prize. How would you select this ticket stub so that the prize will be awarded fairly?

Solution

If the prize is to be awarded fairly, it seems reasonable to require that each ticket stub have the same probability of being drawn. That is, each stub should have a probability of $1/10$ of being selected. A method to achieve the objective of equal selection probabilities is to *mix* the 10 stubs thoroughly and *blindly* pick one of the stubs. If this procedure were repeatedly used, each time replacing the selected stub, a particular stub should be chosen approximately $1/10$ of the time in a long series of draws. This method of sampling is known as **random sampling**.

If a population is not too large and the elements can be numbered on slips of paper, poker chips, etc., you can physically mix the slips of paper or chips and remove n elements from the total. The numbers that appear on the chips selected would indicate the population elements to be included in the sample. Such a procedure will not guarantee a random sample because it is often difficult to achieve a thorough mix (see Case Study 3.3), but it usually provides a reasonably good approximation to random sampling.

Sampling can be thought of as an experiment consisting of drawing n elements from a population of N elements, with each different sample representing a simple event of the experiment. Thus, in Example 3.21, in which a lottery consisted of drawing one of 10 tickets, there are 10 different samples, or simple events. If the drawing is held so that each simple event is equally likely with probability $1/10$, then the result of the experiment is a *random sample*.

Of course, for most applications the population will consist of more than $N = 10$ elements, and the sample will consist of more than $n = 1$ element. Often, the total number of possible samples will not be easy to visualize, so a method for counting the number of samples is needed. For example, suppose the lottery consists of drawing two tickets from ten. We can list the possible samples as in Table 3.7 on page 146, where

*Strictly speaking, this is a **simple random sample.** There are many different types of random samples. The simple random sample is the most common.

T_1 represents ticket 1, T_2 ticket 2, . . . , and T_{10} ticket 10. The systematic listing in Table 3.7 shows the 45 possible samples, the simple events of the experiment of sampling two elements from 10. However, the listing is tedious and only gets more so as the values of the population size N and the sample size n are increased.

TABLE 3.7 Listing of All Possible Samples of Two Tickets Drawn from 10 Tickets

T_1, T_2	T_2, T_3	T_3, T_4	T_4, T_5	T_5, T_6	T_6, T_7	T_7, T_8	T_8, T_9	T_9, T_{10}
T_1, T_3	T_2, T_4	T_3, T_5	T_4, T_6	T_5, T_7	T_6, T_8	T_7, T_9	T_8, T_{10}	
T_1, T_4	T_2, T_5	T_3, T_6	T_4, T_7	T_5, T_8	T_6, T_9	T_7, T_{10}		
T_1, T_5	T_2, T_6	T_3, T_7	T_4, T_8	T_5, T_9	T_6, T_{10}			
T_1, T_6	T_2, T_7	T_3, T_8	T_4, T_9	T_5, T_{10}				
T_1, T_7	T_2, T_8	T_3, T_9	T_4, T_{10}					
T_1, T_8	T_2, T_9	T_3, T_{10}						
T_1, T_9	T_2, T_{10}							
T_1, T_{10}								

A second method of determining the number of samples is to use **combinatorial mathematics**. The combinatorial symbol for the number of different ways of selecting n elements from N elements is $\binom{N}{n}$, which is read "the number of combinations of N elements taken n at a time." The formula* for calculating the number is

$$\binom{N}{n} = \frac{N!}{n!(N-n)!}$$

where "!" is the factorial symbol and is a shorthand for the following multiplication:

$$n! = n(n-1)(n-2) \cdot \cdots \cdot (3)(2)(1)$$

Thus, for example, $5! = 5 \cdot 4 \cdot 3 \cdot 2 \cdot 1 = 120$. (The quantity 0! is defined to be equal to 1.)

EXAMPLE 3.22

a. Use the combinatorial formula to count the number of different ways of drawing one lottery ticket from a total of 10 tickets.

b. Use the combinatorial formula to count the number of ways of drawing two lottery tickets from a total of 10 tickets.

*For a more thorough discussion of the reasoning behind this and other counting rules, consult the references at the end of this chapter.

Solution

a. Substituting $n = 1$ and $N = 10$ into the formula, we find

$$\binom{N}{n} = \binom{10}{1} = \frac{10!}{1!(10-1)!} = \frac{10!}{1!9!}$$

$$= \frac{10 \cdot 9 \cdot 8 \cdot 7 \cdot 6 \cdot 5 \cdot 4 \cdot 3 \cdot 2 \cdot 1}{(1)(9 \cdot 8 \cdot 7 \cdot 6 \cdot 5 \cdot 4 \cdot 3 \cdot 2 \cdot 1)} = 10$$

Thus, as we know intuitively, there are 10 different samples that can be selected when drawing one ticket from 10.

b. Substituting $n = 2$ and $N = 10$ into the formula, we find

$$\binom{N}{n} = \binom{10}{2} = \frac{10!}{2!(10-2)!} = \frac{10!}{2!8!}$$

$$= \frac{10 \cdot 9 \cdot 8 \cdot 7 \cdot 6 \cdot 5 \cdot 4 \cdot 3 \cdot 2 \cdot 1}{(2 \cdot 1)(8 \cdot 7 \cdot 6 \cdot 5 \cdot 4 \cdot 3 \cdot 2 \cdot 1)} = \frac{10 \cdot 9}{2 \cdot 1} = 45$$

This agrees with the number obtained by listing all possible samples in Table 3.7 but requires much less effort. And, as the next example illustrates, the combinatorial formula works long after listing has ceased to be a viable option.

EXAMPLE 3.23

Suppose you wish to randomly sample five households from a population of 100,000 households.

a. How many different samples can be selected?
b. Give a procedure for selecting a random sample.

Solution

a. Using the combinatorial rule, we find

$$\binom{100{,}000}{5} = \frac{100{,}000!}{5!99{,}995!}$$

$$= \frac{100{,}000 \cdot 99{,}999 \cdot 99{,}998 \cdot 99{,}997 \cdot 99{,}996}{5 \cdot 4 \cdot 3 \cdot 2 \cdot 1}$$

$$= 8.33 \times 10^{22}$$

Thus, there are 83.3 billion trillion different samples of five households that can be selected from 100,000.

b. How can we ensure that each of the possible samples has an equal chance of being selected, as required for random sampling, when there are so many? Generally, we will use a **random number table**, such as the one in Table I of Appendix A.

First, we number the households in the population from 1 to 100,000. Then, we turn to a page of Table I, say the first page. (A partial reproduction of the first page of Table I is shown in Table 3.8.) Now, randomly select a starting number, say the random number appearing in the third row, second column. This number is 48360. Proceed down the second column to obtain the remaining four random numbers. The five selected random numbers are shaded in Table 3.8. Using the first five digits to represent the households from 1 to 99,999 and the number 00000 to represent household 100,000, you can see that the households numbered

48,360 93,093 39,975 6,907 72,905

should be included in your sample.

TABLE 3.8 Partial Reproduction of Table I in Appendix A

Row \ Column	1	2	3	4	5	6
1	10480	15011	01536	02011	81647	91646
2	22368	46573	25595	85393	30995	89198
3	24130	48360	22527	97265	76393	64809
4	42167	93093	06243	61680	07856	16376
5	37570	39975	81837	16656	06121	91782
6	77921	06907	11008	42751	27756	53498
7	99562	72905	56420	69994	98872	31016
8	96301	91977	05463	07972	18876	20922
9	89579	14342	63661	10281	17453	18103
10	85475	36857	53342	53988	53060	59533
11	28918	69578	88231	33276	70997	79936
12	63553	40961	48235	03427	49626	69445
13	09429	93969	52636	92737	88974	33488

Can we be sure that all 83.3 billion trillion samples have an equal chance of being selected? We cannot, but to the extent that the random number table contains truly random sequences of digits, the sample should be very close to random.

Table I in Appendix A is just one example of a *table of random numbers*. Most samplers use such a table to obtain random samples. Random number tables are constructed in such a way that every number occurs with (approximately) equal probability. Further, the occurrence of any one number in a position is independent of any of the other numbers that appear in the table. To use a table of random numbers, number the N elements in the population from 1 to N. Then turn to Table I and select a starting number in the table. Proceeding from this number either across the row or down the column, remove and record n numbers from the table. Use only the necessary number of digits in each random number to identify the element to be included in the sample. If, in the course of recording the n numbers from the table, you select a number that has already been selected, simply discard the duplicate and select a replacement at the end of the sequence. Thus, you may have to record more than n numbers from the table to obtain a sample of n unique numbers.

CASE STUDY 3.3 / The 1970 Draft Lottery

From 1948 through the early years of the Vietnam War, the Selective Service System drafted men into military service by age—oldest first, starting with 25-year-olds. A network of local draft boards was used to implement the selection process. Then, on the evening of December 1, 1969, the Selective Service System conducted a lottery to determine the order of selection for 1970 in an attempt to overcome what many believed were inequities in the system. (Such lotteries had been used during World Wars I and II, but it had been 27 years since the last one.)

The objective of the lottery was to randomly order the induction sequence of men between the ages of 19 and 26. To do this, the 366 possible days in a year were written on slips of paper and placed in egg-shaped capsules that were stored in monthly lots. The monthly lots were placed one by one into a wooden box that was ". . . turned end over end several times to mix the numbers" ("Random or Not? Judge Studies Lottery Protest," *The National Observer*, January 12, 1970, p. 2). The capsules were then dumped into a large glass bowl and drawn one by one to obtain the order of induction. All men born on the first day drawn would be inducted first; those born on the second day drawn would be inducted next, etc. Thus, the lottery assigned a rank to each of the 366 birthdays. The results of the lottery are shown in Table 3.9 on page 150.

To generate a random sequence of numbers with this procedure, it is necessary for each (remaining) capsule in the bowl to have an equal probability of being selected on each draw. That is, by means of thorough mixing, each capsule must have an equal opportunity to come to rest precisely where the sampler's hand closes within the bowl. Although a mixing procedure with this property is almost impossible to achieve, the ideal can be closely approximated. Unfortunately, this was apparently not the case in the 1970 lottery. Even though the sequence of dates in Table 3.9 may appear to be random, there is ample statistical evidence to indicate a nonrandom selection of induction dates.*

To obtain an understanding of the problem with the 1970 lottery, we calculated the median rank for each month and plotted them in Figure 3.16(a) on page 151. If the sequence of ranks were randomly generated, there should be no relationship between the size of the median ranks and the months of the year. The medians should vary randomly above and below a horizontal line with intercept 183.5 (the median of the integers 1 through 366). However, Figure 3.16(a) reveals a general downward trend in the medians. Men born later in the year were more likely to be drafted before men born early in the year. Furthermore, since not all men between 19 and 25 years old would be drafted in 1970, men born earlier in the year were more likely not to be drafted at all. Although it is possible to observe such a sequence of medians when random sampling is employed, it is highly unlikely. It is more likely that the capsules were not mixed thoroughly enough to give every capsule an equal chance of selection on each draw. The graph indicates that capsules tended to be drawn in monthly groups, and thus suggests that monthly lots of capsules in the wooden boxes were not mixed thoroughly enough after they were dumped into the large glass bowl (Williams, 1978).

The following year, the Selective Service System used more sophisticated mixing techniques to guard

*Formal statistical tests to detect nonrandomness are beyond the scope of this text.

TABLE 3.9 1970 Draft Lottery Results

1	Sept. 14	54	Aug. 5	107	Nov. 16	160	Sept 22	213	Mar. 8	266	Nov. 4	319	May 23
2	April 24	55	May 16	108	Mar. 1	161	Sept. 2	214	Feb. 5	267	Mar. 3	320	Dec. 15
3	Dec. 30	56	Dec. 5	109	June 23	162	Dec. 23	215	Jan. 4	268	Mar. 27	321	May 8
4	Feb. 14	57	Feb. 23	110	June 6	163	Dec. 13	216	Feb. 10	269	April 5	322	July 15
5	Oct. 18	58	Jan. 19	111	Aug. 1	164	Jan. 30	217	Mar. 30	270	July 29	323	Mar. 10
6	Sept. 6	59	Jan. 24	112	May 17	165	Dec. 4	218	April 10	271	April 2	324	Aug. 11
7	Oct. 26	60	June 21	113	Sept. 15	166	Mar. 16	219	April 9	272	June 12	325	Jan. 10
8	Sept. 7	61	Aug. 29	114	Aug. 6	167	Aug. 28	220	Oct. 10	273	April 15	326	May 22
9	Nov. 22	62	April 21	115	July 3	168	Aug. 7	221	Jan. 12	274	June 16	327	July 6
10	Dec. 6	63	Sept. 20	116	Aug. 23	169	Mar. 15	222	Jan. 28	275	Mar. 4	328	Dec. 2
11	Aug. 31	64	June 27	117	Oct. 22	170	Mar. 26	223	Mar. 28	276	May 4	329	Jan. 11
12	Dec. 7	65	May 10	118	Jan. 23	171	Oct. 15	224	Jan. 6	277	July 9	330	May 1
13	July 8	66	Nov. 12	119	Sept. 23	172	July 23	225	Sept. 1	278	May 18	331	July 14
14	April 11	67	July 25	120	July 16	173	Dec. 26	226	May 29	279	July 4	332	Mar. 18
15	July 12	68	Feb. 12	121	Jan. 16	174	Nov. 30	227	July 19	280	Jan. 20	333	Aug. 30
16	Dec. 29	69	June 13	122	Mar. 7	175	Sept. 13	228	June 2	281	Nov. 28	334	Mar. 21
17	Jan. 15	70	Dec. 21	123	Dec. 28	176	Oct. 25	229	Oct. 29	282	Nov. 10	335	June 9
18	Sept. 26	71	Sept. 10	124	April 13	177	Sept. 19	230	Nov. 24	283	Oct. 8	336	April 19
19	Nov. 1	72	Oct. 12	125	Oct. 2	178	May 14	231	April 14	284	July 10	337	Jan. 22
20	June 4	73	June 17	126	Nov. 13	179	Feb. 25	232	Sept. 4	285	Feb. 29	338	Feb. 9
21	Aug. 10	74	April 27	127	Nov. 14	180	June 15	233	Sept. 27	286	Aug. 25	339	Aug. 22
22	June 26	75	May 19	128	Dec. 18	181	Feb. 8	234	Oct. 7	287	July 30	340	April 26
23	July 24	76	Nov. 6	129	Dec. 1	182	Nov. 23	235	Jan. 17	288	Oct. 17	341	June 18
24	Oct. 5	77	Jan. 28	130	May 15	183	May 20	236	Feb. 24	289	July 27	342	Oct. 9
25	Feb. 19	78	Dec. 27	131	Nov. 15	184	Sept. 8	237	Oct. 11	290	Feb. 22	343	Mar. 25
26	Dec. 14	79	Oct. 31	132	Nov. 25	185	Nov. 20	238	Jan. 14	291	Aug. 21	344	Aug. 20
27	July 21	80	Nov. 9	133	May 12	186	Jan. 21	239	Mar. 20	292	Feb. 18	345	April 20
28	June 5	81	April 4	134	June 11	187	July 20	240	Dec. 19	293	Mar. 5	346	April 12
29	Mar. 2	82	Sept. 5	135	Dec. 20	188	July 5	241	Oct. 19	294	Oct. 14	347	Feb. 6
30	Mar. 31	83	April 3	136	Mar. 11	189	Feb. 17	242	Sept. 12	295	May 13	348	Nov. 3
31	May 24	84	Dec. 25	137	June 25	190	July 18	243	Oct. 21	296	May 27	349	Jan. 29
32	April 1	85	June 7	138	Oct. 13	191	April 29	244	Oct. 3	297	Feb. 3	350	July 2
33	Mar. 17	86	Feb. 1	139	Mar. 6	192	Oct. 20	245	Aug. 26	298	May 2	351	April 25
34	Nov. 2	87	Oct. 6	140	Jan. 18	193	July 31	246	Sept. 18	299	Feb. 28	352	Aug. 27
35	May 7	88	July 28	141	Aug. 18	194	Jan. 9	247	June 22	300	Mar. 12	353	June 29
36	Aug. 24	89	Feb. 15	142	Aug. 12	195	Sept. 24	248	July 11	301	June 3	354	Mar. 14
37	May 11	90	April 18	143	Nov. 17	196	Oct. 24	249	June 1	302	Feb. 20	355	Jan. 27
38	Oct. 30	91	Feb. 7	144	Feb. 2	197	May 9	250	May 21	303	July 26	356	June 14
39	Dec. 11	92	Jan. 26	145	Aug. 4	198	Aug. 14	251	Jan. 3	304	Dec. 17	357	May 26
40	May 3	93	July 1	146	Nov. 18	199	Jan. 8	252	April 23	305	Jan. 1	358	June 24
41	Dec. 10	94	Oct. 28	147	April 7	200	Mar. 19	253	April 6	306	Jan. 7	359	Oct. 1
42	July 13	95	Dec. 24	148	April 16	201	Oct. 23	254	Oct. 16	307	Aug. 13	360	June 20
43	Dec. 9	96	Dec. 16	149	Sept. 25	202	Oct. 4	255	Sept. 17	308	May 28	361	May 25
44	Aug. 16	97	Nov. 8	150	Feb. 11	203	Nov. 19	256	Mar. 23	309	Nov. 26	362	Mar. 29
45	Aug. 2	98	July 17	151	Sept. 29	204	Sept. 21	257	Sept. 28	310	Nov. 5	363	Feb. 21
46	Nov. 11	99	Nov. 29	152	Feb. 13	205	Feb. 27	258	Mar. 24	311	Aug. 19	364	May 5
47	Nov. 27	100	Dec. 31	153	July 22	206	June 10	259	Mar. 13	312	April 8	365	Feb. 26
48	Aug. 8	101	Jan. 5	154	Aug. 17	207	Sept. 16	260	April 17	313	May 31	366	June 8
49	Sept. 3	102	Aug. 15	155	May 6	208	April 30	261	Aug. 3	314	Dec. 12		
50	July 7	103	May 30	156	Nov. 21	209	June 30	262	April 28	315	Sept. 30		
51	Nov. 7	104	June 19	157	Dec. 3	210	Feb. 4	263	Sept. 9	316	April 22		
52	Jan. 25	105	Dec. 8	158	Sept. 11	211	Jan. 31	264	Oct. 27	317	Mar. 9		
53	Dec. 22	106	Aug. 9	159	Jan. 2	212	Feb. 16	265	Mar. 22	318	Jan. 13		

against the monthly clustering of selections. The median plot for the 1971 lottery is shown in Figure 3.16(b). [*Note:* The 1971 lottery involved only men born in 1951. Thus, only 365 birthdays were ranked, and the sequence of monthly ranked medians should be compared to 183 instead of 183.5.] Notice that no apparent trend remains; the new mixing technique appears to have been successful.

This case study emphasizes the value of using a random number table or a random number generator of a statistical software package in the selection of a random sample. Most important, it points out the problems that may be encountered when attempting to acquire a random sample by a mechanical selection process.

You can now better understand what we mean when we use the phrases "assume the selection is made at random," or simply "at random" in probability examples and exercises. In most cases we are indicating that each simple event, or sample, has an equal probability of occurring.

FIGURE 3.16 ▶
Median plots for lottery results: 1970 and 1971

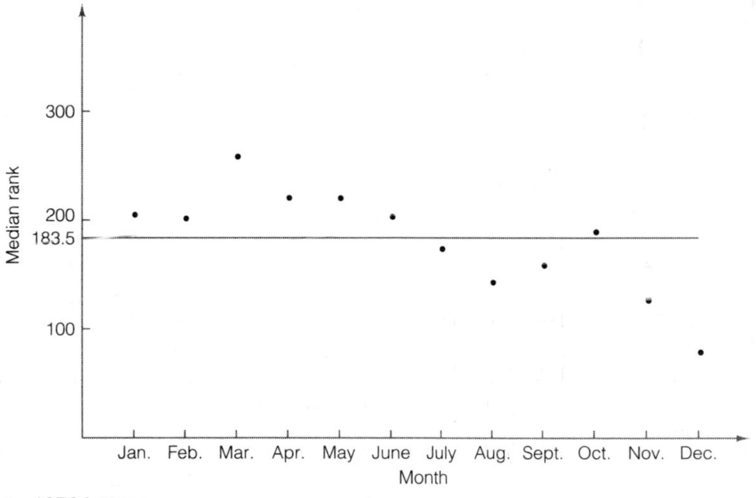

a. 1970 lottery

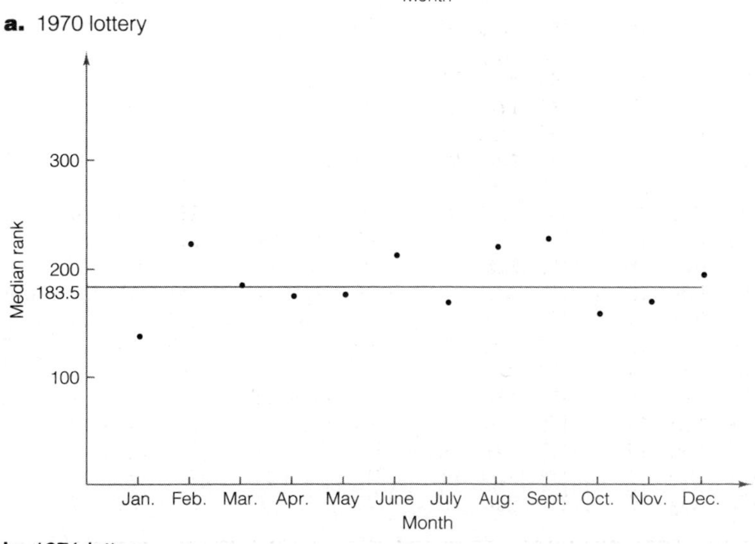

b. 1971 lottery

Exercises 3.61 – 3.65

Learning the Mechanics

3.61 Suppose you wish to sample $n = 3$ elements from a total of $N = 6$ elements.

a. Count the number of different samples that can be drawn, first by listing them, and then by using combinatorial mathematics.

b. If random sampling is to be employed, what is the probability that any particular sample will be selected?

c. Show how to use the random number table, Table I in Appendix A, to select a random sample of 3 elements from a population of 6 elements. Perform the sampling procedure 20 times. Do any two of the samples contain the same three elements? Given your answer to part **b**, did you expect repeated samples?

3.62 Suppose you wish to sample $n = 3$ elements from a total of $N = 600$ elements.

a. Count the number of different samples by using combinatorial mathematics.

b. If random sampling is to be employed, what is the probability that any particular sample will be selected?

c. Show how to use the random number table, Table I in Appendix A, to select a random sample of 3 elements from a population of 600 elements. Perform the sampling procedure 20 times. Do any two of the samples contain the same three elements? Given your answer to part **b**, did you expect repeated samples?

Applying the Concepts

3.63 Archaeologists plan to perform test digs at a location believed to have been inhabited several thousand years ago. The location is approximately 10,000 meters long by 5,000 meters wide. They first draw rectangular grids over the area, consisting of lines every 100 meters, creating a total of $100 \cdot 50 = 5{,}000$ intersections (not counting one of the outer boundaries). The plan is to randomly sample 50 intersection points and dig at the sampled intersections. Explain how you could use the random number table to obtain a random sample of 50 intersections. Develop at least two plans: one that numbers the intersections from 1 to 5,000 prior to selection and another that selects the row and column of each sampled intersection (from the total of 100 rows and 50 columns).

3.64 To ascertain the effectiveness of their advertising campaigns, firms frequently conduct telephone interviews with consumers. Random samples of telephone numbers may be randomly or systematically selected from telephone directories, or an innovation called *random-digit dialing* may be employed. This approach involves using a random number generator to mechanically create the sample of phone numbers to be called. An advantage of random-digit dialing is that it makes it possible to obtain a representative sample from the population of all households with telephones, whereas with telephone directory sampling, it is only possible to obtain a sample from the population of households that have *listed* telephone numbers.

a. Explain how the random number table (Table I of Appendix A) could be used to generate a sample of 7-digit telephone numbers.

b. Use the procedure you described in part **a** to generate a sample of ten 7-digit telephone numbers.

c. Use the procedure you described in part **a** to generate five 7-digit telephone numbers whose first three digits are 373.

3.65 In addition to its decennial enumeration of the population, the U.S. Bureau of the Census regularly samples the population to estimate level of and changes in a number of other attributes, such as income, family size, employment, and marital status. Suppose the Bureau plans to sample 1,000 households in a city that has a total of 534,322 households. Although it would use computer-generated random numbers to obtain such a sample, show how the random number table in Appendix A could be used to generate the sample. Select the first 10 households to be included in the sample.

Summary

We have developed some of the basic tools of probability to enable us to assess the probability of various sample outcomes given a specific population structure. Although many of the examples we presented were of no practical importance, they accomplished their purpose if you now understand the basic concepts and definitions of probability.

The basic understanding of probability presented in this chapter includes the following concepts: **Experiments** are the basis for the generation of data, and the most basic outcomes of experiments, called **simple events**, cannot be predicted with certainty. Therefore, we assign **probabilities** to the simple events, such that they obey two rules: (1) all probabilities must be between 0 and 1, and (2) the simple event probabilities must sum to 1. The collection of all the simple events is called the **sample space** of the experiment.

Unions and **intersections** are useful combinations of events. **Unions** are inclusive: either or both events occur. **Intersections** are limiting: both events occur. The **additive rule** is useful for calculating the probability of unions, and is particularly simple when the events have no intersection, in which case they are said to be **mutually exclusive**.

The **complement** of an event occurs when the event does not occur. Thus the probability of an event plus the probability of its complement sum to 1, a useful relationship for calculating the probabilities of complicated events.

Conditional probability applies when we have some knowledge about what has occurred, reducing the size of the sample space. The **multiplicative rule** is useful for relating the probability of an intersection to conditional probability. When an intersection probability is equal to the product of the two (unconditional) event probabilities, we say the events are **independent**.

Our primary reason for studying the basics of probability is to enable the quantitative evaluation of alternative explanations (or models) of real data. To be able to rule out some explanations as "unlikely" and to accept others as "probable," we must first be able to quantify "unlikely" and "probable," and the theory of probability will provide the foundation for this endeavor.

In the next several chapters, we will present probability models that can be used to solve practical problems. You will see that for most applications, we will need to make inferences about unknown aspects of these probability models, i.e., we will need to apply inferential statistics to the problem.

Supplementary Exercises 3.66–3.88

Learning the Mechanics

3.66 A fair die is tossed and the up face is noted. If the number is even, the die is tossed again; if the number is odd, a fair coin is tossed. Define the events:

A: {A head appears on the coin}

B: {The die is tossed only one time}

a. List the simple events in the sample space.
b. Give the probability for each of the simple events.
c. Find $P(A)$ and $P(B)$.
d. Identify the simple events in A', B', $A \cap B$, and $A \cup B$.
e. Find $P(A')$, $P(B')$, $P(A \cap B)$, $P(A \cup B)$, $P(A \mid B)$, and $P(B \mid A)$.
f. Are A and B mutually exclusive events? Independent events? Why?

3.67 The Venn diagram illustrates a sample space containing six simple events and three events, A, B, and C. The probabilities of the simple events are: $P(1) = .3$, $P(2) = .2$, $P(3) = .1$, $P(4) = .1$, $P(5) = .1$, and $P(6) = .2$.

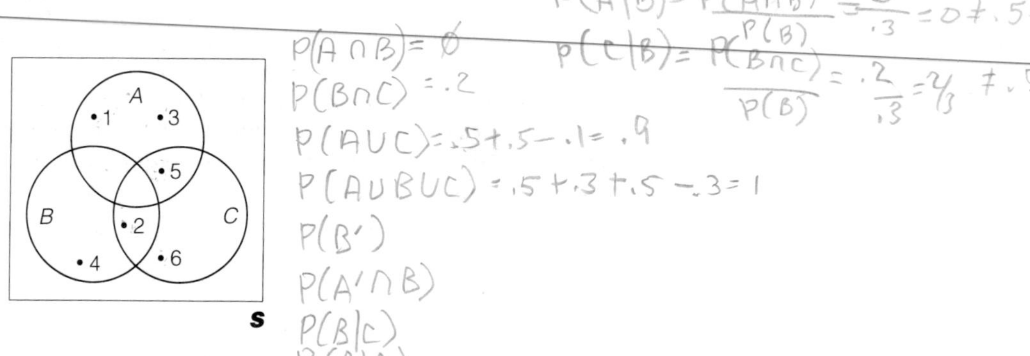

a. Find $P(A \cap B)$, $P(B \cap C)$, $P(A \cup C)$, $P(A \cup B \cup C)$, $P(B')$, $P(A' \cap B)$, $P(B \mid C)$, and $P(B \mid A)$.
b. Are A and B independent? Mutually exclusive? Why?
c. Are B and C independent? Mutually exclusive? Why?

3.68 Which of the following pairs of events are mutually exclusive? Justify your response.

a. The Dow Jones Industrial Average increases on Monday.
   A large New York bank decreases its prime interest rate on Monday.
b. An IBM microcomputer is purchased.
   An Apple microcomputer is purchased.
c. Subject A in a psychology experiment responds to a stimulus within 5 seconds.
   Subject A in a psychology experiment records the fastest response to a stimulus (2.3 seconds).

3.69 Two events, A and B, are independent, with $P(A) = .3$ and $P(B) = .1$.

a. Are A and B mutually exclusive? Why?

b. Find $P(A \mid B)$ and $P(B \mid A)$.
c. Find $P(A \cup B)$.

3.70 A balanced die is thrown once. If a 4 appears, a ball is drawn from urn 1; otherwise a ball is drawn from urn 2. Urn 1 contains four red, three white, and three black balls. Urn 2 contains six red and four white balls.
a. Find the probability that a red ball is drawn.
b. Find the probability that urn 1 was used given that a red ball was drawn.

Applying the Concepts

3.71 In college basketball games a player may be afforded the opportunity to shoot two consecutive foul shots (free throws).
a. Suppose a player who scores on 80% of his foul shots has been awarded two free throws. If the two throws are considered independent, what is the probability that the player scores on both shots? Exactly one? Neither shot?
b. Suppose a player who scores on 80% of his first attempted foul shots has been awarded two free throws, and the outcome on the second shot is dependent on the outcome of the first shot. In fact, if this player makes the first shot he makes 90% of the second shots, and if he misses the first shot, he makes 70% of the second shots. In this case, what is the probability that the player scores on both shots? Exactly one? Neither shot?
c. In parts **a** and **b**, we considered two ways of *modeling* the probability a basketball player scores on two consecutive foul shots. Which model do you think is a more realistic attempt to explain the outcome of shooting foul shots, i.e., do you think two consecutive foul shots are independent or dependent? Explain.

3.72 Psychologists tend to believe that there is a relationship between aggressiveness and order of birth. To test this belief, a psychologist chose 500 elementary school students at random and administered each a test designed to measure the student's aggressiveness. Each student was classified according to one of four categories. The percentages of students falling in the four categories are shown here.

	Firstborn	Not Firstborn
Aggressive	15%	15%
Not Aggressive	25%	45%

a. If one student is chosen at random from the 500, what is the probability that the student is firstborn?
b. What is the probability that the student is aggressive?
c. What is the probability that the student is aggressive, given the student was firstborn?
d. If

A: {Student chosen is aggressive}
B: {Student chosen is firstborn}

are A and B independent events? Explain.

3.73 Two companies, A and B, package and market a chemical substance and claim .15 of the total weight of the substance is sodium. However, a careful survey of 4,000 packages (half from each company) indicates the proportion varies around .15, with the results shown on page 156.

		Proportion of Sodium			
		<.100	.100–.149	.150–.199	>.200
Chemical Brand	A	25%	10%	10%	5%
	B	5%	5%	10%	30%

Suppose a package is chosen at random from the 4,000 packages. If

A: {Package chosen is brand A}
B: {Package chosen is brand B}
C: {Package chosen contains less than .100 sodium}
D: {Package chosen contains between .100 and .149 sodium}
E: {Package chosen contains between .150 and .199 sodium}
F: {Package chosen contains over .200 sodium}

then describe the characteristics of a package portrayed by the following events:
a. $A \cup B$ b. $B \cup F$ c. $A \cap D$ d. $E \cap B$ e. $(A \cap C) \cup (A \cap D)$

3.74 Use your intuitive understanding of independence to form an opinion about whether each of the following scenarios represents independent events.
a. The results of consecutive tosses of a coin
b. The opinions of randomly selected individuals in a pre-election poll
c. A major league baseball player's results in two consecutive at-bats
d. The amount of gain or loss associated with investments in different stocks that are bought on the same day and sold on the same day 1 month later
e. The amount of gain or loss associated with investments in different stocks that are bought and sold in different time periods, 5 years apart
f. The responses of two different subjects to the same stimulus in a psychology experiment

3.75 A local country club has a membership of 600 and operates facilities that include an 18-hole championship golf course and 12 tennis courts. Before deciding whether to accept new members, the club president would like to know how many members regularly use each facility. A survey of the membership indicates that 70% regularly use the golf course, 50% regularly use the tennis courts, and 5% use neither of these facilities regularly.
a. Construct a Venn diagram to describe the results of the survey.
b. If one member is chosen at random, what is the probability that the member uses the golf course or the tennis courts or both?
c. If one member is chosen at random, what is the probability that the member uses both the golf and the tennis facilities?
d. A member is chosen at random from among those known to use the tennis courts regularly. What is the probability that the member also uses the golf facilities regularly?

3.76 Entomologists are often interested in studying the effect of chemical attractants (pheromones) on insects. One common technique is to release several insects equidistant from the pheromone being studied and from a control substance. If the pheromone has an effect, more insects will travel toward it rather than toward the control. Otherwise, the insects are equally likely to travel in either direction. Suppose the pheromone under study has no effect so that it is equally likely that an insect will move toward either the pheromone or the control.

a. If five insects are released, what is the probability that all five travel toward the pheromone?
b. Exactly four?

3.77 All-terrain vehicles (ATVs) came under fire in the 1980s due to the high number of injuries and deaths caused by the machines. In 1988, manufacturers agreed to provide extensive safety warnings to owners, to develop a media safety-awareness program, and to implement a nationwide training program. In an article in the *Journal of Risk and Uncertainty*, Rubinfeld and Rodgers (1992) investigate the relationship of injury rate to a variety of factors. One of the more interesting factors studied, age of the driver, was found to have a strong relationship to injury rate. The article reports that in 1985, 14% of the ATV drivers were under age 12; another 13% were 12–15, and 48% were under age 25.

a. Find the probability a randomly selected ATV driver is 15 years old or younger.
b. Find the probability a randomly selected ATV driver is 25 years old or older.
c. Given that an ATV driver is under age 25, what is the probability the driver is under age 12?
d. Are the events Under age 25 and Under age 12 mutually exclusive? Why or why not?
e. Are the events Under age 25 and Under age 12 independent? Why or why not?

3.78 The performance of quality inspectors affects both the quality of outgoing products and the cost of the products. A product that passes inspection is assumed to meet quality standards; a product that fails inspection may be reworked, scrapped, or reinspected. Quality engineers at Westinghouse Electric Corporation evaluated performances of inspectors in judging the quality of solder joints by comparing each inspector's classifications of a set of 153 joints with the consensus evaluation of a panel of experts. The results for a particular inspector are shown in the table.

		Inspector's Judgment	
		Joint acceptable	Joint rejectable
Committee's Judgment	Joint acceptable	101	10
	Joint rejectable	23	19

Source: Meagher, J. J. and Scazzero, J. A. "Measuring inspector variability," *39th Annual Quality Congress Transactions*, May 1985, pp. 75–81. © 1985 American Society for Quality Control. Reprinted with permission.

One of the 153 solder joints is to be selected at random.

a. What is the probability that the inspector judges the joint to be acceptable? That the committee judges the joint to be acceptable?
b. What is the probability that both the inspector and the committee judge the joint to be acceptable? That neither judge the joint to be acceptable?
c. What is the probability that the inspector and the committee disagree? Agree?

3.79 A manufacturer of 35-mm cameras knows that a shipment of 30 cameras sent to a large discount store contains six defective cameras. The manufacturer also knows that the store will choose two of the cameras at random, test them, and accept the shipment if neither is defective.

a. What is the probability that the first camera chosen by the store will be defective?
b. Given that the first camera chosen passed inspection, what is the probability that the second camera chosen will fail inspection?
c. What is the probability that the shipment will be accepted?

3.80 The probability that a microcomputer salesperson sells a computer to a prospective customer on the first visit to the customer is .4. If the salesperson fails to make the sale on the first visit, the probability that the sale will be made on the second visit is .65. The salesperson never visits a prospective customer more than twice. What is the probability that the salesperson will make a sale to a particular customer?

3.81 Seventy-five percent of all women who submit to pregnancy tests are really pregnant. A certain pregnancy test gives a false positive result with probability .02 and a valid positive result with probability .99. If a particular woman's test is positive, what is the probability that she really is pregnant? [*Hint:* If A is the event that a woman is pregnant and B is the event that the pregnancy test is positive, then B is the union of the two mutually exclusive events $A \cap B$ and $A' \cap B$. Also, the probability of a "false positive result" may be written as $P(B \mid A') = .02$.]

3.82 Blackjack, a favorite game of gamblers, is played by a dealer and at least one opponent and uses a 52-card bridge deck. At the outset of the game, two cards are dealt to the player and two cards to the dealer. Drawing an ace and a face card is called *blackjack*. If the dealer draws it, he or she automatically wins. If the dealer does not draw a blackjack and the player does, the player wins.
a. What is the probability that the dealer will draw a blackjack?
b. What is the probability that the player wins with a blackjack?

3.83 The figure shown here is a schematic representation of a system comprised of three components. The system operates properly only if all three components operate properly. The three components are said to operate *in series*. The components could be mechanical or electrical; they could be work stations in an assembly process; or they could represent the functions of three different departments in an organization. The probability of failure for each component is listed in the table. Assume the components operate independently of each other.

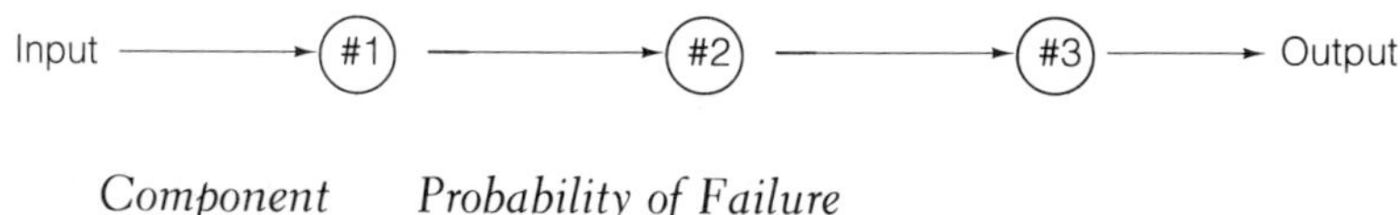

Component	*Probability of Failure*
1	.12
2	.09
3	.11

a. Find the probability that the system operates properly.
b. What is the probability that at least one of the components will fail and therefore that the system will fail?

3.84 The accompanying figure is a representation of a system comprised of two subsystems that are said to operate *in parallel*. Each subsystem has two components that operate in series (refer to Exercise 3.83). The system will operate properly as long as at least one of the subsystems functions properly. The probability of failure for each component in the system is .1. Assume the components operate independently of each other.

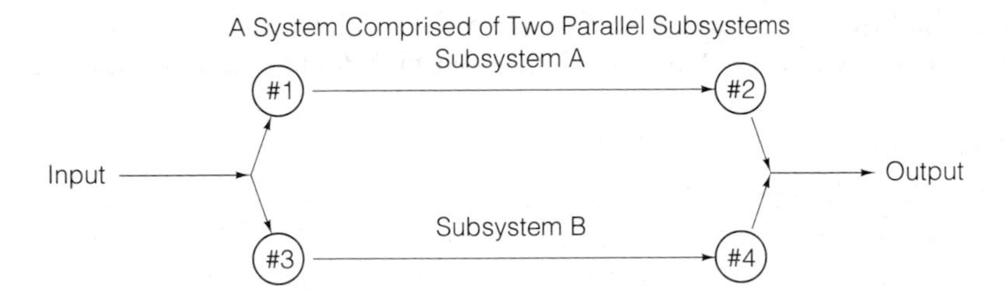

a. Find the probability that the system operates properly.
b. Find the probability that exactly one subsystem fails.
c. Find the probability that the system fails to operate properly.
d. How many parallel subsystems like the two shown here would be required to guarantee that the system would operate properly at least 99% of the time?

3.85 Refer to Exercise 3.37, in which an article by R. Shah and K. Pease in *The Howard Journal of Criminal Justice* was utilized to understand the relationship between the race of the attacker, the race of the victim, and the degree of injury sustained in reported crimes. The tabled information is reprinted below:

		Attacker/Victim				
		White/ White	White/ Nonwhite	Nonwhite/ Nonwhite	Nonwhite/ White	Totals
Degree of Injury	Fatal	183	18	18	18	237
	Serious	580	49	111	141	881
	Slight	4,336	440	594	1,656	7,026
	None	1,422	136	214	1,801	3,573
	Totals	6,521	643	937	3,616	11,717

a. What is the probability that a reported crime involved a nonwhite attacker?
b. What is the probability that a reported crime involved no injury?
c. What is the probability that a reported crime involved both a nonwhite attacker and no injury?
d. Are the race of the attacker and the degree of injury independent?

3.86 A small brewery has two bottling machines. Machine A produces 75% of the bottles and machine B produces 25%. One out of every 20 bottles filled by A is rejected for some reason, while one out of every 30 bottles from B is rejected. What proportion of bottles is rejected? What is the probability that a bottle comes from machine A, given that it is accepted?

3.87 A county welfare agency employs 25 welfare workers who interview prospective food stamp recipients. Periodically, the supervisor selects, at random, the forms completed by two workers to audit for illegal deductions. Unknown to the supervisor, six of the workers have regularly been giving illegal deductions to applicants.
a. What is the probability that the first worker chosen has been giving illegal deductions?
b. Given that the first worker chosen has been giving illegal deductions, what is the probability that the second worker chosen has also been giving illegal deductions?
c. What is the probability that neither of the two workers chosen has been giving illegal deductions?

3.88 A clinical psychologist is asked to view tapes in which each of six experimental subjects is discussing his or her recent dreams. Three of the six subjects have previously been classified as "high-anxiety" individuals and the other three as "low-anxiety." The psychologist is told only that there are three of each type and is asked to select the three high-anxiety subjects.

a. List all possible outcomes (simple events) for this experiment.

b. Assuming that the psychologist guesses at the classifications of the subjects, assign probabilities to the simple events.

c. Find the probability that the psychologist guesses all classifications correctly.

d. Find the probability that the psychologist guesses at least two of the three high-anxiety subjects correctly.

On Your Own

Obtain a standard deck of 52 playing cards (the kind commonly used for bridge, poker, or solitaire). An experiment will consist of drawing 1 card at random from the deck of cards and recording which card was observed. This will be simulated by shuffling the deck thoroughly and observing the top card. Consider the following two events:

A: {Card observed is a heart}

B: {Card observed is an ace, king, queen, or jack}

a. Find $P(A)$, $P(B)$, $P(A \cap B)$, and $P(A \cup B)$.

b. Conduct the experiment 10 times and record the observed card each time. Be sure to return the observed card each time and thoroughly shuffle the deck before making the draw. After 10 cards have been observed, calculate the proportion of observations that satisfy event A, event B, event $A \cap B$, and event $A \cup B$. Compare the observed proportions with the true probabilities calculated in part **a**.

c. Conduct the experiment 40 more times to obtain a total of 50 observed cards. Now calculate the proportion of observations that satisfy event A, event B, event $A \cap B$, and event $A \cup B$. Compare these proportions with those found in part **b** and the true probabilities found in part **a**.

d. Conduct the experiment 50 more times to obtain a total of 100 observations. Compare the observed proportions for the 100 trials with those found previously. What comments do you have concerning the different proportions found in parts **b**, **c**, and **d** as compared to the true probabilities found in part **a**? How do you think the observed proportions and true probabilities would compare if the experiment were conducted 1,000 times? 1 million times?

Using the Computer

Suppose a sociologist is studying income patterns across the United States. She will use the demographic data set described in Appendix B to determine factors that affect the median income for zip code areas.

a. If one of the 1,000 zip codes were to be randomly selected, what is the probability that the selected zone is one for which the median income exceeds $35,000?

b. Suppose the zip code were to be selected from the Northeast census region (from those in the sample). What is the probability that the selected zone is one for which the median income exceeds $35,000?

c. Are the events described in parts **a** and **b** independent? Why or why not? What are the practical implications of the events' independence, or lack thereof?

References

Feller, W. *An Introduction to Probability Theory and Its Applications*, 3d ed., Vol. I. New York: Wiley, 1968, Chapters 1, 3, 4, and 5.

Hains, L. "One-shot wonders." *Golf Digest*, Mar. 1990.

Kuehn, A. A. "An analysis of the dynamics of consumer behavior and its implications for marketing management." Unpublished doctoral dissertation, Graduate School of Industrial Administration, Carnegie Institute of Technology, 1958.

Lindley, D. V. *Making Decisions*, 2d ed. London: Wiley, 1985.

Parzen, E. *Modern Probability Theory and Its Applications*. New York: Wiley, 1960, Chapters 1 and 2.

Preston, M. G., and Baratta, P. "An experimental study of the auction-value of an uncertain outcome." *American Journal of Psychology*, 1948, Vol. 61, pp. 183–193.

Rosenblatt, J. R., and Filliben, J. J. "Randomization and the draft lottery." *Science*, 171, pp. 306–308.

Rubinfeld, D. L., and Rodgers, G. B. "Evaluating the injury risk associated with all-terrain vehicles: An application of Bayes' rule." *Journal of Risk and Uncertainty*, May 1992, Vol. 5, No. 2.

Scheaffer, R. L., and Mendenhall, W. *Introduction to Probability: Theory and Applications*. North Scituate, Mass.: Duxbury, 1975, Chapters 1 and 2.

Williams, B. *A Sampler on Sampling*. New York: Wiley, 1978, pp. 5–8.

Winkler, R. L. *An Introduction to Bayesian Inference and Decision*. New York: Holt, Rinehart and Winston, 1972, Chapter 2.

CHAPTER FOUR

Random Variables and Probability Distributions

Contents

Case Studies

Where We've Been

By illustration we indicated in Chapter 3 how probability would be used to make an inference about a population from data contained in an observed sample. We also noted that probability would be used to measure the reliability of the inference.

Where We're Going

Most experimental events in Chapter 3 were events described in words and denoted by capital letters. In real life, most sample observations are numerical—in other words, numerical data. In this chapter, we learn that data are observed values of random variables. We study three important random variables and learn how to find the probabilities of specific numerical outcomes. One of these, the **normal** random variable, has applications to many of the statistical methods encountered in this text.

4

You may have noticed that many of the examples of experiments in Chapter 3 generated quantitative (numerical) observations. The unemployment rate, the percentage of voters favoring a particular candidate, the cost of textbooks for a school term, and the amount of pesticide in the discharge waters of a chemical plant are all examples of numerical measurements of some phenomenon. Thus, most experiments have simple events that correspond to values of some numerical variable.

Definition 4.1

A **random variable** is a rule that assigns one (and only one) numerical value to each simple event of an experiment.*

The term **random variable** is more meaningful than just the term *variable* because the adjective *random* indicates that the experiment may result in one of the several possible values of the variable, according to the *random* outcome of the experiment. For example, if the experiment is to count the number of customers who use the drive-up window of a bank each day, the random variable (the number of customers) will vary from day to day, partly because of the random phenomena that influence whether customers use the drive-up window. Thus, the possible values of this random variable range from 0 to the maximum number of customers the window could possibly serve in a day.

We define two different types of random variables, **discrete** and **continuous**, in Section 4.1. Then we spend the remainder of this chapter discussing three specific types of discrete random variables and the aspects that make them important to the statistician.

4.1 Two Types of Random Variables

Recall that the simple event probabilities corresponding to an experiment must sum to 1. Dividing one unit of probability among the simple events in a sample space and consequently assigning probabilities to the values of a random variable is not always as easy as the examples in Chapter 3 might lead you to believe. If the number of simple events is finite, that is, if they can be completely listed, the job is relatively easy. However, some experiments result in an infinite number of sample points, in which case assignment of probabilities is more difficult. In fact, we have to use different probability models depending on the number of values that a random variable can assume.

*By *experiment*, we mean an experiment that yields random outcomes (as defined in Chapter 3).

EXAMPLE 4.1

A panel of 10 wine experts is asked to taste a new white wine and assign a rating of 0, 1, 2, or 3. A score is then obtained by adding together the ratings of the 10 experts. How many values can this random variable assume?

Solution

A simple event is a sequence of 10 numbers associated with the rating of each expert. The random variable assigns a score to each one of these simple events by adding the 10 numbers together. Thus, the smallest score is 0 (if all 10 ratings are 0) and the largest score is 30 (if all 10 ratings are 3). Since every integer between 0 and 30 is a possible score, the random variable x can assume 31 values.

This is an example of a **discrete random variable**, since there is a finite number of distinct possible values. Whenever all the possible values a random variable can assume can be listed (or *counted*), the random variable is **discrete**.

EXAMPLE 4.2

Suppose the Environmental Protection Agency (EPA) takes readings once a month on the amount of pesticide in the discharge water of a chemical company. If the amount of pesticide exceeds the maximum level set by the EPA, the company is forced to take corrective action and may be subject to penalty. Consider the following random variable:

Number, x, of months before the company's discharge exceeds the EPA's maximum level

What values can x assume?

Solution

The company's discharge of pesticide may exceed the maximum allowable level on the first month of testing, the second month of testing, etc. It is possible that the company's discharge will *never* exceed the maximum level. Thus, the set of possible values for the number of months until the level is first exceeded is the set of all positive integers:

1, 2, 3, 4, . . .

If we can list the values of a random variable x, even though the list is never-ending, we call the list **countable** and the corresponding random variable *discrete*. Thus, the number of months until the company's discharge first exceeds the limit is a *discrete random variable*.

EXAMPLE 4.3

Refer to Example 4.2. A second random variable of interest is the amount x of pesticide (in milligrams per liter) found in the monthly sample of discharge waters from the chemical company. What values can this random variable assume?

Solution

Unlike the *number* of months before the company's discharge exceeds the EPA's maximum level, the set of all possible values for the *amount* of discharge *cannot* be listed—i.e., is not countable. The possible values for the amount x of pesticide would correspond to the points on the interval between 0 and the largest possible value the amount of the discharge could attain, the maximum number of milligrams that could occupy 1 liter of volume. (Practically, the interval would be much smaller, say, between 0 and 500 milligrams per liter.) When the values of a random variable are not countable but instead correspond to the points on some interval, we call it a **continuous random variable**. Thus, the *amount* of pesticide in the chemical plant's discharge waters is a *continuous random variable.*

Definition 4.2

Random variables that can assume a *countable* number of values are called **discrete**.

Definition 4.3

Random variables that can assume values corresponding to any of the points contained in one or more intervals are called **continuous**.

The following are examples of discrete random variables:

1. The number of sales made by a salesperson in a given week: $x = 0, 1, 2, \ldots$
2. The number of students in a sample of 500 who favor an increase in student activities and, correspondingly, an increase in student activity fees: $x = 0, 1, 2, \ldots, 500$
3. The number of students applying to medical schools this year: $x = 0, 1, 2, \ldots$
4. The number of errors on a page of an accountant's ledger: $x = 0, 1, 2, \ldots$
5. The number of customers waiting to be served in a restaurant at a particular time: $x = 0, 1, 2, \ldots$

Note that each of the examples of discrete random variables begins with the words "The number of . . . ." This wording is very common, since the discrete random variables most frequently observed are counts. The following are examples of continuous random variables:

1. The length of time between arrivals at a hospital clinic: $0 \le x < \infty$ (infinity)
2. For a new apartment complex, the length of time from completion until a specified number of apartments are rented: $0 \le x < \infty$

3. The amount of carbonated beverage loaded into a 12-ounce can in a can-filling operation: $0 \le x \le 12$
4. The depth at which a successful oil drilling venture first strikes oil: $0 \le x \le c$, where c is the maximum depth obtainable
5. The weight of a food item bought in a supermarket: $0 \le x \le 500$
[*Note:* Theoretically, there is no upper limit on x, but it is unlikely that it would exceed 500 pounds.]

In the succeeding sections, we will explain how to construct probability models for both discrete and continuous random variables. Then we will describe the properties of three random variables often encountered in the real world, and see how we can apply their probability distributions to solve practical problems.

Exercises 4.1 – 4.5

Applying the Concepts

4.1 What is a random variable?

4.2 How do discrete and continuous random variables differ?

4.3 Classify the following random variables according to whether they are discrete or continuous:
a. The number of words spelled correctly by a student on a spelling test
b. The amount of water flowing through the Hoover Dam in a day
c. The length of time an employee is late for work
d. The number of bacteria per cubic centimeter of drinking water
e. The amount of carbon monoxide produced per gallon of unleaded gas
f. Your weight

4.4 Identify the following random variables as discrete or continuous:
a. The amount of flu vaccine in a syringe
b. The heart rate (number of beats per minute) of an American male
c. The time it takes a student to complete an examination
d. The barometric pressure at a given location
e. The number of registered voters who vote in a national election
f. Your score on the Scholastic Aptitude Test (SAT)

4.5 Identify the following variables as discrete or continuous:
a. The reaction time difference to the same stimulus before and after training
b. The number of violent crimes committed per month in your community
c. The number of commercial aircraft near-misses per month
d. The number of winners each week in a state lottery
e. The number of free throws made per game by a basketball team
f. The distance traveled by a school bus each day

4.2 Probability Distributions for Discrete Random Variables

A complete description of a discrete random variable requires that we *specify the possible values the random variable can assume* and *the probability associated with each value*. To illustrate, consider Example 4.4.

EXAMPLE 4.4

Recall the experiment of tossing two coins (Chapter 3), and let x be the number of heads observed. Find the probability associated with each value of the random variable x, assuming the two coins are fair.

Solution

Recall from Chapter 3 that the sample space and simple events for this experiment are as shown in Figure 4.1, and the probability associated with each of the four simple events is ¼. The random variable x can assume values 0, 1, 2. Then, identifying the probabilities of the simple events associated with each of these values of x, we have

HH	HT
$x = 2$	$x = 1$
TH	TT
$x = 1$	$x = 0$

S

FIGURE 4.1 ▲
Venn diagram for the two-coin-toss experiment

$$P(x = 0) = P(TT) = \frac{1}{4}$$

$$P(x = 1) = P(TH) + P(HT) = \frac{1}{4} + \frac{1}{4} = \frac{1}{2}$$

$$P(x = 2) = P(HH) = \frac{1}{4}$$

Thus, we now know the values the random variable can assume (0, 1, 2) and how the probability is *distributed over* these values (¼, ½, ¼). This completely describes the random variable and is referred to as the **probability distribution**, denoted by the symbol $\boldsymbol{p(x)}$. The probability distribution for the coin-toss example is shown in tabular form in Table 4.1 and in graphic form in Figure 4.2. Since the probability distribution for a discrete random variable is concentrated at specific points (values of x), the graph in Figure 4.2(a) represents the probabilities as the heights of vertical lines over the corresponding values of x. Although the representation of the probability distribution

TABLE 4.1 Probability Distribution for Coin-Toss Experiment: Tabular Form

x	$p(x)$
0	¼
1	½
2	¼

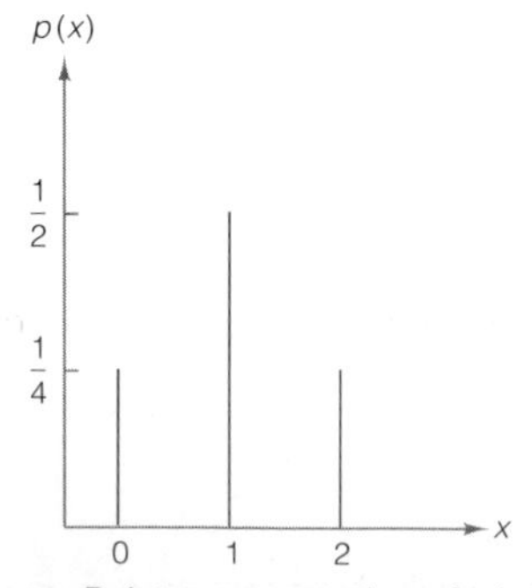

a. Point representation of $p(x)$

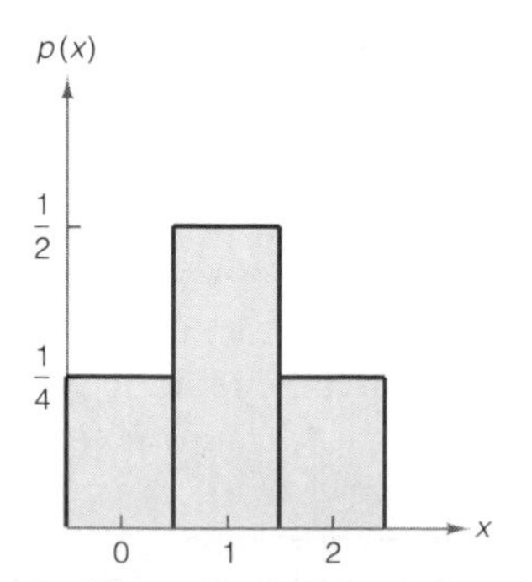

b. Histogram representation of $p(x)$

FIGURE 4.2 ►
Probability distribution for coin-toss experiment: Graphic form

as a histogram, as in Figure 4.2(b), is less precise (since the probability is spread over a unit interval), the histogram representation will prove useful when we approximate probabilities of certain discrete random variables in Section 4.3.

We could also present the probability distribution for x as a formula, but this would unnecessarily complicate a very simple example. We give the formulas for the probability distributions of some common discrete random variables later in this chapter.

Definition 4.4

The **probability distribution** of a discrete random variable is a graph, table, or formula that specifies the probability associated with each possible value the random variable can assume.

Two requirements must be satisfied by all probability distributions for discrete random variables.

Requirements for the Probability Distribution of a Discrete Random Variable x

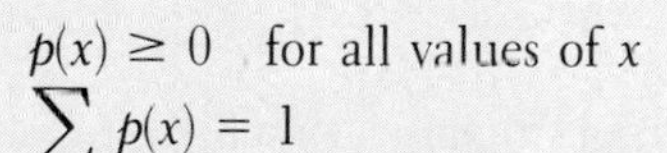

$p(x) \geq 0$ for all values of x

$\sum p(x) = 1$

where the summation of $p(x)$ is over all possible values of x.*

Example 4.4 illustrates how the probability distribution for a discrete random variable can be derived, but for many practical situations the task is much more difficult. Fortunately, many experiments and associated discrete random variables observed in nature possess identical characteristics. Thus, you might observe a random variable in a psychology experiment that would possess the same probability distribution as a random variable observed in an engineering experiment or a social sample survey. We classify random variables according to type of experiment, derive the probability distribution for each of the different types, and then use the appropriate probability distribution when a particular type of random variable is observed in a practical situation. The probability distributions for most commonly occurring discrete random variables have already been derived. (We describe two of these in Sections 4.3 and 4.4.) This fact simplifies the problem of finding the probability distributions for random variables.

---

*Unless otherwise indicated, summations will always be over all possible values of x.

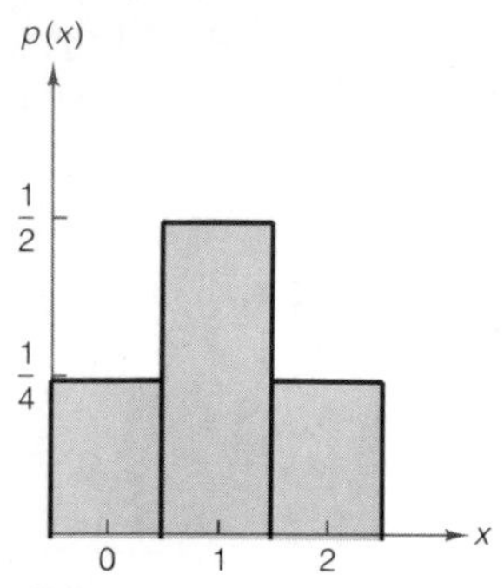

FIGURE 4.3 ▲
Probability distribution for a two-coin toss

Since probability distributions are analogous to the relative frequency distributions of Chapter 2, it should be no surprise that the mean and standard deviation are useful descriptive measures. For example, if a discrete random variable x were observed a very large number of times and the data generated were arranged in a relative frequency distribution, the relative frequency distribution would be indistinguishable from the probability distribution for the random variable. Thus, the probability distribution for a random variable is a theoretical model for the relative frequency distribution of a population. To the extent that the two distributions are equivalent (and we will assume they are), the probability distribution for x possesses a mean μ and a variance σ^2 that are identical to the corresponding descriptive measures for the population.

To illustrate, examine the probability distribution for x (the number of heads observed in the toss of two fair coins) in Figure 4.3. Try to locate the mean of the distribution intuitively. We may reason that the mean μ of this distribution is equal to 1 as follows: In a large number of tosses, ¼ should result in $x = 0$, ½ in $x = 1$, and ¼ in $x = 2$ heads. Therefore, the average number of heads is

$$\mu = 0(1/4) + 1(1/2) + 2(1/4) = 0 + 1/2 + 1/2 = 1$$

Note that to get the population mean of the random variable x, we multiply each possible value of x by its probability $p(x)$, and then we sum this product over all possible values of x. The **mean of x** is also referred to as the **expected value of x**, denoted $E(x)$.

Definition 4.5

The **mean,** or **expected value,** of a discrete random variable x is

$$\mu = E(x) = \sum xp(x)$$

The term *expected* is a mathematical term and should not be interpreted as it is typically used. Specifically, a random variable might never be equal to its "expected value." Rather, the expected value is the mean of the probability distribution, or a measure of its central tendency. You can think of μ as the mean value of x in a *very large* (actually, infinite) number of repetitions of the experiment.

EXAMPLE 4.5

Suppose you work for an insurance company, and you sell a \$10,000 whole-life insurance policy at an annual premium of \$290. Actuarial tables show that the probability of death during the next year for a person of your customer's age, sex, health, etc., is .001. What is the expected gain (amount of money made by the company) for a policy of this type?

Solution

The experiment is to observe whether the customer survives the upcoming year. The probabilities associated with the two simple events, Live and Die, are .999 and .001, respectively. The random variable you are interested in is the gain x, which can assume the values shown in the table.

Gain x	*Simple Event*	*Probability*
\$290	Customer lives	.999
\$290 − \$10,000	Customer dies	.001

If the customer lives, the company gains the \$290 premium as profit. If the customer dies, the gain is negative because the company must pay \$10,000, for a net "gain" of \$(290 − 10,000). The expected gain is therefore

$$\begin{aligned}\mu = E(x) &= \sum xp(x)\\ &= (290)(.999) + (290 - 10{,}000)(.001) = \$280\end{aligned}$$

In other words, if the company were to sell a very large number of 1-year \$10,000 policies to customers possessing the characteristics described above, it would (on the average) net \$280 per sale in the next year.

Example 4.5 illustrates that the expected value of a random variable x need not equal a possible value of x. That is, the expected value is \$280, but x will equal either \$290 or −\$9,710 each time the experiment is performed (a policy is sold and a year elapses). The expected value is a measure of central tendency, and in this case represents the average over a very large number of 1-year policies—but is not a possible value of x.

We learned in Chapter 2 that the mean and other measures of central tendency tell only part of the story about a set of data. The same is true about probability distributions. We need to measure variability as well. Since a probability distribution can be viewed as a representation of a population, we will use the population variance to measure its variability.

The **population variance σ^2** is defined as the average of the squared distance of x from the population mean μ. Since x is a random variable, the squared distance, $(x - \mu)^2$, is also a random variable. Using the same logic used to find the mean value of x, we find the mean value of $(x - \mu)^2$ by multiplying all possible values of $(x - \mu)^2$ by $p(x)$ and then summing over all possible x values.* This quantity,

$$E[(x - \mu)^2] = \sum_{\text{all } x} (x - \mu)^2 p(x)$$

is also called the **expected value of the squared distance from the mean**; that is, $\sigma^2 = E[(x - \mu)^2]$. The standard deviation of x is defined as the square root of the variance σ^2.

*It can be shown that $E[(x - \mu)^2] = E(x^2) - \mu^2$, where $E(x^2) = \Sigma\, x^2 p(x)$. Note the similarity between this expression and the shortcut formula $\Sigma\,(x - \bar{x})^2 = \Sigma\, x^2 - (\Sigma\, x)^2/n$ given in Chapter 2.

Definition 4.6

The **variance** of a random variable x is

$$\sigma^2 = E[(x - \mu)^2] = \sum (x - \mu)^2 p(x)$$

Definition 4.7

The **standard deviation** of a discrete random variable is equal to the square root of the variance, i.e., to $\sigma = \sqrt{\sigma^2}$.

EXAMPLE 4.6

Medical research has shown that a certain type of chemotherapy is successful 70% of the time when used to treat skin cancer. Suppose five skin cancer patients are treated with this type of chemotherapy and let x equal the number of successful cures out of the five. The probability distribution for the number x of successful cures out of five is given in the table:

x	0	1	2	3	4	5
$p(x)$	.002	.029	.132	.309	.360	.168

a. Find $\mu = E(x)$.

b. Find $\sigma = \sqrt{E[(x - \mu)^2]}$.

c. Graph $p(x)$. Locate μ and the interval $\mu \pm 2\sigma$ on the graph. Explain how μ and σ can be used to describe $p(x)$.

Solution

a. Applying the formula,

$$\begin{aligned}\mu = E(x) &= \sum xp(x) \\ &= 0(.002) + 1(.029) + 2(.132) + 3(.309) + 4(.360) + 5(.168) \\ &= 3.50\end{aligned}$$

b. Now we calculate the variance of x:

$$\begin{aligned}\sigma^2 = E[(x - \mu)^2] &= \sum (x - \mu)^2 p(x) \\ &= (0 - 3.5)^2(.002) + (1 - 3.5)^2(.029) + (2 - 3.5)^2(.132) \\ &\quad + (3 - 3.5)^2(.309) + (4 - 3.5)^2(.360) + (5 - 3.5)^2(.168) \\ &= 1.05\end{aligned}$$

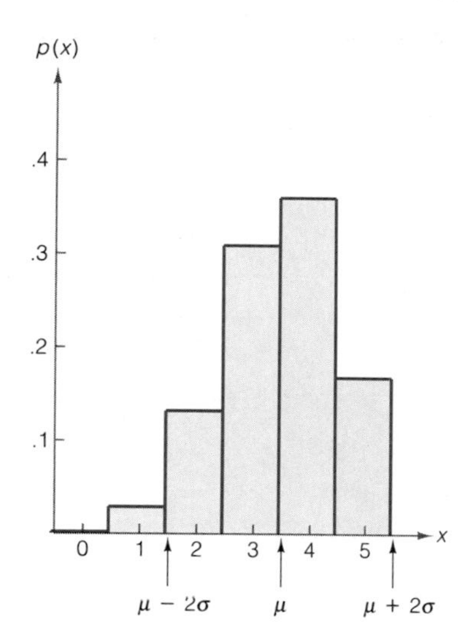

FIGURE 4.4 ▶ Graph of $p(x)$ for Example 4.6

Thus, the standard deviation is

$$\sigma = \sqrt{\sigma^2} = \sqrt{1.05} = 1.02$$

c. The graph of $p(x)$ is shown in Figure 4.4. Note that the mean μ and the interval $\mu \pm 2\sigma$ are shown on the graph. We can use μ and σ to describe the probability distribution $p(x)$ in the same way that we used $\bar{x}$ and s to describe a relative frequency distribution in Chapter 2. Note particularly that $\mu = 3.5$ locates the center of the probability distribution. If five skin cancer patients receive the chemotherapy treatment, we expect the number x that are cured to be near 3.5. Similarly, $\sigma = 1.02$ measures the spread of the probability distribution $p(x)$. Since this distribution is a theoretical relative frequency distribution that is moderately mound-shaped (see Figure 4.4), we expect (see Tables 2.5 and 2.6) at least 75% and, more likely, near 95% of observed x values to fall in the interval $\mu \pm 2\sigma$—that is, between 1.46 and 5.54. Compare this result with the actual probability that x falls in the interval $\mu \pm 2\sigma$. From Figure 4.4 you can see that this probability includes the sum of $p(x)$ for all values of x except $p(0) = .002$ and $p(1) = .029$. Therefore, 96.9% of the probability distribution lies within 2 standard deviations of the mean. This percentage is consistent with Tables 2.5 and 2.6.

CASE STUDY 4.1 / A Restaurant Chain Fights Sales Tax Claim

The June 1, 1977, business section of the Orlando, Florida, *Sentinel Star* featured the following headline: "Red Lobster to Fight Tax Claim." According to the *Sentinel Star*, the Red Lobster Inns of America, a national seafood chain, had decided to take the state of Florida to court. The dispute concerned the 4% sales tax levied on most purchases in the state and focused mainly on the state's "bracket collection system." According to the bracket system, a merchant must collect 1¢ for sales between 10¢ and 25¢, 2¢ for sales from 26¢ to 50¢, 3¢ for sales between 51¢ and 75¢, and 4¢ for sales between 76¢ and 99¢. Red Lobster contended that if this system is followed, merchants will always collect more than 4%. That is, if a sale were made for $10.41, 4% will be collected on the $10, but more than 4% will be collected on the 41¢. This, they contend, would amount to more than 4% on the total sale and therefore is not consistent with the 4% tax required by law.

Concrete evidence supplied by the state of Florida tax records does indeed support the contention that the amount of tax collected using the bracket system exceeds the 4% specified by law. It appears that the sales tax receipts exceeded expected revenue (based on 4%) by $9.5 million in a single year.

What percent sales tax should the state expect to receive using the bracket system for computing the tax? (As noted, the tax on the whole dollar portion of the sale will be 4%.) Using the formula for calculating expected values, it can be shown that the expected percent tax paid on the cents portion of a sale is 4.6%.

Exercises 4.6–4.25

Learning the Mechanics

4.6 Consider the following probability distribution:

x	−4	0	1	3
$p(x)$	.1	.2	.4	.3

a. List the values that x may assume.
b. What value of x is most probable?
c. What is the probability that x is greater than 0?
d. What is the probability that $x = -2$?

4.7 Explain why each of the following is or is not a valid probability distribution for a discrete random variable x:

a.

x	0	1	2	3
$p(x)$	.2	.3	.3	.2

b.

x	−2	−1	0
$p(x)$	.25	.50	.20

c.

x	4	9	20
$p(x)$	−.3	1.0	.3

d.

x	2	3	5	6
$p(x)$	.15	.20	.40	.35

4.8 The random variable x has the following discrete probability distribution:

x	10	11	12	13	14
$p(x)$	.2	.3	.2	.1	.2

Since the values that x can assume are mutually exclusive events, the event $\{x \leq 12\}$ is the union of three mutually exclusive events:

$$\{x = 10\} \cup \{x = 11\} \cup \{x = 12\}$$

a. Find $P(x \leq 12)$. b. Find $P(x > 12)$.
c. Find $P(x \leq 14)$. d. Find $P(x = 14)$.
e. Find $P(x \leq 11 \text{ or } x > 12)$.

4.9 Toss three fair coins and let x equal the number of heads observed.
a. Identify the simple events associated with this experiment and assign a value of x to each simple event.
b. Calculate $p(x)$ for each value of x.
c. Construct a probability histogram for $p(x)$.
d. What is $P(x = 2 \text{ or } x = 3)$?
e. Find $\mu = E(x)$ and σ^2.

4.10 Consider the probability distribution for the random variable x shown here.

x	1	2	3	4	5
$p(x)$	.10	.20	.35	.25	.10

a. Find μ, σ^2, and σ. b. Graph $p(x)$.
c. Locate μ and the interval $\mu \pm \sigma$ on your graph. What is the probability that x will fall within the interval $\mu \pm \sigma$?
d. Locate the interval $\mu \pm 3\sigma$ on your graph. What is the probability that x falls within this interval?

4.11 Consider the probability distribution shown here.

x	1	2	4	10
$p(x)$	.2	.4	.2	.2

a. Find $\mu = E(x)$. b. Find $\sigma^2 = E[(x - \mu)^2]$. c. Find σ.
d. Interpret the value you obtained for μ.
e. In this case, can the random variable x ever assume the value μ? Explain.
f. In general, can a random variable ever assume a value equal to its expected value? Explain.

4.12 Consider the probability distribution shown at the top of page 176.

x	-4	-3	-2	-1	0	1	2	3	4
$p(x)$	.02	.07	.10	.15	.30	.18	.10	.06	.02

a Find $P(x \geq 0)$.
b. Calculate μ, σ^2, and σ.
c. Graph $p(x)$. Locate μ, $\mu - 2\sigma$, and $\mu + 2\sigma$ on the graph.
d. What is the probability that x is in the interval $\mu \pm 2\sigma$?

4.13 Consider the probability distributions shown here.

x	0	1	2
$p(x)$	.3	.4	.3

y	0	1	2
$p(y)$	.1	.8	.1

a. Use your intuition to find the mean for each distribution. How did you arrive at your choice?
b. Which distribution appears to be more variable? Why?
c. Calculate μ and σ^2 for each distribution. Compare these answers to your answers in parts **a** and **b**.

Applying the Concepts

4.14 A patient complaining of severe stomach pains checked into a local hospital. After a series of tests the doctors narrowed their diagnosis to four possible ailments. They believe there is a 40% chance that the patient has hepatitis; a 10% chance that she has cirrhosis; a 45% chance of gallstones; and a 5% chance of cancer of the pancreas. The doctors are certain that the patient has only one of the diseases, but will not know which disease until further tests are performed. The cost associated with treating each disease is given in the table:

Disease	Hepatitis	Cirrhosis	Gallstones	Pancreatic Cancer
Cost	\$700	\$1,110	\$3,320	\$16,450

a. Construct the probability distribution for the cost of treating the patient.
b. Calculate the mean of the probability distribution you constructed in part **a**. What does this number represent?
c. Further testing reveals that the patient has either hepatitis or cirrhosis. Given this information, construct the new probability distribution for the cost of treating the patient.
d. Calculate the mean of the probability distribution you constructed in part **c**. What does this number represent?

4.15 Every human possesses two sex chromosomes. A copy of one or the other (equally likely) is contributed to an offspring. Males have one X chromosome and one Y chromosome. Females have two X chromosomes. If a couple has three children, what is the probability that they have at least one boy? [*Hint:* Define the random variable z as the number of male offspring, and find the probability distribution of z.]

4.16 The number of training units that must be passed before a complex computer software program is mastered varies from one to five, depending on the student. After much experience, the software manufacturer has determined the probability distribution that describes the fraction of users mastering the software after each number of training units:

Number of Units	1	2	3	4	5
Probability of Mastery	.1	.25	.4	.15	.1

a. Calculate the mean number of training units necessary to master the program. Calculate the median. Interpret each.
b. If the firm wants to ensure that at least 75% of the students master the program, what is the minimum number of training units that must be administered? At least 90%?
c. Suppose a new training program is developed that increases the probability that only one unit of training is needed from .1 to .25, increases the probability that only two units are needed to .35, leaves the probability that three units are needed at .4, and completely eliminates the need for four or five units. How do your answers to parts **a** and **b** change for this new program?

4.17 Coach "Bear" Bryant of the University of Alabama, a legendary figure in college football, was known for his winning seasons. He consistently won nine or more games per season. Suppose x equals the number of games won up to the halfway mark (six games) in a 12-game season. If Coach Bryant and his team had a probability $p = .70$ of winning any one game (and the winning or losing of one game was independent of another), then the probability distribution of the number x of winning games in a series of six games (we show how to calculate these probabilities in Section 4.3) is

x	0	1	2	3	4	5	6
$p(x)$	.001	.010	.060	.185	.324	.302	.118

Find the probability that the number of games won by Coach Bryant in the first half of a randomly selected season is

a. 6 b. 5 c. Less than or equal to 4

4.18 Refer to Exercise 4.17.

a. Find the expected number of games that Alabama would win in the first half of the season.
b. Find σ.
c. Find the probability that x is in the interval $\mu \pm 2\sigma$.

4.19 In Exercise 3.60 we noted that the drug cyclosporine appears to retard a body's immune system from rejecting transplanted organs. The drug's developer, Sandoz, reports that kidney transplant patients who receive the drug have an 80% chance of living through the first year (*Newsweek*, August 29, 1983). Suppose four kidney transplant patients are given cyclosporine, and let x equal the number living through the first year.

a. Calculate $P(x)$ for $x = 0, 1, 2, 3, 4$. b. Graph $p(x)$. c. Find $P(x < 2)$.

4.20 President Clinton encountered difficulties early in his new administration in appointing an attorney general. One troublesome issue involved the nominees' hiring of illegal aliens and/or the failure to pay Social Security taxes for domestic help, both of which are against the law. Suppose that a study reveals that 10% of households with incomes exceeding $50,000 annually have hired an illegal alien and/or failed to pay Social Security taxes. Let x be the number of households with incomes in that range who are contacted before one that has not broken either law is found.

a. What is the range of possible values for x?
b. Find $P(x < 3)$.
c. Find $P(x > 2)$.
d. Find the probability distribution for values of x from 1 to 10 and graph it over that domain. Can x exceed 10?

4.21 On one busy holiday weekend, a national airline has many requests for standby flights at half of the usual one-way air fare. However, past experience has shown that these passengers have only about a 1 in 5 chance of getting on the standby flight. When they fail to get on a flight as a standby, their only other choice is to fly first class on the next flight out. Suppose that the usual one-way air fare to a certain city is $70 and the cost of flying first class is $90. Should a passenger who wishes to fly to this city opt to fly as a standby? [*Hint:* Find the expected cost of the trip for a person flying standby.]

4.22 Odds makers try to predict which football teams will win and by how much (the *spread*). If the odds makers do this accurately, adding the spread to the underdog's score should make the final score a tie. Suppose a bookie will give you $6 for every $1 you risk if you pick the winners in three ballgames (adjusted by the spread). Thus, for every $1 bet you will either lose $1 or gain $5. What are the bookie's expected earnings per dollar wagered?

4.23 A rock concert producer has scheduled an outdoor concert for Saturday, May 24. If it does not rain, the producer expects to make $20,000 profit from the concert. If it does rain, the producer will be forced to cancel the concert and will lose $12,000 (rock star's fee, advertising costs, stadium rental, administrative costs, etc.). The producer has learned from the National Weather Service that the probability of rain on May 24 is .4.

a. Find the producer's expected profit from the concert.
b. For a fee of $1,000, an insurance company has offered to insure the producer against all losses resulting from a rained-out concert. If the producer buys the insurance, what is her expected profit from the concert?
c. Assuming the National Weather Service's forecast is accurate, do you believe the insurance company has charged too much or too little for the policy? Explain.

4.24 An automobile insurance company estimates the following loss probabilities for the next year on a $25,000 sports car:

Total loss: .001
50% loss: .01
25% loss: .05
10% loss: .10

Assuming the company will sell only a $500 deductible policy for this model (i.e., the owner covers the first $500 damage), how much annual premium should the company charge in order to average $250 profit per policy sold?

4.25 Many states are approaching fiscal crises in the 1990s, with revenue falling far short of projected expenditures. One governor ordered a survey of the state's voters regarding a proposed increase in the sales tax vs. an increase in the state income tax. The results, with responses ranging from −2 to +2, are summarized below:

	Favor Sales Tax			Favor Income Tax	
	Strongly	Slightly	Undecided	Slightly	Strongly
Response	−2	−1	0	+1	+2
Percent	10%	30%	20%	35%	5%

a. Can the governor report that more of the state's voters prefer the sales tax than prefer the income tax? Explain.

b. Treating the response percentages as probabilities associated with the scores, what is the mean score? Interpret the mean.

4.3 The Binomial Distribution

Many experiments result in dichotomous responses—i.e., responses for which there exist two possible alternatives, such as Yes–No, Pass–Fail, Defective–Nondefective, or Male–Female. A simple example of such an experiment is the coin-toss experiment. A coin is tossed a number of times, say 10. Each toss results in one of two outcomes, Head or Tail, and the probability of observing each of these two outcomes remains the same for each of the 10 tosses. Ultimately, we are interested in the probability distribution of x, the number of heads observed. Many other experiments are equivalent to tossing a coin (either balanced or unbalanced) a fixed number n of times and observing the number x of times that one of the two possible outcomes occurs. Random variables that possess these characteristics are called **binomial random variables**.

Public opinion and consumer preference polls (e.g., the Gallup and Harris polls) frequently yield observations on binomial random variables. For example, suppose a sample of 100 students is selected from a large student body and each person is asked whether he or she favors (a Head) or opposes (a Tail) a certain campus issue. Ultimately, we are interested in x, the number of people in the sample who favor the issue. If each student is randomly selected from the student body and if the (unknown) proportion of students favoring the issue is p, then observing whether a student favors or is opposed to the issue is analogous to tossing an unbalanced coin. The chance that any randomly selected student favors the issue is p; the probability that he or she opposes the issue is $(1 - p)$. Sampling 100 students is analogous to tossing the coin 100 times. Thus, you can see that opinion polls which record the number of people who favor a certain issue are real-life equivalents of coin-toss experiments.

The experiment we have been describing is called a **binomial experiment** and is identified by the following characteristics:

Characteristics of a Binomial Random Variable

1. The experiment consists of n identical trials.
2. There are only two possible outcomes on each trial. We will denote one outcome by S (for Success) and the other by F (for Failure).
3. The probability of S remains the same from trial to trial. This probability is denoted by p, and the probability of F is denoted by q. Note that $q = 1 - p$.
4. The trials are independent.
5. The binomial random variable x is the number of S's in n trials.

EXAMPLE 4.7

For each of the following examples, decide whether x is a binomial random variable.

a. Suppose a university scholarship committee must select two students to receive a scholarship for the next academic year. The committee receives 10 applications for the scholarships–six from male students and four from female students. Suppose the applicants are all equally qualified, so that the selections are randomly made. Let x be the number of female students who receive a scholarship.

b. Before marketing a new product on a large scale, many companies will conduct a consumer-preference survey to determine whether the product is likely to be successful. Suppose a company develops a new diet soda and then conducts a taste-preference survey with 100 randomly chosen consumers stating their preference among the new soda and the two leading sellers. Let x be the number of the 100 who choose the new brand over the two others.

c. Some surveys are conducted by using a method of sampling other than simple random sampling (defined in Chapter 3). For example, suppose a television cable company plans to conduct a survey to determine the fraction of households in the city that would use the cable television service. The sampling method is to choose a city block at random and then survey every household on that block. This sampling technique is called **cluster sampling**. Suppose 10 blocks are so sampled, producing a total of 124 household responses. Let x be the number of the 124 households that would use the television cable service.

Solution

a. In checking the binomial characteristics, a problem arises with independence (characteristic 4 in the preceding box). Given that the first student selected is female, the probability that the second chosen is female is $3/9$. On the other hand, given that the first selection is a male student, the probability that the second is female is 4/9. Thus, the conditional probability of a Success (choosing a female student to receive a scholarship) on the second trial (selection) depends on the outcome of the first trial, and the trials are therefore dependent. Since the trials are *not independent*, this is not a binomial random variable.

b. Surveys that produce dichotomous responses and use random sampling techniques are classic examples of binomial experiments. In our example, each randomly selected consumer either states a preference for the new diet soda or does not. The sample of 100 consumers is a very small proportion of the total number of potential consumers, so the response of one would be, for all practical purposes, independent of another. Thus, x is a binomial random variable.

c. This example is a survey with dichotomous responses (Yes or No to the cable service), but the sampling method is not simple random sampling. Again, the binomial characteristic of independent trials would probably not be satisfied. The responses of households within a particular block almost surely would be dependent, since households within a block tend to be similar with respect to income, level of education, and general interests. Thus, the binomial model would not be satisfactory for x if the cluster sampling technique were employed.

EXAMPLE 4.8

The Heart Association claims that only 10% of adults over 30 years of age in the United States can pass the minimum requirements established by the president's Physical Fitness Commission. Suppose four adults are randomly selected, and each is given the fitness test. Let x be the number of the four who pass the minimum requirements. Find the probability distribution for x, assuming that the Heart Association's claim is true.

Solution

Recall that a probability distribution describes a discrete random variable by assigning probabilities to each of its values. In this example, the possible values of x, the number of four adults who pass the minimum requirements, are 0, 1, 2, 3, 4. Furthermore, there are four identical trials (the sampling and observation of the four adults), each with two possible outcomes (pass or fail minimum requirements). We are assuming that the probability that each adult passes is .1, and the result for each adult will be (at least approximately) independent of that for the others. Thus, the number, x, of adults who pass the minimum requirements is a binomial random variable.

Let us first consider the event $x = 0$, that is, the event that none of the four tested adults passes the test. You can see that the event $x = 0$ is equivalent to the simple event

$$FFFF$$

where F in the first position implies that adult 1 fails, F in the second position implies that adult 2 fails, etc. Since the trials are independent in this binomial experiment (knowing whether adult 1 passes should not affect the probability that adult 2 passes), we can find the probability of an intersection by multiplying the probabilities of the events. Thus,

$$P(x = 0) = P(FFFF) = P(F)P(F)P(F)P(F)$$
$$= (.9)(.9)(.9)(.9) = (.9)^4 = .6561$$

The event $x = 1$ implies that one of the four adults passes the physical fitness test and three fail it. The following list of simple events contains all simple events that imply $x = 1$:

SFFF *FSFF* *FFSF* *FFFS*

where *S* in the first position corresponds to adult 1 passing the test, *S* in the second position corresponds to adult 2 passing the test, etc. Note that each of these simple events will have the same probability, $(.1)(.9)^3$, where .1 corresponds to the one adult who passes the test and $(.9)^3$ corresponds to the three who fail it. Remembering from Chapter 3 that we obtain the probability of an event by summing the probabilities of the simple events of which it is composed, we get

$$P(x = 1) = 4\,[(.1)(.9)^3] = .2916$$

The event $x = 2$ implies that two adults pass the test and two fail it; this event consists of the following six simple events:

SSFF *SFSF* *SFFS* *FSSF* *FSFS* *FFSS*

Each of these simple events has probability $(.1)^2(.9)^2$, so that

$$P(x = 2) = 6[(.1)^2(.9)^2] = .0486$$

Similarly,

$$P(x = 3) = 4[(.1)^3(.9)] = .0036$$
$$P(x = 4) = (.1)^4 = .0001$$

The complete probability distribution is shown in Figure 4.5 and listed in Table 4.2.

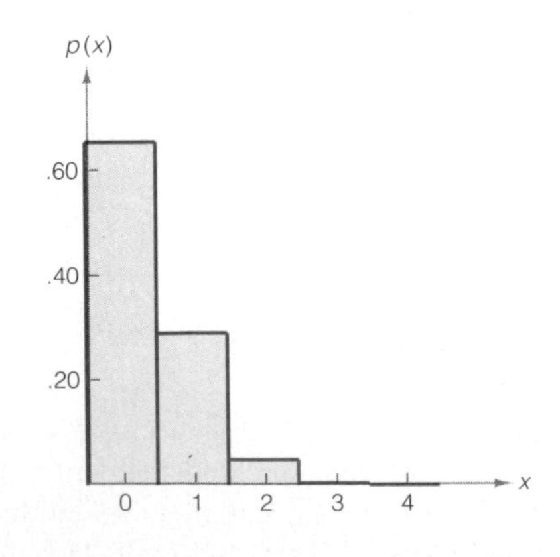

FIGURE 4.5 ▶ Probability distribution for physical fitness example: Graphic form

TABLE 4.2 Probability Distribution for Physical Fitness Example: Tabular Form

x	$p(x)$
0	.6561
1	.2916
2	.0486
3	.0036
4	.0001

Before we give a formula for $p(x)$, we will refresh your memory on factorial notation. Particularly, the symbol $n!$ is to be read "n factorial" and is calculated by

$$n! = n(n-1)(n-2) \cdot \cdots \cdot 3 \cdot 2 \cdot 1$$

We define $0! = 1$. Thus, for example, $4! = 4 \cdot 3 \cdot 2 \cdot 1 = 24$.

Using factorial notation, we write the formula for a binomial probability distribution with $n = 4$ and $p = .1$:

$$p(x) = \frac{4!}{x!(4-x)!}(.1)^x(.9)^{4-x}$$

Then, for $x = 2$, we have

$$p(2) = \frac{4!}{2!(4-2)!}(.1)^2(.9)^{4-2} = \frac{4 \cdot 3 \cdot 2 \cdot 1}{(2 \cdot 1)(2 \cdot 1)}(.1)^2(.9)^2$$
$$= 6(.1)^2(.9)^2 = .0486$$

which agrees with our simple event calculation. Note that the first part of the formula, $4!/[x!(4-x)!]$, counts the number of simple events that result in x adults passing the physical fitness test. The second part of the formula, $(.1)^x(.9)^{4-x}$, is the probability assigned to each simple event that has x adults passing and $(4-x)$ failing. When we multiply the *number* of simple events by the *probability* assigned to each simple event, we get the probability that x adults pass the test. We will see that this formula can be generalized to give the probability distribution of any binomial random variable.

Note that $\binom{n}{x}$, shorthand for $n!/[x!(n-x)!]$, is the number of simple events that have x successes and $(n-x)$ failures, and p^xq^{n-x} is the probability assigned to each simple event that has x successes and $(n-x)$ failures. The product of these two quantities, $\binom{n}{x}p^xq^{n-x}$, is the probability that x successes and $(n-x)$ failures are observed.

The Binomial Probability Distribution

$$p(x) = \binom{n}{x}p^xq^{n-x} \quad (x = 0, 1, 2, \ldots, n)$$

where p = Probability of a success on a single trial
$q = 1 - p$
n = Number of trials
x = Number of successes in n trials

$$\binom{n}{x} = \frac{n!}{x!(n-x)!}$$

The binomial probability distribution is so named because the probabilities, $p(x)$, $x = 0, 1, \ldots, n$, are terms of the binomial expansion, $(q + p)^n$.

CASE STUDY 4.2 / The Space Shuttle *Challenger*: Catastrophe in Space

On January 28, 1986, at 11:39.13 A.M., while traveling at Mach 1.92 at an altitude of 46,000 feet, the space shuttle *Challenger* was totally enveloped in an explosive burn that destroyed the shuttle and resulted in the deaths of all seven astronauts aboard. What happened? What was the cause of this catastrophe? This was the 25th shuttle mission. The preceding 24 missions had all been successful.

The report of the Presidential Commission assigned to investigate the accident concluded that the explosion was caused by the failure of the O-ring seal in the joint between the two lower segments of the right solid rocket booster. The seal is supposed to prevent superhot gases from leaking through the joint during the propellant burn of the booster rocket. The failure of the seal permitted a jet of white-hot gases to escape and to ignite the liquid fuel of the external fuel tank. The fuel tank fireburst destroyed the *Challenger*.

What were the chances of this event occurring? In a 1985 report, the National Aeronautics and Space Administration (NASA) claimed that the probability of such a failure was about $1/60{,}000$, or about once in every 60,000 flights. But a 1983 risk-assessment study conducted for the Air Force assessed the probability of a shuttle catastrophe due to booster rocket "burn-through" to be $1/35$, or about once in every 35 missions.

If it is assumed that (1) p, the probability of shuttle catastrophe due to booster failure, remains the same from mission to mission, and (2) the performance of the booster rockets on one mission is independent of the performance of the boosters on other missions, then the number, x, of shuttle catastrophes due to booster failure in n missions can be treated as a binomial random variable. Accordingly, the probability that no disasters would have occurred during 25 missions is

$$P(x = 0) = \binom{25}{0} p^0(1 - p)^{25-0}$$

$$= \frac{25!}{0!25!} p^0(1 - p)^{25} = (1 - p)^{25}$$

If we use NASA's probability of shuttle catastrophe ($p = 1/60{,}000 = .0000167$), the probability of no catastrophes in 25 missions is approximately .9996. If we use the probability of catastrophe from the study prepared for the Air Force ($p = 1/35 = .02857$), the probability of no catastrophes in 25 missions is .4845. Or, if we consider the complementary event that at least one catastrophe occurs in 25 missions, the chances are .0004, or about 4 in 10,000, given NASA's assumptions. On the other hand, the probability of at least one catastrophe under the Air Force's assumptions is .5155, or slightly more than 50–50. Given the events of January 28, 1986, which risk assessment—NASA's or the Air Force's—appears to be more appropriate? The probability of one or more disasters in 25 missions is so remote using NASA's assessment that it casts serious doubt on the risk assessment practices used by NASA prior to the *Challenger*'s fatal mission (McKean, 1986; Biddle, 1986; Robinson, 1986; *Minneapolis Star and Tribune*, February 11, 1986).

EXAMPLE 4.9 Refer to Example 4.8. Calculate μ and σ, the mean and standard deviation, respectively, of the number of the four adults who pass the test.

Solution From Section 4.2 we know that the mean of a discrete probability distribution is

$$\mu = \sum xp(x)$$

Referring to Table 4.2, the probability distribution for the number x who pass the fitness test, we find

$$\mu = 0(.6561) + 1(.2916) + 2(.0486) + 3(.0036) + 4(.0001)$$
$$= .4 = 4(.1) = np$$

The relationship $\mu = np$ holds in general for a binomial random variable.

The variance is

$$\sigma^2 = \sum (x - \mu)^2 p(x) = \sum (x - .4)^2 p(x)$$
$$= (0 - .4)^2(.6561) + (1 - .4)^2(.2916) + (2 - .4)^2(.0486)$$
$$+ (3 - .4)^2(.0036) + (4 - .4)^2(.0001)$$
$$= .104976 + .104976 + .124416 + .024336 + .001296$$
$$= .36 = 4(.1)(.9) = npq$$

The relationship $\sigma^2 = npq$ holds in general for a binomial random variable.

Finally, the standard deviation of the number who pass the fitness test is

$$\sigma = \sqrt{\sigma^2} = \sqrt{.36} = .6$$

We emphasize that you need not use the expectation summation rules to calculate μ and σ^2 for a binomial random variable. You can find them easily using the formulas $\mu = np$ and $\sigma^2 = npq$.

Mean, Variance, and Standard Deviation for a Binomial Random Variable

Mean: $\mu = np$

Variance: $\sigma^2 = npq$

Standard deviation: $\sigma = \sqrt{npq}$

As we demonstrated in Chapter 2, the mean and standard deviation provide measures of the central tendency and variability, respectively, of a distribution. Thus, we can use μ and σ to obtain a rough visualization of the probability distribution for x when the calculation of the probabilities is too tedious. To illustrate the use of the binomial probability distribution, consider Example 4.10.

EXAMPLE 4.10

A poll of 20 voters is taken in a large city. The purpose is to determine x, the number in favor of a certain candidate for mayor. Suppose that (unknown to us) 60% of all the city's voters favor this candidate.

a. Find the mean and standard deviation of x.

b. Find the probability that x is less than or equal to 10 ($x \leq 10$).

c. Find the probability that x exceeds 12 ($x > 12$).

d. Find the probability that x equals 11 ($x = 11$).

e. Graph the probability distribution of x and locate the interval $\mu - 2\sigma$ to $\mu + 2\sigma$ on the graph.

Solution

a. Given that the sample of 20 was randomly selected from a large number of voters, x, the number of the 20 who favor the candidate, is (approximately) a binomial random variable. The value of p is the fraction of the total voters who favor the candidate; i.e., $p = .6$. Therefore, we calculate the mean and variance:

$$\mu = np = 20(.6) = 12 \qquad \sigma^2 = npq = 20(.6)(.4) = 4.8$$

The standard deviation is then

$$\sigma = \sqrt{4.8} = 2.2$$

b. Calculating binomial probabilities when n is large is a formidable task. For example, to find the probability that $x \leq 10$, we would calculate*

$$P(x \leq 10) = p(0) + p(1) + p(2) + \cdots + p(10)$$

$$= \sum_{x=0}^{10} p(x) = \sum_{x=0}^{10} \binom{20}{x} (.6)^x (.4)^{20-x}$$

We can avoid these tedious calculations by making use of cumulative binomial probability tables (Table II in Appendix A). Part of Table II is shown in Table 4.3. The columns correspond to values of p, and the rows correspond to values of the random variable x.

The entries in Table II are the cumulative sums

$$P(x \leq k) = p(0) + p(1) + p(2) + \cdots + p(k)$$

for values of $k = 0, 1, 2, \ldots, (n - 1)$. Observe that the bottom row of the table, the one corresponding to $k = n$, is omitted. This is because the sum of $p(x)$ from $x = 0$ to $x = n$ is always equal to 1; i.e., $P(x \leq n) = 1$ for any binomial random variable.

To find $P(x \leq 10)$ for $n = 20$ and $p = .6$, we first find the column corresponding to $p = .6$ and then the row corresponding to $k = 10$. The recorded value, shaded in Table 4.3 and in Figure 4.6, is

$$P(x \leq 10) = .245$$

*The value of x below the Σ symbol, $x = 0$, is the **first member**, or **lower limit**, of the summation. The value of x above the Σ symbol, $x = 10$, is the **last member**, or **upper limit**, of the summation. Thus, $\sum_{x=0}^{10} p(x) = p(0) + p(1) + \cdots + p(9) + p(10)$. We include these limits when the summation extends over only some of the possible values of x.

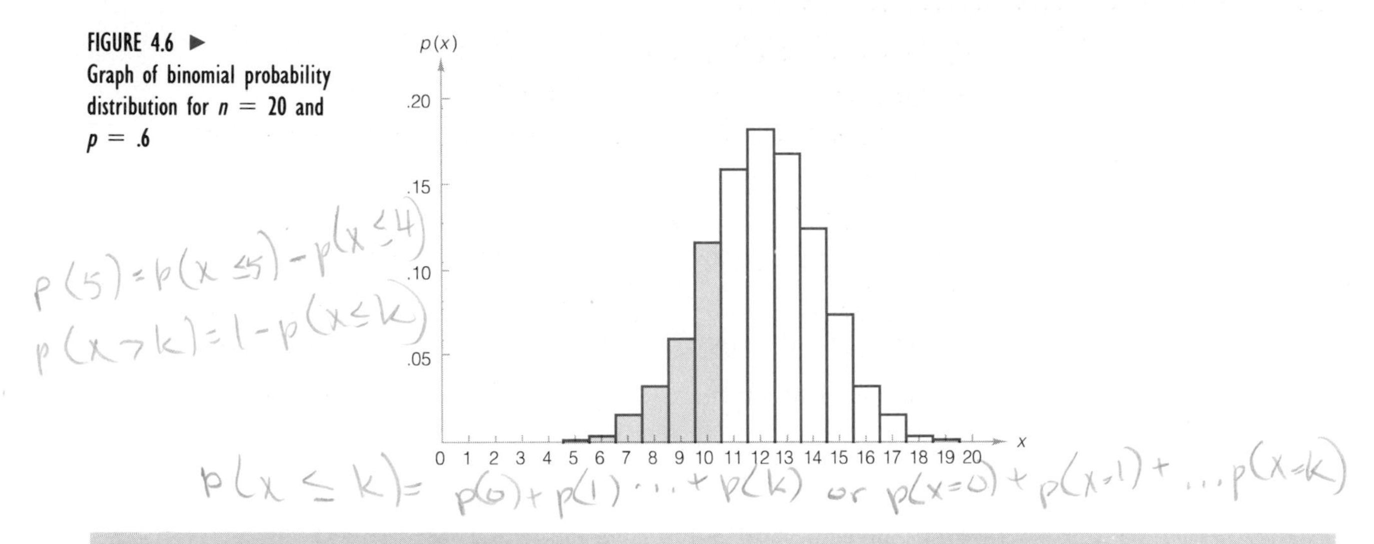

FIGURE 4.6 ▶ Graph of binomial probability distribution for $n = 20$ and $p = .6$

TABLE 4.3 Reproduction of Part of Table II in Appendix A

k \ p	.01	.05	.10	.20	.30	.40	.50	.60	.70	.80	.90	.95	.99
0	.818	.358	.122	.012	.001	.000	.000	.000	.000	.000	.000	.000	.000
1	.983	.736	.392	.069	.008	.001	.000	.000	.000	.000	.000	.000	.000
2	.999	.925	.677	.206	.035	.004	.000	.000	.000	.000	.000	.000	.000
3	1.000	.984	.867	.411	.107	.016	.001	.000	.000	.000	.000	.000	.000
4	1.000	.997	.957	.630	.238	.051	.006	.000	.000	.000	.000	.000	.000
5	1.000	1.000	.989	.804	.416	.126	.021	.002	.000	.000	.000	.000	.000
6	1.000	1.000	.998	.913	.608	.250	.058	.006	.000	.000	.000	.000	.000
7	1.000	1.000	1.000	.968	.772	.416	.132	.021	.001	.000	.000	.000	.000
8	1.000	1.000	1.000	.990	.887	.596	.252	.057	.005	.000	.000	.000	.000
9	1.000	1.000	1.000	.997	.952	.755	.412	.128	.017	.001	.000	.000	.000
10	1.000	1.000	1.000	.999	.983	.872	.588	.245	.048	.003	.000	.000	.000
11	1.000	1.000	1.000	1.000	.995	.943	.748	.404	.113	.010	.000	.000	.000
12	1.000	1.000	1.000	1.000	.999	.979	.868	.584	.228	.032	.000	.000	.000
13	1.000	1.000	1.000	1.000	1.000	.994	.942	.750	.392	.087	.002	.000	.000
14	1.000	1.000	1.000	1.000	1.000	.998	.979	.874	.584	.196	.011	.000	.000
15	1.000	1.000	1.000	1.000	1.000	1.000	.994	.949	.762	.370	.043	.003	.000
16	1.000	1.000	1.000	1.000	1.000	1.000	.999	.984	.893	.589	.133	.016	.000
17	1.000	1.000	1.000	1.000	1.000	1.000	1.000	.996	.965	.794	.323	.075	.001
18	1.000	1.000	1.000	1.000	1.000	1.000	1.000	.999	.992	.931	.608	.264	.017
19	1.000	1.000	1.000	1.000	1.000	1.000	1.000	1.000	.999	.988	.878	.642	.182

c. To find the probability

$$P(x > 12) = p(13) + p(14) + \cdots + p(19) + p(20) = \sum_{x=13}^{20} p(x)$$

we use the fact that for all probability distributions, $\Sigma\ p(x) = 1$. Therefore, using the complementary event, we have

$$P(x > 12) = 1 - [p(0) + p(1) + \cdots + p(12)]$$
$$= 1 - P(x \leq 12) = 1 - \sum_{x=0}^{12} p(x)$$

Consulting Table II, we find the entry in row $k = 12$, column $p = .6$ to be .584. Thus,

$$P(x > 12) = 1 - .584 = .416$$

d. To find the probability that exactly 11 voters favor the candidate, recall that the entries in Table II are cumulative probabilities and use the relationship

$$P(x = 11) = [p(0) + p(1) + \cdots + p(10) + p(11)] - [p(0) + p(1) + \cdots + p(9) + p(10)]$$
$$= P(x \leq 11) - P(x \leq 10)$$

Then

$$P(x = 11) = .404 - .245 = .159$$

e. The probability distribution for x is shown in Figure 4.7. Note that

$$\mu - 2\sigma = 12 - 2(2.2) = 7.6 \qquad \mu + 2\sigma = 12 + 2(2.2) = 16.4$$

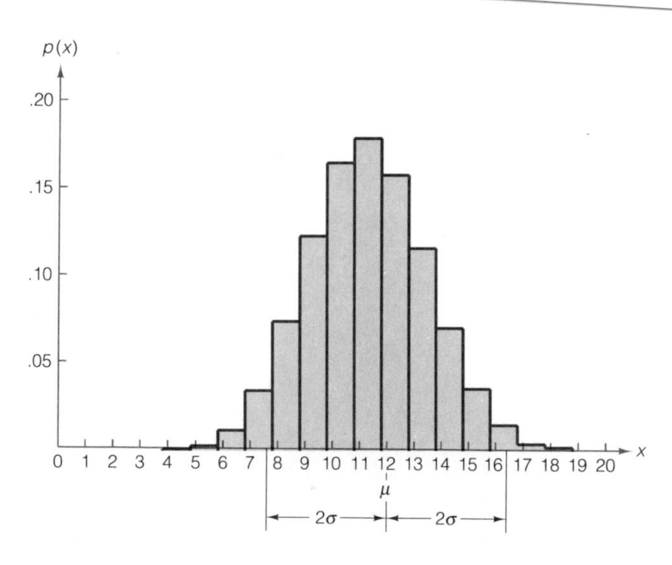

FIGURE 4.7 ▶ The binomial probability distribution for x in Example 4.10: $n = 20$ and $p = .6$

The interval $\mu - 2\sigma$ to $\mu + 2\sigma$ is shown in Figure 4.7. The probability that x falls in the interval, $\mu \pm 2\sigma$, i.e., $P(x = 8, 9, 10, \ldots, 16) = P(x \leq 16) - P(x \leq 7) = .984 - .021 = .963$. Note that this probability is very close to the .95 given by the Empirical Rule.

Cumulative Binomial Tables

Let x be a binomial random variable with n trials and probability of success p. Entries in Table II of Appendix A represent the **cumulative probability**

$$P(x \le k) = P(x = 1) + P(x = 2) + \cdots + P(x = k)$$

where k is a specified value of the binomial random variable.

CASE STUDY 4.3 / A Survey of Children's Political Knowledge

Children's images of political leaders in Britain, France, and the United States were studied by Fred I. Greenstein (1975). Data were collected by means of interviews with small samples of children in the three countries "in order to examine various standard assumptions about political culture and socialization among the three nations, as well as black–white differences in the United States." During one phase of the study, 25 black children from the United States were asked to name the president of their country. This phase represents a binomial experiment with $n = 25$ trials and p equal to the proportion of all black children who could correctly name the president at that time (1969–1970). One objective of the experiment was to obtain an estimate of the value of p.

Of the sample of 25 black children, 24 correctly identified Richard Nixon as president. The implication of this result is that the proportion of all black children who could have made a correct identification must have been quite high. In fact, if the true proportion were equal to .8, the probability that at least 24 out of 25 would correctly identify the president is only .027. (You can verify this result by using Table II in Appendix A.) Thus, unless the observed outcome represents a rare event, the proportion of all black children who could have correctly identified the president was probably in excess of .8 at that time. In Chapter 6 we will develop a systematic approach for making inferences about proportions.

Exercises 4.26–4.44

Learning the Mechanics

4.26 Compute the following:

a. $\dfrac{6!}{2!(6-2)!}$ b. $\dbinom{5}{2}$ c. $\dbinom{7}{0}$ d. $\dbinom{6}{6}$ e. $\dbinom{4}{3}$

4.27 Consider the following probability distribution:

$$p(x) = \binom{5}{x}(.7)^x(.3)^{5-x} \quad (x = 0, 1, 2, \ldots, 5)$$

a. Is x a discrete or a continuous random variable? Explain.
b. What is the name of this probability distribution?

c. Graph the probability distribution.
d. Find the mean and standard deviation of x.
e. Show the mean and the 2-standard-deviation interval on each side of the mean on the graph you drew in part **c**.

4.28 If x is a binomial random variable, compute $p(x)$ for each of the following cases:
a. $n = 5, \quad x = 1, \quad p = .2$ b. $n = 4, \quad x = 2, \quad q = .4$
c. $n = 3, \quad x = 0, \quad p = .7$ d. $n = 5, \quad x = 3, \quad p = .1$
e. $n = 4, \quad x = 2, \quad q = .6$ f. $n = 3, \quad x = 1, \quad p = .9$

4.29 Suppose x is a binomial random variable with $n = 3$ and $p = .3$.
a. Calculate the value of $p(x)$, $x = 0, 1, 2, 3$, using the formula for a binomial probability distribution.
b. Using your answers to part **a**, give the probability distribution for x in tabular form.

4.30 If x is a binomial random variable, calculate μ, σ^2, and σ for each of the following:
a. $n = 25, \quad p = .5$ b. $n = 80, \quad p = .2$
c. $n = 100, \quad p = .6$ d. $n = 70, \quad p = .9$
e. $n = 60, \quad p = .8$ f. $n = 1{,}000, \quad p = .04$

4.31 If x is a binomial random variable, use Table II in Appendix A to find the following probabilities:
a. $P(x = 2)$ for $n = 10, \quad p = .4$
b. $P(x \leq 5)$ for $n = 15, \quad p = .6$
c. $P(x > 1)$ for $n = 5, \quad p = .1$
d. $P(x < 10)$ for $n = 25, \quad p = .7$
e. $P(x \geq 10)$ for $n = 15, \quad p = .9$
f. $P(x = 2)$ for $n = 20, \quad p = .2$

4.32 Suppose x is a binomial random variable with $n = 5$ and $p = .5$. Compute $p(x)$ for $x = 0, 1, 2, 3, 4$, and 5 using the following three methods:
a. List the simple events (using S for a Success and F for a Failure on each trial) corresponding to each value of x, assign probabilities to each simple event, and obtain $p(x)$ by adding simple event probabilities.
b. Use the formula for the binomial probability distribution to obtain $p(x)$.
c. Use Table II to obtain $p(x)$.

4.33 The binomial probability distribution is a family of probability distributions with each single distribution depending on the values of n and p. Assume that x is a binomial random variable with $n = 4$.
a. Determine a value of p such that the probability distribution of x is symmetric.
b. Determine a value of p such that the probability distribution of x is skewed to the right.
c. Determine a value of p such that the probability distribution of x is skewed to the left.
d. Graph each of the binomial distributions you obtained in parts **a**, **b**, and **c**. Locate the mean for each distribution on its graph.
e. In general, for what values of p will a binomial distribution be symmetric? Skewed to the right? Skewed to the left?

Applying the Concepts

4.34 Refer to Exercise 3.53, in which the probability that a male professional golfer makes a hole-in-one is reported to be 1/2,780. Suppose 36 professional male golfers play the sixth hole during a round of golf. Let x be the number of the 36 who make a hole-in-one.

a. Can x be reasonably treated as a binomial random variable? Check the characteristics.
b. Calculate the probability that none of the 36 golfers makes a hole-in-one on the sixth hole.
c. Calculate the probability that exactly four of the 36 golfers make a hole-in-one on the sixth hole—as actually happened during the 1989 U.S. Open.

4.35 A large southern university has determined from past records that the probability a student who registers for fall classes will have his or her schedule rejected (due to overfilled classrooms, clerical error, etc.) is .2.
a. Suppose 25,000 students register for fall classes, and x is the number of students who have their schedules rejected. Is x a binomial random variable? Explain.
b. Suppose a random sample of 20 students is selected from the total of 25,000, and x is the number of these students who have their schedules rejected. Is x a binomial random variable? Explain.
c. Suppose you sample the results of the first 1,000 students who register next fall and record x, the number of rejected registrations. Is x a binomial random variable? Explain.
d. For the random variables in parts **a**, **b**, and **c** that you identified as being binomial, find μ, σ^2, and σ.

4.36 An automobile manufacturer has determined that 30% of all gas tanks that were installed on its 1988 compact model are defective.
a. If 15 of the cars are recalled by a particular dealer, what is the probability that more than 10 of the 15 will need new gas tanks?
b. If 10,000 of the cars are recalled by the manufacturer, what is the probability that fewer than 3,000 will need new gas tanks? Set up the solution but do not perform the calculations. [*Note:* In Section 4.7, we discuss a procedure that can be used to obtain an approximate answer to this question without having to perform the tedious calculations required by the binomial distribution.] Calculate the mean and standard deviation of the distribution, and give an "intuitive" approximation to this probability.

4.37 A problem of considerable impact on the economy is the burgeoning cost of Medicare and other public-funded medical services. One aspect of this problem, reported in the "Behavior" section of *Time* (April 18, 1983), concerns the high percentage of people seeking medical treatment who, in fact, have no physical basis for their ailment. One conservative estimate is that the percentage of people who seek medical assistance but have no real physical ailment is 10%, and some doctors believe that it may be as high as 40%. Suppose we randomly sample the records of a doctor and find that five of 15 patients seeking medical assistance are physically healthy.
a. What is the probability of observing five or more physically healthy patients in a sample of 15 if the proportion p that the doctor normally sees is 10%?
b. What is the probability of observing five or more physically healthy patients in a sample of 15 if the proportion p that the doctor normally sees is 40%?
c. Why might your answer to part **a** make you believe that p is larger than .1?

4.38 The *Wall Street Journal* (March 8, 1984) noted that a "blood clot dissolver that researchers hope can stop heart attacks passed its first test in humans but researchers said the substance must still undergo extensive trials before any life-saving potential can be determined." One aspect of the research by Frans Van de Werf, M.D., and colleagues (1984), reported in the *New England Journal of Medicine*, involved actual tests on seven humans aged 50 or more who suffered heart attacks and were treated with the new drug, *t*-PA. After the treatment, the blood clots in six of the seven patients were dissolved. The blood clot did not dissolve in the seventh patient.

Assume that the drug is ineffective in dissolving blood clots and that without the drug the probability p that a heart attack patient's blood clot would dissolve of its own accord, in a short time, is very small, say less than .1. Let x equal the number of heart attack patients in the sample of seven whose blood clots dissolved in a short time.

a. If the drug is ineffective and $p = .1$, what is the probability that blood clots would have dissolved of their own accord in at least six of seven heart attack patients?

b. If p really is equal to .1 and if the drug *t*-PA is ineffective in treating heart attacks, would you conclude that $x \geq 6$ is a rare event? Using the logic of Section 2.6, what do you think about the utility of *t*-PA in dissolving blood clots in heart attack patients?

4.39 A particular system in a space vehicle must work properly in order for it to gain reentry into the earth's atmosphere. One component of the system operates successfully only 85% of the time. To increase the reliability of the system, four of the components are installed in such a way that the system will operate successfully if at least one component is working. What is the probability that the system will fail? Assume the components operate independently.

4.40 Suppose you are a purchasing officer for a large company. You have purchased 5 million electrical switches and have been guaranteed by the supplier that the shipment will contain no more than .1% defectives. To check the shipment, you randomly sample 500 switches, test them, and find that four are defective. If the switches are as represented, calculate μ and σ for this sample of 500. Based on this evidence, do you think the supplier has complied with the guarantee? Explain. [*Hint:* Calculate μ and σ for this binomial random variable with $p = .001$ to see if a value of x as large as 4 is probable.]

4.41 An experiment is to be conducted to see whether an acclaimed psychic has extrasensory perception (ESP). Five different cards are shuffled, and one is chosen at random. The psychic will then try to identify which card was drawn without seeing it. The experiment is to be repeated 20 times and x, the number of correct decisions, is recorded. (Assume that the 20 trials are independent.)

a. If the psychic is guessing—i.e., if the psychic does not possess ESP—what is the value of p, the probability of a correct decision on each trial?

b. If the psychic is guessing, what is the expected number of correct decisions in 20 trials?

c. If the psychic is guessing, what is the probability of six or more correct decisions in 20 trials?

d. Suppose that the psychic makes six correct decisions in 20 trials. Is there evidence to indicate that the psychic is *not* guessing and actually has ESP? Explain.

4.42 A literature professor decides to give a 20-question true–false quiz to determine who has read an assigned novel. She wants to choose the passing grade such that the probability of passing a student who guesses on every question is less than .05. What score should she set as the lowest passing grade?

4.43 A new drug has been synthesized that is designed to reduce a person's blood pressure. Twenty randomly selected hypertensive patients receive the new drug. Suppose 18 or more of the patients' blood pressures drop.

a. Suppose the probability that a hypertensive patient's blood pressure drops if he or she is *untreated* is .5. Then what is the probability of observing 18 or more blood pressure drops in a random sample of 20 treated patients if the new drug is in fact ineffective in reducing blood pressure?

b. Considering this probability (part **a**), do you think you have observed a rare event, or do you conclude that the drug is effective in reducing hypertension?

4.44 A local newspaper claims that 65% of the items advertised in its classified advertisement section are sold within 1 week of the first appearance of the ad. To check the validity of the claim, the newspaper randomly selected $n = 800$ advertisements from last year's classified advertisements and contacted the people who placed the ads. They found that $x = 472$ of the 800 items sold within a week.

a. Compute μ and σ for the random variable x, the number of items sold within a week.
b. Based on a sample of 800, is it likely that you would observe $x \leq 472$ if the newspaper's claim were true? Explain.
c. Do the results of the newspaper's survey support the claim? Explain.

4.4 The Poisson Distribution (Optional)

A type of probability distribution that is often useful in describing the number of events that will occur in a specific period of time or in a specific area or volume is the **Poisson distribution** (named after the 18th-century physicist and mathematician, Siméon Poisson). Typical examples of random variables for which the Poisson probability distribution provides a good model are

1. The number of traffic accidents per month at a busy intersection
2. The number of noticeable surface defects (scratches, dents, etc.) found by quality inspectors on a new automobile
3. The parts per million of some toxicant found in the water or air emission from a manufacturing plant
4. The number of diseased trees per acre of a certain woodland
5. The number of death claims received per day by an insurance company
6. The number of unscheduled admissions per day to a hospital

Characteristics of a Poisson Random Variable

1. The experiment consists of counting the number of times a certain event occurs during a given unit of time or in a given area or volume (or weight, distance, or any other unit of measurement).
2. The probability that an event occurs in a given unit of time, area, or volume is the same for all the units.
3. The number of events that occur in one unit of time, area, or volume is independent of the number that occur in other units.
4. The mean (or expected) number of events in each unit is denoted by the Greek letter lambda, λ.

The characteristics of the Poisson random variable are usually difficult to verify for practical examples. The examples given satisfy them well enough that the Poisson distribution provides a good model in many instances. As with all probability models, the real test of the adequacy of the Poisson model is in whether it provides a reasonable approximation to reality—that is, whether empirical data support it.

The Poisson probability distribution also provides a good approximation to a binomial probability distribution with mean

$$\lambda = np$$

when n is large, p is small, and $np \le 7$. To illustrate with a relatively small value of n, if $n = 25$ and $p = .05$, then the exact value of the binomial probability $p(2)$ is .231. The Poisson approximation is .224. The approximations are better for $n \ge 100$.

The probability distribution, mean, and variance for a Poisson random variable are shown in the next box.

Probability Distribution, Mean, and Variance for a Poisson Random Variable

$$p(x) = \frac{\lambda^x e^{-\lambda}}{x!} \quad (x = 0, 1, 2, \ldots)$$

$$\mu = \lambda \qquad \sigma^2 = \lambda$$

where λ = Mean number of events during given unit of time, area, volume, etc.

$e = 2.71828\ldots$

The calculation of Poisson probabilities is made easier by the use of Table III in Appendix A, which gives the cumulative probabilities $P(x \le k)$ for various values of λ. The use of Table III is illustrated in Example 4.11.

EXAMPLE 4.11

Ecologists often use the number of reported sightings of a rare species of animal to estimate the remaining population size. For example, suppose the number, x, of reported sightings per week of blue whales is recorded. Assume that x has (approximately) a Poisson probability distribution. Furthermore, assume that the average number of weekly sightings is 2.6.

a. Find the mean and standard deviation of x, the number of blue whale sightings per week.

b. Use Table III to find the probability that fewer than two sightings are made during a given week.

c. Use Table III to find the probability that more than five sightings are made during a given week.

d. Use Table III to find the probability that exactly five sightings are made during a given week.

Solution

a. The mean and variance of a Poisson random variable are both equal to λ. Thus, for this example,

$$\mu = \lambda = 2.6 \qquad \sigma^2 = \lambda = 2.6$$

Then the standard deviation of x is

$$\sigma = \sqrt{2.6} = 1.61$$

Remember that the mean measures the central tendency of the distribution and does not necessarily equal a possible value of x. In this example, the mean is 2.6 sightings, and although there cannot be 2.6 sightings during a given week, the average number of weekly sightings is 2.6. Similarly, the standard deviation of 1.61 measures the variability of the number of sightings per week. Perhaps a more helpful measure is the interval $\mu \pm 2\sigma$, which in this case stretches from $-.62$ to 5.82. We expect the number of sightings to fall in this interval most of the time—with at least 75% relative frequency and probably with more than 90% relative frequency. The mean and the 2-standard-deviation interval around it are shown in Figure 4.8.

FIGURE 4.8 ▶
Probability distribution for number of blue whale sightings

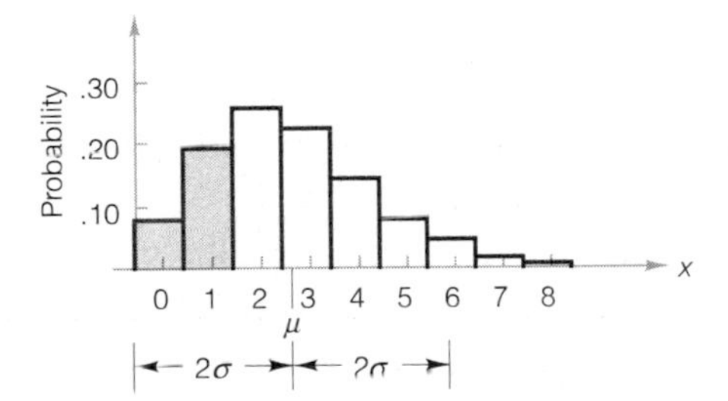

b. A partial reproduction of Table III is shown in Table 4.4. The rows of the table correspond to different values of λ, and the columns correspond to different values of the Poisson random variable x. The entries in the table are cumulative probabilities (much like the binomial probabilities in Table II). To find the probability that fewer than two sightings are made during a given week, we first note that

$$P(x < 2) = P(x \leq 1)$$

This probability is a cumulative probability and therefore is the entry in Table III in the row corresponding to $\lambda = 2.6$ and the column corresponding to $x = 1$. The entry is .267, shown shaded in Table 4.4. This probability corresponds to the shaded area in Figure 4.8 and may be interpreted as meaning that there is a 26.7% chance that fewer than two sightings will be made during a given week.

c. To find the probability that more than five sightings are made during a given week, we consider the complementary event:

$$P(x > 5) = 1 - P(x \leq 5) = 1 - .951 = .049$$

where .951 is the entry in Table III corresponding to $\lambda = 2.6$ and $x = 5$ (see Table 4.4 on page 196). Note from Figure 4.8 that this is the area in the interval $\mu \pm 2\sigma$, or $-.62$ to 5.82. Then the number of sightings should exceed 5—or, equivalently, should be more than 2 standard deviations from the mean—during only about

TABLE 4.4 Reproduction of Part of Table III in Appendix A

λ \ x	0	1	2	3	4	5	6	7	8	9
2.2	.111	.355	.623	.819	.928	.975	.993	.998	1.000	1.000
2.4	.091	.308	.570	.779	.904	.964	.988	.997	.999	1.000
2.6	.074	.267	.518	.736	.877	.951	.983	.995	.999	1.000
2.8	.061	.231	.469	.692	.848	.935	.976	.992	.998	.999
3.0	.050	.199	.423	.647	.815	.916	.966	.988	.996	.999
3.2	.041	.171	.380	.603	.781	.895	.955	.983	.994	.998
3.4	.033	.147	.340	.558	.744	.871	.942	.977	.992	.997
3.6	.027	.126	.303	.515	.706	.844	.927	.969	.988	.996
3.8	.022	.107	.269	.473	.668	.816	.909	.960	.984	.994
4.0	.018	.092	.238	.433	.629	.785	.889	.949	.979	.992
4.2	.015	.078	.210	.395	.590	.753	.867	.936	.972	.989
4.4	.012	.066	.185	.359	.551	.720	.844	.921	.964	.985
4.6	.010	.056	.163	.326	.513	.686	.818	.905	.955	.980
4.8	.008	.048	.143	.294	.476	.651	.791	.887	.944	.975
5.0	.007	.040	.125	.265	.440	.616	.762	.867	.932	.968
5.2	.006	.034	.109	.238	.406	.581	.732	.845	.918	.960
5.4	.005	.029	.095	.213	.373	.546	.702	.822	.903	.951
5.6	.004	.024	.082	.191	.342	.512	.670	.797	.886	.941
5.8	.003	.021	.072	.170	.313	.478	.638	.771	.867	.929
6.0	.002	.017	.062	.151	.285	.446	.606	.744	.847	.916

4.9% of all weeks. Note that this percentage agrees remarkably well with that given by the Empirical Rule for mound-shaped distributions, which tells us to expect approximately 5% of the measurements (values of the random variable) to lie further than 2 standard deviations from the mean.

d. To use Table III to find the probability that *exactly* five sightings are made during a given week, we must write the probability as the difference between two cumulative probabilities:

$$P(x = 5) = P(x \le 5) - P(x \le 4) = .951 - .877 = .074$$

Note that the probabilities in Table III are all rounded to three decimal places. Thus, although in theory a Poisson random variable can assume infinitely large values, the values of x in Table III are extended only until the cumulative probability is 1.000. This does not mean that x *cannot* assume larger values but only that the likelihood is less than .001 (in fact, less than .0005) that it will do so.

Finally, you may need to calculate Poisson probabilities for values of λ not found in Table III. You may be able to obtain an adequate approximation by interpolation, but if not, consult more extensive tables for the Poisson distribution.

Exercises 4.45–4.58

Learning the Mechanics

4.45 Consider the probability distribution shown here:

$$p(x) = \frac{3^x e^{-3}}{x!} \quad (x = 0, 1, 2, \ldots)$$

a. Is x a discrete or continuous random variable? Explain.
b. What is the name of this probability distribution?
c. Graph the probability distribution.
d. Find the mean and standard deviation of x.
e. Find the mean and standard deviation of the probability distribution.

4.46 Given that x is a random variable for which a Poisson probability distribution provides a good approximation, use Table III to compute the following:
a. $P(x \leq 2)$ when $\lambda = 1$ b. $P(x \leq 2)$ when $\lambda = 2$
c. $P(x \leq 2)$ when $\lambda = 3$
d. What happens to the probability of the event $\{x \leq 2\}$ as λ increases from 1 to 3? Is this intuitively reasonable?

4.47 Assume that x is a random variable having a Poisson probability distribution with a mean of 1.5. Use Table III to find the following probabilities:
a. $P(x \leq 3)$ b. $P(x \geq 3)$ c. $P(x = 3)$
d. $P(x = 0)$ e. $P(x > 0)$ f. $P(x > 6)$

4.48 Suppose x is a random variable for which a Poisson probability distribution with $\lambda = 1$ provides a good characterization.
a. Graph $p(x)$ for $x = 0, 1, 2, \ldots, 9$.
b. Find μ and σ for x, and locate μ and the interval $\mu \pm 2\sigma$ on the graph.
c. What is the probability that x will fall within the interval $\mu \pm 2\sigma$?

4.49 Suppose x is a random variable for which a Poisson probability distribution with $\lambda = 3$ provides a good characterization.
a. Graph $p(x)$ for $x = 0, 1, 2, \ldots, 9$.
b. Find μ and σ for x, and locate μ and the interval $\mu \pm 2\sigma$ on the graph.
c. What is the probability that x will fall within the interval $\mu \pm 2\sigma$?

4.50 As mentioned in Section 4.4, when n is large, p is small, and $np \leq 7$, the Poisson probability distribution provides a good approximation to the binomial probability distribution. Since we provide exact binomial probabilities (Table II in Appendix A) for relatively small values of n, you can investigate the adequacy of the approximation for $n = 25$. Use Table II to find $p(0)$, $p(1)$, and $p(2)$ for $n = 25$ and $p = .05$. Calculate the corresponding Poisson approximations using $\lambda = \mu = np$. [*Note:* These approximations are reasonably good for n as small as 25, but to use the approximation in a practical situation we would prefer to have $n \geq 100$.]

Applying the Concepts

4.51 The mean number of patients admitted per day to the emergency room of a small hospital is 2.5. If, on a given day, there are only four beds available for new patients, what is the probability the hospital will not have enough beds to accommodate its newly admitted patients?

4.52 The Environmental Protection Agency (EPA) requires manufacturers of vinyl chloride and similar compounds to limit the amount of these chemicals in plant air emissions to no more than 10 parts per million. Suppose the mean emission of vinyl chloride for a particular plant is 4 parts per million. Assume that the number of parts per million of vinyl chloride in air samples, x, follows a Poisson probability distribution.

a. What is the standard deviation of x for the plant?
b. Is it likely that a sample of air from the plant would yield a value of x that would exceed the EPA limit? Explain.
c. Discuss conditions that would make the Poisson assumption plausible.

4.53 A can company reports that the number of breakdowns per 8-hour shift on its machine-operated assembly line follows a Poisson distribution with a mean of 1.5.

a. What is the probability of exactly two breakdowns on the midnight shift?
b. What is the probability of fewer than two breakdowns on the afternoon shift?
c. What is the probability that more than two breakdowns occur on the midnight shift?
d. What is the probability of no breakdowns during three consecutive 8-hour shifts? (Assume that the machine operates independently across shifts.)

4.54 A certain automatic car wash takes exactly 5 minutes to wash a car. On the average, 10 cars per hour arrive at the car wash. Suppose that, 30 minutes before closing time, five cars are in line. If the car wash is in continuous use until closing time, what is the probability that no one will be in line at closing time?

4.55 The safety supervisor at a large manufacturing plant believes the expected number of industrial accidents per month is 3.4.

a. What is the probability of exactly two accidents occurring next month?
b. What is the probability of three or more accidents occurring next month?
c. What assumptions do you need to make to solve this problem using the methodology of this chapter?

4.56 U.S. airlines fly approximately 26 billion passenger-miles per month and average about 11.8 fatalities per month (*Statistical Abstract of the United States: 1991*, pp. 627, 629). Assume the probability distribution for x, the number of fatalities per month, can be approximated by a Poisson probability distribution.

a. What is the probability that no fatalities will occur during any given month? [*Hint:* Either use Table III of Appendix A and interpolate to approximate the probability, or use a calculator or computer to calculate the probability exactly.]
b. Find $E(x)$ and the standard deviation of x.
c. Use your answers to part **b** to describe the probability that as many as 20 fatalities will occur in any given month.
d. Discuss conditions that would make the Poisson assumption plausible.

4.57 The number x of people who arrive at a cashier's counter in a bank during a specified period of time often possesses (approximately) a Poisson probability distribution. If we know the mean arrival rate λ, the Poisson probability distribution can be used to aid in the design of the customer service facility. Suppose you estimate that the mean number of arrivals per minute for cashier service at a bank is one person per minute.

a. What is the probability that in a given minute the number of arrivals will equal three or more?
b. Can you tell the bank manager that the number of arrivals will rarely exceed two per minute?

4.58 In many cities, neighborhood Crime Watch organizations are formed in an attempt to reduce the amount of criminal activity. Suppose one neighborhood that has experienced an average of 10 crimes per year organizes a Crime Watch group. During the first year following the creation of the group, three crimes are committed in the neighborhood.

a. Use the Poisson distribution to calculate the probability that three or fewer crimes are committed in a year assuming the average number is still 10 crimes per year.
b. Do you think this event provides some evidence that the Crime Watch group has been effective in this neighborhood?

4.5 Probability Distributions for Continuous Random Variables

The graphic form of the probability distribution for a continuous random variable x is a smooth curve that might appear as shown in Figure 4.9. This curve, a function of x, is denoted by the symbol $f(x)$ and is variously called a **probability density function**, a **frequency function**, or a **probability distribution**.

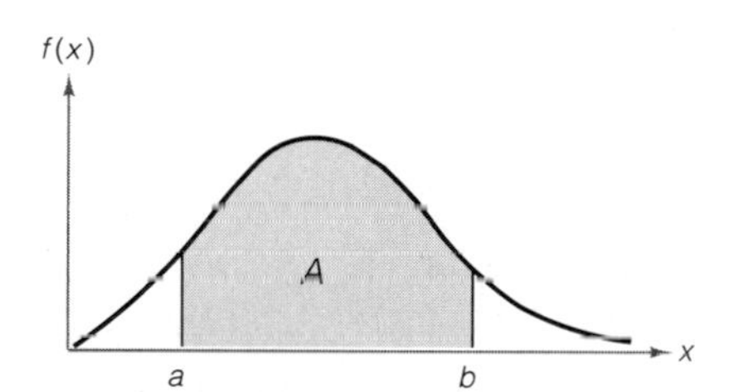

FIGURE 4.9 ► **A probability distribution $f(x)$ for a continuous random variable x**

The areas under a probability distribution correspond to probabilities for x. For example, the area A beneath the curve between the two points a and b, as shown in Figure 4.9, is the probability that x assumes a value between a and b ($a < x < b$). Because there is no area over a point, say $x = a$, it follows that (according to our model) the probability associated with a particular value of x is equal to 0; that is, $P(x = a) = 0$ and hence $P(a < x < b) = P(a \leq x \leq b)$. In other words, the probability is the same regardless of whether you include the endpoints of the interval. Also, because areas over intervals represent probabilities, it follows that the total area under a probability distribution, the probability assigned to all values of x, should equal 1. Note that probability distributions for continuous random variables possess different shapes depending on the relative frequency distributions of real data that the probability distributions are supposed to model.

The areas under most probability distributions are obtained by use of the calculus or numerical methods.* Because this is often a difficult procedure, we will give the areas for some of the most common probability distributions in tabular form in Appendix A. Then to find the area between two values of x, say $x = a$ and $x = b$, you simply have to consult the appropriate table.

For the continuous random variable presented in this chapter, we will give the formula for the probability distribution along with its mean and standard deviation. These two numbers, μ and σ, will enable you to make some approximate probability statements about a random variable even when you do not have access to a table of areas under the probability distribution.

4.6 The Normal Distribution

One of the most commonly observed continuous random variables has a **bell-shaped** probability distribution as shown in Figure 4.10. It is known as a **normal random variable** and its probability distribution is called a **normal distribution**.

FIGURE 4.10 ▶
A normal probability distribution

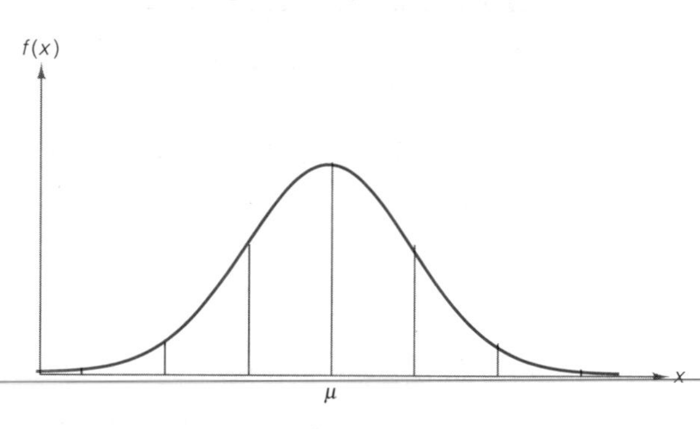

You will see during the remainder of this text that the normal distribution plays a very important role in the science of statistical inference. Moreover, many phenomena generate random variables with probability distributions that are very well approximated by a normal distribution. For example, the error made in measuring a person's blood pressure may be a normal random variable, and the probability distribution for the yearly rainfall in a certain region might be approximated by a normal probability distribution. The normal distribution might also provide an accurate model for the distribution of the scores on an aptitude test. You can determine the adequacy of the normal approximation to an existing population of data by comparing the relative frequency distribution of a large sample of the data to the normal probability distri-

---

*Students with knowledge of calculus should note that the probability that x assumes a value in the interval $a < x < b$ is $P(a < x < b) = \int_a^b f(x)\,dx$, assuming the integral exists. Similar to the requirements for a discrete probability distribution, we require $f(x) \geq 0$ and $\int_{-\infty}^{\infty} f(x)\,dx = 1$.

bution. Tests to detect disagreement between a set of data and the assumption of normality are available, but they are beyond the scope of this book.

The normal distribution is perfectly symmetric about its mean μ, as can be seen in the examples in Figure 4.11. Its spread is determined by the value of its standard deviation σ.

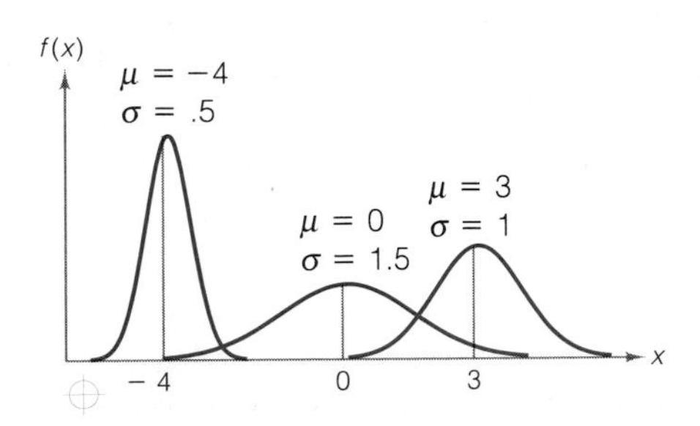

FIGURE 4.11 ► **Several normal distributions with different means and standard deviations**

The formula for the normal probability distribution is shown in the box. When plotted, this formula yields a curve like that shown in Figure 4.10.

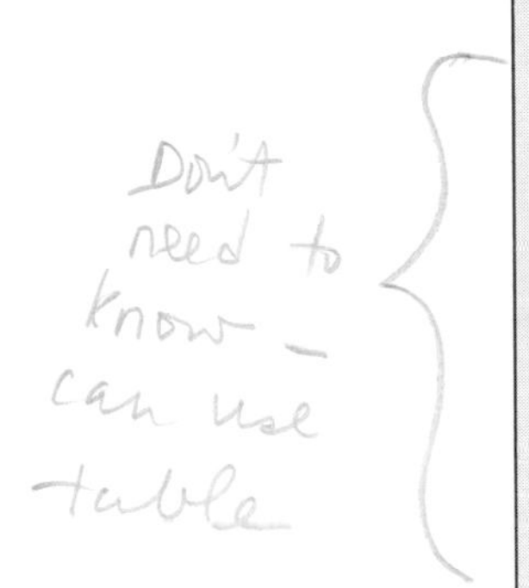

Probability Distribution for a Normal Random Variable *x*

$$f(x) = \frac{1}{\sigma\sqrt{2\pi}} e^{-(1/2)[(x-\mu)/\sigma]^2}$$

where μ = Mean of the normal random variable x
σ = Standard deviation
π = 3.1416. . .
e = 2.71828. . .

Note that the mean μ and standard deviation σ appear in this formula, so that no separate formulas for μ and σ are necessary. To graph the normal curve we have to know the numerical values of μ and σ.

Computing the area over intervals under the normal probability distribution is a difficult task.* Consequently, we will use the computed areas listed in Table IV of Appendix A (and inside the front cover). Although there are an infinitely large number of normal curves—one for each pair of values for μ and σ—we have formed a single table that will apply to any normal curve. This is done by constructing the table of areas

*The student with knowledge of calculus should note that there is not a closed-form expression for $P(a < x < b) = \int_a^b f(x)\,dx$ for the normal probability distribution. The value of this definite integral can be obtained to any desired degree of accuracy by numerical approximation procedures. For this reason, it is tabulated for the user.

as a function of the z-score (presented in Section 2.6). The population z-score for a measurement was defined as the *distance* between the measurement and the population mean, divided by the population standard deviation. Thus, the z-score gives the distance between a measurement and the mean in units equal to the standard deviation. In symbolic form, the z-score for the measurement x is

$$z = \frac{x - \mu}{\sigma}$$

Note that when $x = \mu$, we obtain $z = 0$.

To illustrate the use of Table IV, suppose we know that the length of time x between charges of a pocket calculator has a normal distribution with a mean of 50 hours and a standard deviation of 15 hours. If we were to observe the length of time that elapses before the need for the next charge, what is the probability that this measurement will assume a value between 50 and 70 hours? This probability is the area under the normal probability distribution between 50 and 70, as shown in the shaded area A of Figure 4.12.

FIGURE 4.12 ▶
Normal distribution: $\mu = 50$, $\sigma = 15$

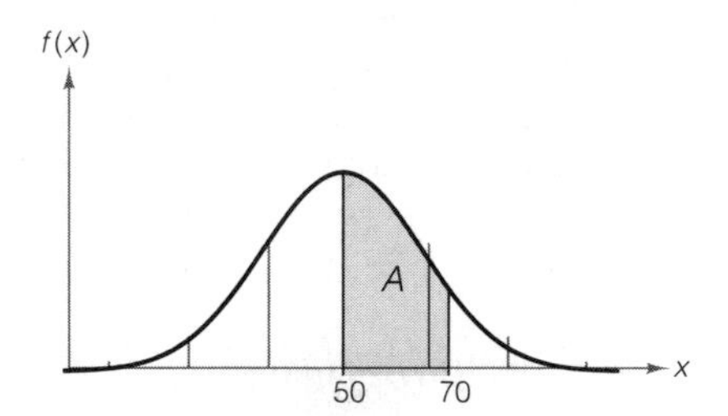

The first step in finding the area A is to calculate the z-score corresponding to the measurement 70. We calculate

$$z = \frac{x - \mu}{\sigma} = \frac{70 - 50}{15} = \frac{20}{15} = 1.33$$

Thus, the measurement 70 is 1.33 standard deviations above the mean, $\mu = 50$. The second step is to refer to Table IV (a partial reproduction of this table is shown in Table 4.5). Note that z-scores are listed in the left-hand column of the table. To find the area corresponding to a z-score of 1.33, we first locate the value 1.3 in the left-hand column. Since this column lists z-values to one decimal place only, we refer to the top row of the table to get the second decimal place, .03. Finally, we locate the number where the row labeled $z = 1.3$ and the column labeled .03 meet. This number represents the area between the mean μ and the measurement that has a z-score of 1.33:

$$A = .4082$$

Hence, the probability that the calculator operates between 50 and 70 hours before needing a charge is .4082.

The use of the z-score simplifies the calculation of normal probabilities because if x is normally distributed with any mean and standard deviation, z is *always* a normal

TABLE 4.5 Reproduction of Part of Table IV in Appendix A

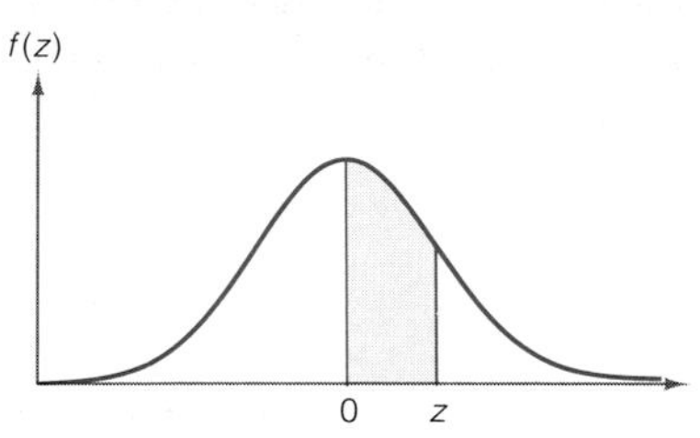

z	.00	.01	.02	.03	.04	.05	.06	.07	.08	.09
.0	.0000	.0040	.0080	.0120	.0160	.0199	.0239	.0279	.0319	.0359
.1	.0398	.0438	.0478	.0517	.0557	.0596	.0636	.0675	.0714	.0753
.2	.0793	.0832	.0871	.0910	.0948	.0987	.1026	.1064	.1103	.1141
.3	.1179	.1217	.1255	.1293	.1331	.1368	.1406	.1443	.1480	.1517
.4	.1554	.1591	.1628	.1664	.1700	.1736	.1772	.1808	.1844	.1879
.5	.1915	.1950	.1985	.2019	.2054	.2088	.2123	.2157	.2190	.2224
.6	.2257	.2291	.2324	.2357	.2389	.2422	.2454	.2486	.2517	.2549
.7	.2580	.2611	.2642	.2673	.2704	.2734	.2764	.2794	.2823	.2852
.8	.2881	.2910	.2939	.2967	.2995	.3023	.3051	.3078	.3106	.3133
.9	.3159	.3186	.3212	.3238	.3264	.3289	.3315	.3340	.3365	.3389
1.0	.3413	.3438	.3461	.3485	.3508	.3531	.3554	.3577	.3599	.3621
1.1	.3643	.3665	.3686	.3708	.3729	.3749	.3770	.3790	.3810	.3830
1.2	.3849	.3869	.3888	.3907	.3925	.3944	.3962	.3980	.3997	.4015
1.3	.4032	.4049	.4066	.4082	.4099	.4115	.4131	.4147	.4162	.4177
1.4	.4192	.4207	.4222	.4236	.4251	.4265	.4279	.4292	.4306	.4319
1.5	.4332	.4345	.4357	.4370	.4382	.4394	.4406	.4418	.4429	.4441

random variable with a mean of 0 and a standard deviation of 1. For this reason z is often referred to as a **standard normal random variable**.

Definition 4.8

The **standard normal random variable** z is defined by the formula

$$z = \frac{x - \mu}{\sigma}$$

where x is a normal random variable with mean μ and standard deviation σ. The standard normal random variable z is normally distributed with mean 0 and standard deviation 1 and can be described as the number of standard deviations between x and μ.

Since we will convert all normal random variables to standard normal in order to use Table IV to find probabilities, it is important that you learn to use Table IV well. The following examples illustrate the use of Table IV.

EXAMPLE 4.12

Find the probability that the standard normal random variable z falls between -1.33 and $+1.33$.

Solution

The standard normal distribution is shown in Figure 4.13. Since all probabilities associated with standard normal random variables can be depicted as areas under the standard normal curve, you should always draw the curve and then equate the desired probability to an area.

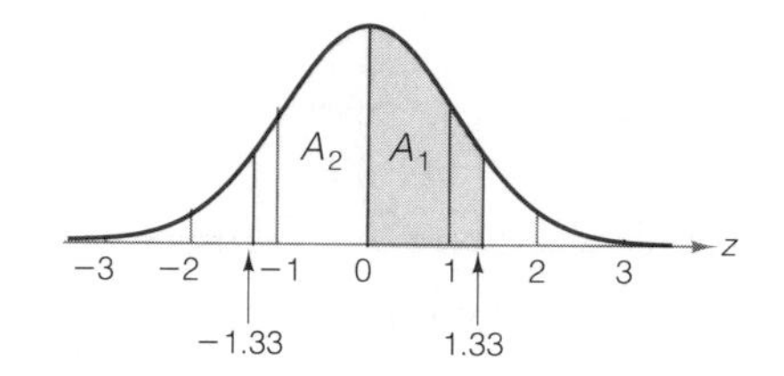

FIGURE 4.13 ▶ A distribution of z-scores (a standard normal distribution)

In this example we want to find the probability that z falls between -1.33 and $+1.33$, which is equivalent to the area between -1.33 and $+1.33$, shown shaded in Figure 4.13. Table IV provides the area between $z = 0$ and any value of z, so that if we look up $z = 1.33$, we find that the area between $z = 0$ and $z = 1.33$ is .4082. This is the area labeled A_1 in Figure 4.13. To find the area A_2 between $z = 0$ and $z = -1.33$, we note that the symmetry of the normal distribution implies that the area between $z = 0$ and any point to the left is equal to the area between $z = 0$ and the point equidistant to the right. Thus, in this example the area between $z = 0$ and $z = -1.33$ is equal to the area between $z = 0$ and $z = +1.33$. That is,

$$A_1 = A_2 = .4082$$

The probability that z falls between -1.33 and $+1.33$ is the sum of the areas A_1 and A_2. We summarize in probabilistic notation:

$$\begin{aligned} P(-1.33 < z < +1.33) &= P(-1.33 < z < 0) + P(0 \leq z < 1.33) \\ &= A_1 + A_2 \\ &= .4082 + .4082 = .8164 \end{aligned}$$

Remember that "$<$" and "$\leq$" are equivalent in events involving z, because the inclusion (or exclusion) of a single point does not alter the probability of an event involving a continuous random variable.

EXAMPLE 4.13

Find the probability that a standard normal random variable exceeds 1.64; i.e., find $P(z > 1.64)$.

Solution

The area under the standard normal distribution to the right of 1.64 is the shaded area labeled A_1 in Figure 4.14. This area represents the desired probability that z exceeds 1.64. However, when we look up $z = 1.64$ in Table IV, we must remember that the probability given in the table corresponds to the area between $z = 0$ and $z = 1.64$ (the area labeled A_2 in Figure 4.14). From Table IV we find that $A_2 = .4495$. To find the area A_1 to the right of 1.64, we make use of two facts:

FIGURE 4.14 ►
Standard normal distribution:
$\mu = 0, \sigma = 1$

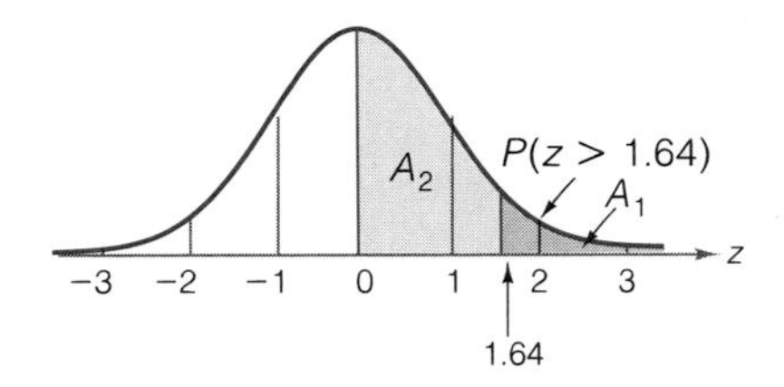

1. The standard normal distribution is symmetric about its mean, $z = 0$.
2. The total area under the standard normal probability distribution equals 1.

Taken together, these two facts imply that the areas on either side of the mean $z = 0$ equal .5; thus, the area to the right of $z = 0$ in Figure 4.14 is $A_1 + A_2 = .5$. Then

$$P(z > 1.64) = A_1 = .5 - A_2 = .5 - .4495 = .0505$$

To attach some practical significance to this probability, note that the implication is that the chance of a standard normal random variable exceeding 1.64 is approximately .05. Or, since z represents the number of standard deviations between *any* normal random variable and its mean, a normal random variable will exceed its mean by more than 1.64 standard deviations only about 5% of the time.

EXAMPLE 4.14

Find the probability that a normal random variable lies to the right of a point −.74 standard deviation from its mean.

Solution

We first must interpret the event in terms of a standard normal random variable, so that we can use Table IV to find the event's probability. Since the standard normal random variable z is simply the number of standard deviations between an observed value of a normal random variable and its mean, the event that a normal random variable lies to the right of a point −.74 standard deviation from the mean is equivalent to the event that the standard normal random variable z exceeds −.74. The event is shown as the shaded area in Figure 4.15 on page 206, and we want to find $P(z > -.74)$.

We divide the shaded area into two parts: the area A_1 between $z = -.74$ and $z = 0$, and the area A_2 to the right of $z = 0$. We must always make such a division when the desired area lies on both sides of the mean ($z = 0$) because Table IV contains areas between $z = 0$ and the point you look up. To find A_1, we remember that the sign of

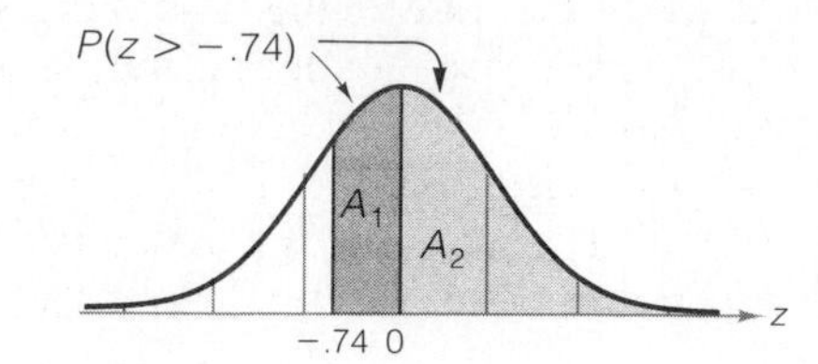

FIGURE 4.15 ▶ Standard normal distribution: $\mu = 0, \sigma = 1$

z is unimportant when determining the area, because the standard normal distribution is symmetric about its mean. We look up $z = .74$ in Table IV to find that $A_1 = .2704$. The symmetry also implies that half the distribution lies on each side of the mean, so the area A_2 to the right of $z = 0$ is .5. Then,

$$P(z > -.74) = A_1 + A_2 = .2704 + .5 = .7704$$

EXAMPLE 4.15

Find the probability that a normal random variable lies more than 1.96 standard deviations from its mean *in either direction*.

Solution

The event that a normal random variable exceeds 1.96 standard deviations in either direction is equivalent to the event that the standard normal random variable z exceeds 1.96 in absolute value. That is, we want to find

$$P(z > |1.96|) = P(z < -1.96 \quad \text{or} \quad z > 1.96)$$

This probability is the shaded area in Figure 4.16. Note that the total shaded area is the sum of two areas, A_1 and A_2—areas that are equal because of the symmetry of the normal distribution.

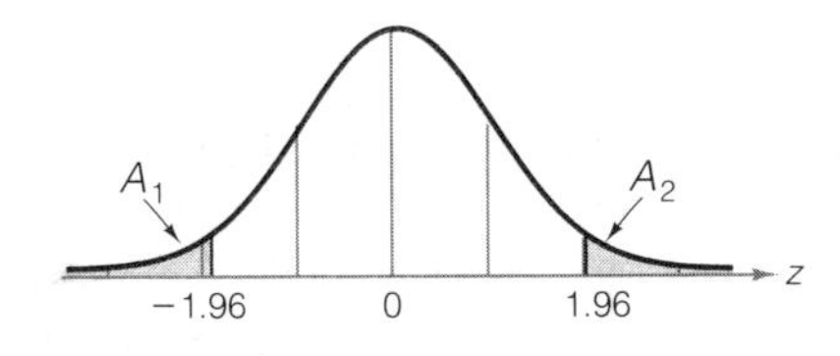

FIGURE 4.16 ▶ Standard normal distribution: $\mu = 0, \sigma = 1$

We look up $z = 1.96$ and find the area between $z = 0$ and $z = 1.96$ to be .4750. Then the area to the right of 1.96, A_2, is $.5 - .4750 = .0250$, so that

$$P(z > |1.96|) = A_1 + A_2 = .0250 + .0250 = .05$$

The implication is that any normal random variable lies more than 1.96 standard deviations from its mean 5% of the time. Recall from Chapter 2 that the Empirical

Rule tells us that about 5% of the measurements in mound-shaped distributions will lie beyond 2 standard deviations from the mean; the normal distribution, which is certainly mound-shaped, has 5% of its area beyond 1.96 standard deviations. In fact, the normal distribution provides the model on which the Empirical Rule is based, along with much "empirical" experience with real data that often approximately obey the rule, whether drawn from a normal distribution or not.

EXAMPLE 4.16

Assume that the length of time, x, between charges of a pocket calculator is normally distributed with a mean of 50 hours and a standard deviation of 15 hours. Find the probability that the calculator will last between 30 and 70 hours between charges.

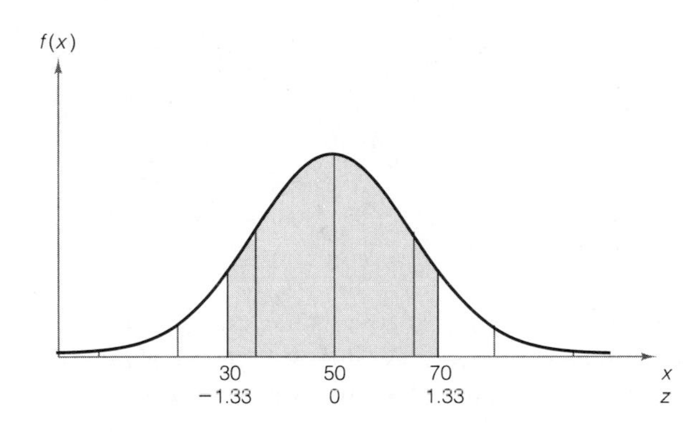

FIGURE 4.17 ► Normal probability distribution: $\mu = 50, \sigma = 15$

Solution

The normal distribution with mean $\mu = 50$ and $\sigma = 15$ is shown in Figure 4.17. The desired probability that the calculator lasts between 30 and 70 hours is shaded. In order to find the probability, we must first convert the distribution to standard normal, which we do by calculating the z-score:

$$z = \frac{x - \mu}{\sigma}$$

The z-scores corresponding to the important values of x are shown beneath the x values on the horizontal axis in Figure 4.17. Note that $z = 0$ corresponds to the mean of $\mu = 50$ hours, whereas the x values 30 and 70 yield z-scores of -1.33 and $+1.33$, respectively. Thus, the event that the calculator lasts between 30 and 70 hours is equivalent to the event that a standard normal random variable lies between -1.33 and $+1.33$. We found this probability in Example 4.12 (see Figure 4.13) by doubling the area corresponding to $z = 1.33$ in Table IV. That is,

$$P(30 \le x \le 70) = P(-1.33 \le z \le 1.33) = 2(.4082) = .8164$$

The steps to follow when calculating a probability corresponding to a normal random variable are shown in the box at the top of page 208.

Know but won't have to state

*

Steps for Finding a Probability Corresponding to a Normal Random Variable

1. Sketch the normal distribution and indicate the mean of the random variable x. Then shade the area corresponding to the probability you want to find.
2. Convert the boundaries of the shaded area from x values to standard normal random variable z values using the formula

$$z = \frac{x - \mu}{\sigma}$$

   Show the z values under the corresponding x values on your sketch.
3. Use Table IV in Appendix A (and inside the front cover) to find the areas corresponding to the z values. If necessary, use the symmetry of the normal distribution to find areas corresponding to negative z values and the fact that the total area on each side of the mean equals .5 to convert the areas from Table IV to the probabilities of the event you have shaded.

EXAMPLE 4.17

Suppose an automobile manufacturer introduces a new model that has an advertised mean in-city mileage of 27 miles per gallon. Although such advertisements seldom report any measure of variability, suppose you write the manufacturer for the details of the tests, and you find that the standard deviation is 3 miles per gallon. This information leads you to formulate a probability model for the random variable x, the in-city mileage for this car model. You believe that the probability distribution of x can be approximated by a normal distribution with a mean of 27 and a standard deviation of 3.

a. If you were to buy this model of automobile, what is the probability that you would purchase one that averages less than 20 miles per gallon for in-city driving? In other words, find $P(x < 20)$.

b. Suppose you purchase one of these new models and it does get less than 20 miles per gallon for in-city driving. Should you conclude that your probability model is incorrect?

Solution

a. The probability model proposed for x, the in-city mileage, is shown in Figure 4.18. We are interested in finding the area A to the left of 20 since this area corresponds to the probability that a measurement chosen from this distribution falls below 20. In other words, if this model is correct, the area A represents the fraction of cars that can be expected to get less than 20 miles per gallon for in-city driving. To find A, we first calculate the z value corresponding to $x = 20$. That is,

$$z = \frac{x - \mu}{\sigma} = \frac{20 - 27}{3} = -\frac{7}{3} = -2.33$$

Then

$$P(x < 20) = P(z < -2.33)$$

as indicated by the shaded area in Figure 4.18. Since Table IV gives only areas to the right of the mean (and because the normal distribution is symmetric about its mean), we look up 2.33 in Table IV and find that the corresponding area is .4901. This is equal to the area between $z = 0$ and $z = -2.33$, so we find

$$P(x < 20) = A = .5 - .4901 = .0099 \approx .01$$

According to this probability model, you should have only about a 1% chance of purchasing a car of this make with an in-city mileage under 20 miles per gallon.

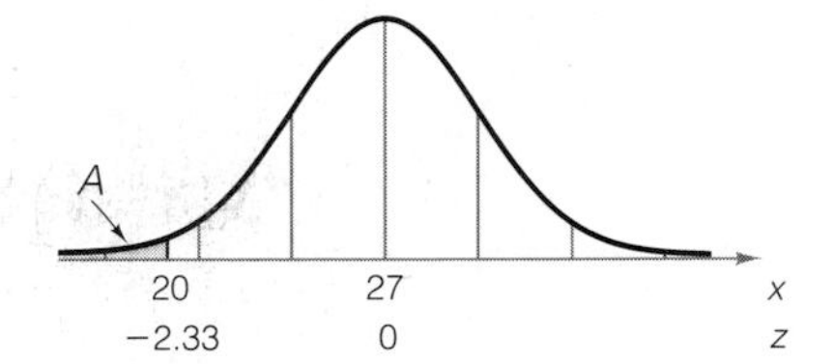

FIGURE 4.18 ▶ Normal probability distribution for x in Example 4.17: $\mu = 27$ miles per gallon, $\sigma = 3$ miles per gallon

b. Now you are asked to make an inference based on a sample—the car you purchased. You are getting less than 20 miles per gallon for in-city driving. What do you infer? We think you will agree that one of two possibilities is true:

(1) The probability model is correct. You simply were unfortunate to have purchased one of the cars in the 1% that get less than 20 miles per gallon in the city.

(2) The probability model is incorrect. Perhaps the assumption of a normal distribution is unwarranted, or the mean of 27 is an overestimate, or the standard deviation of 3 is an underestimate, or some combination of these errors was made. At any rate, the form of the actual probability model certainly merits further investigation.

You have no way of knowing with certainty which possibility is correct, but the evidence points to the second one. We are again relying on the rare event approach to statistical inference that we introduced earlier. The sample (one measurement in this case) was so unlikely to have been drawn from the proposed probability model that it casts serious doubt on the model. We would be inclined to believe that the model is somehow in error.

Occasionally you will be given a probability and will want to find the values of the normal random variable that correspond to the probability. For example, suppose the scores on a college entrance examination are known to be normally distributed, and a certain prestigious university will consider for admission only those applicants whose scores exceed the 90th percentile of the test score distribution. To determine the minimum score for admission consideration, you will need to be able to use Table IV in reverse, as demonstrated in the following example.

EXAMPLE 4.18

Find the value of z, call it z_0, in the standard normal distribution that will be exceeded only 10% of the time. That is, find z_0 such that $P(z \geq z_0) = .10$.

Solution

In this case we are given a probability, or an area, and asked to find the value of the standard normal random variable that corresponds to the area. Specifically, we want to find the value z_0 such that only 10% of the standard normal distribution exceeds z_0 (see Figure 4.19).

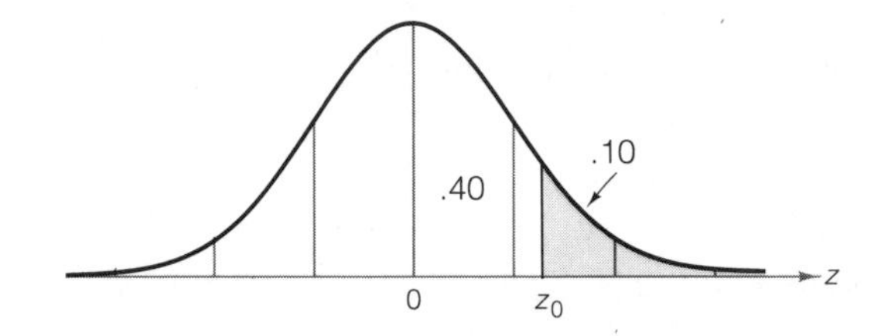

FIGURE 4.19 ► Standard normal distribution for Example 4.18

We know that the total area to the right of the mean $z = 0$ is .5, which implies that z_0 must lie to the right of (above) 0. To pinpoint the value, we use the fact that the area to the right of z_0 is .10, which implies that the area between $z = 0$ and z_0 is $.5 - .1 = .4$. But areas between $z = 0$ and some other z value are exactly the types given in Table IV. Therefore, we look up the area .4000 in the body of Table IV and find that the corresponding z value is (to the closest approximation) $z_0 = 1.28$. The implication is that the point 1.28 standard deviations above the mean is the 90th percentile of a normal distribution.

EXAMPLE 4.19

Find the value of z_0 such that 95% of the standard normal z values lie between $-z_0$ and $+z_0$—i.e., $P(-z_0 \leq z \leq z_0) = .95$.

Solution

Here we wish to move an equal distance z_0 in the positive and negative direction from the mean $z = 0$ until 95% of the standard normal distribution is enclosed. This means that the area on each side of the mean will be equal to $\frac{1}{2}(.95) = .475$, as shown in Figure 4.20. Since the area between $z = 0$ and z_0 is .475, we look up .475 in the body of Table IV to find the value $z_0 = 1.96$. Thus, as we found in the reverse order in

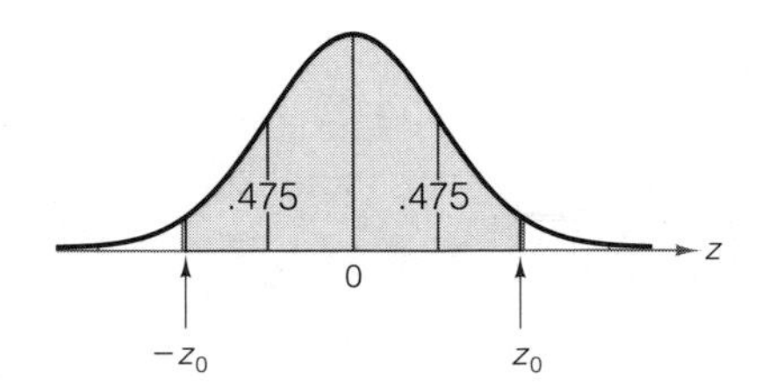

FIGURE 4.20 ► Standard normal distribution: $\mu = 0, \sigma = 1$

Example 4.15, 95% of a normal distribution lies between plus and minus 1.96 standard deviations of the mean.

Now that you have learned to use Table IV to find a standard normal z value that corresponds to a specified probability, we demonstrate a practical application in Example 4.20.

EXAMPLE 4.20

Suppose the scores, x, on a college entrance examination are normally distributed with a mean of 550 and a standard deviation of 100. A certain prestigious university will consider for admission only those applicants whose scores exceed the 90th percentile of the distribution. Find the minimum score an applicant must achieve in order to receive consideration for admission to the university.

Solution

First, we find the 90th percentile in the standard normal distribution. We did this in Example 4.18 (see Figure 4.19) and found that $z = 1.28$ is the standard normal value that will be exceeded only 10% of the time.

Next, we want to convert the standard normal value to an x value, a test score in this example. We know that the minimum test score will be $z = 1.28$ standard deviations above the mean score. To determine the minimum score, remember that

$$z = \frac{x - \mu}{\sigma}$$

If we solve this equation for x, we find

$$x = \mu + z\sigma$$

In this example, $\mu = 550$, $z = 1.28$, and $\sigma = 100$, so

$$x = 550 + 1.28(100) = 550 + 128 = 678$$

This x value is shown in Figure 4.21 corresponding to the z value of 1.28. Thus, the 90th percentile of the test score distribution is 678. An applicant must score at least 678 on the entrance exam to receive consideration for admission by the university.

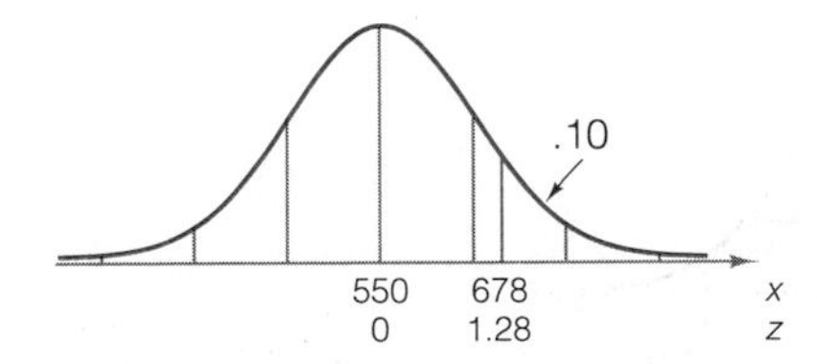

FIGURE 4.21 ► Normal distribution for Example 4.20

CASE STUDY 4.4 / Grading on the Curve

How did teachers ever suppose that the statistician's bell-shaped curve ought somehow to be imposed on the results of their work? The famous curve was developed to describe the distribution of natural phenomena. If we weigh 10,000 grains of corn, or measure the heights of men and women, or their ability to learn nonsense syllables, the findings will cluster around some central value and taper at both ends. The bell-shaped curve is descriptive of raw or unselected phenomena, and then only if vast numbers of cases are used. When the teacher receives his pupils, their distribution with respect to some characteristics may follow the "normal curve." But, having received his charges, the skillful teacher sets out as fast as he can to destroy the "natural" state of affairs.

This statement is the introduction to Clyde W. Bresee's article (1976) about "grading on the curve." His main point is that many teachers who consistently grade on the curve are assuming that the measures of learning (test scores and the like) will always form a normal, or bell-shaped, distribution. These teachers may even use z-scores to determine grades and the corresponding areas under the normal curve (those in Table IV) to obtain the percentages of students who will receive each grade. The implication is that a student's performance will be measured only in relation to the other students in the class. Bresee relates the following anecdote:

After the completion of a particularly successful unit, a teacher was heard to say, "I don't know how I'll grade this thing because they all did so well." This is a sad and dangerous statement because this teacher is on the verge of undoing a month's or even a year's work. Here is a teacher so indoctrinated by an erroneous concept that he questions his own accomplishment even in the face of clear evidence. But question it he must, if he has been trained to grade "on the curve." In a class of 20, for example, the only possible explanation for 15 A's is that he is a weak teacher or a "soft grader," or both.

The caveat presented by Bresee can be more generally applied. The normal distribution is widely used, and in many situations it provides an adequate approximation to reality. However, before each new application of the normal distribution, the situation should be carefully studied, and the user must be confident that all attendant assumptions are satisfied. When using the normal distribution to "grade on a curve," Bresee worries that the day may come when a supervisor admonishes a teacher, "Last year we sent you twenty children to teach. We spent a year's time and upwards of $20,000 on them, and all we have to show for it is the same old bell-shaped curve!"

Exercises 4.59–4.79

Learning the Mechanics

4.59 Find the area under the standard normal probability distribution between the following pairs of z-scores.

a. $z = 0$ and $z = 2.00$
b. $z = 0$ and $z = 1.00$
c. $z = 0$ and $z = 3.00$
d. $z = 0$ and $z = .58$
e. $z = -2.00$ and $z = 0$
f. $z = -1.00$ and $z = 0$
g. $z = -1.69$ and $z = 0$
h. $z = -.58$ and $z = 0$

4.60 Find the following probabilities for the standard normal random variable z.
a. $P(-1 \le z \le 1)$ b. $P(-2 \le z \le 2)$ c. $P(-2.16 \le z \le .55)$
d. $P(-.42 < z < 1.96)$ e. $P(z \ge -2.33)$ f. $P(z < 2.33)$

4.61 Find the following probabilities for the standard normal random variable z.
a. $P(z > 1.46)$ b. $P(z < -1.56)$ c. $P(.67 \le z \le 2.41)$
d. $P(-1.96 \le z < -.33)$ e. $P(z \ge 0)$ f. $P(-2.33 < z < 1.50)$

4.62 Find a value of the standard normal random variable z, call it z_0, such that
a. $P(z \ge z_0) = .05$ b. $P(z \ge z_0) = .025$
c. $P(z \le z_0) = .025$ d. $P(z \ge z_0) = .10$
e. $P(z > z_0) = .10$

4.63 Find a z-score, call it z_0, such that
a. $P(z \ge z_0) = .5$ b. $P(z \ge z_0) = .0057$
c. $P(0 \le z \le z_0) = .4713$ d. $P(z < z_0) = .0392$

4.64 Give the z-score for a measurement from a normal distribution for the following.
a. 1 standard deviation above the mean
b. 1 standard deviation below the mean
c. Equal to the mean
d. 2.5 standard deviations below the mean
e. 3 standard deviations above the mean

4.65 Suppose x is a normally distributed random variable with $\mu = 40$ and $\sigma = 10$. Find each of the following:
a. $P(x \le 50)$ b. $P(x \le 35.6)$ c. $P(35 \le x \le 56.8)$
d. $P(22.9 \le x \le 33.2)$ e. $P(x \ge 25.3)$ f. $P(x \le 25.3)$

4.66 Suppose x is a normally distributed random variable with $\mu = 30$ and $\sigma = 8$. Find a value of the random variable, call it x_0, such that
a. $P(x \ge x_0) = .5$ b. $P(x < x_0) = .025$
c. $P(x > x_0) = .10$ d. $P(x > x_0) = .95$
e. 10% of the values of x are less than x_0
f. 80% of the values of x are less than x_0
g. 1% of the values of x are greater than x_0

4.67 Suppose x is a normally distributed random variable with mean 100 and standard deviation 8. Draw a rough graph of the distribution of x. Locate μ and the interval $\mu \pm 2\sigma$ on the graph. Find the following probabilities:
a. $P(\mu - 2\sigma \le x \le \mu + 2\sigma)$ b. $P(x \ge \mu + 2\sigma)$ c. $P(x \le 92)$
d. $P(92 \le x \le 116)$ e. $P(92 \le x \le 96)$ f. $P(76 \le x \le 124)$

4.68 The random variable x has a normal distribution with standard deviation 25. It is known that the probability that x exceeds 150 is .90. Find the mean μ of the probability distribution.

Applying the Concepts

4.69 Personnel tests are designed to test a job applicant's cognitive and/or physical abilities. An IQ test is an example of the former; a speed test involving the arrangement of pegs on a peg board is an example of the latter. During the 1970s, the proportion of employers using personnel tests dropped from more than 90% to less than 50%, in part due to concerns with equal-rights laws. As a result of improved testing procedures and the realization that such tests could hold down the costs associated with hiring the wrong person, the use of personnel tests has increased dramatically during the 1980s (Dessler, 1986).

A particular dexterity test is administered nationwide by a private testing service. It is known that for all tests administered last year the distribution of scores was approximately normal with mean 75 and standard deviation 7.5.

a. A particular employer requires job candidates to score at least 80 on the dexterity test. Approximately what percentage of the test scores during the past year exceeded 80?
b. The testing service reported to a particular employer that one of its job candidate's scores fell at the 98th percentile of the distribution (i.e., approximately 98% of the scores were lower than the candidate's, and only 2% were higher). What was the candidate's score?

4.70 The tread life of a particular brand of tire is a random variable best described by a normal distribution with a mean of 60,000 miles and a standard deviation of 8,300 miles. If the manufacturer guarantees the tread life of the tires for the first 45,000 miles, what proportion of the tires will need to be replaced under warranty? What if the warranty is for the first 40,000 miles?

4.71 The average salary for a major league baseball player has risen steadily from $19,000 per year in 1967 to $1,100,000 in 1993.

a. If the 1993 distribution of salaries is normally distributed with a standard deviation equal to $400,000, what percentage of major league baseball players are making $1,500,000 per year or more?
b. Can you give a reason why it is unlikely that the distribution of major league baseball salaries is normally distributed?

4.72 The amount of oxygen dissolved in rivers and streams depends on the water temperature and on the amounts of decaying organic matter from natural processes or human disturbances that are present in the water. The Council on Environmental Quality (CEQ) considers a dissolved oxygen content of less than 5 milligrams per liter of water to be undesirable because it is unlikely to support aquatic life. Suppose an industrial plant discharges its waste into a river and the downstream daily oxygen content measurements are normally distributed with a mean equal to 6.3 milligrams per liter and a standard deviation of .6 milligram per liter.

a. What percentage of the days would the dissolved oxygen content in the river be considered undesirable by the CEQ?
b. Within what limits would we expect the dissolved oxygen content to fall?

4.73 The pulse rate per minute of the adult male population between 18 and 25 years of age in the United States is known to have a normal distribution with a mean of 72 beats per minute and a standard deviation of 9.7. If the requirements for military service state that anyone with a pulse rate over 100 is medically unsuitable for service, what proportion of the males between 18 and 25 years of age would be declared unfit because their pulse rates are too high?

4.74 Do security analysts do a good job of forecasting corporate earnings growth and advising their clientele? David Dreman, a *Forbes* columnist, addresses this question in an article titled "Astrology Might Be Better" (*Forbes*, March 26, 1984). The basis of Dreman's article is a study by Professors Michael Sandretto of Harvard and Sudhir Milkrishnamurthi of MIT. The study surveys security analysts' forecasts of annual earnings for the (then) current year for more than 769 companies with five or more forecasts per company per year. The average forecast error for this large number of forecasts was plus or minus 31.3%. To apply this information to a practical situation, suppose the population of analysts' forecast errors is normally distributed with a mean of 31.3% and a standard deviation of 10%.

a. If you obtain a security analyst's forecast for a certain company, what is the probability that it will be in error by more than 50%?

b. If three analysts make the forecast, what is the probability that at least one of the analysts will err by more than 50%?

4.75 A machine used to regulate the amount of dye dispensed for mixing shades of paint can be set so that it discharges an average of μ milliliters of dye per can of paint. The amount of dye discharged is known to have a normal distribution with a standard deviation of .4 milliliter. If more than 6 milliliters of dye are discharged when making a certain shade of blue paint, the shade is unacceptable. Determine the setting for μ so that only 1% of the cans of paint will be unacceptable.

4.76 A physical-fitness association is including the mile run in their secondary-school fitness test for boys. The time for this event for boys in secondary school is approximately normally distributed with a mean of 450 seconds and a standard deviation of 40 seconds. If the association wants to designate the fastest 10% as "excellent," what time should the association set for this criterion?

4.77 The board of examiners that administers the real estate brokers' examination in a certain state found that the mean score on the test was 435 and the standard deviation was 72. If the board wants to set the passing score so that only the best 30% of all applicants pass, what is the passing score? Assume that the scores are normally distributed.

4.78 The distribution of the demand (in number of units per unit time) for a product can often be approximated by a normal probability distribution. For example, a bakery has determined that the number of loaves of its white bread demanded daily has a normal distribution with mean 7,200 loaves and standard deviation 300 loaves. Based on cost considerations, the company has decided that its best strategy is to produce a sufficient number of loaves so that it will fully supply demand on 94% of all days.

a. How many loaves of bread should the company produce?

b. Based on the production in part **a**, on what percentage of days will the company be left with more than 500 loaves of unsold bread?

4.79 What relationship exists between the standard normal distribution and the box-plot methodology (Section 2.7) for describing distributions of data using quartiles? The answer depends on the true underlying probability distribution of the data. Assume for the remainder of this exercise that the distribution is normal.

a. Calculate the values of the standard normal random variable z, call them z_L and z_U, that correspond to the hinges of the box plot—i.e., the lower and upper quartiles, Q_L and Q_U—of the probability distribution.

b. Calculate the z values that correspond to the inner fences of the box plot for a normal probability distribution.

c. Calculate the z values that correspond to the outer fences of the box plot for a normal probability distribution.

d. What is the probability that an observation lies beyond the inner fences of a normal probability distribution? The outer fences?

e. Can you better understand why the inner and outer fences of a box plot are used to detect outliers in a distribution? Explain.

4.7 Approximating a Binomial Distribution with a Normal Distribution

When a binomial random variable can assume a large number of values, the calculation of its probabilities may become very tedious. To contend with this problem, we provide tables in Appendix A to give the probabilities for some values of n and p, but these tables are by necessity incomplete. In particular, the binomial probability table (Table II) can be used only for $n = 5, 6, 7, 8, 9, 10, 15, 20$, or 25. To deal with this limitation, we seek approximation procedures for calculating the probabilities associated with a binomial probability distribution.

When n is large, a normal probability distribution may be used to provide a good approximation to the probability distribution of a binomial random variable. To show how this approximation works, we refer to Example 4.10, in which we used the binomial distribution to model the number x of 20 voters who favor a candidate. We assumed that 60% of all the eligible voters favored the candidate. The mean and standard deviation of x were found to be $\mu = 12$ and $\sigma = 2.2$. The binomial distribution for $n = 20$ and $p = .6$ is shown in Figure 4.22, and the approximating normal distribution with mean $\mu = 12$ and standard deviation $\sigma = 2.2$ is superimposed.

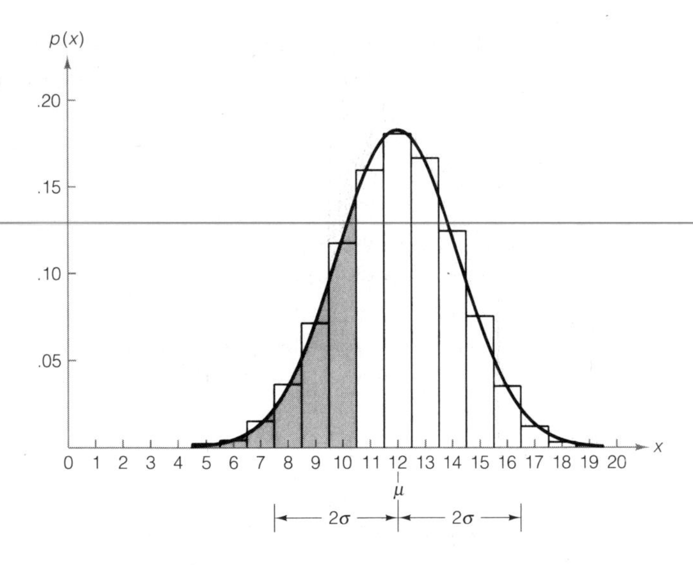

FIGURE 4.22 ▶ Binomial distribution for $n = 20$, $p = .6$ and normal distribution with $\mu = 12$, $\sigma = 2.2$

As part of Example 4.10, we used Table II to find the probability that $x \leq 10$. This probability, which is equal to the sum of the areas contained in the rectangles (shown in Figure 4.22) that correspond to $p(0), p(1), p(2), \ldots, p(10)$, was found to equal .245. The portion of the approximating normal curve that would be used to approximate the area $p(0) + p(1) + \cdots + p(10)$ is shaded in Figure 4.22. Note that this shaded area lies to the left of 10.5 (not 10), so we may include all of the probability in the rectangle corresponding to $p(10)$. Because we are approximating a discrete distribution (the binomial) with a continuous distribution (the normal), we call the use of 10.5 (instead of 10 or 11) a **correction for continuity**. That is, we are correcting the discrete distribution so that it can be approximated by the continuous one. The use of the correction for continuity leads to the calculation of the following standard normal z value:

$$z = \frac{x - \mu}{\sigma} = \frac{10.5 - 12}{2.2} = -.68$$

Using Table IV, we find the area between $z = 0$ and $z = .68$ to be .2517. Then the probability that x is less than or equal to 10 is approximated by the area under the normal distribution to the left of 10.5, shown shaded in Figure 4.22. That is,

$$\begin{aligned} P(x \leq 10) &\approx P(z \leq -.68) = .5 - P(-.68 < z \leq 0) \\ &= .5 - .2517 = .2438 \end{aligned}$$

The approximation differs only slightly from the exact binomial probability, .245. Of course, when tables of exact binomial probabilities are available, we will use the exact value rather than a normal approximation.

Use of the normal distribution will not always provide a good approximation for binomial probabilities. The following is a useful rule of thumb to determine when n is large enough for the approximation to be effective: the interval $\mu \pm 3\sigma$ should lie within the range of the binomial random variable x (i.e., 0 to n) in order for the normal approximation to be adequate. The rule works well because almost all of the normal distribution falls within 3 standard deviations of the mean, so if this interval is contained within the range of x values, there is "room" for the normal approximation to work.

As shown in Figure 4.23(a) (page 218) for the preceding example with $n = 20$ and $p = .6$, the interval $\mu \pm 3\sigma = 12 \pm 3(2.19) = (5.43, 18.57)$ lies within the range 0 to 20. However, if we were to try to use the normal approximation with $n = 10$ and $p = .1$, the interval $\mu \pm 3\sigma$ is $1 \pm 3(.95)$, or $(-1.85, 3.85)$. As shown in Figure 4.23(b), this interval is not contained within the range of x since $x = 0$ is the lower bound for a binomial random variable. Note in Figure 4.23(b) that the normal distribution will not "fit" in the range of x, and therefore it will not provide a good approximation to the binomial probabilities.

EXAMPLE 4.21

The pocket calculator has become relatively inexpensive because its solid-state circuitry is stamped by machine, thus making mass production feasible. One problem with anything that is mass-produced is quality control. The process must somehow be monitored or audited to be sure the output of the process conforms to requirements.

One method of dealing with this problem is **lot acceptance sampling**, in which items being produced are sampled at various stages of the production process and are carefully inspected. The lot of items from which the sample is drawn is then accepted or rejected, based on the number of defectives in the sample. Lots that are accepted may be sent forward for further processing or may be shipped to customers; lots that are

FIGURE 4.23 ▶
Rule of thumb for normal approximation to binomial probabilities

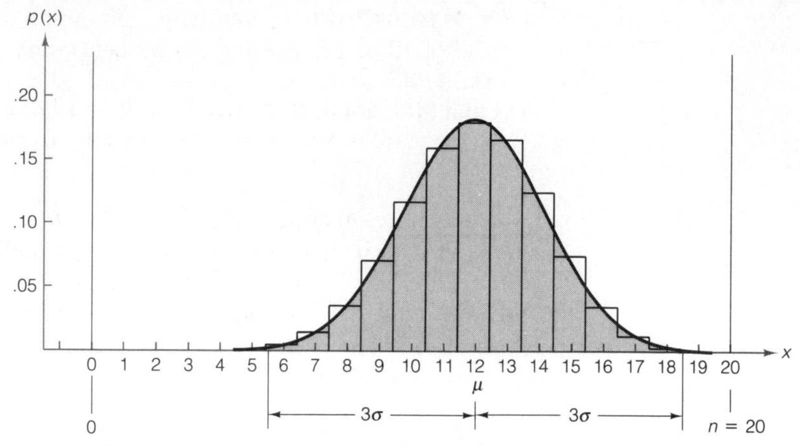

a. $n = 20$, $p = .6$: Normal approximation is good

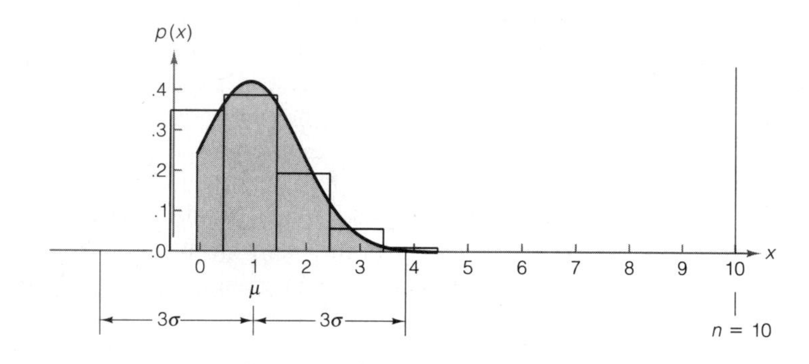

b. $n = 10$, $p = .1$: Normal approximation is poor

rejected may be reworked or scrapped. For example, suppose a manufacturer of calculators chooses 200 stamped circuits from the day's production and determines x, the number of defective circuits in the sample. Suppose that up to a 6% rate of defectives is considered acceptable for the process.

a. Find the mean and standard deviation of x, assuming the defective rate is 6%.

b. Use the normal approximation to determine the probability that 20 or more defectives are observed in the sample of 200 circuits (i.e., find the approximate probability that $x \geq 20$).

Solution

a. The random variable x is binomial with $n = 200$ and the fraction defective $p = .06$. Thus,

$$\mu = np = 200(.06) = 12$$
$$\sigma = \sqrt{npq} = \sqrt{200(.06)(.94)} = \sqrt{11.28} = 3.36$$

We first note that

$$\mu \pm 3\sigma = 12 \pm 3(3.36) = 12 \pm 10.08 = (1.92, 22.08)$$

lies completely within the range from 0 to 200. Therefore, a normal probability distribution should provide an adequate approximation to this binomial distribution.

b. To find the approximating area corresponding to $x \geq 20$, refer to Figure 4.24. Note that we want to include all the binomial probability histogram from 20 to 200, inclusive. But in order to include the entire rectangle corresponding to $x = 20$, we must begin the approximating area at $20 - .5 = 19.5$. In other words, since the event is of the form $x \geq a$, with $a = 20$, the correction for continuity is $a - .5 = 20 - .5 = 19.5$. Thus, the z value is

$$z = \frac{(a - .5) - \mu}{\sigma} = \frac{19.5 - 12}{3.36} = \frac{7.5}{3.36} = 2.23$$

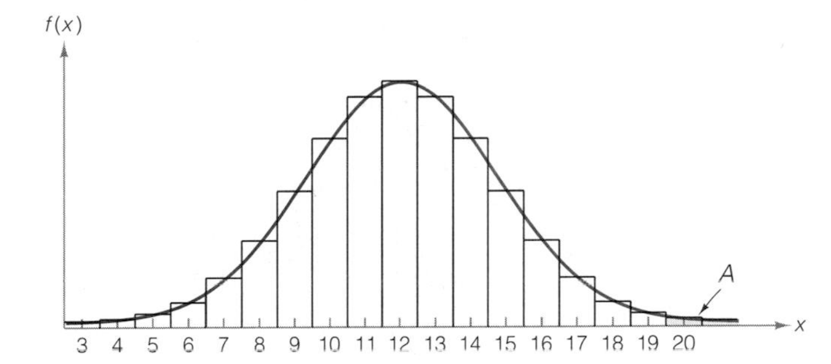

FIGURE 4.24 ▶ Normal approximation to the binomial distribution with $n = 200$, $p = .06$

Referring to Table IV in Appendix A, we find that the area to the right of the mean corresponding to $z = 2.23$ (see Figure 4.25) is .4871. So the area A is

$$A = .5 - .4871 = .0129$$

Thus, the normal approximation to the binomial probability is

$$P(x \geq 20) \approx .0129$$

In other words, the probability is extremely small that 20 or more defectives will be observed in a sample of 200 circuits—*if in fact the true fraction of defectives is* .06. If the manufacturer observes $x \geq 20$, the likely reason is that the process is producing more than the acceptable 6% defectives. The lot acceptance sampling procedure is another example of using the rare event approach to make inferences.

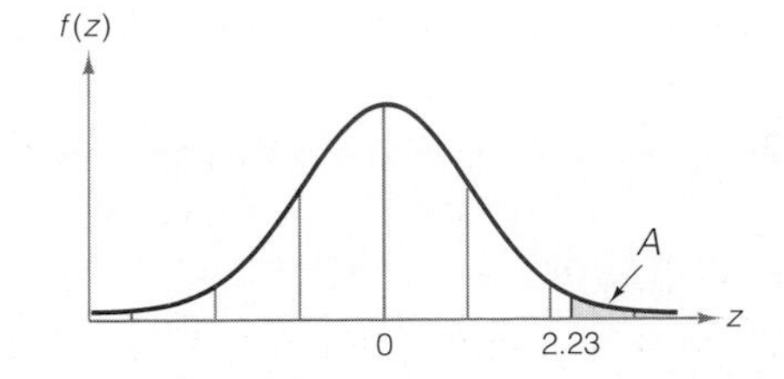

FIGURE 4.25 ▶ Standard normal distribution

The steps for approximating a binomial probability by a normal probability are given in the accompanying box.

Using a Normal Distribution to Approximate Binomial Probabilities

1. After you have determined n and p for the binomial distribution, calculate the interval

$$\mu \pm 3\sigma = np \pm 3\sqrt{npq}$$

If the interval lies in the range 0 to n, the normal distribution will provide a reasonable approximation to the probabilities of most binomial events.

2. Write the binomial probability of interest in the form $P(x \leq a)$ or $P(x \geq a)$ or $P(a \leq x \leq b)$.
3. If the binomial probability to be approximated is of the form $P(x \leq a)$, the correction for continuity is $(a + .5)$, and the approximating standard normal z value is

$$z = \frac{(a + .5) - \mu}{\sigma} \quad \text{See Figure 4.26(a).}$$

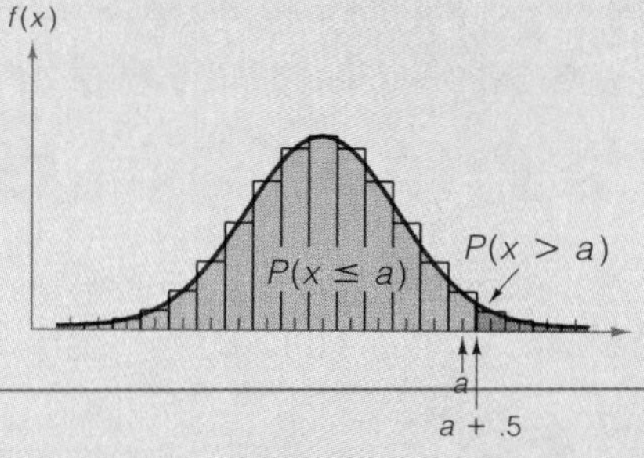

a. $P(x \leq a)$

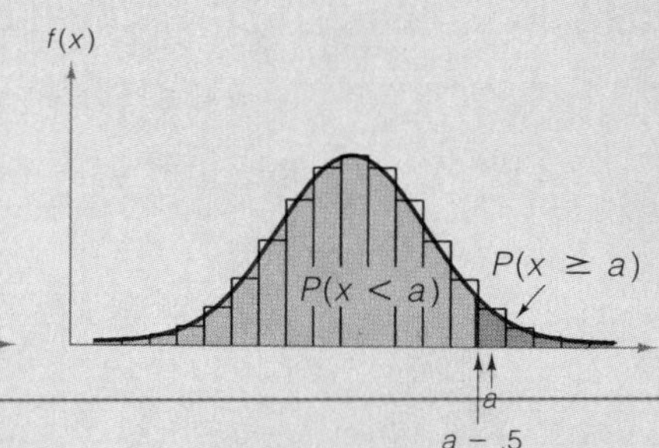

b. $P(x < a)$

FIGURE 4.26 ► Approximating binomial probabilities by normal probabilities

4. If the binomial probability to be approximated is of the form $P(x \geq a)$, the correction for continuity is $(a - .5)$, and the approximating standard normal z value is

$$z = \frac{(a - .5) - \mu}{\sigma} \quad \text{See Figure 4.26(b).}$$

5. If the binomial probability to be approximated is of the form $P(a \leq x \leq b)$, treat the ends of the interval separately, calculating two distinct z values according to step 3. [*Note:* $P(a \leq x \leq b) = P(x \leq b) - P(x \leq a - 1)$]
6. Sketch the approximating normal distribution and shade the area corresponding to the probability of the event of interest, as in Figure 4.26. Verify that the rectangles you have included in the shaded area correspond to the event probability you wish to approximate. Using Table IV and the z value(s) you calculated in steps 3–5, find the shaded area. This is the approximate probability of the binomial event.

Exercises 4.80–4.97

Learning the Mechanics

4.80 Why might you want to use a normal distribution to approximate a binomial distribution?

4.81 What conditions must be satisfied in order for the normal distribution to provide a good approximation to the binomial probability distribution?

4.82 Assume that x is a binomial random variable with n and p as specified in parts **a–f**. For which cases would it be appropriate to use a normal distribution to approximate the binomial distribution?
a. $n = 100, \quad p = .01$ b. $n = 20, \quad p = .6$
c. $n = 10, \quad p = .4$ d. $n = 1{,}000, \quad p = .05$
e. $n = 100, \quad p = .8$ f. $n = 35, \quad p = .7$

4.83 Suppose x is a binomial random variable with $p = .4$ and $n = 25$.
a. Would it be appropriate to approximate the probability distribution of x with a normal distribution? Explain.
b. Assuming that a normal distribution provides an adequate approximation to the distribution of x, what are the mean and variance of the approximating normal distribution?
c. Use Table II of Appendix A to find the exact value of $P(x \geq 9)$.
d. Use the normal approximation to find $P(x \geq 9)$.

4.84 Assume that x is a binomial random variable with $n = 25$ and $p = .5$. Use Table II of Appendix A and the normal approximation to find the exact and approximate values, respectively, for the following probabilities:
a. $P(x \leq 11)$ b. $P(x \geq 16)$ c. $P(8 \leq x \leq 16)$

4.85 Assume that x is a binomial random variable with $n = 100$ and $p = .40$. Use a normal approximation to find the following:
a. $P(x \leq 35)$ b. $P(40 \leq x \leq 50)$ c. $P(x \geq 38)$

4.86 Assume that x is a binomial random variable with $n = 1{,}000$ and $p = .50$. Find each of the following probabilities.
a. $P(x > 500)$ b. $P(490 \leq x < 500)$ c. $P(x > 550)$

Applying the Concepts

4.87 A recent article in *The International Journal of Sports Psychology* evaluates the relationship between physical fitness and stress. Assume that white-collar workers in good physical condition have only a 10% probability of developing a stress-related health problem. What is the probability that more than 60 in a random sample of 400 white-collar employees in good physical condition will develop stress-related illnesses?

4.88 The *Statistical Abstract of the United States: 1989* reports that 24% of the country's 91,061,000 households are inhabited by one person. If 1,000 randomly selected homes are to participate in a Nielsen survey to determine television ratings, find the approximate probability that no more than 250 of these homes are inhabited by one person.

4.89 In Case Study 4.2, the number of shuttle catastrophes due to booster failure in n missions was treated as a binomial random variable. Using the binomial distribution and the probability of catastrophe determined by the Air Force's risk assessment study (1/35), we determined the probability of at least 1 shuttle catastrophe in 25 missions to be .5155.

a. Based on the guidelines presented in this section, would it have been advisable to approximate this probability using the normal approximation to the binomial distribution? Explain.
b. Regardless of your answer to part **a**, use the normal distribution to approximate the binomial probability. Comment on the difference between the exact and approximate probabilities.
c. Refer to part **a**. Would the normal approximation be advisable if $n = 100$? If $n = 500$? If $n = 1{,}000$?
d. Approximate the probability that more than 25 catastrophes occur in 1,000 flights, assuming that the probability of a catastrophe in any given flight remains 1/35.

4.90 If you are attracted to lotteries, one of the best is run by the U.S. government. Furthermore, with some research you can improve your chances of winning. We refer to the Interior Department's lottery for oil and gas leases. The *Wall Street Journal* (March 29, 1984) reports on abuses in the lottery—particularly the tendency of the Interior Department to include valuable oil leases, some worth millions of dollars, among the many included in the lottery. The Interior Department claims that only 5% to 10% of the leases included in the $75 per ticket lottery should be salable to an oil company. Yet 184 of 328 winners in the July 1980 lottery of Wyoming lands were able to sell their leases to oil companies.

Suppose as many as 10% of all leases awarded by the Interior Department are salable to oil companies to answer the following questions:

a. What is the probability that as many as 50 leases (i.e., 50 or more) would be salable to oil companies?
b. If 184 of the 328 leases awarded in the July 1980 Wyoming lottery were in fact salable to oil companies, would you regard this as a rare event? Explain.
c. Given the outcome of the Wyoming lottery and considering your answer to part **b**, do you believe the Interior Department's claim? Explain.

4.91 Melanoma, a malignant form of skin cancer, strikes more than 10,000 Americans per year and kills 60% of this number (*Time*, May 30, 1983).

a. What are the expected value and variance of x, the number of the 10,000 annual melanoma patients who die of the affliction?
b. Find the probability that x will exceed 6,100 patients per year.
c. Would you expect the number x dying of melanoma to exceed 6,500 in any single year? Explain.

4.92 It is against the law to discriminate against job applicants because of race, religion, sex, or age. Of the individuals who apply for an accountant's position in a large corporation, 40% are over 45 years of age. If the company decides to choose 50 of a very large number of applicants for closer credential screening, claiming that the selection will be random and not age-biased, what is the approximate probability that fewer than 15 of those chosen are over 45 years of age? (Assume that the applicant pool is large enough so that x, the number in the sample over 45 years of age, has a binomial probability distribution.)

4.93 Recent market research reveals that an estimated 30% of U.S. households own one or more personal computers (PCs). Suppose that in a sample of 500 households in a large high-tech community, 178 own PCs.

a. Approximate the probability that 178 or more households in the sample would own PCs if in fact 30% of all households in the community own PCs.
b. Would you conclude on the basis of your answer to part **a** that more than 30% of this community's households own PCs? Explain.

4.94 A recent study involving attrition rates at a major university has shown that 43% of all incoming freshmen do not graduate within 4 years of entrance.

a. If 200 freshmen are randomly sampled this year and their progress through college is followed, what is the approximate probability that no less than half will graduate within the next 4 years?
b. What is the approximate probability that the number of sampled freshmen graduating within 4 years will be between 40 and 80?

4.95 A computer disk manufacturer claims that 99.4% of its disks are defect free. A large software company that buys and uses large numbers of the disks wants to verify this claim, and selects 1,600 disks to be tested. The tests reveal 12 disks to be defective. Assuming that the disk manufacturer's claim is correct, what is the probability of finding 12 or more defective disks in a sample of 1,600? Does your answer cast doubt on the manufacturer's claim? Explain.

4.96 The median time a patient waits to see a doctor in a large clinic is 20 minutes. On a day when 150 patients visit the clinic, what is the approximate probability that
a. More than half will have to wait more than 20 minutes?
b. More than 85 will have to wait more than 20 minutes?
c. More than 60 but less than 90 will have to wait more than 20 minutes?

4.97 The percentage of fat in the bodies of American men is an approximate normal random variable with mean equal to 15% and standard deviation equal to 2%.
a. If these values were used to describe the body fat of men in the United States Army and if 20% or more body fat is characterized as obese, what is the approximate probability that a random sample of 10,000 soldiers will contain fewer than 50 who would be characterized as obese?
b. If the army actually were to check the percentage of body fat for a random sample of 10,000 men and if only 30 contained 20% (or higher) body fat, would you conclude that the army was successful in reducing the percentage of obese men below the percentage in the general population? Explain your reasoning.

Summary

Random variables are rules that assign numerical values to the outcomes of an experiment. **Discrete random variables** have countable numbers of possible values, and **continuous random variables** can assume an uncountable number of values corresponding to the points in one or more intervals. For purposes of distinguishing the two, the values of a discrete random variable can be listed, whereas those of a continuous random variable cannot.

The **probability distribution** of a discrete random variable is a table, graph, or formula that specifies the probability associated with each possible value the random variable can assume. The probability distribution of a continuous random variable is a smooth curve that models a population relative frequency distribution. The mean and standard deviation of the probability distribution provide numerical descriptive measures of the distribution. Many applications of statistics involve the estimation of the mean and standard deviation of a probability distribution based on sample data.

The **binomial** and **Poisson** probability distributions describe two specific types of discrete random variables that have many applications in the real world. The **normal** probability distribution describes an important continuous random variable. Many relative frequency distributions of data can be approximated by this distribution. You will learn in the following chapters that the normal distribution plays a primary role in statistical inference.

The characteristics of these important random variables were presented in this chapter, along with many examples of data for which each would be an appropriate model. The formulas for the probability distributions, means, and variances were also presented, along with tables to assist you in calculating probabilities associated with each of these distributions.

Supplementary Exercises 4.98–4.132

Note: Starred (*) exercises refer to the optional section in this chapter.

Learning the Mechanics

4.98 Which of the following describe discrete random variables, and which describe continuous random variables?

a. The length of time that an exercise physiologist's program takes to elevate her client's heart rate to 140 beats per minute
b. The number of crimes committed on a college campus per year
c. The number of square feet of vacant office space in a large city
d. The number of voters who favor a new tax proposal

4.99 Suppose x is a binomial random variable. Find $p(x)$ for each of the following combinations of x, n, and p.

a. $x = 1, \quad n = 3, \quad p = .1$
b. $x = 4, \quad n = 20, \quad p = .3$
c. $x = 0, \quad n = 2, \quad p = .4$
d. $x = 4, \quad n = 5, \quad p = .5$
e. $n = 15, \quad x = 12, \quad p = .9$
f. $n = 10, \quad x = 8, \quad p = .6$

4.100 For each of the following examples, decide whether x is a binomial random variable and explain your decision:

a. A manufacturer of computer chips randomly selects 100 chips from each hour's production in order to estimate the proportion defective. Let x represent the number of defectives in the 100 sampled chips.
b. Of five applicants for a job, two will be selected. Although all applicants appear to be equally qualified, only three have the ability to fulfill the expectations of the company. Suppose that the two selections are made at random from the five applicants, and let x be the number of qualified applicants selected.
c. A software developer establishes a support hotline for customers to call in with questions regarding use of the software. Let x represent the number of calls received on the support hotline during a specified workday.
d. Florida is one of a minority of states with no state income tax. A poll of 1,000 registered voters is conducted to determine how many would favor a state income tax in light of the state's current fiscal condition. Let x be the number in the sample who would favor the tax.

4.101 Consider the discrete probability distribution shown at the top of the next page.

x	10	12	18	20
$p(x)$	.2	.3	.1	.4

a. Calculate μ, σ^2, and σ.
b. What is $P(x < 15)$?
c. Calculate $\mu \pm 2\sigma$.
d. What is the probability that x is in the interval $\mu \pm 2\sigma$?

4.102 Calculate the area under the standard normal probability distribution between the following pairs of z-scores.
a. 0 and 1.96 b. -1.96 and 1.96 c. -1.50 and 1.50
d. -2.25 and -1.25 e. .60 and 2.01 f. -1.43 and 1.95

4.103 Find the following probabilities for the standard normal random variable z.
a. $P(z \leq 2.1)$ b. $P(z \geq 2.1)$ c. $P(z \geq -1.65)$
d. $P(-2.13 \leq z \leq -.41)$ e. $P(-1.45 \leq z \leq 2.15)$ f. $P(z \leq -1.43)$

4.104 Find a z-score, call it z_0, such that
a. $P(z \leq z_0) = .8708$ b. $P(z \geq z_0) = .0526$ c. $P(z \leq z_0) = .5$
d. $P(-z_0 \leq z \leq z_0) = .8164$ e. $P(z \geq z_0) = .8023$ f. $P(z \geq z_0) = .0041$

4.105 The random variable x has a normal distribution with $\mu = 70$ and $\sigma = 10$. Find the following probabilities.
a. $P(x \leq 75)$ b. $P(x \geq 90)$ c. $P(60 \leq x \leq 75)$
d. $P(x > 75)$ e. $P(x = 75)$ f. $P(x \leq 95)$

4.106 The random variable x has a normal distribution with $\mu = 40$ and $\sigma^2 = 36$. Find a value of x, call it x_0, such that
a. $P(x \geq x_0) = .5$ b. $P(x \leq x_0) = .9911$ c. $P(x \leq x_0) = .0028$
d. $P(x > x_0) = .0228$ e. $P(x \leq x_0) = .1003$ f. $P(x \geq x_0) = .7995$

4.107 Suppose x is a binomial random variable with $n = 20$ and $p = .7$.
a. Find $P(x = 14)$. b. Find $P(x \leq 12)$. c. Find $P(x > 12)$.
d. Find $P(9 \leq x \leq 18)$. e. Find $P(8 < x < 18)$. f. Find μ, σ^2, and σ.
g. What is the probability that x is in the interval $\mu \pm 2\sigma$?

***4.108** Suppose x is a Poisson random variable. Compute $p(x)$ for each of the following cases:
a. $\lambda = 2, \quad x = 3$ b. $\lambda = 1, \quad x = 4$ c. $\lambda = .5, \quad x = 2$

4.109 Assume that x is a binomial random variable with $n = 100$ and $p = .5$. Use the normal probability distribution to approximate the following probabilities.
a. $P(x \leq 48)$ b. $P(50 \leq x \leq 65)$ c. $P(x \geq 70)$
d. $P(55 \leq x \leq 58)$ e. $P(x = 62)$ f. $P(x \leq 49 \quad \text{or} \quad x \geq 72)$

Applying the Concepts

4.110 A recent study in *The International Journal of Sports Psychology* (Aggleton and Wood, 1990) investigated the role that *handedness* plays in the sport of golf. Handedness refers to whether a player is right- or left-handed. The authors report that approximately 10.4% of the population of golfers are left-handed.

a. Suppose a random sample of 20 professional golfers is selected. Find the probability that none of the 20 professional golfers is left-handed.
b. The article reports that none of the top 100 professional golfers on the American tour in 1985 was left-handed. Assuming $p = .104$, find the probability of this occurrence.
c. Based on the probability found in part **b**, can you reach any conclusion about the proportion of all professional golfers who are left-handed? Be sure to state any assumptions you make in reaching your conclusion.

4.111 Robert M. Sellers (1992) examined the SAT scores of male student-athletes at NCAA Division I institutions. The SAT score is used to determine whether athletes are eligible to participate in athletics their freshman year, with the NCAA requiring a minimum score of 700. Suppose that SAT scores of athletes on scholarship have an average of 950 and a standard deviation of 200. Assuming the distribution of SAT scores is normal, what percentage of the athletes will not be eligible their freshman year?

4.112 Anticipating a substantial growth in sales over the next 5 years, a printing company is planning today for the warehouse space it will need 5 years hence. It obviously cannot be certain exactly how many square feet of storage space, x, it will need in 5 years, but the company can project its needs by using a probability distribution such as the following:

x	10,000	15,000	20,000	25,000	30,000	35,000
$p(x)$	.05	.15	.35	.25	.15	.05

What is the expected number of square feet of storage space the printing company will need in 5 years?

4.113 Many minor operations at a hospital can be performed the same day the patient is admitted. A hospital serving a large metropolitan area has found that 20% of newly admitted patients needing an operation are scheduled for same-day surgery. Suppose that 10 patients are randomly selected from those admitted to the hospital for surgery over the past year. If x, the number in the sample of 10 who receive same-day surgery, possesses a binomial probability distribution:
a. What is the probability that exactly five of these patients have same-day surgery?
b. What is the probability that at most one has same-day surgery?
c. If the 10 patients are selected from the admissions on a single given day, is it reasonable to expect x to possess the characteristics of a binomial random variable? That is, is this a binomial experiment? Explain.

4.114 Farmers often sell fruits and vegetables at roadside stands during the summer. One such roadside stand has a daily demand for tomatoes that is approximately normally distributed with a mean equal to 125 tomatoes per day and a standard deviation equal to 30 tomatoes per day.
a. If there are 90 tomatoes available to be sold at the roadside stand at the beginning of a day, what is the probability that they will all be sold?
b. If there are 200 tomatoes available to be sold, what is the probability that 50 or more will not be sold that day?
c. How many tomatoes must be available on any given day so that there will be only a 10% chance that all tomatoes will be sold?

4.115 For a student to graduate, a high school requires that each student demonstrate competence in mathematics by scoring 70% or above on a mathematics achievement test. The scores of those students taking the test for the first time are normally distributed with a mean of 77% and a standard deviation of 7.3%. What percentage of students who take the test for the first time will pass the test?

4.116 In a 1986 article in the *Journal of Applied Psychology*, J. Near and M. Miceli report the results of an extensive survey conducted to determine the extent of "whistle blowing" among federal employees and to study the factors that are associated with retaliation against whistle blowers. Whistle blowing refers to an employee's reporting of wrongdoing by coworkers. Among other things, the survey found that about 5% of employees contacted had reported wrongdoing during the past 12 months. Assume that a sample of 25 employees in one agency are contacted, and let x be the number who have observed and reported wrongdoing in the last 12 months. Assume that the probability of whistle blowing is .05 for any federal employee over the past 12 months.

a. Find the mean and standard deviation of x. Can x be equal to its expected value? Explain.
b. Write the event that at least five of the employees are whistle blowers in terms of x. Find the probability of the event.
c. If five of the 25 contacted have been whistle blowers over the past 12 months, what would you conclude about the applicability of the 5% assumption to this agency? Use your answer to part **b** to justify your conclusion.

4.117 On the average, the main chute fails in one of every 1,000 parachutes. Suppose during a lifetime a professional parachutist makes 4,000 jumps, and let x equal the number of times the main chute fails. What is the approximate probability that the parachutist's main chute fails on at least one jump? [*Note:* Because n is large and p is so small, the Poisson probability distribution will also provide a good approximation to this probability. (See Exercise 4.50.) If you covered Section 4.4, find the Poisson approximation to $P(x > 0)$.]

4.118 The owner of construction company A makes bids on jobs so that if awarded the job, company A will make a $10,000 profit. The owner of construction company B makes bids on jobs so that if awarded the job, company B will make a $15,000 profit. Each company describes the probability distribution of the number of jobs the company is awarded per year as shown in the table.

Company A		*Company B*	
2	.05	2	.15
3	.15	3	.30
4	.20	4	.30
5	.35	5	.20
6	.25	6	.05

a. Find the expected number of jobs each will be awarded in a year.
b. What is the expected profit for each company?
c. Find the variance and standard deviation of the distribution of number of jobs awarded per year for each company.
d. Graph $p(x)$ for both companies A and B. For each company, what proportion of the time will x fall in the interval $\mu \pm 2\sigma$?

4.119 The length of time required to assemble a photoelectric cell is normally distributed with $\mu = 18.1$ minutes and $\sigma = 1.3$ minutes. What is the probability that it will require more than 20 minutes to assemble a cell?

4.120 A local track club has decided to sponsor a 10,000-meter road race. From past results of races across the state, it is known that the length of time to complete the race has an approximately normal distribution with a mean of 49 minutes and a standard deviation of 8 minutes. The club has decided that everyone who completes the race will receive a T-shirt. Those who run between 34 and 60 minutes will also receive a medal, and those finishing under 34 minutes will receive a plaque.

a. What proportion of racers would you expect to receive a medal and a T-shirt?
b. What proportion of racers would you expect to receive a plaque and a T-shirt?

4.121 In recent years, the use of the telephone as a data collection instrument for public opinion polls has been steadily increasing. However, one of the major factors bearing on the extent to which the telephone will become an acceptable data collection tool in the future is the refusal rate, i.e., the percentage of the eligible subjects actually contacted who refuse to take part in the poll. Suppose that past records indicate a refusal rate of 20% in a large city. A poll of 25 residents is to be taken and x is the number of residents contacted by telephone who refuse to take part in the poll.

a. Find the mean and variance of x.
b. Find $P(x \leq 5)$.
c. Find $P(x > 10)$.

4.122 The efficacy of insecticides is often measured by the dose necessary to kill a certain percentage of insects. Suppose a certain dose of a new insecticide is supposed to kill 80% of the exposed insects. To test the claim, 25 insects are put in contact with the insecticide.

a. If the insecticide really kills 80% of the exposed insects, what is the probability that fewer than 15 die?
b. If you observed such a result, what would you conclude about the new insecticide? Explain your logic.

4.123 The net weight per package of a certain brand of corn chips is listed as 10 ounces. The weight actually delivered to each package by an automated machine is a normal random variable with mean 10.5 ounces and standard deviation .2 ounce. Suppose 100 packages are chosen at random and the net weights are ascertained. Let x be the number of the 100 selected packages that contain at least 10 ounces of corn chips. Then x is a binomial random variable with $n = 100$ and $p =$ probability that a randomly selected package contains at least 10 ounces. What is the probability that they all contain at least 10 ounces of corn chips? What is the probability that at least 90% of the packages contain 10 ounces or more?

4.124 Sixteen percent of the American black population is known to suffer from sickle-cell anemia. If 1,000 American black people are sampled at random, what is the approximate probability that:

a. More than 175 have the disease?
b. Fewer than 140 have the disease?
c. The number of people in the sample with the disease is between 130 and 180 inclusive?

4.125 A physical fitness specialist claims that the probability is greater than .5 that an average adult male can improve his physical condition by spending 5 minutes per day on a certain exercise program. To test the claim, 15 randomly selected adult males follow the program for a specified amount of time. Maximal oxygen uptake is measured before and after the program for each male and serves as the criterion for assessing physical condition. If the program is not really beneficial (i.e., the probability of improvement is only .5), what is the probability that 11 or more of the 15 men have improved maximal oxygen uptake?

Suppose 11 or more showed increased maximal oxygen uptake. Assuming that the specialist's exercise program is ineffective, would you regard $x \geq 11$ as a rare event, or would you conclude that, in actuality, the probability of improvement exceeds $p = \frac{1}{2}$ and that the program is effective?

4.126 A company has a lump-sum incentive plan for salespeople that is dependent on their level of sales. If they sell less than \$100,000 per year, they receive a \$1,000 bonus; from \$100,000 to \$200,000, they receive

\$5,000; and above \$200,000, they receive \$10,000. Suppose the annual sales per salesperson has approximately a normal distribution with $\mu = \$180,000$ and $\sigma = \$50,000$.

a. Find p_1, the proportion of salespeople who receive a \$1,000 bonus.
b. Find p_2, the proportion of salespeople who receive a \$5,000 bonus.
c. Find p_3, the proportion of salespeople who receive a \$10,000 bonus.
d. What is the mean value of the bonus payout for the company? [*Hint:* See the definition for the expected value of a random variable.]

***4.127** A small life insurance company has determined that on the average it receives five death claims per day.

a. What is the probability that the company will receive three claims or less on a particular day?
b. What is the probability that the company will receive exactly five claims on a particular day?
c. What assumptions must you make to find these probabilities?

4.128 It is quite common for the standard deviation of a random variable to increase proportionally as the mean increases. When this occurs, the **coefficient of variation,**

$$\text{CV} = \frac{\sigma}{\mu}$$

which is the ratio of σ to μ, is the **proportionality constant**. To illustrate, the error (in dollars) in assessing the value of a house increases as the house increases in value. Suppose that long experience with assessors in a given region has shown that the coefficient of variation is .08 and that the probability distribution of assessed valuations on the same house by many different assessors is approximately normal with a mean we will call the "true value" of the house. Suppose the true value of your house is \$50,000 and it is being assessed for taxation purposes. What is the probability the assessor will assess your house in excess of \$55,000?

4.129 An admissions officer for a law school indicates that 35% of the applicants meet all 10 requirements and 95% meet at least eight of the 10 requirements.

a. If a random sample of 300 applicants is taken, what is the approximate probability that fewer than 250 will fail to meet all 10 requirements?
b. What is the approximate probability that more than 280 will meet at least eight of the 10 requirements?

4.130 Contrary to our intuition, very reliable decisions concerning the proportion of a large group of consumers favoring a certain product or a certain social issue can be based on relatively small samples. For example, suppose the target population of consumers contains 50 million people and we wish to decide whether the proportion of consumers, p, in the population that favor some product (or issue) is as large as some value, say .2. Suppose you randomly select a sample as small as 1,600 from the 50 million and you observe the number, x, of consumers in the sample who favor the new product. Assuming that $p = .2$, find the mean and standard deviation of x. Suppose that 400 or (25%) of the sample of 1,600 consumers favor the new product. Why might this sample result lead you to conclude that p (the proportion of consumers favoring the product in the population of 50 million) is at least as large as .2? [*Hint:* Find the values of μ and σ for $p = .2$, and use them to decide whether the observed value of x is unusually large.]

***4.131** An emergency rescue vehicle is used an average of 1.3 times daily.

a. What is the probability that the vehicle will be used exactly twice tomorrow?
b. What is the probability that it will be used more than twice?
c. Exactly three times?

***4.132** The Poisson probability distribution, like the binomial, can be approximated by a normal probability distribution. This approximation, using $\mu = \lambda$ and $\sigma = \sqrt{\lambda}$, will be good when λ is large (large enough so that the distance between $x = 0$ and λ is at least $3\sigma = 3\sqrt{\lambda}$, or $\lambda \geq 9$). The number of union complaints per month at a certain manufacturing plant has a Poisson probability distribution with $\lambda = 40$ complaints per month. Use the normal approximation to the Poisson probability distribution.

a. Approximate the probability that the number of complaints in a given month will be less than 35.

b. Approximate the probability that the number of complaints in a given month exceeds 40.

c. If the mean number of complaints per month remains constant and if the number of complaints in 1 month is independent of the number in any other, what is the probability that in each of 3 successive months, the number of complaints exceeds 40?

On Your Own

For large values of n the computational effort involved in working with the binomial probability distribution is considerable. Fortunately, in many instances the normal distribution provides a good approximation to the binomial distribution. This exercise was designed to enable you to demonstrate to yourself how well the normal distribution approximates the binomial distribution.

a. Let the random variable x have a binomial probability with $n = 10$ and $p = .5$. Using the binomial distribution, find the probability that x takes on a value in each of the following intervals: $\mu \pm \sigma$, $\mu \pm 2\sigma$, and $\mu \pm 3\sigma$.

b. Find the probabilities requested in part **a** by using a normal approximation to the given binomial distribution.

c. Determine the magnitude of the difference between each of the three probabilities as determined by the binomial distribution and by the normal approximation.

d. Letting x have a binomial distribution with $n = 20$ and $p = .5$, repeat parts **a**, **b**, and **c**. Notice that the probability estimates provided by the normal distribution are more accurate for $n = 20$ than for $n = 10$.

e. Letting x have a binomial distribution with $n = 20$ and $p = .01$, repeat parts **a**, **b**, and **c**. Notice that the probability estimates provided by the normal distribution are very poor in this case. Explain why this occurs.

Using the Computer

Calculate the mean percentage of college graduates over all the zip codes for one of the census regions listed in Appendix B. Round the mean to the nearest integer, and use the result as an estimate of the percentage of college graduates among all people at least 25 years old in the region.

a. Suppose 20 individuals from the region respond to an advertisement by an employment agency. If these people represent a random sample of all individuals age 25 or older in the region, what is the probability that more than half of them have college degrees?

b. If 2,000 individuals respond, find the mean μ and standard deviation σ of the number of them who have college degrees. Calculate $\mu \pm 2\sigma$ and $\mu \pm 3\sigma$, and use Chebyshev's rule and the Empirical Rule to estimate the probability that the number of college degree responses falls in each of the intervals. Based on your answers, assess the likelihood that more than half the applicants will have college degrees.

c. What are the potential problems with using the mean percentage of college graduates as an estimate of the percentage for the region? How could the estimate be improved?

d. Use the normal approximation to the binomial to estimate the probability that fewer than 10% of 20 applicants (still assuming they represent a random sample) have a college education. Compare your answer with the exact binomial probability of the same event.

e. Use the normal approximation to the binomial to estimate the probability that fewer than 10% of 200 applicants have a college education. If your statistical software package has a function that calculates exact binomial probabilities, use it to calculate the same probability you approximated, and compare the results.

References

Aggleton, J. P., and Wood, C. J. "Is there a left-handed advantage in ballistic sports?" *International Journal of Sports Psychology*, Jan.–Mar. 1990, 21, pp. 46–57.

Biddle, W. "What destroyed *Challenger*?" *Discover*, Apr. 1986, pp. 40–47.

Bresee, C. W. "On 'grading on the curve'." *Clearing House*, Nov. 1976, 50, pp. 108–110.

Dessler, G. "Personnel tests gain in popularity," *St. Paul Pioneer Press and Dispatch*, March 17, 1986.

"'83 report put booster accident as most likely." *Minneapolis Star and Tribune*, Feb. 11, 1986.

Greenstein, F. I. "The benevolent leader revisited: Children's images of political leaders in three democracies." *American Political Science Review*, Dec. 1975, 69, pp. 1371–1398.

Hogg, R. V., and Craig, A. T. *Introduction to Mathematical Statistics*, 4th ed. New York: Macmillan, 1978, Chapter 1.

McKean, K. "They fly in the face of danger." *Discover*, Apr. 1986, pp. 48–58.

Mendenhall, W., Scheaffer, R. L., and Wackerly, D. *Mathematical Statistics with Applications*, 4th ed. Boston: PWS-Kent, 1990.

Near, J. P., and Miceli, M. P. "Retaliation against whistle blowers: Predictors and effects." *Journal of Applied Psychology*, 1986, Vol. 71, No. 1, pp. 137–145.

Robinson, W. V. "NASA blamed for shuttle disaster." *Boston Globe*, June 10, 1986.

Sellers, R. M. "Racial differences in the predictors for academic achievement of student-athletes in Division I revenue-producing sports." *Sociology of Sport Journal*, 1992, 9, pp. 48–59.

Tucker, L. A. "Physical fitness and psychological distress." *The International Journal of Sports Psychology*, July–Sept. 1990, 21, pp. 185–201.

Van de Werf, F., et al. "Coronary thrombosis with tissue-type plasminogen activator in patients with evolving myocardial infarction." *New England Journal of Medicine*, 1984, p. 310.

CHAPTER FIVE

Sampling Distributions

Contents

Where We've Been

Because sample measurements are observed values of random variables, the value for a sample statistic will vary in a random manner from sample to sample. In other words, since sample statistics are random variables, they therefore possess probability distributions that are either discrete or continuous. These probability distributions, called *sampling distributions* because they characterize the distribution of values of the various statistics over a very large number of samples, are the topic of this chapter. In particular, you will learn why many sampling distributions tend to be approximately normal, and you will see how sampling distributions can be used to evaluate the reliability of inferences made using the statistics. Then in subsequent chapters we will show how to use sampling distributions to make inferences about populations.

Where We're Going

We have learned in earlier chapters that the objective of most statistical investigations is inference—that is, making decisions or predictions about a population based on information in a sample. To actually make the decision, we use the sample data to compute sample statistics, such as the sample mean or variance. The knowledge of random variables and their probability distributions enables us to construct theoretical models of populations.

5

In Chapter 4 we assumed that we knew the probability distribution of a random variable, and using this knowledge we were able to compute the mean, variance, and probabilities associated with the random variable. However, in most practical applications, this information is not available. To illustrate, in Example 4.10 we calculated the probability that the binomial random variable x, the number of 20 polled voters who favor a certain mayoral candidate, assumed specific values. To do this, it was necessary to assume some value for p, the proportion of all voters who favor the candidate. Thus, for the purposes of illustration, we assumed $p = .6$ when, in all likelihood, the exact value of p would be unknown. In fact, the probable purpose of taking the poll is to estimate p. Similarly, when we modeled the in-city gas mileage of a certain automobile model, we used the normal probability distribution with an *assumed* mean and standard deviation of 27 and 3 miles per gallon, respectively. In most situations, the true mean and standard deviation are unknown quantities that would have to be estimated. Numerical quantities that describe probability distributions are called **parameters**. Thus, p, the probability of a success in a binomial experiment, and μ and σ, the mean and standard deviation of a normal distribution, are examples of parameters.

Definition 5.1

A **parameter** is a numerical descriptive measure of a population. Because it is based on the observations in the population, its value is almost always unknown.

We have also discussed the sample mean, $\bar{x}$, sample variance s^2, sample standard deviation s, etc., which are numerical descriptive measures calculated from the sample. We will often use the information contained in these **sample statistics** to make inferences about the parameters of a population.

Definition 5.2

A **sample statistic** is a numerical descriptive measure of a sample. It is calculated from the observations in the sample.

Note that the term *statistic* refers to a *sample* quantity and the term *parameter* refers to a *population* quantity.

Before we can show you how to use sample statistics to make inferences about population parameters, we need to be able to evaluate their properties. Does one sample statistic contain more information than another about a population parameter? On what basis should we choose the "best" statistic for making inferences about a parameter? The purpose of this chapter is to answer these questions.

5.1 What Is a Sampling Distribution?

If we want to estimate a parameter of a population—say, the population mean μ—there are a number of sample statistics that could be used for the estimate. Two possibilities are the sample mean $\bar{x}$ and the sample median m. Which of these do you think will provide a better estimate of μ?

Before answering this question, consider the following example: Toss a fair die, and let x equal the number of dots showing on the up face. Suppose the die is tossed three times, producing the sample measurements 2, 2, 6. The sample mean is $\bar{x} = 3.33$ and the sample median is $m = 2$. Since the population mean of x is $\mu = 3.5$, you can see that for this sample of three measurements, the sample mean $\bar{x}$ provides an estimate that falls closer to μ than does the sample median [see Figure 5.1(a)]. Now suppose we toss the die three more times and obtain the sample measurements 3, 4, 6. The mean and median of this sample are $\bar{x} = 4.33$ and $m = 4$, respectively. This time m is closer to μ [see Figure 5.1(b)].

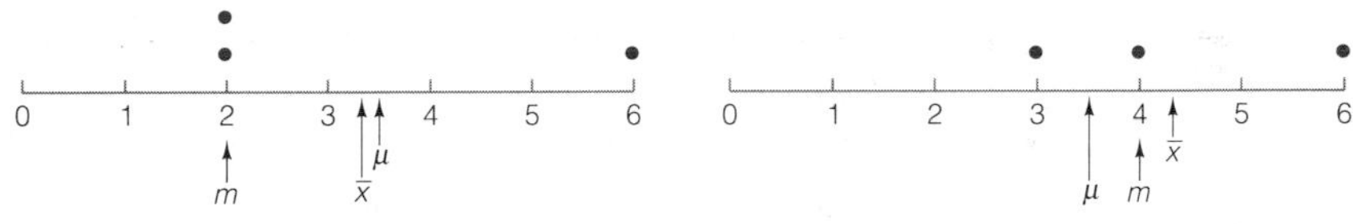

FIGURE 5.1 ► Comparing the sample mean ($\bar{x}$) and sample median (m) as estimators of the population mean (μ)

This simple example illustrates an important point: Neither the sample mean nor the sample median will *always* fall closer to the population mean. Consequently, we cannot compare these two sample statistics, or, in general, any two sample statistics, on the basis of their performance for a single sample. Instead, we need to recognize that sample statistics are themselves random variables, because different samples can lead to different values for the sample statistics. As random variables, sample statistics must be judged and compared on the basis of their probability distributions, i.e., the *collection* of values and associated probabilities of each statistic that would be obtained if the sampling experiment were repeated a *very large number of times*. We will illustrate this concept with an example.

Suppose it is known that in a certain part of Canada the daily high temperature recorded for all past months of January has a mean of $\mu = 10°F$ and a standard deviation of $\sigma = 5°F$. Consider an experiment consisting of randomly selecting 25 daily high temperatures from the records of past months of January and calculating the sample mean $\bar{x}$. If this experiment were repeated a very large number of times, the value of $\bar{x}$ would vary from sample to sample. For example, the first sample of 25 temperature measurements might have a mean $\bar{x} = 9.8$, the second sample a mean $\bar{x} = 11.4$, the third sample a mean $\bar{x} = 10.5$, etc. If the sampling experiment were repeated a very large number of times, the resulting histogram of sample means would be approximately the probability distribution of $\bar{x}$. If $\bar{x}$ is a good estimator of μ, we would expect the values of $\bar{x}$ to cluster around μ as shown in Figure 5.2. This probability distribution is called a **sampling distribution** because it is generated by repeating a sampling experiment a very large number of times.

FIGURE 5.2 ►
Sampling distribution for $\bar{x}$ based on a sample of $n = 25$ measurements

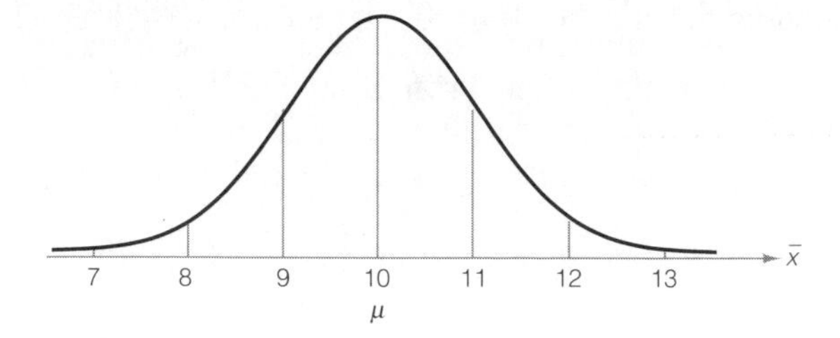

Definition 5.3

The **sampling distribution** of a sample statistic calculated from a sample of n measurements is the probability distribution of the statistic.

In actual practice, the sampling distribution of a statistic is obtained mathematically or (at least approximately) by simulating the sample on a computer using a procedure similar to that just described.

If $\bar{x}$ has been calculated from a sample of $n = 25$ measurements selected from a population with mean $\mu = 10$ and standard deviation $\sigma = 5$, the sampling distribution (Figure 5.2) provides information about the behavior of $\bar{x}$ in repeated sampling. For example, the probability that you will draw a sample of 25 measurements and obtain a value of $\bar{x}$ in the interval $9 \leq \bar{x} \leq 10$ will be the area under the sampling distribution over that interval.

Since the properties of a statistic are typified by its sampling distribution, it follows that to compare two sample statistics you compare their sampling distributions. For example, if you have two statistics, A and B, for estimating the same parameter (for purposes of illustration, suppose the parameter is the population variance σ^2) and if their sampling distributions are as shown in Figure 5.3, you would choose statistic A in preference to statistic B. You would make this choice because the sampling distribution for statistic A centers over σ^2 and has less spread (variation) than the sampling distribution for statistic B. When you draw a single sample in a practical sampling situation, the probability is higher that statistic A will fall nearer σ^2.

FIGURE 5.3 ►
Two sampling distributions for estimating the population variance, σ^2

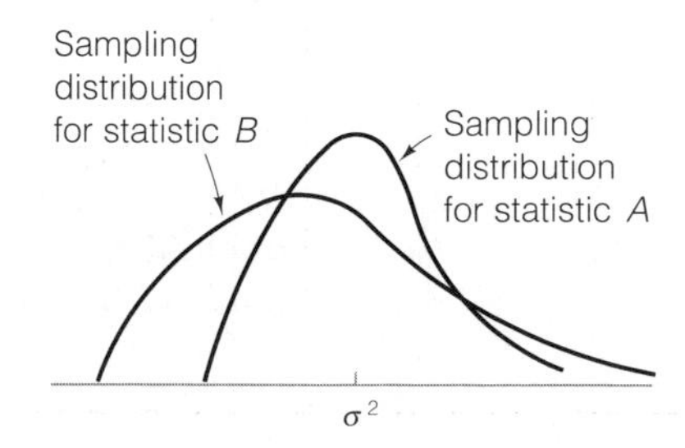

Remember that in practice we will not know the numerical value of the unknown parameter σ^2, so we will not know whether statistic A or statistic B is closer to σ^2 for a sample. We have to rely on our knowledge of the theoretical sampling distributions to choose the best sample statistic and then use it sample after sample. The procedure for finding the sampling distribution for a statistic is demonstrated in Example 5.1.

EXAMPLE 5.1

Consider a population consisting of the measurements 0, 3, and 12 and described by the probability distribution shown here. A random sample of $n = 3$ measurements is selected from the population.

x	0	3	12
$p(x)$	1/3	1/3	1/3

a. Find the sampling distribution of the sample mean $\bar{x}$.

b. Find the sampling distribution of the sample median m.

Solution

Every possible sample of $n = 3$ measurements is listed in Table 5.1 along with the sample mean and median. Also, because any one sample is as likely to be selected as any other (random sampling), the probability of observing any particular sample is 1/27. The probability is also listed in Table 5.1.

TABLE 5.1

Possible Samples	$\bar{x}$	m	Probability	Possible Samples	$\bar{x}$	m	Probability
0, 0, 0	0	0	1/27	3, 3, 12	6	3	1/27
0, 0, 3	1	0	1/27	3, 12, 0	5	3	1/27
0, 0, 12	4	0	1/27	3, 12, 3	6	3	1/27
0, 3, 0	1	0	1/27	3, 12, 12	9	12	1/27
0, 3, 3	2	3	1/27	12, 0, 0	4	0	1/27
0, 3, 12	5	3	1/27	12, 0, 3	5	3	1/27
0, 12, 0	4	0	1/27	12, 0, 12	8	12	1/27
0, 12, 3	5	3	1/27	12, 3, 0	5	3	1/27
0, 12, 12	8	12	1/27	12, 3, 3	6	3	1/27
3, 0, 0	1	0	1/27	12, 3, 12	9	12	1/27
3, 0, 3	2	3	1/27	12, 12, 0	8	12	1/27
3, 0, 12	5	3	1/27	12, 12, 3	9	12	1/27
3, 3, 0	2	3	1/27	12, 12, 12	12	12	1/27
3, 3, 3	3	3	1/27				

a. From Table 5.1 you can see that $\bar{x}$ can assume the values 0, 1, 2, 3, 4, 5, 6, 8, 9, and 12. Because $\bar{x} = 0$ occurs in only one sample, $P(\bar{x} = 0) = 1/27$. Similarly, $\bar{x} = 1$ occurs in three samples: (0, 0, 3), (0, 3, 0), and (3, 0, 0). Therefore, $P(\bar{x} = 1) = 3/27 = 1/9$. Calculating the probabilities of the remaining values of $\bar{x}$ and arranging them in a table, we obtain the probability distribution shown here.

$\bar{x}$	0	1	2	3	4	5	6	8	9	12
$p(\bar{x})$	1/27	3/27	3/27	1/27	3/27	6/27	3/27	3/27	3/27	1/27

This is the sampling distribution for $\bar{x}$ because it specifies the probability associated with each possible value of $\bar{x}$.

b. In Table 5.1 you can see that the median m can assume one of the three values 0, 3, or 12. The value $m = 0$ occurs in seven different samples. Therefore, $P(m = 0) = 7/27$. Similarly, $m = 3$ occurs in 13 samples and $m = 12$ occurs in seven samples. Therefore, the probability distribution (i.e., the sampling distribution) for the median m is as shown below.

m	0	3	12
$p(m)$	7/27	13/27	7/27

Example 5.1 demonstrates the procedure for finding the exact sampling distribution of a statistic when the number of different samples that could be selected from the population is relatively small. In the real world, populations often consist of a large number of different values, making samples difficult (or impossible) to enumerate. When this situation occurs, we may choose to obtain the approximate sampling distribution for a statistic by simulating the sampling over and over again and recording the proportion of times different values of the statistic occur. Example 5.2 illustrates this procedure.

EXAMPLE 5.2

Suppose we perform the following experiment over and over again: Take a sample of 11 measurements from the probability distribution shown in Figure 5.4. [This distribution is known as a *uniform* distribution with mean $\mu = .5$.] Calculate the two sample statistics

$$\bar{x} = \text{Sample mean} = \frac{\sum x}{11}$$

m = Median = Sixth sample measurement when the 11 measurements are arranged in ascending order

Obtain approximations to the sampling distributions of $\bar{x}$ and m.

FIGURE 5.4 ►
Uniform distribution from 0 to 1

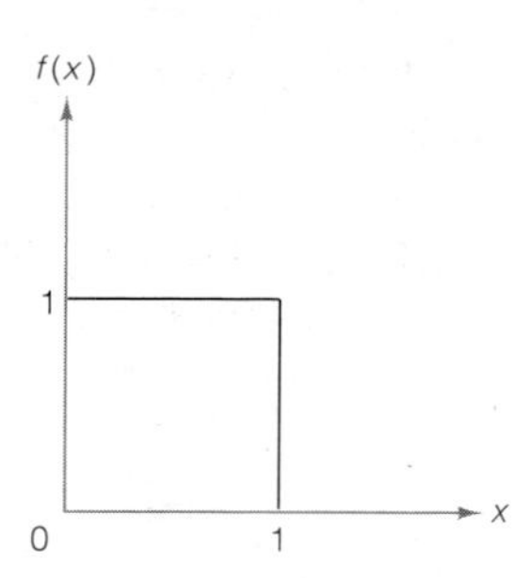

Solution

We use a computer to generate 1,000 samples, each with $n = 11$ observations. Then we compute $\bar{x}$ and m for each sample. Our goal is to obtain approximations to the sampling distributions of $\bar{x}$ and m and to find out which sample statistic ($\bar{x}$ or m) contains more information about μ. (Note that, in this particular example, we *know* the population mean is $\mu = .5$.) The first 10 of the 1,000 samples generated are presented in Table 5.2. For example, the first computer-generated sample from the uniform distribution (arranged in ascending order) contained the following measurements: .125, .138, .139, .217, .419, .506, .516, .757, .771, .786, and .919. The sample mean $\bar{x}$ and median m computed for this sample are:

$$\bar{x} = \frac{.125 + .138 + \cdots + .919}{11} = .481$$

$$m = \text{Sixth ordered measurement} = .506$$

TABLE 5.2 First 10 Samples of $n = 11$ Measurements from a Uniform Distribution

Sample	Measurements										
1	.217	.786	.757	.125	.139	.919	.506	.771	.138	.516	.119
2	.303	.703	.812	.650	.848	.392	.988	.469	.632	.012	.065
3	.383	.547	.383	.584	.098	.676	.091	.535	.256	.163	.390
4	.218	.376	.248	.606	.610	.055	.095	.311	.086	.165	.665
5	.144	.069	.485	.739	.491	.054	.953	.179	.865	.429	.648
6	.426	.563	.186	.896	.628	.075	.283	.549	.295	.522	.674
7	.643	.828	.465	.672	.074	.300	.319	.254	.708	.384	.534
8	.616	.049	.324	.700	.803	.399	.557	.975	.569	.023	.072
9	.093	.835	.534	.212	.201	.041	.889	.728	.466	.142	.574
10	.957	.253	.983	.904	.696	.766	.880	.485	.035	.881	.732

The relative frequency histograms for $\bar{x}$ and m for the 1,000 samples of size $n = 11$ are shown in Figure 5.5 (page 240).

You can see that the values of $\bar{x}$ tend to cluster around μ to a greater extent than do the values of m. Thus, on the basis of the observed sampling distributions, we conclude that $\bar{x}$ contains more information about μ than m does—at least for samples of $n = 11$ measurements from the uniform distribution.

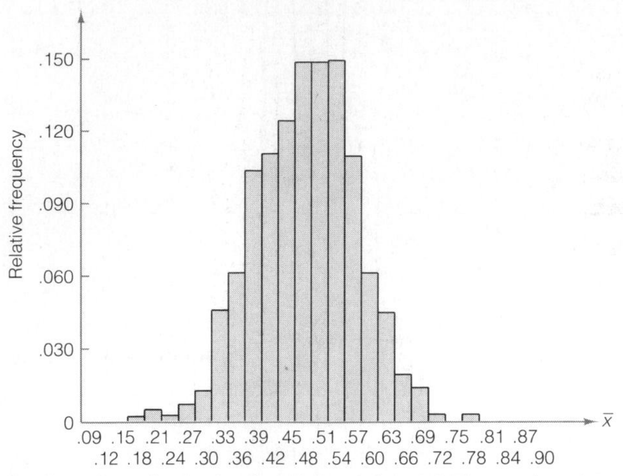

a. Sampling distribution for $\bar{x}$ (based on 1,000 samples of $n = 11$ measurements)

.090
.060
.030
0
.09 .15 .21 .27 .33 .39 .45 .51 .57 .63 .69 .75 .81 .87
.12 .18 .24 .30 .36 .42 .48 .54 .60 .66 .72 .78 .84 .90
m

b. Sampling distribution for m (based on 1,000 samples of $n = 11$ measurements)

FIGURE 5.5 ▲ Relative frequency histograms for $\bar{x}$ and m, Example 5.2

As noted earlier, many sampling distributions can be derived mathematically, but the theory necessary to do this is beyond the scope of this text. Consequently, when we need to know the properties of a statistic, we will present its sampling distribution and simply describe its properties. Several of the important properties we look for in sampling distributions are discussed in the next section.

Exercises 5.1 – 5.7

Note: Starred (*) exercises require the use of a computer.

5.1 The probability distribution shown here describes a population of measurements that can assume values of 0, 2, 4, and 6, each of which occurs with the same relative frequency:

x	0	2	4	6
$p(x)$	1/4	1/4	1/4	1/4

a. List all the different samples of $n = 2$ measurements that can be selected from this population.
b. Calculate the mean of each different sample listed in part **a**.
c. If a sample of $n = 2$ measurements is randomly selected from the population, what is the probability that a specific sample will be selected?
d. Assume that a random sample of $n = 2$ measurements is selected from the population. List the different values of $\bar{x}$ found in part **b**, and find the probability of each. Then give the sampling distribution of the sample mean $\bar{x}$ in tabular form.
e. Construct a probability histogram for the sampling distribution of $\bar{x}$.

5.2 Simulate sampling from the population (Exercise 5.1) by marking the values of x, one on each of four identical coins (or poker chips, etc.). Place the coins (marked 0, 2, 4, and 6) into a bag, randomly select one, and observe its value. Replace this coin, draw a second coin, and observe its value. Finally, calculate the mean $\bar{x}$ for this sample of $n = 2$ observations randomly selected from the population (Exercise 5.1). Replace the coins, mix, and using the same procedure, select a sample of $n = 2$ observations from the population. Record the numbers and calculate $\bar{x}$ for this sample. Repeat this sampling process until you acquire 100 values of $\bar{x}$. Construct a relative frequency distribution for these 100 sample means. This distribution will be an approximation to the exact sampling distribution of $\bar{x}$ found in part **e** of Exercise 5.1. Compare the two distributions. The distribution obtained in this exercise will not be exactly the same as the exact sampling distribution (Exercise 5.1, part **e**).

If you were to repeat the sampling procedure, drawing two coins not 100 times but 10,000 times, the relative frequency distribution for the 10,000 sample means would be almost identical to the sampling distribution of $\bar{x}$ found in Exercise 5.1, part **e**.

5.3 Consider the population described by the probability distribution shown here.

x	1	2	3	4	5
$p(x)$	.2	.3	.2	.2	.1

The random variable x is observed twice. If these observations are independent, verify that the different samples of size 2 and their probabilities are as shown here.

Sample	*Probability*	*Sample*	*Probability*	*Sample*	*Probability*
1, 1	.04	3, 1	.04	5, 1	.02
1, 2	.06	3, 2	.06	5, 2	.03
1, 3	.04	3, 3	.04	5, 3	.02
1, 4	.04	3, 4	.04	5, 4	.02
1, 5	.02	3, 5	.02	5, 5	.01
2, 1	.06	4, 1	.04		
2, 2	.09	4, 2	.06		
2, 3	.06	4, 3	.04		
2, 4	.06	4, 4	.04		
2, 5	.03	4, 5	.02		

a. Find the sampling distribution of the sample mean $\bar{x}$.
b. Construct a probability histogram for the sampling distribution of $\bar{x}$.
c. What is the probability that $\bar{x}$ is 4.5 or larger?
d. Would you expect to observe a value of $\bar{x}$ equal to 4.5 or larger? Explain.

5.4 Refer to Exercise 5.3 and find $E(x) = \mu$. Then use the sampling distribution of $\bar{x}$ found in Exercise 5.3 to find the expected value of $\bar{x}$. Note that $E(\bar{x}) = \mu$.

5.5 Refer to Exercise 5.3. Assume that a random sample of $n = 2$ measurements is randomly selected from the population.

a. List the different values that the sample median m may assume and find the probability of each. Then give the sampling distribution of the sample median.
b. Construct a probability histogram for the sampling distribution of the sample median and compare it with the probability histogram for the sample mean (Exercise 5.3, part **b**).

*5.6 In Example 5.2 we use the computer to generate 1,000 samples, each containing $n = 11$ observations, from a uniform distribution over the interval from 0 to 1. For this exercise, generate 500 samples, each containing $n = 15$ observations, from this population.
a. Calculate the sample mean for each sample. To approximate the sampling distribution of $\bar{x}$, construct a relative frequency histogram for the 500 values of $\bar{x}$.
b. Repeat part **a** for the sample median. Compare this approximate sampling distribution with the approximate sampling distribution of $\bar{x}$ found in part **a**.

*5.7 Consider a population that contains values of x equal to 00, 01, 02, 03, . . . , 96, 97, 98, 99. Assume that these values of x occur with equal probability. Generate 500 samples, each containing $n = 25$ measurements, from this population. Calculate the sample mean $\bar{x}$ and sample variance s^2 for each of the 500 samples.
a. To approximate the sampling distribution of $\bar{x}$, construct a relative frequency histogram for the 500 values of $\bar{x}$.
b. Repeat part **a** for the 500 values of s^2.

5.2 Properties of Sampling Distributions: Unbiasedness and Minimum Variance (Optional)

The simplest type of statistic used to make inferences about a population parameter is a **point estimator**. A point estimator is a rule or formula that tells us how to use the sample data to calculate a single number that is intended to estimate the value of some population parameter. For example, the sample mean $\bar{x}$ is a point estimator of the population mean μ. Similarly, the sample variance s^2 is a point estimator of the population variance σ^2.

Definition 5.4

A **point estimator** of a population parameter is a rule or formula that tells us how to use the sample data to calculate a single number that can be used as an *estimate* of the population parameter.

Often, many different point estimators can be found to estimate the same parameter. Each will have a sampling distribution that provides information about the point estimator. By examining the sampling distribution, we can determine how large the difference between an estimate and the true value of the parameter (called the **error of estimation**) is likely to be. We can also tell whether an estimator is more likely to overestimate or to underestimate a parameter.

EXAMPLE 5.3

Suppose two statistics, A and B, exist to estimate the same population parameter, θ (theta). (Note that θ could be any parameter, μ, σ^2, σ, etc.) Suppose the two statistics have sampling distributions as shown in Figure 5.6. Based on these sampling distributions, which statistic is more attractive as an estimator of θ?

FIGURE 5.6 ▶
Sampling distributions of unbiased and biased estimators

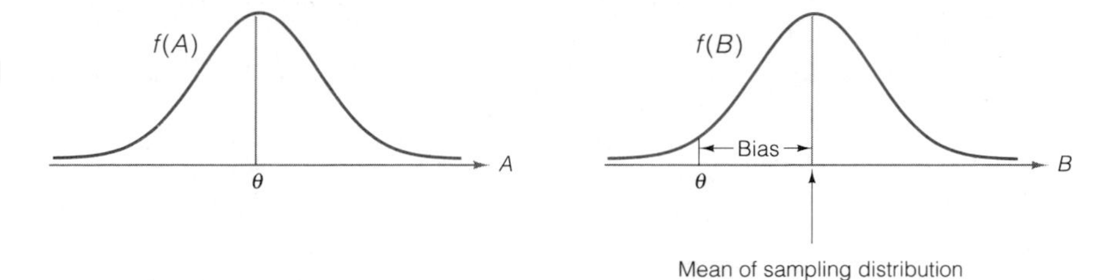

a. Unbiased sample statistic for the parameter θ

b. Biased sample statistic for the parameter θ

Solution

As a first consideration, we would like the sampling distribution to center over the value of the parameter we wish to estimate. One way to characterize this property is in terms of the mean of the sampling distribution. Consequently, we say that a statistic is **unbiased** if the mean of the sampling distribution is equal to the parameter it is intended to estimate. This situation is shown in Figure 5.6(a), where the mean μ_A of statistic A is equal to θ. If the mean of a sampling distribution is not equal to the parameter it is intended to estimate, the statistic is said to be **biased**. The sampling distribution for a biased statistic is shown in Figure 5.6(b). The mean μ_B of the sampling distribution for statistic B is not equal to θ; in fact, it is shifted to the right of θ.

You can see that biased statistics tend either to overestimate or to underestimate a parameter. Consequently, when other properties of statistics tend to be equivalent, we will choose an unbiased statistic to estimate a parameter of interest.*

Definition 5.5

If the sampling distribution of a sample statistic has a mean equal to the population parameter the statistic is intended to estimate, the statistic is said to be an **unbiased** estimate of the parameter.

If the mean of the sampling distribution is not equal to the parameter, the statistic is said to be a **biased** estimate of the parameter.

*Unbiased statistics do not exist for all parameters of interest, but they do exist for the parameters considered in this text.

The standard deviation of a sampling distribution measures another important property of statistics—the spread of these estimates generated by repeated sampling. Suppose two statistics, A and B, are both unbiased estimators of the population parameter. Since the means of the two sampling distributions are the same, we turn to their standard deviations in order to decide which will provide estimates that fall closer to the unknown population parameter we are estimating. Naturally, we will choose the sample statistic that has the smaller standard deviation. Figure 5.7 depicts sampling distributions for A and B. Note that the standard deviation of the distribution of A is smaller than the standard deviation for B, indicating that over a large number of samples, the values of A cluster more closely around the unknown population parameter than do the values of B. Stated differently, the probability that A is close to the parameter value is higher than the probability that B is close to the parameter value.

FIGURE 5.7 ▶ Sampling distributions for two unbiased estimators

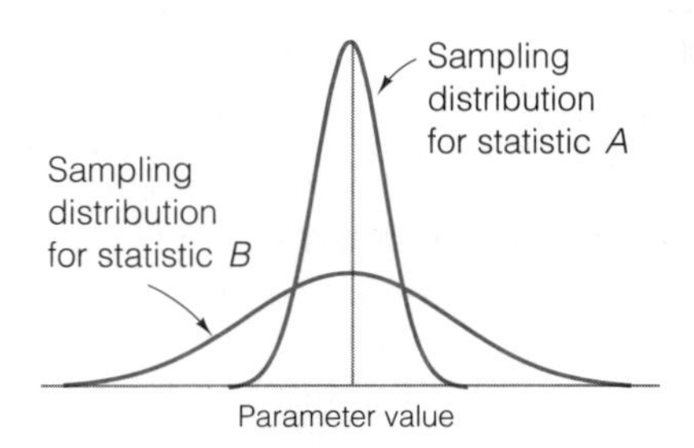

In summary, to make an inference about a population parameter, we use the sample statistic with a sampling distribution that is unbiased and has a small standard deviation (usually smaller than the standard deviation of other unbiased sample statistics). The derivation of this sample statistic will not concern us, because the "best" statistic for estimating specific parameters is a matter of record. We will simply present an unbiased estimator with its standard deviation for each population parameter we consider. [*Note:* The standard deviation of the sampling distribution of a statistic is also called the **standard error of the statistic.**]

EXAMPLE 5.4

In Example 5.1, we found the sampling distributions of the sample mean $\bar{x}$ and the sample median m for random samples of $n = 3$ measurements from a population defined by the probability distribution shown here.

x	0	3	12
$p(x)$	1/3	1/3	1/3

The sampling distributions of $\bar{x}$ and m were found to be as shown at the top of the next page.

$\bar{x}$	0	1	2	3	4	5	6	8	9	12
$p(\bar{x})$	$1/27$	$3/27$	$3/27$	$1/27$	$3/27$	$6/27$	$3/27$	$3/27$	$3/27$	$1/27$

m	0	3	12
$p(m)$	$7/27$	$13/27$	$7/27$

a. Show that $\bar{x}$ is an unbiased estimator of μ in this situation.
b. Show that m is a biased estimator of μ in this situation.

Solution

a. The expected value of a discrete random variable x (see Section 4.2) is defined to be $E(x) = \Sigma\, xp(x)$, where the summation is over all values of x. Then

$$E(x) = \mu = \sum xp(x) = (0)(1/3) + (3)(1/3) + (12)(1/3) = 5$$

The expected value of the discrete random variable $\bar{x}$ is

$$E(\bar{x}) = \sum (\bar{x})p(\bar{x})$$

summed over all values of $\bar{x}$. Or

$$E(\bar{x}) = (0)(1/27) + (1)(3/27) + 2(3/27) + \cdots + (12)(1/27) = 5$$

Since $E(\bar{x}) = \mu$, we see that $\bar{x}$ is an unbiased estimator of μ.

b. The expected value of the sample median m is

$$E(m) = \sum mp(m) = (0)(7/27) + (3)(13/27) + (12)(7/27) = 4.56$$

Since the expected value of m is not equal to μ ($\mu = 5$), the sample median m is a biased estimator of μ.

EXAMPLE 5.5

Refer to Example 5.4 and find the standard deviations of the sampling distributions of $\bar{x}$ and m. Which statistic would appear to be a better estimator of μ?

Solution

The variance of the sampling distribution of $\bar{x}$ (we denote it by the symbol $\sigma_{\bar{x}}^2$) is found to be

$$\sigma_{\bar{x}}^2 = E\{[\bar{x} - E(\bar{x})]^2\} = \sum (\bar{x} - \mu)^2 p(\bar{x})$$

where, from Example 5.4,

$$E(\bar{x}) = \mu = 5$$

Then

$$\sigma_{\bar{x}}^2 = (0 - 5)^2(1/27) + (1 - 5)^2(3/27) + (2 - 5)^2(3/27) + \cdots + (12 - 5)^2(1/27)$$
$$= 8.6667$$

and

$$\sigma_{\bar{x}} = \sqrt{8.6667} = 2.94$$

Similarly, the variance of the sampling distribution of m (we denote it by σ_m^2) is

$$\sigma_m^2 = E\{[m - E(m)]^2\}$$

where, from Example 5.4, the expected value of m is $E(m) = 4.56$. Then

$$\sigma_m^2 = E\{[m - E(m)]^2\} = \sum [m - E(m)]^2 p(m)$$
$$= (0 - 4.56)^2(7/27) + (3 - 4.56)^2(13/27) + (12 - 4.56)^2(7/27) = 20.9136$$

and

$$\sigma_m = \sqrt{20.9136} = 4.57$$

Which statistic appears to be the better estimator for the population mean μ: the sample mean $\bar{x}$ or the median m? To answer this question, we compare the sampling distributions of the two statistics. The sampling distribution of the sample median m is biased (i.e., it is located to the left of the mean μ) and its standard deviation $\sigma_m = 4.57$ is much larger than the standard deviation of the sampling distribution of $\bar{x}$, $\sigma_{\bar{x}} = 2.94$. Consequently, the sample mean $\bar{x}$ would be a better estimator of the population mean μ, for the population in question, than would the sample median m.

Exercises 5.8–5.14

Note: *Starred (*) exercises require the use of a computer.*

5.8 Consider the probability distribution:

x	0	1	4
$p(x)$	1/3	1/3	1/3

a. Find μ and σ^2.

b. Find the sampling distribution of the sample mean $\bar{x}$ for a random sample of $n = 2$ measurements from this distribution.

c. Show that $\bar{x}$ is an unbiased estimator for μ. [*Hint:* Show that $E(\bar{x}) = \Sigma\, \bar{x}p(\bar{x}) = \mu$.]

d. Find the sampling distribution of the sample variance s^2 for a random sample of $n = 2$ measurements from this distribution.
e. Show that s^2 is an unbiased estimator for σ^2.

5.9 Consider the probability distribution shown here.

x	2	4	9
$p(x)$	1/3	1/3	1/3

a. Calculate μ for this distribution.
b. Find the sampling distribution of the sample mean $\bar{x}$ for a random sample of $n = 3$ measurements from this distribution, and show that $\bar{x}$ is an unbiased estimator of μ.
c. Find the sampling distribution of the sample median m for a random sample of $n = 3$ measurements from this distribution, and show that the median is a biased estimator of μ.
d. If you wanted to estimate μ using a sample of three measurements from this population, which estimator would you use? Why?

5.10 Consider the probability distribution shown here.

x	0	1	2
$p(x)$	1/3	1/3	1/3

a. Find μ.
b. For a random sample of $n = 3$ observations from this distribution, find the sampling distribution of the sample mean.
c. Find the sampling distribution of the median of a sample of $n = 3$ observations from this population.
d. Refer to parts **b** and **c** and show that both the mean and median are unbiased estimators of μ for this population.
e. Find the variances of the sampling distributions of the sample mean and the sample median.
f. Which estimator would you use to estimate μ? Why?

***5.11** Generate 500 samples, each containing $n = 25$ measurements, from a population that contains values of x equal to 01, 02, . . . , 48, 49, 50. Assume that these values of x are equally likely. Calculate the sample mean $\bar{x}$ and median m for each sample. Construct relative frequency histograms for the 500 values of $\bar{x}$ and the 500 values of m. Use these approximations to the sampling distributions of $\bar{x}$ and m to answer the following questions:
a. Does it appear that $\bar{x}$ and m are unbiased estimators of the population mean? [*Note:* $\mu = 25.5$.]
b. Which sampling distribution displays greater variation?

5.12 Refer to Exercise 5.3.
a. Show that $\bar{x}$ is an unbiased estimator of μ.
b. Find $\sigma^2_{\bar{x}}$.
c. Find the probability that $\bar{x}$ will fall within $2\sigma_{\bar{x}}$ of μ.

5.13 Refer to Exercise 5.3.

a. Find the sampling distribution of s^2.
b. Find the population variance σ^2.
c. Show that s^2 is an unbiased estimator of σ^2.
d. Find the sampling distribution of the sample standard deviation s.
e. Show that s is a biased estimator of σ.

5.14 Refer to Exercise 5.5, where we found the sampling distribution of the sample median. Is the median an unbiased estimator of the population mean μ?

5.3 The Central Limit Theorem

Estimating the mean useful life of automobiles, the mean number of crimes per month in a large city, and the mean yield per acre of a new soybean hybrid are practical problems with something in common. In each case we are interested in making an inference about the mean μ of some population. As we mentioned in Chapter 2, the sample mean $\bar{x}$ is, in general, a good estimator of μ. We now develop pertinent information about the sampling distribution for this useful statistic.

EXAMPLE 5.6

Suppose a population has the uniform probability distribution given in Figure 5.8. The mean and standard deviation of this probability distribution are $\mu = .5$ and $\sigma = .29$. Now suppose a sample of 11 measurements is selected from this population. Describe the sampling distribution of the sample mean $\bar{x}$ based on the 1,000 sampling experiments discussed in Example 5.2.

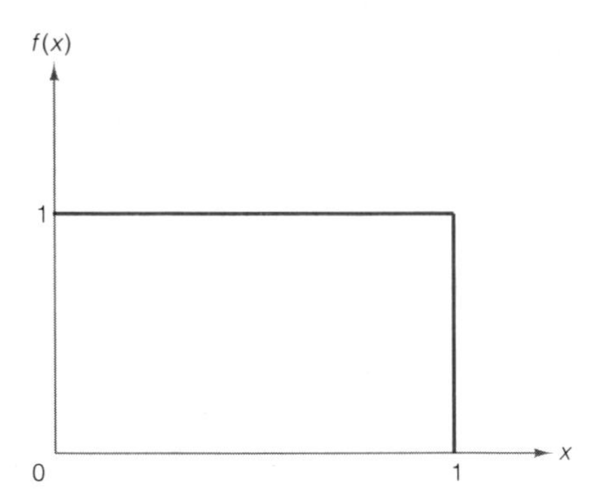

FIGURE 5.8 ▶ Sampled uniform population

Solution

You will recall that in Example 5.2 we generated 1,000 samples of $n = 11$ measurements each. The relative frequency histogram for the 1,000 sample means is shown in Figure 5.9 with a normal probability distribution superimposed. You can see that this normal probability distribution approximates the computer-generated sampling distribution very well.

FIGURE 5.9 ▶
Relative frequency histogram for $\bar{x}$ in 1,000 samples of $n = 11$ measurements with normal distribution superimposed

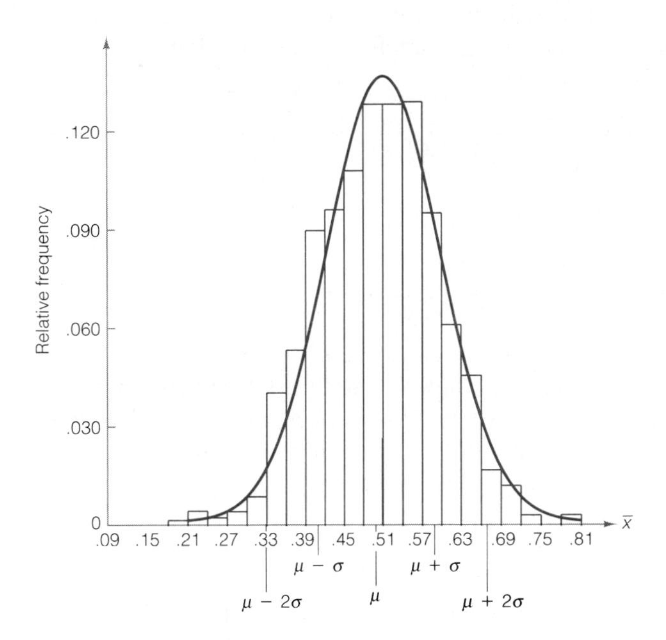

To describe fully a normal probability distribution, it is necessary to know its mean and standard deviation. Inspection of Figure 5.9 indicates that the mean of the distribution of $\bar{x}$, $\mu_{\bar{x}}$, appears to be very close to .5, the mean of the sampled uniform population. Furthermore, for a mound-shaped distribution such as that shown in Figure 5.9, almost all the measurements should fall within 3 standard deviations of the mean. Since the number of values of $\bar{x}$ is very large (1,000), the range of the observed $\bar{x}$'s divided by 6 (rather than 4) should give a reasonable approximation to the standard deviation of the sample means, $\sigma_{\bar{x}}$. The values of $\bar{x}$ range from about .2 to .8, so we calculate

$$\sigma_{\bar{x}} \approx \frac{\text{Range of } \bar{x}\text{'s}}{6} = \frac{.8 - .2}{6} = .1$$

To summarize our findings based on 1,000 samples, each consisting of 11 measurements from a uniform population, the sampling distribution of $\bar{x}$ appears to be approximately normal with a mean of about .5 and a standard deviation of about .1.

The sampling distribution of $\bar{x}$ has the properties given in the box on page 250, assuming only that a random sample of n observations has been selected from *any* population.

Properties of the Sampling Distribution of $\bar{x}$

1. Mean of sampling distribution = Mean of sampled population
   That is, $\mu_{\bar{x}} = E(\bar{x}) = \mu$.
2. Standard deviation of sampling distribution equals

$$\frac{\begin{pmatrix}\text{Standard deviation of}\\ \text{sampled population}\end{pmatrix}}{\text{Square root of sample size}}$$

   That is, $\sigma_{\bar{x}} = \sigma/\sqrt{n}$

   The standard deviation $\sigma_{\bar{x}}$ is often referred to as the **standard error of the mean.**
3. The sampling distribution of $\bar{x}$ is approximately normal for large sample sizes.

You can see that our approximation to $\mu_{\bar{x}}$ in Example 5.6 was precise, since property 1 assures us that the mean is the same as that of the sampled population: .5. Property 2 tells us how to calculate the standard deviation of the sampling distribution of $\bar{x}$. Substituting $\sigma = .29$, the standard deviation of the sampled uniform distribution, and the sample size $n = 11$ into the formula for $\sigma_{\bar{x}}$, we find

$$\sigma_{\bar{x}} = \frac{\sigma}{\sqrt{n}} = \frac{.29}{\sqrt{11}} = .09$$

Thus, the approximation we obtained in Example 5.6, $\sigma_{\bar{x}} \approx .1$, is very close to the exact value, $\sigma_{\bar{x}} = .09$.

The justification for property 3 is contained in one of the most important theoretical results in statistics, the **Central Limit Theorem.**

Central Limit Theorem

If a random sample of n observations is selected from a population (*any* population), then when n is sufficiently large, the sampling distribution of $\bar{x}$ will be approximately a normal distribution. The larger the sample size, the better will be the normal approximation to the sampling distribution of $\bar{x}$.*

*Moreover, because of the Central Limit Theorem the sum of a random sample of n observations, Σx, will possess a sampling distribution that is approximately normal for large samples. This distribution will have a mean equal to $n\mu$ and a variance equal to $n\sigma^2$. Proof of the Central Limit Theorem is beyond the scope of this book, but it can be found in many mathematical statistics texts.

Thus, for sufficiently large samples the sampling distribution of $\bar{x}$ is approximately normal. How large must the sample size n be so that the normal distribution provides a good approximation for the sampling distribution of $\bar{x}$? The answer depends on the shape of the distribution of the sampled population, as shown by Figure 5.10. Generally speaking, the greater the skewness of the sampled population distribution, the larger the sample size must be before the normal distribution is an adequate approximation for the sampling distribution of $\bar{x}$. For most sampled populations, sample sizes of $n \geq 30$ will suffice for the normal approximation to be reasonable. We will use the normal approximation for the sampling distribution of $\bar{x}$ when the sample size is at least 30.

FIGURE 5.10 ►
Sampling distributions of $\bar{x}$ for different populations and different sample sizes

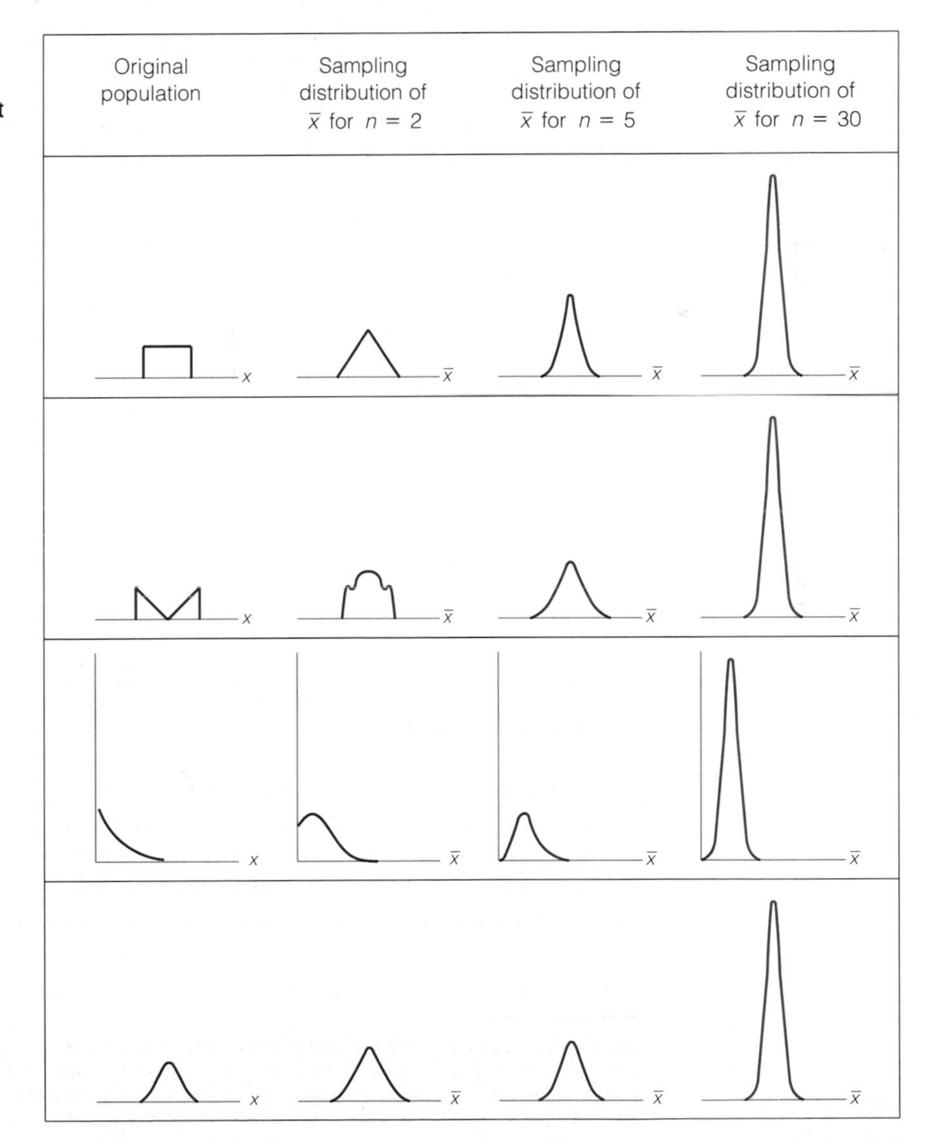

EXAMPLE 5.7

Suppose we have selected a random sample of $n = 25$ observations from a population with mean equal to 80 and standard deviation equal to 5. It is known that the population is not extremely skewed.

a. Sketch the relative frequency distributions for the population and for the sampling distribution of the sample mean, $\bar{x}$.
b. Find the probability that $\bar{x}$ will be larger than 82.

Solution

a. We do not know the exact shape of the population relative frequency distribution, but we do know that it should be centered about $\mu = 80$, its spread should be measured by $\sigma = 5$, and it is not highly skewed. One possibility is shown in Figure 5.11(a). From the Central Limit Theorem, we know that the sampling distribution of $\bar{x}$ will be approximately normal since the sampled population distribution is not extremely skewed. We also know that the sampling distribution will have mean and standard deviation

$$\mu_{\bar{x}} = \mu = 80 \quad \text{and} \quad \sigma_{\bar{x}} = \frac{\sigma}{\sqrt{n}} = \frac{5}{\sqrt{25}} = 1$$

The sampling distribution of $\bar{x}$ is shown in Figure 5.11(b).

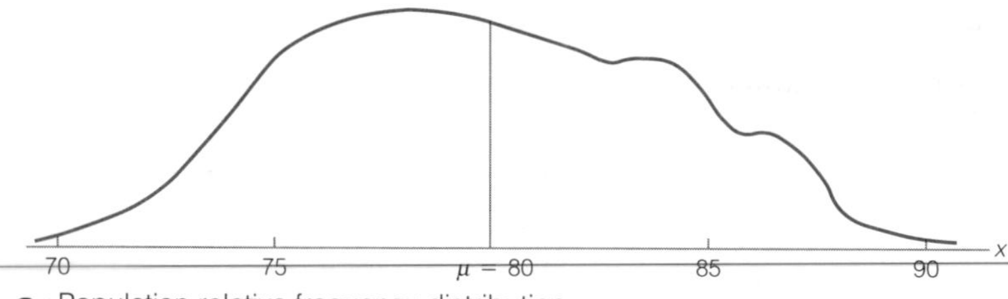

a. Population relative frequency distribution

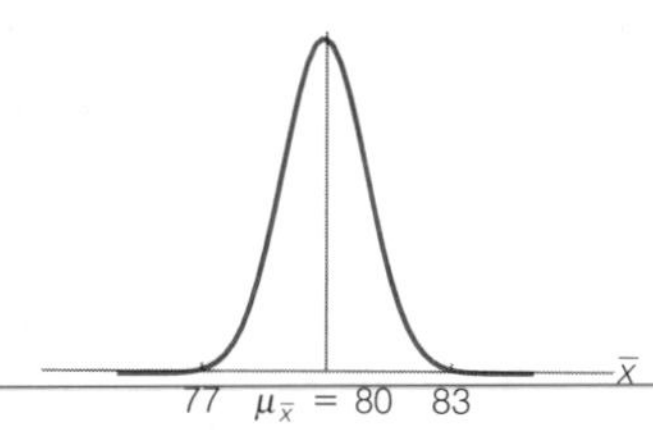

b. Sampling distribution of $\bar{x}$

FIGURE 5.11 ▲
A population relative frequency distribution and the sampling distribution for $\bar{x}$

b. The probability that $\bar{x}$ will exceed 82 is equal to the lightly shaded area in Figure 5.12. To find this area, we need to find the z value corresponding to $\bar{x} = 82$. Recall that the standard normal random variable z is the difference between any normally distributed random variable and its mean, expressed in units of its standard deviation. Since $\bar{x}$ is a normally distributed random variable with mean $\mu_{\bar{x}} = \mu$ and standard deviation $\sigma_{\bar{x}} = \sigma/\sqrt{n}$, it follows that the standard normal z value corresponding to the sample mean, $\bar{x}$, is

$$z = \frac{\text{(Normal random variable)} - \text{(Mean)}}{\text{Standard deviation}} = \frac{\bar{x} - \mu_{\bar{x}}}{\sigma_{\bar{x}}}$$

Therefore, for $\bar{x} = 82$, we have

$$z = \frac{\bar{x} - \mu_{\bar{x}}}{\sigma_{\bar{x}}} = \frac{82 - 80}{1} = 2$$

FIGURE 5.12 ►
The sampling distribution of $\bar{x}$

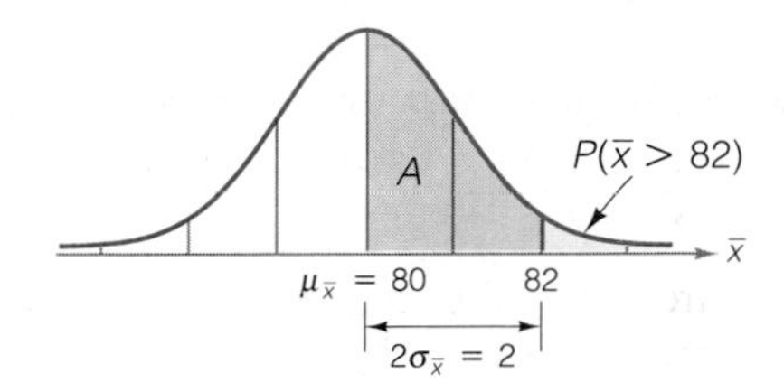

The area A in Figure 5.12 corresponding to $z = 2$ is given in the table of areas under the normal curve (see Table IV of Appendix A) as .4772. Therefore, the tail area corresponding to the probability that $\bar{x}$ exceeds 82 is

$$P(\bar{x} > 82) = P(z > 2) = .5 - .4772 = .0228$$

EXAMPLE 5.8

A manufacturer of automobile batteries claims that the distribution of the lengths of life of its best battery has a mean of 54 months and a standard deviation of 6 months. Suppose a consumer group decides to check the claim by purchasing a sample of 50 of these batteries and subjecting them to tests that determine their lives.

a. Assuming that the manufacturer's claim is true, describe the sampling distribution of the mean lifetime of a sample of 50 batteries.
b. Assuming that the manufacturer's claim is true, what is the probability the consumer group's sample has a mean life of 52 or fewer months?

Solution

a. Even though we have no information about the shape of the probability distribution of the lives of the batteries, we can use the Central Limit Theorem to deduce that the sampling distribution for a sample mean lifetime of 50 batteries is approximately normally distributed. Furthermore, the mean of this sampling distribution is the same as the mean of the sampled population, which is $\mu = 54$ months according to the manufacturer's claim. Finally, the standard deviation of the sampling distribution is given by

$$\sigma_{\bar{x}} = \frac{\sigma}{\sqrt{n}} = \frac{6}{\sqrt{50}} = .85 \text{ month}$$

Note that we used the claimed standard deviation of the sampled population, $\sigma = 6$ months. Thus, if we assume that the claim is true, the sampling distribution for the mean life of the 50 batteries sampled is as shown in Figure 5.13 (page 254).

b. If the manufacturer's claim is true, the probability that the consumer group observes a mean battery life of 52 or fewer months for their sample of 50 batteries, $P(\bar{x} \leq 52)$, is equivalent to the lightly shaded area in Figure 5.13. Since the

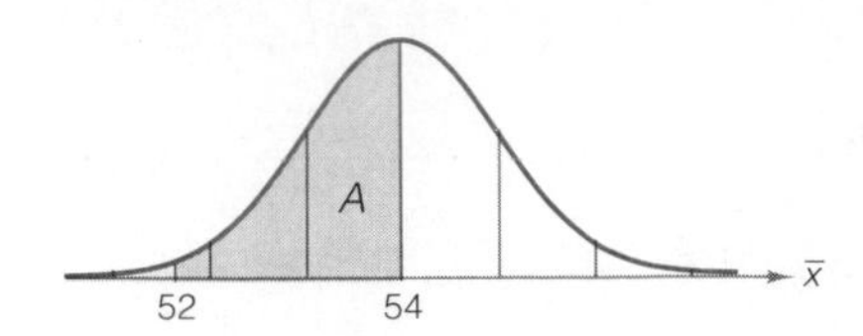

FIGURE 5.13 ▶ Sampling distribution of $\bar{x}$ in Example 5.8 for $n = 50$

sampling distribution is approximately normal, we can find this area by computing the standard normal z value:

$$z = \frac{\bar{x} - \mu_{\bar{x}}}{\sigma_{\bar{x}}} = \frac{\bar{x} - \mu}{\sigma_{\bar{x}}} = \frac{52 - 54}{.85} = -2.35$$

where $\mu_{\bar{x}}$, the mean of the sampling distribution of $\bar{x}$, is equal to μ, the mean of the lives of the sampled population, and $\sigma_{\bar{x}}$ is the standard deviation of the sampling distribution of $\bar{x}$. Note that z is the familiar standardized distance (z-score) of Section 2.6 and, since $\bar{x}$ is approximately normally distributed, it will possess the standard normal distribution of Section 4.6.

The area A shown in Figure 5.13 between $\bar{x} = 52$ and $\bar{x} = 54$ (corresponding to $z = -2.35$) is found in Table IV of Appendix A to be .4906. Therefore, the area to the left of $\bar{x} = 52$ is

$$P(\bar{x} \leq 52) = .5 - A = .5 - .4906 = .0094$$

Thus, the probability the consumer group will observe a sample mean of 52 or less is only .0094 if the manufacturer's claim is true. If the 50 tested batteries do result in a mean of 52 or fewer months, the consumer group will have strong evidence that the manufacturer's claim is untrue, because such an event is very unlikely to occur if the claim is true. (This is still another application of the *rare event approach* to statistical inference.)

In addition to providing a very useful approximation for the sampling distribution of a sample mean, the Central Limit Theorem offers an explanation for the fact that many relative frequency distributions of data possess mound-shaped distributions. Many of the macroscopic measurements we take in various areas of research are really means or sums of many microscopic phenomena. For example, a year's growth of a pine seedling is the total of the many individual components that affect the plant's growth. Similarly, the length of time a construction company takes to complete a house might be viewed as the total of the times taken to complete each of the distinct jobs necessary to build the house. The monthly demand for blood at a hospital may be viewed as the total of the individual patients' needs. Whether the observations entering into these sums satisfy the assumptions basic to the Central Limit Theorem is questionable, but it is a fact that many distributions of data are mound-shaped and possess the appearance of normal distributions. Thus, the Central Limit Theorem offers one explanation for the frequent occurrence of mound-shaped distributions in nature.

Exercises 5.15–5.29

Note: Starred (*) exercises require the use of a computer.

Learning the Mechanics

5.15 Suppose a random sample of n measurements is selected from a population with mean $\mu = 100$ and variance $\sigma^2 = 100$. For each of the following values of n, give the mean and standard deviation of the sampling distribution of the sample mean $\bar{x}$.

a. $n = 4$ b. $n = 25$ c. $n = 100$ d. $n = 50$ e. $n = 500$ f. $n = 1{,}000$

5.16 Suppose a random sample of $n = 25$ measurements is selected from a population with mean μ and standard deviation σ. For each of the following values of μ and σ, give the values of $\mu_{\bar{x}}$ and $\sigma_{\bar{x}}$.

a. $\mu = 10, \quad \sigma = 3$ b. $\mu = 100, \quad \sigma = 25$ c. $\mu = 20, \quad \sigma = 40$ d. $\mu = 10, \quad \sigma = 100$

5.17 Consider the probability distribution shown here.

x	1	2	3	8
$p(x)$	.1	.4	.4	.1

a. Find μ, σ^2, and σ.
b. Find the sampling distribution of $\bar{x}$ for random samples of $n = 2$ measurements from this distribution by listing all possible values of $\bar{x}$, and find the probability associated with each.
c. Use the results of part b to calculate $\mu_{\bar{x}}$ and $\sigma_{\bar{x}}$. Confirm that $\mu_{\bar{x}} = \mu$ and that $\sigma_{\bar{x}} = \sigma/\sqrt{n} = \sigma/\sqrt{2}$.

5.18 Will the sampling distribution of $\bar{x}$ always be approximately normally distributed? Explain. ONLY IF LARGE SAMPLE SIZE

5.19 A random sample of $n = 64$ observations is drawn from a population with a mean equal to 20 and standard deviation equal to 16.

a. Give the mean and standard deviation of the (repeated) sampling distribution of $\bar{x}$.
b. Describe the shape of the sampling distribution of $\bar{x}$. Does your answer depend on the sample size?
c. Calculate the standard normal z-score corresponding to a value of $\bar{x} = 15.5$.
d. Calculate the standard normal z-score corresponding to $\bar{x} = 23$.

5.20 Refer to Exercise 5.19. Find the probability that

a. $\bar{x}$ is less than 16 b. $\bar{x}$ is greater than 23
c. $\bar{x}$ is greater than 25 d. $\bar{x}$ falls between 16 and 22
e. $\bar{x}$ is less than 14

5.21 A random sample of $n = 100$ observations is selected from a population with $\mu = 30$ and $\sigma = 16$. Approximate the following probabilities:

a. $P(\bar{x} \geq 28)$ b. $P(22.1 \leq \bar{x} \leq 26.8)$ c. $P(\bar{x} \leq 28.2)$ d. $P(\bar{x} \geq 27.0)$

5.22 A random sample of $n = 900$ observations is selected from a population with $\mu = 100$ and $\sigma = 10$.

a. What are the largest and smallest values of $\bar{x}$ that you would expect to see?

b. How far, at the most, would you expect $\bar{x}$ to deviate from μ?

c. Did you have to know μ to answer part **b**? Explain.

***5.23** Consider a population that contains values of x equal to 00, 01, 02, . . . , 97, 98, 99. Assume that the values of x are equally likely. For each of the following values of n, generate 500 random samples and calculate $\bar{x}$ for each sample. For each sample size, construct a relative frequency histogram of the 500 values of $\bar{x}$. What changes occur in the histograms as the value of n increases? What similarities exist? Use $n = 2$, $n = 5$, $n = 10$, $n = 30$, and $n = 50$.

Applying the Concepts

5.24 A recent study in the *College Student Journal* (Sheehan et al., 1992) investigated differences in traditional and nontraditional students, where nontraditional students are generally defined as those 25 years or older. Suppose that a random sample of $n = 100$ nontraditional students is selected from the 1993 population of nontraditional students, and the GPA of each student is determined. Assume that the population mean and standard deviation for the GPA of all nontraditional students is $\mu = 3.5$ and $\sigma = .5$. Then $\bar{x}$, the sample mean, will be approximately normally distributed (because of the Central Limit Theorem).

a. Calculate $\mu_{\bar{x}}$ and $\sigma_{\bar{x}}$.

b. What is the approximate probability that the nontraditional student sample has a mean GPA between 3.40 and 3.60?

c. What is the approximate probability that the sample of 100 nontraditional students has a mean GPA that exceeds 3.62?

d. How would the sampling distribution of $\bar{x}$ change if the sample size n were doubled from 100 to 200? How do your answers to parts **b** and **c** change when the sample size is doubled?

5.25 The relationship among socioeconomic status, IQ, and juvenile delinquency has long been the subject of psychological research. In one study (Moffitt et al., 1981), researchers examined the relationship between delinquent behavior and poor verbal abilities. Assume that scores on the verbal IQ test administered during this research have a population mean $\mu = 107$ and a population standard deviation $\sigma = 15$.

a. What shape would you expect the (repeated) sampling distribution of $\bar{x}$ for $n = 84$ juveniles to have? Does your answer depend on the shape of the distribution of verbal IQ scores for all juveniles?

b. Assuming that the population mean and standard deviation for juveniles with no record of delinquency are the same as that for all juveniles, approximate the probability that the sample mean verbal IQ for $n = 84$ juveniles will be 110 or more. State any assumptions you make.

c. As part of the cited study, the researchers found that a sample of $n = 84$ juveniles with no record of delinquency had a mean verbal IQ of $\bar{x} = 110$. Considering your answer to part **b**, do you think that the population mean and standard deviation for nondelinquent juveniles are the same as those for all juveniles? Explain.

5.26 Refer to Exercise 5.25. As part of the study cited there, the researchers reported the sample mean verbal IQ for $n = 22$ juveniles with records of two or more offenses.

a. What shape would you expect the (repeated) sampling distribution of $\bar{x}$ for $n = 22$ juveniles to have? Does your answer depend on the shape of the distribution of verbal IQ scores for all juveniles?

b. Assuming that the population mean and standard deviation for juveniles with no record of delinquency are the same as those for all juveniles, approximate the probability that the sample mean verbal IQ for $n = 22$ juveniles will be 98 or less. State any assumptions you make.

c. As part of the cited study, the researchers found that a sample of $n = 22$ juveniles with no record of delinquency had a mean verbal IQ of $\bar{x} = 98$. Considering your answering to part **b**, do you think that the population mean and standard deviation for nondelinquent juveniles are the same as for all juveniles? Explain.

5.27 The number of violent crimes per day in a certain city possesses a mean equal to 1.3 and a standard deviation equal to 1.7. A random sample of 50 days is observed, and the daily mean number of crimes for this sample, $\bar{x}$, is calculated.

a. Give the mean and standard deviation of the sampling distribution of $\bar{x}$.

b. Will the sampling distribution of $\bar{x}$ be approximately normal? Explain.

c. Find an approximate value of $P(\bar{x} < 1)$.

d. Find an approximate value of $P(\bar{x} > 1.9)$.

5.28 Last year a company initiated a program to compensate its employees for unused sick days, paying each employee a bonus of one-half the usual wage earned for each unused sick day. The question that naturally arises is: "Did this policy motivate employees to use fewer allotted sick days?" *Before* last year, the number of sick days used by employees had a distribution with a mean of 7 days and a standard deviation of 2 days.

a. Assuming that these parameters did not change last year, find the approximate probability that the sample mean number of sick days used by 100 employees chosen at random was less than or equal to 6.4 last year.

b. Suppose the sample mean for the 100 employees was, in fact, 6.4. How would you interpret this result?

5.29 Purchased materials, parts, and services account for over 50% of the manufacturing costs of many companies. As a result, it is important for a company's overall quality program to extend to the vendors (i.e., suppliers) from whom the purchases are made. Many firms include vendor surveillance as part of their quality programs (Juran and Gryna, 1980).

For example, suppose a soft drink bottler requires bottles with an internal pressure strength of at least 150 pounds per square inch (psi). A prospective bottle vendor claims that its production process yields bottles with a mean internal pressure strength of 157 psi and a standard deviation of 3 psi. As part of its vendor surveillance program, the bottler strikes an agreement with the vendor that permits the bottler to sample from the vendor's production process to verify the vendor's claim. The bottler randomly selects 40 bottles from the last 10,000 produced, measures the internal pressure strength of each, and finds the mean strength for the sample to be 1.3 psi below the process mean cited by the vendor.

a. Assuming the vendor's claim to be true, what is the probability of obtaining a sample mean this far or farther below the process mean? What does your answer suggest about the validity of the vendor's claim?

b. If the process standard deviation were 3 psi as claimed by the vendor, but the mean were 156 psi, would the observed sample result be more or less likely than in part **a**? What if the mean were 158 psi?

c. If the process mean were 157 psi as claimed, but the process standard deviation were 2 psi, would the sample result be more or less likely than in part **a**? What if instead the standard deviation were 6 psi?

$\mu = 7$ $\bar{x} =$

$\sigma = 2$

5.4 The Relation Between Sample Size and a Sampling Distribution (Optional)

Sample 1	*Sample 2*
$n = 5$	$n = 10$
$\bar{x} = 12.6$	$\bar{x} = 13.1$

Suppose you draw two random samples, one containing $n = 5$ and the second $n = 10$ observations, from a population with mean μ and standard deviation σ. If you compute the mean $\bar{x}$ for each sample and obtain the results shown in the margin, which estimate do you think contains more information about μ?

Intuitively, it would seem that the sample mean based on 10 measurements contains more information than the mean based on five, but to answer the question correctly we need to compare the sampling distributions of these two statistics.

From Section 5.3 we know that the expected value of the sample means in repeated sampling is μ, regardless of the sample size. That is, both sample means are unbiased estimators of μ. The main difference in the sampling distributions lies in their standard deviations: The standard deviation of the mean based on $n = 5$ is $\sigma/\sqrt{5}$, whereas that based on $n = 10$ is $\sigma/\sqrt{10}$. Since the second standard deviation is smaller, we expect the sample means based on 10 measurements to cluster more closely around μ in repeated sampling than those based on five measurements. Thus, our intuitive feeling that $\bar{x}$ for $n = 10$ contains more information about μ is justified.

For most of the statistics you will encounter in this text, the variance of a statistic's sampling distribution is inversely proportional to the sample size. Or, since the standard deviation of $\bar{x}$ is equal to $\sigma/\sqrt{n}$, you can say that the standard deviation of the sampling distribution is proportional to $1/\sqrt{n}$. To reduce the standard deviation of the sampling distribution of a statistic by one-half, you will therefore need four times as many observations in your sample. To reduce the standard deviation to one-third its original value, you will need nine times as many observations.

The sampling distributions for the sample mean $\bar{x}$, based on random samplings from a normally distributed population, are shown in Figure 5.14 for $n = 1$, 4, and 16 observations. The curve for $n = 1$ represents the probability distribution for the population. Those for $n = 4$ and $n = 16$ are sampling distributions for $\bar{x}$. Note how the distributions contract (variation decreases) as n increases from 1 to 4 to 16. The standard deviation for $\bar{x}$ based on $n = 16$ measurements is one-half the corresponding standard deviation for the distribution based on $n = 4$ measurements.

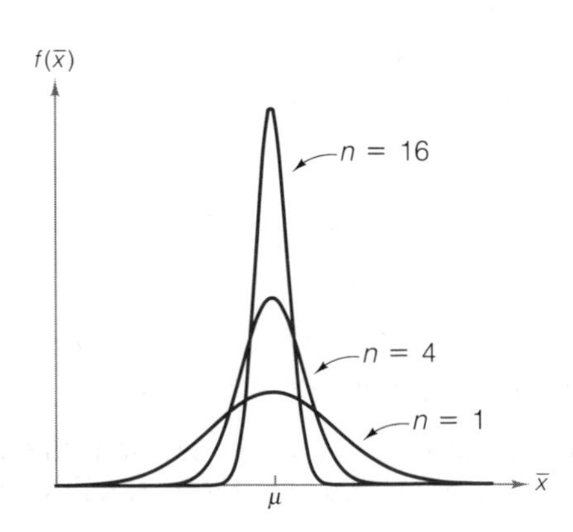

FIGURE 5.14 ► Three sampling distributions for $\bar{x}$

EXAMPLE 5.9

Consider the **Bernoulli random variable** x that can assume the values 1 or 0 with probabilities p and $q = 1 - p$, respectively. The distribution is summarized in Table 5.3, where we have associated the term "Success" with the outcome $x = 1$ and "Failure" with the outcome $x = 0$. The Bernoulli random variable is just a binomial random variable with $n = 1$ trial.

TABLE 5.3 Bernoulli Distribution

Outcome	x	$p(x)$
Failure	0	q
Success	1	p

Suppose a random sample of n measurements is drawn from this Bernoulli distribution, and the sample mean $\bar{x}$ is calculated. Note that

$$\bar{x} = \frac{\sum_{i=1}^{n} x_i}{n}$$

and that $\sum_{i=1}^{n} x_i$ is the total number of Successes (the number of 1's in the sample) in the random sample of n measurements (or "trials"). This sum is therefore a binomial random variable, with n trials and probability of Success p. For example, the Bernoulli random variable might be the *status* of a single computer microchip (nondefective or defective), x the *number* of n such chips that are nondefective, and $\bar{x}$ the *fraction* of nondefective chips in a set of n.

a. Find the mean and standard deviation of the sampling distribution of $\bar{x}$.
b. Simulate the distribution of $\bar{x}$ using $p = .8$, and $n = 1$, 10, 25, and 100 by generating 1,000 samples for each sample size and creating a histogram of the 1,000 sample means.

Solution

a. The mean and variance of the Bernoulli random variable are:

$$\mu = E(x) = 0(q) + 1(p) = p$$
$$\sigma^2 = E[(x - \mu)^2] = (0 - p)^2(q) + (1 - p)^2(p)$$
$$= p^2q + q^2p = pq(p + q) = pq(1) = pq$$

Note that these are the mean and variance of a binomial random variable with $n = 1$, which is another description of a Bernoulli random variable.

We know that the mean $\bar{x}$ of a random sample is unbiased, so that

$$E(\bar{x}) = \mu = p$$

and the standard error is

$$\sigma_{\bar{x}} = \frac{\sigma}{\sqrt{n}} = \frac{\sqrt{pq}}{\sqrt{n}} = \sqrt{\frac{pq}{n}}$$

Because $\bar{x}$ provides an unbiased estimate of the probability of Success p and has a standard error that decreases as the sample size increases, we will use it to estimate p in subsequent chapters, where we will refer to it as the sample fraction of Successes, $\hat{p}$.

b. The mean and standard error of $\bar{x}$ when $p = .8$ are

$$\mu_{\bar{x}} = p = .8 \qquad \sigma_{\bar{x}} = \sqrt{\frac{pq}{n}} = \sqrt{\frac{(.8)(.2)}{n}} = \frac{.4}{\sqrt{n}}$$

You can see how the standard error of $\bar{x}$ decreases as n increases in Table 5.4.

TABLE 5.4 Mean and Standard Error of $\bar{x}$

n	$\mu_{\bar{x}}$	$\sigma_{\bar{x}}$
1	.8	$.4/\sqrt{1} = .4000$
10	.8	$.4/\sqrt{10} = .1265$
25	.8	$.4/\sqrt{25} = .0800$
100	.8	$.4/\sqrt{100} = .0400$

The simulation of 1,000 sample means with each of the sample sizes given in Table 5.4 resulted in the sampling distributions shown in Figure 5.15, where a relative frequency histogram is used to display each sampling distribution. Note that each sampling distribution more closely resembles a normal distribution than the previous one. As the Central Limit Theorem promises, the distribution of $\bar{x}$ becomes approximately normal for large n, and the approximation improves as n increases. Note, too, that the values of $\bar{x}$ cluster more closely around the mean (.8) as n is increased. We make use of these properties to estimate the probability of success p for binomial random variables in Chapter 6.

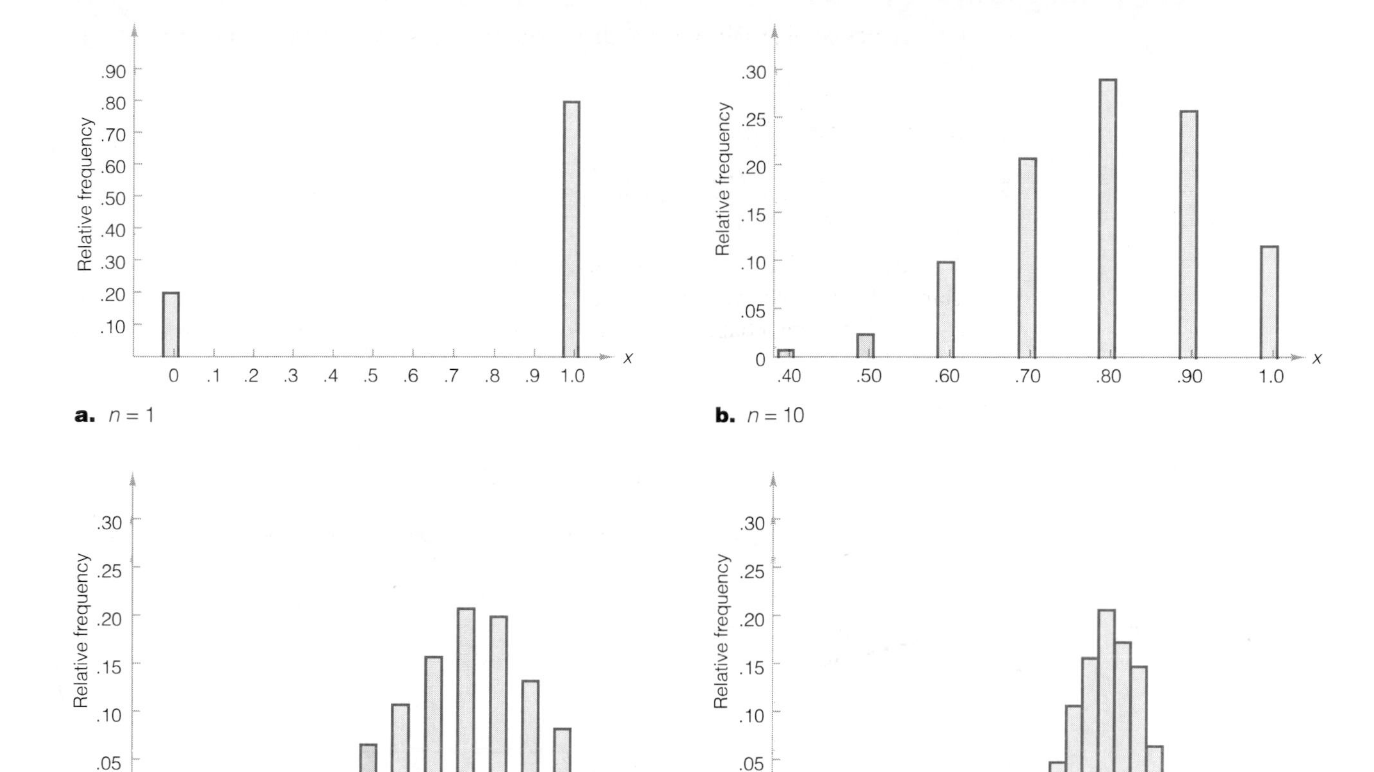

FIGURE 5.15 ▲ Sampling distributions for Bernoulli sample means, Example 5.9

EXAMPLE 5.10

Refer to Example 5.9. Suppose $n = 400$ microchips are to be sampled from a very large batch, of which a proportion p are good. The proportion of nondefective chips in the sample will be determined by assigning a 1 to each nondefective chip, a 0 to each defective chip, and calculating the sample mean of the 400 Bernoulli observations. Assuming that the approximate value of p is .8, what is the probability that $\bar{x}$ will fall within .03 of the exact value of p?

Solution

We first note (see Example 5.9) that the mean and standard deviation of $\bar{x}$ are

$$E(\bar{x}) = p \qquad \sigma_{\bar{x}} = \sqrt{\frac{pq}{n}} = \sqrt{\frac{pq}{400}}$$

Using the approximate value of p to obtain an approximation for the standard deviation of the sampling distribution of $\bar{x}$, we find

$$\sigma_{\bar{x}} \approx \sqrt{\frac{(.8)(.2)}{400}} = .02$$

Next, the Central Limit Theorem implies that the distribution of $\bar{x}$ based on a sample of size 400 is approximately normal. The properties are summarized in Figure 5.16. Note that the mean of the sampling distribution is p, so that the standard normal z value is given by the formula

$$z = \frac{\bar{x} - \mu_{\bar{x}}}{\sigma_{\bar{x}}} = \frac{\bar{x} - p}{.02}$$

FIGURE 5.16 ▶
Sampling distribution of $\bar{x}$

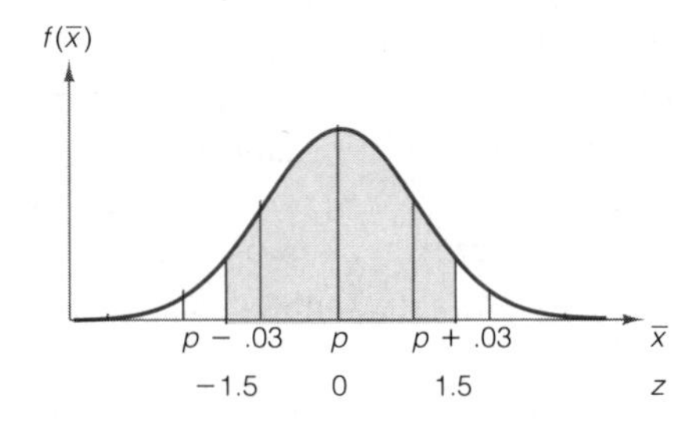

Even though the value of p is unknown, we can find the probability that $\bar{x}$ is within .03 of p by first calculating the z value corresponding to the $\bar{x}$ value that is .03 greater than p:

$$z = \frac{(p + .03) - p}{.02} = \frac{.03}{.02} = 1.5$$

Similarly, the value .03 less than p is 1.5 standard deviations *below* the mean, and has a z value of -1.5. Thus, the event that $\bar{x}$ falls within .03 of p is equivalent to the event that the standard normal random variable z is between -1.5 and 1.5. Using Table IV in Appendix A, we find the probability that $\bar{x}$ falls within 1.5 standard deviations of p is $2(.4332) = .8664$, the area shown shaded in Figure 5.16. The interpretation of this probability is that there is about an 87% chance that the proportion of nondefective chips in the sample of 400 will fall within .03 of the exact proportion of nondefective chips in the entire batch.

For most sampling distributions, the standard deviation of the distribution decreases as the sample size increases. We will use this result in Chapter 6 to help us determine the sample size needed to obtain a specified accuracy of estimation.

Summary

Many practical problems require that an inference be made about some population **parameter** (for example, the mean or the standard deviation). If we want to make this inference on the basis of sample information, we need to compute a **sample statistic** that contains information about the population parameter. The amount of such information contained in a sample statistic is reflected in its **sampling distribution**, the probability distribution of the sample statistic. The sampling distribution describes the behavior of the statistic in repeated sampling. In particular, we want a sample statistic that is an **unbiased** estimator of the population parameter and has a smaller variance than any other unbiased sample statistic.

When the population parameter of interest is the mean μ, the sample mean provides an unbiased estimator with a standard error of $\sigma/\sqrt{n}$. Moreover, the **Central Limit Theorem** assures us that the sampling distribution for the mean of a large sample is approximately normally distributed, no matter what the shape of the relative frequency distribution of the sampled population.

The amount of information in a sample that is relevant to some population parameter is related to the sample size. Another way of saying this is that as n gets larger, the standard error of most sample statistics gets smaller.

The sampling distributions for all the many statistics that can be computed from sample data could be discussed in detail, but this would delay discussion of the practical objective of this course—the role of statistical inference in decision-making. Consequently, we will comment further on the sampling distributions of statistics when we use them as estimators or decision-makers in the following chapters.

Supplementary Exercises 5.30–5.49

Note: Starred (*) exercises require the use of a computer.

Learning the Mechanics

5.30 Consider a sample statistic A. As with all sample statistics, A is computed by utilizing a specified function (formula) of the sample measurements. (For example, if A were the sample mean, the specified formula would be to sum the measurements and divide by the number of measurements.)

a. Describe what we mean by the phrase "the sampling distribution of the sample statistic A."

b. Suppose A is to be used to estimate a population parameter α. What is meant by the assertion that A is an unbiased estimator of α?

c. Consider another sample statistic, B. Assume that B is also an unbiased estimator of the population parameter α. How can we use the sampling distributions of A and B to decide which is the better estimator of α?

d. If the sample sizes on which A and B are based are large, can we apply the Central Limit Theorem and assert that the sampling distributions of A and B are approximately normal? Why or why not?

5.31 The standard deviation (or, as it is usually called, the *standard error*) of the sampling distribution for the sample mean, $\bar{x}$, is equal to the standard deviation of the population from which the sample was selected divided by the square root of the sample size. That is,

$$\sigma_{\bar{x}} = \frac{\sigma}{\sqrt{n}}$$

a. As the sample size is increased, what happens to the standard error of $\bar{x}$? Why is this property considered important?

b. Suppose that a sample statistic has a standard error that is not a function of the sample size. In other words, the standard error remains constant as n changes. What would this imply about the statistic as an estimator of a population parameter?

c. Suppose another unbiased estimator (call it A) of the population mean is a sample statistic with a standard error equal to

$$\sigma_{\text{A}} = \frac{\sigma}{\sqrt[3]{n}}$$

Which of the sample statistics, $\bar{x}$ or A, is preferable as an estimator of the population mean? Why?

d. Suppose that the population standard deviation σ is equal to 10 and that the sample size is 64. Calculate the standard errors of $\bar{x}$ and A. Assuming that the sampling distribution of A is approximately normal, interpret the standard errors. Why is the assumption of (approximate) normality unnecessary for the sampling distribution of $\bar{x}$?

5.32 A random sample of 40 observations is to be drawn from a large population of measurements. It is known that 30% of the measurements in the population are 1's, 20% are 2's, 20% are 3's, and 30% are 4's.

a. Give the mean and standard deviation of the (repeated) sampling distribution of $\bar{x}$, the sample mean of the 40 observations.

b. Describe the shape of the sampling distribution of $\bar{x}$. Does your answer depend on the sample size?

5.33 A random sample of $n = 68$ observations is selected from a population with $\mu = 19.6$ and $\sigma = 3.2$. Approximate each of the following probabilities.

a. $P(\bar{x} \leq 19.6)$ b. $P(\bar{x} \leq 19)$ c. $P(\bar{x} \geq 20.1)$ d. $P(19.2 \leq \bar{x} \leq 20.6)$

5.34 Suppose x equals the number of heads observed when a single coin is tossed; that is, $x = 0$ or $x = 1$. The population corresponding to x is the set of 0's and 1's generated when the coin is tossed repeatedly a large number of times. Suppose we select $n = 2$ observations from this population. (That is, we toss the coin twice and observe two values of x.)

a. List the three different samples (combinations of 0's and 1's) that could be obtained.

b. Calculate the value of $\bar{x}$ for each of the samples.

c. List the values that $\bar{x}$ can assume, and find the probabilities of observing these values.

d. Construct a graph of the sampling distribution of $\bar{x}$.

***5.35** Use a statistical software package to generate 100 random samples of size $n = 2$ from the population described in Exercise 5.32. Compute $\bar{x}$ for each sample, and plot a frequency distribution for the 100 $\bar{x}$ values. Repeat this process for $n = 5$, 10, 30, and 50. Explain how your plots illustrate the Central Limit Theorem.

*5.36 Use a statistical software package to generate 100 random samples of size $n = 2$ from a population characterized by a normal probability distribution with mean 100 and standard deviation 10. Compute $\bar{x}$ for each sample and plot a frequency distribution for the 100 values of $\bar{x}$. Repeat this process for $n = 5$, 10, 30, and 50. How does the fact that the sampled population is normal affect the sampling distribution of $\bar{x}$?

5.37 A random sample of size n is to be drawn from a large population with mean 100 and standard deviation 10, and the sample mean $\bar{x}$ is to be calculated. To see the effect of different sample sizes on the standard deviation of the sampling distribution of $\bar{x}$, plot $\sigma/\sqrt{n}$ against n for $n = 1$, 5, 10, 20, 30, 40, and 50.

5.38 A random sample of size $n = 30$ is to be drawn from a population with $\mu = 500$ and $\sigma = 200$.

a. What is the standard deviation of the sampling distribution of $\bar{x}$?

b. To reduce the standard deviation of $\bar{x}$ to 50% of the value in part **a**, how much larger would n need to be?

c. To reduce $\sigma_{\bar{x}}$ to 75% of the value in part **a**, how much larger would n need to be?

Applying the Concepts

5.39 The fourth *Annual Report: Florida Employer Opinion Survey* (1992) gives the results of an extensive survey of employer opinions in Florida. Each employer was asked to rate his or her satisfaction with the preparation of employees by the public education system. Responses were 1, 1.5, or 2, representing very dissatisfied, neither satisfied nor dissatisfied, and very satisfied, respectively. A sample of 651 employers was selected. Assume that the mean for all employers in Florida is 1.50 (the "dividing" line between satisfied and dissatisfied) and the standard deviation is .45.

a. Which type of distribution describes the individual survey responses, continuous or discrete?

b. Describe the distribution that best approximates the sample mean response of 651 employers. What are the mean and standard deviation of this distribution? What assumptions, if any, are necessary to assure the validity of your answers?

c. What is the approximate probability that the sample mean will be 1.45 or less?

d. The mean of the sample of 651 employers surveyed in 1992 was 1.36. Given this result, do you think it is likely that all Florida employers' opinions were evenly divided on the effectiveness of public education? That is, do you think the assumption that the population mean is 1.50 is correct? Why or why not?

5.40 The distribution of the number of barrels of oil produced by a certain oil well each day for the past 3 years has a mean of 400 and a standard deviation of 75.

a. Describe the sampling distribution of the mean number of barrels produced per day for samples of 40 production days drawn from the past 3 years.

b. What is the approximate probability that the sample mean will be greater than 425?

c. What is the approximate probability that the sample mean will be less than 400?

d. What assumptions did you have to make in order to answer parts **a–c**? Justify the assumptions.

5.41 Water availability is of prime importance in the life cycle of most reptiles. To determine the rate of evaporative water loss of a certain species of lizard at a particular desert site, 34 such lizards were randomly collected, weighed, and placed under the appropriate experimental conditions. After 24 hours, each lizard was removed, reweighed, and its total water loss was calculated as the difference between initial body weight and body weight after treatment. Previous studies have shown that the distribution of water loss for the lizards has a mean of

3.1 grams and a standard deviation of .8 gram. Find the approximate probability that the 34 lizards have a mean water loss of

a. Less than 3.0 grams
b. Between 3.15 and 3.25 grams

5.42 Electric power plants that use water for cooling their condensers sometimes discharge heated water into rivers, lakes, or oceans. It is known that water heated above certain temperatures has a detrimental effect on the plant and animal life in the water. Suppose it is known that the increased temperature of the heated water discharged by a certain power plant on any given day has a distribution with a mean of 5°C and a standard deviation of .5°C.

a. For 50 randomly selected days, what is the approximate probability that the average increase in temperature of the discharged water is greater than 5.0°C?
b. Less than 4.8°C?
c. What assumptions must be made for you to answer the questions?

5.43 *Random-number generators* have many uses in statistics.* One type is designed to produce a sequence of numbers between 0 and 1. A number x can assume any value in the interval from 0 to 1 with equal probability, and any value of x is independent of the values of previous numbers that appear in the sequence. Furthermore, the probability distribution of x has a mean $\mu = .5$ and a standard deviation $\sigma = .29$. Let y be the average of n such random numbers.

a. Graph the probability distribution for x.
b. Give the mean and standard deviation of the sampling distribution of y.
c. What is the approximate form of the sampling distribution of y when n is large?
d. Sketch the sampling distribution of y and compare it with your graph from part **a**.

5.44 To determine whether a metal lathe that produces machine bearings is properly adjusted, a random sample of 25 bearings is collected and the diameter of each is measured.

a. If the standard deviation of the diameters of the bearings measured over a long period of time is .001 inch, what is the approximate probability that the mean diameter $\bar{x}$ of the sample of 25 bearings will lie within .0001 inch of the population mean diameter of the bearings?
b. If the population of diameters has an extremely skewed distribution, how will your approximation in part **a** be affected?

5.45 Refer to Exercise 5.44. The mean diameter of the bearings produced by the machine is supposed to be .5 inch. The company decides to use the sample mean (from Exercise 5.44) to decide whether the process is in control; i.e., whether it is producing bearings with a mean diameter of .5 inch. The machine will be considered out of control if the mean of the sample of $n = 25$ diameters is less than .4994 inch or larger than .5006 inch. If the true mean diameter of the bearings produced by the machine is .501 inch, what is the approximate probability that the test will imply that the process is out of control?

5.46 This past year, an elementary school began using a new method to teach arithmetic to first graders. A standardized test, administered at the end of the year, was used to measure the effectiveness of the new method. The distribution of past scores on the standardized test produced a mean of 75 and a standard deviation of 10.

*Random-number generators are used to produce random numbers such as those that appear in Table 1 of Appendix A.

a. If the new method is no different from the old method, what is the approximate probability that the mean score $\bar{x}$ of a random sample of 36 students will be greater than 79?
b. What assumptions must be satisfied to make your answer valid?

5.47 Much emphasis has recently been placed on preventative health behavior. In one study at a health fair (Price, O'Connell, and Kukulka, 1985), 100 attendees were administered a questionnaire on preventative health behaviors. Assume that the population mean and standard deviation of the questionnaire scores in the area of preventing heart disease are 38 and 5, respectively, when administered to the general public.
a. Describe the sampling distribution of the sample mean questionnaire score of 100 health-fair attendees, assuming that they are a random sample of the general public.
b. What is the probability that the sample mean score of the 100 health-fair attendees will exceed 39.1, if they represent a random sample of the general public?
c. The scores of the 100 health-fair attendees on the preventative behavior questionnaire had a mean of 39.1. Considering your answer to part **b**, do you think the attendees represent a random sample of the general public with regard to the heart disease questionnaire?

5.48 Refer to Exercise 5.47. The 100 health-fair attendees were also administered a questionnaire on malignancies. Assume that the mean and standard deviation of scores of the general public are 42 and 6.5, respectively.
a. Describe the sampling distribution of the sample mean score of 100 attendees, assuming that they are a random sample of the general public.
b. What is the probability that the sample mean questionnaire score of the 100 attendees will exceed 43.6, if they represent a random sample of the general public?
c. The scores of the 100 health-fair attendees on the preventative behavior questionnaire related to malignancies had a mean of 43.6. Considering your answer to part **b**, do you think the attendees represent a random sample of the general public with regard to the malignancy questionnaire?

***5.49** [*Note:* This exercise refers to an optional section in Chapter 4.] A building contractor has decided to purchase a load of factory-reject aluminum siding as long as the average number of flaws per piece of siding in a sample of size 35 from the factory's reject pile is 2.1 or less. If it is known that the number of flaws per piece of siding in the factory's reject pile has a Poisson probability distribution with a mean of 2.5, find the approximate probability that the contractor will not purchase a load of siding. [*Hint:* If x is a Poisson random variable with mean λ, then σ_x^2 also equals λ.]

On Your Own

To understand the Central Limit Theorem and sampling distribution, consider the following experiment: Toss four identical coins, and record the number of heads observed. Then repeat this experiment four more times, so that you end up with a total of five observations for the random variable x, the number of heads when four coins are tossed.

Now derive and graph the probability distribution for x, assuming the coins are balanced. Note that the mean of this distribution is $\mu = 2$ and the standard deviation is $\sigma = 1$. This probability distribution represents the one from which you are drawing a random sample of five measurements.

Next, calculate the mean $\bar{x}$ of the five measurements—i.e., calculate the mean number of heads you observed in five repetitions of the experiment. Although you have repeated the basic experiment five times, you have only one observed value of $\bar{x}$. To derive the probability distribution or sampling distribution of $\bar{x}$ empirically, you have to repeat the entire process (of tossing four coins five times) many times. Do it 100 times.

The approximate sampling distribution of $\bar{x}$ can be derived theoretically by making use of the Central Limit Theorem. We expect at least an approximate normal probability distribution with a mean $\mu = 2$ and a standard deviation

$$\sigma_{\bar{x}} = \frac{\sigma}{\sqrt{n}} = \frac{1}{\sqrt{5}} = .45$$

Count the number of your 100 $\bar{x}$'s that fall in each of the intervals in the figure below. Use the normal probability distribution with $\mu = 2$ and $\sigma_{\bar{x}} = .45$ to calculate the expected number of the 100 $\bar{x}$'s in each of the intervals. How closely does the theory describe your experimental results?

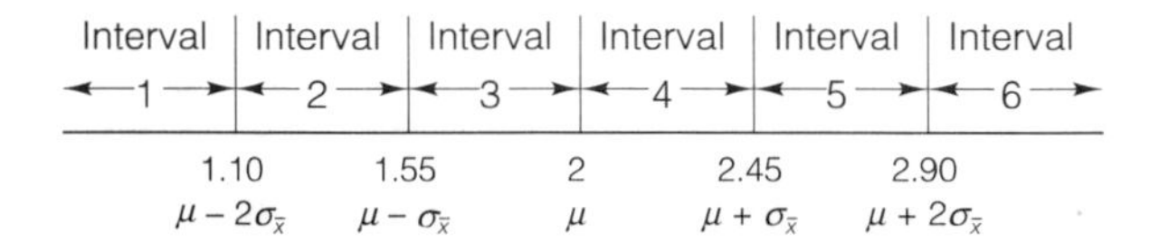

Using the Computer

Calculate the mean and standard deviation for the 1,000 zip codes' median household incomes. We will treat these quantities as the population mean μ and the population standard deviation σ for this exercise.

a. Draw 100 random samples of $n = 20$ observations from the 1,000 zip codes' median incomes. Select the samples with replacement—i.e., replace each measurement before selecting the next.* Calculate the 100 sample means. Generate a stem-and-leaf display or a relative frequency histogram for the 100 means. Then count the number of the 100 sample means that fall in the intervals $\mu \pm \sigma/\sqrt{n}$, $\mu \pm 2\sigma/\sqrt{n}$, and $\mu \pm 3\sigma/\sqrt{n}$. How do the graphic description and the percentage of means falling in the intervals agree with a normal distribution having mean μ and standard deviation $\sigma/\sqrt{n}$?

b. Repeat part **a** using a sample size of $n = 50$. Is the sampling distribution of the sample means closer to normal for the larger sample size?

*In this and future exercises we will specify sampling with replacement to simulate the sampling from very large or infinite populations.

References

Annual Report: Florida Employer Opinion Survey. Florida Education and Training Placement Information Program, State of Florida Department of Education, June 1992.

Hogg, R. V., and Craig, A. T. *Introduction to Mathematical Statistics*, 4th ed., New York: Macmillan, 1978, Chapter 4.

Juran, J. M., and Gryna, F. M., Jr. *Quality Planning and Analysis*. New York: McGraw-Hill Book Company, 1980, Chapter 9.

Lindgren, B. W. *Statistical Theory*, 3d ed. New York: Macmillan, 1976, Chapter 2.

Mendenhall, W., Scheaffer, R. L., and Wackerly, D. *Mathematical Statistics with Applications*, 4th ed. Boston: PWS-Kent, 1990.

Moffitt, T. E., et al. "Socioeconomic status, IQ, and delinquency." *Journal of Abnormal Psychology*, 1981, 90, pp. 152–156.

Price, J. H., O'Connell, J., and Kukulka, G. "Preventative health behaviors related to the ten leading causes of mortality of health-fair attenders and nonattenders." *Psychological Reports*, 1985, 56, pp. 131–135.

Sheehan, E. P., McMenamin, N., and McDevitt, T. M. "Learning styles of traditional and nontraditional university students." *College Student Journal*, Dec. 1992, Vol. 26, No. 4.

CHAPTER SIX

Inferences Based on a Single Sample: Estimation

Contents

Case Study

Where We've Been

In the preceding chapters we learned that populations are characterized by numerical descriptive measures (called *parameters*) and that decisions about their values are based on sample statistics computed from sample data. Since statistics vary in a random manner from sample to sample, inferences based on them will be subject to uncertainty. This property is reflected in the sampling (probability) distribution of a statistic.

Where We're Going

This chapter puts all the preceding material into practice, that is, we estimate population means and proportions based on a single sample selected from the population of interest. Most important, we use the sampling distribution of a sample statistic to assess the reliability of an estimate.

The estimation of the mean gas mileage for a new car model, the estimation of the expected life of a computer monitor, and the estimation of the mean yearly sales for companies in the steel industry are problems with a common element. In each case, we are interested in estimating the mean of a population of measurements. This important problem constitutes the primary topic of this chapter.

You will see that different techniques are used for estimating a mean, depending on whether a sample contains a large or small number of measurements. Regardless, our objectives remain the same: We want to use the sample information to estimate the mean and to assess the reliability of the estimate.

In Sections 6.1 and 6.2 we consider a method of estimating a population mean using a large sample and develop a formula to determine just how large the sample must be to achieve a specified degree of reliability. In Section 6.3 we show how a mean can be estimated when only a small sample is available. Finally, estimation of binomial probabilities and the determination of sample sizes necessary to make reliable estimates are covered in Sections 6.4 and 6.5, respectively.

6.1 Large-Sample Estimation of a Population Mean

We illustrate the **large-sample method** of estimating a population mean with an example. Suppose a large hospital wants to estimate the average length of time patients remain in the hospital. To accomplish this objective, the hospital administrators plan to sample 100 of all previous patients' records and to use the sample mean, $\bar{x}$, of the lengths of stay to estimate the mean stay, μ, of *all* patients' visits. The sample mean $\bar{x}$ represents a *point estimator* of the population mean μ (Definition 5.4). How can we assess the accuracy of this point estimator?

According to the Central Limit Theorem, the sampling distribution of the sample mean is approximately normal for large samples, as shown in Figure 6.1. Let us calculate the interval

$$\bar{x} \pm 2\sigma_{\bar{x}} = \bar{x} \pm \frac{2\sigma}{\sqrt{n}}$$

That is, we will form an interval 4 standard deviations wide—from 2 standard deviations below the sample mean to 2 standard deviations above the mean. What are the chances (answer before we have drawn a sample) that this interval will enclose μ, the population mean?

To answer this question, refer to Figure 6.1. If the 100 measurements yield a value of $\bar{x}$ that falls between the two lines on either side of μ—i.e., within 2 standard deviations of μ—then the interval $\bar{x} \pm 2\sigma_{\bar{x}}$ will contain μ; if $\bar{x}$ falls outside these boundaries, the interval $\bar{x} \pm 2\sigma_{\bar{x}}$ will not contain μ. Since the area under the normal curve (the sampling distribution of $\bar{x}$) between these boundaries is about .95 (more precisely, from Table IV in Appendix A the area is .9544), we know that the interval $\bar{x} \pm 2\sigma_{\bar{x}}$ will contain μ with a probability approximately equal to .95.

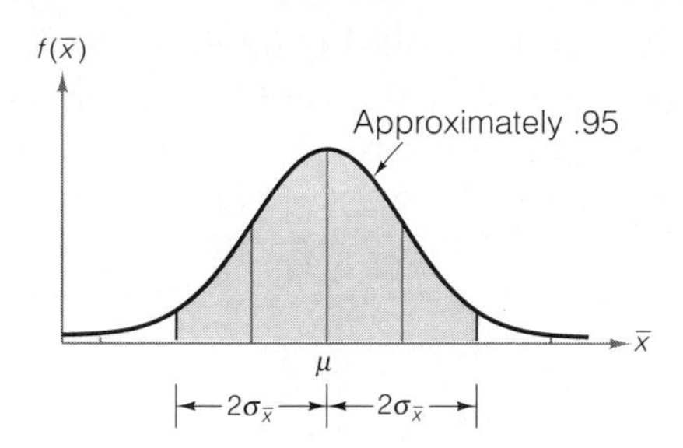

FIGURE 6.1 ▶
Sampling distribution of $\bar{x}$

To illustrate, suppose the sum and the sum of squared deviations for the sample of 100 lengths of time spent in the hospital are

$$\sum x = 465 \text{ days} \quad \text{and} \quad \sum (x - \bar{x})^2 = 2,387$$

Then

$$\bar{x} = \frac{\sum x}{n} = \frac{465}{100} = 4.65$$

$$s^2 = \frac{\sum (x - \bar{x})^2}{n - 1} = \frac{2,387}{99} = 24.11 \quad \text{and} \quad s = 4.9$$

Then, we consider the interval

$$\bar{x} \pm 2\sigma_{\bar{x}} = 4.65 \pm 2\frac{\sigma}{\sqrt{100}}$$

But now we face a problem. You can see that without knowing the standard deviation σ of the original population—i.e., the standard deviation of the lengths of stay of *all* patients—we cannot calculate this interval. However, since we have a large sample ($n = 100$ measurements), we can approximate the interval by using the sample standard deviation s to approximate σ. Thus,

$$\bar{x} \pm 2\frac{\sigma}{\sqrt{100}} \approx \bar{x} \pm 2\frac{s}{\sqrt{100}} = 4.65 \pm 2\left(\frac{4.9}{10}\right) = 4.56 \pm .98$$

That is, we estimate the mean length of stay in the hospital for all patients to fall in the interval 3.67 to 5.63 days.

Can we be sure that μ, the true mean, is in the interval 3.67 to 5.63? We cannot be certain, but we can be reasonably confident that it is. This confidence is derived from the knowledge that if we were to draw repeated random samples of 100 measurements from this population and form the interval $\bar{x} \pm 2\sigma_{\bar{x}}$ each time, approximately 95% of the intervals would contain μ. We have no way of knowing (without looking at all the patients' records) whether our sample interval is one of the 95% that contain μ or one of the 5% that do not, but the odds certainly favor its containing μ. Consequently, the interval 3.67 to 5.63 provides an estimate of the mean length of patient stay in the hospital. The formula that tells us how to calculate an interval estimate based on sample data is called an **interval estimator**. The probability, .95, that measures the confidence we can place in the interval estimate is called a **confidence coefficient**. The percentage, 95%, is called the **confidence level** for the interval estimate. It is not usually possible to assess precisely the reliability of point estimators because they are single points rather than intervals. Since we prefer to use estimators for which a measure of reliability can be calculated, interval estimators will generally be used.

Definition 6.1

An **interval estimator** is a formula that tells us how to use sample data to calculate an interval that estimates a population parameter.

Definition 6.2

The **confidence coefficient** is the probability that an interval estimator encloses the population parameter—i.e., the relative frequency with which the interval estimator encloses the population parameter when the estimator is used repeatedly a very large number of times.

The **confidence level** is the confidence coefficient expressed as a percentage.

Now we have seen how an interval can be used to estimate a population mean. When we use an interval estimator, we can usually calculate the probability that the estimation *process* will result in an interval that contains the true value of the population mean. That is, the probability that the interval contains the parameter in repeated usage is usually known. Figure 6.2 shows what happens when 10 different samples are drawn from a population, and a confidence interval for μ is calculated from each. The location of μ is indicated by the vertical line in the figure. Ten confidence intervals, each based on one of 10 samples, are shown as horizontal line segments. Note that the confidence intervals move from sample to sample—sometimes containing μ and other times missing μ. If our confidence level is 95%, then in the long run, 95% of our sample confidence intervals will contain μ.

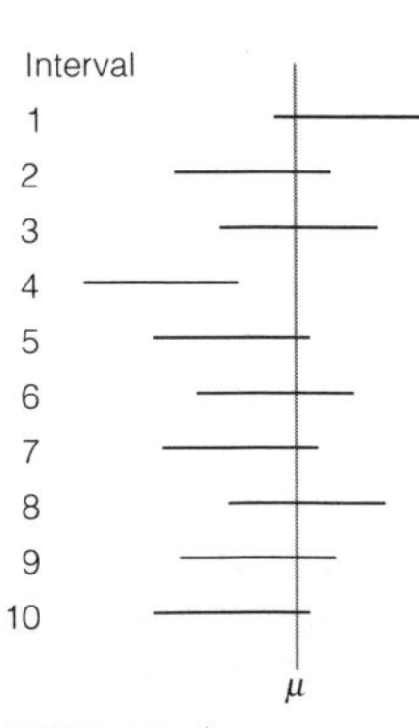

FIGURE 6.2 ▲ Interval estimators for μ: 10 samples

Suppose you wish to choose a confidence coefficient other than .95. Notice in Figure 6.1 that the confidence coefficient .95 is equal to the total area under the sampling distribution, less .05 of the area, which is divided equally between the two tails. Using this idea, we can construct a confidence interval with any desired confidence coefficient by increasing or decreasing the area (call it α) assigned to the tails of the sampling distribution (see Figure 6.3). For example, if we place area $\alpha/2$ in each tail and if $z_{\alpha/2}$ is the z value such that the area $\alpha/2$ will lie to its right, then the confidence interval with confidence coefficient $(1 - \alpha)$ is

$$\bar{x} \pm z_{\alpha/2}\sigma_{\bar{x}}$$

FIGURE 6.3 ▶
Locating $z_{\alpha/2}$ on the standard normal curve

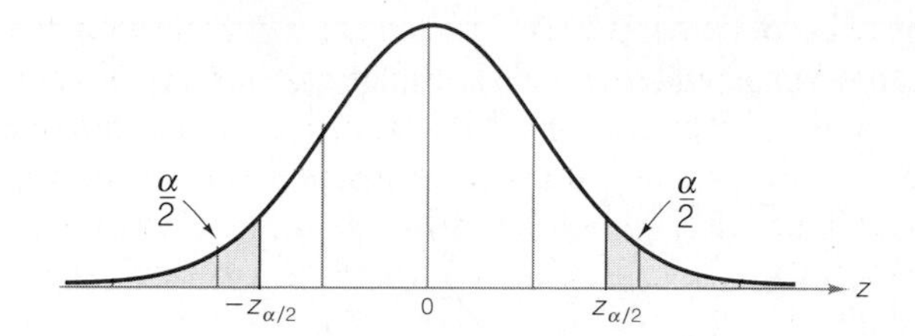

To illustrate, for a confidence coefficient of .90 we have $(1 - \alpha) = .90$, $\alpha = .10$, and $\alpha/2 = .05$; $z_{.05}$ is the z value that locates area .05 in the upper tail of the sampling distribution. Recall that Table IV in Appendix A (and inside the front cover of the book) gives the areas between the mean and a specified z value. Since the total area to the right of the mean is .5, we find that $z_{.05}$ will be the z value corresponding to an area of $.5 - .05 = .45$ to the right of the mean (see Figure 6.4). This z value is $z_{.05} = 1.645$. Confidence coefficients used in practice usually range from .90 to .99. The most commonly used confidence coefficients with corresponding values of α and $z_{\alpha/2}$ are shown in Table 6.1.

FIGURE 6.4 ▶
The z value ($z_{.05}$) corresponding to an area equal to .05 in the upper tail of the z-distribution

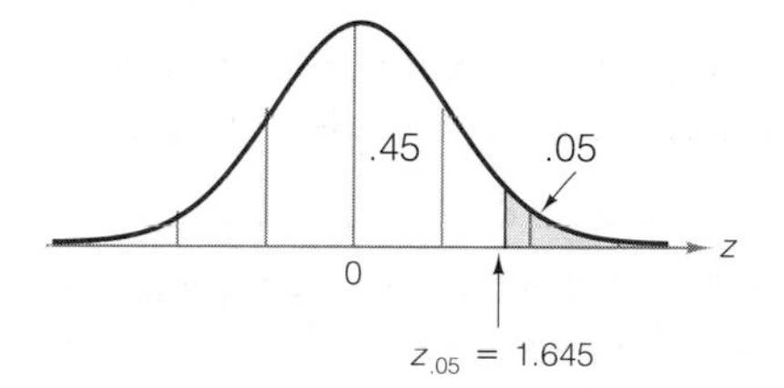

TABLE 6.1 Commonly Used Values of $z_{\alpha/2}$

Confidence Level $100(1 - \alpha)$	α	$\alpha/2$	$z_{\alpha/2}$
90%	.10	.05	1.645
95%	.05	.025	1.96
99%	.01	.005	2.58

Large-Sample 100(1 − α)% Confidence Interval for μ

$$\bar{x} \pm z_{\alpha/2}\sigma_{\bar{x}} = \bar{x} \pm z_{\alpha/2}\frac{\sigma}{\sqrt{n}}$$

where $z_{\alpha/2}$ is the z value with an area $\alpha/2$ to its right (see Figure 6.3) and $\sigma_{\bar{x}} = \sigma/\sqrt{n}$. The parameter σ is the standard deviation of the sampled population and n is the sample size.

When n is equal to 30 or more, the confidence interval is approximately equal to

$$\bar{x} \pm z_{\alpha/2}\left(\frac{s}{\sqrt{n}}\right)$$

where s is the sample standard deviation.

EXAMPLE 6.1

Unoccupied seats on flights cause airlines to lose revenue. Suppose a large airline wants to estimate its average number of unoccupied seats per flight over the past year. To accomplish this, the records of 225 flights are randomly selected, and the number of unoccupied seats is noted for each of the sampled flights. The sample mean and standard deviation are

$$\bar{x} = 11.6 \text{ seats} \qquad s = 4.1 \text{ seats}$$

Estimate μ, the mean number of unoccupied seats per flight during the past year, using a 90% confidence interval.

Solution

The general form of the 90% confidence interval for a population mean is

$$\bar{x} \pm z_{\alpha/2}\sigma_{\bar{x}} = \bar{x} \pm z_{.05}\sigma_{\bar{x}} = \bar{x} \pm 1.645\left(\frac{\sigma}{\sqrt{n}}\right)$$

For the 225 records sampled, we have

$$11.6 \pm 1.645\left(\frac{\sigma}{\sqrt{225}}\right)$$

Since we do not know the value of σ (the standard deviation of the number of unoccupied seats per flight for all flights of the year), we use our best approximation—the sample standard deviation s. Then the 90% confidence interval is, approximately,

$$11.6 \pm 1.645\left(\frac{4.1}{\sqrt{225}}\right) = 11.6 \pm .45$$

or from 11.15 to 12.05. That is, at the 90% confidence level, we estimate the mean number of unoccupied seats per flight to be between 11.15 and 12.05 during the sampled year. We stress that the confidence level refers to the procedure used. If we were to apply this procedure repeatedly to different samples, approximately 90% of the intervals would contain μ. We do not know whether this particular interval (11.15, 12.05) is one of the 90% that contain μ or one of the 10% that do not.

The interpretation of confidence intervals for a population mean is summarized in the accompanying box.

Interpretation of a Confidence Interval for a Population Mean

When we form a $100(1 - \alpha)\%$ confidence interval for μ, we usually express our confidence in the interval with a statement such as, "We can be $100(1 - \alpha)\%$ confident that μ lies between the lower and upper bounds of the confidence interval," where for a particular application, we substitute the appropriate numerical values for the confidence, and the lower and upper bounds. *The statement reflects our confidence in the estimation process rather than in the particular interval that is calculated from the sample data.* We know that repeated application of the same procedure will result in different lower and upper bounds on the interval. Furthermore, we know that $100(1 - \alpha)\%$ of the resulting intervals will contain μ. There is (usually) no way to determine whether any particular interval is one of those that contain μ, or one that does not. However, unlike point estimators, confidence intervals have some measure of reliability, the confidence coefficient, associated with them. For that reason they are generally preferred to point estimators.

CASE STUDY 6.1 / Dancing to the Customer's Tune: The Need to Assess Customer Preferences

The following quotations have been extracted from the December 13, 1976, issue of *Business Week*:*

"We're dancing to the tune of the customer as never before," says J. Janvier Wetzel, vice-president for sales promotion at Los Angeles-based Broadway Department Stores. "With population growth down to a trickle compared with its previous level, we're no longer spoiled with instant success every time we open a new store. Traditional department stores are locked in the biggest competitive battle in their history."

The nation's retailers are becoming uncomfortably aware that today's operating environment is vastly different from that of the 1960s. Population growth is slowing, a growing singles market is emerging, family formations are coming at later ages, and more women are embarking on careers. Of the 71 million households in the U.S. today, the dominant consumer buying segment is

*Reprinted by special permission. All rights reserved.

families headed by persons over 45. But by 1980 this group will have lost its majority status to the 25 to 40-year-old group. Merchants must now reposition their stores to attract these new customers.

To do so retailers are using market research to ferret out new purchasing attitudes and lifestyles and then translating this into customer buying segments. . . . Department stores are taking a hard look at some of the basics of their business by . . . spending heavily for far more elaborate market research. Data on demographics, psychographics (measurement of attitudes), and lifestyle are being fed into retailers' computers so they can make marketing decisions based on actual spending patterns and estimate their inventory needs with less risk.

In order to stock its various departments with the type and style of goods that appeal to its potential group of customers, a downtown department store should be interested in estimating the average age of downtown shoppers, not shoppers in general. Suppose a downtown department store questions 49 downtown shoppers concerning their age (the offer of a small gift certificate may help persuade shoppers to respond to such questions). The sample mean and standard deviation are found to be 40.1 years and 8.6 years, respectively. The store could then estimate the mean age μ of all downtown shoppers with a 95% confidence interval as follows:

$$\bar{x} \pm 1.96\left(\frac{s}{\sqrt{n}}\right) = 40.1 \pm 1.96\left(\frac{8.6}{\sqrt{49}}\right)$$
$$= 40.1 \pm 2.4$$

Thus, the department store should gear its sales to consumers with average age between 37.7 and 42.5.

Exercises 6.1 – 6.17

Learning the Mechanics

6.1 Find $z_{\alpha/2}$ for each of the following:
a. $\alpha = .10$ b. $\alpha = .01$ c. $\alpha = .05$ d. $\alpha = .20$

6.2 What is the confidence level of each of the following confidence intervals for μ?

a. $\bar{x} \pm 1.96\left(\frac{\sigma}{\sqrt{n}}\right)$ b. $\bar{x} \pm 1.645\left(\frac{\sigma}{\sqrt{n}}\right)$ c. $\bar{x} \pm 2.575\left(\frac{\sigma}{\sqrt{n}}\right)$

d. $\bar{x} \pm 1.282\left(\frac{\sigma}{\sqrt{n}}\right)$ e. $\bar{x} \pm .99\left(\frac{\sigma}{\sqrt{n}}\right)$

6.3 A random sample of 90 observations produced a mean $\bar{x} = 25.9$ and a standard deviation $s = 2.7$.
a. Find a 95% confidence interval for the population mean μ.
b. Find a 90% confidence interval for μ.
c. Find a 99% confidence interval for μ.

6.4 A random sample of 100 observations from a normally distributed population possesses a mean equal to 83.2 and a standard deviation equal to 6.4.
a. Find a 95% confidence interval for μ.
b. What is meant when you say that a confidence coefficient is .95?
c. Find a 99% confidence interval for μ.

d. What happens to the width of a confidence interval as the value of the confidence coefficient is increased while the sample size is held fixed?

e. Would your confidence intervals of parts **a** and **c** be valid if the distribution of the original population was not normal? Explain.

6.5 A random sample of n measurements was selected from a population with unknown mean μ and standard deviation σ. Calculate a 95% confidence interval for μ for each of the following situations.

a. $n = 75$, $\bar{x} = 28$, $s^2 = 12$ b. $n = 200$, $\bar{x} = 102$, $s^2 = 22$

c. $n = 100$, $\bar{x} = 15$, $s = .3$ d. $n = 100$, $\bar{x} = 4.05$, $s = .83$

e. Is the assumption that the underlying population of measurements is normally distributed necessary to assure the validity of the confidence intervals in parts **a–d**? Explain.

6.6 Explain the difference between an interval estimator and a point estimator for μ.

6.7 Explain what is meant by the statement, "We are 95% confident that an interval estimate contains μ."

6.8 Will a large-sample confidence interval be valid if the population from which the sample is taken is not normally distributed? Explain.

6.9 The mean and standard deviation of a random sample of n measurements are equal to 33.9 and 3.3, respectively.

a. Find a 95% confidence interval for μ if $n = 100$.

b. Find a 95% confidence interval for μ if $n = 400$.

c. Find the widths of the confidence intervals found in parts **a** and **b**. What is the effect on the width of a confidence interval of quadrupling the sample size while holding the confidence coefficient fixed?

Applying the Concepts

6.10 Research indicates that bicycle helmets save lives. A recent study reported in *Public Health Reports* (Otis et al., 1992) was intended to identify ways of encouraging helmet use in children. One of the variables measured was the children's perception of the risk involved in bicycling. A 4-point scale was used, with scores ranging from 1 (no risk) to 4 (very high risk). A sample of 797 children in grades 4–6 yielded the following results on the perception of risk variable: $\bar{x} = 3.39$, $s = .80$.

a. Calculate a 90% confidence interval for the average perception of risk for all students in grades 4–6. What assumptions did you make to assure the validity of the confidence interval?

b. If the population mean perception of risk exceeds 2.50, the researchers will conclude that students in these grades exhibit an awareness of the risk involved with bicycling. Interpret the confidence interval constructed in part **a** in this context.

6.11 A fact long known but little understood is that twins, in their early years, tend to have lower IQ's and pick up language more slowly than nontwins. Recently, psychologists have found that the slower intellectual growth of most twins may be caused by benign parental neglect. Suppose it is desired to estimate the mean attention time given to twins per week by their parents. A sample of 46 sets of 2½-year-old twin boys is taken, and at the end of 1 week the attention time given to each pair is recorded. The results are as follows: $\bar{x} = 22$ hours, $s = 16$ hours. Using the data, find a 90% confidence interval for the mean attention time given to all twin boys by their parents. Interpret the confidence interval.

6.12 A process has been developed that can transform ordinary iron into a kind of super-iron called *metallic glass* ("One Answer to Imports," 1981). Metallic glass is three to four times as strong as the toughest steel alloys, but it becomes brittle at very high temperatures. To estimate the mean temperature, μ, at which a particular type of metallic glass becomes brittle, 36 pieces of the metallic glass were randomly sampled from a recent production run. Each piece was independently subjected to higher and higher temperatures until it became brittle. The temperature at which brittleness was first noticed was recorded for each piece in the sample. The following results were obtained: $\bar{x} = 480°F$, $s = 11°F$. Use a 90% confidence interval to estimate μ. Interpret your confidence interval.

6.13 Automotive engineers are continually improving their products. Suppose a new type of brake light has been developed by General Motors. As part of a product safety evaluation program, General Motors' engineers wish to estimate the mean driver response time to the new brake light. (Response time is the length of time from the point that the brake is applied until the driver in the following car takes some corrective action.) Fifty drivers are selected at random and the response time (in seconds) for each driver is recorded, yielding the following results: $\bar{x} = .72$, $s^2 = .022$. Estimate the mean driver response time to the new brake light using a 99% confidence interval. Interpret the confidence interval.

6.14 Named for the section of the 1978 Internal Revenue Code that authorized them, 401(k) plans permit employees to shift part of their before-tax salaries into investments such as mutual funds. Employers typically match 50% of the employee's contribution up to about 6% of salary (Pare, 1992). One company, concerned with what it believed was a low employee participation rate in its 401(k) plan, sampled 30 other companies with similar plans and asked for their 401(k) participation rates. The following rates (in percentages) were obtained:

80	76	81	77	82	80	85	60	80	79
82	70	88	85	80	79	83	75	87	78
80	84	72	75	90	84	82	77	75	86

a. Use a 95% confidence interval to estimate the mean participation rate for all companies that have 401(k) plans.
b. Interpret the interval in the context of this problem.
c. What assumption is necessary to ensure the validity of this confidence interval?
d. If the company that conducted the sample has a 71% participation rate, can it safely conclude that its rate is below the population mean rate for all companies with 401(k) plans? Explain.
e. If in the data set the 60% had been 80%, how would the center and width of the confidence interval you constructed in part **a** be affected?

6.15 As an aid in the establishment of personnel requirements, the director of a hospital wishes to estimate the mean number of people who are admitted to the emergency room during a 24-hour period. The director randomly selects 64 different 24-hour periods and determines the number of admissions for each. For this sample, $\bar{x} = 19.8$ and $s^2 = 25$. Estimate the mean number of admissions per 24-hour period with a 95% confidence interval. Interpret the result.

6.16 An article titled "Scientists Seek Upper Hand in Insect Wars" (*Wall Street Journal*, March 6, 1984) notes that the cockroach has had 300 million years to develop a resistance to destruction. That they are "reproductively brilliant" is evidenced by a study conducted by researchers for S. C. Johnson & Son, Inc. (manufacturers of Raid and Off). Five thousand roaches, the expected number in a roach-infested house, were released in the Raid test kitchen. One week later the kitchen was fumigated and 16,298 dead roaches were counted, a gain of 11,298 roaches for the 1-week period. Assume that none of the original roaches died during the 1-week period and that the standard deviation of x, the number of roaches produced per roach in a 1-week period, is 1.5. Use the number of roaches produced by the sample of 5,000 roaches to find a 95% confidence interval for the mean number of roaches produced per week for each roach in a typical roach-infested house.

6.17 Recent research conducted by Peter Warr (1992) and reported in the *Journal of Psychology and Aging* studied the role that the age of workers has in determining their level of job satisfaction. The author hypothesized that both younger and older workers would have a higher job satisfaction rating than middle-age workers. Each of a sample of 1,686 adults was given a job satisfaction score based on answers to a series of questions. Higher job satisfaction scores indicate higher levels of job satisfaction. The data are given below arranged by age group.

	Age Group		
	Younger 18–24	Middle-Age 25–44	Older 45–64
$\bar{x}$	4.17	4.04	4.31
s	.75	.81	.82
n	241	768	677

a. Construct 95% confidence intervals for the mean job satisfaction scores of each age group. Carefully interpret each interval.

b. In the construction of three 95% confidence intervals, is it more or less likely that at least one of them will *not* contain the population mean it is intended to estimate than it is for a single confidence interval to miss the population mean? [*Hint:* Assume the three intervals are independent, and calculate the probability that at least one of them will not contain the population mean it estimates. Compare this probability to the probability that a single interval fails to enclose the mean.]

c. Based on these intervals, does it appear that the author's hypothesis is supported? [*Caution:* We will learn how to use sample information to compare population means in Chapter 8. Here, simply base your opinion on the individual confidence intervals you constructed in part **a**.]

6.2 Determining the Sample Size Necessary to Estimate a Population Mean

In many practical applications of statistics, the analyst observes data that have been collected by an outside agency not under his or her direction. Examples are data collected by federal agencies such as the Bureau of the Census or the Bureau of Labor Statistics, survey data collected by a national pollster, and data published by another researcher. Such data are called **observational**, since they are observed rather than collected by the analyst.

Sometimes the analyst must plan the sampling experiment that generates the data used to make inferences about the population. Such data are generated by **designed experiments**, and perhaps the most important design decision faced by the analyst is to determine the size of the sample. We will show in this section that the appropriate sample size for making an inference about a population mean depends on the desired reliability.

To see this, consider the example from Section 6.1 in which we estimated the mean length of stay for patients in a large hospital. A sample of 100 patients' records produced an estimate $\bar{x}$ that was within .98 day of the true mean length of stay, μ, for all the hospital's patients at the 95% confidence level. That is, the 95% confidence interval for μ was $2(.98) = 1.96$ days wide when 100 accounts were sampled. This is illustrated in Figure 6.5(a).

Now suppose we want to estimate μ to within .25 day with 95% confidence. That is, we want to narrow the width of the confidence interval from 1.96 days to .50 day, as shown in Figure 6.5(b). How much will the sample size have to be increased to accomplish this? If we want the estimator $\bar{x}$ to be within .25 day of μ, we must have

$$2\sigma_{\bar{x}} = .25 \quad \text{or, equivalently,} \quad 2\left(\frac{\sigma}{\sqrt{n}}\right) = .25$$

The necessary sample size is obtained by solving this equation for n. To do this we need an approximation for σ. We have an approximation from the initial sample of 100 patients' records—namely, the sample standard deviation, $s = 4.9$. Thus,

$$2\left(\frac{\sigma}{\sqrt{n}}\right) \approx 2\left(\frac{s}{\sqrt{n}}\right) = 2\left(\frac{4.9}{\sqrt{n}}\right) = .25$$

$$\sqrt{n} = \frac{2(4.9)}{.25} = 39.2$$

$$n = (39.2)^2 = 1{,}536.64$$

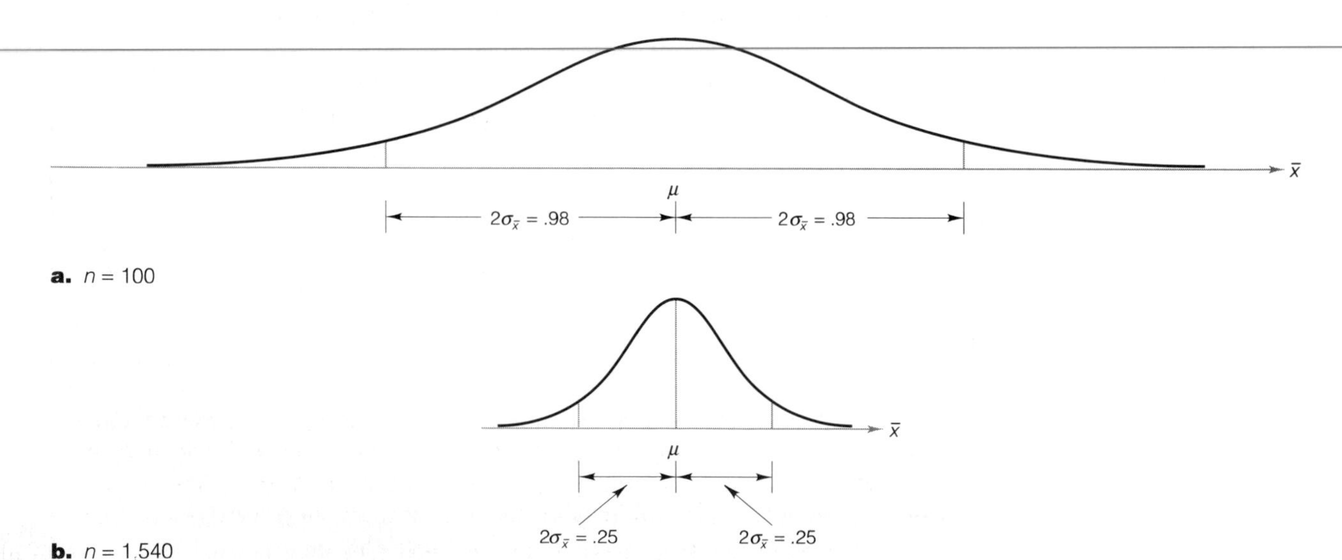

FIGURE 6.5 ▲ Relationship between sample size and width of confidence interval: Hospital stay example

Approximately 1,540 patients' records will have to be sampled to estimate the mean length of stay μ to within .25 day with (approximately) 95% confidence. The confidence interval resulting from a sample of this size will be approximately .50 day wide [see Figure 6.5(b)].

In general, we can express the reliability associated with a confidence interval for the population mean μ in one of two equivalent ways. We can specify the bound, B, within which we want to estimate μ with $100(1 - \alpha)\%$ confidence. The bound B then is equal to the half-width of the confidence interval, as shown in Figure 6.6. Equivalently, we can specify the total width, W, of the confidence interval for μ, also shown in Figure 6.6. Note that $W = 2B$.

FIGURE 6.6 ► Specifying the bound B or total width W for a confidence interval

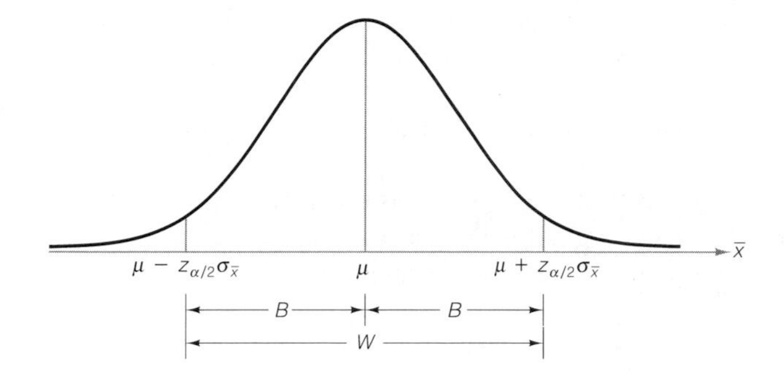

The procedure for finding the sample size necessary to estimate μ to within a given bound B or with a total interval width W is given in the following box.

Sample Size Determination for $100(1 - \alpha)\%$ Confidence Intervals for μ

In order to estimate μ to within a bound B or, equivalently, with a confidence interval of total width W with $100(1 - \alpha)\%$ confidence, the required sample size is found as follows:

$$z_{\alpha/2}\left(\frac{\sigma}{\sqrt{n}}\right) = B \quad \text{or} \quad z_{\alpha/2}\left(\frac{\sigma}{\sqrt{n}}\right) = \frac{W}{2}$$

The solution can be written in terms of either B or W as follows:

$$n = \frac{(z_{\alpha/2})^2\sigma^2}{B^2} \quad \text{or} \quad n = \frac{4(z_{\alpha/2})^2\sigma^2}{W^2}$$

The value of σ is usually unknown. It can be estimated by the standard deviation, s, from a prior sample. Alternatively, we may approximate the range R of observations in the population, and (conservatively) estimate $\sigma \approx R/4$. In any case, you should round the value of n obtained *upward* to ensure that the sample size will be sufficient to achieve the specified reliability.

EXAMPLE 6.2

Suppose the manufacturer of official NFL footballs uses a machine to inflate the new balls to a pressure of 13.5 pounds. When the machine is properly calibrated, the mean inflation pressure is 13.5 pounds, but uncontrollable factors cause the pressures of individual footballs to vary randomly from about 13.3 to 13.7 pounds. For quality control purposes, the manufacturer wishes to estimate the true mean inflation pressure with a 99% confidence interval that is only .05 pound wide. What sample size should be specified for the experiment?

Solution

For a 99% confidence interval, we have $z_{\alpha/2} = z_{.005} = 2.575$. To estimate σ, we note that the range of observations is $R = 13.7 - 13.3 = .4$ and use $\sigma \approx R/4 = .1$. Thus, in order to have a confidence interval of width $W = .05$, we use the formula derived in the box to find the sample size n:

$$n = \frac{4(z_{\alpha/2})^2\sigma^2}{W^2} \approx \frac{4(2.575)^2(.1)^2}{(.05)^2} = 106.09$$

We round this up to $n = 107$. Realizing that σ was approximated by $R/4$, we might even advise that the sample size be specified as $n = 110$ to be more certain of attaining the objective of a 99% confidence interval with width $W = .05$ pound or less.

Sometimes the formulas will lead to a solution that indicates a small sample size is sufficient to achieve the confidence interval goal. As we will see in Section 6.3, the procedures and assumptions for small samples differ from those for large samples. Therefore, if the formulas yield a small sample size ($n < 30$), one simple strategy is to select a sample size $n \geq 30$. Of course, the cost of sampling must also be considered when the sample size is being determined. Although more complex formulas can be derived to take sampling costs into account, these are beyond the scope of this text. For our purposes, it is sufficient to realize that a sampling budget may prove to be a restriction on the sample size, and therefore on the reliability of the confidence interval.

Exercises 6.18–6.28

Learning the Mechanics

6.18 If you wish to estimate a population mean correct to within a bound $B = .2$ with probability .95 and you know from prior sampling that σ^2 is approximately equal to 5.4, how many observations would have to be included in your sample?

6.19 If you wish to estimate a population mean with a 95% confidence interval of width $W = .2$ and you know from prior sampling that σ^2 is approximately equal to 5.4, how many observations would have to be included in your sample? Compare your answer to that for Exercise 6.18. Explain the difference.

6.20 Suppose you wish to estimate the mean of a normal population using a 95% confidence interval, and you know from prior information that $\sigma^2 \approx 1$.

a. To see the effect of the sample size on the width of the confidence interval, calculate the width W of the confidence interval for $n = 16, 25, 49, 100$, and 400.

b. Plot the width as a function of sample size n on graph paper. Connect the points by a smooth curve and note how the width decreases as n increases.

6.21 Suppose you wish to estimate a population mean correct to within a bound $B = .15$ with probability equal to .90. You do not know σ^2, but you know that the observations will range in value between 31 and 39.

a. Find the approximate sample size that will produce the desired accuracy of the estimate. You wish to be conservative to ensure that the sample size will be ample to achieve the desired accuracy of the estimate. [*Hint:* Using your knowledge of data variation from Section 2.5, assume that the range of the observations will equal 4σ.]

b. Calculate the approximate sample size making the less conservative assumption that the range of the observations is equal to 6σ.

6.22 It costs you $10 to draw a sample of size $n = 1$ and measure the attribute of interest. You have a budget of $1,200.

a. Do you have sufficient funds to estimate the population mean for the attribute of interest with a 95% confidence interval 4 units in width? Assume $\sigma = 12$.

b. If a 90% confidence level were used, would your answer to part **a** change? Explain.

Applying the Concepts

6.23 Refer to Exercise 6.17, in which workers' level of job satisfaction was related to age. Suppose that we want to estimate the mean job level satisfaction for younger workers (age 18–24) to within .04 with 95% confidence. How large a sample should be selected? Recall that the standard deviation for this age group was .75.

6.24 The EPA standard on the amount of suspended solids that can be discharged into rivers and streams is a maximum of 60 milligrams per liter daily, with a maximum monthly average of 30 milligrams per liter. Suppose you want to test a randomly selected sample of n water specimens and estimate the mean daily rate of pollution produced by a mining operation. If you want a 95% confidence interval estimate of width 2 milligrams, how many water specimens would you have to include in your sample? Assume prior knowledge indicates that pollution readings in water samples taken during a day are approximately normally distributed with a standard deviation equal to 5 milligrams.

6.25 Suppose a department store wants to estimate μ, the average age of the customers in its contemporary apparel department, correct to within 2 years with probability equal to .95. Approximately how large a sample would be required? [*Note:* Management does not know the standard deviation σ but guesses that the ages of its customers range from 15 to 45. Use a conservative approximation for σ to calculate n.]

6.26 According to a Food and Drug Administration (FDA) study, a cup of coffee contains an average of 115 milligrams of caffeine, with the amount per cup ranging from 60 to 180 milligrams. In contrast, sugar-free Mr. Pibb tested at 58.8 milligrams of caffeine per 12-ounce serving, Coca-Cola and Diet Coke at 45.6 milligrams, and Pepsi at 38.4 milligrams. Suppose you want to repeat the FDA experiment to obtain an estimate of the mean caffeine content in a cup of coffee correct to within 5 milligrams with 95% confidence. How many cups of coffee would have to be included in your sample?

6.27 The United States Golf Association (USGA) tests all new brands of golf balls to assure that they meet USGA specifications. One test conducted is intended to measure the average distance traveled when the ball is hit by a machine called "Iron Byron," a name inspired by the swing of the famous golfer Byron Nelson. Suppose the USGA wishes to estimate the mean distance for a new brand with a 90% confidence interval of width 2 yards. Assume that past tests have indicated that the standard deviation of the distances Iron Byron hits golf balls is approximately 10 yards. How many golf balls should be hit by Iron Byron to achieve the desired accuracy in estimating the mean?

6.28 It costs more to produce defective items—since they must be scrapped or reworked—than it does to produce nondefective items. This simple fact suggests that manufacturers should ensure the quality of their products by perfecting their production processes rather than through inspection of finished products (Deming, 1986). In order to better understand a particular metal stamping process, a manufacturer wishes to estimate the mean length of items produced by the process during the past 24 hours.

a. How many parts should be sampled in order to estimate the population mean to within .1 mm with 90% confidence? Previous studies of this machine have indicated that the standard deviation of lengths produced by the stamping operation is about 2 mm.

b. Time permits the use of a sample size no larger than 100. If a 90% confidence interval for μ is constructed using $n = 100$, will it be wider or narrower than would have been obtained using the sample size determined in part **a**? Explain.

c. If management requires that μ be estimated to within .1 mm and that a sample size of no more than 100 be used, what is (approximately) the maximum confidence level that could be attained for a confidence interval that meets management's specifications?

6.3 Small-Sample Estimation of a Population Mean

Federal legislation requires pharmaceutical companies to perform extensive tests on new drugs before they can be marketed. Initially, a new drug is tested on animals. If the drug is deemed safe after this first phase of testing, the pharmaceutical company is then permitted to begin human testing on a limited basis. During this second phase, inferences must be made about the safety of the drug based on information in very small samples.

Suppose a pharmaceutical company must estimate the average increase in blood pressure of patients who take a certain new drug. Assume that only six patients can be used in the initial phase of human testing. The use of a **small sample** in making an inference about μ presents two immediate problems when we attempt to use the standard normal z as a test statistic.

PROBLEM 1

The shape of the sampling distribution of the sample mean $\bar{x}$ (and the z statistic) now depends on the shape of the population that is sampled. We can no longer assume that the sampling distribution of $\bar{x}$ is approximately normal, because the Central Limit Theorem assures normality only for samples that are sufficiently large.

PROBLEM 2

The population standard deviation σ is almost always unknown. Although it is still true that $\sigma_{\bar{x}} = \sigma/\sqrt{n}$, the sample standard deviation s may provide a poor approximation for σ when the sample size is small.

Solution to Problem 1

The sampling distribution of $\bar{x}$ (and z) is exactly normal even for relatively small samples if the sampled population is normal. It is approximately normal if the sampled population is approximately normal.

Solution to Problem 2

Instead of using the standard normal statistic

$$z = \frac{\bar{x} - \mu}{\sigma_{\bar{x}}} = \frac{\bar{x} - \mu}{\sigma/\sqrt{n}}$$

which requires knowledge of or a good approximation to σ, we define and use the statistic

$$t = \frac{\bar{x} - \mu}{s/\sqrt{n}}$$

in which the sample standard deviation, s, replaces the population standard deviation, σ.

The distribution of the ***t* statistic** in repeated sampling was discovered by W. S. Gosset, a scientist in the Guinness brewery in Ireland, who published his discovery in 1908 under the pen name of Student. The main result of Gosset's work is that if we are sampling from a normal distribution, the t statistic has a sampling distribution very much like that of the z statistic: mound-shaped, symmetric, with mean 0. The primary difference between the sampling distributions of t and z is that the t statistic is more variable than the z, which follows intuitively when you realize that t contains two random quantities ($\bar{x}$ and s), whereas z contains only one ($\bar{x}$).

The actual amount of variability in the sampling distribution of t depends on the sample size n. A convenient way of expressing this dependence is to say that the t statistic has $(n - 1)$ **degrees of freedom (df)**. Recall that the quantity $(n - 1)$ is the divisor that appears in the formula for s^2. This number plays a key role in the sampling distribution of s^2 and appears in discussions of other statistics in later chapters. Particularly, the smaller the number of degrees of freedom associated with the t statistic, the more variable will be its sampling distribution.

In Figure 6.7 (page 288) we show both the sampling distribution of z and the sampling distribution of a t statistic with 4 df. You can see that the increased variability of the t statistic means that the t value, t_α, that locates an area α in the upper tail of the t-distribution is larger than the corresponding value z_α. For any given value of α, the t value t_α increases as the number of degrees of freedom (df) decreases. Values of t that will be used in forming small-sample confidence intervals for μ are given in Table V of Appendix A and inside the front cover of the text. A partial reproduction of this table is shown in Table 6.2 on page 288.

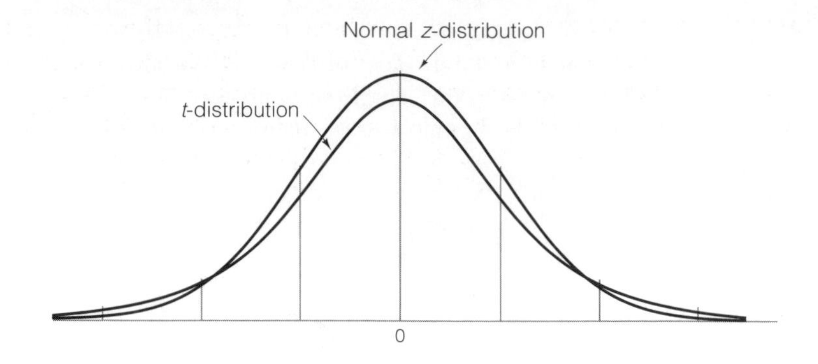

FIGURE 6.7 ▶ Standard normal (z) distribution and t-distribution with 4 df

TABLE 6.2 Reproduction of Part of Table V in Appendix A

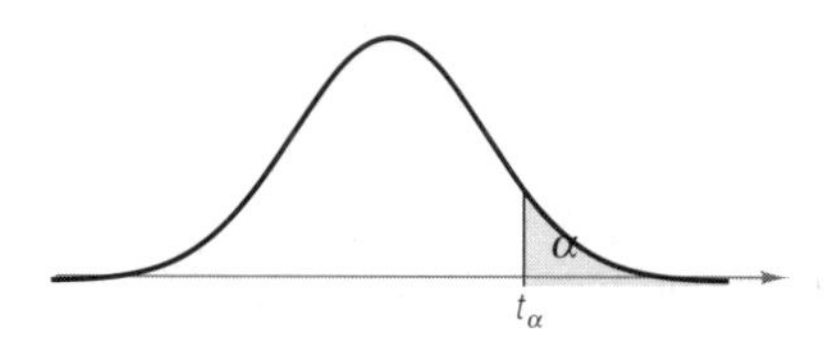

Degrees of Freedom	$t_{.100}$	$t_{.050}$	$t_{.025}$	$t_{.010}$	$t_{.005}$	$t_{.001}$	$t_{.0005}$
1	3.078	6.314	12.706	31.821	63.657	318.31	636.62
2	1.886	2.920	4.303	6.965	9.925	22.326	31.598
3	1.638	2.353	3.182	4.541	5.841	10.213	12.924
4	1.533	2.132	2.776	3.747	4.604	7.173	8.610
5	1.476	2.015	2.571	3.365	4.032	5.893	6.869
6	1.440	1.943	2.447	3.143	3.707	5.208	5.959
7	1.415	1.895	2.365	2.998	3.499	4.785	5.408
8	1.397	1.860	2.306	2.896	3.355	4.501	5.041
9	1.383	1.833	2.262	2.821	3.250	4.297	4.781
10	1.372	1.812	2.228	2.764	3.169	4.144	4.587
11	1.363	1.796	2.201	2.718	3.106	4.025	4.437
12	1.356	1.782	2.179	2.681	3.055	3.930	4.318
13	1.350	1.771	2.160	2.650	3.012	3.852	4.221
14	1.345	1.761	2.145	2.624	2.977	3.787	4.140
15	1.341	1.753	2.131	2.602	2.947	3.733	4.073
⋮	⋮	⋮	⋮	⋮	⋮	⋮	⋮
∞	1.282	1.645	1.960	2.326	2.576	3.090	3.291

Note that t_α values are listed for degrees of freedom from 1 to 29, where α refers to the tail area under the t-distribution to the right of t_α. For example, if we want the t value with an area of .025 to its right and 4 df, we look in the table under the column $t_{.025}$ for the entry in the row corresponding to 4 df. This entry is $t_{.025} = 2.776$, as shown in Figure 6.8. The corresponding standard normal z-score is $z_{.025} = 1.96$.

FIGURE 6.8 ▶ The $t_{.025}$ value in a t-distribution with 4 df and the corresponding $z_{.025}$ value

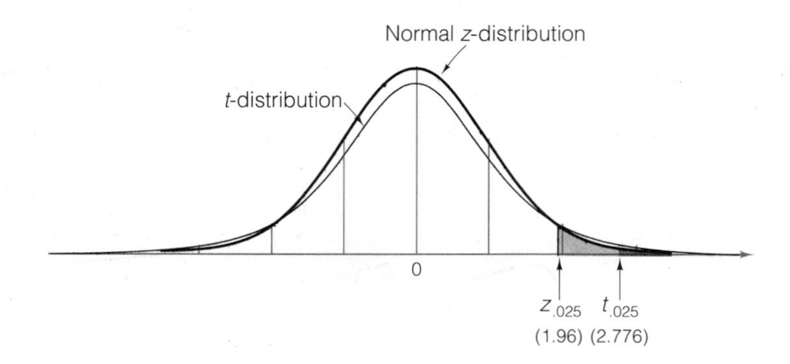

Note that the last row of Table V, where df = infinity, contains the standard normal z values. This follows from the fact that as the sample size n grows very large, s becomes closer to σ and thus t becomes closer in distribution to z. In fact, when df = 29, there is little difference between corresponding tabulated values of z and t. Thus, we choose the arbitrary cutoff of $n = 30$ (df = 29) to distinguish between the large-sample and small-sample inferential techniques.

Returning to the example of testing a new drug, suppose that the six test patients have blood pressure increases of 1.7, 3.0, .8, 3.4, 2.7, and 2.1 points. We calculate

$$\bar{x} = \frac{\sum x}{n} = \frac{13.7}{6} = 2.28$$

$$s^2 = \frac{\sum (x - \bar{x})^2}{n - 1} = \frac{\sum x^2 - \frac{\left(\sum x\right)^2}{n}}{n - 1} = \frac{35.79 - \frac{(13.7)^2}{6}}{5} = .9017$$

$$s = \sqrt{s^2} = .950$$

How can we use this information to construct a 95% confidence interval for μ, the mean increase in blood pressure associated with the new drug for all patients in the population?

First, we know that we are dealing with a sample too small to assume that the sample mean $\bar{x}$ is approximately normally distributed by the Central Limit Theorem.

That is, we do not get the normal distribution of $\bar{x}$ "automatically" from the Central Limit Theorem when the sample size is small. Instead, we must assume that the measured variable, in this case the increase in blood pressure, is normally distributed in order for the distribution of $\bar{x}$ to be normal.

Second, unless we are fortunate enough to know the population standard deviation σ, which in this case represents the standard deviation of *all* the patients' increase in blood pressure when they take the new drug, we cannot use the standard normal z statistic to form our confidence interval for μ. Instead, we must use the t-distribution, with $(n - 1)$ degrees of freedom.

In this case, $n - 1 = 5$ df, and the t value is found in Table 6.2 (or inside front cover) to be

$$t_{.025} = 2.571 \qquad \text{with 5 df}$$

Recall that the large-sample confidence interval would have been of the form

$$\bar{x} \pm z_{\alpha/2}\sigma_{\bar{x}} = \bar{x} \pm z_{\alpha/2}\frac{\sigma}{\sqrt{n}} = \bar{x} \pm z_{.025}\frac{\sigma}{\sqrt{n}}$$

where 95% is the desired confidence level. To form the interval for a small sample *from a normal distribution, we simply substitute t for z and s for σ in the preceding formula*:

$$\bar{x} \pm t_{\alpha/2}\frac{s}{\sqrt{n}}$$

Substituting the numerical values, we get

$$2.28 \pm (2.571)\left(\frac{.95}{\sqrt{6}}\right) = 2.28 \pm 1.00$$

or 1.28 to 3.28 points. That is, we can be 95% confident that the mean increase in blood pressure associated with taking this new drug is between 1.28 and 3.28 points. As with our large-sample interval estimates, our confidence is in the process, not in this particular interval. We know that the procedure we used will produce an interval that contains the true mean μ 95% of the time we utilize it, *assuming that the probability distribution of changes in blood pressure from which our sample was selected is normal.* The latter assumption is necessary for the small-sample interval to be valid.

What price did we pay for having to utilize a small sample to make the inference? First, we had to assume the underlying population is normally distributed, and if the assumption is invalid, our interval might also be invalid.* Second, we had to form the interval using a t value of 2.571 rather than a z value of 1.96, resulting in a wider interval to achieve the same 95% level of confidence. If the interval from 1.28 to 3.28 is too wide to be of much use, then we know how to remedy the situation: increase the number of patients sampled in order to decrease the interval width (on average).

---

*By *invalid*, we mean that the probability that the procedure will yield an interval that contains μ is not equal to $(1 - \alpha)$. Generally, if the underlying population is approximately normal, then the confidence coefficient will approximate the probability that the interval contains μ.

The procedure for forming a small-sample confidence interval is summarized in the accompanying box.

Small-Sample Confidence Interval for μ*

$$\bar{x} \pm t_{\alpha/2}\left(\frac{s}{\sqrt{n}}\right)$$

where $t_{\alpha/2}$ is based on $(n - 1)$ degrees of freedom

Assumption: A random sample is selected from a population with a relative frequency distribution that is approximately normal.

EXAMPLE 6.3

Some quality control experiments require *destructive sampling* (i.e., the test to determine whether the item is defective destroys the product) in order to measure some particular characteristic of the product. For example, suppose a manufacturer of printers for personal computers wishes to estimate the mean number of characters printed before the printhead fails. The cost of destructive sampling often dictates small samples. Suppose the printer manufacturer tests $n = 15$ printheads and calculates the following statistics:

$$\bar{x} = 1.23 \text{ million characters} \qquad s = .27 \text{ million characters}$$

Form a 99% confidence interval for the mean number of characters printed before the printhead fails.

Solution

If we assume that the number of characters printed before printhead failure is normally distributed, we can use the t statistic to form the confidence interval. We use a confidence coefficient of .99 and $n - 1 = 14$ degrees of freedom to find in Table V:

$$t_{\alpha/2} = t_{.005} = 2.977$$

Thus, the small sample forces us to assume normality and extend the interval almost 3 standard deviations (of $\bar{x}$) on each side of the sample mean in order to form the 99% confidence interval. For these data, the interval is

$$\bar{x} \pm t_{.005}\left(\frac{s}{\sqrt{n}}\right) = 1.23 \pm 2.977\left(\frac{.27}{\sqrt{15}}\right)$$
$$= 1.23 \pm .21 \quad \text{or} \quad (1.02, 1.44)$$

*The procedure given in the box assumes that the population standard deviation σ is unknown, which is almost always the case. If σ is known, we can form the small-sample confidence interval just as we would a large-sample confidence interval using a standard normal z value instead of t. However, we must still assume that the underlying population is approximately normal.

Thus, the manufacturer can be 99% confident that the printhead has a mean life between 1.02 and 1.44 million characters. If the manufacturer were to advertise that the mean life of its printheads is (at least) 1 million characters, the interval would support such a claim. Our confidence is derived from the fact that 99% of the intervals formed in repeated applications of this procedure will contain μ.

We have emphasized throughout this section that an assumption that the population is approximately normally distributed is necessary for making small-sample inferences about μ when using the t statistic. Although many phenomena do have approximately normal distributions, it is also true that many random phenomena have distributions that are not normal or even mound-shaped. Empirical evidence acquired over the years has shown that the t-distribution is rather insensitive to moderate departures from normality. That is, use of the t statistic when sampling from mound-shaped populations generally produces credible results; however, for cases in which the distribution is distinctly nonnormal, either take a large sample or use a *nonparametric method*. A simple nonparametric procedure for making inferences about a population mean is described in Chapter 7.

What Do You Do When the Population Relative Frequency Distribution Departs Greatly from Normality?

Answer: Use the nonparametric statistical methods of optional Section 7.5.

Exercises 6.29–6.40

Learning the Mechanics

6.29 Explain the differences in the sampling distributions of $\bar{x}$ for large and small samples under the following assumptions.

a. The variable of interest, x, is normally distributed.
b. Nothing is known about the distribution of the variable x.

6.30 Suppose you have selected a random sample of $n = 7$ measurements from a normal distribution. Compare the standard normal z values with the corresponding t values if you were forming the following confidence intervals.

a. 80% confidence interval
b. 90% confidence interval
c. 95% confidence interval
d. 98% confidence interval
e. 99% confidence interval
f. Use the table values you obtained in parts **a–e** to sketch the z- and t-distributions. What are the similarities and differences?

6.31 Let t_0 be a specific value of t. Use Table V in Appendix A to find t_0 values such that the following statements are true.

a. $P(t \geq t_0) = .025$ where df = 10
b. $P(t \geq t_0) = .01$ where df = 17
c. $P(t \leq t_0) = .005$ where df = 6
d. $P(t \leq t_0) = .05$ where df = 13

6.32 The following random sample was selected from a normal distribution: 4, 6, 3, 5, 9, 3.

a. Construct a 90% confidence interval for the population mean μ.
b. Construct a 95% confidence interval for the population mean μ.
c. Construct a 99% confidence interval for the population mean μ.
d. Assume that the sample mean $\bar{x}$ and sample standard deviation s remain exactly the same as those you just calculated but that they are based on a sample of $n = 25$ observations rather than $n = 6$ observations. Repeat parts **a–c**. What is the effect of increasing the sample size on the width of the confidence intervals?

6.33 The following sample of 16 measurements was selected from a population that is approximately normally distributed:

91	80	99	110	95	106	78	121
106	100	97	82	100	83	115	104

a. Construct an 80% confidence interval for the population mean.
b. Construct a 95% confidence interval for the population mean and compare the width of this interval with that of part **a**.
c. Carefully interpret each of the confidence intervals, and explain why the 80% confidence interval is narrower.

Applying the Concepts

6.34 Pulse rate is an important measure of the fitness of a person's cardiovascular system. The mean pulse rate for American adult males is approximately 72 heart beats per minute. A random sample of 21 American adult males who jog at least 15 miles per week had a mean pulse rate of 52.6 beats per minute and a standard deviation of 3.22 beats per minute.

a. Find a 95% confidence interval for the mean pulse rate of all American adult males who jog at least 15 miles per week.
b. Interpret the interval found in part **a**.
c. What assumptions are required for the validity of the confidence interval?

6.35 Health insurers and the federal government are both putting pressure on hospitals to shorten the average length of stay (LOS) of their patients. A random sample of 20 hospitals in one state had a mean LOS in 1990 of 3.8 days and a standard deviation of 1.2 days.

a. Use a 90% confidence interval to estimate the population mean of the LOS for the state's hospitals in 1990.
b. Interpret the interval in terms of this application.
c. What is meant by the phrase "90% confidence interval"?

6.36 United States firms are increasingly looking to the international market for expansion and growth. A random sample of 15 of the *Fortune* 1,000 firms is selected, and the percentage of their 1990 revenues from foreign sales is recorded for each. The mean is 23.8% and the standard deviation is 9.4%.

a. Use a 99% confidence interval to estimate the mean percentage of 1990 foreign sales for all large U.S. firms.

b. Interpret the interval in terms of this application.
c. What assumption is necessary to ensure the validity of this confidence interval?

6.37 A company purchases large quantities of naphtha in 50-gallon drums. Because the purchases are ongoing, small shortages in the drums can represent a sizable loss to the company. The weights of the drums vary slightly from drum to drum, so the weight of the naphtha is determined by removing it from the drums and measuring it. Suppose the company samples the contents of 20 drums, measures the naphtha in each, and calculates $\bar{x} = 49.70$ gallons and $s = .32$ gallon. Find a 95% confidence interval for the mean number of gallons of naphtha per drum. What assumptions are necessary to assure the validity of the confidence interval?

6.38 A study indicated that the cost of hiring an employee (excluding salary) ranges from about $1,500 for a secretary to more than $40,000 for a manager (Dessler, 1986). To estimate its mean cost of hiring an entry-level secretary, a large corporation randomly selected eight of the entry-level secretaries it had hired during the last 2 years and determined the costs (in dollars) involved in hiring each. The following data were obtained:

2,100 1,650 1,315 2,035 2,245 1,980 1,700 2,190

Assume that the population from which these data were sampled is approximately normally distributed.
a. Describe the population from which the corporation collected the sample data.
b. Use a 90% confidence interval to estimate the mean of interest to the corporation.
c. How wide is the confidence interval you constructed in part b? Would a 95% confidence interval be wider or narrower? Explain.

6.39 A *mortgage* is a type of loan that is secured by a designated piece of property. If the borrower defaults on the loan, the lender can sell the property to recover the outstanding debt. A federal bank examiner is interested in estimating the mean outstanding principal balance of all home mortgages foreclosed by the bank due to default by the borrower during the last 3 years. A random sample of 12 foreclosed mortgages yielded the following data (in dollars):

95,982	81,422	39,888	46,836	66,899	69,110
59,200	62,331	105,812	55,545	56,635	72,123

a. Describe the population from which the bank examiner collected the sample data. What characteristic must this population possess to enable us to construct a confidence interval for the mean outstanding principal balance using the method described in this section?
b. Construct a 90% confidence interval for the mean of interest.
c. Carefully interpret your confidence interval in the context of the problem.

6.40 The Clinton administration considers the increasing cost of health care a primary issue. Suppose that a random sample of 23 small companies (companies with less than $10 million in annual revenues) that offer paid health insurance as a benefit was selected. The mean health insurance cost per worker per month was $135, and the standard deviation was $32.

a. Use a 95% confidence interval to estimate the mean cost per worker per month for all small companies.
b. What assumption is necessary to ensure the validity of the confidence interval?
c. What is meant by the phrase "95% confidence interval"?

6.4 Large-Sample Estimation of a Binomial Probability

In recent years the number of public opinion polls has grown at an astounding rate. Almost daily, the news media report the results of some poll. Pollsters regularly determine the percentage of people in favor of the president's energy program, the fraction of voters in favor of a certain candidate, the fraction of customers who favor a particular brand of wine, and the proportion of people who smoke cigarettes. In each case, we are interested in estimating the percentage (or proportion) of some group with a certain characteristic. In this section we will consider methods for making inferences about population proportions.

EXAMPLE 6.4

The mid-1970s may well be remembered for political unrest across the country. Since the days of Watergate, public opinion polls have been conducted to estimate the fraction of Americans who trust the president. Suppose 1,000 people are randomly chosen and 637 answer that they trust the president. How would you estimate the true fraction of *all* American people who trust the president?

Solution

What we have really asked is how would you estimate the probability p of success in a binomial experiment, where p is the probability that a person chosen trusts the president. One logical method of estimating p for the population is to use the proportion of successes in the sample. That is, we can estimate p by calculating

$$\hat{p} = \frac{\text{Number of people sampled who trust the president}}{\text{Number of people sampled}}$$

where $\hat{p}$ is read "p hat." Thus, in this case,

$$\hat{p} = \frac{637}{1{,}000} = .637$$

To determine the reliability of the estimator $\hat{p}$, we need to know its sampling distribution. That is, if we were to draw samples of 1,000 people over and over again, each time calculating a new estimate $\hat{p}$, what would be the frequency distribution of all the $\hat{p}$ values? The answer lies in viewing $\hat{p}$ as the average, or mean, number of successes per trial over the n trials. If each success is assigned a value equal to 1 and a failure is assigned a value of 0, then the sum of all n sample observations is x, the total number of successes, and $\hat{p} = x/n$ is the average, or mean, number of successes per trial in the n trials. The Central Limit Theorem tells us that the relative frequency distribution of the sample mean for any population is approximately normal for sufficiently large samples.

We demonstrated the properties of $\hat{p}$ in Example 5.9. The repeated sampling distribution of $\hat{p}$ has the characteristics listed in the accompanying box and shown in Figure 6.9 on page 296.

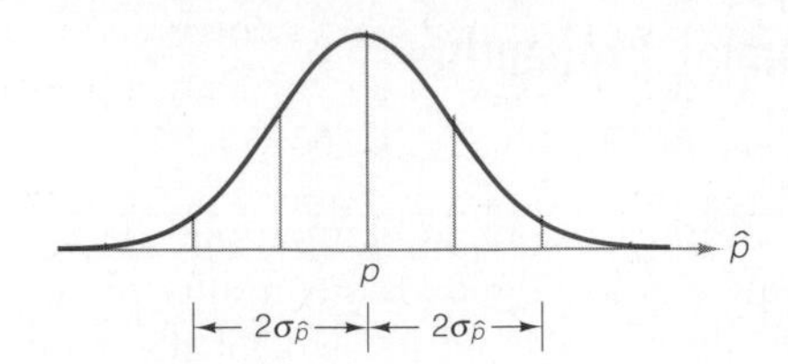

FIGURE 6.9 ► Sampling distribution of $\hat{p}$

Sampling Distribution of $\hat{p}$

1. The mean of the sampling distribution of $\hat{p}$ is p; that is, $\hat{p}$ is an unbiased estimator of p.
2. The standard deviation of the sampling distribution of $\hat{p}$ is $\sqrt{pq/n}$; that is, $\sigma_{\hat{p}} = \sqrt{pq/n}$, where $q = 1 - p$.
3. For large samples, the sampling distribution of $\hat{p}$ is approximately normal. A sample size is considered large if the interval $\hat{p} \pm 3\sigma_{\hat{p}}$ does not include 0 or 1. [*Note:* This requirement is almost equivalent to that given in Section 4.7 for approximating a binomial distribution with a normal one. The difference is that we assumed p to be known in Section 4.7; now we are trying to make inferences about an unknown p, so we use $\hat{p}$ to estimate p in checking the adequacy of the normal approximation.]

The fact that $\hat{p}$ is a "sample mean fraction of successes" allows us to form confidence intervals about p in a manner that is completely analogous to that used for large-sample estimation of μ.

Large-Sample Confidence Interval for p

$$\hat{p} \pm z_{\alpha/2}\sigma_{\hat{p}} = \hat{p} \pm z_{\alpha/2}\sqrt{\frac{pq}{n}} \approx \hat{p} \pm z_{\alpha/2}\sqrt{\frac{\hat{p}\hat{q}}{n}}$$

where $\hat{p} = \frac{x}{n}$ and $\hat{q} = 1 - \hat{p}$

Note: When n is large, we can use $\hat{p}$ to approximate the value of p in the formula for $\sigma_{\hat{p}}$.

Thus, if 637 of 1,000 Americans say they trust the president, a 95% confidence interval for the proportion of *all* Americans who trust the president is

$$\hat{p} \pm z_{\alpha/2}\sigma_{\hat{p}} = .637 \pm 1.96\sqrt{\frac{pq}{1{,}000}}$$

TABLE 6.3 Values of pq for Several Different Values of p

p	pq	$\sqrt{pq}$
.5	.25	.50
.6 or .4	.24	.49
.7 or .3	.21	.46
.8 or .2	.16	.40
.9 or .1	.09	.30

where $q = 1 - p$. Just as we needed an approximation for σ in calculating a large-sample confidence interval for μ, we now need an approximation for p. As Table 6.3 shows, the approximation for p does not have to be especially accurate, because the value of $\sqrt{pq}$ needed for the confidence interval is relatively insensitive to changes in p. Therefore, we can use $\hat{p}$ to approximate p. Keeping in mind that $\hat{q} = 1 - \hat{p}$, we substitute these values into the formula for the confidence interval:

$$\hat{p} \pm 1.96\sqrt{pq/1{,}000} \approx \hat{p} \pm 1.96\sqrt{\hat{p}\hat{q}/1{,}000}$$
$$= .637 \pm 1.96\sqrt{(.637)(.363)/1{,}000} = .637 \pm .030$$
$$= (.607, .667)$$

Then we can be 95% confident that the interval from 60.7% to 66.7% contains the true percentage of *all* Americans who trust the president. That is, in repeated construction of confidence intervals, approximately 95% of all samples would produce confidence intervals that enclose p. Note that the guidelines for interpreting a confidence interval about μ also apply to interpreting a confidence interval for p because p is the "population mean fraction of successes" in a binomial experiment.

EXAMPLE 6.5

United States law requires companies to treat minorities fairly in their hiring, promotion, and pay practices. The Equal Employment Opportunity Commission (EEOC) is the agency charged with the responsibility of monitoring the treatment of minorities in the workplace. Suppose the EEOC samples 135 recent hires of one large company and determines that 12 are minorities. Use a 90% confidence interval to estimate the proportion of all new hires of the company that are minorities.

Solution

The number, x, of the 135 sampled hires who are minorities is a binomial random variable if we can assume that the sample was randomly selected from all the company's hires and that the process by which the company hires is relatively stable during the period of interest (so that the probability of a minority hire remains constant over the period).

Then the point estimate of the proportion of minority hires for this company is

$$\hat{p} = \frac{x}{n} = \frac{12}{135} = .089$$

We first check to be sure that the sample size is sufficiently large that the normal distribution provides a reasonable approximation for this binomial proportion. We check the 3-standard-deviation interval around $\hat{p}$:

$$\hat{p} \pm 3\sigma_{\hat{p}} \approx \hat{p} \pm 3\sqrt{\frac{\hat{p}\hat{q}}{n}}$$
$$= .089 \pm 3\sqrt{\frac{(.089)(.911)}{135}} = .089 \pm .074 = (.015, .163)$$

Since this interval is wholly contained in the interval (0, 1), we may conclude that the normal approximation is reasonable.

We now form the 90% confidence interval for p, the true proportion of minority hires for the company:

$$\hat{p} \pm z_{\alpha/2}\sigma_{\hat{p}} = \hat{p} \pm z_{\alpha/2}\sqrt{\frac{pq}{n}} \approx \hat{p} \pm z_{\alpha/2}\sqrt{\frac{\hat{p}\hat{q}}{n}}$$

$$= .089 \pm 1.645\sqrt{\frac{(.089)(.911)}{135}} = .089 \pm .040 = (.049, .129)$$

Thus, we can be 90% confident that the proportion of all the company's hires (under the current policy) who are minorities is between .049 and .129. As always, our confidence stems from the fact that 90% of all similarly formed intervals will contain the true proportion p and not from any knowledge about whether this particular interval does.

Suppose the EEOC knows that the minority proportion of the qualified work force for jobs in the industry represented by this company is 15%. Can we conclude that the company is hiring minorities at a rate below the 15% qualification rate? Since the 90% confidence interval lies completely below .15 (the upper confidence limit is only .129), we would conclude based on this interval that the company's minority hire rate is below the 15% qualification rate. In fact, we would conclude, with 90% confidence, that the minority hire rate is between 4.9% and 12.9%.

In Example 6.5 we used the confidence interval to make an inference about whether the true value of p is less than .15. That is, we used the sample to **test** whether p is less than .15. When we want to use sample information to test the value of a population parameter, we usually conduct a *test of hypothesis*, the subject of Chapter 7.

We conclude this chapter by showing how to determine the sample size necessary to estimate a binomial proportion with a specified precision.

Exercises 6.41–6.53

Learning the Mechanics

6.41 Describe the sampling distribution of $\hat{p}$ based on large samples of size n. That is, give the mean, the standard deviation, and the (approximate) shape of the distribution of $\hat{p}$ when large samples of size n are (repeatedly) selected from the binomial distribution with probability of success p.

6.42 Explain the meaning of the phrase "$\hat{p}$ is an unbiased estimator of p."

6.43 A random sample of size $n = 196$ yielded $\hat{p} = .64$.

a. Is the sample size large enough to use the methods of this section to construct a confidence interval for p? Explain.

b. Construct a 95% confidence interval for p.

c. Interpret the 95% confidence interval.

d. Explain what is meant by the phrase "95% confidence interval."

6.44 For the binomial sample information summarized in each part, indicate whether the sample size is large enough to use the methods of this chapter to construct a confidence interval for p.

a. $n = 500$, $\hat{p} = .05$ **b.** $n = 100$, $\hat{p} = .05$
c. $n = 10$, $\hat{p} = .5$ **d.** $n = 10$, $\hat{p} = .3$

6.45 A random sample of 50 consumers taste-tested a new snack food. Their responses were coded (0: do not like; 1: like; 2: indifferent) and recorded as follows:

1	0	0	1	2	0	1	1	0	0
0	1	0	2	0	2	2	0	0	1
1	0	0	0	0	1	0	2	0	0
0	1	0	0	1	0	0	1	0	1
0	2	0	0	1	1	0	0	0	1

a. Use an 80% confidence interval to estimate the proportion of consumers who like the snack food.
b. Provide a statistical interpretation for the confidence interval you constructed in part **a**.

Applying the Concepts

6.46 Past research has clearly indicated that the stress produced by today's lifestyles results in health problems for a large proportion of society. A recent article in *The International Journal of Sports Psychology* (Tucker, 1992) evaluates the relationship between physical fitness and stress. Employees of companies that participate in the Health Examination Program offered by Health Advancement Services (HAS) were classified into three fitness levels: poor, average, and good. Each person was tested for signs of stress. The results are reported for the three groups below:

Fitness Level	*Sample Size*	*Proportion with Signs of Stress*
Poor	242	.155
Average	212	.133
Good	95	.108

a. Check to see whether each of these samples is large enough to construct a confidence interval for the true proportion of all employees at each fitness level exhibiting signs of stress.
b. Assuming each sample represents a random sample from its corresponding population, calculate and interpret a 95% confidence interval for the proportion of people with signs of stress within each of the three fitness levels.
c. Interpret each of the confidence intervals constructed in part **b** using the terminology of this exercise.

6.47 A random sample of 122 Illinois law firms was selected to determine their degree of computer usage (Wentling, 1988). She found that 76 of the firms used microcomputers (PCs).

a. Use a 95% confidence interval to estimate the proportion of all Illinois law firms that used microcomputers at the time of the survey.
b. Interpret the interval in the terms of this application.
c. What is meant by the phrase "95% confidence interval"?
d. Do you think this interval provides an estimate for the proportion of all U.S. law firms that were using microcomputers at the time of the survey? Why or why not?

6.48 Refer to Exercise 6.47. In a similar sample survey of 71 Illinois banks, 55 used microcomputers (Wentling and Wentling, 1988).

a. Use a 95% confidence interval to estimate the proportion of all Illinois banks that used microcomputers at the time of the survey.

b. Interpret the interval in terms of this application.

c. How does this interval compare to the one you constructed for the proportion of Illinois law firms that used microcomputers at the time of the survey (Exercise 6.47)?

6.49 The U.S. Commission on Crime wishes to estimate the fraction of crimes related to firearms in an area with one of the highest crime rates in the country. The commission randomly selects 600 files of recently committed crimes in the area and finds 380 in which a firearm was reportedly used. Find a 99% confidence interval for p, the true fraction of crimes in the area in which some type of firearm was reportedly used.

6.50 An article titled "Searching for a Forever Home" (*Time*, May 2, 1983) reports on how television programs aid in the adoption of orphans who have physical or mental handicaps. One method of stimulating the adoption program is to present television profiles of the children, one or more per program. The article documents the success of these programs and notes that Oklahoma City's television station KOCO helped to place 92 of the 119 it profiled, New York's WCBS placed 21 of 35, and Atlanta's WXIA placed 79 of 177. How effective were these three television stations in promoting the adoption of children?

a. Check to see whether each of the sample sizes is large enough to use the normal approximation for the sampling distributions of the sample proportions.

b. Construct a 95% confidence interval for each television station's placement success rate.

c. Suppose the total of 331 children profiled by the three stations could be regarded as a random sample from the population of all similar profiles that might be presented by television stations throughout the country. Construct a 90% confidence interval for the national placement success rate.

6.51 According to *U.S. News and World Report* (May 10, 1982), major infections are the fifth leading cause of death in the United States and approximately 2 million of these cases occur in the nation's hospitals. A powerful antibiotic, piperacillin, was developed by Lederle Laboratories and approved by the Food and Drug Administration in 1982. Tests on 600 patients showed the drug to be 92% effective in curing serious bacterial infections.

a. Find a 99% confidence interval for the probability p that a patient with a serious bacterial infection will be cured after treatment with piperacillin.

b. If all 2 million hospital patients with serious bacterial infections were treated with piperacillin, what is the expected value of the number x that would be cured?

c. Find the standard deviation of x.

d. Give an upper limit on the number x that would be cured.

6.52 An article in the *Gainesville Sun* (March 14, 1984) reports on the topics that teenagers most want to discuss with their parents. The findings, the results of a Gallup Youth Poll, showed that 46% would like more discussion about the family's financial situation, 37% would like to talk about school, and 30% would like to talk about religion. These and other percentages were based on a national sampling of 505 teenagers. What "margin of error" (using the language of the media) would you attach to these findings? Explain.

6.53 Americans over age 50 represent 25% of the population, yet they control 70% of the wealth. Research indicates the highest priority of retirees is travel. A recent study in the *Annals of Tourism Research* (M. Blazey, 1992) investigates the relationship of retirement status (pre- and post-retirement) to various items of interest to the travel industry. As one part of the study, a sample of 323 post-retirees was selected, and the number of nights each typically stayed away from home on trips was determined. One hundred seventy-two (172) responded that their typical stays ranged from 4 to 7 nights. Use a 90% confidence interval to estimate the true proportion of post-retirement travelers who stay between 4 and 7 nights on a typical trip. Interpret the interval.

6.5 Determining the Sample Size Necessary to Estimate a Binomial Probability

We showed in Section 6.2 that sampling experiments can be designed to estimate a population mean μ with a specified degree of reliability. An analogous situation exists for estimating a binomial probability p. For example, in Section 6.4 a pollster used a sample of 1,000 Americans to calculate a 95% confidence interval for the proportion who trust the president, obtaining the interval .607 to .667. Note that the total width of the interval is $.667 - .607 = .06$. Suppose the pollster wishes to estimate more precisely the proportion who trust the president, say with a 95% confidence interval having a width of .03.

The pollster wants a confidence interval width W of .03. This corresponds to a half-width, or bound B on the estimate of p, of $B = W/2 = .015$. The sample size n to generate such an interval is found by solving the following equation for n:

$$z_{\alpha/2}\sigma_{\hat{p}} = B \quad \text{or} \quad z_{\alpha/2}\sqrt{\frac{pq}{n}} = .015 \qquad \text{(see Figure 6.10)}$$

FIGURE 6.10 ▶ Specifying the total width W (or bound B) of a confidence interval for a binomial probability p

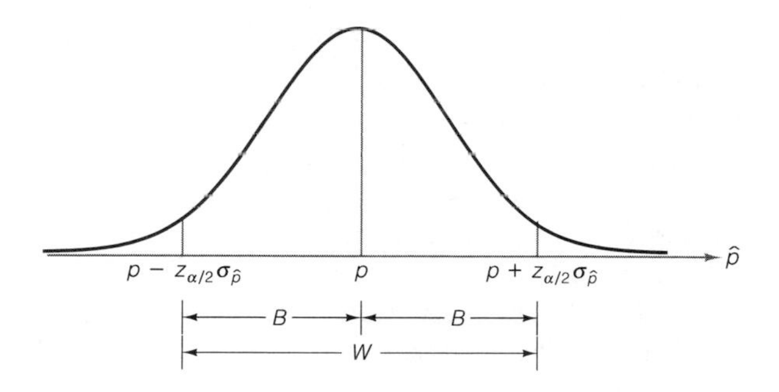

Since a 95% confidence interval is desired, the appropriate z value is $z_{\alpha/2} = z_{.025} = 1.96$. We must approximate the value of the product pq before we can solve the equation for n. As shown in Table 6.3, the closer the values of p and q to .5, the larger the product pq. Thus, to find a conservatively large sample size that will generate a confidence interval with the specified reliability, we generally choose an approximation of p close to .5. In the case of the proportion of Americans who trust the president, however, we have an initial sample estimate of $\hat{p} = .637$. A conservatively large estimate of pq can therefore be obtained by using $p = .60$. We now substitute into the equation and solve for n:

$$1.96\sqrt{\frac{(.60)(.40)}{n}} = .015$$

$$n = \frac{(1.96)^2(.60)(.40)}{(.015)^2} = 4{,}097.7 \approx 4{,}098$$

The pollster must sample about 4,098 Americans to estimate the percentage who trust the president with a confidence interval of width .03.

The procedure for finding the sample size necessary to estimate a binomial probability p to within a given bound B or with a total interval width W is given in the box.

Sample Size Determination for $100(1 - \alpha)\%$ Confidence Interval for p

In order to estimate a binomial probability p to within a bound B or, equivalently, with a confidence interval of total width W with $100(1 - \alpha)\%$ confidence, the required sample size is found by solving one of the following equations for n:

$$z_{\alpha/2}\sqrt{\frac{pq}{n}} = B \quad \text{or} \quad z_{\alpha/2}\sqrt{\frac{pq}{n}} = \frac{W}{2}$$

The solution can be written in terms of either B or W:

$$n = \frac{(z_{\alpha/2})^2(pq)}{B^2} \quad \text{or} \quad n = \frac{4(z_{\alpha/2})^2(pq)}{W^2}$$

The value of the product pq is usually unknown. It can be estimated by using the sample fraction of successes, $\hat{p}$, from a prior sample. Remember (Table 6.3) that the value of pq is at its maximum when p equals .5, so that you can obtain conservatively large values of n by approximating p by .5 or values close to .5. In any case, you should round the value of n obtained *upward* to ensure that the sample size will be sufficient to achieve the specified reliability.

EXAMPLE 6.6

Since the deregulation of the telephonic communications industry, many firms have begun manufacturing telephones. Suppose one large manufacturer that entered the market quickly has an initial problem with excessive customer complaints and consequent returns of the phones for repair or replacement. The manufacturer wants to estimate the magnitude of the problem in order to design a quality control program. How many telephones should be sampled and checked in order to estimate the fraction defective p to within .01 with 90% confidence?

Solution

In order to estimate p to within a bound of .01, we set the half-width of the confidence interval equal to $B = .01$, as shown in Figure 6.11.

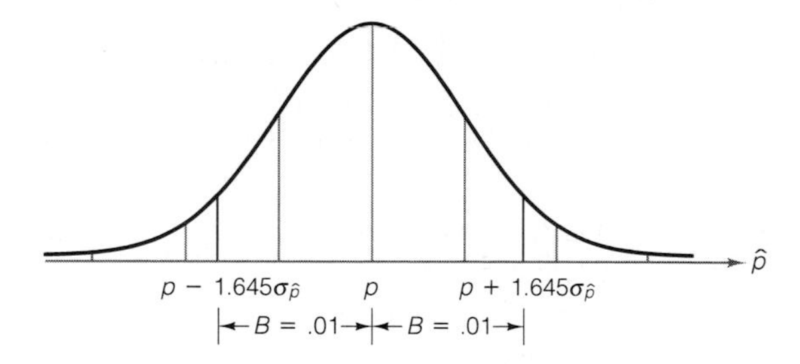

FIGURE 6.11 ► Specified reliability for estimate of fraction defective in Example 6.6

The equation for the sample size n requires an estimate of the product pq. We could most conservatively estimate $pq = .25$ (i.e., use $p = .5$), but this may be overly conservative when estimating a fraction defective. A value of .1, corresponding to 10% defective, will probably be conservatively large for this application. The solution is therefore

$$n = \frac{(z_{\alpha/2})^2(pq)}{B^2} = \frac{(1.645)^2(.1)(.9)}{(.01)^2} = 2{,}435.4 \approx 2{,}436$$

Thus, the manufacturer should sample 2,436 telephones in order to estimate the fraction defective p to within .01 with 90% confidence. Remember that this answer depends on our approximation for pq, where we used .09. If the fraction defective is closer to .05 than .10, we can use a sample of 1,286 telephones (check this) to estimate p to within .01 with 90% confidence.

The cost of sampling will play an important role in the final determination of the sample size to be selected to estimate a binomial probability p. Although more complex formulas can be derived to balance the reliability and cost considerations, we will solve for the necessary sample size and note that the sampling budget may be a limiting factor. Consult the references for a more complete treatment of this problem.

Exercises 6.54–6.64

Learning the Mechanics

6.54 If nothing is known about p, .5 can be substituted for p in the sample-size formula. But when this is done, the resulting sample size may be larger than needed. Under what circumstances will using $p = .5$ in the sample-size formula yield a sample size larger than needed to construct a confidence interval for p with a specified bound and a specified confidence level?

6.55 In each case, find the approximate sample size necessary to estimate a binomial proportion p correct to within .02 with probability equal to .90.

a. Assume you know p is near .8.
b. Assume you have no knowledge of the value of p, but you wish to be certain that your sample is large enough to achieve the specified accuracy for the estimate.

6.56 In each case, find the approximate sample size required to construct a 95% confidence interval for p that has width .12.
a. Assume p is near .3.
b. Assume you have no prior knowledge about p.

6.57 The following is a 90% confidence interval for p: (.26, .54). How large was the sample used to construct this interval?

Applying the Concepts

6.58 According to estimates made by the General Accounting Office, the Internal Revenue Service (IRS) answered 18.3 million telephone inquiries during a recent tax season, and 17% of the IRS offices provided answers that were wrong. These estimates were based on data collected from sample calls to numerous IRS offices. How many IRS offices should be randomly selected and contacted in order to estimate the proportion of IRS offices that fail to correctly answer questions about gift taxes with a 90% confidence interval of width .06?

6.59 Although corporate executives are probably not as highly stressed as air-traffic controllers or inner-city police, research has indicated that they are among the more highly pressured work groups. In order to estimate p, the proportion of managers who perceive themselves to be frequently under stress, Hall and Savery (1986) sampled 532 managers in western Australian corporations. One hundred ninety of these managers fell into the high-stress group. Assume that random sampling was used in this study. Was the sample size large enough to estimate p to within .03 with 95% confidence? Explain.

6.60 If you want to estimate the proportion of operating automobiles that are equipped with air bags, approximately how large a sample would be required to estimate p to within .02 with probability equal to .95?

6.61 A marketing research organization wishes to estimate the proportion of television viewers who watch a particular prime-time situation comedy on May 24. The proportion is expected to be approximately .30. At a minimum, how many viewers should be randomly selected to ensure that a 95% confidence interval for the true proportion of viewers will have a width of .01 or less?

6.62 Before a bill to increase the income tax rate for the wealthy comes before the U.S. Congress, a congressperson would like to know how her wealthy constituents feel about the issue. Approximately how many wealthy constituents should the congressperson survey to estimate the true proportion favoring the tax bill to within .05 with 90% confidence?

6.63 Some quality control testing involves destructive sampling—i.e., the test to determine whether the item is defective destroys the product. This type of sampling is generally expensive, and high costs often prohibit large sample sizes. For example, suppose the National Highway Safety Administration (NHSA) wishes to determine the proportion of new tires that will fail when subjected to hard braking at a speed of 60 miles per hour. NHSA can obtain the tires for \$25 (wholesale price) each. Suppose the budget for the experiment is \$10,000, and NHSA wishes to estimate the percentage that will fail to within .02 with 95% confidence. If the entire \$10,000 can be spent on tires (ignoring other costs) and assuming that the true fraction that will fail is approximately .05, can NHSA attain its goal while staying within the budget? Explain.

6.64 Refer to Exercise 6.46, where a sample of 95 workers in good physical condition was used to estimate the proportion of all workers in good condition who showed signs of stress. Recall that the point estimate of the proportion was .108. How many workers from this category must be sampled to estimate the true proportion who are stressed to within .01 with 95% confidence?

Summary

From the beginning we have stressed that the theme of this text is the use of sample information to make inferences about the parameters of a population. For the first six chapters we developed the tools necessary to perform inference-making procedures. In Chapter 6 we discussed the inferential procedure called **estimation**.

Estimation consists of both **point estimates** and **interval estimates**. Point estimates are single points calculated from a sample to estimate a population parameter. Because we cannot assess the reliability of a point estimate, we prefer to use intervals to estimate the unknown population parameters. Interval estimates utilize **confidence coefficients** to express the reliability of the estimate. The confidence coefficient is the probability that the estimation procedure will generate an interval that contains the parameter being estimated. Thus, confidence is expressed in terms of the long-run performance of the estimation procedure.

We used the standard normal z-distribution to develop large-sample interval estimates for the population mean μ and the binomial probability p. The Student t-distribution was used to form confidence intervals for the mean of normal distributions when n is small.

Supplementary Exercises 6.65–6.82

Note: List the assumptions necessary for the valid implementation of the statistical procedures you use in solving all these exercises.

Learning the Mechanics

6.65 In each of the following instances, determine whether you would use a z or t statistic (or neither) to form a 95% confidence interval, and then look up the appropriate z or t value.

a. Random sample of size $n = 21$ from a normal distribution with unknown mean μ and standard deviation σ
b. Random sample of size $n = 175$ from a normal distribution with unknown mean μ and standard deviation σ
c. Random sample of size $n = 12$ from a normal distribution with unknown mean μ and standard deviation $\sigma = 5$
d. Random sample of size $n = 65$ from a distribution about which nothing is known
e. Random sample of size $n = 8$ from a distribution about which nothing is known

6.66 A random sample of 225 measurements is selected from a population, and the sample mean and standard deviation are $\bar{x} = 32.5$ and $s = 30.0$, respectively.
a. Use a 99% confidence interval to estimate the mean of the population, μ.
b. How large a sample would be needed to estimate μ to within .5 with 99% confidence?
c. What is meant by the phrase "99% confidence" as it is used in this exercise?

6.67 A random sample of 900 college students is selected, and 670 are found to receive some form of financial aid.
a. Use a 90% confidence interval to estimate the proportion of all college students, p, who receive some form of financial aid.
b. How large a sample would be needed to estimate p to within .01 with 90% confidence?
c. What is meant by the phrase "90% confidence" as it is used in this exercise?

Applying the Concepts

6.68 Research reported in *The Professional Geographer* (Johnston-Anumonwo, 1992) investigates the hypothesis that the disproportionate housework responsibility of women in two-income households is a major factor in determining the proximity of a woman's place of employment. The researcher studied the distance (in miles) to work for both men and women in two-income households. Random samples of men and women yielded the following results:

	Central City Residence		Suburban Residence	
	Men	Women	Men	Women
Sample Size	159	119	138	93
Mean	7.4	4.5	9.3	6.6
Std. Deviation	6.3	4.2	7.1	5.6

a. For central city residences, calculate a 95% confidence interval for the average distance to work for men and women in two-income households. Interpret the intervals.
b. Repeat part **a** for suburban residents.
[*Note:* We will show how to use statistical techniques to compare two population means in Chapter 8.]

6.69 A *USA Today* article (July 17, 1992) reveals that both men and women are, on average, waiting longer to marry. Suppose that a random sample of 500 men age 30–34 is selected, and 134 have never married. Use a 90% confidence interval to estimate the true percentage of all men age 30–34 who have never married. Suppose that a similar sample taken in 1970 resulted in only 42 of 290 men age 30–34 who had never married. Calculate the 90% confidence interval for the 1970 percentage of never-married age 30–34 men. How do the 1970 and 1992 intervals compare? [*Note:* We will show how to use statistical techniques to compare two population percentages in Chapter 8.]

6.70 A company is interested in estimating μ, the mean number of days of sick leave taken by all its employees. The firm's statistician selects at random 100 personnel files and notes the number of sick days taken by each employee. The following sample statistics are computed: $\bar{x} = 12.2$ days, $s = 10$ days.
a. Estimate μ using a 90% confidence interval.
b. How many personnel files would the statistician have to select in order to estimate μ with a 99% confidence interval of width 4 days?

6.71 A meteorologist wishes to estimate the mean amount of snowfall per year in Spokane, Washington. A random sample of the recorded snowfalls for 20 years produces a sample mean equal to 54 inches and a standard deviation of 9.59 inches.

a. Estimate the true mean amount of snowfall in Spokane using a 99% confidence interval.

b. If you were purchasing snow-removal equipment for a city, what numerical descriptive measure of the distribution of depth of snowfall would be of most interest to you? Would it be the mean?

6.72 Before approval is given for the use of a new insecticide, the United States Department of Agriculture (USDA) requires that several tests be performed to see how the substance will affect wildlife. In particular, the USDA would like to know the proportion of starlings that will die after being exposed to the insecticide. A random sample of 80 starlings were caught and fed their regular food, which had been treated with the substance. After 10 days, 10 starlings had died. Use a 99% confidence interval to estimate the true proportion of starlings that will be killed by the substance.

6.73 Many people think that a national lobby's successful fight against gun control legislation is reflecting the will of a minority of Americans. A random sample of 4,000 citizens yielded 2,250 who are in favor of gun control legislation. Use a 99% confidence interval to estimate the true proportion of Americans who favor gun control legislation. Interpret the result.

6.74 To help consumers assess the risks they are taking, the Food and Drug Administration (FDA) publishes the amount of nicotine found in all commercial brands of cigarettes. A new cigarette has recently been marketed. The FDA tests on this cigarette gave a mean nicotine content of 26.4 milligrams and standard deviation of 2.0 milligrams for a sample of $n = 9$ cigarettes. Using a 95% confidence interval, estimate the true mean nicotine content per cigarette for the brand. Interpret the result.

6.75 A health researcher wishes to estimate the mean number of cavities per child for children under the age of 12 who live in a specified environment. The number of cavities per child for a random sample of 35 children under the age of 12 has a mean of 2 and a standard deviation of 1.7. Construct a 90% confidence interval for the mean number of cavities per child under the age of 12 who lives in the sampled environment.

6.76 Recycling is receiving increased emphasis among environmental concerns as a means of dealing with the significant growth in the trash and garbage of our "throw-away" society. To estimate the degree of awareness about recycling in one major city, a random sample of 346 households was selected, and 212 were found to use available recycling facilities. Use a 95% confidence interval to estimate the proportion of households in the city that are using available recycling facilities.

6.77 In 1988 tire sales in the United States reached a historic high (*Chemical Week*, 1989). Tires made of synthetic rubber accounted for much of this growth. Suppose one manufacturer was testing a new synthetic rubber design. Twenty tires of the new design were produced and subjected to wear tests. The results indicate that the mean wear for the test tires was 42,250 miles, and the standard deviation was 4,355 miles.

a. What is the point estimate of the true mean wear for the new design?

b. Construct a 90% confidence interval estimate for the mean wear associated with the new design.

c. Which method of estimation is better, point estimation or interval estimation? Why?

6.78 Refer to Exercise 6.77. Suppose that 200 tires rather than 20 were tested, and assume that the sample mean and standard deviation for the 200 tires remain the same: $\bar{x} = 42{,}250$ miles, $s = 4{,}355$ miles. Repeat parts **a–c** of Exercise 6.77 and comment on the similarities and differences in your answers.

6.79 In 1987 a case of salmonella (bacterial) poisoning was traced to a particular brand of ice cream bar, and the manufacturer removed the bars from the market. Despite this response, many consumers refused to purchase any brand of ice cream bars for some period of time after the event (McClave, personal consulting). One manufacturer conducted a survey of consumers 6 months after the poisoning. A sample of 244 ice cream bar consumers were contacted, and 23 of them indicated that they would not purchase ice cream bars because of the potential for food poisoning.

a. What is the point estimate of the true fraction of the entire market who refuse to purchase bars 6 months after the poisoning?
b. Is the sample size large enough to use the normal approximation for the sampling distribution of the estimator of the binomial probability? Justify your response.
c. Construct a 95% confidence interval for the true proportion of the market who still refuse to purchase ice cream bars 6 months after the event.
d. Interpret both the point estimate and confidence interval in terms of this application.

6.80 Refer to Exercise 6.79. Suppose it is now 1 year after the poisoning was traced to ice cream bars. The manufacturer wishes to estimate the proportion who still will not purchase bars using a 95% confidence interval of width .04. How many consumers should be sampled?

6.81 Medicaid health-assistance programs are administered by the individual states, even though part of the funding is federal. The federal government requires that the states perform regular audits in order to assure that payments are accurate. One Florida hospital was audited by the Florida Department of Health and Rehabilitative Services (HRS) in 1989, and a random sample of 25 Medicaid claims was selected. The sample mean of the claims was \$34.76 and the standard deviation was \$11.34.

a. Use a 99% confidence interval to estimate the mean of all claims submitted by this hospital.
b. What assumptions are necessary to assure the validity of this confidence interval?

6.82 Refer to Exercise 6.81.

a. How many claims must be sampled if the HRS wants to estimate the mean size of the hospital's claims to within \$1.00 using a 99% confidence interval?
b. If a sample of this size were to be selected and a 99% confidence interval constructed, what assumptions would be necessary to assure the validity of the interval?

On Your Own

Choose a population pertinent to your major area of interest that has an unknown mean (or, if the population is binomial, that has an unknown probability of success). For example, a marketing major may be interested in the proportion of consumers who prefer a certain product. A sociology major may be interested in estimating the proportion of people in a certain socioeconomic group or the mean income of people living in a certain part of a city. A political scientist may wish to estimate the proportion of an electorate in favor of a certain candidate, a certain amendment, or a certain presidential policy. A person interested in medicine might want to find the average length of time patients stay in the hospital or the average number of people treated daily in the emergency room. We could continue with examples, but the point should be clear—choose something of interest to you.

Define the parameter you want to estimate and conduct a *pilot study* to obtain an initial estimate of the parameter of interest and, more importantly, an estimate of the variability associated with the estimator. A pilot study is a small experiment (perhaps 20 to 30 observations) used to gain some information about the population of interest. The purpose is to help plan more elaborate future experiments. Using the results of your pilot study, determine the sample size necessary to estimate the parameter to within a reasonable bound (of your choice) with a 95% confidence interval.

Using the Computer

Refer to **Using the Computer** in Chapter 5. Recall the values of the "population" mean μ and standard deviation σ for the 1,000 zip code income measurements. Suppose our objective is to sample from this population and to estimate the mean μ using a 95% confidence interval.

a. Determine the sample size n_1 necessary to estimate μ to within \$2,000 with 95% confidence. Then generate one hundred 95% confidence intervals by repeatedly drawing samples of size n_1 (with replacement) from the 1,000 measurements, and using the sample statistics to form a confidence interval. Treat σ as unknown when forming the confidence intervals. What percentage of confidence intervals contain μ?

b. Determine the sample size n_2 necessary to estimate μ to within \$500 with 95% confidence. Then generate one hundred 95% confidence intervals by repeatedly drawing samples of size n_2 (with replacement) from the 1,000 measurements, and using the sample statistics to form a confidence interval. Treat σ as unknown when forming the confidence intervals. What percentage of confidence intervals contain μ?

c. Repeat part **a**, but this time use an 80% confidence interval.

References

Blazey, M. "Travel and retirement status." *Annals of Tourism Research*, 1992, Vol. 19, No. 4.

"Bounce for synthetic rubber: Tires are back in the fast lane." *Chemical Week*, Apr. 19, 1989, pp. 40–44.

Deming, W. E. *Out of the Crisis*. Cambridge, Mass.: M.I.T. Center for Advanced Study of Engineering, 1986.

Dessler, G. "Personnel tests gain in popularity." *St. Paul Pioneer Press and Dispatch*, Mar. 17, 1986.

Hall, K., and Savery, L. K. "Tight reign, more stress." *Harvard Business Review*, Jan.–Feb. 1986, pp. 160–164.

Johnston-Anumonwo, I. "The influence of household type on gender differences in work trip distance." *The Professional Geographer*, May 1992, Vol. 44, No. 2, p. 161.

Mendenhall, W., and Beaver, B. *Introduction to Probability and Statistics*, 8th ed. Boston: PWS-Kent, 1991.

"One answer to imports: Wonder-iron." *Fortune*, Feb. 9, 1981, p. 71.

Otis, J., et al. "Predicting and reinforcing children's intention to wear protective helmets while bicycling." *Public Health Reports*, May–June 1992, Vol. 107, No. 3, p. 283.

Pare, T. P. "Is your 401(k) plan good enough?" *Fortune*, Dec. 28, 1992, pp. 78–83.

Tucker, L. A. "Physical fitness and psychological distress." *International Journal of Sports Psychology*, July–Sept. 1990, 21, pp. 185–201.

Warr, P. "Age and occupational well-being." *Journal of Psychology and Aging*, Mar. 1992, Vol. 7, No. 1, p. 27.

Wentling, R. M. "Master the software explosion." *Legal Assistant Today*, Mar.–Apr. 1988, Vol 5, No. 4, pp. 44–51.

Wentling, R. M., and Wentling, T. L. "Computer technology in banks: What Illinois institutions are using." *Banking Administration*, Jan. 1988, Vol. 64, No. 1, pp. 42–45.

CHAPTER SEVEN

Inferences Based on a Single Sample: Tests of Hypotheses

Contents

Case Studies

Where We've Been

We showed how to use sample information to estimate population parameters in Chapter 6. The sampling distribution of a statistic was used to assess the reliability of an estimate, which was expressed in terms of a confidence interval.

Where We're Going

We now show how to utilize sample information to test whether a population parameter is less than, equal to, or greater than a specified value. This type of inference is called a *test of hypothesis*. We show how to conduct a test of hypothesis about a population mean μ and a binomial probability p. But just as with estimation, we stress the measurement of the reliability of the inference. An inference without a measure of reliability is little more than a guess.

7

Suppose you wanted to determine whether the mean level of blood alcohol exceeds the legal limit after two drinks, or whether the mean breaking strength of sewer pipe exceeds 2,400 pounds per foot, or whether the majority of registered voters approve of the president's performance. In each case you are interested in making an inference about how the value of a parameter relates to a specific numerical value: Is it less than, equal to, or greater than the specified number? This type of inference is called a **test of hypothesis**, and testing hypotheses is the subject of this chapter.

We introduce the elements of a test of hypothesis in Section 7.1. We then show how to conduct a large-sample test of hypothesis about a population mean in Sections 7.2 and 7.3. In Section 7.4 we utilize small samples to conduct tests about means, and in optional Section 7.5 we consider an alternative nonparametric test. Large-sample tests about binomial probabilities are the subject of Section 7.6, and some advanced methods for determining the reliability of a test are covered in optional Section 7.7.

7.1 The Elements of a Test of Hypothesis

Suppose building specifications in a certain city require that the average breaking strength of residential sewer pipe be more than 2,400 pounds per foot of length (i.e., per lineal foot). Each manufacturer who wants to sell pipe in this city must demonstrate that its product meets the specification. Note that we are again interested in making an inference about the mean μ of a population. However, in this example we are less interested in estimating the value of μ than we are in testing a **hypothesis** about its value. That is, we want to decide whether the mean breaking strength of the pipe exceeds 2,400 pounds per lineal foot.

The method used to reach a decision is based on the rare event concept explained in earlier chapters. We define two hypotheses: (1) The **null hypothesis** is that which represents the status quo to the party performing the sampling experiment—the hypothesis that will be accepted unless the data provide convincing evidence that it is false. (2) The **research**, or **alternative**, **hypothesis** is that which will be accepted only if the data provide convincing evidence of its truth. From the point of view of the city conducting the tests, the null hypothesis is that the manufacturer's pipe does *not* meet specifications unless the tests provide convincing evidence otherwise. The null and alternative hypotheses are therefore

Null hypothesis (H_0): $\mu \le 2{,}400$ (i.e., the manufacturer's pipe does not meet specifications)

Alternative (*research*) *hypothesis* (H_a): $\mu > 2{,}400$ (i.e., the manufacturer's pipe meets specifications)

How can the city decide when enough evidence exists to conclude that the manufacturer's pipe meets specifications? Since the hypotheses concern the value of the population mean μ, it is reasonable to use the sample mean $\bar{x}$ to make the inference, just as we did when forming confidence intervals for μ in Sections 6.1 and 6.2. The city will conclude that the pipe meets specifications only when the sample mean $\bar{x}$ convincingly indicates that the population mean exceeds 2,400 pounds per lineal foot.

"Convincing" evidence in favor of the alternative hypothesis will exist when the value of $\bar{x}$ exceeds 2,400 by an amount that cannot be readily attributed to sampling variability. To decide, we compute a **test statistic**, which is the z value that measures the distance between the value of $\bar{x}$ and the value of μ specified in the null hypothesis. When the null hypothesis contains more than one value of μ, as in this case (H_0: $\mu \leq 2{,}400$), we use the value of μ closest to the values specified in the alternative hypothesis. The idea is that if the hypothesis that μ *equals* 2,400 can be rejected in favor of $\mu > 2{,}400$, then μ *less than or equal to* 2,400 can certainly be rejected. Thus, the test statistic is

$$z = \frac{\bar{x} - 2{,}400}{\sigma_{\bar{x}}} = \frac{\bar{x} - 2{,}400}{\sigma/\sqrt{n}}$$

Note that a value of $z = 1$ means that $\bar{x}$ is 1 standard deviation above $\mu = 2{,}400$; a value of $z = 1.5$ means that $\bar{x}$ is 1.5 standard deviations above $\mu = 2{,}400$, etc. How large must z be before the city can be convinced that the null hypothesis can be rejected in favor of the alternative and conclude that the pipe meets specifications?

If you examine Figure 7.1, you will note that the chance of observing $\bar{x}$ more than 1.645 standard deviations above 2,400 is only .05—*if in fact the true mean* μ is 2,400. Thus, if the sample mean is more than 1.645 standard deviations above 2,400, either H_0 is true and a relatively rare event has occurred (.05 probability) or H_a is true and the population mean exceeds 2,400. Since we would most likely reject the notion that a rare event has occurred, we would reject the null hypothesis ($\mu \leq 2{,}400$) and conclude that the alternative hypothesis ($\mu > 2{,}400$) is true. What is the probability that this procedure will lead us to an incorrect decision?

Deciding that the null hypothesis is false when in fact it is true is called a **Type I decision error**. As indicated in Figure 7.1, the risk of making a Type I error—that is, deciding in favor of the research hypothesis when in fact the null hypothesis is true—is denoted by the symbol α. That is,

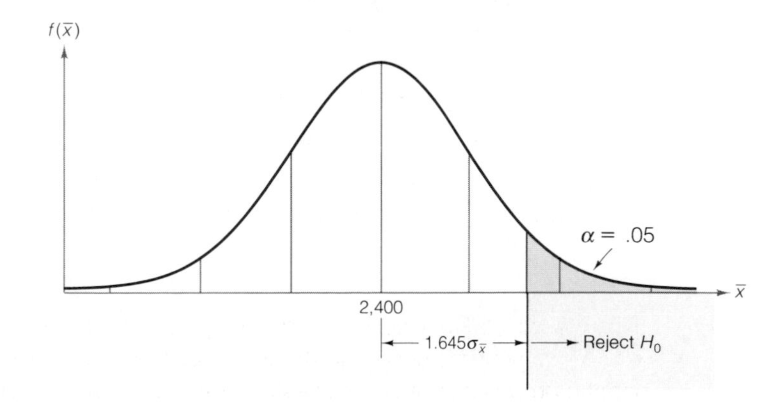

FIGURE 7.1 ▶ The sampling distribution of $\bar{x}$, assuming $\mu = 2{,}400$

$$\alpha = P(\text{Type I error})$$
$$= P(\text{Rejecting the null hypothesis when in fact the null hypothesis is true})$$

In our example

$$\alpha = P(z > 1.645 \text{ when in fact } \mu = 2{,}400) = .05$$

We now summarize the elements of the test:

H_0: $\mu = 2{,}400$

H_a: $\mu > 2{,}400$

Test statistic : $z = \dfrac{\bar{x} - 2{,}400}{\sigma_{\bar{x}}}$

Rejection region: $z > 1.645$, which corresponds to $\alpha = .05$

Note that the **rejection region** refers to the values of the test statistic for which we will **reject the null hypothesis**.

To illustrate the use of the test, suppose we test 50 sections of sewer pipe and find the mean and standard deviation for these 50 measurements to be

$$\bar{x} = 2{,}460 \text{ pounds per lineal foot} \qquad s = 200 \text{ pounds per lineal foot}$$

As in the case of estimation, we can use s to approximate σ when s is calculated from a large set of sample measurements.

The test statistic is

$$z = \frac{\bar{x} - 2{,}400}{\sigma_{\bar{x}}} = \frac{\bar{x} - 2{,}400}{\sigma/\sqrt{n}} \approx \frac{\bar{x} - 2{,}400}{s/\sqrt{n}}$$

Substituting $\bar{x} = 2{,}460$, $n = 50$, and $s = 200$, we have

$$z \approx \frac{2{,}460 - 2{,}400}{200/\sqrt{50}} = \frac{60}{28.28} = 2.12$$

Therefore, the sample mean lies $2.12\sigma_{\bar{x}}$ above the hypothesized value of μ, 2,400, as shown in Figure 7.2. Since this value of z exceeds 1.645, it falls in the rejection

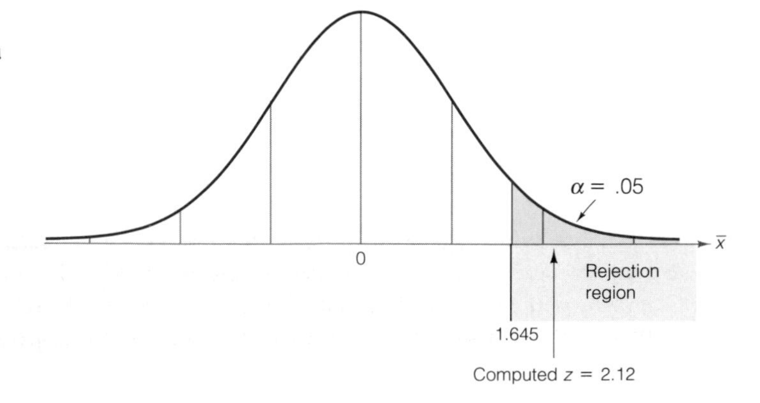

FIGURE 7.2 ▶ Location of the test statistic for a test of the hypothesis H_0: $\mu = 2{,}400$

region. That is, we reject the null hypothesis that $\mu = 2{,}400$ and conclude that $\mu > 2{,}400$. Thus, it appears that the company's pipe has a mean strength that exceeds 2,400 pounds per lineal foot.

How much faith can be placed in this conclusion? What is the probability that our statistical test could lead us to reject the null hypothesis (and conclude that the company's pipe meets the city's specifications) when in fact the null hypothesis is true? The answer is $\alpha = .05$. That is, we selected the level of risk, α, of making a Type I error when we constructed the test. Thus, the chance is only 1 in 20 that our test would lead us to conclude the manufacturer's pipe satisfies the city's specifications when in fact the pipe does *not* meet specifications.

Now, suppose the sample mean breaking strength for the 50 sections of sewer pipe turned out to be $\bar{x} = 2{,}430$ pounds per lineal foot. Assuming that the sample standard deviation is still $s = 200$, the test statistic is

$$z = \frac{2{,}430 - 2{,}400}{200/\sqrt{50}} = \frac{30}{28.28} = 1.06$$

Therefore, the sample mean $\bar{x} = 2{,}430$ is only 1.06 standard deviations above the null hypothesized value of $\mu = 2{,}400$. As shown in Figure 7.3, this value does not fall into the rejection region ($z > 1.645$). Therefore, we know that we cannot reject H_0 using $\alpha = .05$. Even though the sample mean exceeds the city's specification of 2,400 by 30 pounds per lineal foot, it does not exceed the specification by enough to provide *convincing* evidence that the *population mean* exceeds 2,400.

FIGURE 7.3 ▶ Location of test statistic when $\bar{x} = 2{,}430$

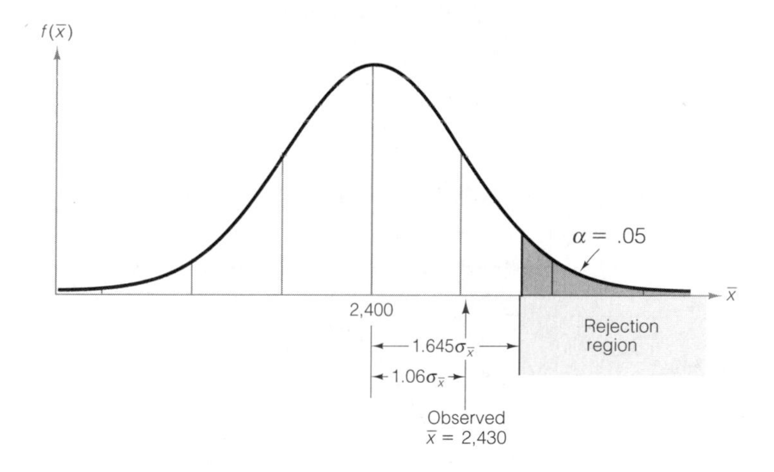

Should we accept the null hypothesis H_0: $\mu \leq 2{,}400$ and conclude that the manufacturer's pipe does not meet specifications? To do so would be to risk a **Type II error**—that of concluding that the null hypothesis is true (the pipe does not meet specifications) when in fact it is false (the pipe does meet specifications). We denote the probability of committing a Type II error by β, and we show in optional Section 7.7 that β is often difficult to determine precisely. Rather than make a decision (accept H_0) for which the probability of error (β) is unknown, we avoid the potential Type II error by avoiding the conclusion that the null hypothesis is true. Instead, we will simply state that *the sample evidence is insufficient to reject* H_0 *at* $\alpha = .05$. Since the null hypothesis is the "status-quo" hypothesis, the effect of not rejecting H_0 is to maintain the status quo. In our pipe-testing example, the effect of having insufficient evidence to reject the null hypothesis that the pipe does not meet specifications is probably to prohibit the utilization of the manufacturer's pipe unless and until there is sufficient evidence that the pipe does meet specifications. That is, until the data indicate convincingly that the null hypothesis is false, we usually maintain the status quo implied by its truth.

Table 7.1 summarizes the four possible outcomes of a test of hypothesis. The "true state of nature" columns in Table 7.1 refer to the fact that either the null hypothesis H_0 is true or the alternative hypothesis H_a is true. Note that the true state of nature is unknown to the researcher conducting the test. The "decision" rows in Table 7.1 refer to the action of the researcher, assuming that he or she will either conclude that H_0 is true or that H_a is true, based on the results of the sampling experiment. Note that a Type I error can be made *only* when the alternative hypothesis is accepted (equivalently, when the null hypothesis is rejected), and a Type II error can be made *only* when the null hypothesis is accepted. Our policy will be to make a decision only when we know the probability of making the error that corresponds to that decision. Since α is usually specified by the analyst, we will generally be able to reject H_0 (accept H_a) when the sample evidence supports that decision. However, since β is usually not specified, we will generally avoid the decision to accept H_0, preferring instead to state that the sample evidence is insufficient to reject H_0 when the test statistic is not in the rejection region.

TABLE 7.1 Conclusions and Consequences for a Test of Hypothesis

		True State of Nature	
		H_0 True	H_a True
Conclusion	H_0 True	Correct decision	Type II error (probability β)
	H_a True	Type I error (probability α)	Correct decision

The elements of a test of hypothesis are summarized in the box. Note that the first four elements are all specified *before* the sampling experiment is performed. In no case will the results of the sample be used to determine the hypotheses—the data are collected to test the predetermined hypotheses, not to formulate them.

Elements of a Test of Hypothesis

1. *Null hypothesis* (H_0): A theory about the values of one or more population parameters. The theory generally represents the status quo, which we accept until proven false.
2. *Alternative (research) hypothesis* (H_a): A theory that contradicts the null hypothesis. The theory generally represents that which we will accept only when sufficient evidence exists to establish its truth.
3. *Test statistic:* A sample statistic used to decide whether to reject the null hypothesis.
4. *Rejection region:* The numerical values of the test statistic for which the null hypothesis will be rejected. The rejection region is chosen so that the probability is α that it will contain the test statistic when the null hypothesis is true, thereby leading to a Type I error. The value of α is usually chosen to be small (e.g., .01, .05, or .10), and is referred to as the **level of significance** of the test.
5. *Assumptions:* Any assumptions made about the population(s) being sampled should be clearly stated.
6. *Experiment and calculation of test statistic:* The sampling experiment is performed, and the numerical value of the test statistic is determined.
7. *Conclusion:*
   a. If the numerical value of the test statistic falls in the rejection region, we reject the null hypothesis and conclude that the alternative hypothesis is true. We know that the hypothesis-testing process will lead to this conclusion incorrectly (Type I error) only $100\alpha\%$ of the time when H_0 is true.
   b. If the test statistic does not fall in the rejection region, we do not reject H_0. Thus, we reserve judgment about which hypothesis is true. We do not conclude that the null hypothesis is true because we do not (in general) know the probability β that our test procedure will lead to an incorrect acceptance of H_0 (Type II error).

CASE STUDY 7.1 / Statistics Is Murder!

The jury trial of an accused murderer is analogous to the statistical hypothesis-testing process. Each of the elements of a test of hypothesis applies to the jury system of deciding the guilt or innocence of the accused:

1. H_0: The null hypothesis in a jury trial is that the accused is innocent. The status-quo hypothesis in the American system of justice is innocence, which is assumed to be true until proven otherwise.
2. H_a: The alternative hypothesis is guilt, which is accepted only when sufficient evidence exists to establish its truth.

3. *Test statistic:* The test statistic in a trial is the final vote of the jury—i.e., the number of the jury members who vote "guilty."
4. *Rejection region:* In a murder trial the jury vote must be unanimous in favor of guilt before the null hypothesis of innocence is rejected in favor of the alternative hypothesis of guilt. Thus, for a 12-member jury trial, the rejection region is $x = 12$, where x is the number of "guilty" votes.
5. *Assumption:* The primary assumption made in trials concerns the method of selecting the jury. The jury is assumed to represent a random sample of citizens who have no prejudice concerning the case.
6. *Experiment and calculation of the test statistic:* The sampling experiment is analogous to the jury selection, the trial, and the jury deliberations. The final vote of the jury is analogous to the calculation of the test statistic.
7. *Conclusion:*
   a. If the vote of the jury is unanimous in favor of guilt, the null hypothesis of innocence is rejected and the court concludes that the accused murderer is guilty. Although the court does not, in general, know the probability α that the conclusion is in error, the system relies on the belief that the value is made very small by requiring a unanimous vote before guilt is concluded.
   b. Any vote other than a unanimous one for guilt results in the court reserving judgment about the hypotheses, either by declaring the accused "not guilty," or by declaring a mistrial and repeating the "test" with a new jury. (The latter is analogous to collecting more data and repeating a statistical test of hypothesis.) The court never accepts the null hypothesis by declaring the accused "innocent," perhaps recognizing both that innocence is the status-quo hypothesis and does not need to be proved and that the probability β of incorrectly concluding innocence may not be as small as α.

As in the case of tests of statistical hypotheses, we may never know whether the verdict in a murder trial is correct. Instead, we rely on the knowledge that the trial procedure will lead to incorrect conclusions (especially, guilt when the accused is in fact innocent) in only a very small percentage of trials.

Exercises 7.1–7.9

Learning the Mechanics

7.1 Which hypothesis, the null or the alternative, is the status-quo hypothesis? Which is the research hypothesis?

7.2 Which element of a test of hypothesis is used to decide whether to reject the null hypothesis in favor of the alternative hypothesis?

7.3 What is the level of significance of a test of hypothesis?

7.4 What is the difference between Type I and Type II errors in hypothesis testing? How do α and β relate to Type I and Type II errors?

7.5 List the four possible results of the combinations of decisions and true states of nature for a test of hypothesis.

7.6 Why do we (generally) reject the null hypothesis when the test statistic falls in the rejection region, but do not accept the null hypothesis when the test statistic does not fall in the rejection region?

7.7 If you test a hypothesis and reject the null hypothesis in favor of the alternative hypothesis, does your test prove that the alternative hypothesis is correct? Explain.

Applying the Concepts

7.8 In 1895 an Italian criminologist, Cesare Lombroso, proposed that blood pressure be used to test for truthfulness. In the 1930s, William Marston added the measurements of respiration and perspiration to the process and called his machine the *polygraph*—or *lie detector*. Today, the federal court system will not consider polygraph results as evidence, but nearly half of the state courts do permit polygraph tests under certain circumstances. In addition, its use in screening job applicants is on the rise (Dujack, 1986). Physicians Michael Phillips, Allan Brett, and John Beary subjected the polygraph to the same careful testing given to medical diagnostic tests. They found that if 1,000 people were subjected to the polygraph and 500 told the truth and 500 lied, the polygraph would indicate that approximately 185 of the truth-tellers were liars and that approximately 120 of the liars were truth-tellers ("Lie Detectors Can Make a Liar of You," *Discover*, June 1986, p. 7).

a. In the application of a polygraph test, an individual is presumed to be a truth-teller (H_0) until "proven" a liar (H_a). In this context, what is a Type I error? A Type II error?

b. According to Phillips, Brett, and Beary, what is the probability (approximately) that a polygraph test will result in a Type I error? A Type II error?

7.9 When a new drug is formulated, the pharmaceutical company must subject it to lengthy and involved testing before receiving the necessary permission from the Food and Drug Administration (FDA) to market the drug. The FDA's policy is that the pharmaceutical company must provide substantial evidence that a new drug is safe prior to receiving FDA approval, so that the FDA can confidently certify the safety of the drug to potential consumers.

a. If the new drug testing were to be placed in a test of hypothesis framework, would the null hypothesis be that the drug is safe or unsafe? The alternative hypothesis?

b. Given the choice of null and alternative hypotheses in part **a**, describe Type I and Type II errors in terms of this application. Define α and β in terms of this application.

c. If the FDA wants to be very confident that the drug is safe before permitting it to be marketed, is it more important that α or β be small? Explain.

7.2 Large-Sample Test of Hypothesis About a Population Mean

In Section 7.1 we learned that the null and alternative hypotheses form the basis for a test of hypothesis inference. The null and alternative hypotheses may take one of several forms. In the sewer pipe example we tested the null hypothesis that the population mean strength of the pipe is less than or equal to 2,400 pounds per lineal foot against the alternative hypothesis that the mean strength exceeds 2,400. That is, we tested

$$H_0: \ \mu \le 2{,}400$$
$$H_a: \ \mu > 2{,}400$$

This is a **one-tailed** (or **one-sided**) statistical test because the alternative hypothesis specifies that the population parameter (the population mean μ, in this example) is strictly greater than a specified value (2,400, in this example). If the null hypothesis had been H_0: $\mu \ge 2{,}400$ and the alternative hypothesis had been H_a: $\mu < 2{,}400$, the test would still be one-sided, because the parameter is still specified to be on "one side" of the null hypothesis value. Some statistical investigations seek to show that the population parameter is *either larger or smaller* than some specified value. Such an alternative hypothesis is called a **two-tailed** (or **two-sided**) hypothesis.

While alternative hypotheses are always specified as strict inequalities, such as $\mu < 2{,}400$, $\mu > 2{,}400$, or $\mu \neq 2{,}400$, null hypotheses are usually specified as equalities, such as $\mu = 2{,}400$. Even when the null hypothesis is an inequality, such as $\mu \leq 2{,}400$, we specify H_0: $\mu = 2{,}400$, reasoning that if sufficient evidence exists to show that H_a: $\mu > 2{,}400$ is true when tested against H_0: $\mu = 2{,}400$, then surely sufficient evidence exists to reject $\mu < 2{,}400$ as well. Therefore, the null hypothesis is specified as the value of μ closest to a one-sided alternative hypothesis and as the only value *not* specified in a two-tailed alternative hypothesis. The steps for selecting the null and alternative hypotheses are summarized in the box.

Steps for Selecting the Null and Alternative Hypotheses

1. Select the *alternative hypothesis* as that which the sampling experiment is intended to establish. The alternative hypothesis will assume one of three forms:
   a. One-tailed, upper-tailed *Example:* H_a: $\mu > 2{,}400$
   b. One-tailed, lower-tailed *Example:* H_a: $\mu < 2{,}400$
   c. Two-tailed *Example:* H_a: $\mu \neq 2{,}400$
2. Select the *null hypothesis* as the status quo, that which will be presumed true unless the sampling experiment conclusively establishes the alternative hypothesis. The null hypothesis will be specified as that parameter value closest to the alternative in one-tailed tests, and as the complementary (or only unspecified) value in two-tailed tests.

   Example: H_0: $\mu = 2{,}400$

The rejection region for a two-tailed test differs from that for a one-tailed test. When we are trying to detect departure from the null hypothesis in *either* direction, we must establish a rejection region in both tails of the sampling distribution of the test statistic. Figures 7.4(a) and (b) show the one-tailed rejection regions for lower- and

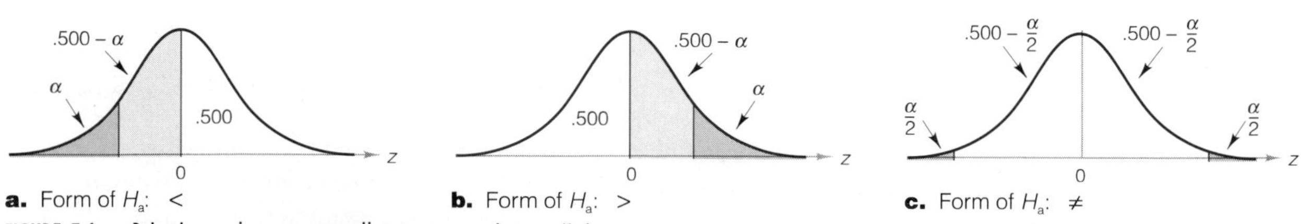

FIGURE 7.4 ▲ Rejection regions corresponding to one- and two-tailed tests

upper-tailed tests, respectively. The two-tailed rejection region is illustrated in Figure 7.4(c). Note that a rejection region is established in each tail of the sampling distribution for a two-tailed test.

The rejection regions corresponding to typical values selected for α are shown in Table 7.2 for one- and two-tailed tests. Note that the smaller α you select, the more evidence (the larger z) you will need before you can reject H_0.

TABLE 7.2 Rejection Regions for Common Values of α

	Alternative Hypotheses		
	Lower-Tailed	Upper-Tailed	Two-Tailed
$\alpha = .10$	$z < -1.28$	$z > 1.28$	$z < -1.645$ or $z > 1.645$
$\alpha = .05$	$z < -1.645$	$z > 1.645$	$z < -1.96$ or $z > 1.96$
$\alpha = .01$	$z < -2.33$	$z > 2.33$	$z < -2.575$ or $z > 2.575$

EXAMPLE 7.1

The effects of drugs and alcohol on the nervous system have been the subject of considerable research recently. Suppose a research neurologist is testing the effects of a drug on response time by injecting 100 rats with a unit dose of the drug, subjecting each to a neurological stimulus, and recording its response time. The neurologist knows that the mean response time for rats not injected with the drug (the "control" mean) is 1.2 seconds. She wishes to test whether the mean response time for drug-injected rats differs from 1.2 seconds. Set up the test of hypothesis for this experiment, using $\alpha = .01$.

Solution

Since the neurologist wishes to detect whether the mean response time, μ, for drug-injected rats differs from the control mean of 1.2 seconds in *either* direction—i.e., $\mu < 1.2$ or $\mu > 1.2$—we conduct a two-tailed statistical test. Following the procedure for selecting the null and alternative hypotheses, we specify as the alternative hypothesis that the mean differs from 1.2 seconds, since determining whether the drug-injected mean differs from the control mean is the purpose of the experiment. The null hypothesis is the presumption that drug-injected rats have the same mean response time as control rats unless the research indicates otherwise. Thus,

$$H_0: \quad \mu = 1.2$$
$$H_a: \quad \mu \neq 1.2 \qquad (\text{i.e., } \mu < 1.2 \quad \text{or} \quad \mu > 1.2)$$

The test statistic measures the number of standard deviations between the observed value of $\bar{x}$ and the null hypothesized value $\mu = 1.2$:

$$\textit{Test statistic:} \quad z = \frac{\bar{x} - 1.2}{\sigma_{\bar{x}}}$$

The rejection region must be designated to detect a departure from $\mu = 1.2$ in *either* direction, so we will reject H_0 for values of z that are either too small (negative) or too large (positive). To determine the precise values of z that comprise the rejection region, we first select α, the probability that the test will lead to incorrect rejection of the null hypothesis. Then we divide α equally between the lower and upper tail of the distribution of z, as shown in Figure 7.5. In this example, $\alpha = .01$, so $\alpha/2 = .005$ is placed in each tail. The areas in the tails correspond to $z = -2.575$ and $z = 2.575$, respectively (from Table 7.2):

Rejection region: $z < -2.575$ or $z > 2.575$ (see Figure 7.5)

Assumptions: Since the sample size of the experiment is large enough ($n > 30$), the Central Limit Theorem will apply, and no assumptions need be made about the population of response time measurements. The sampling distribution of the sample mean response of 100 rats will be approximately normal regardless of the distribution of the individual rats' response times.

FIGURE 7.5 ▶
Two-tailed rejection region:
$\alpha = .01$

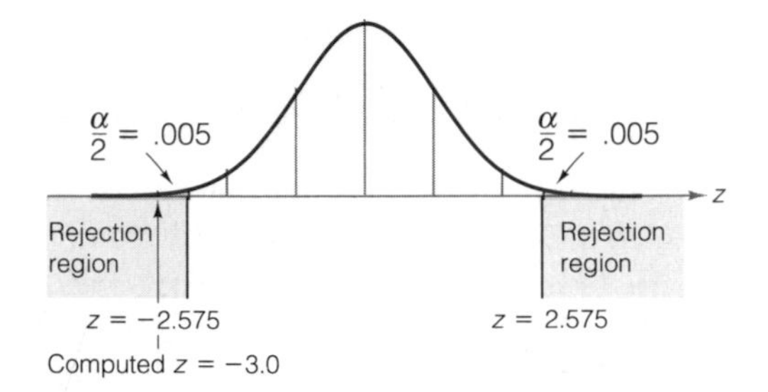

Note that the test in Example 7.1 is set up *before* the sampling experiment is conducted. The data are not used to develop the test. Evidently, the neurologist wants to conclude that the mean response time for the drug-injected rats differs from the control mean only when the evidence is very convincing, because the value of α has been set quite low at .01. If the experiment results in the rejection of H_0, she can be 99% confident that the mean response time of the drug-injected rats differs from the control mean.

Once the test is set up, she is ready to perform the sampling experiment and conduct the test. The test is performed in Example 7.2.

EXAMPLE 7.2

Refer to the neurological response-time test set up in Example 7.1. Suppose the sampling experiment is conducted with the following results:

$n = 100$ response times for drug-injected rats

$\bar{x} = 1.05$ seconds $\qquad s = .5$ second

Use the results of the sampling experiment to conduct the test of hypothesis.

Solution

Since the test is completely specified in Example 7.1, we simply substitute the sample statistics into the test statistic:

$$z = \frac{\bar{x} - 1.2}{\sigma_{\bar{x}}} = \frac{\bar{x} - 1.2}{\sigma/\sqrt{n}} = \frac{1.05 - 1.2}{\sigma/\sqrt{100}}$$

$$\approx \frac{1.05 - 1.2}{s/10} = \frac{-.15}{.5/10} = -3.0$$

The implication is that the sample mean, 1.05, is (approximately) 3 standard deviations below the null hypothesized value of 1.2 in the sampling distribution of $\bar{x}$. You can see in Figure 7.5 that this value of z is in the lower-tail rejection region, which consists of all values of $z < -2.575$. This sampling experiment provides sufficient evidence to reject H_0 and conclude, at the $\alpha = .01$ level of significance, that the mean response time for drug-injected rats differs from the control mean of 1.2 seconds. It appears that the rats receiving an injection of this drug have a mean response time that is less than 1.2 seconds.

Two final points about the test of hypothesis in Example 7.2 apply to all statistical tests:

1. Since z is less than -2.575, it is tempting to state our conclusion at a significance level lower than $\alpha = .01$. We resist the temptation because the level of α is determined *before* the sampling experiment is performed. If we decide that we are willing to tolerate a 1% Type I error rate, the result of the sampling experiment should have no effect on that decision. *In general, the same data should not be used both to set up and to conduct the test.*
2. When we state our conclusion at the .01 level of significance, we are referring to the failure rate of the *procedure*, not the result of this particular test. We know that the test procedure will lead to the rejection of the null hypothesis only 1% of the time when in fact $\mu = 1.2$. *Therefore, when the test statistic falls in the rejection region, we infer that the alternative $\mu \neq 1.2$ is true and express our confidence in the procedure by quoting the α level of significance, or the $100(1 - \alpha)\%$ confidence level.*

The setup of a large-sample test of hypothesis about a population mean is summarized in the next box at the top of page 324. Both the one- and two-tailed tests are shown.

Large-Sample Test of Hypothesis About μ

One-Tailed Test	*Two-Tailed Test*
H_0: $\mu = \mu_0$*	H_0: $\mu = \mu_0$*
H_a: $\mu < \mu_0$ (or H_a: $\mu > \mu_0$)	H_a: $\mu \neq \mu_0$
Test statistic: $z = \dfrac{\bar{x} - \mu_0}{\sigma_{\bar{x}}}$	*Test statistic:* $z = \dfrac{\bar{x} - \mu_0}{\sigma_{\bar{x}}}$
Rejection region: $z < -z_\alpha$ (or $z > z_\alpha$ when H_a: $\mu > \mu_0$)	*Rejection region:* $z < -z_{\alpha/2}$ or $z > z_{\alpha/2}$
where z_α is chosen so that $P(z > z_\alpha) = \alpha$	where $z_{\alpha/2}$ is chosen so that $P(z > z_{\alpha/2}) = \alpha/2$

Assumptions: No assumptions need to be made about the probability distribution of the population because the Central Limit Theorem assures us that, for large samples, the test statistic will be approximately normally distributed regardless of the shape of the underlying probability distribution of the population.

**Note:* μ_0 is the symbol for the numerical value assigned to μ under the null hypothesis.

Once the test has been set up, the sampling experiment is performed and the test statistic is calculated. The next box contains possible conclusions for a test of hypothesis, depending on the result of the sampling experiment.

Possible Conclusions for a Test of Hypothesis

1. If the calculated test statistic falls in the rejection region, reject H_0 and conclude that the alternative hypothesis H_a is true. State that you are rejecting H_0 at the α level of significance. Remember that the confidence is in the testing *process*, not the particular result of a single test.
2. If the test statistic does not fall in the rejection region, conclude that the sampling experiment does not provide sufficient evidence to reject H_0 at the α level of significance. [Generally, we will not "accept" the null hypothesis unless the probability β of a Type II error has been calculated (see optional Section 7.7).]

CASE STUDY 7.2 / Hypothesis Tests in Computer Security Systems

Hackers are able to crack password codes and wander the phone lines, trespassing through data banks. Industrial spies raid secret data files, snatching new product designs. Computer criminals enrich themselves, plundering electronic fund transfer networks. The Defense Department is positive—almost—that its top-security computer systems are invulnerable to penetration by outsiders or, worse, insiders with evil intentions.

This paragraph introduces an article by Daniel Kagan ("Locking Up Data," *Omni*, 1984) on one of the most pressing problems in high-technology industries today: computer security. The movie *War Games* is a fictional account of a young man's successful attempt to crack the security codes of Defense Department computers, gaining access to highly confidential, sensitive, and, as it turns out in the movie, potentially dangerous information. Perhaps more disturbing, however, are the many factual reports of computer thieves obtaining access to bank accounts, government data bases, and university research computers. Several proposed solutions to the growing computer security problem are presented in Kagan's article.

The objective of computer security is to allow only authorized personnel access to the computer's data files. This goal is typically achieved by use of a *password*—a collection of symbols (usually letters and numbers) that must be supplied by the user before the computer permits access to the account. The problem is that persistent computer thieves can program their computers to generate millions of combinations of letters and numbers and to enter them into the computer into which access is desired until the correct password is found. Some accounts are doubly protected with an unlisted phone number providing a first level of protection, followed by the password. This measure simply means that gaining illegal access may take longer, since the thief must now wait until his computer cracks both levels of security. The problem is augmented by the more common, less intriguing, but equally damaging illegal access by former employees and friends of current employees who steal the necessary passwords.

Omni reports on several innovative new proposals for solving the computer security problem. One school of thought consists of "a move away from I.D. verification based on what the operator *has* (like a magnetic stripe card) and what he *knows* (like a password)—toward using what the operator *is*. . . . Authorized users are identified by unique body characteristics, 'You can't leave your body at home like a card or key; and no one can steal it,' quips Tom Catto of Palmguard, Inc., in Beaverton, Oregon." Palmguard's security system consists of computer identification of the user's palm before access is permitted. The system tests the hypothesis

H_0: The proposed user is authorized

versus

H_a: The proposed user is unauthorized

by checking characteristics of the proposed user's palm against those stored in the authorized users' data bank. Palmguard reports that the Type I error rate is less than 1%, whereas the Type II error rate is .00025%.

Since the Type I error refers to the rejection of H_0 when H_0 is true, the implication is that an authorized user will be denied access by the computer less than 1% of the time. This is primarily an inconvenience, since the authorized user can try to gain access again with reasonable assurance of success. Of more importance is the Type II error rate, which refers to the acceptance of H_0 when H_0 is false, or the probability that an unauthorized user will be granted access to the account. Palmguard's claim implies that a Type II error will occur only 25 times in 10 million.

Another new security system, the EyeDentifyer, "spots authorized computer users by reading the one-of-a-kind patterns formed by the network of minute blood vessels across the retina at the back of the eye."

The Type I and II error rates are reported to be .01% (1 in 10,000) and .005% (5 in 100,000), respectively.

As new security systems are developed, the Type I and Type II error rates provide objective standards with which to compare them. Of course, the ultimate security system would have a Type II error rate of 0, but as with tests of statistical hypotheses, this is probably an impractical (and unachievable) goal.

Exercises 7.10–7.20

Learning the Mechanics

7.10 For each of the following rejection regions, sketch the sampling distribution for z and indicate the location of the rejection region.

a. $z > 1.96$ b. $z > 1.645$ c. $z > 2.58$ d. $z < -1.28$
e. $z < -1.645$ or $z > 1.645$ f. $z < -2.58$ or $z > 2.58$
g. For each of the rejection regions specified in parts **a–f**, what is the probability that a Type I error will be made?

7.11 Suppose you are interested in conducting the statistical test of H_0: $\mu = 200$ against H_a: $\mu > 200$, and you have decided to use the following decision rule: Reject H_0 if the sample mean of a random sample of 100 items is more than 215. Assume that the standard deviation of the population is 80.

a. Express the decision rule in terms of z.
b. Find α, the probability of making a Type I error, by using this decision rule.

7.12 A random sample of 100 observations from a population with standard deviation 60 yielded a sample mean of 110.

a. Test the null hypothesis that $\mu = 100$ against the alternative hypothesis that $\mu > 100$ using $\alpha = .05$. Interpret the results of the test.
b. Test the null hypothesis that $\mu = 100$ against the alternative hypothesis that $\mu \neq 100$ using $\alpha = .05$. Interpret the results of the test.
c. Compare the results of the two tests you conducted. Explain why the results differ.

7.13 A random sample of 64 observations produced the following sums:

$$\sum x = 20.7 \qquad \sum (x_i - \bar{x})^2 = 2.155$$

a. Test the null hypothesis that $\mu = .36$ against the alternative hypothesis that $\mu < .36$ using $\alpha = .10$. Interpret the result of the test.
b. Test the null hypothesis that $\mu = .36$ against the alternative hypothesis that $\mu \neq .36$ using $\alpha = .10$. Interpret the result.

Applying the Concepts

7.14 A recent study reported in the *Journal of Occupational and Organizational Psychology* (Harding and Sewel, 1992) investigated the relationship of employment status to mental health. A sample of 49 unemployed men were given a mental health examination using the General Health Questionnaire (GHQ). The GHQ is a widely recognized measure of present mental health, with lower values indicating better mental health. The mean and standard deviation of the GHQ scores were $\bar{x} = 10.94$ and $s = 5.10$, respectively.

a. Specify the appropriate null and alternative hypotheses if we wish to test the research hypothesis that the mean GHQ score for all unemployed men exceeds 10. Is the test one-tailed or two-tailed? Why?
b. If we specify $\alpha = .05$, what is the appropriate rejection region for this test?
c. Conduct the test, and state your conclusion clearly and in the language of this exercise.

7.15 The Environmental Protection Agency (EPA) estimated that the 1991 G-car obtains a mean of 35 miles per gallon on the highway, and the company that manufactures the car claims that it exceeds the EPA estimate in highway driving. To support its assertion, the company randomly selects 36 1991 G-cars and records the mileage obtained for each car over a driving course similar to that used by the EPA. The following data resulted: $\bar{x} = 36.8$ miles per gallon, $s = 6.0$ miles per gallon.
a. If the auto manufacturer wishes to show that the mean miles per gallon for 1991 G-cars is greater than 35 miles per gallon, what should it choose for the alternative hypothesis? The null hypothesis?
b. Do the data provide sufficient evidence to support the auto manufacturer's claim? Test using $\alpha = .05$.

7.16 Refer to Exercise 7.15. Repeat the test using $\alpha = .10$. How does your conclusion differ from that in Exercise 7.15? Why?

7.17 A pain reliever currently being used in a hospital is known to bring relief to patients in a mean time of 3.5 minutes. To compare a new pain reliever with the one currently being used, the new drug is administered to a random sample of 50 patients. The mean time to relief for the sample of patients is 2.8 minutes and the standard deviation is 1.14 minutes. Do the data provide sufficient evidence to conclude that the new drug was effective in reducing the mean time until a patient receives relief from pain? Test using $\alpha = .10$.

7.18 The University of Minnesota uses thousands of fluorescent light bulbs each year. The brand of bulb it currently uses has a mean life of 900 hours. A manufacturer claims that its new brand of bulbs, which cost the same as the brand the university currently uses, has a mean life of more than 900 hours. The university has decided to purchase the new brand if, when tested, the test evidence supports the manufacturer's claim at the .05 significance level. Suppose 64 bulbs were tested with the following results: $\bar{x} = 920$ hours, $s = 80$ hours. Will the University of Minnesota purchase the new brand of fluorescent bulbs?

7.19 The introduction of printed circuit boards (PCBs) in the 1950s revolutionized the electronics industry. However, solder-joint defects on PCBs have plagued electronics manufacturers since the introduction of the PCB. A single PCB may contain thousands of solder joints. Until the 1980s, the only means of checking the quality of solder joints was by visual inspection. Because of the low reliability of visual inspection, some manufacturers required each of their joints to be inspected four times by four different people. Now both X-ray and laser technologies are available for use in inspection (Streeter, 1986). A particular manufacturer of laser-based inspection equipment claims that its product can inspect on average at least 10 solder joints per second when the joints are spaced .1 inch apart. The equipment was tested by a potential buyer on 48 different PCBs. In each case, the equipment was operated for exactly 1 second. The numbers of solder joints inspected on each run are as follows:

10	9	10	10	11	9	12	8	8	9	6	10
7	10	11	9	9	13	9	10	11	10	12	8
9	9	9	7	12	6	9	10	10	8	7	9
11	12	10	0	10	11	12	9	7	9	9	10

a. The potential buyer wants to know whether the sample data refute the manufacturer's claim. Specify the null and alternative hypotheses that the buyer should test.

b. In the context of this exercise, what is a Type I error? A Type II error?
c. Conduct the hypothesis test you described in part **a**, and interpret the test's result in the context of this exercise. Use $\alpha = .05$.

7.20 Nutritionists stress that weight control generally requires significant reductions in the intake of fat. A random sample of 64 middle-age men on weight-control programs is selected to determine whether their mean intake of fat exceeds the recommended 30 grams per day. The sample mean and standard deviation are $\bar{x} = 37$ and $s = 32$, respectively.
a. Considering the sample mean and standard deviation, would you expect the distribution for fat intake per day to be symmetric or skewed? Explain.
b. Do the sample results indicate that the mean intake for middle-age men on weight-control programs exceeds 30 grams? Test using $\alpha = .10$.
c. Would you reach the same conclusion as in part **b** using $\alpha = .05$? Using $\alpha = .01$? Why can the conclusion of a test change when the value of α is changed?

7.3 Observed Significance Levels: *p*-Values

According to the statistical test procedure described in Section 7.2, the rejection region and, correspondingly, the value of α are selected prior to conducting the test, and the conclusions are stated in terms of rejecting or not rejecting the null hypothesis. A second method of presenting the results of a statistical test is one that reports the extent to which the test statistic disagrees with the null hypothesis and leaves to the reader the task of deciding whether to reject the null hypothesis. This measure of disagreement is called the **observed significance level** (or ***p*-value**) for the test.

Definition 7.1

The **observed significance level**, or ***p*-value**, for a specific statistical test is the probability (assuming H_0 is true) of observing a value of the test statistic that is at least as contradictory to the null hypothesis, and supportive of the alternative hypothesis, as the actual one computed from the sample data.

For example, the value of the test statistic computed for the sample of $n = 50$ sections of sewer pipe was $z = 2.12$. Since the test is one-tailed—i.e., the alternative (research) hypothesis of interest is H_a: $\mu > 2{,}400$—values of the test statistic even more contradictory to H_0 than the one observed would be values larger than $z = 2.12$. Therefore, the observed significance level (p-value) for this test is

$$p\text{-value} = P(z \geq 2.12)$$

or, equivalently, the area under the standard normal curve to the right of $z = 2.12$ (see Figure 7.6).

FIGURE 7.6 ▶
Finding the p-value for an upper-tailed test when $z = 2.12$

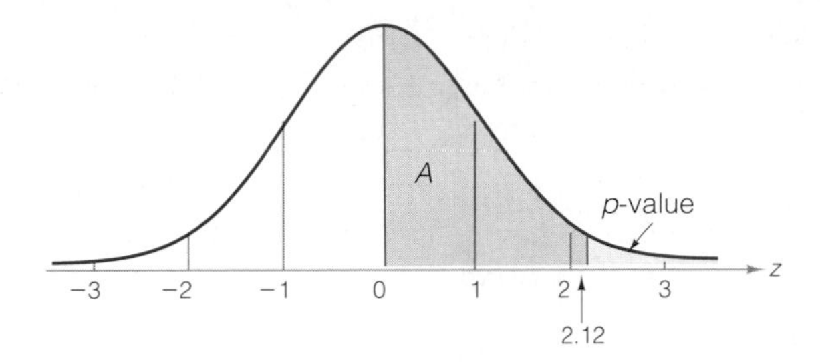

The area A in Figure 7.6 is given in Table IV in Appendix A as .4830. Therefore, the upper-tail area corresponding to $z = 2.12$ is

$$p\text{-value} = .5 - .4830 = .0170$$

Consequently, we say that these test results are "very significant"; i.e., they disagree rather strongly with the null hypothesis, H_0: $\mu = 2{,}400$, and favor H_a: $\mu > 2{,}400$. The probability of observing a z value as large as 2.12 is only .0170, if in fact the true value of μ is 2,400.

If you are inclined to select $\alpha = .05$ for this test, then you would reject the null hypothesis because the p-value for the test, .0170, is less than .05. In contrast, if you choose $\alpha = .01$ you would not reject the null hypothesis because the p-value for the test is larger than .01. Thus, the use of the observed significance level is identical to the test procedure described in the preceding sections except that the choice of α is left to you.

The steps for calculating the p-value corresponding to a test statistic for a population mean are given in the box.

Steps for Calculating the p-Value for a Test of Hypothesis

1. Determine the value of the test statistic z corresponding to the result of the sampling experiment.
2. a. If the test is one-tailed, the p-value is equal to the tail area beyond z in the same direction as the alternative hypothesis. Thus, if the alternative hypothesis is of the form $>$, the p-value is the area to the right of, or above, the observed z value. Conversely, if the alternative is of the form $<$, the p-value is the area to the left of, or below, the observed z value.
   b. If the test is two-tailed, the p-value is equal to twice the tail area beyond the observed z value in the direction of the sign of z. That is, if z is positive, the p-value is twice the area to the right of, or above, the observed z value. Conversely, if z is negative, the p-value is twice the area to the left of, or below, the observed z value.

EXAMPLE 7.3 Find the observed significance level for the test of the mean response time for drug-injected rats in Examples 7.1 and 7.2.

Solution Example 7.1 presented a two-tailed test of the hypothesis

$$H_0: \quad \mu = 1.2 \text{ seconds}$$

against the alternative hypothesis

$$H_a: \quad \mu \neq 1.2 \text{ seconds}$$

The observed value of the test statistic in Example 7.2 was $z = -3.0$, and any value of z less than -3.0 or greater than $+3.0$ (because this is a two-tailed test) would be even more contradictory to H_0. Therefore, the observed significance level for the test is

$$p\text{-value} = P(z < -3.0 \text{ or } z > +3.0)$$

Thus, we calculate the area below the observed z value, $z = -3.0$, and double it. Consulting Table IV in Appendix A, we find that $P(z < -3.0) = .5 - .4987 = .0013$. Therefore, the p-value for this two-tailed test is

$$2P(z < -3.0) = 2(.0013) = .0026$$

We can interpret this p-value as a strong indication that the mean reaction time of drug-injected rats differs from the control mean ($\mu \neq 1.2$), since we would observe a test statistic this extreme or more extreme only 26 in 10,000 times if the drug-injected mean were equal to the control mean ($\mu = 1.2$). The extent to which the mean differs from 1.2 could be better determined by calculating a confidence interval for μ.

When publishing the results of a statistical test of hypothesis in journals, case studies, reports, etc., many researchers make use of p-values. Instead of selecting α beforehand and then conducting a test, as outlined in this chapter, the researcher computes and reports the value of the appropriate test statistic and its associated p-value. It is left to the reader of the report to judge the significance of the result—i.e., the reader must determine whether to reject the null hypothesis in favor of the alternative hypothesis, based on the reported p-value. This p-value is often referred to as the **observed significance level** of the test. Usually, the null hypothesis is rejected if the observed significance level is *less* than the fixed significance level, α, chosen by the reader. The inherent advantages of reporting test results in this manner are twofold: (1) Readers are permitted to select the maximum value of α that they would be willing to tolerate if they actually carried out a standard test of hypothesis in the manner outlined in this chapter, and (2) a measure of the degree of significance of the result (i.e., the p-value) is provided.

Reporting Test Results as p-Values: How to Decide Whether to Reject H_0

1. Choose the maximum value of α that you are willing to tolerate.
2. If the observed significance level (p-value) of the test is less than the chosen value of α, reject the null hypothesis. Otherwise, do not reject the null hypothesis.

Exercises 7.21–7.35

Learning the Mechanics

7.21 If a hypothesis test were conducted using $\alpha = .05$, for which of the following p-values would the null hypothesis be rejected?

a. .06 b. .10 c. .01
d. .001 e. .251 f. .042

7.22 For each α and observed significance level (p-value) pair, indicate whether the null hypothesis would be rejected.

a. $\alpha = .05$, p-value $= .10$ b. $\alpha = .10$, p-value $= .05$
c. $\alpha = .01$, p-value $= .001$ d. $\alpha = .025$, p-value $= .05$
e. $\alpha = .10$, p-value $= .45$

7.23 An analyst tested the null hypothesis $\mu \geq 20$ against the alternative hypothesis that $\mu < 20$. The analyst reported a p-value of .06. What is the smallest value of α for which the null hypothesis would be rejected?

7.24 In a test of H_0: $\mu = 100$ against H_a: $\mu > 100$, the sample data yielded the test statistic $z = 2.17$. Find the p-value for the test.

7.25 In a test of H_0: $\mu = 100$ against H_a: $\mu \neq 100$, the sample data yielded the test statistic $z = 2.17$. Find the p-value for the test.

7.26 In a test of H_0: $\mu \geq 100$ against H_a: $\mu < 100$, the sample data yielded the test statistic $z = -1.24$. Find the observed significance level of the test.

7.27 In a test of the hypothesis H_0: $\mu = 50$ versus H_a: $\mu > 50$, a sample of $n = 100$ observations possessed mean $\bar{x} = 49.4$ and standard deviation $s = 4.1$. Find and interpret the p-value for this test.

7.28 In a test of the hypothesis H_0: $\mu = 10$ versus H_a: $\mu \neq 10$, a sample of $n = 50$ observations possessed mean $\bar{x} = 10.7$ and standard deviation $s = 3.1$. Find and interpret the p-value for this test.

Applying the Concepts

7.29 A new blood pressure drug is advertised to reduce, after 1 week of medication, a patient's blood pressure an average of 10 units. Blood pressure reductions were recorded for 37 patients after treatment with the drug. The mean and standard deviation for this sample were 8.7 and 6.8, respectively. Do the data appear to contradict the advertising claim? Use the observed significance level for the test to answer the question.

7.30 The manufacturer of an over-the-counter pain reliever claims that its product brings pain relief to headache sufferers in less than 3.5 minutes, on average. In order to be able to make this claim in its television advertisements, the manufacturer was required by a particular television network to present statistical evidence in support of the claim. The manufacturer reported that for a random sample of 50 headache sufferers, the mean time to relief was 3.3 minutes and the standard deviation was 1.1 minutes.

a. Do these data support the manufacturer's claim? Test using $\alpha = .05$.
b. Report the p-value of the test.
c. In general, do large p-values or small p-values support the manufacturer's claim? Explain.

7.31 *The Chronicle of Higher Education Almanac* (Sept. 1990) reported that for the 1989–1990 academic year, 4-year private colleges charged students an average of \$8,446 for tuition and fees, whereas at 4-year public colleges the average was \$1,781. Suppose that for 1990–1991 a random sample of 30 colleges yielded the following data on tuition and fees: $\bar{x} = \$9,392$ and $s = \$1,643$. Assume that \$8,446 is the population mean for 1989–1990.

a. Specify the null and alternative hypotheses you would use to investigate whether the mean amount for tuition and fees in 1990–1991 was larger than it was in 1989–1990.
b. Calculate the p-value for the hypothesis test you described in part **a**, and explain what the p-value indicates about the statistical significance of the test results.

7.32 Refer to Exercise 7.14, in which a random sample of 49 unemployed men were administered the General Health Questionnaire (GHQ). The sample mean and standard deviation were 10.94 and 5.10, respectively. Denoting the population mean GHQ for unemployed workers by μ, we wish to test the null hypothesis H_0: $\mu = 10$ versus the one-tailed alternative H_a: $\mu > 10$.

a. When the data are run through statistical software, the results (in part) are as shown below. Check the program's results for accuracy.

```
Z = 1.29          P-VALUE = .0985
```

b. What conclusion would you reach about the test based on the computer analysis?

7.33 In Exercise 6.10 we examined research about bicycle helmets reported in *Public Health Reports* (Otis et al., 1992). One of the variables measured was the children's perception of the risk involved in bicycling. A random sample of 797 children in grades 4–6 were asked to rate their perception of bicycle risk without wearing a helmet, ranging from 1 (no risk) to 4 (very high risk). The mean and standard deviation of the sample were: $\bar{x} = 3.39$, $s = .80$.

a. Assume that a mean score, μ, of 2.5 is indicative of indifference as to risk, and values of μ exceeding 2.5 indicate a perception that a risk exists. What are the appropriate null and alternative hypotheses for testing the research hypothesis that children in this age group perceive a risk associated with failure to wear helmets?
b. Calculate the p-value for the data collected in this study.
c. Interpret the p-value in the context of this research.

7.34 According to advertisements, a strain of soybeans planted on soil prepared with a specified fertilizer treatment has a mean yield of 500 bushels per acre. Fifty farmers who belong to a cooperative plant the soybeans. Each uses a 40-acre plot and records the mean yield per acre. The mean and variance for the sample of 50 farms are $\bar{x} = 485$ and $s^2 = 10,045$.

a. Use the p-value for this test to determine whether the data provide sufficient evidence to indicate that the mean yield for the soybeans is different from that advertised.
b. In reaching the conclusion in part **a**, would you have to qualify your conclusions because of the manner in which the sample was selected? Explain.

7.35 Golf-course designers are concerned that the new equipment introduced each year is making old courses obsolete because it enables golfers to hit the ball so far that the courses are, in effect, "shrinking." One designer believes that new courses need to be built with the expectation that players will be able to hit the ball more than 250 yards (with their drivers), on average. Suppose a sample of 135 golfers is tested, and their mean driving distance is 256.3 yards, with a standard deviation of 43.4 yards.
a. What are appropriate null and alternative hypotheses to test the designer's research hypothesis?
b. Calculate and interpret the p-value for this test.
c. If you were to make this a two-tailed test, how would your answer to part **b** change?

7.4 Small-Sample Test of Hypothesis About a Population Mean

Most water-treatment facilities monitor the quality of their drinking water on an hourly basis. One variable monitored is pH, which measures the degree of alkalinity or acidity in the water. A pH below 7.0 is acidic, one above 7.0 is alkaline, and a pH of 7.0 is neutral. One water-treatment plant has a target pH of 8.5 (most try to maintain a slightly alkaline level). The mean and standard deviation of 1 hour's test results, based on 17 water samples at this plant, are

$$\bar{x} = 8.42 \qquad s = .16$$

Does this sample provide sufficient evidence that the mean pH level in the water differs from 8.5?

This inference can be placed in a test of hypothesis framework. We establish the target pH as the null hypothesized value and then utilize a two-tailed alternative that the true mean pH differs from target:

$$H_0: \mu = 8.5$$
$$H_a: \mu \neq 8.5$$

Recall from Section 6.3 that when we are faced with making inferences about a population mean using the information in a small sample, two problems emerge:

1. The normality of the sampling distribution for $\bar{x}$ does not follow from the Central Limit Theorem when the sample size is small. We must assume that the distribution of measurements from which the sample was selected is approximately normally distributed in order to ensure the approximate normality of the sampling distribution of $\bar{x}$.
2. If the population standard deviation σ is unknown, as is usually the case, then we cannot assume that s will provide a good approximation for σ when the sample size is small. Instead, we must use the t-distribution rather than the standard normal z-distribution to make inferences about the population mean μ.

Therefore, as the test statistic of a small-sample test of a population mean, we use the t statistic:

$$\textit{Test statistic:} \quad t = \frac{\bar{x} - \mu_0}{s/\sqrt{n}} = \frac{\bar{x} - 8.5}{s/\sqrt{n}}$$

where μ_0 is the null hypothesized value of the population mean, μ. In our example, $\mu_0 = 8.5$.

To find the rejection region, we must specify the value of α, the probability that the test will lead to rejection of the null hypothesis when it is true, and then consult the t table (Table V of Appendix A or inside the front cover of the text). Using $\alpha = .05$, the two-tailed rejection region is

Rejection region: $t_{\alpha/2} = t_{.025} = 2.120$ with $n - 1 = 16$ degrees of freedom
Reject H_0 if $t < -2.120$ or $t > 2.120$

The rejection region is shown in Figure 7.7.

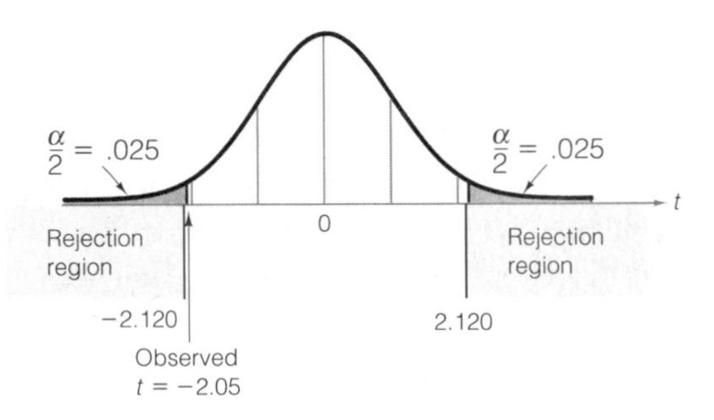

FIGURE 7.7 ▶ Two-tailed rejection region for small-sample t-test

We are now prepared to calculate the test statistic and reach a conclusion:

$$t = \frac{\bar{x} - \mu_0}{s/\sqrt{n}} = \frac{8.42 - 8.50}{.16/\sqrt{17}} = \frac{-.08}{.039} = -2.05$$

Since the calculated value of t does not fall in the rejection region (Figure 7.7), we cannot reject H_0 at the $\alpha = .05$ level of significance. Thus, the water-treatment plant should not conclude that the mean pH differs from the 8.5 target based on the sample evidence.

It is interesting to note that the calculated t value, -2.05, is *less than* the .05 level z value, -1.96. The implication is that if we had *incorrectly* used a z statistic for this test, we would have rejected the null hypothesis at the .05 level, concluding that the mean pH level differs from 8.5. The important point is that the statistical procedure to be used must always be closely scrutinized and all the assumptions understood. Many statistical lies are the result of misapplications of otherwise valid procedures.

The technique for conducting a small-sample test of hypothesis about a population mean is summarized in the following box.

Small-Sample Test of Hypothesis About μ

One-Tailed Test	*Two-Tailed Test*
H_0: $\mu = \mu_0$	H_0: $\mu = \mu_0$
H_a: $\mu < \mu_0$ (or H_a: $\mu > \mu_0$)	H_a: $\mu \neq \mu_0$
Test statistic: $t = \frac{\bar{x} - \mu_0}{s/\sqrt{n}}$	*Test statistic:* $t = \frac{\bar{x} - \mu_0}{s/\sqrt{n}}$
Rejection region: $t < -t_\alpha$ (or $t > t_\alpha$ when H_a: $\mu > \mu_0$)	*Rejection region:* $t < -t_{\alpha/2}$ or $t > t_{\alpha/2}$

where t_α and $t_{\alpha/2}$ are based on $(n - 1)$ degrees of freedom

Assumption: A random sample is selected from a population with a relative frequency distribution that is approximately normal.

EXAMPLE 7.4

A major car manufacturer wants to test a new engine to determine whether it meets new air-pollution standards. The mean emission μ of all engines of this type must be less than 20 parts per million of carbon. Ten engines are manufactured for testing purposes, and the mean and standard deviation of the emissions for this sample of engines are determined to be

$\bar{x} = 17.1$ parts per million $\qquad s = 3.0$ parts per million

Do the data supply sufficient evidence to allow the manufacturer to conclude that this type of engine meets the pollution standard? Assume that the manufacturer is willing to risk a Type I error with probability $\alpha = .01$.

Solution

The manufacturer wants to support the research hypothesis that the mean emission level μ for all engines of this type is less than 20 parts per million. The elements of this small-sample one-tailed test are

H_0: $\mu = 20$

H_a: $\mu < 20$

Test statistic: $t = \frac{\bar{x} - 20}{s/\sqrt{n}}$

Assumption: The relative frequency distribution of the population of emission levels for all engines of this type is approximately normal.

Rejection region: For $\alpha = .01$ and df $= n - 1 = 9$, the one-tailed rejection region (see Figure 7.8 on page 336) is $t < -t_{.01} = -2.821$.

We now calculate the test statistic:

$$t = \frac{\bar{x} - 20}{s/\sqrt{n}} = \frac{17.1 - 20}{3.0/\sqrt{10}} = -3.06$$

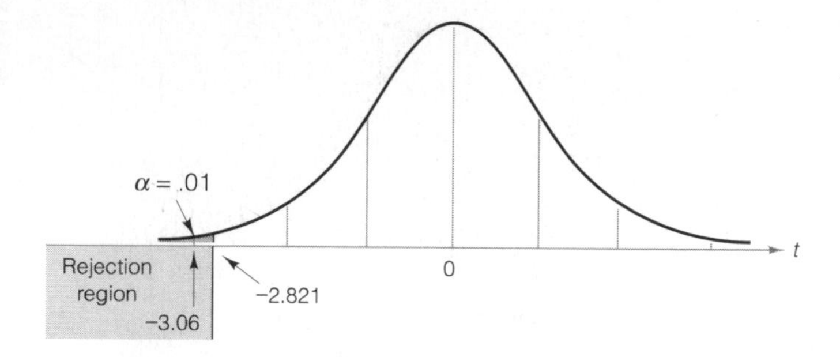

FIGURE 7.8 ▶ A t-distribution with 9 df and the rejection region for Example 7.4

Since the calculated t falls in the rejection region (see Figure 7.8), the manufacturer concludes that $\mu < 20$ parts per million and the new engine type meets the pollution standard. Are you satisfied with the reliability associated with this inference? The probability is only $\alpha = .01$ that the test would support the research hypothesis if in fact it were false.

EXAMPLE 7.5

Find the observed significance level for the test described in Example 7.4.

Solution

The test of Example 7.4 was a lower-tail test: H_0: $\mu = 20$ versus H_a: $\mu < 20$. Since the value of t computed from the sample data was $t = -3.06$, the observed significance level (or p-value) for the test is equal to the probability that t would assume a value less than or equal to -3.06 if in fact H_0 were true. This is equal to the area in the lower tail of the t-distribution (shaded in Figure 7.9). To find this area—i.e., the p-value for the test—we consult the t-table (Table V in Appendix A). Unlike the table of areas under the normal curve, Table V gives only the t values corresponding to the areas .100, .050, .025, .010, .005, .001, and .0005. Therefore, we can only approximate the p-value for the test. Since the observed t value was based on 9 degrees of freedom, we use the df = 9 row in Table V and move across the row until we reach the t values that are closest to the observed $t = -3.06$. [*Note:* We ignore the minus sign.] The t values corresponding to p-values of .010 and .005 are 2.821 and 3.250, respectively. Since the observed t value falls between $t_{.010}$ and $t_{.005}$, the p-value for the test lies between .005 and .010. We could interpolate to locate the p-value for the test more accurately, but it is easier and adequate for our purposes to choose the larger area as the p-value and report it as .010. Thus, we would reject the null hypothesis, H_0: $\mu =$ 20 parts per million, for any value of α larger than .01.

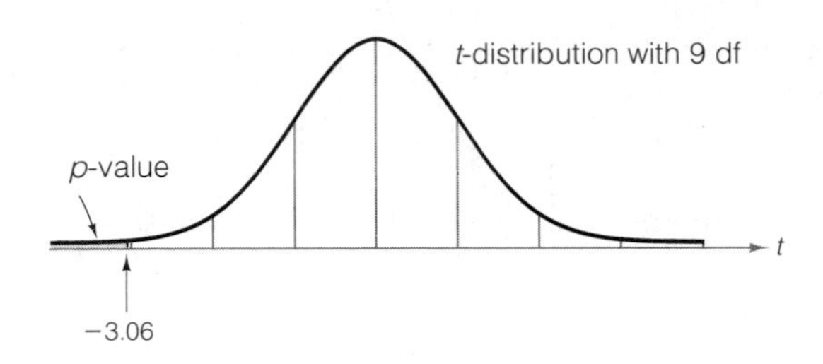

FIGURE 7.9 ▶
The observed significance level for the test of Example 7.4

Small-sample inferences typically require more assumptions and provide less information about the population parameter than do large-sample inferences. Nevertheless, the t-test is a method of testing a hypothesis about a population mean of a normal distribution when only a small number of observations is available.

Exercises 7.36–7.49

Learning the Mechanics

7.36 In what ways are the distributions of the z-test statistic and t-test statistic alike? How do they differ?

7.37 Under what circumstances should you use the t-distribution in testing a hypothesis about a population mean?

7.38 A random sample of n observations is selected from a normal population to test the null hypothesis that $\mu = 10$. Specify the rejection region for each of the following combinations of H_a, α, and n:

a. H_a: $\mu \neq 10$; $\alpha = .05$; $n = 14$ b. H_a: $\mu > 10$; $\alpha = .01$; $n = 24$
c. H_a: $\mu > 10$; $\alpha = .10$; $n = 9$ d. H_a: $\mu < 10$; $\alpha = .01$; $n = 12$
e. H_a: $\mu \neq 10$; $\alpha = .10$; $n = 20$ f. H_a: $\mu < 10$; $\alpha = .05$; $n = 4$

7.39 For each of the following rejection regions, sketch the sampling distribution of t, and indicate the location of the rejection region on your sketch:

a. $t > 1.440$ where df = 6 b. $t < -1.782$ where df = 12
c. $t < -2.060$ or $t > 2.060$ where df = 25

7.40 For each of the rejection regions defined in Exercise 7.39, what is the probability that a Type I error will be made?

7.41 The following sample of five measurements was randomly selected from a normally distributed population: 4, 5, 3, 6, 6.

a. Test the null hypothesis that the mean of the population is 6 against the alternative hypothesis, $\mu < 6$. Use $\alpha = .05$.
b. Test the null hypothesis that the mean of the population is 6 against the alternative hypothesis, $\mu \neq 6$. Use $\alpha = .05$.
c. Find the observed significance level for each test.

7.42 The following sample of six measurements was randomly selected from a normally distributed population: 1, 3, −1, 5, 1, 2.

a. Test the null hypothesis that the mean of the population is 3 against the alternative hypothesis, $\mu < 3$. Use $\alpha = .05$.
b. Test the null hypothesis that the mean of the population is 3 against the alternative hypothesis, $\mu \neq 3$. Use $\alpha = .05$.
c. Find the observed significance level for each test.

7.43 A statistical software package is used to conduct a *t*-test for the null hypothesis H_0: $\mu = 1{,}000$ versus the alternative hypothesis H_a: $\mu > 1{,}000$ based on a sample of 17 observations. The software's output is:

```
T = 1.894          P-VALUE = .0382
```

a. Most software does not indicate what assumptions are necessary for the validity of a statistical procedure. What assumptions are necessary for the validity of this procedure?
b. Interpret the results of the test.
c. Suppose the alternative hypothesis had been the two-tailed H_a: $\mu \neq 1{,}000$. If the t statistic were unchanged, then what would the p-value be for this test? Interpret the p-value for the two-tailed test.

Applying the Concepts

7.44 A consumer protection group is concerned that a catsup manufacturer is filling its 20-ounce family-size containers with less than 20 ounces of catsup. The group purchases 10 family-size bottles of this catsup, weighs the contents of each, and finds that the mean weight is equal to 19.86 ounces, and the standard deviation is equal to .22 ounce.

a. Do the data provide sufficient evidence for the consumer group to conclude that the mean fill per family-size bottle is less than 20 ounces? Test using $\alpha = .05$.
b. If the test in part **a** were conducted on a periodic basis by the company's quality control department, is the consumer group more concerned about making a Type I error or a Type II error? (The probability of making this type of error is called the *consumer's risk*.)
c. The catsup company is also interested in the mean amount of catsup per bottle. It does not wish to overfill them. For the test conducted in part **a**, which type of error is more serious from the company's point of view—a Type I error or a Type II error? (The probability of making this type of error is called the *producer's risk*.)
d. Find a 90% confidence interval for the mean number of ounces of catsup per bottle.

7.45 The Everglades National Park in Florida is considered a national treasure by many, and much environmental research has been conducted to determine how the ecology of the park is changing. Total phosphorus is one water quality parameter of concern in the park. Suppose that the EPA makes 12 measurements in one section of the park, yielding a mean level of total phosphorus at 12.3 parts per billion (ppb) and a standard deviation of 5.4 ppb. The EPA wants to test whether the data support the conclusion that the mean level is less than 15 ppb.

a. What are the null and alternative hypotheses appropriate for the EPA's test?
b. The EPA statistician submitted the data to a statistical software program, with the results shown. Check the computer results: calculate the t statistic, the degrees of freedom, and determine whether the p-value is in the correct range according to Table V of Appendix A.

```
T = -1.732          DF = 11          P-VALUE = .0556
```

c. Interpret the results of the test.

7.46 One of the most feared predators in the ocean is the great white shark. Although it is known that the white shark grows to a mean length of 21 feet, a marine biologist believes that the great white sharks off the Bermuda coast grow much longer due to unusual feeding habits. To test this claim, a number of full-grown great white sharks are captured off the Bermuda coast, measured and then set free. However, because the capture of sharks is difficult, costly, and very dangerous, only three are sampled. Their lengths are 24, 20, and 22 feet.
a. Do the data provide sufficient evidence to support the marine biologist's claim? Use $\alpha = .10$.
b. Give the approximate observed significance level for the test in part **a**, and interpret its value.
c. What assumptions must be made in order to carry out the test?
d. Do you think these assumptions are likely to be satisfied in this sampling situation?

7.47 An important problem facing strawberry growers is the control of nematodes. These organisms compete with the plants for nutrients in the soil, thereby reducing yield. For this reason, fumigation is normally a part of field preparation. In the past, the fumigants used yielded an average of 8 pounds of marketable fruit for a certain standard sized plot. Recently, a new fumigant has been developed. It is applied to six standard plots of strawberries, and the yield of marketable fruit (in pounds) for each plot is 9, 9, 13, 9, 10, and 8.
a. Do the data indicate a significant increase in average yield at the .05 level of significance?
b. What assumptions are necessary for the procedure used to be valid?

7.48 A psychologist was interested in knowing whether male heroin addicts' assessments of self-worth differ from those of the general male population. On a test designed to measure assessment of self-worth, the mean score for males from the general population is 48.6. A random sample of 25 scores achieved by heroin addicts yielded a mean of 44.1 and a standard deviation of 6.2
a. Do the data indicate a difference in assessment of self-worth between male heroin addicts and the general male population? Test using $\alpha = .01$.
b. Give the approximate observed significance level for the test and interpret its value.

7.49 Research reported in the *Journal of Psychology* (Mills, 1991) studied the personality characteristics of obese individuals. One variable, the locus of control (LOC), measures the individual's degree of belief that he or she has control over situations. High scores on the LOC scale indicate less perceived control. For one sample of 19 obese adolescents, the mean LOC score was 10.89 with a standard deviation of 2.48. Suppose we wish to test whether the mean LOC score for all obese adolescents exceeds 10.
a. Specify the null and alternative hypotheses for this test.

b. The data are analyzed by computer, and the following output is obtained. Check the calculation of the t statistic, the degrees of freedom, and determine whether the p-value is in the correct range according to Table V of Appendix A.

```
T = 1.564          DF = 18          P-VALUE = .0676
```

c. Interpret the p-value. What conclusion would you reach regarding this test?

7.5 A Nonparametric Test About a Population Median (Optional)

In Sections 7.2–7.4 we utilized the z and t statistics for testing hypotheses about a population mean. The z statistic is appropriate for large random samples selected from "general" populations—that is, with few limitations on the probability distribution of the underlying population. The t statistic was developed for small-sample tests when the sample is selected at random from a *normal* distribution. The question is: How can we conduct a test of hypothesis when we have a small sample from a *nonnormal* distribution? The answer: Use a procedure that requires fewer or less stringent assumptions about the underlying population, called a **nonparametric method**.

The **sign test** is a relatively simple nonparametric procedure for testing hypotheses about the central tendency of a nonnormal probability distribution. Note that we used the phrase *central tendency* rather than *population mean*. This is because the sign test, like many nonparametric procedures, provides inferences about the population *median*, M, rather than the population mean, μ. Remember (Chapter 2) that the median is the 50th percentile of the distribution (Figure 7.10) and as such is less affected by the skewness of the distribution and the presence of outliers (extreme observations). Since the nonparametric test must be suitable for all distributions, not just the normal, it is reasonable for nonparametric tests to focus on the more robust (less sensitive to extreme values) measure of central tendency, the median.

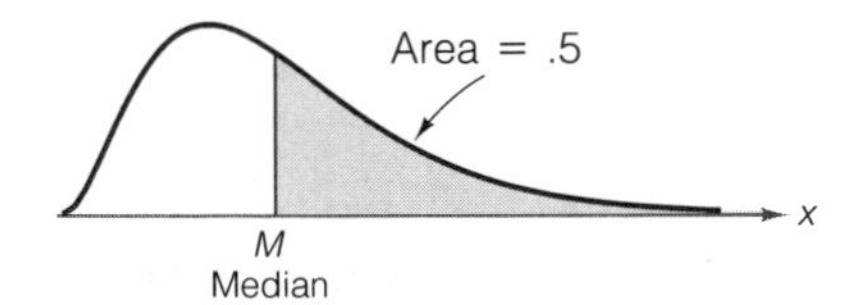

FIGURE 7.10 ▶ Location of the population median, M

For example, increasing numbers of both private and public agencies are requiring their employees to submit to tests for substance abuse. One laboratory that conducts the testing has developed a system with a normalized measurement scale, in which values less than 1.00 indicate "normal" ranges and values equal to or greater than 1.00 are indicative of potential substance abuse. The lab reports a normal result as long as the median level for an individual is less than 1.00. Eight independent measurements of each individual's sample are made; suppose that one individual's results were as follows:

.78 .51 3.79 .23 .77 .98 .96 .89

If the objective is to determine whether the *population* median (that is, the true median level if an indefinitely large number of measurements were made on the same individual sample) is less than 1.00, we establish that as our alternative hypothesis and test

H_0: $M = 1.00$

H_a: $M < 1.00$

The one-tailed sign test is conducted by counting the number of sample measurements that "favor" the alternative hypothesis—in this case, the number that are less than 1.00. If the null hypothesis is true, we expect approximately half of the measurements to fall on each side of the hypothesized median, and if the alternative is true, we expect significantly more than half to favor the alternative—that is, to be less than 1.00. Thus,

Test statistic: S = Number of measurements less than 1.00, the null hypothesized median

If we wish to conduct the test at the $\alpha = .05$ level of significance, then the rejection region can be expressed in terms of the observed significance level, or p-value, of the test:

Rejection region: $p\text{-value} \leq .05$

In this example, $S = 7$ of the 8 measurements are less than 1.00. To determine the observed significance level associated with this outcome, we note that the number of measurements less than 1.00 is a binomial random variable (check the binomial characteristics presented in Chapter 4), and *if H_0 is true*, the binomial probability p that a measurement lies below (or above) the median 1.00 is equal to .5 (Figure 7.10). What is the probability that a result is *as contrary to or more contrary* to H_0 than the one observed? That is, what is the probability that 7 *or more* of 8 binomial measurements will result in Success (be less than 1.00) if the probability of Success is .5? Binomial Table II in Appendix A (using $n = 8$ and $p = .5$) indicates that

$$P(x \geq 7) = 1 - P(x \leq 6) = 1 - .965 = .035$$

Thus, the probability that at least 7 of 8 measurements would be less than 1.00 *if the true median were* 1.00 is only .035. The p-value of the test is therefore .035, which is less than $\alpha = .05$. We therefore conclude that this sample provides sufficient evidence to reject the null hypothesis. The implication of this rejection is that the laboratory can conclude at the $\alpha = .05$ level of significance that the true median level for the tested individual is less than 1.00. However, we note that one of the measurements greatly exceeds the others, with a value of 3.79, and deserves special attention. Note that this large measurement is an outlier that would make the use of a t-test and its concomitant assumption of normality dubious. The only assumption necessary to ensure the validity of the sign test is that the probability distribution of measurements is continuous.

The use of the sign test for testing hypotheses about population medians is summarized in the box.

Sign Test for a Population Median M

One-Tailed Test

H_0: $M = M_0$

H_a: $M > M_0$
[or H_a: $M < M_0$]

Test statistic:

S = Number of sample measurements greater than M_0
[or S = Number of sample measurements less than M_0]

Observed significance level:

p-value $= P(x \geq S)$

Two-Tailed Test

H_0: $M = M_0$

H_a: $M \neq M_0$

Test statistic:

S = Larger of S_1 and S_2, where S_1 is the number of measurements less than M_0 and S_2 is the number of measurements greater than M_0

Observed significance level:

p-value $= 2P(x \geq S)$

where x has a binomial distribution with parameters n and $p = .5$. (Use Table II, Appendix A.)

Rejection region: Reject H_0 if p-value $\leq .05$.

Assumption: The sample is selected randomly from a continuous probability distribution. [*Note:* No assumptions need to be made about the shape of the probability distribution.]

Recall that the normal probability distribution provides a good approximation for the binomial distribution when the sample size is large. For tests about the median of a distribution, the null hypothesis implies that $p = .5$, and the normal distribution provides a good approximation if $n \geq 10$. (Samples with $n \geq 10$ satisfy the condition that $np \pm 3\sqrt{npq}$ is contained in the interval 0 to n.) Thus, we can use the standard normal z-distribution to conduct the sign test for large samples. The large-sample sign test is summarized in the next box.

Large-Sample Sign Test for a Population Median M

One-Tailed Test	*Two-Tailed Test*
H_0: $M = M_0$	H_0: $M = M_0$
H_a: $M > M_0$ [or H_a: $M < M_0$]	H_a: $M \neq M_0$

Test statistic: $z = \dfrac{(S - .5) - .5n}{.5\sqrt{n}}$

[*Note:* S is calculated as shown in the previous box. We subtract .5 from S as the "correction for continuity." The null hypothesized mean value is $np = .5n$, and the standard deviation is

$$\sqrt{npq} = \sqrt{n(.5)(.5)} = .5\sqrt{n}$$

See Chapter 4 for details on the normal approximation to the binomial distribution.]

Rejection region: $z > z_{\alpha}$ *Rejection region:* $z > z_{\alpha/2}$

where tabulated z values can be found inside the front cover.

EXAMPLE 7.6

A manufacturer of compact disk (CD) players has established that the median time to failure for its players is 5,250 hours of utilization. A sample of 20 CDs from a competitor is obtained, and they are continuously tested until each fails. The 20 failure times range from 5 hours (a "defective" player) to 6,575 hours, and 14 of the 20 exceed 5,250 hours. Is there evidence that the median failure time of the competitor differs from 5,250 hours? Use $\alpha = .10$.

Solution

The null and alternative hypotheses of interest are

H_0: $M = 5{,}250$ hours

H_a: $M \neq 5{,}250$ hours

Test statistic: Since $n \geq 10$, we use the standard normal z statistic:

$$z = \frac{(S - .5) - .5n}{.5\sqrt{n}}$$

where S is the maximum of S_1, the number of measurements greater than 5,250, and S_2, the number of measurements less than 5,250.

Rejection region: $z > 1.645$, where $z_{\alpha/2} = z_{.05} = 1.645$

Assumptions: The distribution of the failure times is continuous (time is a continuous variable), but nothing is assumed about the shape of its probability distribution.

Since the number of measurements exceeding 5,250 is $S_2 = 14$ and thus the number of measurements less than 5,250 is $S_1 = 6$, then $S = 14$, the greater of S_1 and S_2. The calculated z statistic is therefore

$$z = \frac{(S - .5) - .5n}{.5\sqrt{n}} = \frac{13.5 - 10}{.5\sqrt{20}} = \frac{3.5}{2.236} = 1.565$$

The value of z is not in the rejection region, so we cannot reject the null hypothesis at the $\alpha = .10$ level of significance. Thus, the CD manufacturer should not conclude, on the basis of this sample, that its competitor's CDs have a median failure time that differs from 5,250 hours.

The one-sample nonparametric sign test for a median provides an alternative to the t-test for small samples from nonnormal distributions. However, if the distribution is approximately normal, the t-test provides a more powerful test about the central tendency of the distribution.

Exercises 7.50–7.57

Learning the Mechanics

7.50 Under what circumstances is the sign test preferred to the t-test for making inferences about the central tendency of a population?

7.51 What is the probability that a randomly selected observation exceeds the
a. Mean of a normal distribution? b. Median of a normal distribution?
c. Mean of a nonnormal distribution? d. Median of a nonnormal distribution?

7.52 Use Table II of Appendix A to calculate the following binomial probabilities:
a. $P(x \geq 6)$ when $n = 7$ and $p = .5$ b. $P(x \geq 5)$ when $n = 9$ and $p = .5$
c. $P(x \geq 8)$ when $n = 8$ and $p = .5$
d. $P(x \geq 10)$ when $n = 15$ and $p = .5$. Also use the normal approximation to calculate this probability, and compare the approximation with the exact value.
e. $P(x \geq 15)$ when $n = 25$ and $p = .5$. Also use the normal approximation to calculate this probability, and compare the approximation with the exact value.

7.53 Consider the following sample of 10 measurements:

8.4 16.9 15.8 12.5 10.3 4.9 12.9 9.8 23.7 7.3

Use these data to conduct each of the following sign tests using the binomial tables (Table II, Appendix A) and $\alpha = .05$:
a. H_0: $M = 9$ versus H_a: $M > 9$ b. H_0: $M = 9$ versus H_a: $M \neq 9$
c. H_0: $M = 20$ versus H_a: $M < 20$ d. H_0: $M = 20$ versus H_a: $M \neq 20$
e. Repeat each of the preceding tests using the normal approximation to the binomial probabilities. Compare the results.
f. What assumptions are necessary to assure the validity of each of the preceding tests?

7.54 Suppose you wish to conduct a test of the research hypothesis that the median of a population is greater than 80. You randomly sample 25 measurements from the population and determine that 16 of them exceed 80. Set up and conduct the appropriate test of hypothesis at the .10 level of significance. Be sure to specify all necessary assumptions.

Applying the Concepts

7.55 The American Cancer Society funds medical research focused on finding treatments for various forms of cancer. One new treatment is being tested to determine whether the average time a particularly virulent form of cancer can be held in remission is extended by the treatment. Assume that current treatments have been shown to provide a median remission time of 4.5 years. Seven patients with this cancer are given the new treatment, and their remission times (in years) are as follows:

5.3 7.3 3.6 5.2 6.1 4.8 8.4

Do these data provide sufficient evidence to indicate that the median remission time is increased by the new treatment?

7.56 Many states now require that mathematics teachers pass a basic math literacy test in order to qualify to teach math in the state. An important part of the process is the development of a fair test. Suppose that one state board of education has developed a test, and it is submitted to a sample of 20 math teachers. The test will be considered too hard if at least 50% of *all* math teachers in the state score less than 60 (on a scale of 0 to 100).

a. Set up the appropriate null and alternative hypotheses to test whether the test is too hard.
b. Establish the appropriate test statistic and rejection region for the test using $\alpha = .05$.
c. What assumptions are necessary to assure the validity of the test?
d. Suppose that 14 of the 20 sampled teachers score less than 60. What is the appropriate conclusion?
e. Calculate the observed significance level of the test. Interpret it.
f. Considering your answers to parts **d** and **e**, what would your recommendation be concerning the adoption of the test statewide?

7.57 A paper company requires that the median height of pine trees exceed 40 feet before they are harvested. A sample of 24 trees in one large plot is selected, and 17 of them are over 40 feet.

a. Test whether the company can conclude at the .05 level of significance that the median height of trees in the plot exceeds 40 feet.
b. Calculate and interpret the p-value of this test.
c. What assumptions must be made about the probability distribution of the tree heights in the plot in order to assure the validity of the test?

7.6 Large-Sample Test of Hypothesis About a Binomial Probability

Inferences about proportions (or percentages) are often made in the context of the probability, p, of "success" for a binomial distribution. We showed how to use large samples from binomial distributions to form confidence intervals for p in Section 6.5. We now consider tests of hypotheses about a binomial probability p.

For example, consider the problem of *insider trading* in the stock market. Insider trading is the buying and selling of stock by an individual privy to inside information in a company, usually a high-level executive in the firm. The Securities and Exchange Commission (SEC) imposes strict guidelines about insider trading so that all investors can have equal access to information that may affect the stock's price. An investor wishing to test the effectiveness of the SEC guidelines monitors the market over a period of a year and records the number of times a stock price increases the day following a significant purchase of stock by an insider. For a total of 576 such transactions, the stock increased the following day 327 times. Does this sample provide evidence that the stock price may be affected by insider trading?

We first view this as a binomial experiment, with the 576 transactions as the trials and Success representing an increase in the stock's price the following day. Let p represent the probability that the stock price will increase following a large insider purchase. If the insider purchase has no effect on the stock price (that is, if the information available to the insider is identical to that available to the general market), then the investor expects the probability of a stock increase to be the same as that of a decrease, or $p = .5$. On the other hand, if insider trading affects the stock price (indicating that the market has not fully accounted for the information known to the insiders), then the investor expects the stock either to decrease or to increase more than half the time following significant insider transactions—i.e., $p \neq .5$.

We can now place the problem in the context of a test of hypothesis:

H_0: $p = .5$

H_a: $p \neq .5$

Recall that the sample proportion, $\hat{p}$, is really just the sample mean of the outcomes of the individual binomial trials and, as such, is approximately normally distributed (for large samples) according to the Central Limit Theorem. Thus, for large samples we can use the standard normal z as the test statistic:

$$\textit{Test statistic:} \quad z = \frac{\text{Sample proportion} - \text{Null hypothesized proportion}}{\text{Standard deviation of sample proportion}} = \frac{\hat{p} - p_0}{\sigma_{\hat{p}}}$$

where we use the symbol p_0 to represent the null hypothesized value of p.

Rejection region: We use the standard normal distribution to find the appropriate rejection region for the specified value of α. Using $\alpha = .05$, the two-tailed rejection region is

$$z < -z_{\alpha/2} = -z_{.025} = -1.96 \quad \text{or} \quad z > z_{\alpha/2} = z_{.025} = 1.96$$

See Figure 7.11.

We are now prepared to calculate the value of the test statistic. Before doing so, we want to be sure that the sample size is large enough to ensure that the normal approximation for the sampling distribution of $\hat{p}$ is reasonable. To check this, we

FIGURE 7.11 ▶ Rejection region for insider trading example

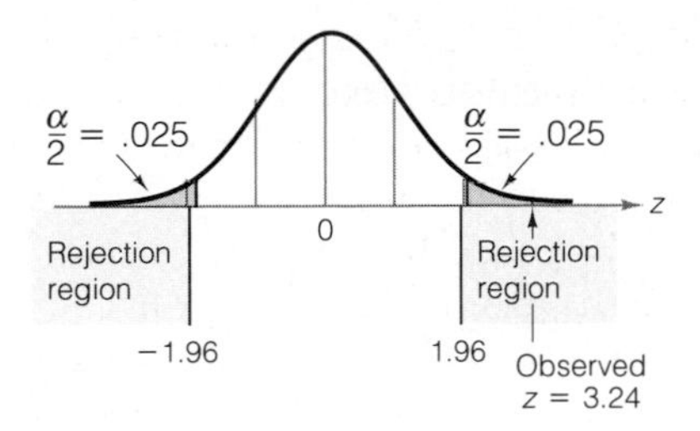

calculate a 3-standard-deviation interval around the null hypothesized value, p_0, which is assumed to be the true value of p until our test procedure proves otherwise. Recall that $\sigma_{\hat{p}} = \sqrt{pq/n}$ and that we need an estimate of the product pq in order to calculate a numerical value of the test statistic z. Since the null hypothesized value is generally the accepted-until-proven-otherwise value, we use the value of p_0q_0 (where $q_0 = 1 - p_0$) to estimate pq in the calculation of z. Thus,

$$\sigma_{\hat{p}} = \sqrt{\frac{pq}{n}} \approx \sqrt{\frac{p_0q_0}{n}} = \sqrt{\frac{(.5)(.5)}{576}} = .021$$

and the 3-standard-deviation interval around p_0 is

$$p_0 \pm 3\sigma_{\hat{p}} \approx .5 \pm 3(.021) = (.437, .563)$$

As long as this interval does not contain 0 or 1 (i.e., is completely contained in the interval 0 to 1), as is the case here, then the normal distribution will provide a reasonable approximation for the sampling distribution of $\hat{p}$.

Returning to the hypothesis test at hand, the proportion of the sampled transactions that resulted in a stock increase is

$$\hat{p} = \frac{327}{576} = .568$$

Finally, we calculate the number of standard deviations (the z value) between the sampled and hypothesized value of the binomial probability:

$$z = \frac{\hat{p} - p_0}{\sigma_{\hat{p}}} \approx \frac{\hat{p} - p_0}{\sqrt{p_0q_0/n}} = \frac{.568 - .5}{.021} = \frac{.068}{.021} = 3.24$$

The implication is that the observed sample proportion is (approximately) 3.24 standard deviations above the null hypothesized probability, .5 (Figure 7.11). Therefore, we reject the null hypothesis, concluding at the .05 level of significance that the true probability of an increase or decrease in a stock's price differs from .5 the day following significant insider purchase of the stock. It appears that an insider purchase significantly *increases* the probability that the stock price will increase the following day.

The test of hypothesis about a binomial probability p is summarized in the box on page 348. Note that the procedure is entirely analogous to that used for conducting large-sample tests about a population mean.

Large-Sample Test of Hypothesis About p

One-Tailed Test	*Two-Tailed Test*
H_0: $p = p_0$ (p_0 = hypothesized value of p)	H_0: $p = p_0$
H_a: $p < p_0$ (or H_a: $p > p_0$)	H_a: $p \neq p_0$
Test statistic: $z = \dfrac{\hat{p} - p_0}{\sigma_{\hat{p}}}$	*Test statistic:* $z = \dfrac{\hat{p} - p_0}{\sigma_{\hat{p}}}$

where, according to H_0, $\sigma_{\hat{p}} = \sqrt{p_0 q_0/n}$ and $q_0 = 1 - p_0$

Rejection region: $z < -z_\alpha$ (or $z > z_\alpha$ when H_a: $p > p_0$)	*Rejection region:* $z < -z_{\alpha/2}$ or $z > z_{\alpha/2}$

Assumption: The experiment is binomial, and the sample size is large enough that the interval $p_0 \pm 3\sigma_{\hat{p}}$ does not include 0 or 1.

EXAMPLE 7.7

The reputations (and hence sales) of many businesses can be severely damaged by shipments of manufactured items that contain a large percentage of defectives. For example, a manufacturer of alkaline batteries may want to be reasonably certain that fewer than 5% of its batteries are defective. Suppose 300 batteries are randomly selected from a very large shipment, each is tested, and 10 defective batteries are found. Does this provide sufficient evidence for the manufacturer to conclude that the fraction defective in the entire shipment is less than .05? Use $\alpha = .01$.

Solution

Before conducting the test of hypothesis, we check to determine whether the sample size is large enough to use the normal approximation for the sampling distribution of $\hat{p}$. The criterion is tested by the interval

$$p_0 \pm 3\sigma_{\hat{p}} = p_0 \pm 3\sqrt{\frac{p_0 q_0}{n}} = .05 \pm 3\sqrt{\frac{(.05)(.95)}{300}}$$
$$= .05 \pm .04 \quad \text{or} \quad (.01, .09)$$

Since the interval lies within the interval (0, 1), the normal approximation will be adequate.

The objective of the sampling is to determine whether there is sufficient evidence to indicate that the fraction defective, p, is less than .05. Consequently, we will test the null hypothesis that $p = .05$ against the alternative hypothesis that $p < .05$. The elements of the test are

H_0: $p = .05$
H_a: $p < .05$

Test statistic: $z = \dfrac{\hat{p} - p_0}{\sigma_{\hat{p}}}$

Rejection region: $z < -z_{.01} = -2.33$ (see Figure 7.12)

FIGURE 7.12 ▶
Rejection region for Example 7.7

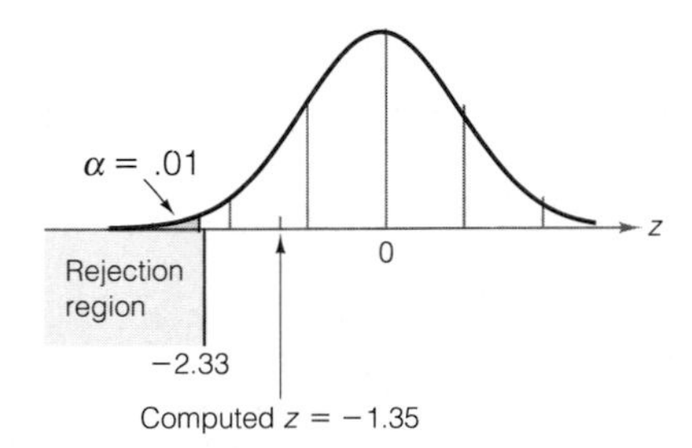

We now calculate the test statistic:

$$z = \frac{\hat{p} - .05}{\sigma_{\hat{p}}} = \frac{(10/300) - .05}{\sqrt{p_0 q_0/n}} = \frac{.033 - .05}{\sqrt{p_0 q_0/300}}$$

Notice that we use p_0 to calculate $\sigma_{\hat{p}}$ because, in contrast to calculating $\sigma_{\hat{p}}$ for a confidence interval, the test statistic is computed on the assumption that the null hypothesis is true—that is, $p = p_0$. Therefore, substituting the values for $\hat{p}$ and p_0 into the z statistic, we obtain

$$z \approx \frac{-.017}{\sqrt{(.05)(.95)/300}} = \frac{-.017}{.0126} = -1.35$$

As shown in Figure 7.12, the calculated z value does not fall in the rejection region. Therefore, there is insufficient evidence at the .01 level of significance to indicate that the shipment contains fewer than 5% defective batteries.

EXAMPLE 7.8

In Example 7.7 we found that we did not have sufficient evidence, at the $\alpha = .01$ level of significance, to indicate that the fraction defective p of alkaline batteries was less than $p = .05$. How strong was the weight of evidence favoring the alternative hypothesis (H_a: $p < .05$)? Find the observed significance level for the test.

Solution

The computed value of the test statistic z was $z = -1.35$. Therefore, for this lower-tail test, the observed significance level is

$$\text{Observed significance level} = P(z \leq -1.35)$$

This lower-tail area is shown in Figure 7.13. The area between $z = 0$ and $z = 1.35$ is given in Table IV in Appendix A as .4115. Therefore, the observed significance level is $.5 - .4115 = .0885$. Note that this probability is quite small. Although we did not reject H_0: $p = .05$ at $\alpha = .01$, the probability of observing a z value as small as or smaller than -1.35 is only .0885 if in fact H_0 is true. Therefore, we would reject H_0 if we choose $\alpha = .10$ (since the observed significance level is less than .10), and we would not reject H_0 (the conclusion of Example 7.7) if we choose $\alpha = .05$ or $\alpha = .01$.

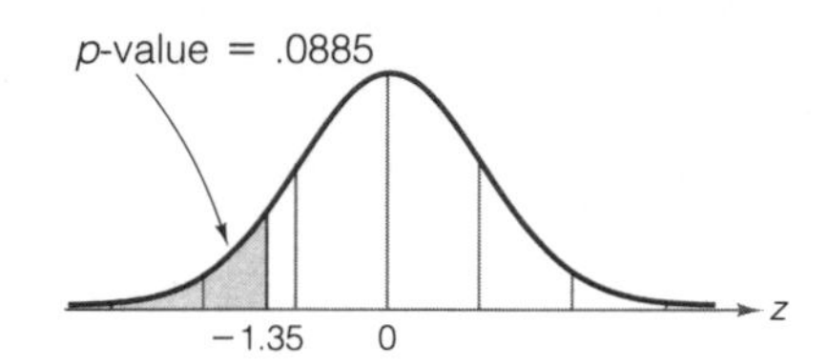

FIGURE 7.13 ▶ The observed significance level for Example 7.8

Small-sample test procedures are also available for p. These are omitted from our discussion because most surveys use samples that are large enough to employ the large-sample tests presented in this section.

Exercises 7.58–7.67

Learning the Mechanics

7.58 For the binomial sample sizes and null hypothesized values of p in each part, determine whether the sample size is large enough to use the normal approximation methodology presented in this section to conduct a test of the null hypothesis H_0: $p = p_0$.

a. $n = 500,\ p_0 = .05$ **b.** $n = 100,\ p_0 = .99$ **c.** $n = 50,\ p_0 = .2$
d. $n = 20,\ p_0 = .2$ **e.** $n = 10,\ p_0 = .4$

7.59 Suppose a random sample of 100 observations from a binomial population gives a value of $\hat{p} = .69$ and you wish to test the null hypothesis that the population parameter p is equal to .75 against the alternative hypothesis that $p < .75$.

a. Noting that $\hat{p} = .69$, what does your intuition tell you? Does the value of $\hat{p}$ appear to contradict the null hypothesis?

b. Use the large-sample z-test to test H_0: $p = .75$ against the alternative hypothesis, H_a: $p < .75$. Use $\alpha = .05$. How do the test results compare with your intuitive decision from part **a**?

c. Find and interpret the observed significance level of the test you conducted in part **b**.

7.60 Suppose the sample in Exercise 7.59 has produced $\hat{p} = .84$ and we wish to test H_0: $p = .9$ against the alternative H_a: $p < .9$.

a. Calculate the value of the z statistic for this test.

b. Note that the numerator of the z statistic ($\hat{p} - p_0 = .84 - .90 = -.06$) is the same as for Exercise 7.59. Considering this, why is the absolute value of z for this exercise larger than that calculated in Exercise 7.59?

c. Complete the test using $\alpha = .05$ and interpret the result.
d. Find the observed significance level for the test and interpret its value.

7.61 A statistics student will use a computer program to test the null hypothesis H_0: $p = .5$ against the one-tailed alternative H_a: $p > .5$. A sample of 500 observations are input into the computer, which returns the following result:

```
Z = .44          P-VALUE = .3300
```

a. The student concludes, based on the p-value, that there is a 33% chance that the alternative hypothesis is true. Do you agree? If not, correct the interpretation.
b. How would the p-value change if the alternative hypothesis were two-tailed, H_a: $p \neq .5$? Interpret this p-value.

Applying the Concepts

7.62 Earthquakes are not uncommon in California. A recent article in the *Annals of the Association of American Geographers* (Palm and Hodgson, 1992) investigated many factors that California residents consider when purchasing earthquake insurance. The survey revealed that only 133 of 337 randomly selected residences in Los Angeles County were protected by earthquake insurance.
a. What are the appropriate null and alternative hypotheses to test the research hypothesis that less than 40% of the residents of Los Angeles County have earthquake insurance?
b. Do the data provide sufficient evidence to support the research hypothesis? Use $\alpha = .10$.
c. Calculate and interpret the p-value for the test.

7.63 A method currently used by doctors to screen women for possible breast cancer fails to detect cancer in 15% of the women who actually have the disease. A new method has been developed that researchers hope will be able to detect cancer more accurately. A random sample of 70 women known to have breast cancer were screened using the new method. Of these, the new method failed to detect cancer in six.
a. Do the data provide sufficient evidence to indicate that the new screening method is better than the one currently in use? Test using $\alpha = .05$.
b. Find the observed significance level for the test and interpret its value.

7.64 Major oil companies have been criticized for raising prices following major oil spills and Mideast crises. In the past, Standard Oil of California has used a sample survey to determine whether people's attitudes toward Standard's corporate image tended to be favorable or unfavorable. The sample results indicated that, for the first time in 30 years, more people had unfavorable than favorable attitudes. Standard Oil responded by initiating an institutional advertising campaign to help improve its image (*Marketing News*, 1976). In 1990, another large oil corporation conducted a similar survey, with the following results:

Unfavorable opinions	3,465
Favorable opinions	2,502
No opinions	821

a. Examine the data. Based on your intuition, would you say that more than 50% of the general public possess an unfavorable attitude toward the company?

b. Do the sample data support the hypothesis that more than 50% of the general public hold unfavorable opinions about the company? Test at $\alpha = .05$.
c. Construct a 90% confidence interval for the proportion of individuals with no opinion.
d. List any assumptions that you made in answering parts **b** and **c**.

7.65 Refer to Exercise 7.64. Find the observed significance level for the test you conducted in part **b** and interpret its value.

7.66 Increasing numbers of businesses are offering child-care benefits for their workers. However, one union claims that at least 90% of firms in the manufacturing sector still do not offer any child-care benefits to their workers. A random sample of 350 manufacturing firms is selected, and only 28 of them offer child-care benefits.
a. Does this sample result support the claim of the union? Test using $\alpha = .10$.
b. Calculate and interpret the p-value associated with this test.

7.67 Health care reform was one of the major platforms of the Clinton campaign in 1992 and one of the top priorities of his new administration in 1993. One of the polls taken after President Clinton's inauguration indicated that 273 of 521 registered voters believed that health care reform should be the leading priority of the new president.
a. What are the appropriate null and alternative hypotheses to test whether these data support the hypothesis that a majority of voters believe that health care reform is the leading priority?
b. When the data are analyzed using a statistical software package, the following results are obtained. Interpret the results.

```
Z = 1.10          P-VALUE = .1357
```

7.7 Calculating Type II Error Probabilities: More About β (Optional)

In our introduction to hypothesis testing in Section 7.1, we showed that the probability of committing a Type I error, α, can be controlled by the selection of the rejection region for the test. Thus, when the test statistic falls in the rejection region and we make the decision to reject the null hypothesis, we do so knowing the error rate for incorrect rejections of H_0. The situation corresponding to accepting the null hypothesis, and thereby risking a Type II error, is not generally as controllable. For that reason, we adopted a policy of nonrejection of H_0 when the test statistic does not fall in the rejection region, rather than risking an error of unknown magnitude.

To see how β, the probability of a Type II error, can be calculated for a test of hypothesis, recall the example in Section 7.1 in which a city tests a manufacturer's pipe to see whether it meets the requirement that the mean strength exceeds 2,400 pounds per lineal foot. The setup for the test is as follows:

$$H_0: \mu = 2{,}400$$
$$H_a: \mu > 2{,}400$$

Test statistic: $z = \dfrac{\bar{x} - 2{,}400}{\sigma/\sqrt{n}}$

Rejection region: $z > 1.645$ for $\alpha = .05$

Figure 7.14(a) (page 354) shows the rejection region for the **null distribution**—that is, the distribution of the test statistic assuming the null hypothesis is true. The area in the rejection region is .05, and this area represents α, the probability that the test statistic leads to rejection of H_0 when in fact H_0 is true.

The Type II error probability β is calculated assuming that the null hypothesis is false, because it is defined as the *probability of accepting* H_0 *when it is false*. Since H_0 is false for any value of μ exceeding 2,400, one value of β exists for each possible value of μ greater than 2,400 (an infinite number of possibilities). Figures 7.14(b), (c), and (d) show three of the possibilities, corresponding to alternative hypothesis values of μ equal to 2,425, 2,450, and 2,475, respectively. Note that β is the area in the *nonrejection* (or *acceptance*) *region* in each of these distributions and that β decreases as the true value of μ moves farther from the null hypothesized value of $\mu = 2{,}400$. This is sensible because the probability of incorrectly accepting the null hypothesis should decrease as the distance between the null and alternative values of μ increases.

In order to calculate the value of β for a specific value of μ in H_a, we proceed as follows:

1. Calculate the value of $\bar{x}$ that corresponds to the border between the acceptance and rejection regions. For the sewer pipe example, this is the value of $\bar{x}$ that lies 1.645 standard deviations above $\mu = 2{,}400$ in the sampling distribution of $\bar{x}$. Denoting this value by $\bar{x}_0$, corresponding to the largest value of $\bar{x}$ that supports the null hypothesis, we find (recalling that $s = 200$ and $n = 50$)

$$\begin{aligned}\bar{x}_0 &= \mu_0 + 1.645\sigma_{\bar{x}} = 2{,}400 + 1.645\left(\frac{\sigma}{\sqrt{n}}\right)\\ &\approx 2{,}400 + 1.645\left(\frac{s}{\sqrt{n}}\right) = 2{,}400 + 1.645\left(\frac{200}{\sqrt{50}}\right)\\ &= 2{,}400 + 1.645(28.28) = 2{,}446.5\end{aligned}$$

2. For a particular alternative distribution corresponding to a value of μ, denoted by μ_a, we calculate the z value corresponding to $\bar{x}_0$, the border between the rejection and acceptance regions. We then use this z value and Table IV of Appendix A to determine the area in the *acceptance* region under the alternative distribution. This area is the value of β corresponding to the particular alternative μ_a. For example, for the alternative $\mu_a = 2{,}425$, we calculate

$$\begin{aligned}z &= \frac{\bar{x}_0 - 2{,}425}{\sigma_{\bar{x}}} = \frac{\bar{x}_0 - 2{,}425}{\sigma/\sqrt{n}}\\ &\approx \frac{\bar{x}_0 - 2{,}425}{s/\sqrt{n}} = \frac{2{,}446.5 - 2{,}425}{28.28} = .76\end{aligned}$$

FIGURE 7.14 ► Values of α and β for various values of μ

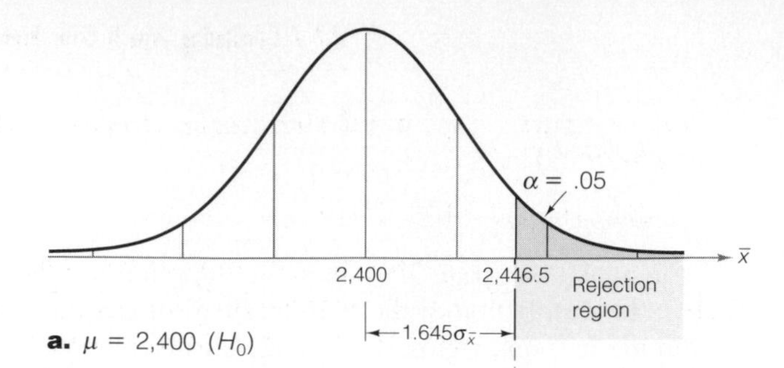

a. μ = 2,400 (H_0)

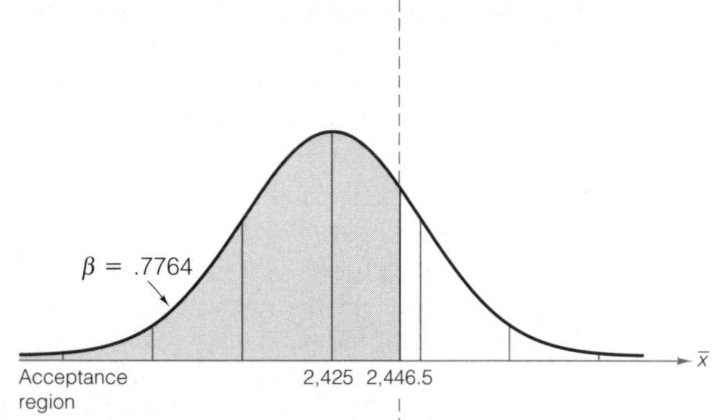

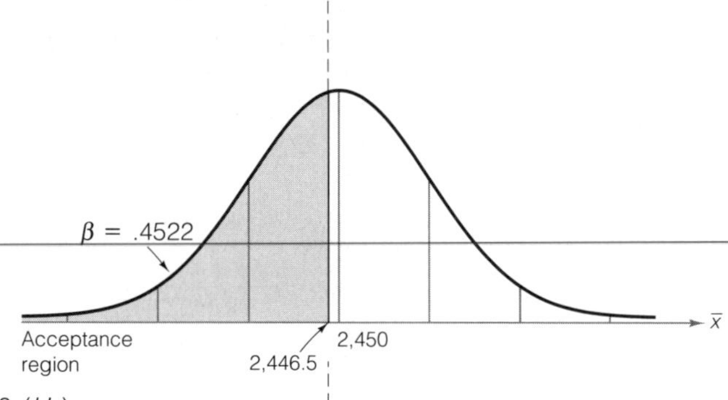

c. μ = 2,450 (H_a)

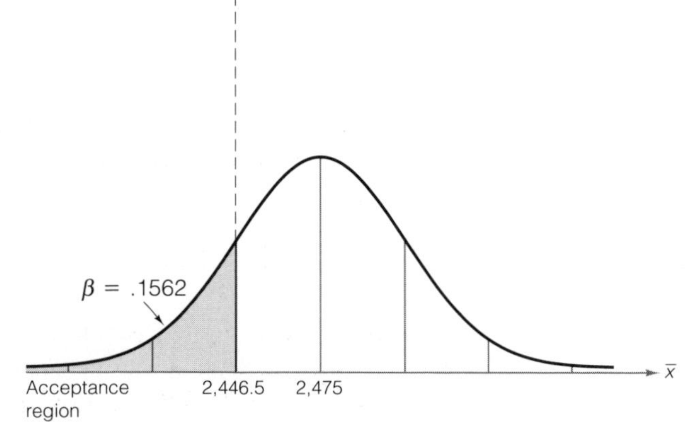

d. μ = 2,475 (H_a)

Note in Figure 7.14(b) that the area in the acceptance region is the area to the left of $z = .76$. This area is

$$\beta = .5 + .2764 = .7764$$

Thus, the probability that the test procedure will lead to an incorrect acceptance of the null hypothesis $\mu = 2{,}400$ when in fact $\mu = 2{,}425$ is about .78. As the average strength of the pipe increases to 2,450, the value of β decreases to .4522 [Figure 7.14(c)]. If the mean strength is further increased to 2,475, the value of β is further decreased to .1562 [Figure 7.14(d)]. Thus, even if the true mean strength of the pipe exceeds the minimum specification by 75 pounds per lineal foot, the test procedure will lead to an incorrect acceptance of the null hypothesis (rejection of the pipe) approximately 16% of the time. The upshot is that the pipe must be manufactured so that the mean strength well exceeds the minimum requirement if the manufacturer wants the probability of its acceptance by the city to be large (i.e., β to be small).

The steps for calculating β for a large-sample test about a population mean are summarized in the box.

Steps for Calculating β for a Large-Sample Test About μ

1. Calculate the value(s) of $\bar{x}$ corresponding to the border(s) of the rejection region. There will be one border value for a one-tailed test, and two for a two-tailed test. The formula is one of the following, corresponding to a test with level of significance α:

Upper-tailed test: $$\bar{x}_0 = \mu_0 + z_\alpha \sigma_{\bar{x}} \approx \mu_0 + z_\alpha\left(\frac{s}{\sqrt{n}}\right)$$

Lower-tailed test: $$\bar{x}_0 = \mu_0 - z_\alpha \sigma_{\bar{x}} \approx \mu_0 - z_\alpha\left(\frac{s}{\sqrt{n}}\right)$$

Two-tailed test: $$\bar{x}_{0,\text{L}} = \mu_0 - z_{\alpha/2} \sigma_{\bar{x}} \approx \mu_0 - z_{\alpha/2}\left(\frac{s}{\sqrt{n}}\right)$$

$$\bar{x}_{0,\text{U}} = \mu_0 + z_{\alpha/2} \sigma_{\bar{x}} \approx \mu_0 + z_{\alpha/2}\left(\frac{s}{\sqrt{n}}\right)$$

2. Specify the value of μ_a in the alternative hypothesis for which the value of β is to be calculated. Then convert the border value(s) of $\bar{x}_0$ to z value(s) using the alternative distribution with mean μ_a. The general formula for the z value is

$$z = \frac{\bar{x}_0 - \mu_a}{\sigma_{\bar{x}}}$$

Sketch the alternative distribution (centered at μ_a), and shade the area in the acceptance (nonrejection) region. Use the z statistic(s) and Table IV of Appendix A to find the shaded area, which is β.

Following the calculation of β for a particular value of μ_a, you should interpret the value in the context of the hypothesis testing application. It is often useful to interpret the value of $1 - \beta$, which is known as the **power of the test** corresponding to a particular alternative, μ_a. Since β is the probability of accepting the null hypothesis when the alternative hypothesis is true with $\mu = \mu_a$, $1 - \beta$ is the probability of the complementary event, or the probability of rejecting the null hypothesis when the alternative H_a: $\mu = \mu_a$ is true. That is, the power $1 - \beta$ measures the likelihood that the test procedure will lead to the correct decision (reject H_0) for a particular value of the mean in the alternative hypothesis.

Definition 7.2

The **power** of a test is the probability that the test will correctly lead to the rejection of the null hypothesis for a particular value of μ in the alternative hypothesis. The power is equal to $1 - \beta$ for the particular alternative considered.

For example, in the sewer pipe example we found that $\beta = .7764$ when $\mu = 2{,}425$. This is the probability that the test leads to the (incorrect) acceptance of the null hypothesis when $\mu = 2{,}425$. Or, equivalently, the power of the test is $1 - .7764 = .2236$, which means that the test will lead to the (correct) rejection of the null hypothesis only 22% of the time when the pipe exceeds specifications by 25 pounds per lineal foot. When the manufacturer's pipe has a mean strength of 2,475 (that is, 75 pounds per lineal foot in excess of specifications), the power of the test increases to $1 - .1562 = .8438$. That is, the test will lead to the acceptance of the manufacturer's pipe 84% of the time if $\mu = 2{,}475$.

EXAMPLE 7.9

Recall the drug experiment in Examples 7.1 and 7.2, in which we tested to determine whether the mean response time for rats injected with a drug differs from the control mean response time of $\mu = 1.2$ seconds. The test setup is repeated here:

H_0: $\mu = 1.2$

H_a: $\mu \neq 1.2$ (i.e., $\mu < 1.2$ or $\mu > 1.2$)

Test statistic: $z = \dfrac{\bar{x} - 1.2}{\sigma_{\bar{x}}}$

Rejection region: $z < -1.96$ or $z > 1.96$ for $\alpha = .05$
$z < -2.575$ or $z > 2.575$ for $\alpha = .01$

Note that two rejection regions have been specified corresponding to values of $\alpha = .05$ and $\alpha = .01$, respectively. Assume that $n = 100$ and $s = .5$.

a. Suppose drug-injected rats have a mean response time of 1.1 seconds, i.e., $\mu = 1.1$. Calculate the values of β corresponding to the two rejection regions. Discuss the relationship between the values of α and β.

b. Calculate the power of the test for each of the rejection regions when $\mu = 1.1$.

Solution

a. We first consider the rejection region corresponding to $\alpha = .05$. The first step is to calculate the border values of $\bar{x}$ corresponding to the two-tailed rejection region, $z < -1.96$ or $z > 1.96$:

$$\bar{x}_{0,\text{L}} = \mu_0 - 1.96\sigma_{\bar{x}} \approx \mu_0 - 1.96\left(\frac{s}{\sqrt{n}}\right) = 1.2 - 1.96\left(\frac{.5}{10}\right) = 1.102$$

$$\bar{x}_{0,\text{U}} = \mu_0 + 1.96\sigma_{\bar{x}} \approx \mu_0 + 1.96\left(\frac{s}{\sqrt{n}}\right) = 1.2 + 1.96\left(\frac{.5}{10}\right) = 1.298$$

These border values are shown in Figure 7.15(a).

FIGURE 7.15 ▶
Calculation of β for drug-injected rats example

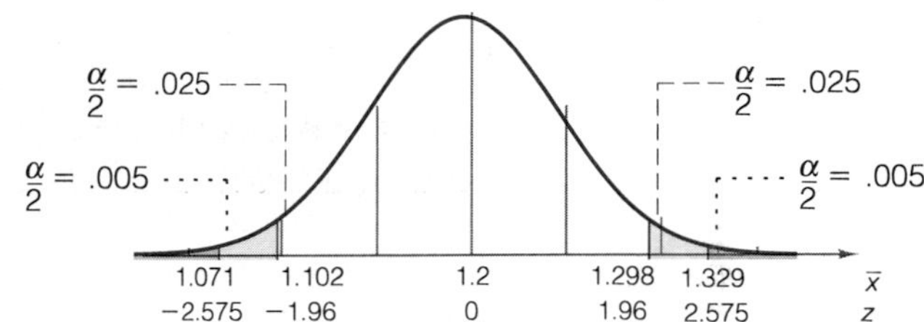

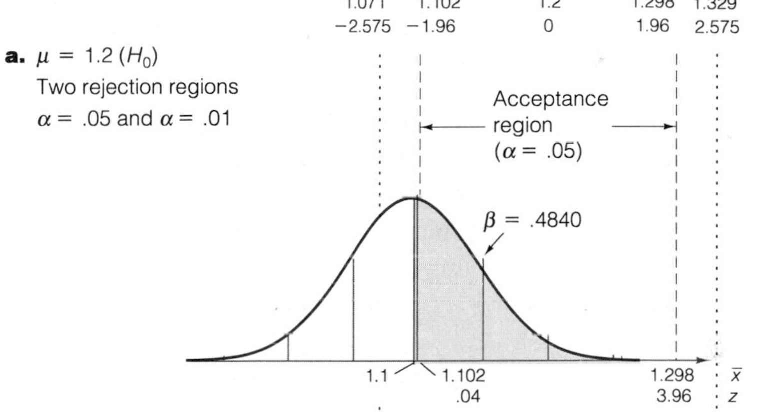

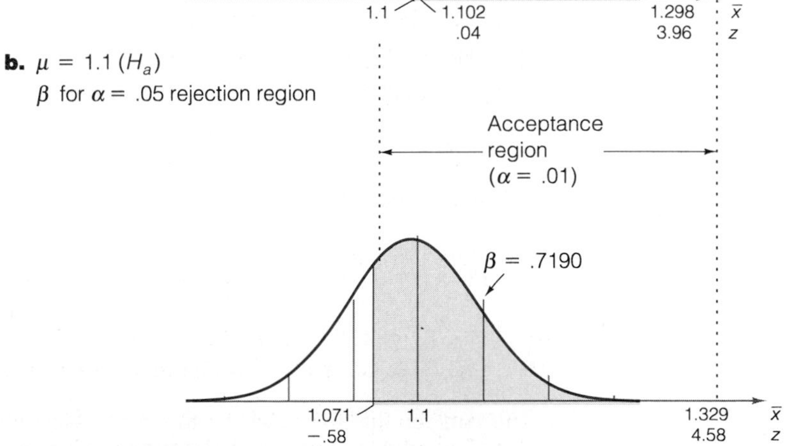

Next, we convert these values to z values in the alternative distribution with $\mu_a = 1.1$:

$$z_L = \frac{\bar{x}_{0,L} - \mu_a}{\sigma_{\bar{x}}} \approx \frac{1.102 - 1.1}{.05} = .04$$

$$z_U = \frac{\bar{x}_{0,U} - \mu_a}{\sigma_{\bar{x}}} \approx \frac{1.298 - 1.1}{.05} = 3.96$$

These z values are shown in Figure 7.15(b). You can see that the acceptance (or nonrejection) region is the area between them. Using Table IV of Appendix A, we find that the area between $z = 0$ and $z = .04$ is .0160, and the area between $z = 0$ and $z = 3.96$ is (approximately) .5 (since $z = 3.96$ is off the scale of Table IV). Then the area between $z = .04$ and $z = 3.96$ is, approximately,

$$\beta = .5 - .0160 = .4840$$

Thus, the test with $\alpha = .05$ will lead to a Type II error about 48% of the time when the mean reaction time for drug-injected rats is .1 second less than the control mean response time.

For the rejection region corresponding to $\alpha = .01$, $z < -2.575$ or $z > 2.575$, we find

$$\bar{x}_{0,L} = 1.2 - 2.575\left(\frac{.5}{10}\right) = 1.0712$$

$$\bar{x}_{0,U} = 1.2 + 2.575\left(\frac{.5}{10}\right) = 1.3288$$

These border values of the rejection region are shown in Figure 7.15(c).

Converting these to z values in the alternative distribution with $\mu_a = 1.1$, we find $z_L = -.58$ and $z_U = 4.58$. The area between these values is, approximately,

$$\beta = .2190 + .5 = .7190$$

Thus, the chance that the test procedure with $\alpha = .01$ will lead to an incorrect acceptance of H_0 is about 72%.

Note that the value of β increases from .4840 to .7190 when we decrease the value of α from .05 to .01. This is a general property of the relationship between α and β: *as α is decreased (increased), β is increased (decreased).*

b. The power is defined to be the probability of (correctly) rejecting the null hypothesis when the alternative is true. When $\mu = 1.1$ and $\alpha = .05$, we find

$$\text{Power} = 1 - \beta = 1 - .4840 = .5160$$

When $\mu = 1.1$ and $\alpha = .01$, we find

$$\text{Power} = 1 - \beta = 1 - .7190 = .2810$$

You can see that the power of the test is decreased as the level of α is decreased. This means that as the probability of incorrectly rejecting the null hypothesis is decreased, the probability of correctly accepting the null hypothesis for a given alternative is also decreased. Thus, the value of α must be selected carefully, with the realization that a test is made less powerful to detect departures from the null hypothesis when the value of α is decreased.

We have shown that the probability of committing a Type II error, β, is inversely related to α (Example 7.9), and that the value of β decreases as the value of μ moves farther from the null hypothesis value (sewer pipe example). The sample size n also affects β. Remember that the standard deviation of the sampling distribution of $\bar{x}$ is inversely proportional to the square root of the sample size ($\sigma_{\bar{x}} = \sigma/\sqrt{n}$). Thus, as illustrated in Figure 7.16, the variability of both the null and alternative sampling distributions is decreased as n is increased. If the value of α is specified and remains fixed, the value of β decreases as n increases, as illustrated in Figure 7.16(b). Conversely, the power of the test for a given alternative hypothesis is increased as the sample size is increased.

FIGURE 7.16 ► **Relationship between α, β, and n**

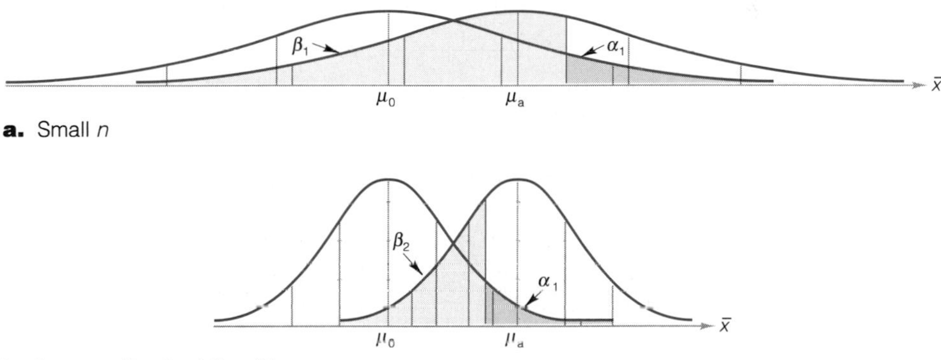

The properties of β and power are summarized in the box.

Properties of β and Power

1. The value of β decreases and the power increases as the distance between the null and alternative values of μ increases (see Figure 7.14).
2. The value of β increases and the power decreases as the value of α is decreased (see Figure 7.15).
3. The value of β decreases and the power increases as the sample size is increased, assuming α remains fixed (see Figure 7.16).

Exercises 7.68–7.79

Learning the Mechanics

7.68 What is the relationship between β, the probability of committing a Type II error, and the power of a test?

7.69 List three factors that will increase the power of a test.

7.70 Suppose you want to test H_0: $\mu = 1{,}000$ against H_a: $\mu > 1{,}000$ using $\alpha = .05$. The population in question is normally distributed with standard deviation 120. A random sample of size $n = 36$ will be used.

a. Sketch the sampling distribution of $\bar{x}$ assuming that H_0 is true.

b. Find the value of $\bar{x}_0$, that value of $\bar{x}$ above which the null hypothesis will be rejected. Indicate the rejection region on your graph of part **a**. Shade the area above the rejection region and label it α.

c. On your graph of part **a**, sketch the sampling distribution of $\bar{x}$ if $\mu = 1{,}020$. Shade the area under this distribution that corresponds to the probability that $\bar{x}$ falls in the nonrejection region when $\mu = 1{,}020$. Label this area β.

d. Find β.

e. Compute the power of this test for detecting the alternative H_a: $\mu = 1{,}020$.

7.71 Refer to Exercise 7.70.

a. If $\mu = 1{,}040$ instead of 1,020, what is the probability that the hypothesis test will incorrectly fail to reject H_0? That is, what is β?

b. If $\mu = 1{,}040$, what is the probability that the test will correctly reject the null hypothesis? That is, what is the power of the test?

c. Compare β and the power of the test when $\mu = 1{,}040$ to the values you obtained in Exercise 7.70 for $\mu = 1{,}020$. Explain the differences.

7.72 It is desired to test H_0: $\mu = 50$ against H_a: $\mu < 50$ using $\alpha = .10$. The population in question is uniformly distributed with standard deviation 20. A random sample of size 64 will be drawn from the population.

a. Describe the (approximate) sampling distribution of $\bar{x}$ under the assumption that H_0 is true.

b. Describe the (approximate) sampling distribution of $\bar{x}$ under the assumption that the population mean is 45.

c. If μ were really equal to 45, what is the probability that the hypothesis test would lead the investigator to commit a Type II error?

d. What is the power of this test for detecting the alternative H_a: $\mu = 45$?

7.73 Refer to Exercise 7.72. If the true value of the population mean is $\mu = 48$, what is the power of the test? How does it compare with the power when $\mu = 45$?

7.74 Suppose you want to conduct the two-tailed test of H_0: $\mu = 10$ against H_a: $\mu \neq 10$ using $\alpha = .05$. A random sample of size 100 will be drawn from the population in question. Assume the population has a standard deviation equal to 1.0.

a. Describe the sampling distribution of $\bar{x}$ under the assumption that H_0 is true.

b. Describe the sampling distribution of $\bar{x}$ under the assumption that $\mu = 9.9$.

c. If μ were really equal to 9.9, find the value of β associated with the test.

d. Find the value of β for the alternative H_a: $\mu = 10.1$.

7.75 Refer to Exercises 7.72 and 7.73.

a. Find β for each of the following values of the population mean: 49, 47, 45, 43, and 41.

b. Plot each value of β you obtained in part **a** against its associated population mean. Show β on the vertical axis and μ on the horizontal axis. Draw a curve through the five points on your graph.

c. Use your graph of part **b** to find the approximate probability that the hypothesis test will lead to a Type II error when $\mu = 48$. Compare your answer to the result you obtained in Exercise 7.73.

d. Convert each of the β values you calculated in part **a** to the power of the test at the specified value of μ. Plot the power on the vertical axis against μ on the horizontal axis. Compare the graph of part **b** to the *power curve* of this part.

e. Examine the graphs of parts **b** and **d**. Explain what they reveal about the relationships among the distance between the true mean μ and the null hypothesized mean μ_0, the value of β, and the power.

Applying the Concepts

7.76 Refer to Exercise 7.19, in which the performance of a particular type of laser-based inspection equipment was investigated. Assume that the standard deviation of the number of solder joints inspected on each run is 1.2. If $\alpha = .05$ is used in conducting the hypothesis test of interest using a sample of 48 circuit boards, and if the true mean number of solder joints that can be inspected is really equal to 9.5, what is the probability that the test will result in a Type II error?

7.77 Refer to Exercise 7.15, in which the alternative hypothesis that the mean miles per gallon achieved by 1991 G-cars exceeds 35 is tested against the null hypothesis that the mean is 35 (or less). A sample of 36 automobiles was tested. Assume that the resulting standard deviation of $s = 6$ is a good estimate of the true standard deviation.

a. Calculate the power of the test for the mean values of 35.5, 36.0, 36.5, 37.0, and 37.5.

b. Plot the power of the test on the vertical axis against the mean on the horizontal axis. Draw a curve through the points.

c. Use the power curve of part **b** to estimate the power for the mean value $\mu = 36.75$. Calculate the power for this value of μ, and compare it to your approximation.

d. Use the power curve to approximate the power of the test when $\mu = 40$. If the true value of the mean miles per gallon for this model is really 40, what (approximately) are the chances that the test will fail to reject the null hypothesis that the mean is 35?

7.78 Refer to Exercise 7.77. Show what happens to the power curve when the sample size is increased from $n = 36$ to $n = 100$. Assume that the standard deviation is $\sigma = 6$.

7.79 If a manufacturer (the vendee) buys all items of a particular type from a particular vendor, the manufacturer is practicing *sole sourcing*. Sole sourcing is a purchasing policy that is generally recognized as an important component of a firm's quality system. One of the major benefits of sole sourcing for the vendee is the improved communication that results from the closer vendee/vendor relationship (Treleven, 1986). As part of a sole sourcing arrangement, a vendor agrees to periodically supply its vendee with sample data from its production process. The vendee uses the data to investigate whether the mean length of rods produced by the vendor's production process is truly 5.0 mm or more, as claimed by the vendor and desired by the vendee.

a. If the production process has a standard deviation of .01 mm, the vendor supplies $n = 100$ items to the vendee, and the vendee uses $\alpha = .05$ in testing H_0: $\mu = 5.0$ mm against H_a: $\mu < 5.0$ mm, what is the probability that the vendee's test will fail to reject the null hypothesis when in fact $\mu = 4.9975$ mm? What is the name given to this type of error?

b. Refer to part **a**. What is the probability that the vendee's test will reject the null hypothesis when in fact $\mu = 5.0$? What is the name given to this type of error?
c. What is the power of the test to detect a departure of .0025 mm below the specified mean rod length of 5.0 mm?

Summary

In this chapter we extended the concept of making inferences from using samples to estimate parameters of a distribution to performing tests of hypotheses about these parameters. The essential elements of a test of hypothesis are the **null** and **alternative hypotheses**, the **test statistic**, the **rejection region**, the **calculation of the test statistic**, and the **conclusion**.

The null hypothesis is the status-quo hypothesis, the hypothesis accepted until contradicted by sufficient sample evidence. The alternative is the research hypothesis, the hypothesis that will not be accepted until convincingly established by sample information. The test statistic is a value calculated from the sample information and utilized to decide whether to reject the null hypothesis. The rejection region is the collection of values of the test statistic that will cause us to reject the null hypothesis.

We design the test to have a specified **Type I error probability**, α, so that we know the probability of rejecting H_0 when H_0 is true. When the test statistic falls in the rejection region, we make the inference that the null hypothesis is false (that is, the alternative is true), at the α level of significance. If the test statistic does not fall in the rejection region, then we fail to reject the null hypothesis. We do not accept the null hypothesis in this case unless we know the **Type II error probability**, β.

We covered large-sample tests involving the population mean μ and the binomial probability p, both using the standard normal z as a test statistic. We also discussed the small-sample test involving the mean of a normal distribution using the Student t statistic. A **nonparametric** procedure is required when a small sample is drawn from a nonnormal population.

Supplementary Exercises 7.80–7.108

Note: List the assumptions necessary for the valid implementation of the statistical procedures you use in solving all these exercises. Starred (*) exercises refer to the optional sections.

Learning the Mechanics

7.80 Which of the elements of a test of hypothesis can and should be specified *prior* to analyzing the data that are to be utilized to conduct the test?

7.81 *Complete the following statement:* The smaller the p-value associated with a test of hypothesis, the stronger the support for the ________ hypothesis. Explain your answer.

7.82 Specify the differences between a large-sample and small-sample test of hypothesis about a population mean μ. Focus on the assumptions and test statistics.

7.83 Medical tests have been developed to detect many serious diseases. A medical test is designed to minimize the probability that it will produce a "false positive" or a "false negative." A false positive refers to a positive test result when in fact the individual does not have the disease, whereas a false negative is a negative test result for an individual who does have the disease.

a. If we treat a medical test for a disease as a statistical test of hypothesis, what are the null and alternative hypotheses for the medical test?
b. What are the Type I and Type II errors for the test? Relate each to false positives and false negatives.
c. Considering which of the errors has more grave consequences, is it more important to minimize α or β? Explain.

7.84 *Complete the following statement:* The larger the p-value associated with a test of hypothesis, the stronger the support for the ________ hypothesis. Explain your answer.

7.85 A random sample of 20 observations selected from a normal population produced $\bar{x} = 72.6$ and $s^2 = 19.4$.

a. Test H_0: $\mu = 80$ against H_a: $\mu < 80$. Use $\alpha = .05$.
b. Test H_0: $\mu = 80$ against H_a: $\mu \neq 80$. Use $\alpha = .01$.

7.86 A random sample of $n = 200$ observations from a binomial population yields $\hat{p} = .29$.

a. Test H_0: $p = .35$ against H_a: $p < .35$. Use $\alpha = .05$.
b. Test H_0: $p = .35$ against H_a: $p \neq .35$. Use $\alpha = .05$.

7.87 A random sample of 175 measurements possessed a mean $\bar{x} = 8.2$ and a standard deviation $s = .79$.

a. Test H_0: $\mu = 8.3$ against H_a: $\mu \neq 8.3$. Use $\alpha = .05$.
b. Test H_0: $\mu = 8.4$ against H_a: $\mu \neq 8.4$. Use $\alpha = .05$.

7.88 A t-test is conducted for the null hypothesis H_0: $\mu = 10$ versus the alternative H_a: $\mu > 10$ for a random sample of $n = 17$ observations. The data are analyzed using a statistical software package, with the following results:

```
T = 1.174          DF = 16          P-VALUE = .1288
```

a. Interpret the p-value.
b. What assumptions are necessary for the validity of this test?
c. Calculate and interpret the p-value assuming the alternative hypothesis was instead H_a: $\mu \neq 10$.

Applying the Concepts

7.89 Failure to meet payments on student loans guaranteed by the United States government has been a major problem for both banks and the government. Approximately 50% of all student loans guaranteed by the government are in default. A random sample of 350 loans to college students in one region of the United States indicates that 147 loans are in default.

a. Do the data indicate that the proportion of student loans in default in this area of the country differs from the proportion of all student loans in the United States that are in default? Use $\alpha = .01$.
b. Find the observed significance level for the test and interpret its value.

7.90 In order to be effective, the mean length of life of a certain mechanical component used in a spacecraft must be longer than 1,100 hours. Due to the prohibitive cost of the components, only three can be tested under simulated space conditions. The lifetimes (hours) of the components were recorded and the following statistics were computed: $\bar{x} = 1{,}173.6$ and $s = 36.3$. These data were analyzed using a statistical software package, with the following results:

```
T = 3.512          DF = 2          P-VALUE = .0362
```

a. Verify that the software has correctly calculated the t statistic, and use Table V of Appendix A to determine whether the p-value is in the appropriate range.
b. Interpret the p-value.
c. What assumptions are necessary for the validity of this test?
d. Which type of error, I or II, is of greater concern for this test? Explain.
e. Would you recommend that this component be passed as meeting specifications?

7.91 The mean score on a Peace Corps application test, based on many tests conducted over a long period of time, is 80. Ten prospective applicants have taken a course designed to improve their scores on the test. The scores of the 10 applicants who completed the course had a mean equal to 86.1 and a standard deviation equal to 12.4.
a. Do the data provide sufficient evidence to conclude that students taking the course will have a higher mean score than those who do not? Test using $\alpha = .10$.
b. Find the approximate observed significance level for the test and interpret its value.
c. What assumptions must be made in order for the procedure that you used in part **a** to be valid?

7.92 A sporting goods manufacturer who produces both white and yellow golf balls claims that more than 75% of all golf balls sold are white. A marketing study of the purchases of white and yellow golf balls at a number of stores showed that of 470 balls sold, 410 were white and 60 were yellow.
a. Is there sufficient evidence to support the manufacturer's claim? Test using $\alpha = .01$.
b. Find the observed significance level for the test and interpret its value.
*c. Calculate the probability, β, of a Type II error if in fact 80% of the golf balls sold are white.

***7.93** Many water treatment facilities add hydrofluosilicic acid to the water to supplement the natural fluoride concentration in order to reach a target concentration of fluoride in the drinking water. Certain levels are thought to enhance dental health, but very high concentrations can be dangerous. Suppose that one such treatment plant targets .75 milligram per liter (mg/L) for their water. The plant tests 25 samples each day to determine whether the median level differs from the target.
a. Set up the null and alternative hypotheses.
b. Set up the test statistic and rejection region using $\alpha = .10$.
c. Explain the implication of a Type I error in the context of this application. A Type II error.
d. Suppose that one day's samples result in 18 values that exceed .75 mg/L. Conduct the test and state the appropriate conclusion in the context of this application.
e. When it was suggested to the plant's supervisor that a t-test should be used to conduct the daily test, she replied that the probability distribution of the fluoride concentrations was "heavily skewed to the right." Show graphically what she meant by this, and explain why this is a reason to prefer the sign test to the t-test.

7.94 During past harvests, a farmer has averaged 68.2 bushels of corn per acre. A new fertilizer has been placed on the market, and after using the new fertilizer the farmer notes the yield of corn for four randomly selected fields of equal size. The mean yield is 72.4 bushels per acre and the standard deviation is 2.2 bushels.

a. If these data truly represent a random sample of corn yields that the farmer might expect (now and in the future) when using the new fertilizer, do they suggest that the mean yield of corn per acre has changed from past years? Test using $\alpha = .05$.

b. Note that the four yield measurements were selected from within the same year. Are these measurements a random sample selected from the population of interest to the farmer? If not, what information do the data provide the farmer?

7.95 The EPA sets a limit of 5 parts per million (ppm) on PCB (a dangerous substance) in water. A major manufacturing firm producing PCB for electrical insulation discharges small amounts from the plant. The company management, attempting to control the PCB in its discharge, has given instructions to halt production if the mean amount of PCB in the effluent exceeds 3 ppm. A random sample of 50 water specimens produced the following statistics: $\bar{x} = 3.1$ ppm and $s = .5$ ppm.

a. Do these statistics provide sufficient evidence to halt the production process? Use $\alpha = .01$.

b. If you were the plant manager, would you want to use a large or a small value for α for the test in part **a**?

***7.96** Refer to Exercise 7.95.

a. In the context of the problem, define a Type II error.

b. Calculate β for the test described in part **a** of Exercise 7.95 assuming that the true mean is $\mu = 3.1$ ppm.

c. What is the power of the test to detect the effluent's departure from the standard of 3.0 ppm when the mean is 3.1 ppm?

d. Repeat parts **b** and **c** assuming that the true mean is 3.2 ppm. What happens to the power of the test as the plant's mean PCB departs further from the standard?

***7.97** Refer to Exercises 7.95 and 7.96.

a. Suppose an α value of .05 is used to conduct the test. Does this change favor the manufacturer? Explain.

b. Determine the value of β and the power for the test when $\alpha = .05$ and $\mu = 3.1$.

c. What happens to the power of the test when α is increased?

7.98 One study (Sauer et al., 1988) of gambling newsletters that purport to improve a bettor's odds of winning bets on NFL football games indicates that the newsletters' betting schemes were not profitable. Suppose a random sample of 50 games is selected to test one gambling newsletter. Following the newsletter's recommendations, 30 of the 50 games produced winning wagers. Test whether the newsletter can be said to significantly increase the odds of winning over what one could expect by selecting the winner at random. Use $\alpha = .05$.

7.99 Refer to Exercise 7.98. Calculate and interpret the p-value for the test.

***7.100** Refer to Exercise 7.98.

a. Describe a Type II error in terms of this application.

b. Calculate the probability β of a Type II error for this test assuming that the newsletter really does increase the probability of winning a wager on an NFL game to $p = .55$.

c. Suppose the number of games sampled is increased from 50 to 100. How does this affect the probability of a Type II error for $p = .55$?

7.101 A large mail-order company has placed an order for 5,000 electric can openers with a supplier on condition that no more than 2% of the devices will be defective. To check the shipment, the company tests a random sample of 400 of the can openers and finds 11 are defective. Does this provide sufficient evidence to indicate that the proportion of defective can openers in the shipment exceeds 2%? Test using $\alpha = .05$.

7.102 Refer to Exercise 7.101. Find and interpret the observed significance level for the hypothesis test.

***7.103** Increasing numbers of private and public agencies are requiring their employees to submit to tests for substance abuse. One laboratory that conducts the testing has developed a system with a normalized measurement scale, in which values less than 1.00 indicate "normal" ranges and values equal to or greater than 1.00 are indicative of potential substance abuse. The lab reports a normal result as long as the median level for an individual is less than 1.00. Nine independent measurements of each individual's sample are made; suppose that one individual's results were as follows:

.78 .51 4.32 .23 .77 .98 .96 .89 1.11

a. Set up the appropriate null and alternative hypotheses if the agency conducting the test wants "proof" beyond a reasonable doubt that the individual is in the normal range.
b. Test to determine whether the laboratory can conclude that the individual's median level is in the normal range using a .05 level of significance.
c. What assumptions are necessary to assure the validity of this test?

7.104 The mean grade-point average (GPA) at a certain university was 3.20 in 1980. To show that grade inflation has been reversed, a dean sets out to show that the mean GPA is now lower than 3.20. A random sample of 100 students yields a mean GPA of 3.05 and a standard deviation equal to .90. Do the data provide sufficient evidence to indicate that grade inflation has been reversed? (Test using $\alpha = .10$.)

7.105 The "beta coefficient" of a stock is a measure of the stock's volatility (or risk) relative to the market as a whole. Stocks with beta coefficients greater than 1 generally bear greater risk (more volatility) than the market, whereas stocks with beta coefficients less than 1 are less risky (less volatile) than the overall market. A random sample of 15 high-technology stocks was selected at the end of 1990, and the mean and standard deviation of the beta coefficients were calculated: $\bar{x} = 1.23$, $s = .37$.

a. Set up the appropriate null and alternative hypotheses to test whether the average high-technology stock is riskier than the market as a whole.
b. Establish the appropriate test statistic and rejection region for the test. Use $\alpha = .10$.
c. What assumptions are necessary to assure the validity of the test?
d. Calculate the test statistic and state your conclusion.
e. What is the approximate p-value associated with this test? Interpret it.

7.106 Refer to Exercise 7.105. A statistical software package was used to analyze the data, and the output was as follows:

```
T = 2.408          P-VALUE = .0304
```

The P-VALUE corresponds to a two-tailed test of the null hypothesis $\mu = 1$.

a. Interpret the p-value on the computer output.
b. If the alternative hypothesis of interest is $\mu > 1$, what is the appropriate p-value of the test?

7.107 Officials from a high school claim that at least 85% of the students who have graduated from the school have received a college degree or are enrolled in a college degree program. A random sample of 60 former graduates indicates that 47 have received or are enrolled in a program to receive a college degree. Do the data contradict the school official's claim? (Use $\alpha = .05$.)

7.108 A random sample of 200 residents in one large city is selected, and each is asked whether a new increase in the property tax would be favored if the income were used for public education. A total of 113 indicate support for the tax.

a. What are the appropriate null and alternative hypotheses to test whether a majority of all residents in the city favor the tax?

b. The data are analyzed using a statistical software package, with the output shown here. Interpret the result.

```
Z = 1.84          P-VALUE = .0329
```

c. What assumptions, if any, are necessary to assure the validity of this test?

On Your Own

The "efficient market" theory postulates that the best predictor of a stock's price at some point in the future is the current price of the stock (with some adjustments for inflation and transaction costs, which we shall assume to be negligible for the purpose of this exercise). To test this theory, select a random sample of 25 stocks on the New York Stock Exchange and record the closing prices on the last days of 2 recent consecutive months. Calculate the increase or decrease in the stock price over the 1-month period.

a. Define μ as the mean change in price of all stocks over a 1-month period. Set up the appropriate null and alternative hypotheses in terms of μ.

b. Use the sample of the 25 stock price differences to conduct the test of hypothesis established in part **a**. Use $\alpha = .05$.

c. Tabulate the number of rejections and nonrejections of the null hypothesis in your class. If the null hypothesis were true, how many rejections of the null hypothesis would you expect among those in your class? How does this expectation compare with the actual number of nonrejections? What does the result of this exercise indicate about the efficient market theory?

Using the Computer

Refer to **Using the Computer** in Chapters 5 and 6. Let μ_0 be the mean of the population of 1,000 zip codes you selected—i.e., μ_0 is the "true" value of the population mean. In each of the following scenarios you will be testing the null hypothesis H_0: $\mu = \mu_0$ versus the two-tailed alternative hypothesis H_a: $\mu \neq \mu_0$. Thus, in this exercise you know that the null hypothesis is true.

a. Select 100 samples of size 50 (with replacement) from the 1,000 zip codes, and conduct the test for each sample using $\alpha = .05$. In how many of the tests did the sample lead to an incorrect rejection of the null hypothesis? How does this compare with what you expected to occur? Repeat the exercise using $\alpha = .10$.

b. Select 100 samples of size 20 (with replacement) from the 1,000 zip codes, and conduct the test for each sample using $\alpha = .05$. In how many of the tests did the sample lead to an incorrect rejection of the null hypothesis? How does this compare with what you expected to occur? Repeat the exercise using $\alpha = .10$.

References

Dujack, S. R. "Science leaves no doubt: The polygraph lies," *Minneapolis Star and Tribune*, July 31, 1986.

Gibbons, J. D. *Nonparametric Statistical Inference*, 2d ed. New York: McGraw-Hill, 1985.

Harding, L., and Sewel, J. "Psychological health and employment status in an island community." *Journal of Occupational and Organizational Psychology*, Dec. 1992, Vol. 65, Part 4.

Hollander, M., and Wolfe, D. A. *Nonparametric Statistical Methods*. New York: Wiley, 1973.

Mills, J. K. "Difference in locus of control between obese adults and adolescent females undergoing weight reduction." *Journal of Psychology*, Mar. 1991, Vol. 125, No. 2, pp. 195–197.

Noether, G. E. *Elements of Nonparametric Statistics*. New York: Wiley, 1967.

Otis, J., Lesage, D., Godin, G., Brown, B., Farley, C., and Lambert, J. "Predicting and reinforcing children's intention to wear protective helmets while bicycling." *Public Health Reports*, May–June 1992, Vol. 107, No. 3, p. 283.

Palm, R., and Hodgson, M., "Earthquake insurance: Mandated disclosure and homeowner response in California." *Annals of the Association of American Geographers*, June 1992, Vol. 82, No. 2.

Sauer, R. D., Brajer, V., Ferris, S. P., and Marr, M. W. "Hold your bets: Another look at the efficiency of the gambling market for National Football League games." *Journal of Political Economy*, Feb. 1988, Vol. 96, No. 1.

Snedecor, G. W., and Cochran, W. G. *Statistical Methods*, 7th ed. Ames: Iowa State University Press, 1980, Chapters 4 and 5.

Streeter, J. P. "Solder joint inspection using a laser inspector." *Quality Congress Transactions*. Milwaukee: American Society for Quality Control, 1986, pp. 507–515.

Treleven, M. "Sole sourcing from the vendor side." *Quality Congress Transactions*. Milwaukee: American Society for Quality Control, 1986, pp. 584–590.

CHAPTER EIGHT

Comparing Two Population Means

Contents

Case Studies

Where We've Been

Two methods for making statistical inferences, estimation and tests of hypotheses based on single samples, were presented in Chapters 6 and 7. In particular, we gave confidence intervals and tests of hypotheses concerning a population mean μ and a binomial proportion p, and we learned how to select the sample size necessary to obtain a specified amount of information concerning a parameter.

Where We're Going

Now that we have learned to make inferences about a single population, we will learn how to compare two populations. Such problems often arise in practice. We may wish to compare the mean gas mileages for two models of automobiles, the mean retirement ages of workers in the public and private sectors, or the mean reaction times of men and women to a visual stimulus. How to decide whether differences exist and how to estimate the differences between population means are the subjects of this chapter. We will learn how to compare population proportions in Chapter 9.

8

Many experiments involve a comparison of two population means. For example, a sales manager for a steel company may want to estimate the difference in mean sales per customer between two different salespeople. A consumer group may want to test whether two major brands of food freezers differ in the mean amount of electricity they use. In this chapter we consider techniques for using two samples to compare the means of the populations from which they were selected.

8.1 Large-Sample Inferences About the Difference Between Two Population Means: Independent Sampling

Many of the same procedures that are used to estimate and test hypotheses about a single parameter can be modified to make inferences about two parameters. Both the z and t statistics may be adapted to make inferences about the difference between two population means.

In this section we develop the large-sample z statistic for comparing two population means. The t statistic for making small-sample inferences about the difference between two population means is introduced in Section 8.2.

We use Example 8.1 to introduce the procedures for making large-sample inferences about the difference between two population means.

EXAMPLE 8.1

A dietitian has developed a diet that is low in fats, carbohydrates, and cholesterol. Although the diet was initially intended to be used by people with heart disease, the dietitian wishes to examine the effect this diet has on the weights of obese people. Two random samples of 100 obese people each are selected, and one group of 100 is placed on the low-fat diet. The other 100 are placed on a diet that contains approximately the same quantity of food but is not as low in fats, carbohydrates, and cholesterol. For each person, the amount of weight lost (or gained) in a 3-week period is recorded. Using the data given in the table, form a 95% confidence interval for the difference between the population mean weight losses for the two diets.

	Low-Fat Diet	Other Diet
Sample size	100	100
Sample mean weight loss	9.3 pounds	3.7 pounds
Sample variance	22.4	16.3

Solution

Recall that the general form of a large-sample confidence interval for a single mean μ is $\bar{x} \pm z_{\alpha/2}\sigma_{\bar{x}}$. That is, we add and subtract $z_{\alpha/2}$ standard deviations of the sample estimate, $\bar{x}$, to the value of the estimate. We employ a similar procedure to form the confidence interval for the difference between two population means.

Let μ_1 represent the mean of the conceptual population of weight losses for all obese people who could be placed on the low-fat diet. Let μ_2 be similarly defined for the other diet. We wish to form a confidence interval for $(\mu_1 - \mu_2)$. An intuitively appealing estimator for $(\mu_1 - \mu_2)$ is the difference between the sample means, $(\bar{x}_1 - \bar{x}_2)$. Thus, we will form the confidence interval of interest by

$$(\bar{x}_1 - \bar{x}_2) \pm z_{\alpha/2}\sigma_{(\bar{x}_1 - \bar{x}_2)}$$

Assuming the two samples are independent, the standard deviation of the difference between the sample means is

$$\sigma_{(\bar{x}_1 - \bar{x}_2)} = \sqrt{\frac{\sigma_1^2}{n_1} + \frac{\sigma_2^2}{n_2}} \approx \sqrt{\frac{s_1^2}{n_1} + \frac{s_2^2}{n_2}}$$

Using the sample data and noting that $\alpha = .05$ and $z_{.025} = 1.96$, we find that the 95% confidence interval is, approximately,

$$(9.3 - 3.7) \pm 1.96\sqrt{\frac{22.4}{100} + \frac{16.3}{100}} = 5.6 \pm (1.96)(.62) = 5.6 \pm 1.22$$

or (4.38, 6.82). Using this estimation procedure over and over again for different samples, we know that approximately 95% of the confidence intervals formed in this manner will enclose the difference in population means $(\mu_1 - \mu_2)$. Therefore, we are reasonably confident that the mean weight loss for the low-fat diet is between 4.38 and 6.82 pounds more than the mean weight loss for the other diet. With this information, the dietitian better understands the potential of the low-fat diet as a weight-reducing diet.

The justification for the procedure used in Example 8.1 to estimate $(\mu_1 - \mu_2)$ relies on the properties of the sampling distribution of $(\bar{x}_1 - \bar{x}_2)$. The performance of the estimator in repeated sampling is pictured in Figure 8.1, and its properties are summarized in the box at the top of page 372.

FIGURE 8.1 ▶ Sampling distribution of $(\bar{x}_1 - \bar{x}_2)$

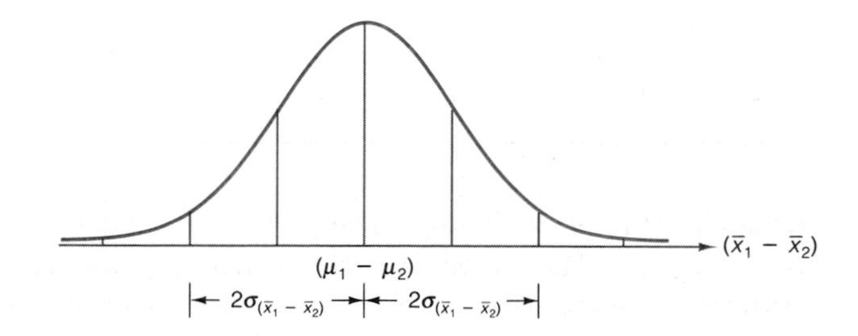

Properties of the Sampling Distribution of $(\bar{x}_1 - \bar{x}_2)$

1. The mean of the sampling distribution of $(\bar{x}_1 - \bar{x}_2)$ is $(\mu_1 - \mu_2)$.
2. If the two samples are independent, the standard deviation of the sampling distribution is

$$\sigma_{(\bar{x}_1-\bar{x}_2)} = \sqrt{\frac{\sigma_1^2}{n_1} + \frac{\sigma_2^2}{n_2}}$$

   where σ_1^2 and σ_2^2 are the variances of the two populations being sampled and n_1 and n_2 are the respective sample sizes. We also refer to $\sigma_{(\bar{x}_1-\bar{x}_2)}$ as the **standard error** of the statistic $(\bar{x}_1 - \bar{x}_2)$.
3. The sampling distribution of $(\bar{x}_1 - \bar{x}_2)$ is approximately normal for *large samples* by the Central Limit Theorem.

In Example 8.1, we noted the similarity in the procedures for forming a large-sample confidence interval for one population mean and a large-sample confidence interval for the difference between two population means. When we are testing hypotheses, the procedures are again very similar. The general large-sample procedures for forming confidence intervals and testing hypotheses about $(\mu_1 - \mu_2)$ are summarized in the next two boxes.

Large Sample Confidence Interval for $(\mu_1 - \mu_2)$

$$(\bar{x}_1 - \bar{x}_2) \pm z_{\alpha/2}\sigma_{(\bar{x}_1-\bar{x}_2)} = (\bar{x}_1 - \bar{x}_2) \pm z_{\alpha/2}\sqrt{\frac{\sigma_1^2}{n_1} + \frac{\sigma_2^2}{n_2}}$$

Assumptions: The two samples are randomly selected in an independent manner from the two populations. The sample sizes, n_1 and n_2, are large enough so that $\bar{x}_1$ and $\bar{x}_2$ each have approximately normal sampling distributions and so that s_1^2 and s_2^2 provide good approximations to σ_1^2 and σ_2^2. This will be true if $n_1 \geq 30$ and $n_2 \geq 30$.

Large-Sample Test of Hypothesis for $(\mu_1 - \mu_2)$

One-Tailed Test	*Two-Tailed Test*
H_0: $(\mu_1 - \mu_2) = D_0$ H_a: $(\mu_1 - \mu_2) < D_0$ [or H_a: $(\mu_1 - \mu_2) > D_0$]	H_0: $(\mu_1 - \mu_2) = D_0$ H_a: $(\mu_1 - \mu_2) \neq D_0$

where D_0 = Hypothesized difference between the means (this difference is often hypothesized to be equal to 0)

Test statistic: $z = \frac{(\bar{x}_1 - \bar{x}_2) - D_0}{\sigma_{(\bar{x}_1 - \bar{x}_2)}}$	*Test statistic:* $z = \frac{(\bar{x}_1 - \bar{x}_2) - D_0}{\sigma_{(\bar{x}_1 - \bar{x}_2)}}$

where $\sigma_{(\bar{x}_1 - \bar{x}_2)} = \sqrt{\frac{\sigma_1^2}{n_1} + \frac{\sigma_2^2}{n_2}}$

Rejection region: $z < -z_\alpha$ [or $z > z_\alpha$ when H_a: $(\mu_1 - \mu_2) > D_0$]	*Rejection region:* $z < -z_{\alpha/2}$ or $z > z_{\alpha/2}$

Assumptions: Same as for the large-sample confidence interval.

EXAMPLE 8.2

In recent years, the United States and Japan have been involved in very intense negotiations regarding restrictions on trade between the two countries. One of the claims made repeatedly by U.S. officials is that many Japanese manufacturers are pricing their goods higher in Japan than they are in the United States, in effect subsidizing low prices in the United States by extremely high prices in Japan. The basis of the U.S. argument is that Japan is able to do that only because they are keeping competitive U.S. goods from reaching the Japanese marketplace.

An economist wishes to test the hypothesis that higher retail prices are being charged for Japanese automobiles in Japan than in the United States. She obtains random samples of 50 retail sales in the United States and 30 retail sales in Japan over the same time period and for the same model of automobile, converts the Japanese sales prices from yen to dollars using current conversion rates, and obtains the summary statistics shown:

U.S. Sales	*Japanese Sales*
$n_1 = 50$	$n_2 = 30$
$\bar{x}_1 = \$11{,}545$	$\bar{x}_2 = \$12{,}243$
$s_1 = \$1{,}989$	$s_2 = \$1{,}843$

Do these data provide sufficient evidence for the economist to conclude that the mean sales price for this model is higher in Japan than in the United States?

Solution

We can best answer this question by performing a test of hypothesis. Defining μ_1 as the mean sales price in the United States and μ_2 as the mean sales price in Japan during this period and for this model of automobile, we want to test whether the data support the alternative (research) hypothesis that $\mu_2 > \mu_1$ [i.e., that $(\mu_1 - \mu_2) < 0$]. Thus, we will test the null hypothesis, $(\mu_1 - \mu_2) = 0$. The evidence necessary to reject this hypothesis in favor of the alternative is a sufficiently large negative value of the difference between the sample means, $(\bar{x}_1 - \bar{x}_2)$.

The elements of the test are as follows:

H_0: $(\mu_1 - \mu_2) = 0$ (i.e., $\mu_1 = \mu_2$; note that $D_0 = 0$ for this hypothesis test)

H_a: $(\mu_1 - \mu_2) < 0$ (i.e., $\mu_1 < \mu_2$)

Test statistic: $z = \dfrac{(\bar{x}_1 - \bar{x}_2) - D_0}{\sigma_{(\bar{x}_1 - \bar{x}_2)}} = \dfrac{(\bar{x}_1 - \bar{x}_2) - 0}{\sigma_{(\bar{x}_1 - \bar{x}_2)}}$

Rejection region: $z < -z_\alpha = -1.645$ (see Figure 8.2)

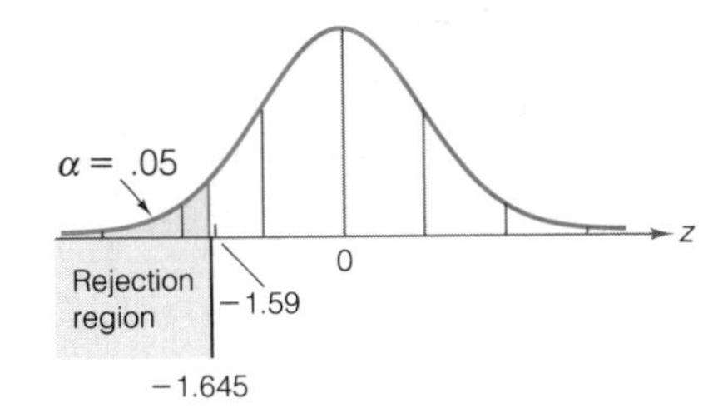

FIGURE 8.2 ▶
Rejection region for Example 8.2

Assuming the samples from the United States and Japan are independent, we now calculate

$$z = \frac{(\bar{x}_1 - \bar{x}_2) - 0}{\sigma_{(\bar{x}_1 - \bar{x}_2)}} = \frac{11{,}545 - 12{,}243}{\sqrt{\dfrac{\sigma_1^2}{n_1} + \dfrac{\sigma_2^2}{n_2}}}$$

$$\approx \frac{-698}{\sqrt{\dfrac{s_1^2}{n_1} + \dfrac{s_2^2}{n_2}}} = \frac{-698}{\sqrt{\dfrac{(1{,}989)^2}{50} + \dfrac{(1{,}843)^2}{30}}} = \frac{-698}{438.57} = -1.59$$

As you can see in Figure 8.2, the calculated z value does not fall in the rejection region. The samples do not provide sufficient evidence at $\alpha = .05$ for the economist to conclude that the mean retail price in Japan exceeds that in the United States.

EXAMPLE 8.3

Find the observed significance level for the test in Example 8.2.

Solution

The alternative hypothesis in Example 8.2, H_a: $\mu_1 - \mu_2 < 0$, required a lower one-tailed test using

$$z = \frac{\bar{x}_1 - \bar{x}_2}{\sigma_{(\bar{x}_1 - \bar{x}_2)}}$$

as a test statistic. Since the value z calculated from the sample data was -1.59, the observed significance level (p-value) for the test is the probability of observing a value of z at least as contradictory to the null hypothesis as $z = -1.59$; i.e.,

$$p\text{-value} = P(z \leq -1.59)$$

This probability is computed assuming H_0 is true and is equal to the shaded area shown in Figure 8.3.

FIGURE 8.3 ▶
The observed significance level for Example 8.2

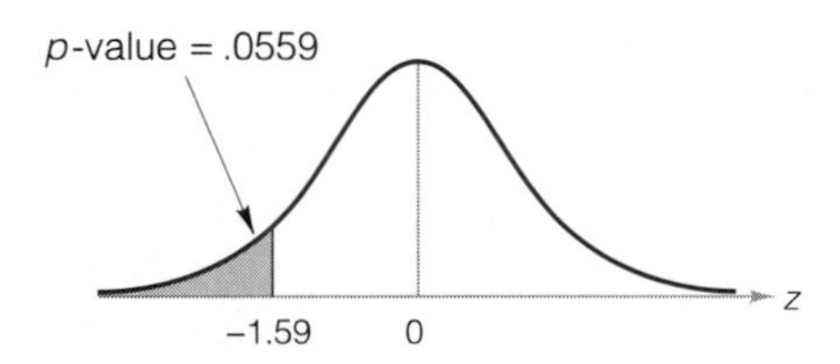

The tabulated area corresponding to $z = 1.59$ in Table IV of Appendix A is .4441. Therefore, the observed significance level for the test is

$$p\text{-value} = .5 - .4441 = .0559$$

You will recall that in Example 8.2, we chose $\alpha = .05$ as the probability of a Type I error and consequently did not reject H_0; that is, we did not find sufficient evidence to indicate that $(\mu_1 - \mu_2) < 0$. If the results of the test had been presented in terms of an observed significance level and left for us to interpret, we might not have reached such an inflexible conclusion. Observing a value of z as small as $z = -1.59$ is an improbable event (rare event) if, in fact, $\mu_1 = \mu_2$. Since the probability is fairly small (.0559)—in fact, quite close to .05—we would conclude that there is some evidence to suggest that $\mu_1 < \mu_2$. Naturally, we would be more certain of this conclusion if the observed significance level were smaller, say .05, .01, or, better yet, .001. However, the practical question to be answered is not whether the test results are statistically significant but whether the difference between μ_1 and μ_2 is large enough to have a practical business significance. To shed light on this question, we wish to estimate the difference, $(\mu_1 - \mu_2)$.

EXAMPLE 8.4

Refer to Example 8.2. Find a 95% confidence interval for the difference in the mean retail prices in the United States and Japan and discuss the implication of the confidence interval.

Solution

The 95% confidence interval for $(\mu_1 - \mu_2)$ is

$$(\bar{x}_1 - \bar{x}_2) \pm z_{\alpha/2}\sqrt{\frac{\sigma_1^2}{n_1} + \frac{\sigma_2^2}{n_2}}$$

Once again, we substitute s_1^2 and s_2^2 for σ_1^2 and σ_2^2, because these quantities provide good approximations to σ_1^2 and σ_2^2 for samples as large as $n_1 = 50$ and $n_2 = 30$. Then, the 95% confidence interval for $(\mu_1 - \mu_2)$ is

$$(11{,}545 - 12{,}243) \pm 1.96\sqrt{\frac{(1{,}989)^2}{50} + \frac{(1{,}843)^2}{30}} = -698 \pm 859.60$$

Thus, we estimate the difference in retail prices to fall in the interval −\$1,557.60 to \$161.60. In other words, we estimate that μ_2, the mean retail price in Japan, could be larger than μ_1, the mean retail price in the United States, by as much as \$1,557.60, or it could be less than μ_1 by as much as \$161.60.

What is the practical interpretation of all this? By all appearances, the Japanese retail price for this model car was indeed higher than the U.S. retail price during the same time period. However, the data collected were insufficient to provide statistical support for this conclusion at the specified level of significance. What can be done to further refine and sharpen the analysis? The first and most obvious improvement would be to increase the sample sizes: Sample sizes of 50 and 30 yield a rather wide confidence interval (total width of \$1,719.20) for estimating the difference between the means. Increasing the sample sizes will decrease the standard error $\sigma_{(\bar{x}_1 - \bar{x}_2)}$, and so the width of the confidence interval will also be decreased.

A second, less obvious, refinement to the analysis would be to decrease somehow the rather large standard deviations associated with the samples, s_1 and s_2. Both exceed \$1,800, indicating that the retail prices contained within each sample vary widely. Perhaps the options available on this model result in significant price discrepancies, or perhaps the dealerships in urban settings have significantly different pricing strategies than those in rural settings. **Regression analysis** is a useful statistical methodology for comparing population means while controlling for other factors affecting the variability of the measurements, such as options and geographic location in this example. Regression analysis is the subject of Chapter 10.

CASE STUDY 8.1 / Productivity and Mobility

One theory regarding the mobility of college and university faculty members is that those who are most productive in the publishing of scholarly articles are also the most mobile. The logic behind this theory is that good researchers and publishers receive more job offers and are therefore more likely to move from one university to another. Since most people in the academic world are aware that a similar strong relationship does not exist for academics who are known to be fine teachers, we might wonder whether it holds for persons employed in industry. George F. Dreher (1982) considered this question by examining the personnel records of a large national oil company. Dreher obtained early career performance records for 529 of the company's employees. Of these, 174 were classified as *stayers,* those who stayed with the company; the other 355, who left the company at varying points during a 15-year period, were classified as *leavers*. Dreher made a number of comparisons between these two groups. Among these were a comparison of measures of the mean initial performances of the stayers and leavers, their rates of career advancement (number of promotions per year), and their final performance appraisals. The means and standard deviations for these sample comparisons are shown in Table 8.1.

Although a large-sample z-test would be appropriate, Dreher tests for differences between the means of stayers and leavers using a small-sample t-test. This test, to be discussed in the following section, is based on assumptions that are not necessary for the large-sample z-test. When the sample sizes are very large, as in Dreher's sampling, the numerical values of the t and z statistics are almost the same and lead to essentially the same results. Dreher's computed t values are shown in Table 8.1.

We have not computed the values of the large-sample z statistics for comparing the three pairs of population means, but it is clear (from examining Dreher's large t values) that there is ample evidence to indicate a difference in means for all three comparisons. (The p-values for all three tests are extremely small.) In fact, it appears that stayers had higher mean measures of initial and final performance and a higher mean rate of career advancement than the leavers.

TABLE 8.1 Differences Between Leavers and Continuing Employees

	Stayers ($n_1 = 174$)		Leavers ($n_2 = 355$)		
Variable	$\bar{x}_1$	s_1	$\bar{x}_2$	s_2	t
Initial performance	3.51	.51	3.24	.52	5.23
Rate of career advancement	.43	.20	.31	.31	4.63
Final performance appraisal	3.78	.62	3.15	.68	9.76

Exercises 8.1–8.17

Learning the Mechanics

8.1 The purpose of this exercise is to compare the variability of $\bar{x}_1$ and $\bar{x}_2$ with the variability of $(\bar{x}_1 - \bar{x}_2)$.

a. Suppose the first sample is selected from a population with mean $\mu_1 = 150$ and variance $\sigma_1^2 = 900$. Within what range should the sample mean vary about 95% of the time in repeated samples of 100 measurements from this distribution? That is, construct an interval extending 2 standard deviations of $\bar{x}_1$ on each side of μ_1.

b. Suppose the second sample is selected independently of the first from a second population with mean $\mu_2 = 150$ and variance $\sigma_2^2 = 1{,}600$. Within what range should the sample mean vary about 95% of the time in repeated samples of 100 measurements from this distribution? That is, construct an interval extending 2 standard deviations of $\bar{x}_2$ on each side of μ_2.

c. Now consider the difference between the two sample means, $(\bar{x}_1 - \bar{x}_2)$. What are the mean and standard deviation of the sampling distribution of $(\bar{x}_1 - \bar{x}_2)$?

d. Within what range should the difference in sample means vary about 95% of the time in repeated independent samples of 100 measurements each from the two populations?

e. What, in general, can be said about the variability of the difference between independent sample means relative to the variability of the individual sample means?

8.2 Independent random samples of 100 observations each are chosen from two normal populations with the following means and standard deviations:

Population 1	*Population 2*
$\mu_1 = 14$	$\mu_2 = 10$
$\sigma_1 = 4$	$\sigma_2 = 3$

Let $\bar{x}_1$ and $\bar{x}_2$ denote the two sample means.
a. Give the mean and standard deviation of the sampling distribution of $\bar{x}_1$.
b. Give the mean and standard deviation of the sampling distribution of $\bar{x}_2$.
c. Suppose you were to calculate the difference $(\bar{x}_1 - \bar{x}_2)$ between the sample means. Find the mean and standard deviation of the sampling distribution of $(\bar{x}_1 - \bar{x}_2)$.
d. Will the statistic $(\bar{x}_1 - \bar{x}_2)$ be normally distributed? Explain.

8.3 Refer to Exercise 8.2.
a. Give the z-score corresponding to $(\bar{x}_1 - \bar{x}_2) = 3.0$.
b. Find the probability that $(\bar{x}_1 - \bar{x}_2)$ is greater than 3.0.
c. Find the probability that $(\bar{x}_1 - \bar{x}_2)$ is less than 3.0.
d. Find the probability that $(\bar{x}_1 - \bar{x}_2)$ is greater than 4.5 or less than 3.

8.4 Two independent random samples have been selected, 100 observations from population 1 and 100 from population 2. Sample means $\bar{x}_1 = 70$ and $\bar{x}_2 = 50$ were obtained. From previous experience with these populations, it is known that the variances are $\sigma_1^2 = 100$ and $\sigma_2^2 = 64$.
a. Find $\sigma_{(\bar{x}_1 - \bar{x}_2)}$.
b. Sketch the approximate sampling distribution for $(\bar{x}_1 - \bar{x}_2)$ assuming $(\mu_1 - \mu_2) = 5$.
c. Locate the observed value of $(\bar{x}_1 - \bar{x}_2)$ on the graph you drew in part **b**. Does it appear that this value contradicts the null hypothesis H_0: $(\mu_1 - \mu_2) = 5$?
d. Use the z table on the inside of the front cover to determine the rejection region for the test of H_0: $(\mu_1 - \mu_2) = 5$ against H_a: $(\mu_1 - \mu_2) \neq 5$. Use $\alpha = .05$.
e. Conduct the hypothesis test of part **d** and interpret your result.

8.5 Refer to Exercise 8.4. Construct a 95% confidence interval for $(\mu_1 - \mu_2)$. Interpret the interval. Which inference provides more information about the value of $(\mu_1 - \mu_2)$—the test of hypothesis in Exercise 8.4 or the confidence interval in this exercise?

8.6 Two independent random samples have been selected, 130 from population 1 and 170 from population 2. Sample means $\bar{x}_1 = 534$ and $\bar{x}_2 = 615$ were obtained. The sample standard deviations are $s_1 = 25$ and $s_2 = 30$.
a. Describe the sampling distribution of $(\bar{x}_1 - \bar{x}_2)$. Assume that $(\mu_1 - \mu_2) = -70$.
b. Test H_0: $(\mu_1 - \mu_2) = -70$ against H_a: $(\mu_1 - \mu_2) < -70$ using $\alpha = .01$. Interpret the result of your test.
c. Report the p-value of your test.

8.7 Independent random samples are selected from two populations, and are used to test the hypothesis H_0: $(\mu_1 - \mu_2) = 0$ against the alternative H_a: $(\mu_1 - \mu_2) \neq 0$. A total of 233 observations from population 1 and 312 from population 2 are analyzed by a statistical software package, with the following result:

```
X̄1 = 473          X̄2 = 485
S1 = 84           S2 = 93
Z = -1.576        P-VALUE = .1150
```

a. Interpret the results of the computer analysis.
b. If the alternative hypothesis had been H_a: $(\mu_1 - \mu_2) < 0$, how would the p-value change? Interpret the p-value for this one-tailed test.

Applying the Concepts

8.8 Lesley E. Tan investigated the relationship between handedness and motor competence in preschool children. Random samples of 41 right-handers and 41 left-handers were administered several tests of motor skills, yielding the means and standard deviations shown in the accompanying table.

	Left-Handed	Right-Handed
n	41	41
$\bar{x}$	97.5	98.1
s	17.5	19.2

Source: Tan, L. E. "Laterality and motor skills in four-year-olds," *Child Development*, 56.

a. Is there evidence of a difference between the motor skills of right- and left-handed preschool children based on this experiment? Use $\alpha = .10$.
b. Use a 90% confidence interval to estimate the true difference in mean motor skill scores between left- and right-handed preschoolers. Does the confidence interval support the result of the test you conducted in part **a**?
c. What assumptions about the distributions of the populations of test scores are necessary to assure the validity of the inferences you made in parts **a** and **b**?
d. What is the observed significance level of the test you conducted in part **a**?
e. Tan concluded that ". . . tests indicated no difference between left-handers and right-handers . . ." Does your analysis support this conclusion?

8.9 An experiment has been conducted at a university to compare the mean number of study hours expended per week by student athletes with the mean number of hours expended by nonathletes. A random sample of 55 athletes produced a mean equal to 20.6 hours studied per week and a standard deviation equal to 5.3 hours. A second random sample of 200 nonathletes produced a mean equal to 23.5 hours per week and a standard deviation equal to 4.1 hours.

a. Describe the two populations involved in the comparison.
b. Do the samples provide sufficient evidence to conclude that there is a difference in the mean number of hours of study per week between athletes and nonathletes? Test using $\alpha = .01$.
c. Construct a 99% confidence interval for $(\mu_1 - \mu_2)$.
d. Would a 95% confidence interval for $(\mu_1 - \mu_2)$ be narrower or wider than the one you found in part **c**? Why?

8.10 A new type of band has been developed by a dental laboratory for children who have to wear braces. The new bands are designed to be more comfortable, look better, and provide more rapid progress in realigning teeth. An experiment was conducted to compare the mean wearing time necessary to correct a specific type of misalignment between the old braces and the new brands. One hundred children were randomly assigned, 50 to each group. A summary of the data is shown in the table.

	Old Braces	New Bands
$\bar{x}$	410 days	380 days
s	45 days	60 days

a. Is there sufficient evidence to conclude that the old braces have to be worn longer than the new bands? Use $\alpha = .01$.
b. Find a 95% confidence interval for the difference in wearing times for the two types of devices. Interpret the interval.

8.11 Suppose it is desired to compare two physical education training programs for preadolescent girls. A total of 80 girls are randomly selected, with 40 assigned to each program. After three 6-week periods on the program, each girl is given a fitness test that yields a score between 0 and 100. The means and variances of the scores for the two groups are shown in the table. Calculate a 99% confidence interval for the true difference in mean fitness scores for girls trained using these two programs.

	n	$\bar{x}$	s^2
Program 1	40	78.7	201.6
Program 2	40	75.3	259.2

8.12 It is often said that economic status is related to the commission of crimes. To test this theory, a sociologist selected a random sample of 70 people (who had no record of criminal conviction) from the census records of a certain city and recorded their annual incomes. Similarly, a random sample of 60 people, each of whom had committed their first crime, was selected from court records and the annual income (prior to arrest) was recorded for each. The means and variances of the annual incomes (in thousands of dollars) for the people in the two groups are shown in the table. Do the data provide sufficient evidence to indicate that the mean income of criminals, prior to committing their first offense, is lower than that for the noncriminal public? Test using $\alpha = .05$.

	$\bar{x}$	s^2
Criminals	13.3	24.2
Noncriminals	15.4	42.6

8.13 A large supermarket chain is interested in determining whether a difference exists between the mean shelf life (in days) of brand S bread and brand H bread. Random samples of 50 freshly baked loaves of each brand were tested, with the results shown in the table.

Brand S	*Brand H*
$\bar{x}_1 = 4.1$	$\bar{x}_2 = 5.2$
$s_1 = 1.2$	$s_2 = 1.4$

a. What are the appropriate null and alternative hypotheses to determine whether these samples provide sufficient evidence to conclude that the mean shelf lives of the two brands differ?

b. The test is conducted using statistical software, with the results shown here. Interpret the results of the test.

```
Z = -4.218          P-VALUE = .0000
```

c. Do you think the *p*-value of the above test is really zero? What is the appropriate interpretation of the computer result, "P-VALUE = .0000"?

d. Let μ_1 and μ_2 represent the mean shelf lives for brands S and H, respectively. Construct a 90% confidence interval for $(\mu_1 - \mu_2)$. Give an interpretation of your confidence interval.

8.14 Recent research reported in *The Professional Geographer* (Johnston-Anumonwo, 1992) examines the hypothesis that the disproportionate housework responsibility of women in two-income households is a major factor in determining the proximity of a woman's place of employment. The distance to work for both men and women in two-income households was reported for random samples of both central city and suburban residences:

	Central City Residence		Suburban Residence	
	Men	Women	Men	Women
n	159	119	138	93
$\bar{x}$	7.4	4.5	9.3	6.6
s	6.3	4.2	7.1	5.6

a. For central city residences, calculate a 99% confidence interval for the difference in average distance to work for men and women in two-income households. Interpret the interval.

b. Repeat part **a** for suburban residences.

c. Interpret the confidence intervals. Do they indicate that women tend to work closer to home than men?

d. What assumptions have you made to assure the validity of the confidence intervals constructed in parts **a** and **b**?

8.15 Two manufacturers of corrugated fiberboard both claim that the strength of their product tests on the average at more than 360 pounds per square inch. As a result of consumer complaints, a consumer products testing firm suspects that firm A's product is more than 5 pounds per square inch stronger than firm B's. To test its suspicion, 100 fiberboards were chosen randomly from firm A's inventory and 100 were chosen from firm B's inventory. The results of tests run on the samples are shown in the table.

A	B
$\bar{x}_1 = 365$	$\bar{x}_2 = 352$
$s_1 = 23$	$s_2 = 41$

a. Set up the appropriate null and alternative hypotheses to test the testing firm's belief. How do they differ from the typical hypotheses used to compare two population means?
b. Does the sample information support the consumer products testing firm's suspicion? Test at the .05 significance level.
c. What assumptions did you make in conducting the test in part **a**? Do you think such assumptions could comfortably be made in practice? Why or why not?
d. Find the observed significance level for the test and interpret its value.

8.16 As part of a study of participative management, George H. Hines sampled workers from two types of New Zealand sociocultural backgrounds: those who believed in the existence of a class system and those who believed that they lived and worked in a classless society. Each worker in the sampling was selected from a work environment with participatory management. Do workers who consider themselves to be social equals with their management superiors possess different levels of job satisfaction than those workers who see themselves as socially different from management? Each worker in the independent random samples was asked to answer this question by rating his or her job satisfaction on a scale of 1 (poor) to 7 (excellent). Using the results of this study (shown in the table), what can you say about the differences in job satisfaction for the two different sociocultural types of workers?

	Belief in Existence of a Class System	
	Yes	No
Sample size	175	277
Mean	5.42	5.19
Standard deviation	1.24	1.17

Source: Hines, G. H. "Influences on employee expectancy and participative management," *Academy of Management Journal*, 1974, 17.

8.17 Do the early questions in an exam affect your performance on later questions? Research reported in *The Journal of Psychology* (Petiprin and Johnson, 1991) investigated this relationship by conducting the following experiment: 140 college students were randomly assigned to two groups. Group A took an exam that contained 15 difficult questions followed by five moderate questions. Group B took an exam that contained 15 easy questions followed by the same five moderate questions. One of the goals of the study was to compare the average score on the moderate questions for the two groups.
a. The authors hypothesized that the students with the easy questions would score better than the students with the difficult questions. Set up the appropriate null and alternative hypotheses to test their research hypothesis, defining any symbols you use.
b. Suppose the p-value for this test were reported as .3248. What conclusion would you reach based on this p-value?
c. What assumptions are necessary to assure the validity of this conclusion?

8.2 Small-Sample Inferences About the Difference Between Two Population Means: Independent Sampling

Suppose a television network wants to determine whether major sports events or first-run movies attract more viewers in the prime-time hours. It selects 28 prime-time evenings; of these, 13 have programs devoted to major sports events and the remaining 15 have first-run movies. The number of viewers (estimated by a television viewer rating firm) is recorded for each program. If μ_1 is the mean number of sports viewers per evening of sports programming and μ_2 is the mean number of movie viewers per evening of movie programming, we want to detect a difference between μ_1 and μ_2—if such a difference exists. Therefore, we want to test the null hypothesis

$$H_0\text{:}\quad (\mu_1 - \mu_2) = 0$$

against the alternative hypothesis

$$H_a\text{:}\quad (\mu_1 - \mu_2) \neq 0 \qquad \text{(i.e., either } \mu_1 > \mu_2 \text{ or } \mu_2 > \mu_1\text{)}$$

Since the sample sizes are small, estimates of σ_1^2 and σ_2^2 are unreliable and the z-test statistic is inappropriate for the test. But as in the case of a single mean (Section 7.4), we can construct a Student t statistic. This statistic (formula to be given subsequently) has the familiar t-distribution described in Chapter 6. *To use the t statistic, both sampled populations must be approximately normally distributed with equal population variances, and the random samples must be selected independently of each other.* The normality and equal variances assumptions imply relative frequency distributions for the populations that would appear as shown in Figure 8.4.

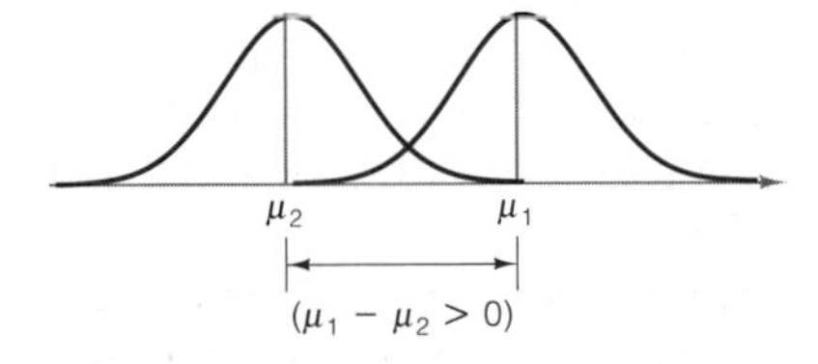

FIGURE 8.4 ► Assumptions for the two-sample t: (1) normal populations, (2) equal variances

Will these assumptions be satisfied for the television-viewing problem? We think that both assumptions will be adequately satisfied in this sampling situation. Since we assume the two populations have equal variances ($\sigma_1^2 = \sigma_2^2 = \sigma^2$), it is reasonable to use the information contained in both samples to construct a **pooled sample estimator** of σ^2 for use in the t statistic. Thus, if s_1^2 and s_2^2 are the two sample variances (both estimating the variance σ^2 common to both populations), the pooled estimator of σ^2, denoted as s_p^2, is

$$s_p^2 = \frac{(n_1 - 1)s_1^2 + (n_2 - 1)s_2^2}{(n_1 - 1) + (n_2 - 1)} = \frac{(n_1 - 1)s_1^2 + (n_2 - 1)s_2^2}{n_1 + n_2 - 2}$$

or

$$s_p^2 = \frac{\overbrace{\sum (x_1 - \bar{x}_1)^2}^{\text{From sample 1}} + \overbrace{\sum (x_2 - \bar{x}_2)^2}^{\text{From sample 2}}}{n_1 + n_2 - 2}$$

where x_1 represents a measurement from sample 1 and x_2 represents a measurement from sample 2. Recall that the term *degrees of freedom* was defined in Section 6.3 as 1 less than the sample size. Thus, in this case, we have $(n_1 - 1)$ degrees of freedom for sample 1 and $(n_2 - 1)$ degrees of freedom for sample 2. Since we are pooling the information on σ^2 obtained from both samples, the degrees of freedom associated with the pooled variance s_p^2 is equal to the sum of the degrees of freedom for the two samples, namely, the denominator of s_p^2; i.e., $(n_1 - 1) + (n_2 - 1) = n_1 + n_2 - 2$.

Note that the second formula given for s_p^2 shows that the pooled variance is simply a **weighted average** of the two sample variances, s_1^2 and s_2^2. The weight given each variance is proportional to its degrees of freedom. If the two variances have the same number of degrees of freedom (i.e., if the sample sizes are equal), then the pooled variance is a simple average of the two sample variances. The result is an average or "pooled" variance that is a better estimate of σ^2 than either s_1^2 or s_2^2 alone.

To obtain the **small-sample test statistic** for testing H_0: $(\mu_1 - \mu_2) = D_0$, substitute the pooled estimate of σ^2 into the formula for the two-sample z statistic (Section 8.1) to obtain

$$t = \frac{(\bar{x}_1 - \bar{x}_2) - D_0}{\sqrt{s_p^2\left(\dfrac{1}{n_1} + \dfrac{1}{n_2}\right)}}$$

It can be shown that this statistic has a t-distribution, but with $(n_1 + n_2 - 2)$ degrees of freedom.

We use the television viewer example to outline the final steps for this t-test: The hypothesized difference in mean number of viewers is $D_0 = 0$. The rejection region will be two-tailed and will be based on a t-distribution with $(n_1 + n_2 - 2)$ or $(13 + 15 - 2) = 26$ df. For $\alpha = .05$, the rejection region for the test would be

$$t < -t_{\alpha/2} \text{ or } t > t_{\alpha/2}$$

The value for $t_{.025}$ given in Table V of Appendix A is 2.056. Thus, the rejection region for the television example is

$$t < -2.056 \text{ or } t > 2.056$$

This rejection region is shown in Figure 8.5.

FIGURE 8.5 ▶
Rejection region for a two-tailed t-test: $\alpha = .05$, df $= 26$

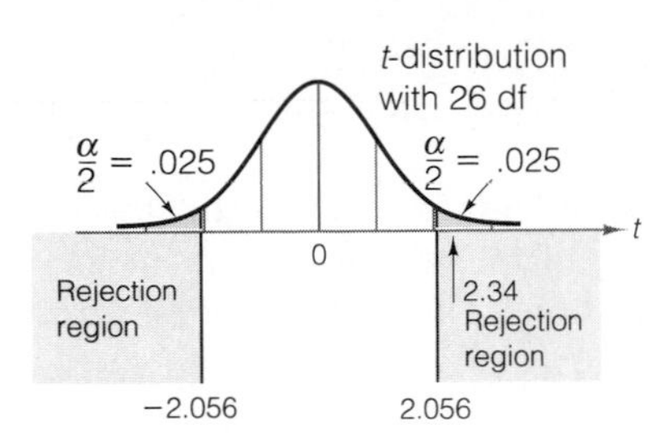

EXAMPLE 8.5

Suppose the television network's samples produced the following results:

Sports	*Movie*
$n_1 = 13$	$n_2 = 15$
$\bar{x}_1 = 6.8$ million	$\bar{x}_2 = 5.3$ million
$s_1 = 1.8$ million	$s_2 = 1.6$ million

Do the data provide sufficient evidence to indicate a difference between the mean number of viewers for major sports events and first-run movies shown in prime time? Test using $\alpha = .05$.

Solution

We must assume that these independent samples were selected from normal distributions (i.e., the number of viewers per evening is normally distributed for each type of program) with equal variances. [*Note:* Although the sample estimates of variance are not equal, the assumption that the population variances are equal may still be valid. Statistical tests of hypotheses are available for comparing population variances. Consult the references at the end of this chapter for details on these statistical tests.] We now calculate

$$s_p^2 = \frac{(n_1 - 1)s_1^2 + (n_2 - 1)s_2^2}{n_1 + n_2 - 2} = \frac{(13 - 1)(1.8)^2 + (15 - 1)(1.6)^2}{13 + 15 - 2}$$

$$= \frac{74.72}{26} = 2.87$$

Then

$$t = \frac{(\bar{x}_1 - \bar{x}_2) - D_0}{\sqrt{s_p^2\left(\frac{1}{n_1} + \frac{1}{n_2}\right)}} = \frac{(6.8 - 5.3) - 0}{\sqrt{2.87\left(\frac{1}{13} + \frac{1}{15}\right)}} = \frac{1.5}{.64} = 2.34$$

Since the observed value of t ($t = 2.34$) falls in the rejection region (see Figure 8.5), the samples provide sufficient evidence to indicate that the mean numbers of viewers differ for major sports events and first-run movies shown in prime time. Or, we can say that the test results are statistically significant at the $\alpha = .05$ level of significance. Because the rejection was in the positive or upper tail of the t-distribution, it appears that the mean number of viewers for sports events exceeds that for movies.

EXAMPLE 8.6 Find the approximate observed significance level for the test in Example 8.5.

Solution The observed significance level (p-value) for the test is the probability of observing a value of the test statistic that is *at least* as contradictory to H_0: $\mu_1 = \mu_2$ as the value observed if in fact H_0 is true. Thus, since the test was two-sided, we have

$$\begin{aligned} p\text{-value} &= P(t < -2.34 \text{ or } t > 2.34) \\ &= 2P(t > 2.34) \quad \text{(because the } t\text{-distribution is symmetric about its mean, 0)} \end{aligned}$$

Turning to Table V in Appendix A for 26 degrees of freedom, you can see that

$$P(t < -2.479 \text{ or } t > 2.479) = 2P(t > 2.479) = 2(.01) = .02$$

and

$$P(t < -2.056 \text{ or } t > 2.056) = 2P(t > 2.056) = 2(.025) = .05$$

Since the observed value of t lies between 2.479 and 2.056, the p-value for the test lies between .02 and .05. We have agreed to choose the larger of these as the approximate observed significance level and would thus give the approximate p-value for the test as .05. A reader of this reported p-value would reject the null hypothesis (and conclude that a difference exists between the mean numbers of sports and movie viewers) for values of α greater than or equal to .05.

The same t statistic can also be used to construct confidence intervals for the difference between population means. Both the confidence interval and the test of hypothesis procedures are summarized in the accompanying boxes.

Small-Sample Confidence Interval for $(\mu_1 - \mu_2)$ (Independent Samples)

$$(\bar{x}_1 - \bar{x}_2) \pm t_{\alpha/2}\sqrt{s_p^2\left(\frac{1}{n_1} + \frac{1}{n_2}\right)}$$

where $s_p^2 = \dfrac{(n_1 - 1)s_1^2 + (n_2 - 1)s_2^2}{n_1 + n_2 - 2}$

and $t_{\alpha/2}$ is based on $(n_1 + n_2 - 2)$ degrees of freedom.

Assumptions:
1. Both sampled populations have relative frequency distributions that are approximately normal.
2. The population variances are equal.
3. The samples are randomly and independently selected from the populations.

Small-Sample Test of Hypothesis for $(\mu_1 - \mu_2)$ (Independent Samples)

One-Tailed Test	*Two-Tailed Test*
H_0: $(\mu_1 - \mu_2) = D_0$ H_a: $(\mu_1 - \mu_2) < D_0$ [or H_a: $(\mu_1 - \mu_2) > D_0$]	H_0: $(\mu_1 - \mu_2) = D_0$ H_a: $(\mu_1 - \mu_2) \neq D_0$
Test statistic: $t = \dfrac{(\bar{x}_1 - \bar{x}_2) - D_0}{\sqrt{s_p^2\left(\dfrac{1}{n_1} + \dfrac{1}{n_2}\right)}}$	*Test statistic:* $t = \dfrac{(\bar{x}_1 - \bar{x}_2) - D_0}{\sqrt{s_p^2\left(\dfrac{1}{n_1} + \dfrac{1}{n_2}\right)}}$
Rejection region: $t < -t_\alpha$ [or $t > t_\alpha$ when H_a: $(\mu_1 - \mu_2) > D_0$]	*Rejection region:* $t < -t_{\alpha/2}$ or $t > t_{\alpha/2}$

where t_α and $t_{\alpha/2}$ are based on $(n_1 + n_2 - 2)$ degrees of freedom.

Assumptions: Same as for the small-sample confidence interval for $(\mu_1 - \mu_2)$ in the previous box.

EXAMPLE 8.7

Suppose you wish to compare a new method of teaching reading to "slow learners" to the current standard method. You decide to base this comparison on the results of a reading test given at the end of a learning period of 6 months. Of a random sample of 20 slow learners, 8 are taught by the new method and 12 are taught by the standard method. All 20 children are taught by qualified instructors under similar conditions for a 6-month period. The results of the reading test at the end of this period are summarized in the accompanying table. Estimate the true mean difference $(\mu_1 - \mu_2)$ between the test scores for the new method and the standard method. Use a 90% confidence interval, and interpret the interval. What assumptions must be made in order that the estimate be valid?

New Method	*Standard Method*
$n_1 = 8$	$n_2 = 12$
$\bar{x}_1 = 76.9$	$\bar{x}_2 = 72.7$
$s_1 = 4.85$	$s_2 = 6.35$

Solution

The objective of this experiment is to obtain a 90% confidence interval for $(\mu_1 - \mu_2)$. To use the small-sample confidence interval for $(\mu_1 - \mu_2)$, the following assumptions must be satisfied:

1. We assume that the populations of test scores are normally distributed for both the new and standard methods of instruction. Since a test score can be viewed as a sum of the results on the various components of the test, the Central Limit Theorem lends credence to this assumption.

2. The variance of the test scores is assumed to be the same for the two populations. Under the circumstances, we might expect the variation in test scores to be approximately the same for both methods.
3. The samples are randomly and independently selected from the two populations. We have randomly chosen 20 different slow learners for the two samples in such a way that the test score for one child is not dependent on the test score for any other child. Therefore, this assumption would probably be valid.

The first step in constructing the confidence interval is to calculate the pooled estimate of variance:

$$s_p^2 = \frac{(n_1 - 1)s_1^2 + (n_2 - 1)s_2^2}{n_1 + n_2 - 2}$$
$$= \frac{(8 - 1)(4.85)^2 + (12 - 1)(6.35)^2}{8 + 12 - 2} = 33.7892$$

where s_p^2 is based on $(n_1 + n_2 - 2) = (8 + 12 - 2) = 18$ degrees of freedom. Then the 90% confidence interval for $(\mu_1 - \mu_2)$, the difference between mean test scores for the two methods, is

$$(\bar{x}_1 - \bar{x}_2) \pm t_{\alpha/2}\sqrt{s_p^2\left(\frac{1}{n_1} + \frac{1}{n_2}\right)} = (76.9 - 72.7) \pm t_{.05}\sqrt{33.7892\left(\frac{1}{8} + \frac{1}{12}\right)}$$
$$= 4.20 \pm 1.734(2.653) = 4.20 \pm 4.60$$

or $(-.40, 8.80)$. This means that with a confidence coefficient equal to .90, we estimate the difference in mean test scores between using the new method of teaching and the standard method to fall in the interval $-.40$ to 8.80. In other words, we estimate the mean test score for the new method to be anywhere from .40 less than to 8.80 more than the mean test score for the standard method. Although the sample means seem to suggest that the new method is associated with a higher mean test score, there is insufficient evidence to indicate that $(\mu_1 - \mu_2)$ differs from 0 because the interval includes 0 as a possible value for $(\mu_1 - \mu_2)$. To show a difference in mean test scores (if it exists), you could increase the sample size and thereby narrow the width of the confidence interval for $(\mu_1 - \mu_2)$. An alternative is to design the experiment differently. This possibility is discussed in the next section.

The two-sample t statistic is a powerful tool for comparing population means when the assumptions are satisfied. It has also been shown to retain its usefulness when the sampled populations are only approximately normally distributed. And when the sample sizes are equal, the assumption of equal population variances can be relaxed. That is, if $n_1 = n_2$, then σ_1^2 and σ_2^2 can be quite different and the test statistic will still possess, approximately, a Student t-distribution. When the assumptions are not satisfied, you can select larger samples from the populations or you can use other available statistical tests (nonparametric statistical tests, which are described in optional Section 8.5).

What Should You Do if the Assumptions Are Not Satisfied?

Answer: If you are concerned that the assumptions are not satisfied, use the nonparametric Wilcoxon rank sum test for independent samples to test for a shift in population distributions. See optional Section 8.5.

Exercises 8.18–8.32

Learning the Mechanics

8.18 To use the t statistic to test for a difference between the means of two populations, what assumptions must be made about the two populations? About the two samples?

8.19 Two populations are described in each of the following cases. In which cases would it be appropriate to apply the small sample t-test to investigate the difference between the population means?

a. Population 1: Normal distribution with variance σ_1^2
   Population 2: Skewed to the right with variance $\sigma_2^2 = \sigma_1^2$

b. Population 1: Normal distribution with variance σ_1^2
   Population 2: Normal distribution with variance $\sigma_2^2 \neq \sigma_1^2$

c. Population 1: Skewed to the left with variance σ_1^2
   Population 2: Skewed to the left with variance $\sigma_2^2 = \sigma_1^2$

d. Population 1: Normal distribution with variance σ_1^2
   Population 2: Normal distribution with variance $\sigma_2^2 = \sigma_1^2$

e. Population 1: Uniform distribution with variance σ_1^2
   Population 2: Uniform distribution with variance $\sigma_2^2 = \sigma_1^2$

8.20 In the t-tests of this section, σ_1^2 and σ_2^2 are assumed to be equal. Thus, we say $\sigma_1^2 = \sigma_2^2 = \sigma^2$. Why is a pooled estimator of σ^2 used instead of either s_1^2 or s_2^2?

8.21 Assume that $\sigma_1^2 = \sigma_2^2 = \sigma^2$. Calculate the pooled estimator of σ^2 for each of the following cases:

a. $s_1^2 = 200, \quad s_2^2 = 180, \quad n_1 = n_2 = 25$

b. $s_1^2 = 25, \quad s_2^2 = 40, \quad n_1 = 20, \quad n_2 = 10$

c. $s_1^2 = .20, \quad s_2^2 = .30, \quad n_1 = 8, \quad n_2 = 12$

d. $s_1^2 = 2{,}500, \quad s_2^2 = 1{,}800, \quad n_1 = 16, \quad n_2 = 17$

e. Note that the pooled estimate is a weighted average of the sample variances. Which of the variances does the pooled estimate fall nearer in each of the above cases?

8.22 Independent random samples from normal populations produced the results shown in the table.

Sample 1	*Sample 2*
1.2	4.2
3.1	2.7
1.7	3.6
2.8	3.9
3.0	

a. Calculate the pooled estimate of σ^2.
b. Do the data provide sufficient evidence to indicate that $\mu_2 > \mu_1$? Test using $\alpha = .10$.
c. Find a 90% confidence interval for $(\mu_1 - \mu_2)$.
d. Which of the two inferential procedures, the test of hypothesis in part **b** or the confidence interval in part c, provides more information about $(\mu_1 - \mu_2)$?

8.23 Independent random samples selected from two normal populations produced the sample means and standard deviations shown in the table.

Sample 1	*Sample 2*
$n_1 = 17$	$n_2 = 12$
$\bar{x}_1 = 5.4$	$\bar{x}_2 = 7.9$
$s_1 = 3.4$	$s_2 = 4.8$

a. The test H_0: $(\mu_1 - \mu_2) = 0$ against H_a: $(\mu_1 - \mu_2) \neq 0$ was conducted using statistical software, with the results shown. Check and interpret the results.

```
T = -1.646          DF = 27          P-VALUE(2-TAILED) = .1114
```

b. Estimate $(\mu_1 - \mu_2)$ using a 95% confidence interval.

8.24 Independent random samples from approximately normal populations produced the results shown in the table.

Sample 1				*Sample 2*			
52	33	42	44	52	43	47	56
41	50	44	51	62	53	61	50
45	38	37	40	56	52	53	60
44	50	43		50	48	60	55

a. Do the data provide sufficient evidence to conclude that $(\mu_2 - \mu_1) > 10$? Test using $\alpha = .01$.
b. Construct a 98% confidence interval for $(\mu_1 - \mu_2)$. Interpret your result.

Applying the Concepts

8.25 Helping smokers kick the habit is big business in today's no-smoking environment. One of the more commonly used treatments according to an article in the *Journal of Imagination, Cognition and Personality* (Spanos et al., 1992/93) is Spiegel's three-point message:

1. For your body, smoking is a poison.
2. You need your body to live.
3. You owe your body this respect and protection.

To determine the effectiveness of this treatment, the authors conducted a study consisting of a sample of 52 smokers placed in two groups, a Spiegel treatment group or a control group (no treatment). Each participant was asked to record the number of cigarettes he or she smoked each week. The results for the study are shown below for the beginning period and four follow-up time periods.

Number of Cigarettes Smoked in Week

	n	$\bar{x}$	s
Beginning			
Treatment	35	165.09	71.20
Control	17	159.00	67.45
First follow-up (2 wks)			
Treatment	35	105.00	69.08
Control	17	157.24	66.80
Second follow-up (4 wks)			
Treatment	35	111.11	69.08
Control	17	159.52	65.73
Third follow-up (8 wks)			
Treatment	35	120.20	67.59
Control	17	157.88	64.41
Fourth follow-up (12 wks)			
Treatment	35	123.63	74.09
Control	17	162.17	67.01

a. Create 95% confidence intervals for the difference in the average number of cigarettes smoked per week for the two groups for the beginning and each follow-up period. Interpret the results.
b. What assumptions are necessary for the validity of these confidence intervals?

8.26 Amid concerns and protests about its potential dangers, nuclear-generated energy is reportedly on the rise, and the growth rate of nuclear power facilities is expected to continue to increase. One reason is that nuclear power is cheaper to produce than conventional sources. The Atomic Industrial Forum, which represents the nuclear industry, reports that coal-produced energy costs 3.5¢ per kilowatt-hour to produce compared to 3.1¢ for nuclear power (*Time*, February 13, 1984).

Assume that these costs are averages and that the results were taken over 9 nuclear power plants and 11 coal-powered plants. Assume also that the variances of the cost estimates for nuclear and coal-produced power were .05 and .04, respectively. Does this information provide sufficient evidence to say that nuclear-generated power is less expensive to produce than coal-produced power? Test using $\alpha = .05$. What assumptions are required for the test to be valid?

8.27 Since the emergence of Japan as an industrial superpower, American businesses have been looking more closely at Japanese management styles and philosophies. Some of the credit for the high quality of Japanese products has been attributed to the Japanese system of permanent employment for their workers. In the

United States, high job turnover rates are common in many industries and are associated with high product defect rates. High turnover rates mean that U.S. plants are more highly populated with inexperienced workers who are unfamiliar with the company's product lines than is the case in Japan. In a study of the room air conditioner industry in Japan and the United States, David Garvin reported that the difference in the average annual turnover rate of workers between American plants and Japanese plants was 3.1%. In a different study, five Japanese and five American plants that manufacture room air conditioners were randomly sampled and their turnover rates were determined as shown in the table.

U.S. Plants	*Japanese Plants*
7.11%	3.52%
6.06%	2.02%
8.00%	4.91%
6.87%	3.22%
4.77%	1.92%

Source: Garvin, D. "Quality on the line," *Harvard Business Review*, September–October, 1983, pp. 65–75.

a. Do these data provide sufficient evidence to contradict the results of Garvin's study? Test using $\alpha = .10$.
b. Report the observed significance level of the test you conducted in part **a**.
c. List any assumptions necessary to assure the validity of the test you conducted. Use the terminology of this exercise.

8.28 An experiment is conducted to investigate the effect of a drug on the time to complete a task. Twenty people are divided at random into two groups of ten each. One group is given a placebo, while the second experimental group is administered a drug thought to increase the ability to complete the task quickly. For the control group, the times required to complete the task had a mean of 14.8 minutes and a variance of 3.9; for the experimental group, the average was 12.3 minutes and the variance was 4.3.
a. Test the null hypothesis that the drug has no effect in reducing the mean length of time to complete the task against the alternative hypothesis that the mean time is less for those subjects who receive the drug. Conduct the test at the $\alpha = .10$ level of significance.
b. Find the approximate observed significance level for the test and interpret its value.

8.29 To compare two methods of teaching reading, randomly selected groups of elementary school children were assigned to each of the two teaching methods for a 6-month period. The criterion for measuring achievement was a reading comprehension test. The results are shown in the accompanying table. Do the data provide sufficient evidence to indicate a difference in mean scores on the comprehension test for the two teaching methods? Test using $\alpha = .05$.

	Number of Children per Group	$\bar{x}$	s^2
Method 1	11	64	52
Method 2	14	69	71

8.30 Suppose your plant purifies its liquid waste and discharges the water into a local river. An EPA inspector has collected water specimens of the discharge of your plant and also water specimens in the river upstream from your plant. Each water specimen is divided into five parts, the bacteria count is read on each, and the mean count for each specimen is reported. The average bacteria counts for each of six specimens are reported in the table for the two locations.

Plant Discharge	*Upstream*
30.1	29.7
36.2	30.3
33.4	26.4
28.2	27.3
29.8	31.7
34.9	32.3

a. Why might the bacteria counts shown here tend to be approximately normally distributed?
b. What are the appropriate null and alternative hypotheses to test whether the mean bacteria count for the plant discharge exceeds that for the upstream location? Be sure to define any symbols you use.
c. When the data are submitted to SPSS, part of the output is the following:

```
T          TWO-TAILED SIGNIFICANCE LEVEL
1.53                   .156
```

Carefully interpret this output.
d. What assumptions are necessary to assure the validity of this test?

8.31 Shorkey, McRoy, and Armendariz examined the relationship between the intensity of parental punishment practices and various demographic characteristics of mothers. The intensity score descriptive statistics are shown in the accompanying table, with higher scores showing a greater intensity of parental punishments.

	Single	Separated
n	8	6
$\bar{x}$	70.75	77.33
s	14.80	13.69

Source: Shorkey, C. T., McRoy, R. G., and Armendariz, J. "Intensity of parental punishments and problem-solving attitudes and behaviors," *Psychological Reports*, 1985, 56.

a. Use a 95% confidence interval to estimate the difference between the mean intensity of parental punishment scores for single and separated mothers. Interpret the interval in terms of this application.
b. Does the confidence interval in part **a** support an inference that the true means for single and separated mothers differ?
c. Conduct a test of hypothesis comparing the two means using $\alpha = .05$. Does your conclusion agree with that in part **b**?
d. What assumptions are necessary to assure the validity of the inferences in parts **a–c**? State them in terms of this application.

8.32 How are physical and mental health related? Research conducted by Jon K. Mills (1991) and reported in the *Journal of Psychology* investigates the relationship between age and perception of control among obese females. Perception of control was measured by the Locus of Control (LOC), a test instrument widely used for determining whether individuals perceive themselves as being in control ("internal" LOC), or as being controlled by others ("external" LOC). Forty-six obese female adults and 19 obese female adolescents were sampled and an LOC score for each was determined. Higher scores represent external LOC and lower scores represent internal LOC. The data reported in the article are shown below:

	Adults	Adolescents
$\bar{x}$	6.45	10.89
s	2.89	2.48
n	46	19

a. What are the appropriate null and alternative hypotheses for determining whether a difference exists between the mean LOC score for obese female adults and adolescents? Define any symbols you use.

b. The data were analyzed using a statistical software package, with the results shown here. Interpret these results. Do you think the *p*-value is exactly zero? Explain.

```
T = -5.858        DF = 63        TWO-TAILED P-VALUE = .0000
```

c. Another part of the computer output is shown here. Interpret this interval.

```
95% CONFIDENCE INTERVAL: (-5.93, -2.95)
```

d. Which do you find more informative, the test of hypothesis or the confidence interval? Explain.

8.3 Inferences About the Difference Between Two Population Means: Paired Difference Experiments

In Example 8.7 we compared two methods of teaching reading to slow learners by means of a 90% confidence interval. Suppose it is possible to measure the slow learners' "reading IQ's" *before* they are subjected to a teaching method. Eight pairs of slow learners with similar reading IQ's are found, and one member of each pair is randomly assigned to the standard teaching method while the other is assigned to the new method. Do the data in Table 8.2 support the hypothesis that the population mean reading test score for slow learners taught by the new method is greater than the mean reading test score for those taught by the standard method?

TABLE 8.2 Reading Test Scores for Eight Pairs of Slow Learners

Pair	New Method	Standard Method
1	77	72
2	74	68
3	82	76
4	73	68
5	87	84
6	69	68
7	66	61
8	80	76
	$\bar{x}_1 = 76.0$	$\bar{x}_2 = 71.625$
	$s_1^2 = 48.0$	$s_2^2 = 49.1$

We want to test

$$H_0: \ (\mu_1 - \mu_2) = 0$$
$$H_a: \ (\mu_1 - \mu_2) > 0$$

If we use the two-sample t statistic (Section 8.2), we first calculate

$$s_p^2 = \frac{(n_1 - 1)s_1^2 + (n_2 - 1)s_2^2}{n_1 + n_2 - 2} = \frac{(8 - 1)(48.0) + (8 - 1)(49.1)}{8 + 8 - 2} = 48.55$$

and then the test statistic

$$t = \frac{(\bar{x}_1 - \bar{x}_2) - 0}{\sqrt{s_p^2\left(\frac{1}{n_1} + \frac{1}{n_2}\right)}} = \frac{76.0 - 71.625}{\sqrt{48.55\left(\frac{1}{8} + \frac{1}{8}\right)}} = \frac{4.375}{3.485} = 1.26$$

This small t value will not lead to rejection of H_0 when compared to the t-distribution with $n_1 + n_2 - 2 = 14$ df, even if α is chosen as large as .10 ($t_{.10} = 1.345$). Thus, from *this* analysis we might conclude that there is insufficient evidence to infer a difference in the mean test scores for the two methods.

If you carefully examine the data in Table 8.2, however, you will find this result difficult to accept. The test score of the new method is larger than the corresponding test score for the standard method *for every one of the eight pairs of slow learners*. This, in itself, seems to provide strong evidence to indicate that μ_1 exceeds μ_2. Why, then, did the t-test fail to detect this difference? The answer is: *The two-sample t is not a valid procedure to use with this set of data.*

The two-sample t is inappropriate because the assumption of independent samples is invalid. We have randomly chosen *pairs of test scores*, and thus, once we have chosen the sample for the new method, we have *not* independently chosen the sample for the standard method. The dependence between observations within pairs can be seen by examining the pairs of test scores, which tend to rise and fall together as we go from pair to pair. This pattern provides strong visual evidence of a violation of the assumption of independence required for the two-sample t-test of Section 8.2. In this situation, you will note the *large variation within samples* (reflected by the large value of s_p^2) in comparison to the relatively *small difference between the sample means*. Because s_p^2 is so large, the t-test of Section 8.2 is unable to detect a difference between μ_1 and μ_2.

We now consider a valid method of analyzing the data of Table 8.2. In Table 8.3 (page 396) we add the column of differences between the test scores of the pairs of slow learners. We can regard these differences in test scores as a random sample of differences for all pairs (matched on reading IQ) of slow learners, past and present. Then we can use this sample to make inferences about the mean of the population of differences, μ_D—which is equal to the difference $(\mu_1 - \mu_2)$. That is, the mean of the population (and sample) of differences equals the difference between the population (and sample) means. Thus, our test becomes

$$H_0: \ \mu_D = 0 \qquad (\mu_1 - \mu_2 = 0)$$
$$H_a: \ \mu_D > 0 \qquad (\mu_1 - \mu_2 > 0)$$

TABLE 8.3

Pair	New Method	Standard Method	Difference (New Method − Standard Method)
1	77	72	5
2	74	68	6
3	82	76	6
4	73	68	5
5	87	84	3
6	69	68	1
7	66	61	5
8	80	76	4
			$\bar{x}_D = 4.375$
			$s_D = 1.69$

The test statistic is a one-sample t (Section 7.4), since we are now analyzing a single sample of differences:

Test statistic: $$t = \frac{\bar{x}_D - 0}{s_D/\sqrt{n_D}}$$

where $\bar{x}_D$ = Sample mean difference

s_D = Sample standard deviation of differences

n_D = Number of differences = Number of pairs

Assumptions: The population of differences in test scores is approximately normally distributed. The sample differences are randomly selected from the population of differences. [*Note:* We do not need to make the assumption that $\sigma_1^2 = \sigma_2^2$.]

Rejection region: At significance level $\alpha = .05$, we will reject H_0 if $t > t_{.05}$, where $t_{.05}$ is based on $(n_D - 1)$ degrees of freedom.

Referring to Table V in Appendix A, we find the t value corresponding to $\alpha = .05$ and $n_D - 1 = 8 - 1 = 7$ df to be $t_{.05} = 1.895$. Then we will reject the null hypothesis if $t > 1.895$. Note that the number of degrees of freedom has decreased from $n_1 + n_2 - 2 = 14$ to 7 by using the paired difference experiment rather than the two independent random samples design. Now calculate

$$t = \frac{\bar{x}_D - 0}{s_D/\sqrt{n_D}} = \frac{4.375}{1.69/\sqrt{8}} = 7.32$$

Because this value of t falls in the rejection region, we conclude that the mean test score for slow learners taught by the new method exceeds the mean score for those taught by the standard method. We have confidence in this conclusion because the probability that this test would lead to the rejection of H_0, given that it is true, is only $\alpha = .05$.

This kind of experiment, in which observations are paired and the differences are analyzed, is called a **paired difference experiment**. In many cases, a paired difference experiment can provide more information about the difference between population means than an independent samples experiment. The idea is to compare population means by comparing the differences between pairs of experimental units (objects, people, etc.) that were very similar prior to the experiment. The differencing removes sources of variation that tend to inflate σ^2. For example, when two children are taught to read by two different methods, the observed difference in achievement may be due to a difference in the effectiveness of the two teaching methods *or* it may be due to differences in the initial reading levels and IQ's of the two children (random error). To reduce the effect of differences in the children on the observed differences in reading achievement, the two methods of reading are imposed on two children who are more likely to possess similar intellectual potentials, namely children with nearly equal IQ's. The effect of this pairing is to remove the larger source of variation that would be present if children with different abilities were randomly assigned to the two samples. Making comparisons within groups of similar experimental units is called **blocking**, and the paired difference experiment is a simple example of a **randomized block experiment**. In our example, pairs of children with matching IQ scores represent the blocks.

Some other examples for which the paired difference experiment might be appropriate are the following:

1. Suppose you want to estimate the difference $(\mu_1 - \mu_2)$ in mean price per gallon between two major brands of premium gasoline. If you choose two independent random samples of stations for each brand, the variability in price due to geographic location may be large. To eliminate this source of variability you could choose pairs of stations of similar size, one station for each brand, in close geographic proximity and use the sample of differences between the prices of the brands to make an inference about $(\mu_1 - \mu_2)$.
2. Suppose a college placement center wants to estimate $(\mu_1 - \mu_2)$, the difference in mean starting salaries for men and women graduates who seek jobs through the center. If it independently samples men and women, the starting salaries may vary because of their different college majors and differences in grade-point averages. To eliminate these sources of variability, the placement center could match male and female job-seekers according to their majors and grade-point averages. Then the differences between the starting salaries of each pair in the sample could be used to make an inference about $(\mu_1 - \mu_2)$.
3. Suppose you wish to estimate the difference $(\mu_1 - \mu_2)$ in mean absorption rate into the bloodstream for two drugs that relieve pain. If you independently sample people, the absorption rates might vary because of age, weight, sex, blood pressure, etc. In fact, there are many possible sources of nuisance variability, and pairing individuals who are similar in all the possible sources would be quite difficult. However, it may be possible to obtain two measurements *on the same person*. First, we administer one of the two drugs and record the time until absorption. After a sufficient amount of time, the other drug is administered and a second measurement on absorption time is obtained. The differences between the measurements

for each person in the sample could then be used to estimate $(\mu_1 - \mu_2)$. This procedure would be advisable only if the amount of time allotted between drugs is sufficient to guarantee little or no carryover effect. Otherwise, it would be better to use different people matched as closely as possible on the factors thought to be most important.

The one-tailed and two-tailed hypothesis-testing procedures and the method of forming confidence intervals for the difference between two means using a paired difference experiment are summarized in the boxes.

Paired Difference Confidence Interval

$$\bar{x}_D \pm t_{\alpha/2} \frac{s_D}{\sqrt{n_D}}$$

where $t_{\alpha/2}$ is based on $(n_D - 1)$ degrees of freedom

Assumptions:
1. The relative frequency distribution of the population of differences is normal.
2. The sample differences are randomly selected from the population of differences.

Paired Difference Test of Hypothesis

One-Tailed Test

H_0: $(\mu_1 - \mu_2) = D_0$; i.e., $\mu_D = D_0$

H_a: $(\mu_1 - \mu_2) < D_0$; i.e., $\mu_D < D_0$

[or H_a: $(\mu_1 - \mu_2) > D_0$; i.e., $\mu_D > D_0$]

Test statistic: $t = \dfrac{\bar{x}_D - D_0}{s_D/\sqrt{n_D}}$

Rejection region: $t < -t_\alpha$ [or $t > t_\alpha$ when H_a: $(\mu_1 - \mu_2) > D_0$]

Two-Tailed Test

H_0: $(\mu_1 - \mu_2) = D_0$; i.e., $\mu_D = D_0$

H_a: $(\mu_1 - \mu_2) \neq D_0$; i.e., $\mu_D \neq D_0$

Test statistic: $t = \dfrac{\bar{x}_D - D_0}{s_D/\sqrt{n_D}}$

Rejection region: $t < -t_{\alpha/2}$ or $t > t_{\alpha/2}$

where t_α and $t_{\alpha/2}$ are based on $(n_D - 1)$ degrees of freedom.

Assumptions:
1. The relative frequency distribution of the population of differences is normal.
2. The differences are randomly selected from the population of differences.

EXAMPLE 8.8

A paired difference experiment is conducted to compare the starting salaries of male and female college graduates who find jobs. Pairs are formed by choosing a male and a female with the same major and similar grade-point averages. Suppose a random sample of 10 pairs is formed in this manner and the starting annual salary of each person is recorded. The results are shown in Table 8.4. Test to see whether there is evidence that the mean starting salary, μ_1, for males exceeds the mean starting salary, μ_2, for females. Use $\alpha = .05$.

TABLE 8.4

Pair	Male	Female	Difference (Male − Female)	Pair	Male	Female	Difference (Male − Female)
1	\$24,300	\$23,800	\$ 500	6	\$22,800	\$23,000	\$−200
2	26,500	26,600	−100	7	24,500	24,200	300
3	25,400	24,800	600	8	26,200	25,100	1,100
4	23,500	23,500	0	9	23,400	23,200	200
5	28,500	27,600	900	10	24,200	23,500	700

Solution

The elements of the paired difference test are

$$H_0: \mu_D = 0 \qquad (\mu_1 - \mu_2 = 0)$$
$$H_a: \mu_D > 0 \qquad (\mu_1 - \mu_2 > 0)$$

Note that we propose a one-sided research hypothesis, since we are interested in determining whether the data indicate that μ_1 exceeds μ_2—that is, male mean starting salary exceeds female mean starting salary.

Test statistic: $t = \dfrac{\bar{x}_D - 0}{s_D/\sqrt{n_D}}$

Assumptions:
1. The relative frequency distribution for the population of differences is normal.
2. The sample differences are randomly selected from the population.

Since the test is upper-tailed, we will reject H_0 if $t > t_\alpha$, where $t_{.05} = 1.833$ is based on $(n_D - 1) = 9$ degrees of freedom. The rejection region is shown in Figure 8.6.

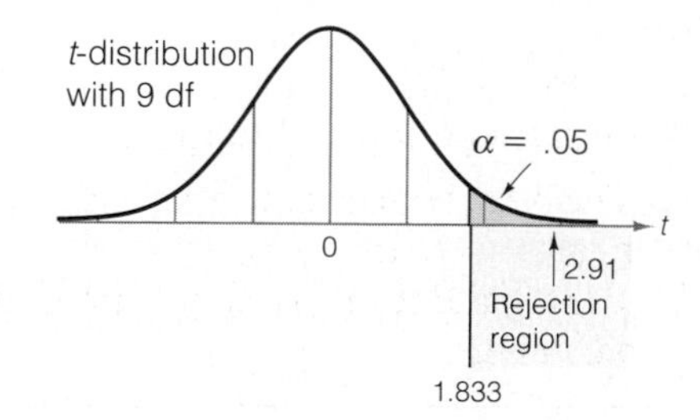

FIGURE 8.6 ► Rejection region for Example 8.8

We now calculate

$$\sum x_D = 500 + (-100) + \cdots + 700 = 4{,}000$$

and

$$\sum x_D^2 = 3{,}300{,}000$$

Then

$$\bar{x}_D = \frac{\sum x_D}{10} = \frac{4{,}000}{10} = 400$$

$$s_D^2 = \frac{\sum (x_D - \bar{x}_D)^2}{n_D - 1} = \frac{\sum x_D^2 - \left(\sum x_D\right)^2/10}{9}$$

$$= \frac{3{,}300{,}000 - (4{,}000)^2/10}{9} = 188{,}888.89$$

$$s_D = \sqrt{s_D^2} = 434.61$$

Substituting these values into the formula for the test statistic, we find that

$$t = \frac{\bar{x}_D - 0}{s_D/\sqrt{n_D}} = \frac{400}{434.61/\sqrt{10}} = \frac{400}{137.44} = 2.91$$

As you can see in Figure 8.6, the calculated t falls in the rejection region. Thus, we conclude at the $\alpha = .05$ level of significance that the mean starting salary for males exceeds the mean starting salary for females.

One measure of the amount of information about $(\mu_1 - \mu_2)$ gained by using a paired difference experiment rather than an independent samples experiment in Example 8.8 is the relative widths of the confidence intervals obtained by the two methods. A 95% confidence interval for $(\mu_1 - \mu_2)$ using the paired difference experiment is

$$\bar{x}_D \pm t_{\alpha/2} \frac{s_D}{\sqrt{n_D}} = 400 \pm t_{.025} \frac{434.61}{\sqrt{10}} = 400 \pm 2.262 \frac{434.61}{\sqrt{10}}$$

$$= 400 \pm 310.88 \approx 400 \pm 311 = (\$89, \$711)$$

If we analyzed the same data as though this were an independent samples experiment,* we would first calculate the following quantities:

*This is done only to provide a measure of the increase in the amount of information obtained by a paired design in comparison to an unpaired design. Actually, if an experiment is designed using pairing, an unpaired analysis would be invalid because the assumption of independent samples would not be satisfied.

Males	*Females*
$n_1 = 10$	$n_2 = 10$
$\bar{x}_1 = \$24{,}930$	$\bar{x}_2 = \$24{,}530$
$s_1^2 = 3{,}009{,}000$	$s_2^2 = 2{,}331{,}222.22$

Then

$$s_p^2 = \frac{(n_1 - 1)s_1^2 + (n_2 - 1)s_2^2}{n_1 + n_2 - 2} = \frac{9(3{,}009{,}000) + 9(2{,}331{,}222.22)}{10 + 10 - 2}$$
$$= 2{,}670{,}111.11$$
$$s_p = \sqrt{s_p^2} = 1{,}634.05$$

The 95% confidence interval is

$$(\bar{x}_1 - \bar{x}_2) \pm t_{.025}\sqrt{s_p^2\left(\frac{1}{n_1} + \frac{1}{n_2}\right)} = 400 \pm (2.101)\sqrt{2{,}670{,}111.11\left(\frac{1}{10} + \frac{1}{10}\right)}$$
$$= 400 \pm 1{,}535.35 \approx 400 \pm 1{,}535$$
$$= (-\$1{,}135,\ \$1{,}935)$$

The confidence interval for the independent sampling experiment is about five times wider than for the corresponding paired difference confidence interval. Blocking out the variability due to differences in majors and grade-point averages significantly increases the information about the difference in male and female mean starting salaries by providing a much more accurate (smaller confidence interval for the same confidence coefficient) estimate of $(\mu_1 - \mu_2)$.

You may wonder whether conducting a paired difference experiment is always superior to an independent samples experiment. The answer is: Most of the time, but not always. We sacrifice half the degrees of freedom in the t statistic when a paired difference design is used instead of an independent samples design. This is a loss of information, and unless this loss is more than compensated for by the reduction in variability obtained by blocking (pairing), the paired difference experiment will result in a net loss of information about $(\mu_1 - \mu_2)$. Thus, we should be convinced that the pairing will significantly reduce variability before performing the paired difference experiment. Most of the time this will happen.

One final note: The pairing of the observations is determined *before* the experiment is performed (that is, by the *design* of the experiment). A paired difference experiment is *never* obtained by pairing the sample observations after the measurements have been acquired.

What Do You Do When the Assumption of a Normal Distribution for the Population of Differences Is Not Satisfied?

Answer: Use the nonparametric Wilcoxon signed rank test for the paired difference design (optional Section 8.6).

CASE STUDY 8.2 / Matched Pairing in Studying the Mentally Retarded

The statistical implications underlying matching procedures have frequently been overlooked in educational research with the mentally retarded population. It is, therefore, the purpose of this paper to point out some of the advantages and disadvantages of different matching procedures.

Stainback and Stainback (1973) describe a number of experimental situations in which blocking (matching) subjects before experimentation might reduce variability and thereby increase the amount of information obtained. One of the matching procedures discussed by the authors can be summarized as follows: Suppose it is desired to form two groups of mentally retarded subjects to compare two methods of educational therapy. Subjects could be randomly selected from an existing (large) group of subjects and ordered (from lowest to highest) on the scores of an appropriate matching variable. Several variables suggested by the authors are "pretest measures of the experimental criterion, measures of learning rate (mental age and Intelligence Quotients), chronological age, personal characteristics (sex, race), environmental conditions (socioeconomic level), or combinations of two or more variables." The two highest-ranking subjects on the matching variables would then form pair 1, the next two pair 2, etc., and one member from each pair would be randomly assigned to each therapy group. This experiment is a practical example of a paired difference experiment, and if the matching variables are correctly chosen, the responses of subjects will be more homogeneous within pairs than between pairs. The authors conclude:

It is important that when comparing two groups on a criterion variable the two groups be as equal as possible on all relevant factors excepting only the independent variables. This consideration deserves particular emphasis in the area of mental retardation since the researchers are constantly dealing with diverse groups. It should be restated, therefore, that researchers in the area of mental retardation should become acutely aware of advantages and disadvantages of matching procedures and their alternatives.

Exercises 8.33–8.47

Learning the Mechanics

8.33 Suppose a paired difference experiment is to be performed in order to test the null hypothesis H_0: $\mu_D = 2$ against H_a: $\mu_D > 2$. Assume that the second observation in each pair will be subtracted from the first. Define μ_D, and explain what is meant by the hypotheses $\mu_D = 2$ and $\mu_D > 2$.

8.34 A paired difference experiment yielded n_D pairs of observations. In each case, what is the rejection region for testing H_0: $\mu_D = 2$ against H_a: $\mu_D > 2$?

a. $n_D = 10$, $\alpha = .05$
b. $n_D = 20$, $\alpha = .10$
c. $n_D = 5$, $\alpha = .025$
d. $n_D = 9$, $\alpha = .01$

8.35 The data for a random sample of six paired observations are shown in the accompanying table.

Pair	Sample from Population 1	Sample from Population 2
	Observation 1	Observation 2
1	7	4
2	3	1
3	9	7
4	6	2
5	4	4
6	8	7

a. Calculate the difference between each pair of observations by subtracting observation 2 from observation 1. Use the differences to calculate $\bar{x}_D$ and s_D^2.
b. If μ_1 and μ_2 are the means of populations 1 and 2, respectively, express μ_D in terms of μ_1 and μ_2.
c. Form a 95% confidence interval for μ_D.
d. Test the null hypothesis H_0: $\mu_D = 0$ against the alternative hypothesis H_a: $\mu_D \neq 0$. Use $\alpha = .05$.

8.36 The data for a random sample of 10 paired observations are shown in the table.

Pair	Sample from Population 1	Sample from Population 2
1	19	24
2	25	27
3	31	36
4	52	53
5	49	55
6	34	34
7	59	66
8	47	51
9	17	20
10	51	55

a. If you wish to test whether these data are sufficient to indicate that the mean for population 2 is larger than that for population 1, what are the appropriate null and alternative hypotheses? Define any symbols you use.
b. The data are analyzed using a statistical software package, with the results shown here. Interpret these results.

```
T = -5.29          DF = 9          P-VALUE = .0003
```

c. The output of the statistical program also included the following. Interpret this output.

```
95% CONFIDENCE INTERVAL: (-5.258, -2.116)
```

d. What assumptions are necessary to assure the validity of this analysis?

8.37 A paired difference experiment produced the following data:

$$n_D = 16 \quad \bar{x}_1 = 143 \quad \bar{x}_2 = 150 \quad \bar{x}_D = -7 \quad s_D^2 = 64$$

a. Determine the values of t for which the null hypothesis, $\mu_1 - \mu_2 = 0$, would be rejected in favor of the alternative hypothesis $\mu_1 - \mu_2 < 0$. Use $\alpha = .10$.
b. Conduct the paired difference test described in part **a**. Draw the appropriate conclusions.
c. What assumptions are necessary so that the paired difference test will be valid?
d. Find a 90% confidence interval for the mean difference μ_D.
e. Which of the two inferential procedures, the confidence interval of part **d** or the test of hypothesis of part b, provides more information about the difference between the population means?

8.38 A paired difference experiment yielded the data shown in the table.

Pair	*x*	*y*	*Pair*	*x*	*y*
1	55	44	5	75	62
2	68	55	6	52	38
3	40	25	7	49	31
4	55	56			

a. Test H_0: $\mu_D = 10$ against H_a: $\mu_D \neq 10$, where $\mu_D = (\mu_1 - \mu_2)$. Use $\alpha = .05$.
b. Report the p-value for the test you conducted in part **a**. Interpret the p-value.

Applying the Concepts

8.39 The data shown in the table are part of a study conducted to compare the abilities of men and women to perform the strenuous tasks of a firefighter. They represent the pulling force (in newtons) that a firefighter was able to exert in pulling the starter cord of a P-250 fire pump. Firefighters were matched in pairs according to weight, thus producing data for a matched pairs (or paired difference) experiment.

Pair	*Female*	*Male*
1	40.03	84.51
2	75.62	80.06
3	53.38	102.30
4	62.27	88.96

Source: Phillips, M. D., and Pepper, R. L. "Shipboard firefighting performance of females and males," *Human Factors*, 1982, 24. Copyright 1982 by The Human Factors Society, Inc. Reproduced by permission.

a. Do the data provide sufficient evidence to indicate a difference in mean pulling force between female and male firefighters? Test using $\alpha = .05$.
b. Find a 95% confidence interval for the difference in mean pulling force between female and male firefighters. Interpret the interval.

8.40 A new weight-reducing technique, consisting of a liquid protein diet, is currently undergoing tests by the Food and Drug Administration (FDA) before its introduction into the market. A typical test performed by the FDA is the following: The weights of a random sample of five people are recorded before they are introduced to the liquid protein diet. The five individuals are then instructed to follow the liquid protein diet for 3 weeks. At the end of this period, their weights (in pounds) are again recorded. The results are listed in the table. Let μ_1 be the true mean weight of individuals before starting the diet and let μ_2 be the true mean weight of individuals after 3 weeks on the diet. Construct a 95% confidence interval for the difference between the true mean weights before and after the diet is used. What assumptions are necessary to ensure the validity of the procedure you used?

Person	*Weight Before Diet*	*Weight After Diet*
1	150	143
2	195	190
3	188	185
4	197	191
5	204	200

8.41 A study reported in the *Journal of Psychology* (Hesse-Biber and Marino, 1991) measures the change in female students' self-concepts as they move from high school to college. A sample of 133 Boston college incoming female freshmen were selected for the study. Each was asked to evaluate several aspects of her life at two points in time: at the end of her senior year of high school, and during her sophomore year of college. Each student was asked to evaluate where she believed she stood on a scale that ranged from top 10% of class (1) to lowest 10% of class (5). The results for three of the traits evaluated are reported below.

Trait		Senior Year of High School	Sophomore Year of College
	n	$\bar{x}$	$\bar{x}$
Leadership	133	2.09	2.33
Popularity	133	2.48	2.69
Intellectual self-confidence	133	2.29	2.55

a. What null and alternative hypotheses would you test to determine whether there is a decrease in the mean self-concept of females between the senior year of high school and the sophomore year of college as measured by each of these three traits?
b. Are these tests more appropriately analyzed using an independent samples test or a paired difference test? Explain.
c. Noting the size of the sample, what assumptions are necessary to assure the validity of the tests?
d. The article reports that the leadership test results in a p-value greater than .05, while the tests for popularity and intellectual self-confidence result in p-values less than .05. Interpret these results.

8.42 The data shown in the table provide information on the relationship between the mean daily air temperature and the cocoon temperature of wooly-bear caterpillars of the High Arctic. You can see from the table that the data indicate that the caterpillar's body temperature (inside the cocoon) is higher than the outside air temperature. Estimate the mean difference in temperature between the cocoon and the outside air. Use a 95% confidence interval.

Day	Temperature (°C)		Day	Temperature (°C)		Day	Temperature (°C)	
	Air	Cocoon[a]		Air	Cocoon[a]		Air	Cocoon[a]
1	10.4	15.1	5	4.1	8.0	9	3.0	7.0
2	9.2	14.6	6	3.7	8.7	10	3.5	7.1
3	2.2	6.8	7	1.7	3.6	11	4.5	9.6
4	2.6	6.8	8	2.0	5.3	12	4.4	9.5

[a]Each cocoon temperature is the average of the temperatures of two cocoons.

Source: Kevan, P. G., Jensen, T. S., and Shorthouse, J. D. "Body temperatures and behavioral thermoregulation of High Arctic wooly-bear caterpillars and pupae (*Gynaephora rossii*, Lymantridae: Lepidoptera) and the importance of sunshine," *Arctic and Alpine Research*, 1982, 14. Reproduced with permission of the Regents of the University of Colorado.

8.43 In the past, many bodily functions were thought to be beyond conscious control. However, recent experimentation suggests that it may be possible for a person to control certain body functions if that person is trained in a program of *biofeedback* exercises. An experiment is conducted to show that blood pressure levels can be consciously reduced in people trained in this program. The blood pressure measurements (in millimeters of mercury) listed in the table represent readings before and after the biofeedback training of six subjects.

Subject	*Before*	*After*
1	136.9	130.2
2	201.4	180.7
3	166.8	149.6
4	150.0	153.2
5	173.2	162.6
6	169.3	160.1

a. If we want to test whether the mean blood pressure decreases after the training, what are the appropriate null and alternative hypotheses? Define any symbols you use.

b. When the test is conducted using a statistical software package, the results are as shown here. Interpret these results.

```
T = 2.98          DF = 5          P-VALUE = .016
```

c. The output of the software package also includes the following confidence interval. Interpret this interval.

```
95% CONFIDENCE INTERVAL: (1.389, 19.011)
```

d. What assumptions are necessary to assure the validity of these results?

8.44 While producing many economic benefits to the state of Florida, gypsum and phosphate mines also produce a harmful byproduct: *radiation*. It has been known for a number of years that the mine tailings (waste) contain radioactive radon 222. In fact, new housing complexes built on top of the leveled piles of residue have shown disturbing radiation levels within the houses. The radiation levels in waste gypsum and phosphate mounds in Polk County, Florida, are regularly monitored by the Eastern Environmental Radiation Facility (EERF) and by the Polk County Health Department (PCHD), Winter Haven, Florida. The table shows measurements of the exhalation rate (a measure of radiation) for 15 soil samples obtained from waste mounds in Polk County, Florida. The exhalation rate was measured for each soil sample by both the PCHD and the EERF. The objective of selecting the paired measurements was to determine whether a bias exists, a difference in the mean readings, between PCHD and EERF. They represent part of the data contained in a report by Thomas R. Horton of EERF.

Charcoal Canister No.	*PCHD*	*EERF*	*Charcoal Canister No.*	*PCHD*	*EERF*
71	1,709.79	1,479.0	85	393.55	187.7
58	357.17	257.8	46	880.84	630.4
84	1,150.94	1,287.0	4	2,996.49	3,707.0
91	1,572.69	1,395.0	20	2,367.40	2,791.0
44	558.33	416.5	36	599.84	706.8
43	4,132.28	3,993.0	42	538.37	618.5
79	1,489.86	1,351.0	55	2,770.23	2,639.0
61	3,017.48	1,813.0			

Source: Horton, T. R. "Preliminary radiological assessment of radon exhalation from phosphate gypsum piles and inactive uranium mill tailings piles," EPA-520/5-79-004. Washington, D. C.: Environmental Protection Agency, 1979.

a. Considering the relative size of the measurements from canister to canister, explain why a paired difference experiment was conducted rather than an independent samples experiment.

b. Given that the mean difference (PCHD − EERF) for the 15 sampled canisters is 84.17 with standard deviation 408.92, do the data provide sufficient evidence to indicate a difference in the mean exhalation rates between PCHD and EERF? Test using $\alpha = .05$.

c. Find a 95% confidence interval for the difference in mean measurements between PCHD and EERF. Interpret the interval. Does it support the result of the test in part a?

8.45 A manufacturer of automobile shock absorbers was interested in comparing the durability of its shocks with that of the biggest competitor. To make the comparison, one of the manufacturer's and one of the competitor's shocks were randomly selected and installed on the rear wheels of six cars. After the cars had been driven 20,000 miles, the strength of each test shock was measured, coded, and recorded. The following are the results of the examination:

Car Number	*Manufacturer's*	*Competitor's*
1	8.8	8.4
2	10.5	10.1
3	12.5	12.0
4	9.7	9.3
5	9.6	9.0
6	13.2	13.0

a. Do the data present sufficient evidence to conclude there is a difference in the mean strength of the two types of shocks after 20,000 miles of use? Use $\alpha = .05$.
b. What assumptions are necessary in order to apply a paired difference analysis to the data?
c. Construct a 95% confidence interval for $(\mu_1 - \mu_2)$. Interpret your confidence interval.

8.46 Suppose the data in Exercise 8.45 are based on independent random samples.
a. Do the data provide sufficient evidence to indicate a difference between the mean strengths for the two types of shocks? Use $\alpha = .05$.
b. Construct a 95% confidence interval for $(\mu_1 - \mu_2)$. Interpret your result.
c. Compare the confidence intervals you obtained in Exercise 8.45 and part **b** of this exercise. Which is larger? To what do you attribute the difference in size? Assuming in each case that the appropriate assumptions are satisfied, which interval provides you with more information about $(\mu_1 - \mu_2)$? Explain.
d. Are the results of an unpaired analysis valid when the data have been collected from a paired experiment?

8.47 An experiment was conducted to measure the effects of fructose and glucose on high-endurance performance of athletes. Six trained female runners were used in the experiment. Each was given 300 milliliters of a liquid 45 minutes prior to running for 85 minutes or until they reached a state of exhaustion, whichever occurred first. Various measures of endurance, such as performance time (time to exhaustion), rating of perceived exertion, etc., were recorded at the end of each run. Four liquids (treatments) were used in the experiment. The first contained fructose, the second contained glucose, the third contained water sweetened with a calcium saccharine solution (a placebo designed to suggest the presence of fructose or glucose), and the fourth contained water alone. Each of the six subjects performed the run for each of the four liquids, which were arranged in random order. The table gives the averages of the six runners' times (in minutes) to exhaustion for only two of the mixtures, glucose and the placebo. The table also gives the sample sizes and standard deviations for the two samples.

	Glucose	Placebo
n	6	6
$\bar{x}$	63.9	52.2
s	20.3	13.5

Source: McMurry, R. G., Wilson, J. R., and Kitchell, B. S. "The effects of fructose and glucose on high-endurance performance," *Research Quarterly for Exercise and Sport*, 1983, 54. Reprinted by permission of the American Alliance for Health, Physical Education, Recreation, and Dance, 1900 Association Drive, Reston, VA 20091.

a. Describe the experiment and explain how and why it does or does not satisfy the assumptions of independent random sampling.
b. Suppose that the data were based on independent random samples. Would the data provide sufficient evidence to indicate a difference in the mean time to exhaustion between runners given the glucose mixture and those given the placebo? Test using $\alpha = .05$. Interpret your result.
c. Consider the manner in which the experiment was actually conducted. What do you gain or lose by analyzing the data using the method of part **b**?

8.4 Determining the Sample Size

You can find the appropriate sample size to estimate the difference between a pair of parameters with a specified degree of reliability by using the method described in Section 6.2. That is, to estimate the difference between a pair of parameters correct to within B units with probability $(1 - \alpha)$, let $z_{\alpha/2}$ standard deviations of the sampling distribution of the estimator equal B. Then solve for the sample size. To do this, you have to solve the problem for a specific ratio between n_1 and n_2. Most often, you will want to have equal sample sizes, that is, $n_1 = n_2 = n$. We will illustrate the procedure with two examples.

EXAMPLE 8.9

New fertilizer compounds are often advertised with the promise of increased yields. Suppose we want to compare the mean yield μ_1 of wheat when a new fertilizer is used to the mean yield μ_2 with a fertilizer in common use. The estimate of the difference in mean yield per acre is to be correct to within .25 bushel with a confidence coefficient of .95. If the sample sizes are to be equal, find $n_1 = n_2 = n$, the number of 1-acre plots of wheat assigned to each fertilizer.

Solution

To solve the problem, you need to know something about the variation in the bushels of yield per acre. Suppose from past records you know the yields of wheat possess a range of approximately 10 bushels per acre. You could then approximate $\sigma_1 = \sigma_2 = \sigma$ by letting the range equal 4σ. Thus,

$$\begin{aligned} 4\sigma &\approx 10 \text{ bushels} \\ \sigma &\approx 2.5 \text{ bushels} \end{aligned}$$

The next step is to solve the equation

$$z_{\alpha/2}\sigma_{(\bar{x}_1 - \bar{x}_2)} = B \quad \text{or} \quad z_{\alpha/2}\sqrt{\frac{\sigma_1^2}{n_1} + \frac{\sigma_2^2}{n_2}} = B$$

for n, where $n = n_1 = n_2$. Since we want the estimate to lie within $B = .25$ of $(\mu_1 - \mu_2)$ with confidence coefficient equal to .95, we have $z_{\alpha/2} = z_{.025} = 1.96$. Then, letting $\sigma_1 = \sigma_2 = 2.5$ and solving for n, we have

$$\begin{aligned} 1.96\sqrt{\frac{(2.5)^2}{n} + \frac{(2.5)^2}{n}} &= .25 \\ 1.96\sqrt{\frac{2(2.5)^2}{n}} &= .25 \\ n = 768.32 &\approx 769 \text{ (rounding up)} \end{aligned}$$

Consequently, you will have to sample 769 acres of wheat for each fertilizer to estimate the difference in mean yield per acre to within .25 bushel. Since this would necessitate extensive and costly experimentation, you might decide to allow a larger bound (say, $B = .50$ or $B = 1$) in order to reduce the sample size, or you might decrease the confidence coefficient. The point is that we can obtain an idea of the experimental effort necessary to achieve a specified precision in our final estimate by determining the approximate sample size *before* the experiment is begun.

EXAMPLE 8.10

A laboratory manager wishes to compare the difference in the mean readings of two instruments, A and B, designed to measure the potency (in parts per million) of an antibiotic. To conduct the experiment, the manager plans to select n_D specimens of the antibiotic from a vat and to measure each specimen with both instruments. The difference $(\mu_A - \mu_B)$ will be estimated based on the n_D paired differences $(x_A - x_B)$ obtained in the experiment. If preliminary measurements suggest that the differences will range between plus or minus 10 parts per million, how many differences will be needed to estimate $(\mu_A - \mu_B)$ correct to within 1 part per million with confidence coefficient equal to .99?

Solution

The estimator for $(\mu_A - \mu_B)$, based on a paired difference experiment, is $\bar{x}_D = (\bar{x}_A - \bar{x}_B)$ and

$$\sigma_{\bar{x}_D} = \frac{\sigma_D}{\sqrt{n_D}}$$

Thus, the number n_D of pairs of measurements needed to estimate $(\mu_A - \mu_B)$ to within 1 part per million can be obtained by solving for n_D in the equation

$$z_{\alpha/2} \frac{\sigma_D}{\sqrt{n_D}} = B$$

where $z_{.005} = 2.58$ and $B = 1$. To solve this equation for n_D, we need to have an approximate value for σ_D.

We are given the information that the differences are expected to range from -10 to 10 parts per million. Letting the range equal $4\sigma_D$, we find

$$\text{Range} = 20 \approx 4\sigma_D$$
$$\sigma_D \approx 5$$

Substituting $\sigma_D = 5$, $B = 1$, and $z_{.005} = 2.58$ into the equation and solving for n_D, we obtain

$$2.58 \frac{5}{\sqrt{n_D}} = 1$$
$$n_D = [(2.58)(5)]^2$$
$$= 166.41$$

Therefore, it will require approximately $n_D = 166$ pairs of measurements to estimate $(\mu_A - \mu_B)$ correct to within 1 part per million using the paired difference experiment.

The box summarizes the procedures for determining the sample sizes necessary for estimating $(\mu_1 - \mu_2)$ for the case $n_1 = n_2 = n$ and for estimating μ_D.

Determination of Sample Size for Comparing Two Means

Independent Random Samples

To estimate $(\mu_1 - \mu_2)$ to within a given bound B with probability $(1 - \alpha)$ or, equivalently, with a $100(1 - \alpha)\%$ confidence interval of width $W = 2B$, use the following formula to solve for equal sample sizes that will achieve the desired reliability:

$$n_1 = n_2 = \frac{(z_{\alpha/2})^2(\sigma_1^2 + \sigma_2^2)}{B^2} = \frac{4(z_{\alpha/2})^2(\sigma_1^2 + \sigma_2^2)}{W^2}$$

You will need to substitute estimates for the values of σ_1^2 and σ_2^2 before solving for the sample size. These estimates might be sample variances s_1^2 and s_2^2 from prior sampling (e.g., a pilot sample), or from an educated (and conservatively large) guess based on the range, i.e., $s \approx R/4$.

Paired Difference Experiment

To estimate μ_D to within a given bound B with probability $(1 - \alpha)$ or, equivalently, with a $100(1 - \alpha)\%$ confidence interval of width $W = 2B$, use the following formula to solve for n:

$$n = \frac{(z_{\alpha/2})^2\sigma_D^2}{B^2} = \frac{4(z_{\alpha/2})^2\sigma_D^2}{W^2}$$

You will need to substitute an estimate of σ_D^2 before solving for the sample size. This estimate might be the sample variance s_D^2 from prior sampling (e.g., a pilot study), or from an educated (and conservatively large) guess based on the range, i.e., $s_D \approx R/4$.

Exercises 8.48–8.55

Learning the Mechanics

8.48 Suppose you want to estimate the difference between two population means correct to within 2.2 with probability .95. If prior information suggests that the population variances are approximately equal to $\sigma_1^2 =$

$\sigma_2^2 = 15$ and you want to select independent random samples of equal size from the populations, how large should the sample sizes, n_1 and n_2, be?

8.49 Find the appropriate values of n_1 and n_2 (assume $n_1 = n_2$) needed to estimate $(\mu_1 - \mu_2)$ with:

a. A bound on the error of estimation equal to 3.2 with 95% confidence. From prior experience it is known that $\sigma_1 \approx 15$ and $\sigma_2 \approx 17$.

b. A bound on the error of estimation equal to 8 with 99% confidence. The range of each population is 60.

c. A 90% confidence interval of width 1.0. Assume that $\sigma_1^2 \approx 5.8$ and $\sigma_2^2 \approx 7.5$.

8.50 If you were to employ a paired difference experiment, determine the number of differences that would be needed to estimate $(\mu_1 - \mu_2)$ to within:

a. A bound on the error of estimation equal to 3 with 95% confidence. From prior experience it is known that $\sigma_D \approx 11$.

b. A bound on the error of estimation equal to 5 with 99% confidence. The range of the differences is approximately 84.

c. A 90% confidence interval with width equal to .8. Assume that $\sigma_D^2 \approx 1.3$.

Applying the Concepts

8.51 Is housework hazardous to your health? A recent study in the *Public Health Reports* (Rogot, Sorlie, and Johnson, 1992) compares the life expectancies of 25-year-old white women in the labor force to those who are housewives. How large a sample would have to be taken from each group in order to be 95% confident that the estimate of difference in life expectancies for the two groups is within 1 year of the true difference in life expectancies? Assume that equal sample sizes will be selected from the two groups, and that the standard deviation for both groups is approximately 15 years.

8.52 One reason high school seniors are encouraged to attend college is that the job opportunities are much better for those with college degrees than for those without. A high school counselor wants to estimate the difference in mean income per day between high school graduates who have a college education and those who have not gone on to college. Suppose it is decided to compare the daily incomes of 30-year-olds, and the range of daily incomes for both groups is approximately \$200 per day. How many people from each group should be sampled in order to estimate the true difference between mean daily incomes correct to within \$10 per day with probability .9? Assume that $n_1 = n_2$.

8.53 In seeking a good professional football running back, a coach is looking for a player with high mean yards gained per carry and a small standard deviation. Suppose the coach wishes to compare the mean yards gained per carry for two major prospects based on independent random samples of their yards gained per carry in the early part of the coming pro football season. Suppose data from last year indicate that $\sigma_1 = \sigma_2 \approx 5$ yards. If the coach wants to estimate the difference in means correct to within 1 yard with probability equal to .9, how many runs would have to be observed for each player? (Assume equal sample sizes.)

8.54 Refer to the liquid protein diet study of Exercise 8.40. How many people must participate in the experiment to estimate the true mean weight loss to within .5 pound with 90% confidence? Use the value of s_D found in Exercise 8.40 in your calculations.

8.55 In Exercise 8.26 we compared the mean costs of electrical energy produced by nuclear and coal-powered electrical power plants. In that exercise, we were given the information that the variances in costs for the nuclear and coal-powered plants were $\sigma_1^2 = .05$ and $\sigma_2^2 = .04$, respectively. How many plants of each type

should be sampled in order to estimate the difference in mean costs per kilowatt-hour with a 95% confid
interval of width .2 cent?

8.5 A Nonparametric Test for Comparing Two Populations: Independent Sampling (Optional)

The independent-samples t-test for comparing two populations (Section 8.2) is unsuitable for some types of data. These data fall into two categories: (1) data sets that do not satisfy the assumptions upon which the t-test is based—namely, that the random variables being measured have normal probability distributions with equal variances; and (2) ordinal data, i.e., data that are not susceptible to measurement but that can be *ranked in order of magnitude*. An example of ordinal data occurs during new product testing. For a food product this may entail taste tests in which consumers rank the new product in order of preference with respect to one or more currently popular brands. A consumer probably has a preference for each product, but the strength of the preference is difficult, if not impossible to measure. Consequently, the best we can do is to have each consumer examine the new product, along with a few established products, and rank them according to preference: 1 for the most preferred, 2 for second, etc.

The *nonparametric* counterpart of the independent-samples test is called the **Wilcoxon rank sum test**, developed by Frank Wilcoxon. The Wilcoxon rank sum test compares the probability distributions of the sampled populations rather than specific parameters (i.e., the means) of these populations. For example, the nonparametric test can be used to compare the probability distribution of the strengths of preferences for a new product to the probability distributions of the strengths of preferences for the currently popular brands. If it can be inferred that the distribution for the new product lies above (to the right of) the others (see Figure 8.7), the implication is that the new product tends to be more preferred than the currently popular products. Such an inference might lead to a decision to market the product nationally. The following numerical example illustrates the use of the Wilcoxon rank sum test.

FIGURE 8.7 ▶ **Probability distributions of strengths of preference measurements (new product is preferred)**

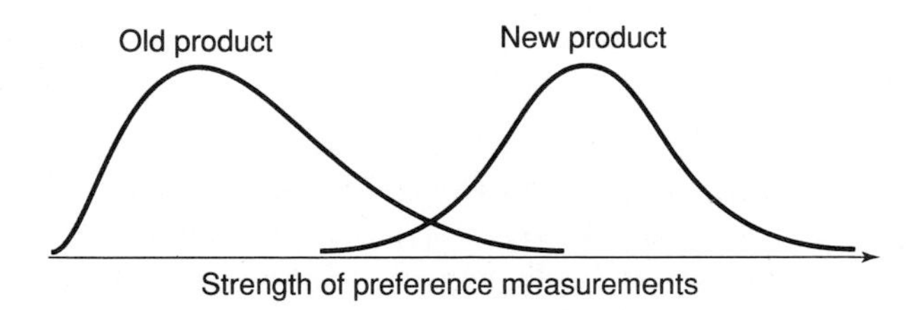

Suppose an experimental psychologist wants to compare reaction times for adult males under the influence of drug A to those under the influence of drug B. Experience has shown that populations of reaction time measurements often possess probability distributions that are skewed to the right, as shown in Figure 8.8. Consequently, a t-test should not be used to compare the mean reaction times for the two drugs because the normality assumption that is required for the t-test may not be valid.

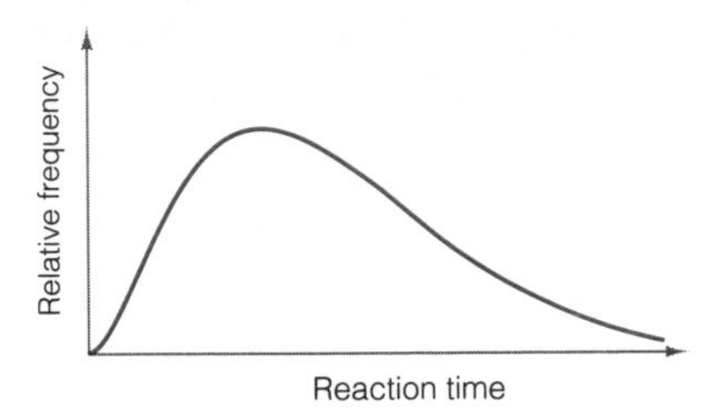

FIGURE 8.8 ▶ Typical probability distribution of reaction times

Suppose the psychologist randomly assigns seven subjects to each of two groups, one group to receive drug A and the other to receive drug B. The reaction time for each subject is measured at the completion of the experiment. These data (with the exception of the measurement for one subject in group A who was eliminated from the experiment for personal reasons) are shown in Table 8.5.

The population of reaction times for either of the drugs, say drug A, is that which could conceptually be obtained by giving drug A to all adult males. To compare the probability distributions for populations A and B, *we first rank the sample observations as though they were all drawn from the same population*. That is, we pool the measurements from both samples and then rank the measurements from the smallest (a rank of 1) to the largest (a rank of 13). The results of this ranking process are also shown in Table 8.5.

TABLE 8.5 Reaction Times of Subjects Under the Influence of Drug A or B

Drug A		Drug B	
Reaction time (seconds)	Rank	Reaction time (seconds)	Rank
1.96	4	2.11	6
2.24	7	2.43	9
1.71	2	2.07	5
2.41	8	2.71	11
1.62	1	2.50	10
1.93	3	2.84	12
		2.88	13

If the two populations were identical, we would expect the ranks to be *randomly mixed* between the two samples. If, on the other hand, one population tends to have longer reaction times than the other, we would expect the larger ranks to be mostly in one sample and the smaller ranks mostly in the other. Thus, the test statistic for the Wilcoxon test is based on the totals of the ranks for each of the two samples—that is, on the **rank sums**. When the sample sizes are equal, for example, the greater the difference in the rank sums, the greater will be the weight of evidence to indicate a difference between the probability distributions for populations A and B. In the reaction times example, we denote the rank sum for drug A by T_A and that for drug B by T_B. Then

$$T_A = 4 + 7 + 2 + 8 + 1 + 3 = 25$$
$$T_B = 6 + 9 + 5 + 11 + 10 + 12 + 13 = 66$$

The sum of T_A and T_B will always equal $n(n + 1)/2$, where $n = n_1 + n_2$. So, for this example, $n_1 = 6$, $n_2 = 7$, and

$$T_A + T_B = \frac{13(13 + 1)}{2} = 91$$

Since $T_A + T_B$ is fixed, a small value for T_A implies a large value for T_B (and vice versa) and a large difference between T_A and T_B. Therefore, the smaller the value of one of the rank sums, the greater the evidence to indicate that the samples were selected from different populations.

The test statistic for this test is the rank sum for the smaller sample or, in the case where $n_1 = n_2$, either rank sum can be used. Values that locate the rejection region for this rank sum are given in Table VI of Appendix A. A partial reproduction of this table is shown in Table 8.6. The columns of the table represent n_1, the first sample size, and the rows represent n_2, the second sample size. *The T_L and T_U entries in the table are the boundaries of the lower and upper regions, respectively, for the rank sum associated with the sample that has fewer measurements.* If the sample sizes n_1 and n_2 are the same, either rank sum may be used as the test statistic. To illustrate, suppose

TABLE 8.6 Reproduction of Part of Table VI in Appendix A: Critical Values for the Wilcoxon Rank Sum Test

a. $\alpha = .025$ one-tailed; $\alpha = .05$ two-tailed

n_2 \ n_1	3		4		5		6		7		8		9		10	
	T_L	T_U	T_L	T_U	T_L	T_U	T_L	T_U	T_L	T_U	T_L	T_U	T_L	T_U	T_L	T_U
3	5	16	6	18	6	21	7	23	7	26	8	28	8	31	9	33
4	6	18	11	25	12	28	12	32	13	35	14	38	15	41	16	44
5	6	21	12	28	18	37	19	41	20	45	21	49	22	53	24	56
6	7	23	12	32	19	41	26	52	28	56	29	61	31	65	32	70
7	7	26	13	35	20	45	28	56	37	68	39	73	41	78	43	83
8	8	28	14	38	21	49	29	61	39	73	49	87	51	93	54	98
9	8	31	15	41	22	53	31	65	41	78	51	93	63	108	66	114
10	9	33	16	44	24	56	32	70	43	83	54	98	66	114	79	131

Wilcoxon Rank Sum Test: Independent Samples*

One-Tailed Test

H_0: Two sampled populations have identical probability distributions

H_a: The probability distribution for population A is shifted to the right of that for B

Test statistic: The rank sum T associated with the sample with fewer measurements (if sample sizes are equal, either rank sum can be used)

Rejection region: Assuming the smaller sample size is associated with distribution A (if sample sizes are equal, we use the rank sum T_A), we reject the null hypothesis if

$$T_A \geq T_U$$

where T_U is the upper value given by Table VI in Appendix A for the chosen *one-tailed* α value.

Two-Tailed Test

H_0: Two sampled populations have identical probability distributions

H_a: The probability distribution for population A is shifted to the left *or* to the right of that for B

Test statistic: The rank sum T associated with the sample with fewer measurements (if the sample sizes are equal, either rank sum can be used)

Rejection region: $T \leq T_L$ or $T \geq T_U$

where T_L is the lower value given by Table VI in Appendix A for the chosen *two-tailed* α value and T_U is the upper value from Table VI.

[*Note:* If the one-sided alternative is that the probability distribution for A is shifted to the *left* of B (and T_A is the test statistic), we reject the null hypothesis if $T_A \leq T_L$.]

Assumptions:
1. The two samples are random and independent.
2. The two probability distributions from which the samples are drawn are continuous.

Ties: Assign tied measurements the average of the ranks they would receive if they were unequal but occurred in successive order. For example, if the third-ranked and fourth-ranked measurements are tied, assign each a rank of $(3 + 4)/2 = 3.5$.

*Another statistic used for comparing two populations based on independent random samples is the **Mann–Whitney *U* statistic**. The U statistic is a simple function of the rank sums. It can be shown that the Wilcoxon rank sum test and the Mann–Whitney U-test are equivalent.

$n_1 = 8$ and $n_2 = 10$. For a two-tailed test with $\alpha = .05$, we consult part **a** of the table and find that the null hypothesis will be rejected if the rank sum of sample 1 (the sample with fewer measurements), T, is less than or equal to $T_L = 54$ *or* greater than or equal to $T_U = 98$. The Wilcoxon rank sum test is summarized in the preceding box.

Note that the assumptions necessary for the validity of the Wilcoxon rank sum test do not specify the shape or type of probability distribution. However, the distributions are assumed to be continuous so that the probability of tied measurements is 0 (see Chapter 4), and each measurement can be assigned a unique rank. In practice, however, rounding of continuous measurements will sometimes produce ties. As long as the number of ties is small relative to the sample sizes, the Wilcoxon test procedure will still have approximate significance level α. The test is not recommended to compare discrete distributions for which many ties are expected.

EXAMPLE 8.11

Do the data given in Table 8.5 provide sufficient evidence to indicate a shift in the probability distributions for drugs A and B, that is, that the probability distribution corresponding to drug A lies either to the right or left of the probability distribution corresponding to drug B? Test at the .05 level of significance.

Solution

H_0: The two populations of reaction times corresponding to drug A and drug B have the same probability distribution

H_a: The probability distribution for drug A is shifted to the right or left of the probability distribution corresponding to drug B*

Test statistic: Since drug A has fewer subjects than drug B, the test statistic is T_A, the rank sum of drug A's reaction times.

Rejection region: Since the test is two-sided, we consult part **a** of Table VI for the rejection region corresponding to $\alpha = .05$. We will reject H_0 for $T_A \leq T_L$ or $T_A \geq T_U$. Thus, we will reject H_0 if $T_A \leq 28$ or $T_A \geq 56$.

Since T_A, the rank sum of drug A's reaction times in Table 8.5, is 25, it is in the rejection region (see Figure 8.9 on page 418).†

We can conclude that the probability distributions for drugs A and B are not identical. In fact, it appears that drug B tends to be associated with reaction times that are larger than those associated with drug A (because T_A fell in the lower tail of the rejection region).

*The alternative hypotheses in this chapter will be stated in terms of a difference in the *location* of the distributions. However, since the shapes of the distributions may also differ under H_a, some of the figures (e.g., Figure 8.9) depicting the alternative hypothesis will show probability distributions with different shapes.

†Figure 8.9 depicts only one side of the two-sided alternative hypothesis. The other would show distribution A shifted to the right of distribution B.

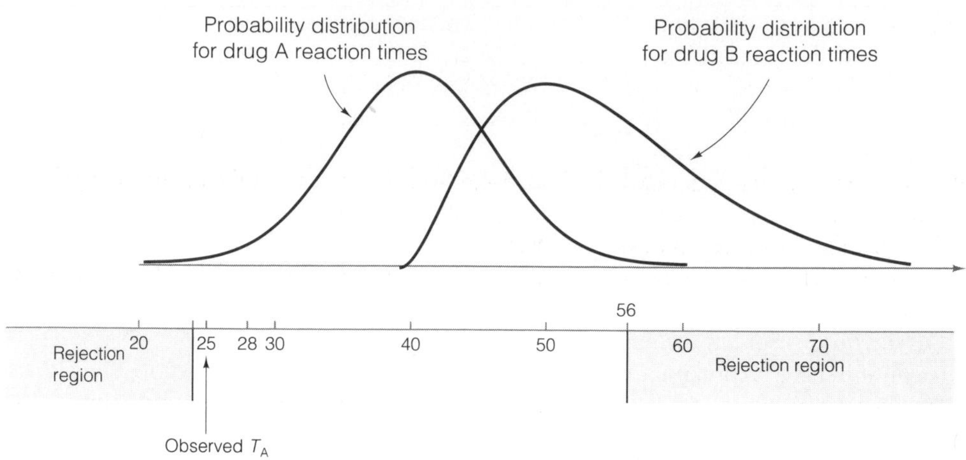

FIGURE 8.9 ▲ **Alternative hypothesis and rejection region for Example 8.11**

Table VI in Appendix A gives values of T_L and T_U for values of n_1 and n_2 less than or equal to 10. When both samples sizes, n_1 and n_2, are 10 or larger, the sampling distribution of T_A can be approximated by a normal distribution with mean and variance

$$E(T_A) = \frac{n_1(n_1 + n_2 + 1)}{2} \quad \text{and} \quad \sigma^2_{T_A} = \frac{n_1 n_2(n_1 + n_2 + 1)}{12}$$

Therefore, for $n_1 \geq 10$ and $n_2 \geq 10$ we can conduct the Wilcoxon rank sum test using the familiar z-test of Section 8.1. The test is summarized in the box.

Wilcoxon Rank Sum Test: Large Independent Samples

One-Tailed Test

H_0: Two sampled populations have identical probability distributions

H_a: The probability distribution for population A is shifted to the right of that for B

Test statistic: $$z = \frac{T_A - \dfrac{n_1(n_1 + n_2 + 1)}{2}}{\sqrt{\dfrac{n_1 n_2(n_1 + n_2 + 1)}{12}}}$$

Rejection region: $z > z_\alpha$

Assumptions: $n_1 \geq 10$ and $n_2 \geq 10$

Two-Tailed Test

H_0: Two sampled populations have identical probability distributions

H_a: The probability distribution for population A is shifted to the left *or* to the right of that for B

Test statistic: $$z = \frac{T_A - \dfrac{n_1(n_1 + n_2 + 1)}{2}}{\sqrt{\dfrac{n_1 n_2(n_1 + n_2 + 1)}{12}}}$$

Rejection region: $z < -z_{\alpha/2}$ or $z > z_{\alpha/2}$

Assumptions: $n_1 \geq 10$ and $n_2 \geq 10$

Exercises 8.56–8.67

Learning the Mechanics

8.56 Specify the test statistic and the rejection region for the Wilcoxon rank sum test for independent samples in each of the following situations:

a. H_0: Two probability distributions, A and B, are identical
H_a: Probability distribution for population A is shifted to the right or left of the probability distribution for population B
$n_A = 7, \quad n_B = 8, \quad \alpha = .10$

b. H_0: Two probability distributions, A and B, are identical
H_a: Probability distribution for population A is shifted to the right of the probability distribution for population B
$n_A = 6, \quad n_B = 6, \quad \alpha = .05$

c. H_0: Two probability distributions, A and B, are identical
H_a: Probability distribution for population A is shifted to the left of the probability distribution for population B
$n_A = 7, \quad n_B = 10, \quad \alpha = .025$

d. H_0: Two probability distributions, A and B, are identical
H_a: Probability distribution for population A is shifted to the right or left of the probability distribution for population B
$n_A = 20, \quad n_B = 20, \quad \alpha = .05$

8.57 Suppose you want to compare two treatments, A and B. In particular, you wish to determine whether the distribution for population B is shifted to the right of the distribution for population A. You plan to use the Wilcoxon rank sum test.

a. Specify the null and alternative hypotheses you would test.

b. Suppose you obtained the following independent random samples of observations on experimental units subjected to the two treatments:

Sample A: 37, 40, 33, 29, 42, 33, 35, 28, 34 *Sample B*: 65, 35, 47, 52

Conduct a test of the hypotheses described in part **a**. Test using $\alpha = .05$.

8.58 Explain the difference between the one-tailed and two-tailed versions of the Wilcoxon rank sum test for independent random samples.

8.59 Random samples of sizes $n_1 = 16$ and $n_2 = 12$ were drawn from populations 1 and 2, respectively. The measurements obtained are listed in the table.

Population 1				*Population 2*		
9.0	15.6	25.6	31.1	10.1	11.1	13.5
21.1	26.9	24.6	20.0	12.0	18.2	10.3
24.8	16.5	26.0	25.1	9.2	7.0	14.2
17.2	30.1	18.7	26.1	15.8	13.6	13.2

a. Conduct a hypothesis test to determine whether the probability distribution for population 2 is shifted to the left of the probability distribution for population 1. Use $\alpha = .05$.

b. What is the approximate p-value of the test of part **a**?

8.60 Suppose you wish to compare two treatments, A and B, based on independent random samples of 15 observations selected from each of the two populations. If $T_A = 173$, do the data provide sufficient evidence to indicate that distribution A is shifted to the left of distribution B? Test using $\alpha = .05$.

Applying the Concepts

8.61 Are tax assessments fair? One measure of "fairness" is the ratio of a property's assessed value to its market value—or a proxy for market value such as a recent sale price (Freedman, 1985). The accompanying table lists assessment ratios for random samples of 10 properties in neighborhood A and 8 properties in neighborhood B.

Neighborhood A		*Neighborhood B*	
.850	.880	.911	.835
1.060	.895	.770	.800
.910	.844	.815	.793
.813	.965	.748	.796
.737	.875		

a. Use the Wilcoxon rank sum test to investigate the fairness of the assessments between the two neighborhoods. Use $\alpha = .05$ and interpret your findings in the context of the problem.
b. Under what circumstances could the two-sample t-test be used to investigate the fairness issue of part **a**?
c. What assumptions are necessary to assure the validity of the test you conducted in part **a**?

8.62 Recall that the variance of a binomial sample proportion $\hat{p}$ depends on the value of the population parameter p. As a consequence, the variance of a sample percentage, $(100\hat{p})\%$, also depends on p. Thus, if you conduct an unpaired t-test (Section 8.2) to compare the means of two populations of percentages, you may be violating the assumption that $\sigma_1^2 = \sigma_2^2$, upon which the t-test is based. In Exercise 8.27, we used a Student t-test to compare the mean annual percentages of labor turnover between U.S. and Japanese manufacturers of air conditioners. The annual percentage turnover rates for five U.S. and five Japanese plants are shown in the table. Do the data provide sufficient evidence to indicate that the mean annual percentage turnover for U.S. plants exceeds the corresponding mean for Japanese plants? Test using the Wilcoxon rank sum test with $\alpha = .05$.

U.S. Plants	*Japanese Plants*
7.11%	3.52%
6.06%	2.02%
8.00%	4.91%
6.87%	3.22%
4.77%	1.92%

8.63 An educational psychologist claims that the order in which test questions are asked affects a student's ability to answer correctly. To investigate this assertion, a professor randomly divides a class of 13 students into two groups—7 in one group and 6 in the other. The professor prepares one set of test questions but arranges the questions in two different orders. On test A the questions are arranged in order of increasing difficulty (that is, from easiest to most difficult), while on test B the order is reversed. One group of students is given test A, the other test B, and the test score is recorded for each student. The results are as follows:

Test A: 90, 71, 83, 82, 75, 91, 65 *Test B*: 66, 78, 50, 68, 80, 60

Do the data provide sufficient evidence to indicate a difference (a shift in location) in the probability distributions of student scores on the two tests? Test using $\alpha = .05$.

8.64 A major razor blade manufacturer advertises that its twin-blade disposable razor will "get you more shaves" than any single-blade disposable razor on the market. A rival blade company that has been very successful in selling single-blade razors wishes to test this claim. Independent random samples of eight single-blade shavers and eight twin-blade shavers are taken, and the number of shaves that each gets before indicating a preference to change blades is recorded. The results are shown in the table.

Twin Blades		*Single Blades*	
8	15	10	13
17	10	6	14
9	6	3	5
11	12	7	7

a. Do the data support the twin-blade manufacturer's claim? Use $\alpha = .05$.

b. Do you think this experiment was designed in the best possible way? If not, what design might have been better?

c. What assumptions are necessary for the validity of the test you performed in part **a**? Do the assumptions seem reasonable for this application?

8.65 Fourteen rats were used in an experiment aimed at comparing two deprivation schedules, A and B, for their effect on hoarding behavior. An independent sampling design was used, with seven rats randomly assigned to each schedule. At the end of the deprivation period, the rats were permitted free access to food pellets and the number of pellets hoarded (taken but not eaten) during a given time period was recorded. The data are given in the table. Is there sufficient evidence to indicate that rats on one of the deprivation schedules have a greater tendency to hoard than those on the other schedule? Test using $\alpha = .05$.

Schedule A		*Schedule B*	
15	4	5	2
10	9	1	6
5	7	2	3
7		8	

8.66 In a comparison of visual acuity of deaf and hearing children, eye movement rates are taken on 10 deaf and 10 hearing children. (See the table.) A clinical psychologist believes that deaf children have greater visual acuity than hearing children. Test the psychologist's claim by using the data in the table. (The larger a child's eye movement rate, the more visual acuity the child possesses.) Use $\alpha = .05$.

Deaf Children		*Hearing Children*	
2.75	1.95	1.15	1.23
3.14	2.17	1.65	2.03
3.23	2.45	1.43	1.64
2.30	1.83	1.83	1.96
2.64	2.23	1.75	1.37

8.67 Conduct the test in Exercise 8.66 by using the large-sample approximation for the Wilcoxon rank sum test. Compare the results with those found in Exercise 8.66.

8.6 A Nonparametric Test for Comparing Two Populations: Paired Difference Experiments (Optional)

Nonparametric techniques may also be employed to compare two probability distributions when a paired difference design is used. For example, consumer preferences for two competing products are often compared by having each of a sample of consumers rate both products. Thus, the ratings have been paired on each consumer. Here is an example of this type of experiment.

For some paper products, softness of the paper is an important consideration in determining consumer acceptance. One method of determining softness is to have judges give a sample of the products a softness rating. Suppose each of ten judges is given a sample of two products that a company wants to compare. Each judge rates the softness of each product on a scale from 1 to 10, with higher ratings implying a softer product. The results of the experiment are shown in Table 8.7.

TABLE 8.7 Softness Ratings of Paper

Judge	Product A	Product B	Difference (A − B)	Absolute Value of Difference	Rank of Absolute Value
1	6	4	2	2	5
2	8	5	3	3	7.5
3	4	5	−1	1	2
4	9	8	1	1	2
5	4	1	3	3	7.5
6	7	9	−2	2	5
7	6	2	4	4	9
8	5	3	2	2	5
9	6	7	−1	1	2
10	8	2	6	6	10

T_+ = Sum of positive ranks = 46
T_- = Sum of negative ranks = 9

Since this is a paired difference experiment, we analyze the differences between the measurements (see Section 8.3). However, the nonparametric approach requires that we calculate the ranks of the absolute values of the differences between the measurements, i.e., the ranks of the differences after removing any minus signs. *Note that tied absolute differences are assigned the average of the ranks they would receive if they were unequal but successive measurements.* After the absolute differences are ranked, the sum of the ranks of the positive differences of the original measurements,

T_+, and the sum of the ranks of the negative differences of the original measurements, T_-, are computed.

We are now prepared to test the nonparametric hypothesis:

H_0: The probability distributions of the ratings for products A and B are identical.

H_a: The probability distributions of the ratings differ (in location) for the two products. (Note that this is a two-sided alternative and that it implies a two-tailed test.)

Test statistic: T = Smaller of the positive and negative rank sums T_+ and T_-

The smaller the value of T, the greater the evidence to indicate that the two probability distributions differ in location. The rejection region for T can be determined by consulting Table VII in Appendix A (part of the table is shown in Table 8.8). This table gives a value T_0 for both one-tailed and two-tailed tests for each value of n, the number of matched pairs. For a two-tailed test with $\alpha = .05$, we will reject H_0 if $T \leq T_0$. You can see in Table 8.8 that the value of T_0 that locates the boundary of the

TABLE 8.8 Reproduction of Part of Table VII of Appendix A: Critical Values for the Wilcoxon Paired Difference Signed Rank Test

One-Tailed	Two-Tailed	$n = 5$	$n = 6$	$n = 7$	$n = 8$	$n = 9$	$n = 10$
$\alpha = .05$	$\alpha = .10$	1	2	4	6	8	11
$\alpha = .025$	$\alpha = .05$		1	2	4	6	8
$\alpha = .01$	$\alpha = .02$			0	2	3	5
$\alpha = .005$	$\alpha = .01$				0	2	3
		$n = 11$	$n = 12$	$n = 13$	$n = 14$	$n = 15$	$n = 16$
$\alpha = .05$	$\alpha = .10$	14	17	21	26	30	36
$\alpha = .025$	$\alpha = .05$	11	14	17	21	25	30
$\alpha = .01$	$\alpha = .02$	7	10	13	16	20	24
$\alpha = .005$	$\alpha = .01$	5	7	10	13	16	19
		$n = 17$	$n = 18$	$n = 19$	$n = 20$	$n = 21$	$n = 22$
$\alpha = .05$	$\alpha = .10$	41	47	54	60	68	75
$\alpha = .025$	$\alpha = .05$	35	40	46	52	59	66
$\alpha = .01$	$\alpha = .02$	28	33	38	43	49	56
$\alpha = .005$	$\alpha = .01$	23	28	32	37	43	49
		$n = 23$	$n = 24$	$n = 25$	$n = 26$	$n = 27$	$n = 28$
$\alpha = .05$	$\alpha = .10$	83	92	101	110	120	130
$\alpha = .025$	$\alpha = .05$	73	81	90	98	107	117
$\alpha = .01$	$\alpha = .02$	62	69	77	85	93	102
$\alpha = .005$	$\alpha = .01$	55	61	68	76	84	92

rejection region for the judges' ratings for $\alpha = .05$ and $n = 10$ pairs of observations is 8. Thus, the rejection region for the test (see Figure 8.10) is

Rejection region: $T \leq 8$ for $\alpha = .05$

Since the smaller rank sum for the paper data, $T_- = 9$, does not fall within the rejection region, the experiment has not provided sufficient evidence to indicate that the two paper products differ with respect to their softness ratings at the $\alpha = .05$ level.

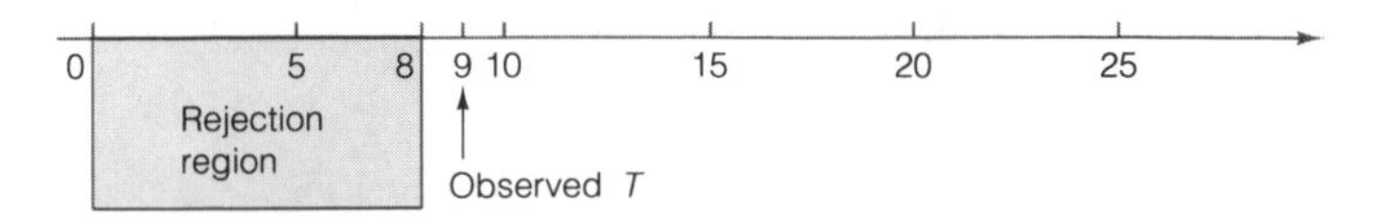

FIGURE 8.10 ▶ **Rejection region for paired difference experiment**

Wilcoxon Signed Rank Test for a Paired Difference Experiment

One-Tailed Test	*Two-Tailed Test*
H_0: Two sampled populations have identical probability distributions	H_0: Two sampled populations have identical probability distributions
H_a: The probability distribution for population A is shifted to the right of that for population B	H_a: The probability distribution for population A is shifted to the right *or* to the left of that for population B
Test statistic: T_-, the negative rank sum (we assume the differences are computed by subtracting each paired B measurement from the corresponding A measurement)	*Test statistic:* T, the smaller of the positive and negative rank sums, T_+ and T_-
Rejection region: $T_- \leq T_0$, where T_0 is found in Table VII (in Appendix A) for the one-tailed significance level α and the number of untied pairs, n.	*Rejection region:* $T \leq T_0$, where T_0 is found in Table VII (in Appendix A) for the two-tailed significance level α and the number of untied pairs, n.

[*Note:* If the alternative hypothesis is that the probability distribution for A is shifted to the left of B, we use T_+ as the test statistic and reject H_0 if $T_+ \leq T_0$.]

Assumptions: 1. The sample of differences is randomly selected from the population of differences.
2. The probability distribution from which the sample of paired differences is drawn is continuous.

Ties: Assign tied absolute differences the average of the ranks they would receive if they were unequal but occurred in successive order. For example, if the third-ranked and fourth-ranked differences are tied, assign both a rank of $(3 + 4)/2 = 3.5$.

Note that if a significance level of $\alpha = .10$ had been used, the rejection region would have been $T \leq 11$ and we would have rejected H_0. In other words, the samples do provide evidence that the probability distributions of the softness ratings differ at the $\alpha = .10$ significance level.

The Wilcoxon signed rank test is summarized in the preceding box. Note that the difference measurements are assumed to have a continuous probability distribution so that the absolute differences will have unique ranks. Although tied (absolute) differences can be assigned ranks by averaging, the number of ties should be small relative to the number of observations to assure the validity of the test.

EXAMPLE 8.12

Suppose the police commissioner in a small community must choose between two plans for patrolling the town's streets. Plan A, the less expensive plan, uses voluntary citizen groups to patrol certain high-risk neighborhoods. In contrast, plan B would utilize police patrols. As an aid in reaching a decision, both plans are examined by 10 trained criminologists, each of whom is asked to rate the plans on a scale from 1 to 10. (High ratings imply a more effective crime prevention plan.) The city will adopt plan B (and hire extra police) only if the data provide sufficient evidence that criminologists tend to rate plan B more effective than plan A.

The results of the survey are shown in Table 8.9. Do the data provide evidence at the $\alpha = .05$ level that the distribution of ratings for plan B lies above that for plan A?

TABLE 8.9 Effectiveness Ratings by 10 Qualified Crime Prevention Experts

Crime Prevention Expert	Plan A	Plan B	Difference (A − B)	Rank of Absolute Difference
1	7	9	−2	4.5
2	4	5	−1	2
3	8	8	0	(Eliminated)
4	9	8	1	2
5	3	6	−3	6
6	6	10	−4	7.5
7	8	9	−1	2
8	10	8	2	4.5
9	9	4	5	9
10	5	9	−4	7.5
			Positive rank sum = T_+ =	15.5

Solution

The null and alternative hypotheses are

H_0: The two probability distributions of effectiveness ratings are identical

H_a: The effectiveness ratings of the more expensive plan (B) tend to exceed those of plan A

Observe that the alternative hypothesis is one-sided (i.e., we only wish to detect a shift in the distribution of the B ratings to the right of the distribution of A ratings) and therefore it implies a one-tailed test of the null hypothesis (see Figure 8.11). If the alternative hypothesis is true, the B ratings will tend to be larger than the paired A ratings, more negative differences in pairs will occur, T_- will be large, and T_+ will be small. Because Table VII is constructed to give lower-tail values of T_0, we will use T_+ as the test statistic and reject H_0 for $T_+ \leq T_0$.

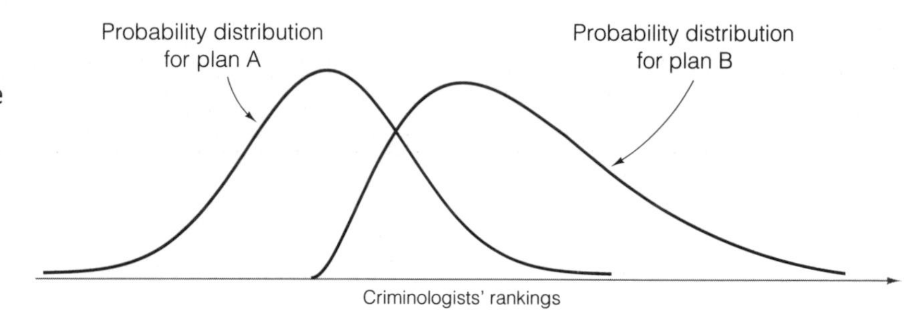

FIGURE 8.11 ► The alternative hypothesis for Example 8.12. We expect T_+ to be small.

The differences in ratings for the pairs (A − B) are shown in Table 8.9. Note that one of the differences equals 0. Consequently, we eliminate this pair from the ranking and reduce the number of pairs to $n = 9$. Looking in Table VII, for a one-tailed test with $\alpha = .05$ and $n = 9$, we have $T_0 = 8$. Therefore, the test statistic and rejection region for the test are:

Test statistic: T_+, the positive rank sum

Rejection region: $T_+ \leq 8$

Summing the ranks of the positive differences from Table 8.9, we find $T_+ = 15.5$. Since this value exceeds the critical value, $T_0 = 8$, we conclude that this sample provides insufficient evidence at the $\alpha = .05$ level to support the alternative hypothesis. The commissioner *cannot* conclude that the plan utilizing police patrols tends to be rated higher than the plan using citizen volunteers. That is, on the basis of this study, extra police will not be hired.

As is the case for the rank sum test for independent samples, the sampling distribution of the signed rank statistic can be approximated by a normal distribution when the number n of paired observations is large (say $n \geq 25$). The large-sample z-test is summarized in the box.

Wilcoxon Signed Rank Test for a Paired Difference Experiment: Large Sample

One-Tailed Test

H_0: Two sampled populations have identical probability distributions

H_a: The probability distribution for population A is shifted to the right of that for population B

Test statistic: $$z = \frac{T_+ - \frac{n(n+1)}{4}}{\sqrt{\frac{n(n+1)(2n+1)}{24}}}$$

Rejection region: $z > z_\alpha$

Assumptions: $n \geq 25$

Two-Tailed Test

H_0: Two sampled populations have identical probability distributions

H_a: The probability distribution for population A is shifted to the right *or* to the left of that for population B

Test statistic: $$z = \frac{T_+ - \frac{n(n+1)}{4}}{\sqrt{\frac{n(n+1)(2n+1)}{24}}}$$

Rejection region: $z < -z_{\alpha/2}$ or $z > z_{\alpha/2}$

Assumptions: $n \geq 25$

Exercises 8.68–8.80

Learning the Mechanics

8.68 Specify the test statistic and the rejection region for the Wilcoxon signed rank test for the paired difference design in each of the following situations:

a. H_0: Two probability distributions, A and B, are identical
   H_a: Probability distribution for population A is shifted to the right or left of probability distribution for population B
   $n = 20, \quad \alpha = .10$

b. H_0: Two probability distributions, A and B, are identical
   H_a: Probability distribution for population A is shifted to the right of the probability distribution for population B
   $n = 39, \quad \alpha = .05$

c. H_0: Two probability distributions, A and B, are identical
   H_a: Probability distribution for population A is shifted to the left of the probability distribution for population B
   $n = 7, \quad \alpha = .005$

8.69 Suppose you want to test a hypothesis that two treatments, A and B, are equivalent against the alternative hypothesis that the responses for A tend to be larger than those for B. You plan to use a paired difference experiment and to analyze the resulting data using the Wilcoxon signed rank test.

a. Specify the null and alternative hypotheses you would test.
b. Suppose the paired difference experiment yielded the data in the table. Conduct the test of part **a**. Test using $\alpha = .025$.

Pair	*Treatment* A	B	*Pair*	*Treatment* A	B
1	54	45	6	77	75
2	60	45	7	74	63
3	98	87	8	29	30
4	43	31	9	63	59
5	82	71	10	80	82

8.70 Explain the difference between the one- and two-tailed versions of the Wilcoxon signed rank test for the paired difference experiment.

8.71 In order to conduct the Wilcoxon signed rank test, why do we need to assume the probability distribution of differences is continuous?

8.72 A random sample of nine pairs of measurements is shown in the table.

Pair	*Sample Data from Population 1*	*Sample Data from Population 2*
1	8	7
2	10	1
3	6	4
4	10	10
5	7	4
6	8	3
7	4	6
8	9	2
9	8	4

a. Use the Wilcoxon signed rank test to determine whether the data provide sufficient evidence to indicate that the probability distribution for population 1 is shifted to the right of the probability distribution for population 2. Test using $\alpha = .05$.
b. Use the Wilcoxon signed rank test to determine whether the data provide sufficient evidence to indicate that the probability distribution for population 1 is shifted either to the right or to the left of the probability distribution for population 2. Test using $\alpha = .05$.

8.73 A paired difference experiment with $n = 30$ pairs yielded $T_+ = 354$.
a. Specify the null and alternative hypotheses that should be used in conducting a hypothesis test to determine whether the probability distribution for population A is located to the right of that for population B.
b. Conduct the test of part **a** using $\alpha = .05$.
c. What is the approximate p-value of the test of part **b**?
d. What assumptions are necessary to assure the validity of the test you performed in part **b**?

Applying the Concepts

8.74 Which is the more effective means of dealing with complex group problem-solving tasks—face-to-face meetings or video teleconferencing? In an experiment similar to the one described in this exercise, Daniel K. Rosetti and Theodore J. Surynt of Stetson University concluded that video teleconferencing may be the more effective method. Ten groups of four people each were randomly assigned both to a specific communication setting (face-to-face or video teleconferencing) and to one of two specific complex problems. Upon completion of the problem-solving task, the same groups were placed in the alternative communication setting and asked to complete the second problem-solving task. The percentage of each problem task correctly completed was recorded for each group, with the results given in the accompanying table.

Group	*Face-to-Face*	*Video Teleconferencing*	*Group*	*Face-to-Face*	*Video Teleconferencing*
1	65%	75%	6	85%	90%
2	82	80	7	98	98
3	54	60	8	35	40
4	69	65	9	85	89
5	40	55	10	70	80

a. What type of experimental design was used in this study?

b. Specify the null and alternative hypotheses that should be used in determining whether the data provide sufficient evidence to conclude (as did Rosetti and Surynt) that the problem-solving performance of video teleconferencing groups is superior to that of groups that interact face-to-face.

c. Conduct the hypothesis test of part **b**. Use $\alpha = .05$. Interpret the results of your test in the context of the problem.

d. What is the p-value of the test in part **c**?

8.75 According to the American Bar Association, in 1982 there were 612,593 lawyers in the United States. About 70% of these lawyers were in private practice; about 15% worked in government as judges, prosecutors, legislators, etc.; and about 9% worked for businesses. Because of mushrooming government regulation, high outside legal fees, and complex litigation, the number of corporate lawyers has been growing at a rapid pace. The data shown in the table are the average salaries for lawyers with 8 years experience for 10 U.S. cities.

City	*Corporate Lawyers*	*Lawyers with Law Firms*
Atlanta	$45,500	$45,500
Chicago	43,000	48,000
Cincinnati	43,500	45,000
Dallas/Ft. Worth	49,500	46,500
Los Angeles	47,000	60,000
Milwaukee	37,500	50,000
Minneapolis/St. Paul	47,500	43,500
New York	43,500	54,000
Pittsburgh	42,000	44,000
San Francisco	47,500	59,500

Source: Reprinted by permission of Avon Books from *The American Almanac of Jobs and Salaries* by John W. Wright. Copyright 1982, 1984, 1987 by John W. Wright.

a. Use the Wilcoxon signed rank test to determine whether the data provide sufficient evidence to conclude that the salaries of corporate lawyers differ from those of lawyers working for law firms. Test using $\alpha = .05$.

b. Under what circumstances would it be appropriate to conduct the test in part **a** using the paired difference *t*-test?

8.76 A 1974 Supreme Court decision (*Milliken v. Bradley*) held that desegregation could not extend beyond the boundary of the school systems that were found to be segregated. One dissenting justice feared that the decision would trigger white flight, the migration of white families out of the inner cities to the suburbs. In discussing this decision, Charles T. Clotfelter (1976) presented data showing the percentages of minority students for 1968, 1970, and 1972 in the school systems of 12 large cities. Examine the last two columns of the table, which give the percentage change in white students over two time periods, 1968–1970 and 1970–1972. Was the percentage change in white students larger in the time period 1970–1972 than for the comparable period, 1968–1970? Test using a Wilcoxon signed rank test with $\alpha = .05$.

Racial Change in Selected City School Systems, 1968–1972

City	Percentage Minority			Percentage Change in White Students[a]	
	1968	1970	1972	1968–1970	1970–1972
Boston	31.5	35.9	40.4	−3.9	−7.4
Philadelphia	61.8	63.6	64.8	−5.7	−2.2
Baltimore	65.1	67.1	69.3	−5.6	−9.3
St. Louis	63.8	65.9	69.1	−9.4	−13.8
Chicago	62.3	65.4	69.2	−9.0	−14.7
Detroit	60.7	65.5	69.5	−15.6	−14.0
Atlanta	61.8	68.7	77.4	−22.3	−34.4
Charlotte	29.5	31.1	32.8	−3.1	−5.6
Jacksonville	28.2	29.4	32.6	−1.8	−11.4
Houston	46.7	50.6	56.4	−9.1	−17.5
San Francisco	58.8	63.1	68.2	−13.5	−22.4
San Diego	23.9	24.6	26.3	−1.1	−5.5

[a]Percentages based on beginning years.
Source: U.S. Dept. of Health, Education, and Welfare, Office for Civil Rights, Directory of Public Elementary and Secondary Schools in Selected Districts, Fall 1968, Fall 1970, and Fall 1972.

8.77 Hypoglycemia is a condition in which blood sugar is below normal limits. To compare two compounds, X and Y, for treating hypoglycemia, each compound is applied to half the diaphragms of each of seven white mice. Blood glucose uptake in milligrams per gram of tissue is measured for each half, producing the results listed in the accompanying table. Do the data provide sufficient evidence to indicate that one of the compounds tends to produce higher blood sugar uptake readings than the other? Test using $\alpha = .10$.

Mouse	*Compound* X	Y	*Mouse*	*Compound* X	Y
1	4.7	5.1	5	7.0	6.1
2	3.3	4.6	6	4.7	4.1
3	8.5	8.7	7	5.2	5.1
4	3.9	3.6			

8.78 Children completing the sixth grade at a school located in a large city have the choice of going to one of two junior high schools, A or B. Members of the school board want to compare the academic effectiveness of the two schools. The parents of six sets of identical twins agree to send one child to school A and the other to school B. Since each set of twins is in the same class at each grade level through the sixth grade, a paired difference design could be employed. Near the end of the ninth grade, an achievement test is given to each child in the experiment. The results are given in the table. Test to determine whether there is evidence of a difference (shift in location) in the probability distributions of achievement test scores at the two schools. Use $\alpha = .10$.

Twin Pair	*School* A	B
1	65	69
2	72	72
3	86	74
4	50	52
5	60	47
6	81	72

8.79 Twelve sets of identical twins are given psychological tests to determine whether the firstborn of the twins tends to be more aggressive than the secondborn. The results are shown in the table, where the higher score indicates greater aggressiveness. Do the data provide sufficient evidence to indicate that the firstborn of a pair of twins is more aggressive than the other? Test using $\alpha = .05$.

Set	*Firstborn*	*Secondborn*	*Set*	*Firstborn*	*Secondborn*
1	86	88	7	77	65
2	71	77	8	91	90
3	77	76	9	70	65
4	68	64	10	71	80
5	91	96	11	88	81
6	72	72	12	87	72

8.80 In Exercise 8.44 we compared matched pairs of measurements on the exhalation rate (a measure of radiation) of 15 soil samples from waste gypsum and phosphate mounds in Polk County, Florida. Each soil sample was measured for exhalation rate by the Polk County Health Department (PCHD) and the Eastern Environmental Radiation Facility (EERF). The data are reproduced in the table at the top of page 432. Do the data provide sufficient evidence to indicate that one of the measuring facilities, PCHD or EERF, tends to read higher or lower than the other? Test using the Wilcoxon signed rank test with $\alpha = .05$.

Charcoal Canister No.	*PCHD*	*EERF*	*Charcoal Canister No.*	*PCHD*	*EERF*
71	1,709.79	1,479.0	85	393.55	187.7
58	357.17	257.8	46	880.84	630.4
84	1,150.94	1,287.0	4	2,996.49	3,707.0
91	1,572.69	1,395.0	20	2,367.40	2,791.0
44	558.33	416.5	36	599.84	706.8
43	4,132.28	3,993.0	42	538.37	618.5
79	1,489.86	1,351.0	55	2,770.23	2,639.0
61	3,017.48	1,813.0			

Source: Horton, T. R. "Preliminary radiological assessment of radon exhalation from phosphate gypsum piles and inactive uranium mill tailings piles." EPA-520/5-79-004. Washington, D. C.: Environmental Protection Agency, 1979.

Summary

We have presented various techniques for using the information in two or more samples to make inferences about the difference between two or more population means. As you would expect, we are able to make reliable inferences with fewer assumptions about the sampled populations when the sample sizes are large. When we cannot take large samples from the populations, the **two-sample *t* statistic** permits us to use the limited sample information to make inferences about the **difference between two means** when the assumptions of normality and equal population variances are at least approximately true. The **paired difference experiment** offers the possibility of increasing the information about $(\mu_1 - \mu_2)$ by pairing similar observational units to control variability. In designing a paired difference experiment, we expect that the reduction in variability will more than compensate for the loss in degrees of freedom.

For situations when the assumptions required to conduct a *t*-test are invalid (e.g., nonnormal populations or unequal variances), **nonparametric procedures** are available. The **Wilcoxon rank sum test** (optional Section 8.5) can be used to compare two populations based on an independent sampling experiment, and the **Wilcoxon signed rank test** (optional Section 8.6) can be used for a paired difference experiment.

Supplementary Exercises 8.81–8.101

Learning the Mechanics

Note: *List the assumptions necessary to ensure the validity of the statistical procedures you use to work these exercises. Starred (*) exercises refer to optional sections.*

8.81 Independent random samples were selected from two normally distributed populations with means μ_1 and μ_2, respectively. The sample sizes, means, and variances are shown in the table.

Sample 1	*Sample 2*
$n_1 = 12$	$n_2 = 14$
$\bar{x}_1 = 17.8$	$\bar{x}_2 = 15.3$
$s_1^2 = 74.2$	$s_2^2 = 60.5$

a. Test H_0: $(\mu_1 - \mu_2) = 0$ against H_a: $(\mu_1 - \mu_2) > 0$. Use $\alpha = .05$.
b. Form a 99% confidence interval for $(\mu_1 - \mu_2)$.
c. How large must n_1 and n_2 be if you wish to estimate $(\mu_1 - \mu_2)$ to within 2 units with 99% confidence? Assume that $n_1 = n_2$.

8.82 Two independent random samples are taken from two populations. The results of these samples are summarized in the table.

Sample 1	*Sample 2*
$n_1 = 135$	$n_2 = 148$
$\bar{x}_1 = 12.2$	$\bar{x}_2 = 8.3$
$s_1^2 = 2.1$	$s_2^2 = 3.0$

a. Form a 90% confidence interval for $(\mu_1 - \mu_2)$.
b. Test H_0: $(\mu_1 - \mu_2) = 0$ against H_a: $(\mu_1 - \mu_2) \neq 0$. Use $\alpha = .01$.
c. What sample sizes would be required if you wish to estimate $(\mu_1 - \mu_2)$ to within .2 with 90% confidence? Assume that $n_1 = n_2$.

8.83 A random sample of five pairs of observations were selected, one of each pair from a population with mean μ_1, the other from a population with mean μ_2. The data are shown in the accompanying table.

Pair	*Value from Population 1*	*Value from Population 2*
1	28	22
2	31	27
3	24	20
4	30	27
5	22	20

a. Test the null hypothesis H_0: $\mu_D = 0$ against H_a: $\mu_D \neq 0$, where $\mu_D = \mu_1 - \mu_2$. Use $\alpha = .05$.
b. Form a 95% confidence interval for μ_D.
c. When are the procedures you used in parts **a** and **b** valid?

8.84 List the assumptions necessary for each of the following inferential techniques:
a. Large-sample inferences about the difference $(\mu_1 - \mu_2)$ between population means using a two-sample z statistic
b. Small-sample inferences about $(\mu_1 - \mu_2)$ using an independent samples design and a two-sample t statistic
c. Small-sample inferences about $(\mu_1 - \mu_2)$ using a paired difference design and a single-sample t statistic to analyze the differences
d. Large-sample inferences about the differences $(p_1 - p_2)$ between binomial proportions using a two-sample z statistic

*8.85 Two independent random samples produced the measurements listed in the table. Do the data provide sufficient evidence to conclude that there is a difference between the locations of the probability distributions for the sampled populations? Test using $\alpha = .05$.

Sample from Population 1		*Sample from Population 2*	
1.2	1.0	1.5	1.9
1.9	1.8	1.3	2.7
.7	1.1	2.9	3.5
2.5			

Applying the Concepts

8.86 Nontraditional university students, generally defined as those at least 25 years old, comprise an increasingly large proportion of undergraduate student bodies at most universities. A recent study reported in the *College Student Journal* (Sheehan, McMenamin, and McDevitt, 1992) compared traditional and nontraditional students on a number of factors, including grade point average (GPA). The table below summarizes the information from the sample.

GPA	*Traditional Students*	*Nontraditional Students*
n	94	73
$\bar{x}$	2.90	3.50
s	.50	.50

a. What are the appropriate null and alternative hypotheses if we want to test whether the mean GPAs of traditional and nontraditional students differ?
b. Conduct the test using $\alpha = .01$, and interpret the result.
c. What assumptions are necessary to assure the validity of the test?

8.87 A pupillometer is a device used to observe changes in an individual's pupil dilations as he or she is exposed to different visual stimuli. Since there is a direct correlation between the amount an individual's pupil dilates and his or her interest in the stimuli, marketing organizations sometimes use pupillometers to help them evaluate potential consumer interest in new products, alternative package designs, and other factors. The Design and Market Research Laboratories of the Container Corporation of America used a pupillometer to evaluate consumer reaction to different silverware patterns for one of its clients. Suppose 15 consumers were chosen at random, and each was shown two different silverware patterns. The pupillometer readings for each consumer, with the means and standard deviations of each sample of observations and their differences, are shown in the tables (in millimeters).

Consumer	*Pattern 1*	*Pattern 2*
1	1.00	.80
2	.97	.66
3	1.45	1.22
4	1.21	1.00
5	.77	.81
6	1.32	1.11
7	1.81	1.30
8	.91	.32
9	.98	.91
10	1.46	1.10
11	1.85	1.60
12	.33	.21
13	1.77	1.50
14	.85	.65
15	.15	.05

	$\bar{x}$	s
Pattern 1	1.12	.50
Pattern 2	.88	.45
Difference (1 − 2)	.24	.16

a. Which type of experiment does this represent—independent samples or paired difference? Explain.
b. Use a 90% confidence interval to estimate the difference in mean pupil dilation per consumer for silverware patterns 1 and 2. Interpret the confidence interval, assuming that the pupillometer indeed measures consumer interest.
c. Test the hypothesis that the mean dilation differs for the two patterns. Use $\alpha = .10$. Does the test conclusion support your interpretation of the confidence interval in part **b**?
d. What assumptions are necessary to assure the validity of the inferences in parts **b** and **c**?

8.88 Lack of motivation is a problem of many students in inner-city schools. To cope with this problem, an experiment was conducted to determine whether motivation could be improved by allowing students greater choice in the structures of their curricula. Two schools with similar student populations were chosen, and 50 students were randomly selected from each to participate in the experiment. School A permitted its 50 students to choose only the courses they wanted to take. School B permitted its students to choose their courses and also to choose when and from which instructors to take the courses. The measure of student motivation was the number of times each student was absent from or late for a class during a 20-day period. The means and variances for the two samples are shown in the table. Do the data provide sufficient evidence to indicate that students from school B were late or absent less frequently than those from school A? (Use $\alpha = .10$.)

School A	*School B*
$\bar{x}_A = 20.5$	$\bar{x}_B = 19.6$
$s_A^2 = 26.2$	$s_B^2 = 24.1$

8.89 Was the average amount spent by firms in the electronics industry on company-sponsored research and development (R&D) higher in 1989 than it was in 1988? The table at the top of page 436 lists R&D expenditures (in millions of dollars) for a sample of firms in the electronics industry.

Firm	*1988*	*1989*
Adams-Russell	6.3	5.1
Harris	116.9	104.0
Aydin	8.5	6.6
Andrew	14.1	17.0
Compudyn	2.3	1.8
Raytheon	271.0	274.7
Varian Associates	80.2	83.1
General Instr.	37.5	46.9

Source: *Business Week*, Special Issue on Innovation in America, 1989 and 1990.

a. A securities analyst who follows the electronic industry believes R&D expenditures have increased. Do the data support the analyst's beliefs? Test using $\alpha = .10$.
b. In the context of this exercise, what are the Type I and Type II errors associated with the hypothesis test of part **a**?
c. What assumptions must hold in order for your test of part **a** to be valid?

8.90 Refer to Exercise 8.89. Use a 95% confidence interval to estimate the mean difference between 1988 and 1989 R&D expenditures. Interpret the interval.

***8.91** An experiment was conducted to compare two print types, A and B, to determine whether type A is easier to read. Ten subjects were randomly divided into two groups of five. Each subject was given the same material to read, one group receiving the material printed with type A, the other group receiving print type B. The times necessary for each subject to read the material (in seconds) are shown below.

Type A: 95, 122, 101, 99, 108 *Type B*: 110, 102, 115, 112, 120

Do the data provide sufficient evidence to indicate that print type A is easier to read? Test using $\alpha = .05$.

***8.92** David J. Teece (1981) used the Wilcoxon signed rank test to examine the differential performance between organizations using an innovative decentralized and divisional organization structure known as the M-form, and their principal competitors. He identified the first firm in each of 20 industries that adopted the M-form structure. The principal competitor of each of these firms was identified. To qualify for inclusion in the study, the competing firm must also have adopted the M-form structure, but at a later date. A total of 14 pairs of firms qualified for the sample. The difference in performance within the pairs of firms was measured over two 3-year time periods—a "before" period in which only the M-form originator in each pair used the M-form, and an "after" period in which both firms had M-form structures. Differential performance in each time period was measured as the average difference in yearly return on stockholders' equity, where the differences were formed by subtracting the competitor's return on equity from the M-form originator's. The performance data appear in the following table.

Industry	*Average Difference in Yearly Return on Equity*	
	Before	*After*
Grocery	25.70%	10.78%
Chemicals	6.26	−.39
Textiles	5.84	1.94
Aluminum	4.16	.56
Meat packing	−2.43	−5.80
Packaged foods	.77	−1.18
Can manufacturing	3.79	2.49
Grain milling	3.86	4.81
Petroleum	3.29	2.38
Tires	−3.36	−2.46
Autos	13.48	14.38
Electrical equipment	−10.85	−10.40
Tobacco	1.50	1.90
Retail department stores	12.32	12.37

Source: Teece, D. J. "Internal organization and economic performance: An empirical analysis of the profitability of principal firms," *The Journal of Industrial Economics*, Vol. 30, No. 2, December 1981, pp. 173–199.

a. Teece concluded that ". . . the M-form innovation has been shown to display a statistically significant impact on firm performance." Do you agree? Test using $\alpha = .05$.

b. What is the approximate p-value of the test of part **a**?

c. What assumptions are necessary to assure the validity of the test procedure you used in part **a**?

8.93 The interocular pressure of glaucoma patients is often reduced by treatment with adrenaline. To compare a new synthetic drug with adrenaline, seven glaucoma patients were treated with both drugs, one eye with adrenaline and one with the synthetic drug. The reduction in pressure in each eye was then recorded, as shown in the table. Do the data provide sufficient evidence to indicate a difference in the mean reductions in eye pressure for the two drugs? Test using $\alpha = .10$.

Patient	*Adrenaline*	*Synthetic*
1	3.5	3.2
2	2.6	2.8
3	3.0	3.1
4	1.9	2.4
5	2.9	2.9
6	2.4	2.2
7	2.0	2.2

8.94 A recent study in *The Journal of Psychology & Marketing* (Jackson et al., 1992) investigates the degree to which American consumers are concerned about product tampering. Random samples of male and female consumers were asked to rate their concern about product tampering on a scale of 1 (little or no concern) to 9 (very concerned).

a. What are the appropriate null and alternative hypotheses to determine whether a difference exists in the mean level of concern about product tampering between men and women? Define any symbols you use.
b. The statistics reported include those shown here. Interpret these results.

```
MEAN SCORES:  MEN   = 3.209
              WOMEN = 3.923

     Z = -2.69   TWO-TAILED P-VALUE = .0072
```

c. What assumptions are necessary to assure the validity of this test?

8.95 Does the time of day during which one works affect job satisfaction? A study in *The Journal of Occupational Psychology* (Barton and Folkard, 1991) examined differences in job satisfaction between day-shift and night-shift nurses. Nurses' satisfaction with their hours of work, free time away from work, and breaks during work were measured. The following table shows the mean scores for each measure of job satisfaction (higher scores indicate greater satisfaction), along with the observed significance level comparing the means for the day-shift and night-shift samples:

	Mean Satisfaction		
	Day Shift	*Night Shift*	*p-Value*
Satisfaction with:			
Hours of work	3.91	3.56	.813
Free time	2.55	1.72	.047
Breaks	2.53	3.75	.0073

a. Specify the null and alternative hypotheses if we wish to test whether a difference in job satisfaction exists between day-shift and night-shift nurses on each of the three measures. Define any symbols you use.
b. Interpret the p-value for each of the tests. (Each of the p-values in the table is two-tailed.)
c. Assume that each of the tests is based on small samples of nurses from each group. What assumptions are necessary for the tests to be valid?

8.96 A physiologist wishes to study the effect of birth-control pills on exercise capacity. Five female subjects who have never taken the pill have their maximal oxygen uptake measured (in milliliters per kilogram of their body weight) during a treadmill session. The five subjects then take the pill for a specified length of time and their uptakes are measured again, as given in the table. Do the data provide sufficient evidence to indicate that the mean maximal oxygen uptake after taking birth-control pills is less than the mean uptake before taking the pill? Use $\alpha = .01$.

Subject	*Maximal Oxygen Uptake*	
	Before	*After*
1	35.0	29.5
2	36.5	33.5
3	36.0	32.0
4	39.0	36.5
5	37.5	35.0

8.97 The National Football League (NFL) Rules committee in 1974 changed the way in which the football was turned over following a missed field goal. Before 1974, the ball was always placed on the 20-yard line for the start of the next possession. Since 1974, the ball has been placed at the line of scrimmage from which the missed field goal was attempted. A recent article in the *Sociology of Sport Journal* (Harris, 1992) hypothesizes that this rule change made accurate placekickers more valuable to NFL teams. To test the theory, salary data were collected for the five statistically most accurate placekickers in the league from 1968 through 1986. The data in the table present the ratio of the average salary of the five most accurate placekickers divided by the entire league mean salary.

Before Rule Change		*After Rule Change*			
1968	.710	1974	.718	1981	.719
1969	.721	1975	.697	1982	.718
1970	.701	1976	.722	1983	.696
1971	.714	1977	.707	1984	.711
1972	.729	1978	.721	1985	.708
1973	.709	1979	.704	1986	.704
		1980	.703		

a. What are the appropriate null and alternative hypotheses for testing that accurate placekickers became more valuable (relative to players in other positions) after the 1974 rule change took effect? Define any symbols you use.
b. Conduct the test and interpret the result. Use $\alpha = .10$.
c. Use a statistical software package to check the result you obtained in part **b**. Find and interpret the p-value in the computer output.
d. What assumptions must be made to assure the validity of the test?

8.98 How does gender affect the type of advertising that proves to be most effective? An article in *The Journal of Advertising Research* (Prokash, 1990) makes reference to numerous studies that conclude males tend to be more competitive with others than with themselves. To apply this to advertising, the author creates two ads promoting a new brand of soft drink:

Ad 1: Four men are shown competing in racquetball
Ad 2: One man is shown competing against himself in racquetball

The author hypothesized that the first ad will be more effective when shown to males. To test this hypothesis, 43 males were shown both ads and asked to measure their attitude toward the advertisement (Aad), their attitude toward the brand of soft-drink (Ab), and their intention to purchase the soft drink (Intention). Each variable was measured using a 7-point scale, with higher scores indicating a more favorable attitude. The results are shown below:

	Sample Means		
	Aad	Ab	Intention
Ad 1	4.465	3.311	4.366
Ad 2	4.150	2.902	3.813
Level of significance	$p = .091$	$p = .032$	$p = .050$

a. What are the appropriate null and alternative hypotheses to test the author's research hypothesis? Define any symbols you use.
b. Based on the information provided about this experiment, do you think this is an independent samples experiment or a paired difference experiment? Explain.
c. Interpret the p-value for each test.
d. What assumptions are necessary for the validity of the tests?

8.99 The state of Florida now requires all high school students to pass a literacy test before they receive a high school diploma. A student who fails the test can enroll in a refresher course and retake the test at a later date. To evaluate the effectiveness of the refresher course, eight students' test scores were compared, before and after, with the results shown in the table. Do the data provide sufficient evidence to conclude that the mean test score has increased? (Use $\alpha = .05$.)

Student	*Before*	*After*
1	45	49
2	52	50
3	63	70
4	68	71
5	57	53
6	55	61
7	60	62
8	59	67

8.100 A number of computer programs are available to conduct two-sample tests of hypotheses to compare the means of two populations, for both independent and paired samples. Most of these report both the test statistic and the observed significance level for the test, but some report only the observed significance level of the test. Suppose you use one of these programs to test the null hypothesis H_0: $(\mu_1 - \mu_2) = 0$ against H_a: $(\mu_1 - \mu_2) \neq 0$ with independent samples of size 12 and 10, respectively. Assuming you are using $\alpha = .05$, what conclusion would you reach in each of the following instances of an observed significance level reported by the program?

a. P-VALUE = .0429 b. P-VALUE = .1984
c. P-VALUE = .0001 d. P-VALUE = .0344
e. P-VALUE = .0545 f. P-VALUE = .9633
g. You should always be sure that the program is performing the calculations correctly, especially programs with which you do not have much experience. Even programs that perform the calculations correctly usually do not remind you of the assumptions necessary for the validity of the procedure. What assumptions are necessary for this test?

8.101 Some power plants are located near rivers or oceans so the water can be used for cooling their condensers. As part of an environmental impact study, suppose a power company wants to estimate the mean difference in water temperature between the discharge of its plant and the offshore waters. How many sample measurements must be taken at each site to obtain a 95% confidence interval of width .4°C? Assume the range in readings will be about 4°C at each site and the same number of readings will be taken at each site.

On Your Own

We have now discussed two methods of collecting data to compare two population means. In many experimental situations a decision must be made either to collect two independent samples or to conduct a paired difference experiment. The importance of this decision cannot be overemphasized, since the amount of information obtained and the cost of the experiment are both directly related to the method of experimentation that is chosen.

Choose two populations (pertinent to your major area) that have unknown means and for which you could both collect two independent samples and collect paired observations. Before conducting the experiment, state which method of sampling you think will provide more information (and why). To compare the two methods, first perform the independent sampling procedure by collecting 10 observations from each population (a total of 20 measurements), and then perform the paired difference experiment by collecting 10 pairs of observations.

Construct two 95% confidence intervals, one for each experiment you conduct. Which method provides the narrower confidence interval and thus more information on this performance of the experiment? Does this result agree with your preliminary expectations?

Using the Computer

Select two of the census regions given in Appendix B, and consider the sports purchasing index. Suppose a marketing firm wants to target one of the two regions for a sports magazine marketing campaign.

a. Treat the sports index measurements for the zip codes in the regions as a random sample of all the zip codes for the regions. Test the null hypothesis that the populations' mean sports purchasing indexes are equal using $\alpha = .01$, and place a 99% confidence interval on the true difference between the mean purchasing index for the two regions.

b. Repeat part **a** using the following pairs of α and confidence levels for the tests and confidence intervals: (.05, 95%), (.10, 90%), and (.20, 80%). Describe what happens to the tests and confidence intervals as α is increased and the confidence level is decreased. Which do you think is more informative—the tests or the confidence intervals?

References

Barton, J., and Folkard, S. "The response of day and night nurses to their work schedules." *Journal of Occupational Psychology*, Sept. 1991, Vol. 64, Part 3, pp. 207–218.

Clotfelter, C. T. "Detroit decision and white flight." *Journal of Legal Studies*, 1976, 5.

Dreher, G. F. "The role of performance in the turnover process." *Academy of Management Journal*, 1982, 25.

Freedman, D., Pisani, R., and Purves, R. *Statistics*. New York: W. W. Norton and Co., 1978.

Gibbons, J. D. *Nonparametric Statistical Inference*, 2d ed. New York: McGraw-Hill, 1985.

Harris, W. T. "Rule changes and the earnings of National Football League field goal kickers." *Sociology of Sport Journal*, Dec. 1992, Vol. 9, No. 4.

Hesse-Biber, S., and Marino, M. "From high school to college: Changes in women's self-concept." *Journal of Psychology*, Mar. 1991, Vol. 125, No. 2, pp. 199–216.

Hollander, M., and Wolfe, D. A. *Nonparametric Statistical Methods*. New York: Wiley, 1973.

Jackson, G. B., Jackson, R. W., and Newmiller, C. E., Jr. "Consumer demographics and reaction to product tampering." *Journal of Psychology & Marketing*, Jan. 1992, Vol. 9, No. 1, pp. 45–57.

Johnston-Anumonwo, I. "The influence of household type on gender differences in work trip distance." *The Professional Geographer*, May 1992, Vol. 44, No. 2, p. 161.

Mendenhall, W. *Introduction to Probability and Statistics*, 8th ed. Boston: PWS-Kent, 1991, Chapters 8 and 9.

Mills, J. K. "Difference in locus of control between obese adults and adolescent females undergoing weight reduction." *Journal of Psychology*, Mar. 1991, Vol. 125, No. 2, pp. 195–197.

Petiprin, G. L., and Johnson, M. E. "Effects of gender, attributional style, and item difficulty on academic performance." *Journal of Psychology*, Jan. 1991, Vol. 125, No. 1, pp. 45–50.

Prokash, V. "Sex roles and advertising preferences." *Journal of Advertising Research*, May/June 1990.

Rogot, E., Sorlie, P. D., and Johnson, N. T. "Life expectancy by employment status, income, and education in the national longitudinal mortality study." *Public Health Reports*, July–Aug. 1992, Vol. 107, No. 4, p. 457.

Rosetti, D. K., and Surynt, T. J. "Video teleconferencing and performance." *Journal of Business Communication*, Fall 1985, Vol. 22, No. 4, pp. 25–31.

Sheehan, E. P., McMenamin, N., and McDevitt, T. M. "Learning styles of traditional and nontraditional university students." *College Student Journal*, Dec. 1992, Vol. 26, No. 4.

Siegel, S. *Nonparametric Statistics for the Behavioral Sciences*. New York: McGraw-Hill, 1956.

Snedecor, G. W., and Cochran, W. *Statistical Methods*, 7th ed. Ames: Iowa State University Press, 1980.

Spanos, N. P., Sims, A., de Faye, B., Mondoux, T. J., and Gabora, N. J. "A comparison of hypnotic and nonhypnotic treatments for smoking." *Journal of Imagination, Cognition and Personality*, 1992/93, Vol. 12, No. 1.

Stainback, S., and Stainback, W. C. "Matched procedures and research in mental retardation." *Training School Bulletin*, May 1973, 70, pp. 33–37.

CHAPTER NINE

Comparing Population Proportions

Contents

Case Studies

Where We've Been

Chapter 8 presented methods for comparing two population means using z and t statistics and for comparing two populations using nonparametric methods.

Where We're Going

In this chapter, we will consider a problem of comparable importance—comparing two or more population proportions. The need to compare population proportions arises because many business and social experiments involve questioning people and classifying their responses. We will learn how to determine whether the proportion of consumers favoring product A differs from the proportion favoring product B, and we will learn how to estimate the difference with a confidence interval.

9

Many experiments are conducted in the biological, physical, and social sciences to compare two or more population proportions. Those conducted in business and the social sciences to sample the opinions of people are called **sample surveys.** For example, a state government might wish to estimate the difference between the proportions of people in two regions of the state who would qualify for a new welfare program. Or, after an innovative process change, an engineer might wish to determine whether the proportion of defective items produced by a manufacturing process was less than the proportion of defectives produced before the change. This chapter will show you how to test hypotheses about the difference between two population proportions based on independent random sampling. We will also show how to find a confidence interval for the difference. Then, in optional Section 9.3 we will compare more than two population proportions, and in optional Section 9.4 we will present a related problem.

9.1 Inferences About the Difference Between Two Population Proportions: Independent Binomial Experiments

Suppose a presidential candidate wants to compare the preference of registered voters in the northeastern United States (NE) to those in the southeastern United States (SE). Such a comparison would help determine where to concentrate campaign efforts. The candidate hires a professional pollster to randomly choose 1,000 registered voters in the northeast and 1,000 in the southeast and interview each to learn her or his voting preference. The objective is to use this sample information to make an inference about the difference $(p_1 - p_2)$ between the proportion p_1 of *all* registered voters in the northeast and the proportion p_2 of *all* registered voters in the southeast who plan to vote for the presidential candidate.

The two samples represent independent binomial experiments. (See Section 4.3 for the characteristics of binomial experiments.) The binomial random variables are the numbers x_1 and x_2 of the 1,000 sampled voters in each area who indicate they will vote for the candidate. The results are summarized in the accompanying table.

NE	*SE*
$n_1 = 1{,}000$	$n_2 = 1{,}000$
$x_1 = 546$	$x_2 = 475$

We can now calculate the sample proportions $\hat{p}_1$ and $\hat{p}_2$ of the voters in favor of the candidate in the northeast and southeast, respectively:

$$\hat{p}_1 = \frac{x_1}{n_1} = \frac{546}{1{,}000} = .546 \qquad \hat{p}_2 = \frac{x_2}{n_2} = \frac{475}{1{,}000} = .475$$

The difference between the sample proportions, $(\hat{p}_1 - \hat{p}_2)$ makes an intuitively appealing point estimator of the difference between the population parameters, $(p_1 - p_2)$. For our example, the estimate is

$$(\hat{p}_1 - \hat{p}_2) = .546 - .475 = .071$$

To judge the reliability of the estimator $(\hat{p}_1 - \hat{p}_2)$, we must observe its performance in repeated sampling from the two populations. That is, we need to know the sampling distribution of $(\hat{p}_1 - \hat{p}_2)$. The properties of the sampling distribution are given in the box. Remember that $\hat{p}_1$ and $\hat{p}_2$ can be viewed as means of the number of successes per trial in the respective samples, so the Central Limit Theorem applies when the sample sizes are large.

Properties of the Sampling Distribution of $(\hat{p}_1 - \hat{p}_2)$

1. The mean of the sampling distribution of $(\hat{p}_1 - \hat{p}_2)$ is $(p_1 - p_2)$, that is,

$$E(\hat{p}_1 - \hat{p}_2) = p_1 - p_2$$

   Thus, $(\hat{p}_1 - \hat{p}_2)$ is an unbiased estimator of $(p_1 - p_2)$.
2. The standard deviation of the sampling distribution of $(\hat{p}_1 - \hat{p}_2)$ is

$$\sigma_{(\hat{p}_1 - \hat{p}_2)} = \sqrt{\frac{p_1 q_1}{n_1} + \frac{p_2 q_2}{n_2}}$$

3. If the sample sizes n_1 and n_2 are large (see Section 6.4 for a guideline), the sampling distribution of $(\hat{p}_1 - \hat{p}_2)$ is approximately normal.

Since the distribution of $(\hat{p}_1 - \hat{p}_2)$ in repeated sampling is approximately normal, we can use the z statistic to derive confidence intervals for $(p_1 - p_2)$ or to test a hypothesis about $(p_1 - p_2)$.

For the voter example, a 95% confidence interval for the difference $(p_1 - p_2)$ is

$$(\hat{p}_1 - \hat{p}_2) \pm 1.96\sigma_{(\hat{p}_1 - \hat{p}_2)} \quad \text{or} \quad (\hat{p}_1 - \hat{p}_2) \pm 1.96\sqrt{\frac{p_1 q_1}{n_1} + \frac{p_2 q_2}{n_2}}$$

The quantities $p_1 q_1$ and $p_2 q_2$ must be estimated in order to complete the calculation of the standard deviation, $\sigma_{(\hat{p}_1 - \hat{p}_2)}$, and hence the calculation of the confidence interval. In Section 6.4 we showed that the value of pq is relatively insensitive to the value chosen to approximate p. Therefore, $\hat{p}_1 \hat{q}_1$ and $\hat{p}_2 \hat{q}_2$ will provide satisfactory estimates to approximate $p_1 q_1$ and $p_2 q_2$, respectively. Then

$$\sqrt{\frac{p_1 q_1}{n_1} + \frac{p_2 q_2}{n_2}} \approx \sqrt{\frac{\hat{p}_1 \hat{q}_1}{n_1} + \frac{\hat{p}_2 \hat{q}_2}{n_2}}$$

and we will approximate the 95% confidence interval by

$$(\hat{p}_1 - \hat{p}_2) \pm 1.96\sqrt{\frac{\hat{p}_1 \hat{q}_1}{n_1} + \frac{\hat{p}_2 \hat{q}_2}{n_2}}$$

Substituting the sample quantities yields

$$(.546 - .475) \pm 1.96\sqrt{\frac{(.546)(.454)}{1,000} + \frac{(.475)(.525)}{1,000}}$$

or $.071 \pm .044$. Thus, we are 95% confident that the interval from .027 to .115 contains $(p_1 - p_2)$.

We infer that there are between 2.7% and 11.5% more registered voters in the northeast than in the southeast who plan to vote for the presidential candidate. It seems that the candidate should direct a greater campaign effort in the southeast compared to the northeast.

The general form of a confidence interval for the difference $(p_1 - p_2)$ between binomial proportions is given in the box.

Large-Sample $100(1 - \alpha)\%$ Confidence Interval for $(p_1 - p_2)$

$$(\hat{p}_1 - \hat{p}_2) \pm z_{\alpha/2}\sigma_{(\hat{p}_1-\hat{p}_2)} = \hat{p}_1 - \hat{p}_2 \pm z_{\alpha/2}\sqrt{\frac{p_1q_1}{n_1} + \frac{p_2q_2}{n_2}}$$

$$\approx (\hat{p}_1 - \hat{p}_2) \pm z_{\alpha/2}\sqrt{\frac{\hat{p}_1\hat{q}_1}{n_1} + \frac{\hat{p}_2\hat{q}_2}{n_2}}$$

Assumptions: The two samples are independent random samples from binomial distributions. Both samples should be large enough that the normal distribution provides an adequate approximation to the sampling distribution of $\hat{p}_1$ and $\hat{p}_2$ (see Section 7.6).

The z statistic,

$$z = \frac{(\hat{p}_1 - \hat{p}_2) - (p_1 - p_2)}{\sigma_{(\hat{p}_1-\hat{p}_2)}}$$

is used to test the null hypothesis that $(p_1 - p_2)$ equals some specified difference, say D_0. For the special case where $D_0 = 0$, i.e., where we want to test the null hypothesis H_0: $(p_1 - p_2) = 0$ (or, equivalently, H_0: $p_1 = p_2$), the best estimate of $p_1 = p_2 = p$ is obtained by dividing the total number of successes $(x_1 + x_2)$ for the two samples by the total number of observations $(n_1 + n_2)$; that is,

$$\hat{p} = \frac{x_1 + x_2}{n_1 + n_2} \quad \text{or} \quad \hat{p} = \frac{n_1\hat{p}_1 + n_2\hat{p}_2}{n_1 + n_2}$$

The second equation shows that $\hat{p}$ is a weighted average of $\hat{p}_1$ and $\hat{p}_2$, with the larger sample receiving more weight. If the sample sizes are equal, then $\hat{p}$ is a simple average of the two sample proportions of successes.

We now substitute the weighted average $\hat{p}$ for both p_1 and p_2 in the formula for the standard deviation of $(\hat{p}_1 - \hat{p}_2)$:

$$\sigma_{(\hat{p}_1-\hat{p}_2)} = \sqrt{\frac{p_1q_1}{n_1} + \frac{p_2q_2}{n_2}} \approx \sqrt{\frac{\hat{p}\hat{q}}{n_1} + \frac{\hat{p}\hat{q}}{n_2}} = \sqrt{\hat{p}\hat{q}\left(\frac{1}{n_1} + \frac{1}{n_2}\right)}$$

The test is summarized in the box.

Large-Sample Test of Hypothesis About $(p_1 - p_2)$

One-Tailed Test

H_0: $(p_1 - p_2) = 0$*
H_a: $(p_1 - p_2) < 0$
[or H_a: $(p_1 - p_2) > 0$]

Test statistic: $z = \dfrac{(\hat{p}_1 - \hat{p}_2)}{\sigma_{(\hat{p}_1-\hat{p}_2)}}$

Rejection region: $z < -z_\alpha$
[or $z > z_\alpha$ when H_a: $(p_1 - p_2) > 0$]

Two-Tailed Test

H_0: $(p_1 - p_2) = 0$
H_a: $(p_1 - p_2) \neq 0$

Test statistic: $z = \dfrac{(\hat{p}_1 - \hat{p}_2)}{\sigma_{(\hat{p}_1-\hat{p}_2)}}$

Rejection region: $z < -z_{\alpha/2}$ or $z > z_{\alpha/2}$

Note: $\sigma_{(\hat{p}_1-\hat{p}_2)} = \sqrt{\dfrac{p_1q_1}{n_1} + \dfrac{p_2q_2}{n_2}} \approx \sqrt{\hat{p}\hat{q}\left(\dfrac{1}{n_1} + \dfrac{1}{n_2}\right)}$ where $\hat{p} = \dfrac{x_1 + x_2}{n_1 + n_2}$

Assumption: Same as for large-sample confidence interval for $(p_1 - p_2)$. (See previous box.)

EXAMPLE 9.1

In the past decade there have been intensive antismoking campaigns sponsored by both federal and private agencies. Suppose the American Cancer Society randomly sampled 1,500 adults in 1985 and then sampled 2,000 adults in 1990 to determine whether there was evidence that the percentage of smokers had decreased. The results of the two sample surveys are shown in the table, where x_1 and x_2 represent the numbers of smokers in the 1985 and 1990 samples, respectively. Do these data indicate that the fraction of smokers decreased over this 5-year period? Use $\alpha = .05$.

1985	*1990*
$n_1 = 1{,}500$	$n_2 = 2{,}000$
$x_1 = 576$	$x_2 = 652$

*The test can be adapted to test for a difference $D_0 \neq 0$. Because most applications call for a comparison of p_1 and p_2, implying $D_0 = 0$, we will confine our attention to this case.

Solution

If we define p_1 and p_2 as the true proportions of adult smokers in 1985 and 1990 the elements of our test are

$$H_0: \quad (p_1 - p_2) = 0$$
$$H_a: \quad (p_1 - p_2) > 0$$

(The test is one-tailed since we are interested only in determining whether the proportion of smokers *decreased*.)

$$\textit{Test statistic:} \quad z = \frac{(\hat{p}_1 - \hat{p}_2) - 0}{\sigma_{(\hat{p}_1 - \hat{p}_2)}}$$

$$\textit{Rejection region:} \quad \alpha = .05$$
$$z > z_\alpha = z_{.05} = 1.645 \quad \text{(see Figure 9.1)}$$

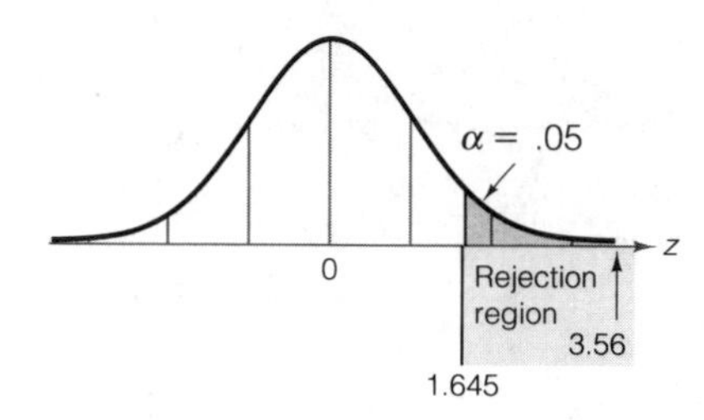

FIGURE 9.1 ▶ Rejection region for Example 9.1

We now calculate the sample proportions of smokers

$$\hat{p}_1 = \frac{576}{1,500} = .384 \qquad \hat{p}_2 = \frac{652}{2,000} = .326$$

Then

$$z = \frac{(\hat{p}_1 - \hat{p}_2) - 0}{\sigma_{(\hat{p}_1 - \hat{p}_2)}} \approx \frac{(\hat{p}_1 - \hat{p}_2)}{\sqrt{\hat{p}\hat{q}\left(\frac{1}{n_1} + \frac{1}{n_2}\right)}}$$

$$\text{where} \quad \hat{p} = \frac{x_1 + x_2}{n_1 + n_2} = \frac{576 + 652}{1,500 + 2,000} = .351$$

Note that $\hat{p}$ is a weighted average of $\hat{p}_1$ and $\hat{p}_2$, with more weight given to the larger (1990) sample.

Thus, the computed value of the test statistic is

$$z = \frac{.384 - .326}{\sqrt{(.351)(.649)\left(\frac{1}{1,500} + \frac{1}{2,000}\right)}} = \frac{.058}{.0164} = 3.56$$

There is sufficient evidence at the $\alpha = .05$ level to conclude that the proportion of adults who smoke has decreased over the 1985–1990 period. We could place a confidence interval on $(p_1 - p_2)$ if we were interested in estimating the extent of the decrease.

Exercises 9.1–9.19

Learning the Mechanics

9.1 What are the characteristics of a binomial experiment?

9.2 Explain why the Central Limit Theorem is important in finding an approximate distribution for $(\hat{p}_1 - \hat{p}_2)$. (See Section 5.3.)

9.3 In each case, determine whether the sample sizes are large enough to conclude that the sampling distribution of $(\hat{p}_1 - \hat{p}_2)$ is approximately normal.
a. $n_1 = 10,\quad n_2 = 12,\quad \hat{p}_1 = .50,\quad \hat{p}_2 = .50$
b. $n_1 = 10,\quad n_2 = 12,\quad \hat{p}_1 = .10,\quad \hat{p}_2 = .08$
c. $n_1 = n_2 = 30,\quad \hat{p}_1 = .20,\quad \hat{p}_2 = .30$
d. $n_1 = 100,\quad n_2 = 200,\quad \hat{p}_1 = .05,\quad \hat{p}_2 = .09$
e. $n_1 = 100,\quad n_2 = 200,\quad \hat{p}_1 = .95,\quad \hat{p}_2 = .91$

9.4 For each of the following values of α, find the values of z for which H_0: $(p_1 - p_2) = 0$ would be rejected in favor of H_a: $(p_1 - p_2) < 0$.
a. $\alpha = .01$ b. $\alpha = .025$ c. $\alpha = .05$ d. $\alpha = .10$

9.5 Independent random samples, each containing 800 observations, were selected from two binomial populations. The samples from populations 1 and 2 produced 320 and 400 successes, respectively.
a. Test H_0: $(p_1 - p_2) = 0$ against H_a: $(p_1 - p_2) \neq 0$. Use $\alpha = .05$.
b. Test H_0: $(p_1 - p_2) = 0$ against H_a: $(p_1 - p_2) \neq 0$. Use $\alpha = .01$.
c. Test H_0: $(p_1 - p_2) = 0$ against H_a: $(p_1 - p_2) < 0$. Use $\alpha = .01$.
d. Form a 90% confidence interval for $(p_1 - p_2)$.

9.6 Construct a 95% confidence interval for $(p_1 - p_2)$ in each of the following situations:
a. $n_1 = 400,\quad \hat{p}_1 = .65;\quad n_2 = 400,\quad \hat{p}_2 = .58$
b. $n_1 = 180,\quad \hat{p}_1 = .31;\quad n_2 = 250,\quad \hat{p}_2 = .25$
c. $n_1 = 100,\quad \hat{p}_1 = .46;\quad n_2 = 120,\quad \hat{p}_2 = .61$

9.7 Sketch the sampling distribution of $(\hat{p}_1 - \hat{p}_2)$ based on independent random samples of $n_1 = 100$ and $n_2 = 200$ observations from two binomial populations with success probabilities $p_1 = .1$ and $p_2 = .5$, respectively.

9.8 Random samples of size $n_1 = 50$ and $n_2 = 60$ were drawn from populations 1 and 2, respectively. The samples yielded $\hat{p}_1 = .4$ and $\hat{p}_2 = .2$. Test H_0: $(p_1 - p_2) = .1$ against H_a: $(p_1 - p_2) > .1$ using $\alpha = .05$.

Applying the Concepts

9.9 Women are filling managerial positions in increasing numbers, although there is much dissension about whether progress has been fast enough. Does marriage act as a hindrance to women's career progression to a greater degree than men's? An article in *The Journal of Applied Psychology* (Stroh, Brett, and Reilly, 1992) investigates this and other questions relating to managerial careers of men and women in today's workforce. In a random sample of 795 male managers and 223 female managers from 20 Fortune 500 corporations, 86% of the male managers and 45% of the female managers were married.

a. Use a 95% confidence interval to estimate the difference between the proportions of men and women managers who are married.

b. Interpret the interval. State your conclusion in terms of the populations from which the samples were drawn.

9.10 Two surgical procedures are widely used to treat a certain type of cancer. To compare the success rates of the two procedures, random samples of the two types of surgical patients were obtained and the numbers of patients who showed no recurrence of the disease after a 1-year period were recorded. The data are shown in the table. Do the data provide sufficient evidence to indicate a difference in the success rates of the two procedures? Use $\alpha = .05$.

	n	Number of Successes
Procedure A	100	78
Procedure B	100	87

9.11 A new insect spray, type A, is to be compared with a spray, type B, that is currently in use. Two rooms of equal size are sprayed with the same amount of spray, one room with A, the other with B. Two hundred insects are released into each room, and after 1 hour the numbers of dead insects are counted. The results are given in the table.

	Spray A	Spray B
Number of insects	200	200
Number of dead insects	120	90

a. Do the data provide sufficient evidence to indicate that spray A is more effective than spray B in controlling the insects? Test using $\alpha = .05$.

b. Find a 90% confidence interval for $(p_1 - p_2)$, the difference in the rates of kill for the two sprays. Interpret this interval.

9.12 Does cohabitation prior to marriage change the probability of the marriage's success? A recent article in *The Journal of Marriage and the Family* (DeMaris and Rao, 1992) compares the divorce rate for couples who cohabited prior to marriage against that for couples who did not cohabit. Suppose that a random sample of 500 couples are selected from each group.

a. What are the appropriate null and alternative hypotheses to be tested? Define any symbols you use.

b. What is the appropriate test statistic? Sketch the probability distribution of this test statistic, assuming the null hypothesis is true.

c. The authors report that cohabitation prior to marriage is associated with a greater risk of divorce, with a p-value $< .05$. In which tail of the distribution you sketched in part **b** did the observed test statistic fall? Show its approximate location on the sketch of the distribution.

d. Based on the observed significance level, what conclusion would you reach about the test you set up in part **a**?

9.13 It is estimated that more than half the votes cast in upcoming elections will be cast by women. Of concern to the prevalent political parties is the possibility of a "gender gap." This hypothesis states that there is a difference between male and female perceptions of which political party best addresses the nation's problems. Particularly, those who favor this hypothesis believe that the Democratic platform may appear to be more in line with philosophies of the majority of women. Suppose a Republican campaign manager wishes to test this theory. Assume that 300 registered voters (150 men, 150 women) were given the platforms of both the Republicans and Democrats for five major issues and then asked to give their preference for one of the parties. Suppose 81 men and 70 women preferred the Republican views.

a. Does this result provide sufficient evidence to indicate that a gender gap favoring the Democratic party exists? Test at $\alpha = .05$.

b. If there were no undecided responses, can we say that women prefer the Democrats' responses to issues? Test at $\alpha = .01$.

9.14 One method of treating a major form of blindness in elderly people uses laser beams to seal abnormal blood vessels in the eye. When the method was tried on 224 patients, only 14% went blind in 1 year. In a control group of the same number of untreated patients, 42% went blind in 1 year. Assume the control group contained the same number of patients as the treated group.

a. If we want to test whether the method reduces the probability that a patient will become blind within 1 year, what are the appropriate null and alternative hypotheses? Define any symbols you use.

b. When the data are analyzed using a statistical software package, the results are as shown. Interpret these results.

```
Z = -6.60          P-VALUE = .0000
```

c. Is the p-value for this test exactly zero? Explain.

9.15 The Reserve Mining Company of Minnesota commissioned a team of physicians to study the breathing patterns of its miners who were exposed to taconite dust. The physicians compared the breathing of 307 miners who had been employed in the Reserve's Babbit, Minnesota, mine for more than 20 years with 35 Duluth area men with no history of exposure to taconite dust. The physicians concluded that "there is no significant difference in respiratory symptoms or breathing ability between the group of men who have worked in the taconite industry for more than 20 years and a group of men of similar smoking habits but without exposure to taconite dust." Using the statistical procedures you have learned in this chapter, design a hypothesis test (give H_0, H_a, test statistic, etc.) that would have been appropriate for use in the physicians' study. [Source: Associated Press, *Minneapolis Tribune*, February 20, 1977.]

9.16 Refer to Exercise 9.15. Suppose the physicians determined that 61 of the 307 miners had breathing irregularities and that five of the 35 Duluth men had breathing irregularities.

a. Test to determine whether these data indicate that a higher proportion of breathing irregularities exists among those who have been exposed to taconite dust than among those who have not been exposed.

b. Find the observed significance level for the test and interpret its value.

9.17 In 1984 the *Orlando Sentinel* (March 8, 1984) suggested that coffee drinking was dropping in favor of soda. The article noted that the "Winter Coffee Drinking Study" showed that 55.2% of all Americans drank coffee as compared with 74.7% in 1962. No sample sizes are given, but let us assume that both surveys were based on samples of 1,000 adult Americans. Use this information and the results just cited to find a 95% confidence interval for the percentage drop in coffee-drinking adult Americans between 1962 and 1984.

9.18 The number of practicing attorneys more than doubled in the decade 1973–1983, and the percentage of female attorneys increased from 5% to 15% of the legal profession. A random sample of 400 female attorneys and 200 male attorneys, from among the approximately 606,000 total number of attorneys in the United States, revealed that 25% of the women finished in the top 10% of their classes versus 18% of the males. The survey also found that women tended to enter law school at a later age than men. (Nearly 33% of women began practicing after age 30 compared with 14% for men.) For older women and men beginning practice, women's starting salaries tended to be higher than men of the same age. Finally, the median salary for different age groups increased with age, peaking for men at age 51–55 and leveling off thereafter (*American Bar Association Journal*, October 1983). Based on the independent random samples of 400 female and 200 male attorneys selected from all attorneys in the United States, do the data provide sufficient evidence to indicate that the probabilities of finishing in the upper 10% of their law school class differ between male and female lawyers?

9.19 Refer to Exercise 6.46, in which we examined data from an article in *The International Journal of Sports Psychology* (Tucker, 1990) studying the relationship between levels of fitness and stress among employees of companies offering health and fitness programs. Random samples of employees were selected from each of three fitness level categories, and each was evaluated for signs of stress. The data are repeated here.

Fitness Level	*Sample Size*	*Proportion with Signs of Stress*
Poor	242	.155
Average	212	.133
Good	95	.108

a. What are the appropriate null and alternative hypotheses to test whether a greater proportion of employees in the poor fitness category show signs of stress than those in the average fitness category? Define any symbols you use.

b. Conduct the test constructed in part **a** using $\alpha = .10$. Interpret the result.

c. How would your null and alternative hypotheses change if you wanted to compare the proportions showing signs of stress in the poor and good fitness categories?

d. To conduct the test in part **c**, the data are analyzed by a statistical software package with the results shown here. Based on these results, state whether you would be willing to support the following statement: "Fitness level has no bearing on whether an individual shows signs of stress." Explain.

```
Z = 1.11        P-VALUE = .1335
```

9.2 Determining the Sample Size

The sample sizes n_1 and n_2 required to compare two population proportions can be found in a manner similar to the method described in Section 8.4 for comparing two population means. We will assume equal sample sizes, i.e., $n_1 = n_2 = n$, and then choose n so that $(\hat{p}_1 - \hat{p}_2)$ will differ from $(p_1 - p_2)$ by no more than a bound B with a specified probability. We will illustrate the procedure with an example.

EXAMPLE 9.2

A production supervisor suspects a difference exists between the proportions p_1 and p_2 of defective items produced by two different machines. Experience has shown that the proportion defective for each of the two machines is in the neighborhood of .03. If the supervisor wants to estimate the difference in the proportions using a 95% confidence interval of width .01, how many items must be randomly sampled from the production of each machine? (Assume that you want $n_1 = n_2 = n$.)

Solution

For the specified level of reliability,

$$
\begin{aligned}
z_{\alpha/2} &= z_{.025} \\
&= 1.96
\end{aligned}
$$

Then, letting $p_1 = p_2 = .03$ and $n_1 = n_2 = n$, we find the required sample size per machine by solving the following equation for n:

$$z_{\alpha/2}\sigma_{(\hat{p}_1-\hat{p}_2)} = B$$

or

$$z_{\alpha/2}\sqrt{\frac{p_1q_1}{n_1} + \frac{p_2q_2}{n_2}} = B$$

$$1.96\sqrt{\frac{(.03)(.97)}{n} + \frac{(.03)(.97)}{n}} = 1.96\sqrt{\frac{2(.03)(.97)}{n}} = .005$$

$$n = 8{,}943.2$$

You can see that this may be a tedious sampling procedure. If the supervisor insists on estimating $(p_1 - p_2)$ correct to within .005 with probability equal to .95, approximately 9,000 items will have to be inspected for each machine.

You can see from the calculations in Example 9.2 that $\sigma_{(\hat{p}_1-\hat{p}_2)}$ (and hence the solution, $n_1 = n_2 = n$) depends on the actual (but unknown) values of p_1 and p_2. In fact, the required sample size $n_1 = n_2 = n$ is largest when $p_1 = p_2 = \frac{1}{2}$. Therefore, if you have no prior information on the approximate values of p_1 and p_2, use $p_1 = p_2 = \frac{1}{2}$ in the formula for $\sigma_{(\hat{p}_1-\hat{p}_2)}$. If p_1 and p_2 are in fact close to $\frac{1}{2}$, then the values of n_1 and n_2 that you have calculated will be correct. If p_1 and p_2 differ substantially from $\frac{1}{2}$, then your solutions for n_1 and n_2 will be larger than needed. Consequently, using $p_1 = p_2 = \frac{1}{2}$ when solving for n_1 and n_2 is a conservative procedure because the sample sizes n_1 and n_2 will be at least as large (and probably larger than) needed.

The procedure for determining sample sizes necessary for estimating $(p_1 - p_2)$ for the case $n_1 = n_2$ is given in the next box.

Determination of Sample Size for Comparing Two Proportions

To estimate $(p_1 - p_2)$ to within a given bound B with probability $(1 - \alpha)$ or, equivalently, with a $100(1 - \alpha)\%$ confidence interval of width $W = 2B$, use the following formula to solve for equal sample sizes that will achieve the desired reliability:

$$n_1 = n_2 = \frac{(z_{\alpha/2})^2(p_1q_1 + p_2q_2)}{B^2} = \frac{4(z_{\alpha/2})^2(p_1q_1 + p_2q_2)}{W^2}$$

You will need to substitute estimates for the values of p_1 and p_2 before solving for the sample size. These estimates might be based on prior samples, obtained from educated guesses or, most conservatively, specified as $p_1 = p_2 = .5$.

Exercises 9.20–9.25

Learning the Mechanics

9.20 Assuming that $n_1 = n_2$, find the sample sizes needed to estimate $(p_1 - p_2)$ for each of the following situations:

a. Bound $= .01$ with 99% confidence. Assume that $p_1 \approx .4$ and $p_2 \approx .7$.

b. A 90% confidence interval of width .05. Assume there is no prior information available to obtain approximate values of p_1 and p_2.

c. Bound $= .03$ with 90% confidence. Assume that $p_1 \approx .2$ and $p_2 \approx .3$.

Applying the Concepts

9.21 A pollster wants to estimate the difference between the proportions of men and women who favor a particular national candidate using a 90% confidence interval of width .04. Suppose the pollster has no prior information about the proportions. If equal numbers of men and women are to be polled, how large should the sample sizes be?

9.22 Nationally televised home shopping was introduced in 1985. Overnight it became the hottest craze in television programming. By December 1986 there were 34 home shopping cable services (Covert, 1986). Today, nearly all cable companies carry at least one home shopping channel. Who uses these home shopping services? Are the shoppers primarily men or women? Suppose you want to estimate the difference in the proportions of men and women who say they have used or expect to use televised home shopping using an 80% confidence interval of width .06 or less.

a. Approximately how many people should be included in your samples?

b. Suppose you want to obtain individual estimates for the two proportions of interest. Will the sample size found in part **a** be large enough to provide estimates of each proportion correct to within .02 with probability equal to .90? Justify your response.

9.23 Rat damage creates a large financial loss in the production of sugarcane. One aspect of the problem that has been investigated by the U.S. Department of Agriculture concerns the optimal place to locate rat poison. To be most effective in reducing rat damage, should the poison be located in the middle of the field or on the ou perimeter? One way to answer this question is to determine where the greater amount of damage occurs. If damage is measured by the proportion of cane stalks that have been damaged by rats, how many stalks from each section of the field should be sampled in order to estimate the true difference between proportions of stalks damaged in the two sections to within .02 with probability .95?

9.24 A television manufacturer wants to compare with a competitor the proportions of its best sets that need repair within 1 year. If it is desired to estimate the difference between proportions to within .05 with 90% confidence, and if the manufacturer plans to sample twice as many buyers (n_1) of its sets as buyers (n_2) of the competitor's sets, how many buyers of each brand must be sampled? Assume that the proportion of sets that need repair will be about .2 for both brands.

9.25 Refer to Exercise 9.19, where we examined Larry A. Tucker's (1990) data relating levels of fitness and stress. Essentially, we failed to reject the null hypothesis that those in poor physical condition exhibit signs of stress in the same proportion as those in good physical condition, even though the sample proportions differed by nearly .05. (That is, of those sampled, almost 5% more of those in poor condition exhibited signs of stress than those in good condition.)

a. How large would the samples have to be to estimate the difference in the proportions showing signs of stress to within .04 with 95% confidence? Assume the sample sizes for the two groups will be equal, and remember that the first samples selected resulted in proportions of .155 and .108 for the poor and good fitness levels, respectively.

b. Suppose samples of the size you calculated in part **a** were selected from each group, and the sample proportions again turned out to be .155 and .108. Test the null hypothesis H_0: $(p_1 - p_2) = 0$ against the alternative H_a: $(p_1 - p_2) > 0$, where p_1 is the proportion of all employees at the poor fitness level who show signs of stress and p_2 is the proportion of all employees at the good fitness level who show signs of stress. Calculate and interpret the observed significance level.

9.3 Comparing Population Proportions from a Multinomial Experiment (Optional)

In this section, we consider a statistical method to compare two or more, say k, population proportions. For example, suppose we wish to compare the percentage of voters favoring each of three political candidates running for the same elective position. The voting preferences of a random sample of 150 eligible voters are obtained, and the resulting count data are classified according to a single criterion: candidate preference. The data are shown in Table 9.1. Do you think these data indicate a voter preference for any of the candidates?

TABLE 9.1 Voter-Preference Survey

Candidate		
1	2	3
61	53	36

To answer this question with a valid statistical analysis, we need to know the underlying probability distribution of these **count data.** This distribution, called the **multinomial probability distribution**, is an extension of the binomial distribution (Section 4.3). The properties of a multinomial experiment are given in the box.

Properties of the Multinomial Experiment

1. The experiment consists of n identical trials.
2. There are k possible outcomes to each trial.
3. The probabilities of the k outcomes, denoted by $p_1, p_2, \ldots, p_k$, remain the same from trial to trial, where $p_1 + p_2 + \cdots + p_k = 1$.
4. The trials are independent.
5. The random variables of interest are the counts $n_1, n_2, \ldots, n_k$ in each of the k cells.

Note that our voter-preference survey satisfies the properties of a multinomial experiment. The experiment consists of randomly sampling $n = 150$ voters from a large population of voters containing an unknown proportion p_1 who favor candidate 1, a proportion p_2 who favor candidate 2, and a proportion p_3 who favor candidate 3. Each voter sampled represents a single trial that can result in one of three outcomes: The voter will favor either candidate 1, 2, or 3 with probabilities p_1, p_2, and p_3, respectively. (Assume that all voters will have a preference.) The voting preference of any single voter in the sample does not affect the preference of another; consequently, the trials are independent. And, finally, you can see that the recorded data are the numbers of voters in each of the three voter-preference categories. Thus, the voter-preference survey satisfies the five properties of a multinomial experiment. You can see that the properties of the multinomial experiment closely resemble those of the binomial experiment and that, in fact, a binomial experiment is a multinomial experiment for the special case where $k = 2$.

In the voter-preference survey, and in most practical applications of the multinomial experiment, the k outcome probabilities $p_1, p_2, \ldots, p_k$ are unknown and we want to use the survey data to make inferences about their values. The unknown probabilities in the voter-preference survey are

$$p_1 = \text{Proportion of all voters who favor candidate 1}$$
$$p_2 = \text{Proportion of all voters who favor candidate 2}$$
$$p_3 = \text{Proportion of all voters who favor candidate 3}$$

To decide whether the voters have a preference for any of the candidates, we will want to test the null hypothesis that the candidates are equally preferred (that is, $p_1 = p_2 = p_3 = 1/3$) against the alternative hypothesis that one candidate is preferred (that is, at least one of the probabilities p_1, p_2, and p_3 exceeds $1/3$). Thus, we want to test

H_0: $p_1 = p_2 = p_3 = 1/3$ (no preference)
H_a: At least one of the proportions exceeds $1/3$ (a preference exists)

If the null hypothesis is true and $p_1 = p_2 = p_3 = 1/3$, the expected value (mean value) of the number of voters who prefer candidate 1 is given by

$$E(n_1) = np_1 = (n)1/3 = (150)1/3 = 50$$

Similarly, $E(n_2) = E(n_3) = 50$ if the null hypothesis is true and no preference exists.

The following test statistic measures the degree of disagreement between the data and the null hypothesis:

$$X^2 = \frac{[n_1 - E(n_1)]^2}{E(n_1)} + \frac{[n_2 - E(n_2)]^2}{E(n_2)} + \frac{[n_3 - E(n_3)]^2}{E(n_3)}$$
$$= \frac{(n_1 - 50)^2}{50} + \frac{(n_2 - 50)^2}{50} + \frac{(n_3 - 50)^2}{50}$$

Note that the farther the observed numbers n_1, n_2, and n_3 are from their expected value (50), the larger X^2 will become. That is, large values of X^2 imply that the null hypothesis is false.

We have to know the distribution of X^2 in repeated sampling before we can decide whether the data indicate that a preference exists. When H_0 is true, X^2 can be shown to have approximately a **chi-square (χ^2) distribution.** The shape of the chi-square distribution will depend on the degrees of freedom (df) associated with Table 9.1. For this application, the number of degrees of freedom will always be calculated as shown in the box.*

Number of Degrees of Freedom for Chi-Square When Comparing k Population Proportions

$$df = k - 1$$

Critical values of χ^2 are shown in Table VIII of Appendix A, a portion of which is shown in Table 9.2 on page 458.

*The derivation of the degrees of freedom for χ^2 involves the number of linear restrictions imposed on the count data. In the present case, the only constraint is that $\Sigma n_i = n$, where n (the sample size) is fixed in advance. Therefore, df $= k - 1$. For other cases, we will give the degrees of freedom for each usage of χ^2 and refer the interested reader to the references at the end of the chapter for more detail.

TABLE 9.2 Reproduction of Part of Table VIII of Appendix A

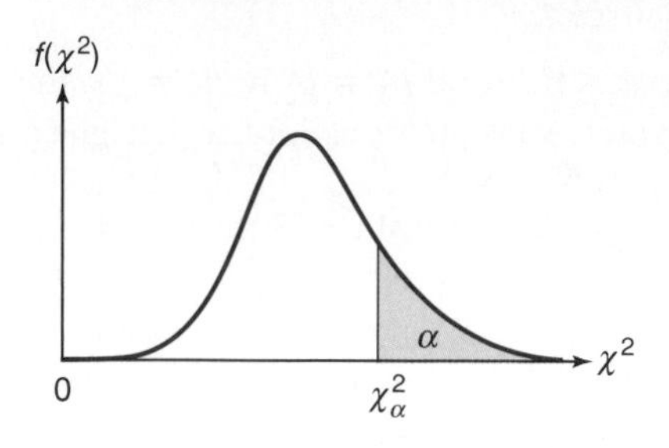

Degrees of Freedom	$\chi^2_{.100}$	$\chi^2_{.050}$	$\chi^2_{.025}$	$\chi^2_{.010}$	$\chi^2_{.005}$
1	2.70554	3.84146	5.02389	6.63490	7.87944
2	4.60517	5.99147	7.37776	9.21034	10.5966
3	6.25139	7.81473	9.34840	11.3449	12.8381
4	7.77944	9.48773	11.1433	13.2767	14.8602
5	9.23635	11.0705	12.8325	15.0863	16.7496
6	10.6446	12.5916	14.4494	16.8119	18.5476
7	12.0170	14.0671	16.0128	18.4753	20.2777

To illustrate, the rejection region for the voter-preference survey for $\alpha = .05$ and $k - 1 = 3 - 1 = 2$ df is

Rejection region: $X^2 > \chi^2_{.05}$

This value of $\chi^2_{.05}$ (found in Table VIII) is 5.99147. (See Figure 9.2.) The computed value of the test statistic is

$$X^2 = \frac{(n_1 - 50)^2}{50} + \frac{(n_2 - 50)^2}{50} + \frac{(n_3 - 50)^2}{50}$$

$$= \frac{(61 - 50)^2}{50} + \frac{(53 - 50)^2}{50} + \frac{(36 - 50)^2}{50} = 6.52$$

Since the computed $X^2 = 6.51$ exceeds the critical value of 5.99147, we conclude at the $\alpha = .05$ level of significance that there does exist a voter preference for one or more of the candidates.

Now that we have evidence to indicate that the proportions p_1, p_2, and p_3 are unequal, we can make inferences concerning their individual values using the methods of Section 6.4. [*Note:* We cannot use the methods of Section 9.1 to compare two proportions because the cell counts are dependent random variables.] The general form for a test of a hypothesis concerning multinomial probabilities is shown in the box.

FIGURE 9.2 ► Rejection region for voter-preference survey

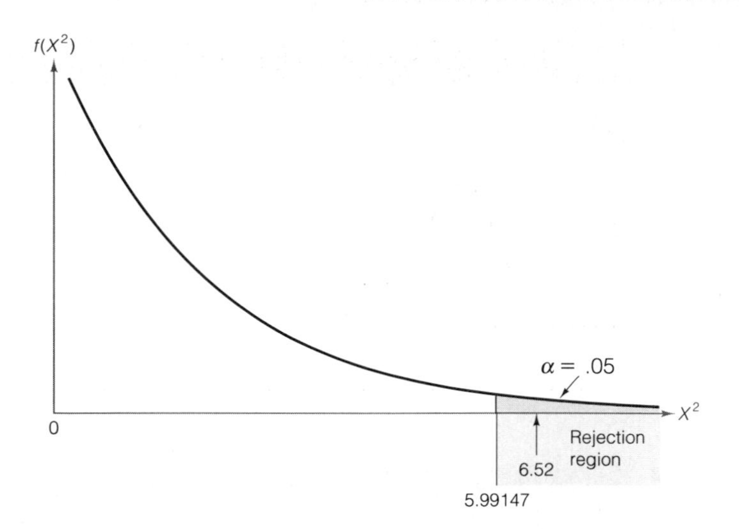

A Test of a Hypothesis About Multinomial Probabilities

H_0: $p_1 = p_{1,0}$, $p_2 = p_{2,0}$, . . . , $p_k = p_{k,0}$

where $p_{1,0}$, $p_{2,0}$, . . . , $p_{k,0}$ represent the hypothesized values of the multinomial probabilities

H_a: At least one of the multinomial probabilities does not equal its hypothesized value

Test statistic: $X^2 = \sum \frac{[n_i - E(n_i)]^2}{E(n_i)}$

where $E(n_i) = np_{i,0}$, the expected number of outcomes of type i assuming that H_0 is true. The total sample size is n.

Rejection region: $X^2 > \chi^2_\alpha$, where χ^2_α has $(k - 1)$ df

Assumptions:
1. A multinomial experiment has been conducted. This is generally satisfied by taking a random sample from the population of interest.
2. The sample size n will be large enough so that for every cell, the expected cell count $E(n_i)$ will be equal to 5 or more.*

*The assumption that all expected cell counts are at least 5 is necessary in order to ensure that the χ^2 approximation is appropriate. Exact methods for conducting the test of a hypothesis exist and may be used for small expected cell counts, but these methods are beyond the scope of this text.

EXAMPLE 9.3

Suppose an educational television station has broadcast a series of programs on the physiological and psychological effects of smoking marijuana. Now that the series is finished, the station wants to see whether the citizens within the viewing area have changed their minds about how possession of marijuana should be considered legally. Before the series was shown, it was determined that 7% of the citizens favored legalization, 18% favored decriminalization, 65% favored the existing law (a person could be fined or imprisoned), and 10% had no opinion.

A summary of the opinions (after the series was shown) of a random sample of 500 people in the viewing area is given in Table 9.3. Test at the $\alpha = .01$ level to see whether these data indicate that the distribution of opinions differs significantly from the proportions that existed before the educational series was aired.

TABLE 9.3 Distribution of Opinions About Marijuana Possession

Legalization	Decriminalization	Existing Laws	No Opinion
39	99	336	26

Solution

Define the proportions after the airing to be

p_1 = Proportion of citizens favoring legalization
p_2 = Proportion of citizens favoring decriminalization
p_3 = Proportion of citizens favoring existing laws
p_4 = Proportion of citizens with no opinion

Then the null hypothesis representing no change in the distribution of percentages is

$$H_0: \quad p_1 = .07, \quad p_2 = .18, \quad p_3 = .65, \quad p_4 = .10$$

and the alternative is

H_a: At least one of the proportions differs from its null hypothesized value

$$\textit{Test statistic:} \quad X^2 = \sum \frac{[n_i - E(n_i)]^2}{E(n_i)}$$

where

$$E(n_1) = np_{1,0} = 500(.07) = 35$$
$$E(n_2) = np_{2,0} = 500(.18) = 90$$
$$E(n_3) = np_{3,0} = 500(.65) = 325$$
$$E(n_4) = np_{4,0} = 500(.10) = 50$$

Since all these values are larger than 5, the χ^2 approximation is appropriate. Also, if the citizens in the sample were randomly selected, the properties of the multinomial probability distribution are satisfied.

Rejection region: For $\alpha = .01$ and df $= k - 1 = 3$, reject H_0 if $X^2 > \chi^2_{.01}$, where (from Table VIII in Appendix A) $\chi^2_{.01} = 11.3449$.

We now calculate the test statistic:

$$X^2 = \frac{(39-35)^2}{35} + \frac{(99-90)^2}{90} + \frac{(336-325)^2}{325} + \frac{(26-50)^2}{50} = 13.249$$

Since this value exceeds the table value of χ^2 (11.3449), the data provide sufficient evidence ($\alpha = .01$) that the opinions on legalization of marijuana have changed since the series was aired.

If we focus on one particular outcome of a multinomial experiment, we can use the methods developed in Section 6.4 for a binomial proportion to establish a confidence interval for any one of the multinomial probabilities.* For example, if we want a 95% confidence interval for the proportion of citizens in the viewing area who have no opinion about the issue, we calculate

$$\hat{p}_4 \pm 1.96\sigma_{\hat{p}_4}$$

where

$$\hat{p}_4 = \frac{n_4}{n} = \frac{26}{500} = .052 \quad \text{and} \quad \sigma_{\hat{p}_4} \approx \sqrt{\frac{\hat{p}_4(1-\hat{p}_4)}{n}}$$

Thus, we get

$$.052 \pm 1.96\sqrt{\frac{(.052)(.948)}{500}} = .052 \pm .019$$

or (.033, .071). Thus, we estimate that between 3.3% and 7.1% of the citizens now have no opinion on the issue of marijuana legalization. The series of programs may have helped citizens who formerly had no opinion on the issue to form an opinion, since it appears that the proportion of "no opinions" is now less than 10%.

Exercises 9.26–9.39

Learning the Mechanics

9.26 Use Table VIII of Appendix A to find each of the following χ^2 values:

a. $\chi^2_{.05}$ for df = 10 **b.** $\chi^2_{.990}$ for df = 50

c. $\chi^2_{.10}$ for df = 16 **d.** $\chi^2_{.005}$ for df = 50

*Note that focusing on one outcome has the effect of lumping the other ($k - 1$) outcomes into a single group. Thus, we obtain, in effect, two outcomes—or a binomial experiment.

9.27 Find the rejection region for a one-dimensional χ^2-test of a null hypothesis concerning $p_1, p_2, \ldots, p_k$ if:
a. $k = 3$; $\alpha = .05$ b. $k = 5$; $\alpha = .10$ c. $k = 4$; $\alpha = .01$

9.28 What are the characteristics of a multinomial experiment? Compare the characteristics to those of a binomial experiment.

9.29 What conditions must n satisfy to make the χ^2-test valid?

9.30 A multinomial experiment with $k = 3$ cells and $n = 320$ produced the data shown in the table. Do these data provide sufficient evidence to contradict the null hypothesis that $p_1 = .25$, $p_2 = .25$, and $p_3 = .50$? Test using $\alpha = .05$.

	Cell		
	1	2	3
n_i	78	60	182

9.31 A multinomial experiment with $k = 4$ cells and $n = 205$ produced the data shown in the table.

	Cell			
	1	2	3	4
n_i	43	56	59	47

a. Do these data provide sufficient evidence to conclude that the multinomial probabilities differ? Test using $\alpha = .05$.
b. What are the Type I and Type II errors associated with the test of part **a**?

9.32 Refer to Exercise 9.31. Construct a 95% confidence interval for the multinomial probability associated with cell 3.

Applying the Concepts

9.33 A 1985 Gallup survey portrays U.S. entrepreneurs as ". . . the mavericks, dreamers, and loners whose rough edges and uncompromising need to do it their own way set them in sharp contrast to senior executives in major American corporations" (Graham, 1985). One of the many questions put to a sample of $n = 100$ entrepreneurs about their job characteristics, work habits, social activities, etc. concerned the origin of the car they personally drive most frequently. The responses given in the table were obtained.

United States	*Europe*	*Japan*
45	46	9

a. Do these data provide evidence of a difference in the preference of entrepreneurs for the cars of the United States, Europe, and Japan? Test using $\alpha = .05$.

b. Do these data provide evidence of a difference in the preference of entrepreneurs for domestic versus foreign cars? Test using $\alpha = .05$.
c. What assumptions must be made in order for your inferences of parts **a** and **b** to be valid?
d. Use a 90% confidence interval to estimate the proportion of U.S. entrepreneurs who drive foreign cars.

9.34 Overweight trucks are responsible for much of the damage sustained by our local, state, and federal highway systems. For a particular week, the proportion of the week's total truck traffic (five-axle tractor truck semi-trailers) was distributed, Monday to Sunday, as follows:

Monday	*Tuesday*	*Wednesday*	*Thursday*	*Friday*	*Saturday*	*Sunday*
.191	.198	.187	.180	.155	.043	.046

Source: Dahlin, C., and Owen, F. "An analysis of data collected at the I-494 weighing-in-motion site." St. Paul: Minnesota Department of Transportation, 1984, p. 10.

During the same week, the number of overweight trucks per day was as follows:

Monday	*Tuesday*	*Wednesday*	*Thursday*	*Friday*	*Saturday*	*Sunday*
90	82	72	70	51	18	31

Source: Dahlin and Owen (1984).

a. A planning agency would like to know whether the number of overweight trucks per week is distributed over the 7 days of the week in direct proportion to the volume of truck traffic. Test using $\alpha = .05$.
b. Find the approximate p-value for the test of part **a**.

9.35 After purchasing a policy from one life insurance company, a person has a certain period of time in which the policy can be canceled without financial obligation. An insurance company is interested in seeing whether those who cancel a policy during this time period are as likely to be in one policy-size category as another. Records for 250 people who canceled policies during this period are selected at random from company files with the results shown in the table. Is there sufficient evidence to conclude that the canceled policies are not distributed equally among the five policy-size categories? Use $\alpha = .05$.

Size of Policy (Thousands of Dollars)	10	15	20	25	30
Number of Policies	31	39	67	54	59

9.36 In the game of chess, the first few moves play a very important role in determining the final outcome. Five different opening strategies are highly favored by chess experts. To determine whether one or more of these strategies is most preferred by grand masters in international competition, a random sample of 100 grand masters is taken, and each is asked which of the strategies he or she would prefer to employ. A summary of their responses is shown here:

Strategy:	A	B	C	D	E
Frequency:	17	27	22	15	19

Do these data present sufficient evidence to indicate a preference for one or more of the strategies? Use $\alpha = .05$.

9.37 There are four standard surgical techniques, A, B, C, and D, presently used in abdominal surgery. To find out whether one method is preferred over any other, each of a random sample of 200 surgeons was asked which technique he or she preferred. A summary of the data is shown here:

A	B	C	D
48	68	45	39

Do the data provide sufficient evidence to indicate differences in preferences for the four techniques? Test using $\alpha = .05$.

9.38 According to a geneticist's theory, a crossing of red and white snapdragons should produce offspring that are 25% red, 50% pink, and 25% white. An experiment conducted to test the theory produces 30 red, 78 pink, and 36 white offspring in 144 crossings. Do the data provide sufficient evidence to contradict the geneticist's theory?

9.39 Supermarket chains often carry products with their own brand labels and usually price them lower than the nationally known brands. A supermarket conducted a taste test to determine whether there was a difference in taste among the four brands of ice cream it carried: its own brand (A) and three national brands (B, C, D). A sample of 200 people participated, and they indicated the preferences shown in the table. Is there evidence of a difference in preference for the four brands? Test at $\alpha = .05$.

Brand			
A	B	C	D
39	57	55	49

9.4 Contingency Table Analysis (Optional)

In optional Section 9.3, we introduced the multinomial probability distribution and considered data classified according to a single criterion. We now consider multinomial experiments in which the data are classified according to two criteria, i.e., *classification with respect to two factors*.

For example, the energy shortage has made many consumers more aware of the size of the automobiles they purchase. Suppose an automobile manufacturer is interested in determining the relationship between the size and manufacturer of newly purchased automobiles. One thousand recent buyers of American-made cars are randomly sampled, and each purchase is classified with respect to the size and manufacturer of the automobile. The data are summarized in the **two-way table** shown in Table 9.4. This table is called a **contingency table**; it presents multinomial count data classified on two scales, or **dimensions**, of classification—namely, automobile size and manufacturer.

TABLE 9.4 Contingency Table for Automobile Size Example

	Manufacturer				Totals
	A	B	C	D	
Small	157	65	181	10	413
Intermediate	126	82	142	46	396
Large	58	45	60	28	191
Totals	341	192	383	84	1,000

The symbols representing the cell counts for the multinomial experiment in Table 9.4 are shown in Table 9.5, part A, and the corresponding cell, row, and column probabilities are shown in Table 9.5, part B. Thus, n_{11} represents the number of buyers who purchase a small car of manufacturer A and p_{11} represents the corresponding cell probability. Note the symbols for the row and column totals and also the symbols for the probability totals. The latter are called **marginal probabilities** for each row and column. The marginal probability p_{r1} is the probability that a small car is purchased; the marginal probability p_{c1} is the probability that a car by manufacturer A is purchased. Thus, $p_{r1} = p_{11} + p_{12} + p_{13} + p_{14}$ and $p_{c1} = p_{11} + p_{21} + p_{31}$.

TABLE 9.5 (A) Observed Counts for Contingency Table 9.4

	Manufacturer				Totals
	A	B	C	D	
Small	n_{11}	n_{12}	n_{13}	n_{14}	r_1
Intermediate	n_{21}	n_{22}	n_{23}	n_{24}	r_2
Large	n_{31}	n_{32}	n_{33}	n_{34}	r_3
Totals	c_1	c_2	c_3	c_4	n

(B) Probabilities for Contingency Table 9.4

	Manufacturer				Totals
	A	B	C	D	
Small	p_{11}	p_{12}	p_{13}	p_{14}	p_{r1}
Intermediate	p_{21}	p_{22}	p_{23}	p_{24}	p_{r2}
Large	p_{31}	p_{32}	p_{33}	p_{34}	p_{r3}
Totals	p_{c1}	p_{c2}	p_{c3}	p_{c4}	1

Thus, we can see that this really is a multinomial experiment with a total of 1,000 trials, (3)(4) = 12 cells or possible outcomes, and probabilities for each cell as shown in Table 9.5, part B. If the 1,000 recent buyers are randomly chosen, the trials are considered independent and the probabilities are viewed as remaining constant from trial to trial.

Suppose we want to know whether the two classifications, manufacturer and size, are dependent. That is, if we know which size car a buyer will choose, does that information give us a clue about the manufacturer of the car the buyer will choose? In a probabilistic sense we know (Chapter 3) that independence of events A and B implies $P(AB) = P(A)P(B)$. Similarly, in the contingency table analysis, if the two classifications are independent, the probability that an item is classified in any particular cell of the table is the product of the corresponding marginal probabilities. Thus, under the hypothesis of independence, in Table 9.5, part B, we must have

$$p_{11} = p_{r1}p_{c1} \qquad p_{12} = p_{r1}p_{c2}$$

and so forth.

To test the hypothesis of independence, we use the same reasoning employed in the chi-square test of Section 9.3. First, we calculate the expected, or mean, count in each cell assuming that the null hypothesis of independence is true. We do this by noting that the expected count in a cell of the table is just the total number of multinomial trials, n, times the cell probability. Recall that n_{ij} represents the observed count in the cell located in the ith row and jth column. Then the expected cell count for the upper-left-hand cell (first row, first column) is

$$E(n_{11}) = np_{11}$$

or, when the null hypothesis (the classifications are independent) is true,

$$E(n_{11}) = np_{r1}p_{c1}$$

Since these true probabilities are not known, we estimate p_{r1} and p_{c1} by the same proportions $\hat{p}_{r1} = r_1/n$ and $\hat{p}_{c1} = c_1/n$. Thus, the estimate of the expected value $E(n_{11})$ is

$$\hat{E}(n_{11}) = n\left(\frac{r_1}{n}\right)\left(\frac{c_1}{n}\right) = \frac{r_1c_1}{n}$$

Similarly, for each i, j,

$$\hat{E}(n_{ij}) = \frac{\text{(Row total)(Column total)}}{\text{Total sample size}}$$

Thus,

$$\hat{E}(n_{12}) = \frac{r_1c_2}{n}$$

$$\vdots \qquad \vdots$$

$$\hat{E}(n_{34}) = \frac{r_3c_4}{n}$$

Using the data in Table 9.4, we find

$$\hat{E}(n_{11}) = \frac{r_1c_1}{n} = \frac{(413)(341)}{1{,}000} = 140.833$$

$$\hat{E}(n_{12}) = \frac{r_1c_2}{n} = \frac{(413)(192)}{1{,}000} = 79.296$$

$$\vdots \qquad \vdots \qquad \vdots \qquad \vdots$$

$$\hat{E}(n_{34}) = \frac{r_3c_4}{n} = \frac{(191)(84)}{1{,}000} = 16.044$$

The observed data and the estimated expected values (in parentheses) are shown in Table 9.6.

We now use the X^2 statistic to compare the observed and expected (estimated) counts in each cell of the contingency table:

$$X^2 = \frac{[n_{11} - \hat{E}(n_{11})]^2}{\hat{E}(n_{11})} + \frac{[n_{12} - \hat{E}(n_{12})]^2}{\hat{E}(n_{12})} + \cdots + \frac{[n_{34} - \hat{E}(n_{34})]^2}{\hat{E}(n_{34})}$$

$$= \sum \frac{[n_{ij} - \hat{E}(n_{ij})]^{2*}}{\hat{E}(n_{ij})}$$

TABLE 9.6 Observed and Estimated Expected (in Parentheses) Counts

	Manufacturer				Totals
	A	B	C	D	
Small	157 (140.833)	65 (79.296)	181 (158.179)	10 (34.692)	413
Intermediate	126 (135.036)	82 (76.032)	142 (151.668)	46 (33.264)	396
Large	58 (65.131)	45 (36.672)	60 (73.153)	28 (16.044)	191
Totals	341	192	383	84	1,000

Substituting the data of Table 9.6 into this expression, we get

$$X^2 = \frac{(157 - 140.833)^2}{140.833} + \frac{(65 - 79.296)^2}{79.296} + \cdots + \frac{(28 - 16.044)^2}{16.044} = 45.81$$

Large values of X^2 imply that the observed counts do not closely agree and therefore imply that the hypothesis of independence is false. To determine how large X^2 must be before it is too large to be attributed to chance, we make use of the fact that the sampling distribution of X^2 is approximately a χ^2 probability distribution when the classifications are independent.

When testing the null hypothesis of independence in a two-way contingency table, the appropriate degrees of freedom will be $(r - 1)(c - 1)$, where r is the number of rows and c is the number of columns in the table.

*The use of Σ in the context of a contingency table analysis refers to a sum over all cells in the table.

For the size and make of automobiles example, the degrees of freedom for χ^2 is $(r - 1)(c - 1) = (3 - 1)(4 - 1) = 6$. Then, for $\alpha = .05$, we reject the hypothesis of independence when

$$X^2 > \chi^2_{.05} = 12.5916$$

Since the computed $X^2 = 45.81$ exceeds the value 12.5916, we conclude that the size and manufacturer of a car selected by a purchaser are dependent events.

The pattern of dependence can be seen more clearly by expressing the data as percentages. We first select one of the two classifications to be used as the base variable. In the automobile size preference example, suppose we select manufacturer as the classificatory variable to be the base. Next, we represent the responses for each level of the second categorical variable (size of automobile in our example) as a percentage of the subtotal for the base variable. For example, from Table 9.6 we convert the response for small car sales for manufacturer A (157) to a percentage of the total sales for manufacturer A (341). That is,

$$(157/341)100\% = 46\%$$

The conversions of all Table 9.6 entries are similarly computed, and the values are shown in Table 9.7. The value shown at the right of each row is the row's total expressed as a percentage of the total number of responses in the entire table. Thus, the small car percentage is $413/1{,}000(100\%) = 41\%$ (rounded to the nearest percent).

TABLE 9.7 Percentage of Car Sizes by Manufacturer

	Manufacturer				All
	A	B	C	D	
Small	46	34	47	12	41
Intermediate	37	43	37	55	40
Large	17	23	16	33	19
Totals	100	100	100	100	100

If the size and manufacturer variables are independent, then the percentages in the cells of the table are expected to be approximately equal to the corresponding row percentages. Thus, we would expect the small car percentages for each of the four manufacturers to be approximately 41% if size and manufacturer were independent.

The extent to which each manufacturer's percentage departs from this value determines the dependence of the two classifications, with greater variability of the row percentages meaning a greater degree of dependence. A plot of the percentages helps summarize the observed pattern. In Figure 9.3 we show the manufacturer (the base variable) on the horizontal axis, and the size percentages on the vertical axis. The "expected" percentages under the assumption of independence are shown as horizontal lines, and each observed value is represented by a symbol indicating the size category.

FIGURE 9.3 ▶
Size as a percentage of manufacturer subtotals

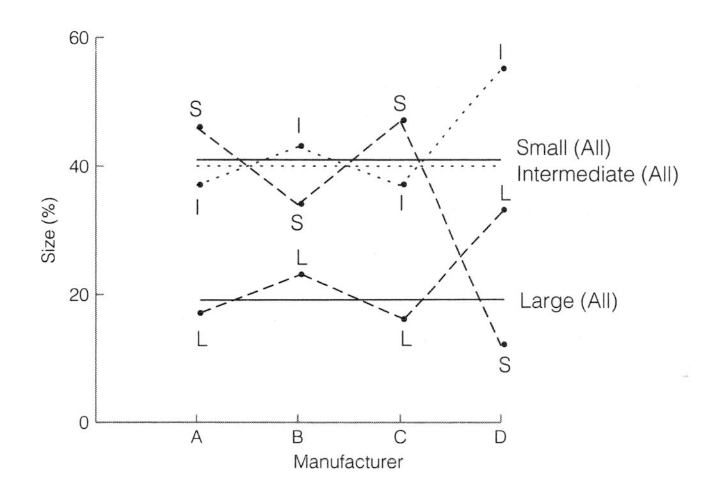

Figure 9.3 clearly indicates the reason that the test resulted in the conclusion that the two classifications in the contingency table are dependent. Note that the sales of manufacturers A, B, and C fall relatively close to the expected percentages under the assumption of independence. However, the sales of manufacturer D deviate significantly from the expected values, with much higher percentages for large and intermediate cars and a much smaller percentage for small cars than expected under independence. Also, manufacturer B deviates slightly from the expected pattern, with a greater percentage of intermediate than small car sales. Statistical measures of the degree of dependence and procedures for making comparisons of pairs of levels for classifications are available. They are beyond the scope of this text, but can be found in the references at the end of the chapter. We will, however, utilize descriptive summaries such as Figure 9.3 to examine the degree of dependence exhibited by the sample data.

The general form of a two-way contingency table containing r rows and c columns (called an $r \times c$ contingency table) is shown in Table 9.8 on page 470. Note that the observed count in the (ij) cell is denoted by n_{ij}, the ith row total is r_i, the jth column total is c_j, and the total sample size is n. Using this notation, we give the general form of the contingency table test for independent classifications in the box.

TABLE 9.8 General $r \times c$ Contingency Table

		Column				Row Totals
		1	2	...	c	
Row	1	n_{11}	n_{12}	...	n_{1c}	r_1
	2	n_{21}	n_{22}	...	n_{2c}	r_2
	$\vdots$	$\vdots$	$\vdots$		$\vdots$	$\vdots$
	r	n_{r1}	n_{r2}	...	n_{rc}	r_r
Column Totals		c_1	c_2	...	c_c	n

General Form of a Contingency Table Analysis: A Test for Independence

H_0: The two classifications are independent
H_a: The two classifications are dependent

Test statistic: $X^2 = \sum \frac{[n_{ij} - \hat{E}(n_{ij})]^2}{\hat{E}(n_{ij})}$

where $\hat{E}(n_{ij}) = \frac{r_i c_j}{n}$

Rejection region: $X^2 > \chi^2_\alpha$, where χ^2_α has $(r - 1)(c - 1)$ df.

Assumptions:
1. The n observed counts are a random sample from the population of interest. We may then consider this to be a multinomial experiment with $r \times c$ possible outcomes.
2. The sample size, n, will be large enough so that, for every cell, the expected count, $E(n_{ij})$, will be equal to 5 or more.

EXAMPLE 9.4

A social scientist wants to determine whether the marital status (divorced or not divorced) of American men is independent of their religious affiliation (or lack thereof). A sample of 500 American men is surveyed and the results are tabulated as shown in Table 9.9.

TABLE 9.9 Observed and Estimated (in Parentheses) Expected Counts, Example 9.4

		Religious Affiliation					Totals
		A	B	C	D	None	
Marital Status	Divorced	39 (48.952)	19 (18.560)	12 (12.992)	28 (22.736)	18 (12.760)	116
	Never divorced	172 (162.048)	61 (61.440)	44 (43.008)	70 (75.264)	37 (42.240)	384
Totals		211	80	56	98	55	500

a. Test to see whether there is sufficient evidence to indicate that the marital status of men who have been or are currently married is dependent on religious affiliation. Test using $\alpha = .01$.

b. Plot the data and describe the patterns revealed. Is the result of the test supported by the plot?

Solution

a. The first step is to calculate estimated expected cell frequencies under the assumption that the classifications are independent. Thus,

$$\hat{E}(n_{11}) = \frac{r_1 c_1}{n} = \frac{(116)(211)}{500} = 48.952$$

$$\hat{E}(n_{12}) = \frac{r_1 c_2}{n} = \frac{(116)(80)}{500} = 18.560$$

and so forth. All the estimated expected cell counts are shown in Table 9.9.

We are now ready to conduct the test for independence:

H_0: The marital status of American men and their religious affiliation are independent

H_a: The marital status of American men and their religious affiliation are dependent

Test statistic: $$X^2 = \sum \frac{[n_{ij} - \hat{E}(n_{ij})]^2}{\hat{E}(n_{ij})}$$

Since all the estimated expected cell frequencies are greater than 5, the χ^2 approximation is appropriate. Assuming the men chosen were randomly selected from all married or previously married American men, the characteristics of the multinomial probability distribution are satisfied.

Rejection region: For $\alpha = .01$ and $(r - 1)(c - 1) = (1)(4) = 4$ df, reject H_0 if $X^2 > \chi^2_{.01}$, where $\chi^2_{.01} = 13.2767$

The calculated value of the test statistic is

$$X^2 = \frac{(39 - 48.952)^2}{48.952} + \frac{(19 - 18.560)^2}{18.560} + \cdots + \frac{(37 - 42.240)^2}{42.240}$$
$$= 7.135$$

Since $X^2 = 7.135$ is less than $\chi^2_{.01} = 13.2767$, we cannot conclude that the marital status of American men depends on their religious affiliation. (Note that we could not reject H_0 even with $\alpha = .10$, since $\chi^2_{.10} = 7.77944$.)

b. The marital status frequencies are expressed as percentages of the number of men in each religious affiliation category in Table 9.10. The expected percentages under the assumption of independence are shown at the right of each row. The plot of the percentage data is shown in Figure 9.4, where horizontal lines represent the expected percentages assuming independence. Note that the response percentages deviate only slightly from those expected under the assumption of independence, supporting the result of the test in part **a**. That is, neither the descriptive plot nor the statistical test provides evidence that the male divorce rate depends on (varies with) religious affiliation.

TABLE 9.10 Marital Status as Percentage of Religious Affiliation

		Religious Affiliation				All
		A	B	C	D	
Marital Status	Divorced	18	24	21	33	23
	Never divorced	82	76	79	67	77
Totals		100	100	100	100	100

FIGURE 9.4 ▶
Plot of marital status–religious affiliation contingency table

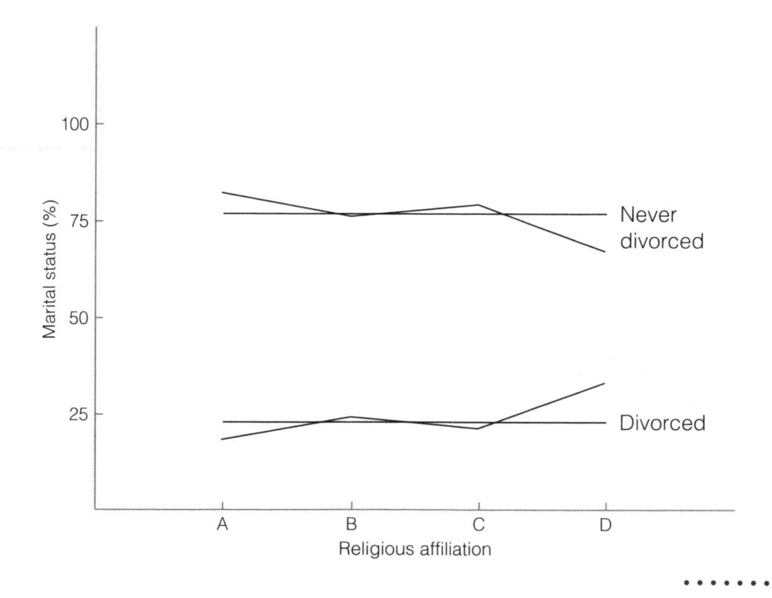

CASE STUDY 9.1 / Evaluating a New Method for Treating Cancer

The typical method of treating moderately advanced cancer of the larynx (voice box) is removal by surgery. This method achieves initial control of the cancer in approximately 75% of the cases; if the cancer recurs, salvage treatment results in approximately 87% of the cancers ultimately controlled. Attempts have been made to treat cancer of the larynx by radiation therapy alone, thereby saving the patient's larynx. W. M. Mendenhall et al. present data on a group of patients treated by this method at the University of Florida's Shands Hospital. Eighteen patients with cancer of the larynx were treated by radiation alone and 23 were treated by surgery alone. Of those treated by radiation, 11 cancers were controlled at the primary site (the larynx). In seven patients the cancer recurred in the larynx; six were treated by surgery and one refused further treatment. Of these, four were controlled so that the number ultimately controlled by radiation therapy alone (11) or by surgical salvage (four) was 15 of 18 patients. Initial control of the cancer was achieved for 18 of the 23 patients treated by surgery alone. Salvage treatment was successful for three of the remaining five. Thus, removal of the larynx by surgery alone achieved ultimate control in 21 of the 23 cases. These data are summarized in Table 9.11.

TABLE 9.11 Comparison of Two Methods for Treating Cancer of the Larynx

	Surgery	Radiation Therapy
Number of Cancers Ultimately Controlled	21	15
Number Not Ultimately Controlled	2	3
Totals	23	18

Source: Mendenhall, W. M., Million, R. R., Sharkey, D. E., and Cassisi, N. J. "Stage T3 squamous cell carcinoma of the larynx treated with surgery and/or radiation therapy." *International Journal of Radiation Oncology, Biology and Physics*, 1984, 10. Copyright 1984, Pergamon Journals, Ltd. Reprinted with permission.

The data of Table 9.11 cannot be analyzed using the chi-square test to detect differences in rates of control for the two methods of treatment. The expected numbers in the "no ultimate control" cells are too small. However, the same test can be conducted by using a small-sample method known as *Fisher's exact test*. This method calculates the exact probability (p-value) of observing sample results at least as contradictory to the hypothesis of independence as those observed for the researchers' data. Mendenhall and coworkers report the p-value for this test as .187, a value that indicates little evidence of a difference in the rates of achieving ultimate control for the two methods of treatment.

Mendenhall et al. clearly intend their research to be a first step in a study that will require the treatment of many more cancer patients using radiation alone. Small-sample methods are available for constructing a confidence interval for the rate of achieving ultimate control using radiation alone, but the interval width would be quite large. As we learned in Chapter 6, it requires a fairly large sample to estimate a binomial proportion p with a small margin of error. This larger sample will be obtained by combining treatment results from other clinics and by collecting data on patients who will be treated in the future.

Exercises 9.40–9.54

Learning the Mechanics

9.40 Find the rejection region for a test of independence of two classifications where the contingency table contains r rows and c columns and:

a. $r = 5, \quad c = 5, \quad \alpha = .05$ **b.** $r = 3, \quad c = 6, \quad \alpha = .10$ **c.** $r = 2, \quad c = 3, \quad \alpha = .01$

9.41 Consider the accompanying 2 × 3 (i.e., $r = 2$ and $c = 3$) contingency table.

		Column		
		1	2	3
Row	1	9	34	53
	2	16	30	25

a. Specify the null and alternative hypotheses that should be used in testing the independence of the row and column classifications.
b. Specify the test statistic and the rejection region that should be used in conducting the hypothesis test of part **a**. Use $\alpha = .01$.
c. Assuming the row classification and the column classification are independent, find estimates for the expected cell counts.
d. Conduct the hypothesis test of part **a**. Interpret your result.

9.42 Refer to Exercise 9.41.
a. Convert the frequency responses to percentages by calculating the percentage of each column total falling in each row. Also convert the row totals to percentages of the total number of responses. Display the percentages in a table.
b. Create a graph with percentage on the vertical axis and column number on the horizontal axis. Show the row total percentages as horizontal lines on the plot, and plot the cell percentages from part **a** using the row number as a plotting symbol.
c. What pattern do you expect to see if the rows and columns are independent? Does the plot support the result of the test of independence in Exercise 9.41?

9.43 Test the null hypothesis of independence of the two classifications, A and B, of the 3 × 3 contingency table shown here. Test using $\alpha = .05$.

		B		
		B_1	B_2	B_3
A	A_1	40	72	42
	A_2	63	53	70
	A_3	31	38	30

9.44 Refer to Exercise 9.43. Convert the responses to percentages by calculating the percentage of each B class total falling into each A classification. Also, calculate the percentage of the total number of responses that constitute each of the A classification totals. Create a graph with percentage on the vertical axis and B classification on the horizontal axis. Show the percentages corresponding to the A classification totals as horizontal lines on the graph, and plot the individual cell percentages using the A class number as a plotting symbol. Does the graph support the result of the test of hypothesis in Exercise 9.43? Explain.

Applying the Concepts

9.45 Many companies use well-known celebrities in their ads, while other companies create their own spokesperson (such as the Maytag repairman). A recent study in the *Journal of Marketing* (Tom et al., 1992) investigated the relationship between the gender of the spokesperson and gender of the viewer and how this relationship affected brand awareness. Three hundred television viewers were asked to identify the products advertised by celebrity spokespersons. The results are shown below:

Male Spokesperson

	Audience Gender		Total
	Male	Female	
Identified Product	95	41	136
Could Not Identify Product	55	109	164
Total	150	150	300

Female Spokesperson

	Audience Gender		Total
	Male	Female	
Identified Product	47	61	108
Could Not Identify Product	103	89	192
Total	150	150	300

a. For the products advertised by male spokespersons, conduct a test to determine whether audience gender and product identification are dependent factors. Test using $\alpha = .05$.
b. Repeat part **a** for the products advertised by female spokespersons.
c. How would you interpret these results?

9.46 Refer to *The International Journal of Sports Psychology* (Tucker, 1990) study of the relationship between physical fitness and stress in Exercise 9.19. Recall that 549 employees of companies that participate in the Health Examination Program offered by Health Advancement Services (HAS) were classified into three groups of fitness levels (good, average, and poor), and each person was tested for signs of stress. The results are reproduced below. [*Note:* The proportions given are the proportions of the entire group that show signs of stress and fall into each particular fitness level.]

Fitness Level	*Sample Size*	*Proportion with Signs of Stress*
Poor	242	.155
Average	212	.133
Good	95	.108

Do the data provide evidence to indicate that the likelihood for stress is dependent on an employee's fitness level? Use $\alpha = .05$.

9.47 Americans over age 50 represent 25% of the population, yet they control 70% of the wealth. Research indicates that the highest priority of retirees is travel. A recent study in the *Annals of Tourism Research* (Blazey, 1992) investigates the relationship of retirement status (pre- and postretirement) to various items related to the travel industry. One part of the study investigated the differences in the length of stay of a trip for pre- and postretirees. A sample of 703 travelers were asked how long they stayed on a typical trip. The results are shown in the table at the top of page 476.

Number of Nights	*Preretirement*	*Postretirement*
4 to 7	247	172
8 to 13	82	67
14 to 21	35	52
22 or more	16	32
Total	380	323

Use the information in the table to determine whether the retirement status of a traveler and the duration of a typical trip are dependent. Test using $\alpha = .05$.

9.48 Researchers conducted a study of 58 children admitted to the child psychiatry ward of the University of Iowa for aggressive conduct. The purpose of the study was to determine the influence of social class, sex, and age on the clinical characteristics of the children. A small portion of their data is shown in the table, which lists the numbers of children exhibiting antisocial behavior for two categories of social class. Classes I to III represent children from middle-class families. Those in Classes IV and V were from poor families.

	Classes I–III	**Classes IV–V**
Numbers Exhibiting Antisocial Behavior	24	17
Sample Size	28	30

Source: Behar, D., and Stewart, M. A. "Aggressive conduct disorder: The influence of social class, sex, and age on the clinical picture." *Journal of Child Psychology and Psychiatry*, 1984, 25.

a. Do the data provide sufficient evidence to indicate a dependence between antisocial behavior and social class? Test using $\alpha = .05$.
b. The authors give the approximate p-value for the test as .015. Calculate the approximate p-value for the test and compare it with the authors' value. Interpret the p-value.

9.49 Refer to Exercise 9.48 and the study by Behar and Stewart (1984) on aggressive behavior in children. Another comparison given in their paper was the numbers of male and female children for whom physical aggressiveness was a presenting complaint. The data are shown in the table.

	Female	**Male**
Number for Which Physical Aggressiveness Is a Presenting Complaint	7	40
Sample Size	12	46

Source: Behar and Stewart (1984).

a. Do the data provide sufficient evidence to indicate a dependence between sex of the child and the proportion for which physical aggressiveness is a presenting complaint? Test using $\alpha = .05$.
b. The authors give the p-value for the test as .025. Find the p-value for your test and compare it with the authors' value. Interpret the p-value.

9.50 Many scientists believe that alcoholism is linked to social isolation. One measure of social isolation is marital status, i.e., whether a person is married or not. To test the notion that alcoholics are socially isolated, 280 adults were randomly selected and each was classified as a diagnosed alcoholic, undiagnosed alcoholic, or not alcoholic and categorized according to his or her marital status. A summary of the responses is shown in the table.

		Alcoholic Classification		
		Diagnosed	Undiagnosed	Not Alcoholic
Marital Status	Married	21	37	58
	Not married	59	63	42

a. Is there evidence of a relationship between marital status and alcoholic classifications? Test using $\alpha = .05$.
b. Construct a graph showing the percentage of married and not married for each alcoholic classification. Compare these to the percentages based on the total over all alcoholic classifications. Does the plot support the result of the test in part **a**?

9.51 One criterion used to evaluate employees in the assembly section of a large factory is the number of defective pieces per 1,000 parts produced. The quality control department wants to find out whether there is a relationship between years of experience and defect rate. Since the job is repetitious, after the initial training period any improvement due to a learning effect might be offset by a decrease in the motivation of a worker. A defect rate is calculated for each worker for a yearly evaluation. The results for 100 workers are given in the table. Is there evidence of a relationship between defect rate and years of experience? Use $\alpha = .05$.

		Years of Experience (After Training Period)		
		1	2–5	6–10
Defect Rate	High	6	9	9
	Average	9	19	23
	Low	7	8	10

9.52 An experimenter wishes to determine whether there is a relationship between hair color and eye color. One hundred people are randomly sampled and the eyes and hair of each person are judged to be light or dark. A summary of the number of people in each of the four categories is shown in the table. Do the data provide sufficient evidence to indicate a relationship between eye and hair color? Test using $\alpha = .10$.

	Light Hair	Dark Hair	Totals
Light Eyes	31	21	52
Dark Eyes	14	34	48
Totals	45	55	100

9.53 As noted earlier in our discussion, the X^2 statistic possesses approximately a chi-square distribution when the sample size n is sufficiently large. The actual sample size n needed to achieve a satisfactory approximation depends on the application. To be safe, we suggested that n be large enough so that all expected cell counts exceed 5. Case Study 9.1 produced data that did not satisfy this criterion. Calculate the X^2 statistic for the data of Case Study 9.1 and find its p-value. Compare this approximate p-value with the exact p-value given by Mendenhall et al. This comparison will enable you to see how well the results for the two tests agree.

9.54 Over the years, pollsters have found that the public's confidence in big business has been closely tied to the economic climate of the country. When businesses are growing and employment is increasing, public confidence is high. When the opposite occurs, public confidence is low. In one study, Harvey Kahalas (1981) explored the relationship between confidence in business and job satisfaction. He hypothesized that there is a relationship between level of confidence and job satisfaction and that this relationship holds true for both union and nonunion workers. To test his hypothesis he used the sample data given in the tables, which were collected by the National Opinion Research Center.

Sample of Union Members

		Job Satisfaction			
		Very Satisfied	Moderately Satisfied	Little Dissatisfied	Very Dissatisfied
Confidence in Major Corporations	A Great Deal	26	15	2	1
	Only Some	95	73	16	5
	Hardly Any	34	28	10	9

Sample of Nonunion Workers

		Job Satisfaction			
		Very Satisfied	Moderately Satisfied	Little Dissatisfied	Very Dissatisfied
Confidence in Major Corporations	A Great Deal	111	52	13	4
	Only Some	246	142	37	18
	Hardly Any	73	51	19	9

a. Kahalas concluded that his hypothesis was not supported by the data. Do you agree? Conduct the appropriate hypothesis tests using $\alpha = .05$. Be sure to specify the null and alternative hypotheses of your tests.
b. Find and interpret the approximate p-values of the tests you conducted in part **a**.
c. Construct graphs to assist in the interpretation of the test results that you obtained.

Summary

This chapter showed you how to compare two population proportions using methodology that is very similar to the methodology used to compare two population means (Chapter 8). When the samples have been randomly and independently selected from two binomial populations and when the sample sizes are large, the difference $(\hat{p}_1 - \hat{p}_2)$ between the sample proportions will be approximately normally distributed (the Central Limit Theorem, Section 5.3). This enables us to form a large-sample standard normal z statistic similar to the one used to compare two population means. This statistic is then used as a test statistic for testing hypotheses about the difference between two population proportions and to construct a confidence interval for $(p_1 - p_2)$. The width of this confidence interval can be controlled by adjusting the sample sizes n_1 and n_2.

Optional Section 9.3 discussed how to compare more than two population proportions in a multinomial distribution using a **chi-square test.** This methodology was applied in optional Section 9.4 to test for a dependence between two methods of classification in a **contingency table.**

Supplementary Exercises 9.55–9.72

[Note: Starred (*) exercises refer to the optional sections in this chapter.]

Learning the Mechanics

9.55 Independent random samples were selected from two binomial populations. The sizes and number of observed successes for each sample are shown in the table.

Sample 1	*Sample 2*
$n_1 = 200$	$n_2 = 200$
$x_1 = 110$	$x_2 = 130$

a. Test H_0: $(p_1 - p_2) = 0$ against H_a: $(p_1 - p_2) < 0$. Use $\alpha = .10$.
b. Form a 95% confidence interval for $(p_1 - p_2)$.
c. What sample sizes would be required if we wish to use a 95% confidence interval of width .01 to estimate $(p_1 - p_2)$?

9.56 Assuming that $n_1 = n_2$, find the sample sizes needed to estimate $(p_1 - p_2)$ for each of the following situations:
a. Bound = .01 with 99% confidence. Assume that $p_1 \approx .2$ and $p_2 \approx .5$.
b. Bound = .05 with 90% confidence. Assume there is no prior information available to obtain approximate values of p_1 and p_2.
c. Bound = .05 with 90% confidence. Assume that $p_1 \approx .1$ and $p_2 \approx .3$.

*9.57 A random sample of 150 observations was classified into the categories shown in the table.

	Category				
	1	2	3	4	5
n_i	28	35	33	25	29

a. Do the data provide sufficient evidence that the categories are not equally likely? Use $\alpha = .10$.
b. Form a 90% confidence interval for p_2, the probability that an observation will fall in category 2.

*9.58 A random sample of 250 observations was classified according to the row and column categories shown in the table.

		Column		
		1	2	3
Row	1	20	20	10
	2	10	20	70
	3	20	50	30

a. Do the data provide sufficient evidence to conclude that the rows and columns are dependent? Test using $\alpha = .05$.
b. Would the analysis change if the row totals were fixed before the data were collected?
c. Do the assumptions required for the analysis to be valid differ according to whether the row (or column) totals are fixed? Explain.
d. Convert the table entries to percentages by using each column total as a base, and calculating each row response as a percentage of the corresponding column total. In addition, calculate the row totals, and convert them to percentages of all 250 observations.
e. Plot the row percentages on the vertical axis against the column number on the horizontal axis. Draw horizontal lines corresponding to the row total percentages. Does the deviation (or lack thereof) of the individual row percentages from the row total percentages support the result of the test conducted in part **a**?

Applying the Concepts

9.59 Radio stations sometimes conduct prize giveaways in an attempt to increase their share of the listening audience. Suppose a station manager calls 300 randomly selected households in a city and finds that 65 have members who regularly listen to the station. The station then conducts a 2-month promotional contest and follows it with a survey of 500 randomly chosen households. The survey shows that 154 households have members who regularly listen to the station.
a. Use a 90% confidence interval to estimate the difference between the proportions of those who regularly listen to the station before and after the promotional contest.
b. Construct a 95% confidence interval for the proportion of those who listen to the station after the promotion is over.

9.60 A consumer protection agency wants to compare the work of two electrical contractors in order to evaluate their safety records. The agency plans to inspect residences in which each of these contractors has done the wiring in order to estimate the difference in the proportions of residences that are electrically deficient. Suppose the proportions of residences with deficient work are expected to be about .10 for both contractors. How many homes should be sampled in order to estimate the difference in proportions using a 90% confidence interval of width .05?

9.61 A large shipment of produce contains Valencia and navel oranges. To determine whether there is a difference in the proportions of nonmarketable fruit between the two varieties, random samples of 850 Valencia oranges and 1,500 navel oranges are independently selected and the number of nonmarketable oranges of each type is counted. It is found that 30 Valencia and 90 navel oranges from these samples are nonmarketble. Do these data provide sufficient evidence to indicate a difference between the proportions of nonmarketable Valencia and navel oranges? Test at the $\alpha = .05$ level of significance.

9.62 The threat of earthquakes is a part of life for homeowners in California. Scientists have been warning about "the big one" for decades. A recent article in the *Annals of the Association of American Geographers* (Palm and Hodgson, 1992) explored some factors that are considered when California homeowners purchase earthquake insurance, including the proximity to a major earthquake fault. Surveys were mailed to residents in four California counties. The data collected are shown in the table below.

	Contra Costa	*Santa Clara*	*Los Angeles*	*San Bernardino*
Sample size	521	556	337	372
Number with earthquake insurance	117	222	133	109

a. Los Angeles County is the closest of the four to a major earthquake fault. Calculate 95% confidence intervals for the difference in the proportions of earthquake-insured residents in Los Angeles County and each of the other counties.

b. Do these results support the contention that closer proximities to major earthquake faults result in higher proportions of earthquake-insured residents?

***9.63** Refer to Exercise 9.62.

a. State the hypothesis necessary to test the theory that the proportion of earthquake-insured residents differs for the four sampled counties.

b. Do the data provide evidence of a difference in the proportion of earthquake-insured residents for the four sampled counties? Test using $\alpha = .05$.

9.64 Two basketball players engage in a foul-shooting contest in which each player takes 100 shots. It is desired to compare the percentages of shots made by each player.

a. What is the parameter of interest?

b. In this contest, player A made 93 shots and player B made 86 shots. Estimate the parameter of interest using a 95% confidence interval.

9.65 Refer to Exercise 9.64. How large would the samples from Los Angeles and San Bernardino counties have to be in order to estimate the difference between earthquake-insured proportions to within .03 with 95% confidence?

***9.66** Despite a good winning percentage, a certain major league baseball team has not drawn as many fans as one would expect. In hopes of finding ways to increase attendance, the management plans to interview fans who come to the games to find out why they come. One thing that the management might want to know is whether there are differences in support for the team among various age groups. Suppose the information in the table was collected during interviews with fans selected at random. Can you conclude that there is a relationship between age and number of games attended per year? Use $\alpha = .05$.

		Number of Games Attended per Year		
		1 or 2	3–5	Over 5
Age of Fan	Under 20	78	107	17
	21–30	147	87	13
	31–40	129	86	19
	41–55	55	103	40
	Over 55	23	74	22

***9.67** If a company can identify times of day when accidents are most likely to occur, extra precautions can be instituted during those times. A random sampling of the accident reports over the last year at a plant gives the frequency of occurrence of accidents during the different hours of the workday. Can it be concluded from the data in the table that the proportions of accidents are different for at least two of the four time periods?

Hours	1–2	3–4	5–6	7–8
Number of Accidents	31	28	45	47

***9.68** The classification of solder joints as acceptable or rejectable is a particularly difficult inspection task due to its subjective nature. Westinghouse Electric Company has experimented with different means of evaluating the performance of solder inspectors. One approach involves comparing an individual inspector's classifications with those of the group of experts that comprise Westinghouse's Work Standards Committee. In an experiment reported by Joseph J. Meagher and Joseph A. Scazzero (1985) of Westinghouse, 153 solder connections were evaluated by the committee and 111 were classified as acceptable. An inspector evaluated the same 153 connections and classified 124 as acceptable. Of the items rejected by the inspector, the committee agreed with 19.

a. Construct a contingency table that summarizes the classifications of the committee and the inspector.
b. Based on a visual examination of the table you constructed in part **a**, does it appear that there is a relationship between the inspector's classifications and the committee's? Explain. (A plot of the percentage rejected by committee and inspector will aid your examination.)
c. Conduct a chi-square test of independence for these data. Use $\alpha = .05$. Carefully interpret the results of your test in the context of the problem.

***9.69** A sociologist was interested in knowing whether sons have a tendency to choose the same occupation as their fathers. To investigate this question, 500 males were polled and questioned concerning their occupation and the occupation of their father. A summary of the numbers of father–son pairs falling in each occupational category is shown in the table. Do the data provide sufficient evidence to indicate a dependence between a son's choice of occupation and his father's occupation? Test using $\alpha = .05$.

		Son			
		Professional or Business	Skilled	Unskilled	Farmer
Father	Professional or Business	55	38	7	0
	Skilled	79	71	25	0
	Unskilled	22	75	38	10
	Farmer	15	23	10	32

***9.70** A study was done on the accuracy of newspaper advertisements by the five types of food stores in a southeastern city. On each of 4 days, items were randomly selected from the advertisements for each type of store and the actual price was compared to the advertised price. Each of the stores in the city was classified as one of the following types: national, regional chain A, regional chain B, regional chain C, or independent. Values in the table represent the number of items that were correctly and incorrectly priced.

Type of Store	*Number Correctly Priced*	*Number Incorrectly Priced*
National chain	89	10
Regional chain A	53	14
Regional chain B	43	12
Regional chain C	32	13
Independent	41	7

a. Determine whether these data provide sufficient evidence to conclude that the proportion of correctly priced items differs for at least two types of stores. Use $\alpha = .10$.

b. Use a 95% confidence interval to estimate the proportion of correctly priced items in the stores in the national chain category.

***9.71** Teenage alcoholism is a big problem in the United States. To discover why teenagers are turning to alcohol, a survey was conducted to find out whether a teenager's family status has any relationship to the amount of alcohol he or she consumes. A random sample of 200 teenagers between the ages of 15 and 19 were questioned concerning their use of alcohol. A summary of the responses is shown in the table.

		Alcohol		
		None	Occasional	Frequent
Family Status	Upper class	4	16	10
	Upper middle class	11	40	24
	Lower middle class	9	47	9
	Lower class	6	17	7

a. Do the data provide sufficient evidence to indicate a relationship between family status and the use of alcohol? Test using $\alpha = .05$.

b. What assumptions are necessary to assure the validity of the inferential procedure you applied in part **a**? Do you think they are satisfied in this application?

c. Calculate the percentages of the total number of teenagers in each family status classification falling into each alcohol use classification. Graph these percentages on the vertical axis against the family status on the horizontal axis. Show the overall percentages for each alcohol use classification as horizontal lines on the graph. Does the graph support the result of the test in part **a**? Explain.

***9.72** Consumers have traditionally viewed products with warranties more favorably than products without warranties. In fact, several studies have demonstrated that, when given the choice between two similar products, one of which is warranted, consumers prefer the warranted product, even at a higher price. Thus, consumers generally perceive warranties as a kind of value added to the product. However, a substantial number of firms have been found to perceive their warranties primarily as legal disclaimers of responsibility and nothing more. As a result of the differences in perceptions by consumers and businesses, Congress passed the Magnuson–Moss Warranty Act, which took effect in 1977. Its purpose was to reform consumer product warranty practices. According to this act, all warranties must be designated as "full" or "limited" and must be clearly written in readily understood language. Further, it specified what was to be contained in warranties (McDaniel and Rao, 1982).

McDaniel and Rao undertook a study to investigate consumer satisfaction with warranty practices since the advent of the Magnuson–Moss Warranty Act. Using a mailed questionnaire, they sampled 237 midwestern consumers who had purchased a major appliance within the past 6 to 18 months. One of the questions they asked the consumers was, "Do most retailers and dealers make a conscientious effort to satisfy their customers' warranty claims?" One hundred fifty-six answered yes, 61 were uncertain, and 20 said no.

The population of consumers from which this sample was drawn had also been investigated 2 years prior to the Magnuson–Moss Warranty Act. At that time, 37.0% of the population answered yes to the same question, 53.3% were uncertain, and 9.7% said no.

a. As reflected in the answers to the above question, have consumer attitudes toward warranties changed since the pre-Magnuson–Moss Act study? Test using $\alpha = .05$.

b. Compare the pre- and post-Magnuson–Moss Warranty Act responses, and describe the changes that have occurred.

On Your Own

Many researchers rely on surveys to estimate the proportions of experimental units in populations that possess certain specified characteristics. A political scientist may want to estimate the proportion of an electorate in favor of a certain legislative bill. A social scientist may be interested in the proportions of people in a geographical region who fall in certain socioeconomic classifications. A psychologist might want to compare the proportions of patients who have different psychological disorders.

Choose a specific topic, similar to those described above, that interests you. Clearly define the population of interest, identify data categories of specific interest, and identify the proportions associated with them. Now *guesstimate* the proportions of the population that you think fall in each of the categories. For example, you might guess that all the proportions are equal, or that the first proportion is twice as large as the second but equal to the third, etc.

You are now ready to collect the data by obtaining a random sample from your population of interest. Select a sample size so that all expected cell counts are at least 5 (preferably larger), and collect the data.

Use the count data you have obtained to test the null hypothesis that the true proportions in the population equal your presampling guesstimates of these actual proportions. Would failure to reject this null hypothesis imply that your guesstimates are correct?

Using the Computer

Are monthly homeowner costs dependent on the region of the country in which the homeowner lives? Use the zip code data set described in Appendix B to investigate this question.

a. Form a contingency table using region as one classification, and the following four levels of homeowner costs as the second classification: \$0–\$200, \$200.01–\$400, \$400.01–\$600, and \$600.01–\$800. The contingency table should contain a count of the total number of zip codes in each region–cost classification combination.
b. Conduct a χ^2 test to determine whether homeowner costs are dependent on region.
c. Calculate the percentage of zip code areas in each cost class using the total number of zip code areas in each region as the base. Then calculate the percentage in each cost class for all 1,000 zip codes. Construct a table like Table 9.10 to summarize the results. Display the percentages in a graph like Figure 9.4. Does the graphical display support the results of the test in part **b**?

References

Agresti, A., and Agresti, B. F. *Statistical Methods for the Social Sciences*, 2d edition. San Francisco: Dellen, 1986, Chapter 8.

Blazey, M. "Travel and retirement status." *Annals of Tourism Research*, 1992, Vol. 19, No. 4.

Cochran, W. G. "The χ^2 test of goodness of fit." *Annals of Mathematical Statistics*, 1952, 23, pp. 315–345.

Covert, C. "Television viewers snapping up Home Shopping Network." *Minneapolis Star and Tribune*, December 16, 1986.

DeMaris, A., and Rao, K. V. "Premarital cohabitation and subsequent marital stability in the United States: A reassessment." *Journal of Marriage and the Family*, Feb. 1992, 54, pp. 178–190.

Graham, E. "The entrepreneurial mystique." *Wall Street Journal*, May 20, 1985, p. 1C.

Kahalas, H. "The relationship between confidence in business and job satisfaction for union and nonunion members." *Baylor Business Studies*, February–April 1981, 127, pp. 45–53.

McDaniel, S. W., and Rao, C. P. "Consumer attitudes toward and satisfaction with warranties and warranty performance—Before and after Magnuson–Moss." *Baylor Business Studies*, November–December 1982, 130, pp. 47–61.

Meagher, J. J., and Scazzero, J. A. "Measuring inspector variability." *1985 ASQC Quality Congress Transaction*, May 1985, 75–81.

Palm, R., and Hodgson, M. "Earthquake insurance: Mandated disclosure and homeowner response in California." *Annals of the Association of American Geographers*, June 1992, Vol. 82, No. 2.

Savage, I. R. "Bibliography of nonparametric statistics and related topics." *Journal of the American Statistical Association*, 1953, 48, pp. 844–906.

Siegel, S. *Nonparametric Statistics for the Behavioral Sciences*. New York: McGraw-Hill, 1956, Chapter 9.

Stroh, L. K., Brett, J. M., and Reilly, A. H. "All the right stuff: A comparison of female and male managers' career progression." *Journal of Applied Psychology*, 1992, Vol. 77, No. 3, pp. 251–260.

Tom, G., Clark, R., Elmer, L., Grech, E., Masetti, J., Jr., and Sandhar, H. "The use of created versus celebrity spokespersons in advertisements." *The Journal of Consumer Marketing*, Fall 1992, Vol. 9, No. 4.

Tucker, L. A. "Physical fitness and psychological distress." *The International Journal of Sports Psychology*, July–Sept., 1990, 21, pp. 185–201.

CHAPTER TEN

Simple Linear Regression

Contents

Case Study

Where We've Been

We have learned how to estimate and test hypotheses about population parameters based on a random sample of observations from the population and have extended these methods to allow for a comparison of parameters from two or more populations.

Where We're Going

For many sampling situations, we have much more information available on a random variable (and the population it generates) than that contained in a single random sample. For example, if we wanted to predict the rainfall at a given location on a given day, we could select a single random sample of n daily rainfalls, use the methods of Chapter 6 to estimate the mean daily rainfall μ, and then use this quantity to predict the rainfall for any given day. But this method fails to utilize scientific information that is available to any forecaster. We know the daily rainfall is related to barometric pressure, cloud cover, etc. By measuring barometric pressure and cloud cover at the same time we sample the daily rainfall, we hope to establish the relationship between these variables and to utilize them for prediction.

This chapter is devoted to the most elementary situation—relating two variables.

In Chapters 6–8 we described methods for making inferences about population means. The mean of a population has been treated as a *constant*, and we have shown how to use sample data to estimate or to test hypotheses about this constant mean. In many applications, the mean of a population is not viewed as a constant but rather as a variable. For example, the mean sale price of residences in a large city during 1990 can be treated as a constant and might be equal to $150,000. But we might also treat the mean sale price as a variable that depends on the square feet of living space in the residence. For example, the relationship might be

Mean sale price = $30,000 + $60(Square feet)

The formula implies that the mean sale price of 1,000-square-foot homes is $90,000, the mean sale price of 2,000-square-foot homes is $150,000, and the mean sale price of 3,000-square-foot homes is $210,000.

What do we gain by treating the mean as a variable rather than a constant? In many practical applications we will be dealing with highly variable data, data for which the standard deviation is so large that a constant mean is almost "lost" in a sea of variability. For example, if the mean residential sale price is $150,000 but the standard deviation is $75,000, then the actual sale prices will vary considerably, and the mean price is not a very meaningful or useful characterization of the price distribution. On the other hand, if the mean sale price is treated as a variable that depends on the square feet of living space, the standard deviation of sale prices for any given size of home might be only $10,000. In this case, the mean price will provide a much better characterization of sale prices when it is treated as a variable rather than a constant.

In this chapter we discuss situations in which the mean of the population is treated as a variable, dependent on the value of another variable. The preceding example of residential sale price depending on the square feet of living space is one illustration. Other examples are: the mean reaction time depending on the amount of a drug in the bloodstream, the mean starting salary of a college graduate depending on the student's GPA, and the mean number of years to which a criminal is sentenced depending on the number of previous convictions.

In this chapter we discuss the simplest of all models relating a population mean to another variable, the *straight-line model*. We show how to use sample data to estimate the straight-line relationship between the mean value of one variable, y, as it relates to a second variable, x. The methodology of estimating and using a straight-line relationship is referred to as *simple linear regression analysis*.

10.1 Probabilistic Models

An important consideration when taking a drug is how it may affect one's perception or general awareness. Suppose you want to model the length of time it takes to respond to a stimulus (a measure of awareness) as a function of the percentage of a certain drug in the bloodstream. The first question to be answered is this: "Do you think an exact relationship exists between these two variables?" That is, do you think it is possible to state the exact length of time it takes an individual (subject) to respond if the amount of the drug in the bloodstream is known? We think you will agree with us that this is *not* possible for several reasons. The reaction time depends on many variables other than the percentage of the drug in the bloodstream—for example, the time of day, the amount of sleep the subject had the night before, the subject's visual acuity, the subject's general reaction time without the drug, and the subject's age would all probably affect reaction time. Even if many variables are included in a model, it is still unlikely that we would be able to predict *exactly* the subject's reaction time. There will almost certainly be some variation in response times due strictly to *random phenomena* that cannot be modeled or explained.

If we were to construct a model that hypothesized an exact relationship between variables, it would be called a **deterministic model**. For example, if we believe that y, the reaction time (in seconds), will be exactly one and one-half times x, the amount of drug in the blood, we write

$$y = 1.5x$$

This represents a **deterministic relationship** between the variables y and x. It implies that y can always be determined exactly when the value of x is known. *There is no allowance for error in this prediction.*

If, on the other hand, we believe there will be unexplained variation in reaction times—perhaps caused by important but unincluded variables or by random phenomena—we discard the deterministic model and use a model that accounts for this **random error**. This **probabilistic model** includes both a deterministic component and a random error component. For example, if we hypothesize that the response time y is related to the percentage of drug x by

$$y = 1.5x + \text{Random error}$$

we are hypothesizing a **probabilistic relationship** between y and x. Note that the deterministic component of this probabilistic model is $1.5x$.

Figure 10.1(a) shows the possible responses for five different values of x, the percentage of drug in the blood, when the model is deterministic. All the responses must fall exactly on the line because a deterministic model leaves no room for error.

FIGURE 10.1 ▶ Possible reaction times, y, for five different drug percentages, x

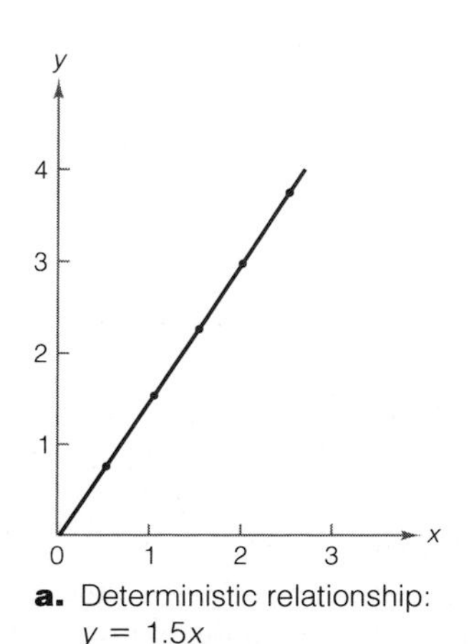

a. Deterministic relationship: $y = 1.5x$

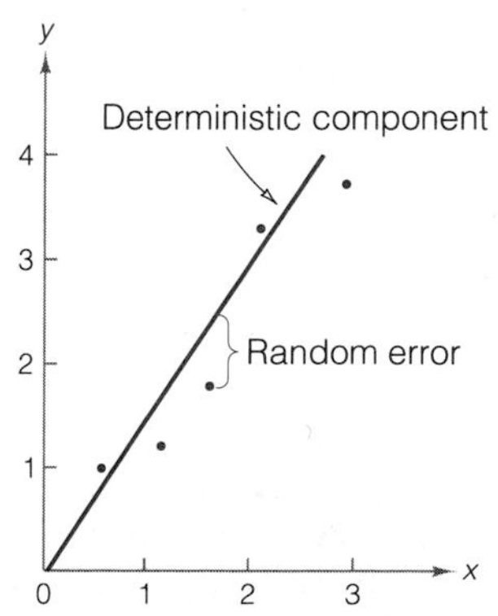

b. Probabilistic relationship: $y = 1.5x + \text{Random error}$

Figure 10.1(b) shows a possible set of responses for the same values of x when we are using a probabilistic model. Note that the deterministic part of the model (the straight line itself) is the same. Now, however, the inclusion of a random error component allows the response times to vary from this line. Since we know that the response time does vary randomly for a given value of x, the probabilistic model provides a more realistic model for y than does the deterministic model.

General Form of Probabilistic Models

$$y = \text{Deterministic component} + \text{Random error}$$

where y is the variable of interest. We always assume that the mean value of the random error equals 0. This is equivalent to assuming that the mean value of y, $E(y)$, equals the deterministic component of the model, i.e.,

$$E(y) = \text{Deterministic component}$$

We begin with the simplest of probabilistic models—the **straight-line model**—which derives its name from the fact that the deterministic portion of the model graphs as a straight line. Fitting this model to a set of data is an example of **regression analysis**, or **regression modeling**. The elements of the straight-line model are summarized in the box.

A First-Order (Straight-Line) Probabilistic Model

$$y = \beta_0 + \beta_1 x + \epsilon$$

where y = *Dependent* or *response* variable (variable to be modeled)

x = *Independent* or *predictor variable* (variable used as a predictor of y)*

ϵ (epsilon) = Random error component

β_0 (beta zero) = y-intercept of the line—i.e., point at which the line intercepts or cuts through the y-axis (see Figure 10.2)

β_1 (beta one) = Slope of the line—i.e., amount of increase (or decrease) in the deterministic component of y for every 1-unit increase in x. You can see (Figure 10.2) that $E(y)$ increases by the amount β_1 as x increases from 2 to 3.

*The word *independent* should not be interpreted in a probabilistic sense, as defined in Chapter 3. The phrase *independent variable* is used in regression analysis to refer to a predictor variable for the response y.

FIGURE 10.2 ►
The straight-line model

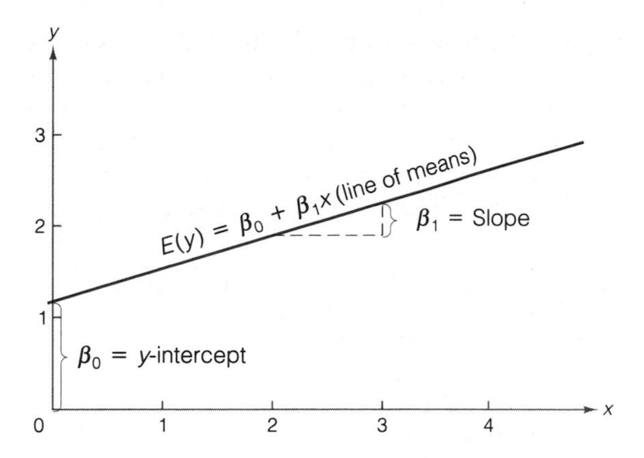

In the probabilistic model, the deterministic component is referred to as the **line of means**, because the mean of y, $E(y)$, is equal to the straight-line component of the model. That is,

$$E(y) = \beta_0 + \beta_1 x$$

Note that the Greek symbols β_0 and β_1 represent the y-intercept and slope of the model. They are population parameters that will be known only if we have access to the entire population of (x, y) measurements. Together with a specific value of the independent variable x, they determine the mean value of y, which is just a specific point on the line of means (Figure 10.2).

The values of β_0 and β_1 will be unknown in almost all practical applications of regression analysis. The process of developing a model, estimating the unknown parameters, and using the model can be viewed as the five-step procedure shown in the box.

Step 1 Hypothesize the deterministic component of the model that relates the mean, $E(y)$, to the independent variable x (Section 10.1).

Step 2 Use the sample data to estimate unknown parameters in the model (Section 10.2).

Step 3 Specify the probability distribution of the random error term, and estimate the standard deviation of this distribution (Sections 10.3 and 10.4).

Step 4 Statistically evaluate the usefulness of the model (Sections 10.5, 10.6, and 10.7).

Step 5 When satisfied that the model is useful, use it for prediction, estimation, and other purposes (Section 10.8).

In this chapter we deal only with the straight-line model. More complex **multiple regression** models are beyond the scope of this text. Consult the references at the end of this chapter if you wish to learn about these models.

Exercises 10.1 – 10.9

Learning the Mechanics

10.1 In each case, graph the line that passes through the given points.
a. (1, 1) and (5, 5) b. (0, 3) and (3, 0) c. (−1, 1) and (4, 2) d. (−6, −3) and 2, 6)

10.2 Give the slope and y-intercept for each of the lines graphed in Exercise 10.1.

10.3 The equation for a straight line (deterministic) is

$$y = \beta_0 + \beta_1 x$$

If the line passes through the point (−2, 4), then $x = -2y = 4$ must satisfy the equation; i.e.,

$$4 = \beta_0 + \beta_1(-2)$$

Similarly, if the line passes through the point (4, 6), then $x = 4$, $y = 6$ must satisfy the equation; i.e.,

$$6 = \beta_0 + \beta_1(4)$$

Use these two equations to solve for β_0 and β_1, and find the equation of the line that passes through the points (−2, 4) and 4, 6).

10.4 Refer to Exercise 10.3. Find the equations of the lines that pass through the points listed in Exercise 10.1.

10.5 Plot the following lines:
a. $y = 4 + x$ b. $y = 5 - 2x$ c. $y = -4 + 3x$
d. $y = -2x$ e. $y = x$ f. $y = .50 + 1.5x$

10.6 Give the slope and y-intercept for each of the lines defined in Exercise 10.5.

10.7 Why do we generally prefer a probabilistic model to a deterministic model? Give examples for which the two types of models might be appropriate.

10.8 What is the line of means?

10.9 If a straight-line probabilistic relationship relates the mean $E(y)$ to an independent variable x, does it imply that every value of the variable y will always fall exactly on the line of means? Why or why not?

10.2 Fitting the Model: The Least Squares Approach

After the straight-line model has been hypothesized to relate the mean $E(y)$ to the independent variable x, the next step is to collect data and to estimate the (unknown) population parameters, the y-intercept β_0 and the slope β_1.

To begin with a simple example, suppose an experiment involving five subjects is conducted to determine the relationship between the percentage of a certain drug in the bloodstream and the length of time it takes to react to a stimulus. The results are shown in Table 10.1. (The number of measurements and the measurements them-

TABLE 10.1 Reaction Time Versus Drug Percentage

Subject	Amount of Drug x (%)	Reaction Time y (seconds)
1	1	1
2	2	1
3	3	2
4	4	2
5	5	4

selves are unrealistically simple in order to avoid arithmetic confusion in this introductory example.) This set of data will be used to demonstrate the five-step procedure of regression modeling given in Section 10.1. In this section we hypothesize the deterministic component of the model and estimate its unknown parameters (steps 1 and 2). Discussion of the model assumptions and the random error component (step 3) are the subjects of Sections 10.3 and 10.4, whereas Sections 10.5–10.7 assess the utility of the model (step 4). Finally, using the model for prediction and estimation (step 5) is the subject of Section 10.8.

Step 1 *Hypothesize the deterministic component of the probabilistic model.* As stated before, we will consider only straight-line models in this chapter, and thus the complete model to relate mean response time $E(y)$ to drug percentage x is given by

$$E(y) = \beta_0 + \beta_1 x$$

Step 2 *Use sample data to estimate unknown parameters in the model.* This step is the subject of this section—namely, how can we best use the information in the sample of five observations in Table 10.1 to estimate the unknown y-intercept β_0 and slope β_1?

To determine whether a linear relationship between y and x is plausible, it is helpful to plot the sample data. Such a plot, called a **scattergram**, locates each of the five data points on a graph, as shown in Figure 10.3. Note that the scattergram suggests a general tendency for y to increase as x increases. If you place a ruler on the scattergram, you will see that a line may be drawn through three of the five points, as

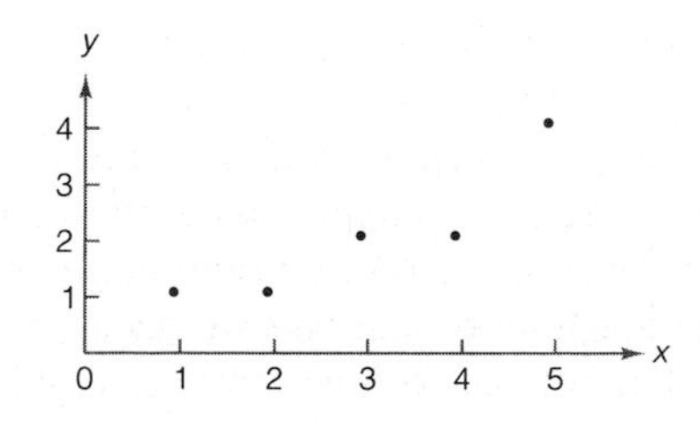

FIGURE 10.3 ► Scattergram for data in Table 10.1

shown in Figure 10.4. To obtain the equation of this visually fitted line, note that the line intersects the y-axis at $y = -1$, so the y-intercept is -1. Also, y increases exactly 1 unit for every 1-unit increase in x, indicating that the slope is $+1$. Therefore, the equation is

$$\tilde{y} = -1 + 1(x) = -1 + x$$

where $\tilde{y}$ is used to denote the predicted y from the visual model.

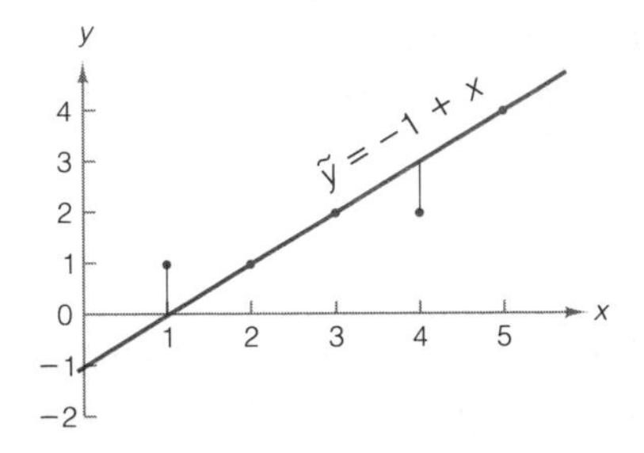

FIGURE 10.4 ▶ Visual straight-line fit to the data

One way to decide quantitatively how well a straight line fits a set of data is to note the extent to which the data points deviate from the line. For example, to evaluate the model in Figure 10.4, we calculate the magnitude of the **deviations**, i.e., the differences between the observed and the predicted values of y. These deviations, or **errors**, are the vertical distances between observed and predicted values (see Figure 10.4). The observed and predicted values of y, their differences, and their squared differences are shown in Table 10.2. Note that the *sum of errors* equals 0 and the *sum of squares of the errors* (SSE), which gives greater emphasis to large deviations of the points from the line, is equal to 2.

TABLE 10.2 Comparing Observed and Predicted Values for the Visual Model

x	y	$\tilde{y} = -1 + x$	$(y - \tilde{y})$	$(y - \tilde{y})^2$
1	1	0	$(1 - 0) = 1$	1
2	1	1	$(1 - 1) = 0$	0
3	2	2	$(2 - 2) = 0$	0
4	2	3	$(2 - 3) = -1$	1
5	4	4	$(4 - 4) = 0$	0
			Sum of errors = 0	Sum of squared errors (SSE) = 2

You can see by shifting the ruler around the graph that it is possible to find many lines for which the sum of the errors is equal to 0, but it can be shown that there is one (and only one) line for which the SSE is a *minimum*. This line is called the **least squares line**, the **regression line**, or **least squares prediction equation**.

To find the least squares prediction equation for a set of data, assume that we have a sample of n data points consisting of pairs of values of x and y, say (x_1, y_1), (x_2, y_2), $\ldots$, (x_n, y_n). For example, the $n = 5$ data points shown in Table 10.2 are (1, 1), (2, 1), (3, 2), (4, 2), and (5, 4). The fitted line, which we will calculate based on the five data points, is written as

$$\hat{y} = \hat{\beta}_0 + \hat{\beta}_1 x$$

The "hats" indicate that the symbols below them are estimates: $\hat{y}$ (y-hat) is an estimator of the mean value of y, $E(y)$, and a predictor of some future value of y; and $\hat{\beta}_0$ and $\hat{\beta}_1$ are estimators of β_0 and β_1, respectively.

For a given data point, say the point (x_i, y_i), the observed value of y is y_i and the predicted value of y would be obtained by substituting x_i into the prediction equation:

$$\hat{y}_i = \hat{\beta}_0 + \hat{\beta}_1 x_i$$

And the deviation of the ith value of y from its predicted value is

$$(y_i - \hat{y}_i) = [y_i - (\hat{\beta}_0 + \hat{\beta}_1 x_i)]$$

Then the sum of squares of the deviations of the y-values about their predicted values for all the n data points is

$$\text{SSE} = \sum [y_i - (\hat{\beta}_0 + \hat{\beta}_1 x_i)]^2$$

The quantities $\hat{\beta}_0$ and $\hat{\beta}_1$ that make the SSE a minimum are called the **least squares estimates** of the population parameters β_0 and β_1, and the prediction equation $\hat{y} = \hat{\beta}_0 + \hat{\beta}_1 x$ is called the **least squares line**.

Definition 10.1

The **least squares line** is one that has a smaller sum of squared errors (SSE) than any other straight-line model.

The values of $\hat{\beta}_0$ and $\hat{\beta}_1$ that minimize the SSE are (proof omitted) given by the formulas in the box at the top of page 496.*

*Students who are familiar with calculus should note that the values of β_0 and β_1 that minimize $\text{SSE} = \Sigma(y_i - \hat{y}_i)^2$ are obtained by setting the two partial derivatives $\partial\text{SSE}/\partial\beta_0$ and $\partial\text{SSE}/\partial\beta_1$ equal to 0. The solutions to these two equations yield the formulas shown in the box. Furthermore, we denote the *sample* solutions to the equations by $\hat{\beta}_0$ and $\hat{\beta}_1$, where the "hat" denotes that these are sample estimates of the true population intercept β_0 and slope β_1.

Formulas for the Least Squares Estimates

Slope: $\hat{\beta}_1 = \frac{SS_{xy}}{SS_{xx}}$

y-intercept: $\hat{\beta}_0 = \bar{y} - \hat{\beta}_1\bar{x}$

where $SS_{xy} = \sum (x_i - \bar{x})(y_i - \bar{y}) = \sum x_i y_i - \frac{\left(\sum x_i\right)\left(\sum y_i\right)}{n}$

$$SS_{xx} = \sum (x_i - \bar{x})^2 = \sum x_i^2 - \frac{\left(\sum x_i\right)^2}{n}$$

n = Sample size

Preliminary computations for finding the least squares line for the drug reaction example are presented in Table 10.3. We can now calculate

$$SS_{xy} = \sum x_i y_i - \frac{\left(\sum x_i\right)\left(\sum y_i\right)}{5} = 37 - \frac{(15)(10)}{5} = 37 - 30 = 7$$

$$SS_{xx} = \sum x_i^2 - \frac{\left(\sum x_i\right)^2}{5} = 55 - \frac{(15)^2}{5} = 55 - 45 = 10$$

TABLE 10.3 Preliminary Computations for the Drug Reaction Example

	x_i	y_i	x_i^2	$x_i y_i$
	1	1	1	1
	2	1	4	2
	3	2	9	6
	4	2	16	8
	5	4	25	20
Totals	$\Sigma x_i = 15$	$\Sigma y_i = 10$	$\Sigma x_i^2 = 55$	$\Sigma x_i y_i = 37$

Then the slope of the least squares line is

$$\hat{\beta}_1 = \frac{SS_{xy}}{SS_{xx}} = \frac{7}{10} = .7$$

and the y-intercept is

$$\hat{\beta}_0 = \bar{y} - \hat{\beta}_1\bar{x} = \frac{\sum y_i}{5} - \hat{\beta}_1\frac{\left(\sum x_i\right)}{5}$$

$$= \frac{10}{5} - (.7)\frac{(15)}{5} = 2 - (.7)(3) = 2 - 2.1 = -.1$$

The least squares line is thus

$$\hat{y} = \hat{\beta}_0 + \hat{\beta}_1 x = -.1 + .7x$$

The graph of this line is shown in Figure 10.5.

FIGURE 10.5 ▶ The line $\hat{y} = -.1 + .7x$ fit to the data

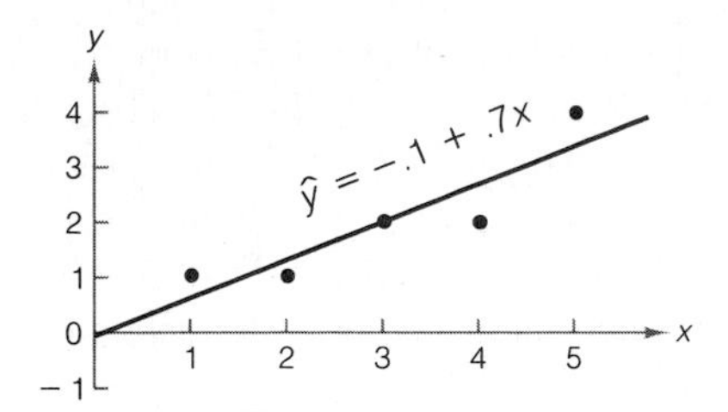

The predicted value of y for a given value of x can be obtained by substituting into the formula for the least squares line. Thus, when $x = 2$ we predict y to be

$$\hat{y} = -.1 + .7x = -.1 + .7(2) = 1.3$$

We show how to find a prediction interval for y in Section 10.8.

The observed and predicted values of y, the deviations of the y values about their predicted values, and the squares of these deviations are shown in Table 10.4. Note that the sum of squares of the deviations, SSE, is 1.10, and (as we would expect) this is less than the SSE = 2.0 obtained in Table 10.2 for the visually fitted line.

TABLE 10.4 Comparing Observed and Predicted Values for the Least Squares Prediction Equation

x	y	$\hat{y} = -.1 + .7x$	$(y - \hat{y})$	$(y - \hat{y})^2$
1	1	.6	$(1 - .6) = .4$	.16
2	1	1.3	$(1 - 1.3) = -.3$	.09
3	2	2.0	$(2 - 2.0) = 0$	.00
4	2	2.7	$(2 - 2.7) = -.7$	.49
5	4	3.4	$(4 - 3.4) = .6$	.36
			Sum of errors = 0	SSE = 1.10

It is important that you be able to interpret the intercept and slope in terms of the data being utilized to fit the model. In the drug reaction example, the estimated y-intercept, $\hat{\beta}_0$, is $-.1$. This value would seem to imply that the estimated mean reaction time is equal to $-.1$ second when the amount of drug, x, is equal to 0%. Since negative reaction times are not possible, this seems to make the model nonsensical. However, *the model parameters should be interpreted only within the sampled range of the independent variable*—in this case, for amounts of drug in the bloodstream between 1% and 5%. Thus, the y-intercept, which is, by definition, at $x = 0$

(0% drug), is not within the range of the sampled values of x and is not subject to meaningful interpretation.

The slope, $\hat{\beta}_1$, of the least squares line was calculated to be .7. The implication is that for every unit increase of x, the mean value of y is estimated to increase by .7 unit. In terms of this example, for every 1% increase in the amount of drug in the bloodstream, the mean reaction time is estimated to increase by .7 second *over the sampled range of drug amounts from 1% to 5%*. Thus, the model does not imply that increasing the drug amount from 5% to 10% will result in an increase in mean reaction time of 3.5 seconds, because the range of x in the sample does not extend to 10% ($x = 10$). In fact, 10% might be such a high concentration of the drug that it would kill the subject! Be careful to interpret the estimated parameters only within the sampled range of x.

Even when the interpretations of the estimated parameters are meaningful, it should be remembered that they are only estimates based on the sample. As such, their values will typically change in repeated sampling. How much confidence do we have that the estimated slope, $\hat{\beta}_1$, accurately estimates the true slope, β_1? This requires statistical inference, in the form of confidence intervals and tests of hypotheses, which we address in Section 10.5.

To summarize, we defined the best-fitting straight line to be the one that minimizes the sum of squared errors around the line, and we called it the least squares line. We should interpret the least squares line only within the sampled range of the independent variable. In subsequent sections we show how to make statistical inferences about the model.

Exercises 10.10–10.21

Learning the Mechanics

10.10 The following table is similar to Table 10.3. It is used for making the preliminary computations for finding the least squares line for the given pairs of x and y values.

	x_i	y_i	x_i^2	x_iy_i
	7	2		
	4	4		
	6	2		
	2	5		
	1	7		
	1	6		
	3	5		
Totals	$\Sigma x_i =$	$\Sigma y_i =$	$\Sigma x_i^2 =$	$\Sigma x_iy_i =$

a. Complete the table. b. Find SS_{xy}. c. Find SS_{xx}.
d. Find $\hat{\beta}_1$. e. Find $\bar{x}$ and $\bar{y}$. f. Find $\hat{\beta}_0$.
g. Find the least squares line.

10.11 Refer to Exercise 10.10. After the least squares line has been obtained, the following table (which is similar to Table 10.4) can be used for (1) comparing the observed and the predicted values of y, and (2) computing SSE.

x	y	$\hat{y} =$	$(y - \hat{y})$	$(y - \hat{y})^2$
7	2			
4	4			
6	2			
2	5			
1	7			
1	6			
3	5			
			$\Sigma (y - \hat{y}) =$	$\text{SSE} = \Sigma (y - \hat{y})^2 =$

a. Complete the table.
b. Plot the least squares line on a scattergram of the data. Plot the following line on the same graph:

$$\hat{y} = 14 - 2.5x$$

c. Show that SSE is larger for the line in part **b** than it is for the least squares line.

10.12 Construct a scattergram for the data in the table.

x	.5	1	1.5
y	2	1	3

a. Plot the following two lines on your scattergram:

$$y = 3 - x \quad \text{and} \quad y = 1 + x$$

b. Which of these lines would you choose to characterize the relationship between x and y? Explain.
c. Show that the sum of errors for both of these lines equals 0.
d. Which of these lines has the smaller SSE?
e. Find the least squares line for the data, and compare it to the two lines described in part **a**.

10.13 Consider the following pairs of measurements:

x	5	3	−1	2	7	6	4
y	4	3	0	1	8	5	3

a. Construct a scattergram for these data.
b. What does the scattergram suggest about the relationship between x and y?
c. Given that $SS_{xx} = 43.4286$, $SS_{xy} = 39.8571$, $\bar{y} = 3.4286$, and $\bar{x} = 3.7143$, calculate the least squares estimates of β_0 and β_1.
d. Plot the least squares line on your scattergram. Does the line appear to fit the data well? Explain.
e. Interpret the y-intercept and slope of the least squares line. Over what range of x are these interpretations meaningful?

10.14 Suppose that $n = 100$ recent residential home sales in a city are used to fit a least squares straight-line model relating the sale price, y, to the square feet of living space, x. Homes in the sample range from 1,500 square feet to 4,000 square feet of living space, and the resulting least squares equation is

$$\hat{y} = -30{,}000 + 70x$$

a. What is the underlying hypothesized probabilistic model for this application? What does it imply about the relationship between the mean sale price and living space?
b. Identify the least squares estimates of the y-intercept and slope of the model.
c. Interpret the least squares estimate of the y-intercept. Is it meaningful for this application? Why?
d. Interpret the least squares estimate of the slope of the model. Over what range of x is the interpretation meaningful?
e. Use the least squares model to estimate the mean sale price of a 3,000-square-foot home. Is the estimate meaningful? Explain.
f. Use the least squares model to estimate the mean sale price of a 5,000-square-foot home. Is the estimate meaningful? Explain.

[*Note:* We show how to measure the statistical reliability of these least squares estimates in subsequent sections.]

Applying the Concepts

10.15 Is the number of games won by a major league baseball team in a season related to the team's batting average? The accompanying table shows the number of games won and the batting averages for the 14 teams in the American League for the 1991 season. [Team batting average is the ratio of the total number of hits to the total number of "at-bats" for the team.]

a. If you were to model the relationship between the mean (or expected) number of games won by a major league team and the team's batting average, x, using a straight line, would you expect the slope of the line to be positive or negative? Explain.
b. Construct a scattergram for the data. Does the pattern revealed by the scattergram agree with your answer to part **a**?
c. Given that $\Sigma\, y = 1{,}134$, $\Sigma\, x = 3.642$, $\Sigma\, y^2 = 93{,}110$, $\Sigma\, x^2 = .948622$, and $\Sigma\, xy = 295.54$, fit a simple linear regression model to the data.
d. Graph the least squares line on your scattergram. Does your least squares line seem to fit the points on your scattergram?
e. Interpret the least squares intercept and slope in terms of this application. Can you explain why the mean (or expected) number of games won does not appear to be strongly related to a team's batting average?

Team	Number of Games Won y	Team Batting Average x
Cleveland	57	.254
New York	71	.256
Boston	84	.269
Toronto	91	.257
Texas	85	.270
Detroit	84	.247
Minnesota	95	.280
Baltimore	67	.254
California	81	.255
Milwaukee	83	.271
Seattle	83	.255
Kansas City	82	.264
Oakland	84	.248
Chicago	87	.262

Source: *Official Major League Baseball 1992 Stat Book*. Major League Baseball Properties, Inc., and the editors of *The Baseball Encyclopedia*, New York.

10.16 Due primarily to the price controls of the Organization of Petroleum Exporting Countries (OPEC), a cartel of crude oil suppliers, the price of crude oil rose dramatically from the mid-1970s to the early 1980s. As a result, motorists were confronted with a similar upward spiral of gasoline prices. The data in the table are typical prices for a gallon of regular leaded gasoline and a barrel of crude oil (refiner acquisition cost) for the indicated years.

Year	Gasoline y (¢ per gallon)	Crude Oil ($ per barrel)	Year	Gasoline y (¢ per gallon)	Crude Oil ($ per barrel)
1975	57	10.38	1983	116	28.99
1976	59	10.89	1984	113	28.63
1977	62	11.96	1985	112	26.75
1978	63	12.46	1986	86	14.55
1979	86	17.72	1987	90	17.90
1980	119	28.07	1988	90	14.67
1981	131	35.24	1989	100	17.97
1982	122	31.87	1990	115	22.23

Source: *Statistical Abstract of the United States: 1982–1992.*

Given that $\sum y = 1{,}521$, $\sum x = 330.28$, $\sum y^2 = 153{,}735$, $\sum x^2 = 7{,}824.1822$, and $\sum xy = 34{,}259.58$:

a. Use the data to calculate the least squares line that describes the relationship between the price of a gallon of gasoline and the price of a barrel of crude oil.

b. Plot your least squares line on a scattergram of the data. Does your least squares line appear to be an appropriate characterization of the relationship between y and x? Explain.
c. If the price of crude oil fell to \$15 per barrel, to what level (approximately) would the price of regular gasoline fall? Justify your response.

10.17 In recent years, physicians have used the "diving reflex" to reduce abnormally rapid heartbeats in humans by briefly submerging the patient's face in cold water. The reflex, triggered by cold water temperatures, is an involuntary neural response that shuts off circulation to the skin, muscles, and internal organs to divert extra oxygen-carrying blood to the heart, lungs, and brain. A research physician conducted an experiment to investigate the effects of various cold water temperatures on the pulse rate of small children. The data for seven 6-year-old children are shown in the accompanying table.

Child	*Temperature of Water* x (°F)	*Decrease in Pulse Rate* y (*beats/minute*)
1	68	2
2	65	5
3	70	1
4	62	10
5	60	9
6	55	13
7	58	10

a. Find the least squares line for the data.
b. Construct a scattergram for the data; then graph the least squares line as a check on your calculations.
c. If the water temperature is 60°F, predict the drop in pulse rate for a 6-year-old child. [*Note:* A measure of the reliability of such predictions is discussed in Section 10.8.]

10.18 To investigate the relationship between yield of potatoes, y, and level of fertilizer application, x, an experimenter divides a field into eight plots of equal size and applies differing amounts of fertilizer to each. The yield of potatoes (in pounds) and the fertilizer application (in pounds) are recorded for each plot. The data are as follows:

x	1	1.5	2	2.5	3	3.5	4	4.5
y	25	31	27	28	36	35	32	34

a. Construct a scattergram for the data.
b. Find the the least squares estimates for β_0 and β_1.
c. According to your least squares line, approximately how many pounds of potatoes would you expect from a plot to which 3.75 pounds of fertilizer has been applied? [*Note:* A measure of the reliability of such predictions is discussed in Section 10.8.]

10.19 In Exercise 8.42, we gave data on the mean daily air temperature and corresponding cocoon temperature of woolly-bear caterpillars of the High Arctic (Kevan et al., 1982). The data, collected over 12 days, are reproduced in the table.

Day	*Temperature (°C)*		*Day*	*Temperature (°C)*	
	Air	*Cocoon*[a]		*Air*	*Cocoon*[a]
1	10.4	15.1	7	1.7	3.6
2	9.2	14.6	8	2.0	5.3
3	2.2	6.8	9	3.0	7.0
4	2.6	6.8	10	3.5	7.1
5	4.1	8.0	11	4.5	9.6
6	3.7	8.7	12	4.4	9.5

[a]Each cocoon temperature is the average of the temperatures of two cocoons.

Given that $SS_{xx} = 83.3425$, $SS_{xy} = 100.0825$, $\bar{x} = 4.2750$, and $\bar{y} = 8.5083$:

a. Fit a least squares line to relate the cocoon temperature y to the outside air temperature x.

b. Plot the data points and graph your least squares line on the same sheet of graph paper. Does your line provide a good fit to the data?

c. Interpret the least squares intercept and slope in terms of this application.

10.20 The sand lance is a small fish that can be found in the Northwest Atlantic from Cape Hatteras to Greenland. In a study of the biological and demographic characteristics of the sand lance, G. H. Winters (1983) includes data on the mean length of specimens collected each year for the period 1969 through 1979 for sand lance ages 2 through 8 years:

Mean Length (Millimeters)	176	194	212	226	236	244	254
Age (Years)	2	3	4	5	6	7	8

a. Find the least squares prediction equation relating the mean length y of the sand lance to its age x.

b. Plot the data points and graph your least squares lines on the same sheet of graph paper. Does your line appear to provide a good fit to the data?

c. Interpret the least squares intercept and slope in terms of this application. Is the intercept's interpretation meaningful in this case?

10.21 The data shown in the table at the top of page 504 are part of a series of experiments conducted to investigate the effect of water temperature on the absorption by rainbow trout of sublethal levels of cyanide (Kovacs and Leduc, 1982). This specific set of data is the result of a preliminary experiment to determine the relationship between mean weight gain (percentage of body weight) over a 20-day period as a function of the ration fed the fish (percentage of body weight). Twenty fish were included in the experiment for each ration level and water temperature combination. The mean percentage weight gain for each sample of 20 fish is shown in the table. (Note that, as expected, a 0 level of rations produced a negative weight gain, i.e., a loss.) Since all the sample sizes are equal, we can fit a simple linear regression model to the data for any water temperature level by fitting the model to the four sample means.

Ration (*% body weight per day*)	*Mean Wet Weight Gain (%)* 6°C	12°C	18°C
.0	−8.14	−10.33	−13.21
.8	12.31		
1.2		14.29	
1.5	28.19		15.51
2.5	29.12	38.05	
3.5			51.13
4.0		60.13	
4.5			65.49
Maintenance ration	.32	.48	.69

a. Find the least squares line relating mean weight gain to ration level for each of the three water temperature levels.

b. Plot the data points and graph the least squares lines on the same sheet of graph paper. Do the least squares lines appear to provide good fits to their respective data sets? Does the relationship between mean weight gain and ration appear to depend on the water temperature?

10.3 Model Assumptions

In Section 10.2 we assumed that the probabilistic model relating drug reaction time y to the percentage of drug x in the bloodstream is

$$y = \beta_0 + \beta_1 x + \epsilon$$

and recall that the least squares estimate of the deterministic component of the model, $\beta_0 + \beta_1 x$, is

$$\hat{y} = \hat{\beta}_0 + \hat{\beta}_1 x = -.1 + .7x$$

Now we turn our attention to the random component ϵ of the probabilistic model and its relation to the errors in estimating β_0 and β_1. We will use a probability distribution to characterize the behavior of ϵ. We will see how the probability distribution of ϵ determines how well the model describes the relationship between the dependent variable y and the independent variable x.

Step 3 in a regression analysis requires us to specify the probability distribution of the random error ϵ. We will make four basic assumptions about the general form of this probability distribution:

Assumption I The mean of the probability distribution of ϵ is 0. That is, the average of the values of ϵ over an infinitely long series of experiments is 0 for each setting of the independent variable x. This assumption implies that the mean value of y, $E(y)$, for a given value of x is $E(y) = \beta_0 + \beta_1 x$.

Assumption 2 The variance of the probability distribution of ϵ is constant for all settings of the independent variable x. For our straight-line model, this assumption means that the variance of ϵ is equal to a constant, say σ^2, for all values of x.

Assumption 3 The probability distribution of ϵ is normal.

Assumption 4 The values of ϵ associated with any two observed values of y are independent. That is, the value of ϵ associated with one value of y has no effect on the values of ϵ associated with other y values.

The implications of the first three assumptions can be seen in Figure 10.6, which shows distributions of errors for three values of x, namely, x_1, x_2, and x_3. Note that the relative frequency distributions of the errors are normal with a mean of 0 and a constant variance σ^2. (All the distributions shown have the same amount of spread or variability.) The straight line shown in Figure 10.6 is the line of means. It indicates the mean value of y for a given value of x. We denote this mean value as $E(y)$. Then, the line of means is given by the equation

$$E(y) = \beta_0 + \beta_1 x$$

FIGURE 10.6 ▶
The probability distribution of ϵ

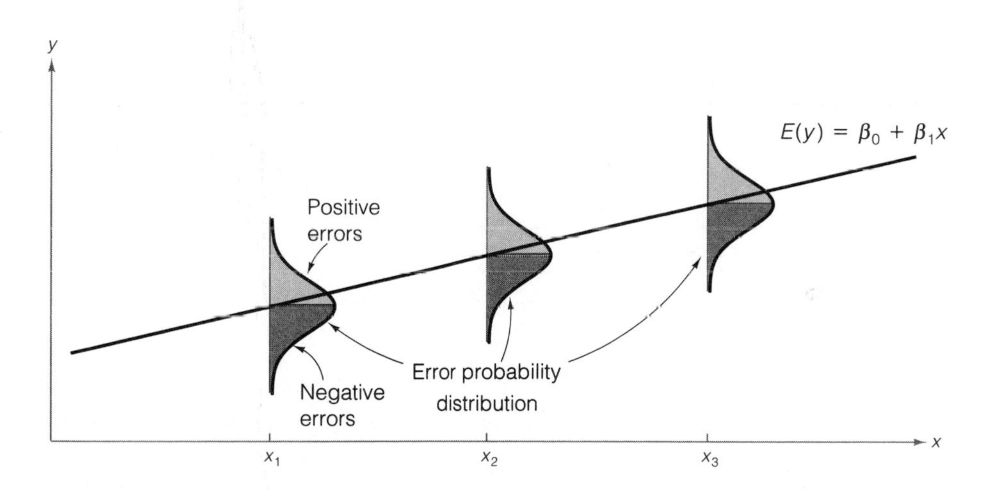

These assumptions make it possible for us to develop measures of reliability for the least squares estimators and to develop hypothesis tests for examining the usefulness of the least squares line. Various techniques exist for checking the validity of these assumptions, and there are remedies to be applied when they appear to be invalid. These topics are beyond the scope of this text, but they are discussed in some of the references listed at the end of the chapter. Fortunately, the assumptions need not hold exactly in order for least squares estimators to be useful. The assumptions will be satisfied adequately for many applications encountered in practice.

10.4 An Estimator of σ^2

It seems reasonable to assume that the greater the variability of the random error ϵ (which is measured by its variance σ^2), the greater will be the errors in the estimation of the model parameters β_0 and β_1 and in the error of prediction when $\hat{y}$ is used to predict y for some value of x. Consequently, you should not be surprised, as we proceed through this chapter, to find that σ^2 appears in the formulas for all confidence intervals and test statistics that we will be using.

Estimation of σ^2 for a (First-Order) Straight-Line Model

$$s^2 = \frac{\text{SSE}}{\text{Degrees of freedom for error}} = \frac{\text{SSE}}{n-2}$$

where $\text{SSE} = \sum (y_i - \hat{y}_i)^2 = \text{SS}_{yy} - \hat{\beta}_1 \text{SS}_{xy}$

$$\text{SS}_{yy} = \sum (y_i - \bar{y})^2 = \sum y_i^2 - \frac{\left(\sum y_i\right)^2}{n}$$

To estimate the standard deviation σ of ϵ, we calculate

$$s = \sqrt{s^2} = \sqrt{\frac{\text{SSE}}{n-2}}$$

We will refer to s as the **estimated standard error of the regression model.**

Warning: When performing these calculations, you may be tempted to round the calculated values of SS_{yy}, $\hat{\beta}_1$, and SS_{xy}. Be certain to carry at least six significant figures for each of these quantities to avoid substantial errors in calculation of the SSE.

In most practical situations, σ^2 is unknown and we must use our data to estimate its value. The best estimate of σ^2, denoted by s^2, is obtained by dividing the sum of squares of the deviations of the y values from the prediction line,

$$\text{SSE} = \sum (y_i - \hat{y}_i)^2$$

by the number of degrees of freedom associated with this quantity. We use 2 df to estimate the two parameters β_0 and β_1 in the straight-line model, leaving $(n - 2)$ df for the error variance estimation.

In the drug reaction example, we previously calculated SSE $= 1.10$ for the least squares line $\hat{y} = -.1 + .7x$. Recalling that there were $n = 5$ data points, we have $n - 2 = 5 - 2 = 3$ df for estimating σ^2. Thus,

$$s^2 = \frac{\text{SSE}}{n-2} = \frac{1.10}{3} = .367$$

is the estimated variance, and

$$s = \sqrt{.367} = .61$$

is the standard error of the regression model.

You may be able to obtain an intuitive feeling for s by recalling the interpretation given to a standard deviation in Chapter 2 and remembering that the least squares line estimates the mean value of y for a given value of x. Since s measures the spread of the distribution of y values about the least squares line, we should not be surprised to find that most of the observations lie within $2s$, or $2(.61) = 1.22$, of the least squares line. For this simple example (only five data points), all five data points fall within $2s$ of the least squares line. In Section 10.8, we use s to evaluate the error of prediction when the least squares line is used to predict a value of y to be observed for a given value of x.

Exercises 10.22–10.29

Learning the Mechanics

10.22 Suppose you fit a least squares line to 12 data points and the calculated value of SSE is .429.

a. Find s^2, the estimator of σ^2 (the variance of the random error term ϵ).

b. What is the largest deviation that you might expect between any one of the 12 points and the least squares line?

10.23 Calculate SSE and s^2 for each of the following cases:

a. $n = 20,\quad SS_{yy} = 95,\quad SS_{xy} = 50,\quad \hat{\beta}_1 = .75$

b. $n = 40,\quad \Sigma y^2 = 860,\quad \Sigma y = 50,\quad SS_{xy} = 2{,}700,\quad \hat{\beta}_1 = .2$

c. $n = 10,\quad \Sigma (y_i - \bar{y})^2 = 58,\quad SS_{xy} = 91,\quad SS_{xx} = 170$

10.24 Refer to Exercises 10.10 and 10.13. Calculate SSE, s^2, and s for the least squares lines obtained in these exercises. Interpret the standard error of the regression model, s, for each.

10.25 Visually compare the following scattergrams. If a least squares line were determined for each data set, which do you think would have the smallest variance, s^2? Explain.

a.

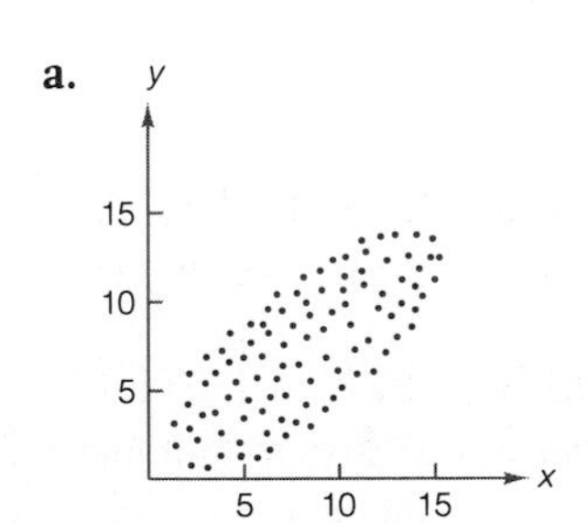

b.

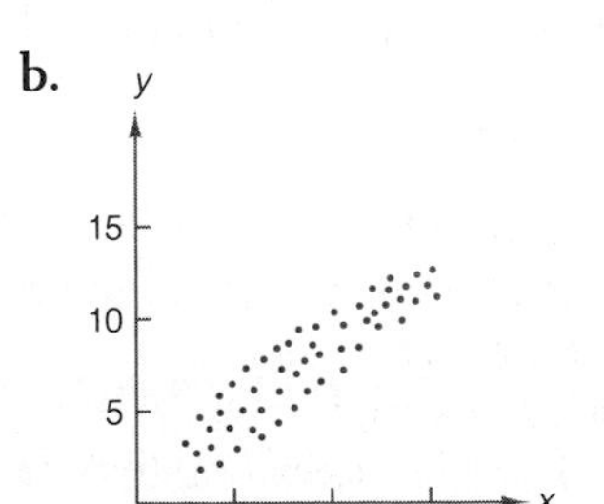

c.

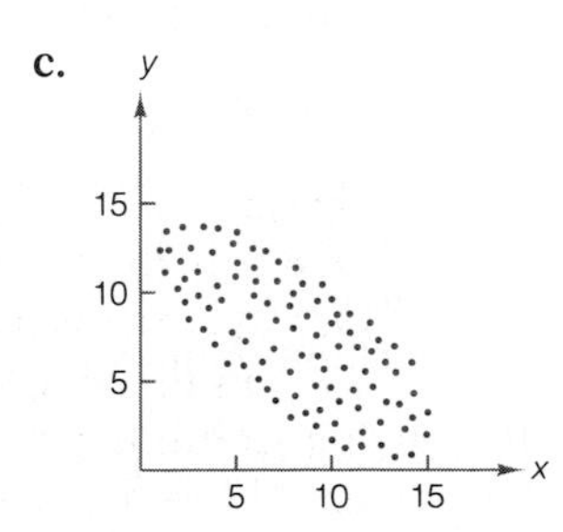

Applying the Concepts

10.26 A much larger proportion of U.S. teenagers are working while attending high school than was the case a decade ago, and this proportion is much larger than that in Japan or Germany or Sweden. The change was fueled by the growth of the service sector after World War II, the rise of the fast-food industry in the 1960s and 1970s, and an increase in the number of teenage girls entering the work force. These heavy workloads often result in underachievement in the classroom and lower grades. As a result, many states are imposing tighter child labor laws. Some business groups have mobilized to block these restrictions. However, a 1991 study of high school students in California and Wisconsin showed that those who worked only a few hours per week had the highest grade point averages (Waldman and Springen, 1992). The table shows grade point averages and number of hours worked per week for a sample of five students.

Grade Point Average, y	2.93	3.00	2.86	3.04	2.66
Hours Worked per Week, x	12	0	17	5	21

a. Given that $SS_{xx} = 294.00000$, $SS_{xy} = -4.55000$, $SS_{yy} = .08968$, $\bar{x} = 11.00000$, and $\bar{y} = 2.89800$, fit a least squares line to the data.
b. Plot the data and graph the least squares line.
c. Predict the grade point average of a high school student who works 10 hours per week. Repeat this for one who works 16 hours per week.
d. Calculate SSE, s^2, and s.
e. Within what approximate distance do you expect your predictions in part **c** to fall from the student's true grade point average? [*Note:* A more precise measure of reliability for these predictions is discussed in Section 10.8.]

10.27 A breeder of thoroughbred horses wishes to model the relationship between the gestation period and the length of life of a horse. The breeder believes that the two variables may follow a linear trend. The information in the table was supplied to the breeder from various thoroughbred stables across the state.

Horse	*Gestation Period* x *(days)*	*Life Length* y *(years)*	*Horse*	*Gestation Period* x *(days)*	*Life Length* y *(years)*
1	416	24	5	356	22
2	279	25.5	6	403	23.5
3	298	20	7	265	21
4	307	21.5			

a. Given that $SS_{xx} = 21{,}752$, $SS_{xy} = 236.5$, $SS_{yy} = 22$, $\bar{x} = 332$, and $\bar{y} = 22.5$, fit a least squares line to the data. Plot the data points and graph the least squares line as a check on your calculations.
b. According to your least squares line, approximately how long would you expect a horse to live whose gestation period was 400 days?
c. Calculate SSE and s^2.
d. Give an interpretation of the standard deviation s in the context of this problem.

10.28 To improve the quality of the output of any production process, it is necessary first to understand the capabilities of the process (Deming, 1982). In a particular manufacturing process, the useful life of a cutting tool is related to the speed at which the tool is operated. It is necessary to understand this relationship in order to predict when the tool should be replaced and how many spare tools should be available. The data in the table were derived from life tests for the two different brands of cutting tools currently used in the production process.

Cutting Speed (meters per minute)	*Useful Life (hours) Brand A*	*Brand B*	*Cutting Speed (meters per minute)*	*Useful Life (hours) Brand A*	*Brand B*
30	4.5	6.0	50	1.0	3.7
30	3.5	6.5	60	4.0	3.8
30	5.2	5.0	60	2.0	3.0
40	5.2	6.0	60	1.1	2.4
40	4.0	4.5	70	1.1	1.5
40	2.5	5.0	70	.5	2.0
50	4.4	4.5	70	3.0	1.0
50	2.8	4.0			

a. Construct a scattergram for each brand of cutting tool.
b. For each brand, use the method of least squares to model the relationship between useful life and cutting speed.
c. Find SSE, s^2, and s for each least squares line.
d. For a cutting speed of 70 meters per minute, find $\hat{y} \pm 2s$ for each least squares line.
e. For which brand would you feel more confident in using the least squares line to predict useful life for a given cutting speed? Explain.

10.29 A company keeps extensive records on its new salespeople on the premise that sales should increase with experience. A random sample of seven new salespeople produced the data on experience and sales shown in the table.

Months on Job x	*Monthly Sales* y *($ thousands)*
2	2.4
4	7.0
8	11.3
12	15.0
1	.8
5	3.7
9	12.0

a. Given that $SS_{xx} = 94.8571$, $SS_{xy} = 124.7571$, $SS_{yy} = 176.5171$, $\bar{x} = 5.8571$, and $\bar{y} = 7.4571$, fit a least squares line to the data.
b. Plot the data and graph the least squares line.
c. Predict the sales that a new salesperson would be expected to generate after 6 months on the job. After 9 months.
d. Calculate SSE, s^2, and s.
e. Within what approximate distance do you expect your predictions in part **c** to fall from the true number of sales generated by the new salesperson? [*Note:* A more precise measure of reliability for these predictions is discussed in Section 10.8.]

10.5 Assessing the Usefulness of the Model: Making Inferences About the Slope β_1

Now that we have specified the probability distribution of ϵ and found an estimate of the variance σ^2, we are ready to make statistical inferences about the model's usefulness for predicting the response y. This is step 4 in our regression modeling procedure.

Refer again to the data of Table 10.1 and suppose the reaction times are *completely unrelated* to the percentage of drug in the bloodstream. What could be said about the values of β_0 and β_1 in the hypothesized probabilistic model

$$y = \beta_0 + \beta_1 x + \epsilon$$

if x contributes no information for the prediction of y? The implication is that the mean of y—that is, the deterministic part of the model $E(y) = \beta_0 + \beta_1 x$—does not change as x changes. In the straight-line model, this means that the true slope, β_1, is equal to 0 (see Figure 10.7). Therefore, to test the null hypothesis that the linear model contributes no information for the prediction of y against the alternative hypothesis that the linear model is useful for predicting y, we test

$$H_0: \beta_1 = 0$$
$$H_a: \beta_1 \neq 0$$

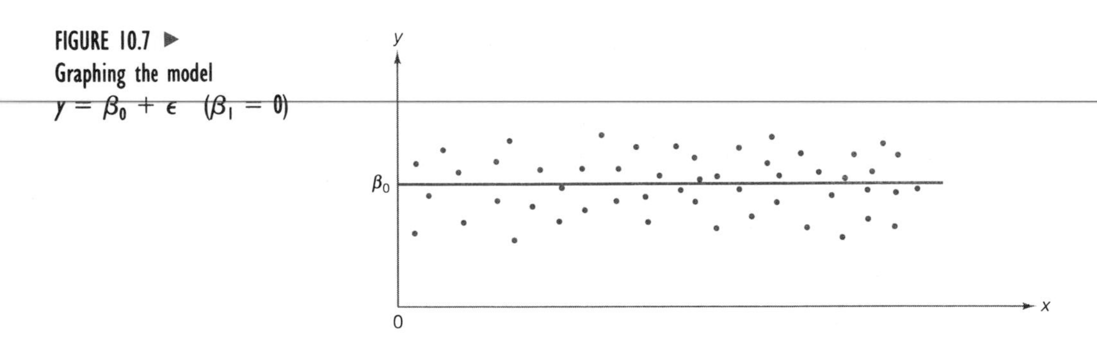

FIGURE 10.7 ▶ Graphing the model $y = \beta_0 + \epsilon$ $(\beta_1 = 0)$

If the data support the alternative hypothesis, we will conclude that x does contribute information for the prediction of y using the straight-line model (although the true relationship between $E(y)$ and x could be more complex than a straight line). Thus, in effect, this is a test of the usefulness of the hypothesized model.

The appropriate test statistic is found by considering the sampling distribution of $\hat{\beta}_1$, the least squares estimator of the slope β_1:

Sampling Distribution of $\hat{\beta}_1$

If we make the four assumptions about ϵ (see Section 10.3), the sampling distribution of the least squares estimator $\hat{\beta}_1$ of the slope will be normal with mean β_1 (the true slope) and standard deviation

$$\sigma_{\hat{\beta}_1} = \frac{\sigma}{\sqrt{SS_{xx}}} \qquad \text{(see Figure 10.8)}$$

We estimate $\sigma_{\hat{\beta}_1}$ by $s_{\hat{\beta}_1} = \frac{s}{\sqrt{SS_{xx}}}$ and refer to this quantity as the estimated standard error of the least squares slope $\hat{\beta}_1$.

FIGURE 10.8 ▶
Sampling distribution of $\hat{\beta}_1$

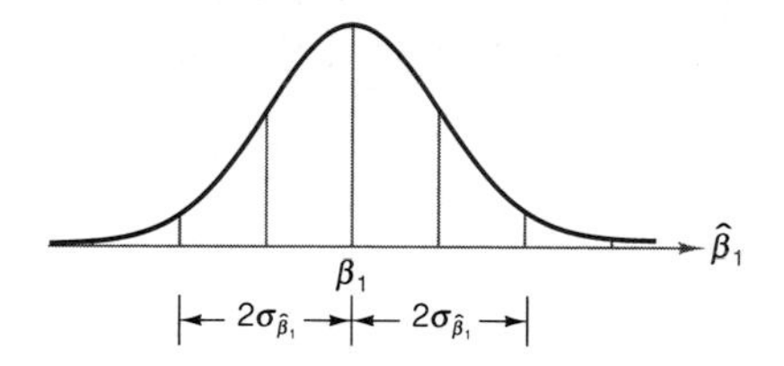

Since σ usually is unknown, the appropriate test statistic is a t statistic, formed as follows:

$$t = \frac{\hat{\beta}_1 - \text{Hypothesized value of } \beta_1}{s_{\hat{\beta}_1}} \qquad \text{where} \quad s_{\hat{\beta}_1} = \frac{s}{\sqrt{SS_{xx}}}$$

Thus,

$$t = \frac{\hat{\beta}_1 - 0}{s/\sqrt{SS_{xx}}}$$

Note that we have substituted the estimator s for σ and then formed the estimated standard error $s_{\hat{\beta}_1}$ by dividing s by $\sqrt{SS_{xx}}$. The number of degrees of freedom associated with this t statistic is the same as the number of degrees of freedom associated with s. Recall that this number is $(n - 2)$ df when the hypothesized model is a straight line (see Section 10.4). The setup of our test of the usefulness of the straight-line model is summarized in the box at the top of page 512.

A Test of Model Usefulness

One-Tailed Test

H_0: $\beta_1 = 0$
H_a: $\beta_1 < 0$
(or H_a: $\beta_1 > 0$)

Test statistic: $t = \dfrac{\hat{\beta}_1}{s_{\hat{\beta}_1}} = \dfrac{\hat{\beta}_1}{s/\sqrt{\text{SS}_{xx}}}$

Rejection region: $t < -t_\alpha$
(or $t > t_\alpha$ when H_a: $\beta_1 > 0$)

Two-Tailed Test

H_0: $\beta_1 = 0$
H_a: $\beta_1 \neq 0$

Test statistic: $t = \dfrac{\hat{\beta}_1}{s_{\hat{\beta}_1}} = \dfrac{\hat{\beta}_1}{s/\sqrt{\text{SS}_{xx}}}$

Rejection region: $t < -t_{\alpha/2}$
or $t > t_{\alpha/2}$

where t_α and $t_{\alpha/2}$ are based on $(n - 2)$ degrees of freedom

Assumptions: The four assumptions about ϵ listed in Section 10.3.

For the drug reaction example, we will choose $\alpha = .05$ and, since $n = 5$, t will be based on $n - 2 = 3$ df and the rejection region will be

$$t < -t_{.025} = -3.182 \quad \text{or} \quad t > t_{.025} = 3.182$$

We previously calculated $\hat{\beta}_1 = .7$, $s = .61$, and $\text{SS}_{xx} = 10$. Thus,

$$t = \frac{\hat{\beta}_1}{s/\sqrt{\text{SS}_{xx}}} = \frac{.7}{.61/\sqrt{10}} = \frac{.7}{.19} = 3.7$$

Since this calculated t value falls in the upper-tail rejection region (see Figure 10.9), we reject the null hypothesis and conclude that the slope β_1 is not 0. The sample evidence indicates that x contributes information for the prediction of y when a linear model is used to characterize the relationship between reaction time and the amount of the drug in the bloodstream.

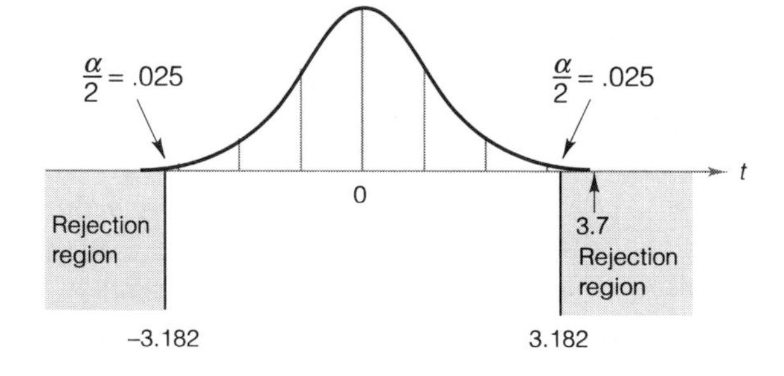

FIGURE 10.9 ▶ Rejection region and calculated t value for testing H_0: $\beta_1 = 0$ versus H_a: $\beta_1 \neq 0$

What conclusion can be drawn if the calculated t value does not fall in the rejection region? We know from previous discussions of the philosophy of hypothesis testing that such a t value does *not* lead us to accept the null hypothesis. That is, we do not conclude that $\beta_1 = 0$. Additional data might indicate that β_1 differs from 0, or a more complex relationship may exist between x and y, requiring the fitting of a model other than the straight-line model.

Another way to make inferences about the slope β_1 is to estimate it using a confidence interval. This interval is formed as shown in the box.

A $100(1 - \alpha)\%$ Confidence Interval for the Slope β_1

$$\hat{\beta}_1 \pm t_{\alpha/2} s_{\hat{\beta}_1}$$

where the estimated standard error of $\hat{\beta}_1$ is calculated by

$$s_{\hat{\beta}_1} = \frac{s}{\sqrt{SS_{xx}}}$$

and $t_{\alpha/2}$ is based on $(n - 2)$ degrees of freedom.

Assumptions: The four assumptions about ϵ listed in Section 10.3.

For the drug reaction example, $t_{\alpha/2}$ is based on $(n - 2) = 3$ degrees of freedom. Therefore, a 95% confidence interval for the slope β_1, the expected change in reaction time for a 1% increase in the amount of drug in the bloodstream, is

$$\hat{\beta}_1 \pm t_{.025} s_{\hat{\beta}_1} = .7 \pm 3.182\left(\frac{s}{\sqrt{SS_{xx}}}\right) = .7 \pm 3.182\left(\frac{.61}{\sqrt{10}}\right) = .7 \pm .61$$

Thus, the interval estimate of the slope parameter β_1 is .09 to 1.31. In terms of this example, the implication is that we can be 95% confident that the *true* mean increase in reaction time per additional 1% of the drug is between .09 and 1.31 seconds. This inference is meaningful only over the sampled range of x—that is, from 1% to 5% of the drug in the bloodstream. Since all the values in this interval are positive, it appears that β_1 is positive and that the mean of y, $E(y)$, increases as x increases. However, the rather large width of the confidence interval reflects the small number of data points (and, consequently, a lack of information) in the experiment. We would expect a narrower interval if the sample size were increased.

Exercises 10.30–10.40

Learning the Mechanics

10.30 Construct both a 95% and a 90% confidence interval for β_1 for each of the following cases:

a. $\hat{\beta}_1 = 31, \quad s = 3, \quad SS_{xx} = 35, \quad n = 12$
b. $\hat{\beta}_1 = 64, \quad SSE = 1{,}960, \quad SS_{xx} = 30, \quad n = 18$
c. $\hat{\beta}_1 = -8.4, \quad SSE = 146, \quad SS_{xx} = 64, \quad n = 24$

10.31 Consider the following pairs of observations:

x	1	5	3	2	6	6	0
y	1	3	3	1	4	5	1

a. Construct a scattergram for the data.
b. Use the method of least squares to fit a straight line to the seven data points in the table.
c. Plot the least squares line on your scattergram of part **a**.
d. Specify the null and alternative hypotheses you would use to test whether the data provide sufficient evidence to indicate that x contributes information for the (linear) prediction of y.
e. What is the test statistic that should be used in conducting the hypothesis test of part **d**? Specify the degrees of freedom associated with the test statistic.
f. Conduct the hypothesis test of part **d** using $\alpha = .05$.

10.32 Refer to Exercise 10.31. Construct an 80% and a 98% confidence interval for β_1.

10.33 Do the accompanying data provide sufficient evidence to conclude that a straight line is useful for characterizing the relationship between x and y?

y	4	2	5	3	2	4
x	1	4	5	3	2	4

10.34 Suppose that $n = 100$ recent residential home sales in a city are used to fit a least squares straight-line model relating the sale price, y, to the square feet of living space, x. Homes in the sample range from 1,500 square feet to 4,000 square feet of living space, and the resulting least squares equation is

$$\hat{y} = -30{,}000 + 70x$$

a. When the null hypothesis that the true slope is zero is tested, the result is a t value of 6.572. Give the appropriate p-value of this test, and interpret the result in the context of this application.
b. The 95% confidence interval for the slope is calculated to be 49.1 to 90.9. Interpret this interval in the context of this application. What can be done to obtain a narrower confidence interval?

Applying the Concepts

10.35 A breeder of thoroughbred horses wishes to model the relationship between the gestation period and the life span of a horse. The breeder believes that the two variables may follow a linear trend. The information in the table was supplied to the breeder from various thoroughbred stables across the state. (Note that the horse has the greatest variation of gestation period of any species, due to seasonal and nutritional factors.)

Horse	Gestation Period x (days)	Life Span y (years)	Horse	Gestation Period x (days)	Life span y (years)
1	416	24	5	356	22
2	279	25.5	6	403	23.5
3	298	20	7	265	21
4	307	21.5			

For these data, $SS_{yy} = 22.00$, $SS_{xx} = 21{,}752.00$, $SS_{xy} = 236.50$, $\bar{y} = 22.50$, and $\bar{x} = 332.00$. The least squares line was obtained in Exercise 10.27.

a. Do the data provide sufficient evidence to support the breeder's hypothesis? Test using $\alpha = .05$.

b. Find a 90% confidence interval for β_1. Interpret this interval.

10.36 A group of children, ranging from 10 to 12 years of age, were administered a verbal test in order to study the relationship between the number of words used and the silence interval before response. The tester believes that a linear relationship exists between the two variables. Each subject was asked a series of questions, and the total number of words used in answering was recorded. The time (in seconds) before the subject responded to each question was also recorded. The data for eight children are given in the table.

Subject	*Total Words* y	*Total Silence Time* x *(seconds)*	*Subject*	*Total Words* y	*Total Silence time* x *(seconds)*
1	61	23	5	91	17
2	70	37	6	63	21
3	42	38	7	71	42
4	52	25	8	55	16

a. Write a simple linear probabilistic model relating total words to total silence time, and use the least squares method to estimate the deterministic part of the model.

b. Does x contribute information for the prediction of y? Test the null hypothesis, slope $\beta_1 = 0$, against the alternative hypothesis, $\beta_1 \neq 0$. Use $\alpha = .05$. Interpret the results of the test.

10.37 In Exercise 10.19, we fit a least squares line to relate the cocoon temperature y of a woolly-bear caterpillar to the outside air temperature x. The data are reproduced in the table.

Day	*Temperature (°C)*		*Day*	*Temperature (°C)*	
	Air	*Cocoon*[a]		*Air*	*Cocoon*[a]
1	10.4	15.1	7	1.7	3.6
2	9.2	14.6	8	2.0	5.3
3	2.2	6.8	9	3.0	7.0
4	2.6	6.8	10	3.5	7.1
5	4.1	8.0	11	4.5	9.6
6	3.7	8.7	12	4.4	9.5

[a]Each cocoon temperature is the average of the temperatures of two cocoons.

Given that $SS_{xx} = 83.3425$, $SS_{xy} = 100.0825$, $SS_{yy} = 127.5092$, $\bar{x} = 4.2750$, and $\bar{y} = 8.5083$:

a. Calculate SSE, s^2, and s.

b. Interpret the standard error of the regression model.

c. Do the data provide sufficient evidence to indicate that the outside air temperature provides information for predicting the woolly-bear caterpillar cocoon temperature? Test using $\alpha = .05$.

d. Find the observed significance level of the test.

10.38 In Exercise 10.20, we fit a least squares line to relate the length of a sand lance to its age. The data are reproduced in the table at the top of page 516.

Mean Length (Millimeters)	176	194	212	226	236	244	254
Age (Years)	2	3	4	5	6	7	8

a. Do the data provide sufficient evidence to indicate that age x contributes information for the prediction of the length y of a sand lance? Test using $\alpha = .05$.
b. Find the p-value for the test and interpret it.

10.39 In Exercise 10.21, we discussed an experiment conducted to determine the effect of ration level on the growth rate of rainbow trout. Twenty fish were fed at each of four ration levels at 6°C water temperature. The experiment was repeated for water temperatures of 12°C and 18°C. The table gives the means and standard deviations (in parentheses) for each sample of 20 fish. Refer only to the data collected on fish fed in water maintained at 12°C.

Ration	*Mean Wet Weight Gain (%)*		
(% body weight per day)	*6°C*	*12°C*	*18°C*
.0	−8.14 (3.27)	−10.33 (2.35)	−13.21 (1.96)
.8	12.31 (3.37)		
1.2		14.29 (5.32)	
1.5	28.19 (5.39)		15.51 (4.48)
2.5	29.12 (5.67)	38.05 (6.99)	
3.5			51.13 (7.14)
4.0		60.13 (12.71)	
4.5			65.49 (13.04)
Maintenance ration	.32	.48	.69

a. Find SSE and s^2 for the $n = 4$ data points.
b. Find a 95% confidence interval for the mean gain (percentage of body weight) for a 1% increase in ration level.
c. State the assumptions required for your inference in part **b** to be valid.
d. Which assumption may not be satisfied? Explain why.

10.40 Based on an observational study of five chief executives, Mintzberg (1973) identified 10 managerial roles that can be found in all managerial jobs: figurehead, leader, liaison, monitor, disseminator, spokesperson, entrepreneur, disturbance handler, resource allocator, and negotiator. In an observational study of 19 managers from a medium-sized manufacturing plant, Luthans, Rosenkrantz, and Hennessey (1985) extended Mintzberg's work by investigating which activities *successful* managers actually perform. Each manager was observed during eighty 10-minute intervals over a 2-week period and their activities were recorded. The researchers used regression analysis to investigate which of the recorded activities were related to managerial success. To measure success, they devised an index based on the manager's length of time in the organization and his or her level within the firm; the higher the index, the more successful the manager. The table presents data (which are representative of the data collected by Luthans, Rosenkrantz, and Hennessey) that can be used to determine whether managerial success can in part be explained by extensiveness of a manager's network-building interactions with people outside the manager's work unit. Such interactions include phone and face-to-face meetings with customers and suppliers, attending external meetings, and doing public relations work.

Manager	*Manager Success Index* y	*Number of Interactions with Outsiders* x	*Manager*	*Manager Success Index* y	*Number of Interactions with Outsiders* x
1	40	12	11	70	20
2	73	71	12	47	81
3	95	70	13	80	40
4	60	81	14	51	33
5	81	43	15	32	45
6	27	50	16	50	10
7	53	42	17	52	65
8	66	18	18	30	20
9	25	35	19	42	21
10	63	82			

a. Construct a scattergram for the data.
b. Given $SS_{yy} = 7{,}006.6316$, $SS_{xx} = 10{,}824.5263$, $SS_{xy} = 2{,}561.2632$, $\bar{y} = 54.5789$, and $\bar{x} = 44.1579$, use the method of least squares to find a prediction equation for managerial success.
c. Find SSE, s^2, and s for your prediction equation. Interpret the standard deviation s in the context of this problem.
d. Plot the least squares line on your scattergram of part **a**. Does it appear that the number of interactions with outsiders contributes information for the prediction of managerial success? Explain.
e. Conduct a formal statistical hypothesis test to answer the question posed in part **d**. Use $\alpha = .05$.
f. Construct a 95% confidence interval for β_1. Interpret the interval in the context of the problem.

10.6 The Coefficient of Correlation

The claim is often made that the number of cigarettes smoked and the incidence of lung cancer are "highly correlated." Another popular belief is that the crime rate and the unemployment rate are "correlated." Some people even believe that the Dow Jones Industrial Average and the lengths of fashionable skirts are "correlated." In this section we will discuss the concept of **correlation**. A numerical descriptive measure of the correlation between two variables x and y is provided by the **Pearson product moment coefficient of correlation, r.**

Definition 10.2

The sample **Pearson product moment coefficient of correlation, *r*,** is defined as

$$r = \frac{SS_{xy}}{\sqrt{SS_{xx}SS_{yy}}}$$

It is a measure of the strength of the linear relationship between two random variables x and y.

Note that the computational formula for the correlation coefficient r given in Definition 10.2 involves the same quantities that were used in computing the least squares prediction equation. In fact, since the numerators of the expressions for $\hat{\beta}_1$ and r are identical, you can see that $r = 0$ when $\hat{\beta}_1 = 0$ (the case where x contributes no information for the prediction of y) and that r is positive when the slope is positive and negative when the slope is negative. Unlike $\hat{\beta}_1$, the correlation coefficient r is *scaleless* and assumes a value between -1 and $+1$, regardless of the units of x and y.

A value of r near or equal to 0 implies little or no linear relationship between y and x. In contrast, the closer r comes to 1 or -1, the stronger the linear relationship between y and x. And if $r = 1$ or $r = -1$, all the sample points fall exactly on the least squares line. Positive values of r imply a positive linear relationship between y and x; that is, y increases as x increases. Negative values of r imply a negative linear relationship between y and x; that is, y decreases as x increases. Each of these situations is portrayed in Figure 10.10.

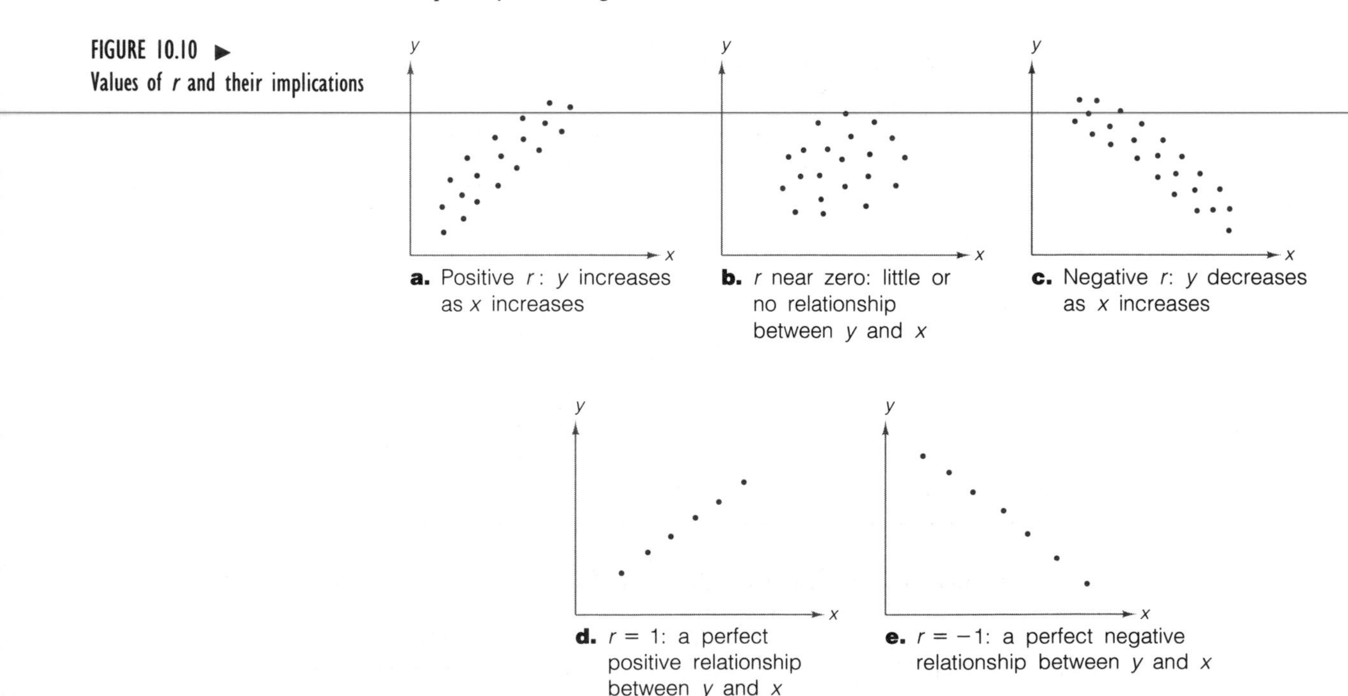

FIGURE 10.10 ▶ Values of r and their implications

We demonstrate how to calculate the coefficient of correlation r using the data in Table 10.1 for the drug reaction example. The quantities needed to calculate r are SS_{xy}, SS_{xx}, and SS_{yy}. The first two quantities have been calculated previously and are repeated here for convenience:

$$SS_{xy} = 7 \qquad SS_{xx} = 10 \qquad SS_{yy} = \sum y^2 - \frac{\left(\sum y\right)^2}{n}$$

$$= 26 - \frac{(10)^2}{5} = 26 - 20 = 6$$

We now find the coefficient of correlation:

$$r = \frac{SS_{xy}}{\sqrt{SS_{xx}SS_{yy}}} = \frac{7}{\sqrt{(10)(6)}} = \frac{7}{\sqrt{60}} = .904$$

The fact that r is positive and near 1 in value indicates that the reaction time tends to increase as the amount of drug in the bloodstream increases—*for this sample of five subjects*. This is the same conclusion we reached when we found the calculated value of the least squares slope to be positive.

EXAMPLE 10.1

A firm wants to know the correlation between the size of its sales force and its yearly sales revenue. The records for the past 10 years are examined, and the results listed in Table 10.5 are obtained. Calculate the coefficient of correlation r for the data.

TABLE 10.5 Sales Versus Number of Salespeople

Year	Number of Salespeople x	Sales y (hundred thousand dollars)
1981	15	1.35
1982	18	1.63
1983	24	2.33
1984	22	2.41
1985	25	2.63
1986	29	2.93
1987	30	3.41
1988	32	3.26
1989	35	3.63
1990	38	4.15

Solution

We need to calculate SS_{xy}, SS_{xx}, and SS_{yy}:

$$SS_{xy} = \sum xy - \frac{\left(\sum x\right)\left(\sum y\right)}{n} = 800.62 - \frac{(268)(27.73)}{10} = 57.456$$

$$SS_{xx} = \sum x^2 - \frac{\left(\sum x\right)^2}{n} = 7{,}668 - \frac{(268)^2}{10} = 485.6$$

$$SS_{yy} = \sum y^2 - \frac{\left(\sum y\right)^2}{n} = 83.8733 - \frac{(27.73)^2}{10} = 6.97801$$

Then, the coefficient of correlation is

$$r = \frac{SS_{xy}}{\sqrt{SS_{xx}SS_{yy}}} = \frac{57.456}{\sqrt{(485.6)(6.97801)}} = \frac{57.456}{58.211} = .99$$

Thus, the size of the sales force and sales revenue are very highly correlated—at least over the past 10 years. The implication is that a strong positive linear relationship exists between these variables (see Figure 10.11). We must be careful, however, not to jump to any unwarranted conclusions. For instance, the firm may be tempted to conclude that the best thing it can do to increase sales is to hire a large number of new salespeople—that is, that there is a *causal relationship* between the two variables. However, high correlation does not imply causality. The fact is, many things have probably contributed both to the increase in the size of the sales force and to the increase in sales revenue. The firm's expertise has undoubtedly grown, the economy has inflated (so that 1990 dollars are not worth as much as 1981 dollars), and perhaps the scope of products and services sold by the firm has widened. *We must be careful not to infer a causal relationship on the basis of high sample correlation. The only safe conclusion when a high correlation is observed in the sample data is that a linear trend may exist between* x *and* y.

FIGURE 10.11 ▶
Scattergram for Example 10.1

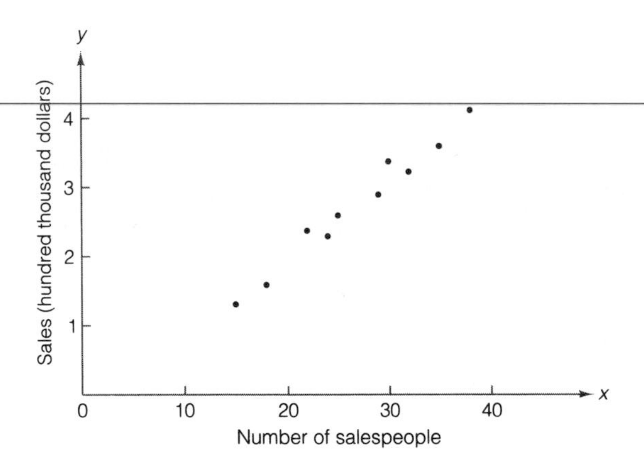

The correlation coefficient r measures the linear correlation between x values and y values in the sample, and a similar linear coefficient of correlation exists for the population from which the data points were selected. The **population correlation coefficient** is denoted by the symbol ρ (rho). As you might expect, ρ is estimated by the corresponding sample statistic, r. Or, rather than estimating ρ, we might want to test the null hypothesis H_0: $\rho = 0$ against H_a: $\rho \neq 0$—i.e., test the hypothesis that x

contributes no information for the prediction of y by using the straight-line model against the alternative that the two variables are at least linearly related.

However, we already performed this *identical* test in Section 10.5 when we tested H_0: $\beta_1 = 0$ against H_a: $\beta_1 \neq 0$. That is, the null hypothesis H_0: $\rho = 0$ is equivalent to the hypothesis H_0: $\beta_1 = 0$.* When we tested the null hypothesis H_0: $\beta_1 = 0$ in connection with the drug reaction example, the data led to a rejection of the null hypothesis at the $\alpha = .05$ level. This rejection implies that the null hypothesis of a 0 linear correlation between the two variables (drug and reaction time) can also be rejected at the $\alpha = .05$ level. The only real difference between the least squares slope $\hat{\beta}_1$ and the coefficient of correlation r is the measurement scale. Therefore, the information they provide about the usefulness of the least squares model is to some extent redundant. For this reason, we will use the slope to make inferences about the existence of a positive or negative linear relationship between two variables.

10.7 The Coefficient of Determination

Another way to measure the usefulness of the model is to measure the contribution of x in predicting y. To accomplish this, we calculate how much the errors of prediction of y were reduced by using the information provided by x. To illustrate, consider the sample shown in the scattergram of Figure 10.12(a) (page 522). If we assume that x contributes no information for the prediction of y, the best prediction for a value of y is the sample mean $\bar{y}$, which is shown as the horizontal line in Figure 10.12(b). The vertical line segments in Figure 10.12(b) are the deviations of the points about the mean $\bar{y}$. Note that the sum of squares of deviations for the prediction equation $\hat{y} = \bar{y}$ is

$$SS_{yy} = \sum (y_i - \bar{y})^2$$

Now suppose you fit a least squares line to the same set of data and locate the deviations of the points about the line as shown in Figure 10.12(c). Compare the deviations about the prediction lines in parts (b) and (c) of Figure 10.12. You can see that:

1. If x contributes little or no information for the prediction of y, the sums of squares of deviations for the two lines,

$$SS_{yy} = \sum (y_i - \bar{y})^2 \quad \text{and} \quad SSE = \sum (y_i - \hat{y}_i)^2$$

will be nearly equal.

2. If x does contribute information for the prediction of y, the SSE will be smaller than SS_{yy}. In fact, if all the points fall on the least squares line, then SSE = 0.

---

*The correlation test statistic that is equivalent to $t = \hat{\beta}_1/s_{\hat{\beta}_1}$ is

$$t = \frac{r}{\sqrt{(1 - r^2)/(n - 2)}}$$

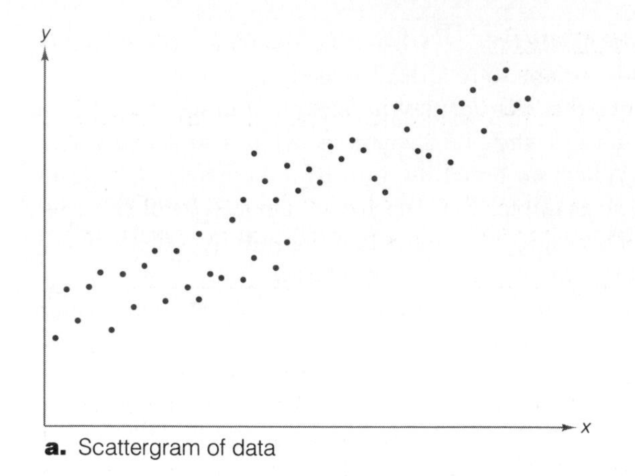

a. Scattergram of data

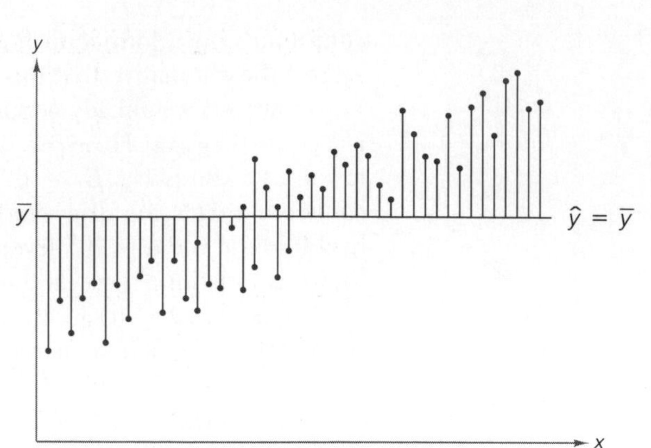

b. Assumption: x contributes no information for predicting y, $\hat{y} = \bar{y}$

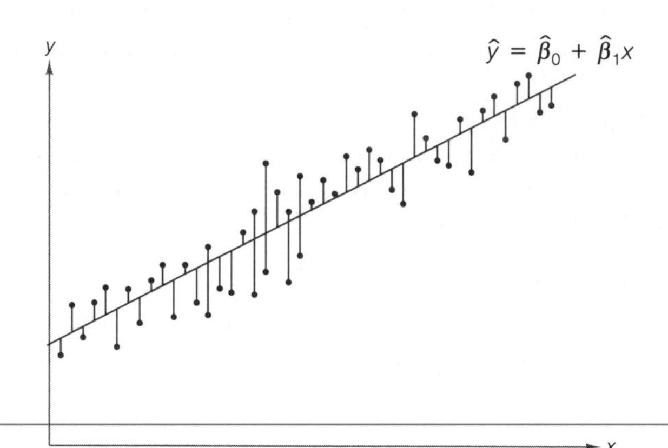

c. Assumption: x contributes information for predicting y, $\hat{y} = \hat{\beta}_0 + \hat{\beta}_1 x$

FIGURE 10.12 ▲ A comparison of the sum of squares of deviations for two models

Then, the reduction in the sum of squares of deviations that can be attributed to x, expressed as a proportion of SS_{yy}, is

$$\frac{SS_{yy} - SSE}{SS_{yy}}$$

Note that SS_{yy} is the "total sample variation" of the observations around the mean $\bar{y}$ and that SSE is the remaining "unexplained sample variability" after fitting the line $\hat{y}$. Thus, the difference (SS_{yy} − SSE) is the "explained sample variability" attributable to the linear relationship with x. Then a verbal description of the proportion is

$$\frac{SS_{yy} - SSE}{SS_{yy}} = \frac{\text{Explained sample variability}}{\text{Total sample variability}}$$

$$= \text{Proportion of total sample variability explained by the linear relationship}$$

It can be shown that this proportion is equal to the square of the simple linear coefficient of correlation r (the Pearson product moment coefficient of correlation).

Definition 10.3

The **coefficient of determination** is the square of the coefficient of correlation. It represents the proportion of the total sample variability around $\bar{y}$ that is explained by the linear relationship between y and x. It can be computed as

$$r^2 = \frac{SS_{yy} - SSE}{SS_{yy}} = 1 - \frac{SSE}{SS_{yy}}$$

Note that r^2 is always between 0 and 1, because r is between -1 and $+1$. Thus, an r^2 of .60 means that the sum of squares of deviations of the y values about their predicted values has been reduced 60% by the use of the least squares equation $\hat{y}$, instead of $\bar{y}$, to predict y.

EXAMPLE 10.2

Calculate the coefficient of determination for the drug reaction example using the formula given with Definition 10.3. The data are repeated in Table 10.6 for convenience.

TABLE 10.6

Amount of Drug x (%)	Reaction Time y (seconds)
1	1
2	1
3	2
4	2
5	4

Solution

From previous calculations,

$$SS_{yy} = 6 \quad \text{and} \quad SSE = \sum (y - \hat{y})^2 = 1.10$$

Then the coefficient of determination is given by

$$r^2 = \frac{SS_{yy} - SSE}{SS_{yy}} = \frac{6.0 - 1.1}{6.0} = \frac{4.9}{6.0} = .82$$

[In Section 10.6, we calculated $r = .904$. Now we have $r^2 = (.904)^2 = .82$.] So we know that using the amount of drug in the blood, x, to predict y with the least squares line

$$\hat{y} = -.1 + .7x$$

accounts for 82% of the total sum of squares of deviations of the five sample y values about their mean.

CASE STUDY 10.1 / Predicting the United States Crime Index

Reporting and analyzing crime rates is an important function of many law enforcement agencies. David Heaukulani (1975) comments: "The simple linear regression analysis remains one of the most useful tools for crime prediction." He demonstrates this point by fitting a straight-line model to predict the annual value of the United States crime index as a function of the United States population for each year. The data for the years 1963–1973 are given in Table 10.7, and the least squares line is shown in Figure 10.13.

Heaukulani reports that the correlation is $r = .94$ ($r^2 = .88$), which provides sufficient evidence at $\alpha = .01$ to indicate that the size of the population is useful for predicting the crime index using the straight-line model. Heaukulani concludes that

most agencies are using too narrow a span of statistical measurement to evaluate the overall crime picture. Year-by-year comparisons and percent analyses do not take average fluctuations into consideration. A lack of knowledge about statistical principles on the part of laymen, especially those in the news media, leads to a distortion or an invalid representation of the crime figures. Any increase is reported either as "bucking a trend" or "in line with the rise in crime."

If we expect to make any progress in solving the crime problem, we must evaluate the crime index objectively. We have to accept the fact that the crime index will probably increase from year

TABLE 10.7 United States Population and Crime Index

Year	Population	Actual Crime Index	Year	Population	Actual Crime Index
1963	188,531,000	2,259,081	1969	201,921,000	4,989,747
1964	191,334,000	2,604,426	1970	203,184,772	5,568,197
1965	193,818,000	2,780,015	1971	206,256,000	5,995,211
1966	195,857,000	3,243,400	1972	208,232,000	5,891,924
1967	197,864,000	3,802,273	1973	209,851,000	8,638,375
1968	199,861,000	4,466,573			

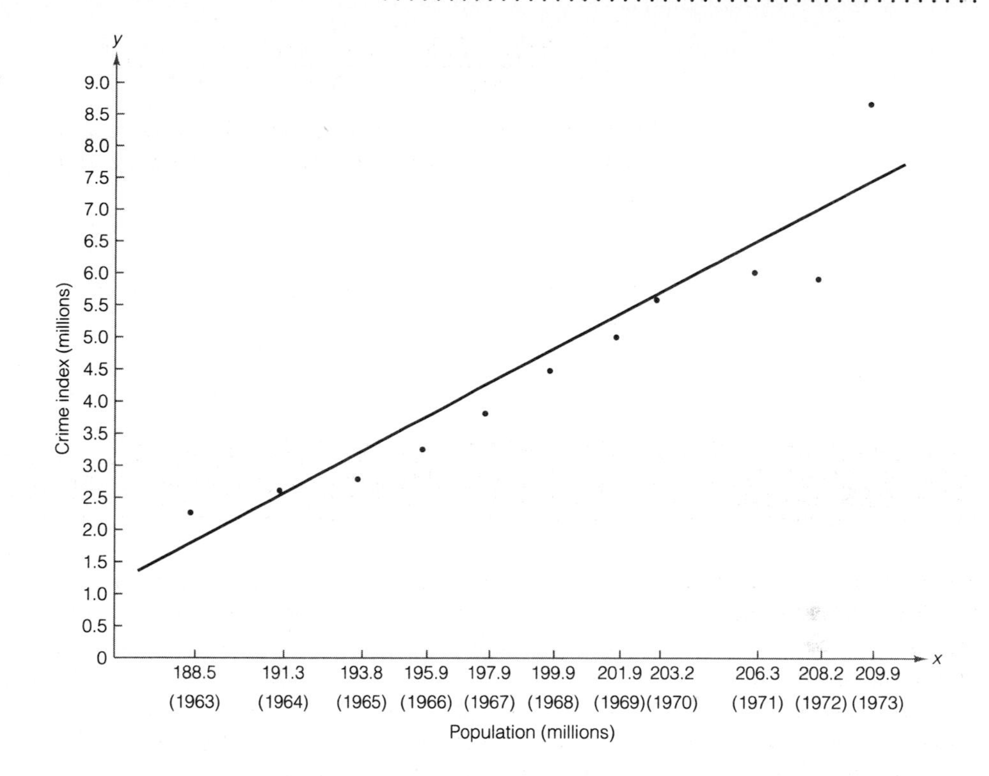

FIGURE 10.13 ▶ Least squares line relating crime index to population size (1963–1973)

to year. We should be willing to accept an average amount of increase and a set maximum limit. If the crime index exceeds the limit, only then should we become concerned and attempt to examine the factors that may have been responsible for the excess deviation.

Note that the 1973 crime rate of 8.6 million greatly exceeded the predicted value of less than 8 million. Perhaps other factors unique to this time period (Vietnam War, Watergate, price controls, and so forth) explain this "excess deviation." In any case, we cannot ascribe causation based only on high correlation between variables.

Exercises 10.41 – 10.53

Learning the Mechanics

10.41 Explain what each of the following sample correlation coefficients tells you about the relationship between the x and y values in the sample:

a. $r = 1$ **b.** $r = -1$ **c.** $r = 0$ **d.** $r = .90$ **e.** $r = .10$ **f.** $r = -.88$

10.42 Describe the slope of the least squares line if:

a. $r = .7$ **b.** $r = -.7$ **c.** $r = 0$ **d.** $r^2 = .64$

10.43 Construct a scattergram for each data set. Then calculate r and r^2 for each data set. Interpret their values.

a.

x	−2	−1	0	1	2
y	−2	1	2	5	6

b.

x	−2	−1	0	1	2
y	6	5	3	2	0

c.

x	1	2	2	3	3	3	4
y	2	1	3	1	2	3	2

d.

x	0	1	3	5	6
y	0	1	2	1	0

10.44 Calculate r^2 for the least squares line in each of the following exercises. Interpret their values.
a. Exercise 10.10 b. Exercise 10.13

Applying the Concepts

10.45 Find the correlation coefficient and the coefficient of determination for the sample data listed in the table and interpret your results; can a causal relationship be inferred?

Year	*Hunting Licenses Sold in U.S. (millions)*	*Divorces and Annulments in U.S. (millions)*
1970	22.2	.71
1975	25.9	1.04
1980	27.0	1.19
1983	28.9	1.16
1984	28.5	1.17
1985	27.7	1.19
1986	27.9	1.18
1987	28.8	1.17
1988	30.0	1.17

Source: U.S. Bureau of the Census, *Statistical Abstract of the United States:* 1992, pp. 90, 238.

10.46 In Exercise 10.15, we gave the number of games won, y, and the batting average, x, for the 14 American League baseball teams at the end of the 1991 season. They are repeated in the table (page 527). As before, $\Sigma y = 1{,}134$, $\Sigma x = 3.642$, $\Sigma y^2 = 93{,}110$, $\Sigma x^2 = .948622$, and $\Sigma xy = 295.54$.

a. Calculate the correlation coefficient, r, and the coefficient of determination, r^2, for the data. Interpret their values.
b. Do the data provide sufficient evidence to conclude that a correlation exists between a team's number of wins and its batting average? Test using $\alpha = .05$. [*Hint:* See the last paragraph of Section 10.6.]

Team	Number of Games Won	Team Batting Average
	y	x
Cleveland	57	.254
New York	71	.256
Boston	84	.269
Toronto	91	.257
Texas	85	.270
Detroit	84	.247
Minnesota	95	.280
Baltimore	67	.254
California	81	.255
Milwaukee	83	.271
Seattle	83	.255
Kansas City	82	.264
Oakland	84	.248
Chicago	87	.262

Source: *Official Major League Baseball 1992 Stat Book*. Major League Baseball Properties, Inc. and the editors of *The Baseball Encyclopedia*, New York.

10.47 Calculate the coefficient of determination and the sample coefficient of correlation for the woolly-bear caterpillar data in Exercise 10.19. Interpret them.

10.48 Is the maximal oxygen uptake, a measure often used by physiologists to indicate an individual's state of cardiovascular fitness, related to the performance of distance runners? Six long-distance runners submitted to treadmill tests for determination of their maximal oxygen uptake. The results, along with each runner's best mile time (in seconds), are shown in the table.

Athlete	Maximal Oxygen Uptake (Milliliters/kilogram)	Mile Time (Seconds)
1	63.3	241.5
2	60.1	249.8
3	53.6	246.1
4	58.8	232.4
5	67.5	237.2
6	62.5	238.4

a. Calculate r and r^2.

b. Do the data provide sufficient evidence to indicate that mile time is negatively correlated with maximal oxygen uptake? Test H_0: $\rho = 0$ against the alternative hypothesis, H_a: $\rho < 0$, using $\alpha = .05$.*

10.49 Is there a correlation between the amount of education received by people living in urban centers and that received by people living in the urban fringes? To determine whether a correlation exists, a sociologist

*Recall (Section 10.6) that the test of H_0: $\rho = 0$ is equivalent to the test of H_0: $\beta_1 = 0$.

compared the percentages of people with 4 years of high school education or more for the two groups. The data are given in the table.

City	Urban Center y	Urban Fringe x	City	Urban Center y	Urban Fringe x
Baltimore	28.2	42.3	Milwaukee	39.7	54.4
Boston	44.6	55.8	New Orleans	33.3	44.6
Chicago	35.3	53.9	New York	36.4	48.7
Cleveland	30.1	55.5	Philadelphia	30.7	48.0
Dallas	48.9	56.4	St. Louis	26.3	43.3
Detroit	34.4	47.5	San Francisco	49.4	57.9
Houston	45.2	50.1	Washington	47.8	67.5
Los Angeles	53.4	53.4			

Source: Computed from U.S. Bureau of the Census, *U.S. Census of Population: 1960, General Social and Economic Characteristics*, and *U.S. Census of Population and Housing: 1960, Census Tracts* (Washington, D.C.: Government Printing Office, 1961).

a. Find the correlation coefficient for the data.
b. Find the coefficient of determination, and explain its meaning in terms of this problem.
c. Is there sufficient evidence to indicate a nonzero correlation between x and y? Test using $\alpha = .05$.*

10.50 In the summer of 1981, the Minnesota Department of Transportation installed a state-of-the-art weigh-in-motion scale in the concrete surface of the eastbound lanes of Interstate 494 in Bloomington, Minnesota. After installation, a study was undertaken to determine whether the scale's readings correspond with the static weights of the vehicles being monitored. (Studies of this type are known as *calibration studies*.) After some preliminary comparisons using a two-axle, six-tire truck carrying different loads (see table), calibration adjustments were made in the software of the weigh-in-motion system and the scales were reevaluated.

Trial Number	Static Weight of Truck x (thousand pounds)	Weigh-in-Motion Reading Prior to Calibration Adjustment y_1 (thousand pounds)	Weigh-in-Motion Reading After Calibration Adjustment y_2 (thousand pounds)
1	27.9	26.0	27.8
2	29.1	29.9	29.1
3	38.0	39.5	37.8
4	27.0	25.1	27.1
5	30.3	31.6	30.6
6	34.5	36.2	34.3
7	27.8	25.1	26.9
8	29.6	31.0	29.6
9	33.1	35.6	33.0
10	35.5	40.2	35.0

Source: Adapted from data in Wright, J. L., Owen, F., and Pena, D. "Status of MN/DOT's weigh-in-motion program." St. Paul: Minnesota Department of Transportation, January 1983.

*Recall (Section 10.6) that the test of H_0: $\rho = 0$ is equivalent to the test of H_0: $\beta_1 = 0$.

a. Construct two scattergrams, one of y_1 versus x and the other of y_2 versus x.
b. Use the scattergram of part **a** to evaluate the performance of the weigh-in-motion scale both before and after the calibration adjustment.
c. Calculate the correlation coefficient for both sets of data, and interpret their values. Explain how these correlation coefficients can be used to evaluate the weigh-in-motion scale.
d. Suppose the sample correlation coefficient for y_2 and x were 1. Could this happen if the static weights and the weigh-in-motion readings disagreed? Explain.

10.51 A problem of economic and social concern in the United States is the importation and sale of illicit drugs. The data shown in the table are part of a larger body of data collected by the Florida attorney general's office in an attempt to relate the incidence of drug seizures and drug arrests to the characteristics of the Florida counties. Given are the number, y, of drug arrests per county in 1982, the density, x_1, of the county (population per square mile), and the number, x_2, of law enforcement employees. In order to simplify the calculations, we show data for only 10 counties.

	County									
	1	2	3	4	5	6	7	8	9	10
Population Density, x_1	169	68	278	842	18	42	112	529	276	613
Number of Law Enforcement Employees, x_2	498	35	772	5,788	18	57	300	1,762	416	520
Number of Arrests in 1982, y	370	44	716	7,416	25	50	189	1,097	256	432

a. Fit a least squares line to relate the number, y, of drug arrests per county in 1982 to the county population density, x_1.
b. We might expect the mean number of arrests to increase as the population density increases. Do the data support this theory? Test using $\alpha = .05$.
c. Calculate the coefficient of determination for this regression analysis and interpret its value.

10.52 Repeat parts **a**, **b**, and **c** of Exercise 10.51 using the number x_2 of county law enforcement employees as the independent variable. Then answer the following questions:
d. Which least squares line has the lower SSE?
e. Which independent variable explains more of the variation in y? Explain.

10.53 Is there a relationship between leisure activities and high school performance? David A. Bergin (1992) presented a list of 43 leisure activities to 159 high school students. The students were asked to identify the activities they participated in each week. Dr. Bergin reports that the correlation between high school GPA and the number of leisure activities is $r = .13$, which has a two-tailed p-value of .0512.
a. What are the appropriate null and alternative hypotheses to test whether the number of leisure activities and GPA are linearly related?
b. Interpret the p-value in terms of this test.

10.8 Using the Model for Estimation and Prediction

If we are satisfied that a useful model has been found to describe the relationship between reaction time and amount of drug in the bloodstream, we are ready for step 5 in our regression modeling procedure: using the model for estimation and prediction.

The most common uses of a probabilistic model for making inferences can be divided into two categories. The first is the use of the model for estimating the mean value of y, E(y), for a specific value of x.

For our drug reaction example, we may want to estimate the mean response time for all people whose blood contains 4% of the drug.

The second use of the model entails predicting a new individual y value for a given x.

That is, we may want to predict the reaction time for a specific person who possesses 4% of the drug in the bloodstream.

In the first case, we are attempting to estimate the mean value of y for a very large number of experiments at the given x value. In the second case, we are trying to predict the outcome of a single experiment at the given x value. Which of these model uses—estimating the mean value of y or predicting an individual new value of y (for the same value of x)—can be accomplished with the greater accuracy?

Before answering this question, we first consider the problem of choosing an estimator (or predictor) of the mean (or a new individual) y value. We will use the least squares prediction equation

$$\hat{y} = \hat{\beta}_0 + \hat{\beta}_1 x$$

both to estimate the mean value of y and to predict a specific new value of y for a given value of x. For our example, we found

$$\hat{y} = -.1 + .7x$$

so that the estimated mean reaction time for all people when $x = 4$ (drug is 4% of blood content) is

$$\hat{y} = -.1 + .7(4) = 2.7 \text{ seconds}$$

The same value is used to predict a new y value when $x = 4$. That is, both the estimated mean and the predicted value of y are $\hat{y} = 2.7$ when $x = 4$, as shown in Figure 10.14.

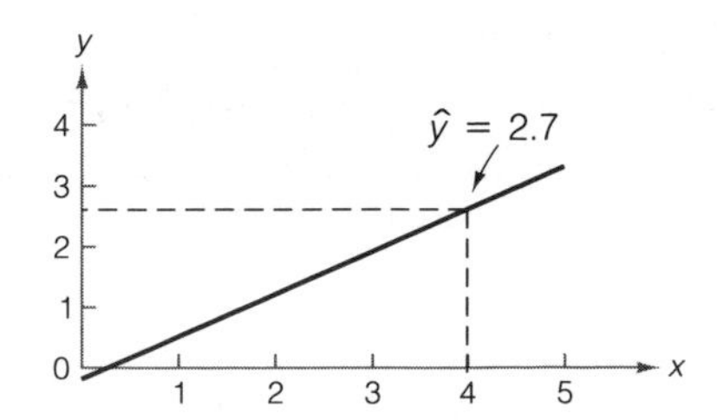

FIGURE 10.14 ▶ Estimated mean value and predicted individual value of reaction time y for $x = 4$

The difference between these two model uses lies in the relative accuracy of the estimate and the prediction. These accuracies are best measured by using the sampling errors of the least squares line when it is used as an estimator and as a predictor, respectively. These errors are reflected in the standard deviations given in the next box.

Sampling Errors for the Estimator of the Mean of y and the Predictor of an Individual New Value of y

1. The standard deviation of the sampling distribution of the estimator $\hat{y}$ of the mean value of y at a specific value of x, say x_p, is

$$\sigma_{\hat{y}} = \sigma\sqrt{\frac{1}{n} + \frac{(x_p - \bar{x})^2}{SS_{xx}}}$$

where σ is the standard deviation of the random error ϵ. We refer to $\sigma_{\hat{y}}$ as the standard error of $\hat{y}$.

2. The standard deviation of the prediction error for the predictor $\hat{y}$ of an individual new y value at a specific value of x is

$$\sigma_{(y-\hat{y})} = \sigma\sqrt{1 + \frac{1}{n} + \frac{(x_p - \bar{x})^2}{SS_{xx}}}$$

where σ is the standard deviation of the random error ϵ. We refer to $\sigma_{(y-\hat{y})}$ as the standard error of prediction.

The true value of σ is rarely known, so we estimate σ by s and calculate the estimation and prediction intervals as shown in the next two boxes.

A $100(1 - \alpha)\%$ Confidence Interval for the Mean Value of y at $x = x_p$

$$\hat{y} \pm t_{\alpha/2}(\text{Estimated standard error of } \hat{y})$$

or

$$\hat{y} \pm t_{\alpha/2}\, s\sqrt{\frac{1}{n} + \frac{(x_p - \bar{x})^2}{SS_{xx}}}$$

where $t_{\alpha/2}$ is based on $(n - 2)$ degrees of freedom.

A 100(1 − α)% Prediction Interval* for an Individual New Value of y at $x = x_p$

$$\hat{y} \pm t_{\alpha/2}(\text{Estimated standard error of prediction})$$

or

$$\hat{y} \pm t_{\alpha/2}s\sqrt{1 + \frac{1}{n} + \frac{(x_p - \bar{x})^2}{SS_{xx}}}$$

where $t_{\alpha/2}$ is based on $(n - 2)$ degrees of freedom.

EXAMPLE 10.3

Find a 95% confidence interval for the mean reaction time when the concentration of the drug in the bloodstream is 4%.

Solution

For a 4% concentration, $x = 4$ and the confidence interval for the mean value of y is

$$\hat{y} \pm t_{\alpha/2}s\sqrt{\frac{1}{n} + \frac{(x_p - \bar{x})^2}{SS_{xx}}} = \hat{y} \pm t_{.025}s\sqrt{\frac{1}{5} + \frac{(4 - \bar{x})^2}{SS_{xx}}}$$

where $t_{.025}$ is based on $n - 2 = 5 - 2 = 3$ degrees of freedom. Recall that $\hat{y} = 2.7$, $s = .61$, $\bar{x} = 3$, and $SS_{xx} = 10$. From Table V in Appendix A, $t_{.025} = 3.182$. Thus, we have

$$\begin{aligned} 2.7 \pm (3.182)(.61)\sqrt{\frac{1}{5} + \frac{(4 - 3)^2}{10}} &= 2.7 \pm (3.182)(.61)(.55) \\ &= 2.7 \pm (3.182)(.34) \\ &= 2.7 \pm 1.1 \end{aligned}$$

Therefore, when the percentage of drug in the bloodstream is 4%, the standard error of $\hat{y}$ is .34 and the corresponding 95% confidence interval for the mean reaction time for all possible subjects is 1.6 to 3.8 seconds. Note that we used a small amount of data (small sample size) for purposes of illustration in fitting the least squares line. The interval would probably be narrower if more information had been obtained from a larger sample.

*The term *prediction interval* is used when the interval formed is intended to enclose the value of a random variable. The term *confidence interval* is reserved for estimation of population parameters (such as the mean).

EXAMPLE 10.4

Predict the reaction time for the next performance of the experiment for a subject with a drug concentration of 4%. Use a 95% prediction interval.

Solution

To predict the response time for an individual new subject for whom $x = 4$, we calculate the 95% prediction interval as

$$\hat{y} \pm t_{\alpha/2} s \sqrt{1 + \frac{1}{n} + \frac{(x_p - \bar{x})^2}{SS_{xx}}} = 2.7 \pm (3.182)(.61)\sqrt{1 + \frac{1}{5} + \frac{(4-3)^2}{10}}$$
$$= 2.7 \pm (3.182)(.61)(1.14)$$
$$= 2.7 \pm (3.182)(.70)$$
$$= 2.7 \pm 2.2$$

Therefore, the standard error of prediction when the drug concentration is 4% is .70, and we predict with 95% confidence that the reaction time for this new individual will fall in the interval from .5 to 4.9 seconds. Like the confidence interval for the mean value of y, the prediction interval for y is quite large. This is because we have chosen a simple example (only five data points) to fit the least squares line. The width of the prediction interval could be reduced by using a larger number of data points.

A comparison of the confidence interval for the mean value of y and the prediction interval for a new value of y for 4% drug concentration ($x = 4$) is illustrated in Figure 10.15. Note that the prediction interval for an individual new value of y is *always* wider than the corresponding confidence interval for the mean value of y. You can see this by examining the formulas for the two intervals and by studying Figure 10.15.

FIGURE 10.15 ▶
A 95% confidence interval for mean sales and a prediction interval for sales when $x = 4$

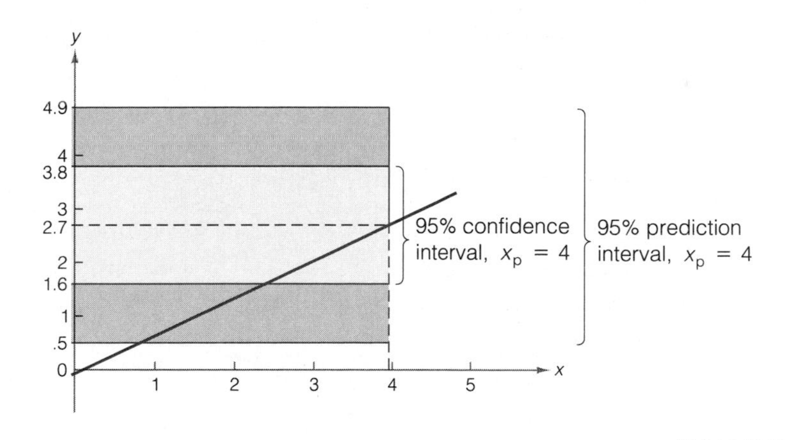

The error in estimating the mean value of y, $E(y)$, for a given value of x, say x_p, is the distance between the least squares line and the true line of means, $E(y) = \beta_0 + \beta_1 x$. This error, $[\hat{y} - E(y)]$, is shown in Figure 10.15. In contrast, *the error* $(y_p - \hat{y})$ *in predicting some future value of y is the sum of two errors*—the error of estimating the mean of y, $E(y)$, shown in Figure 10.16 (page 534), plus the random error that is a

FIGURE 10.16 ▶ Error of estimating the mean value of y for a given value of x

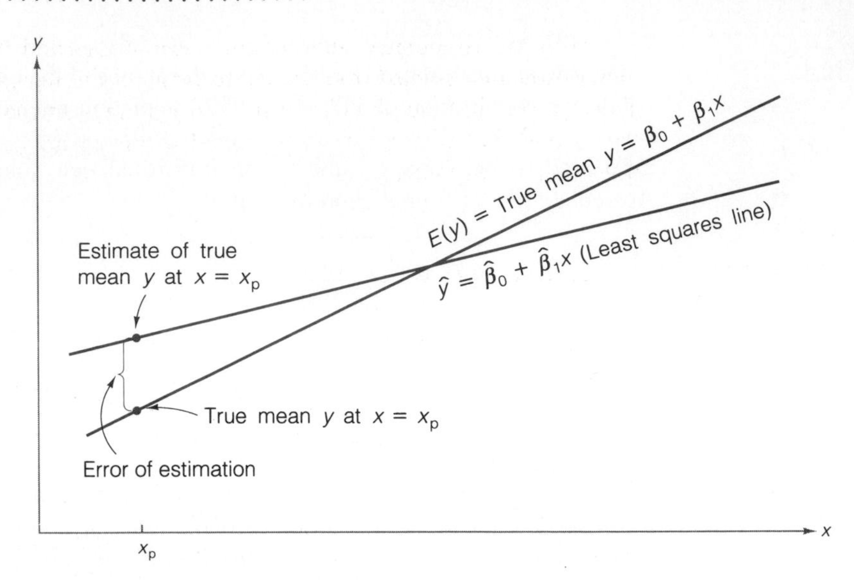

component of the value of y to be predicted (see Figure 10.17). Consequently, the error of predicting a particular value of y will be larger than the error of estimating the mean value of y for a particular value of x. Note from their formulas that both the error of estimation and the error of prediction take their smallest values when $x_p = \bar{x}$. The farther x_p lies from $\bar{x}$, the larger will be the errors of estimation and prediction. You can see why this is true by noting the deviations for different values of x_p between the line of means $E(y) = \beta_0 + \beta_1 x$ and the predicted line of means $\hat{y} = \hat{\beta}_0 + \hat{\beta}_1 x$ shown in Figure 10.17. The deviation is larger at the extremities of the interval where the largest and smallest values of x in the data set occur.

FIGURE 10.17 ▶ Error of predicting a future value of y for a given value of x

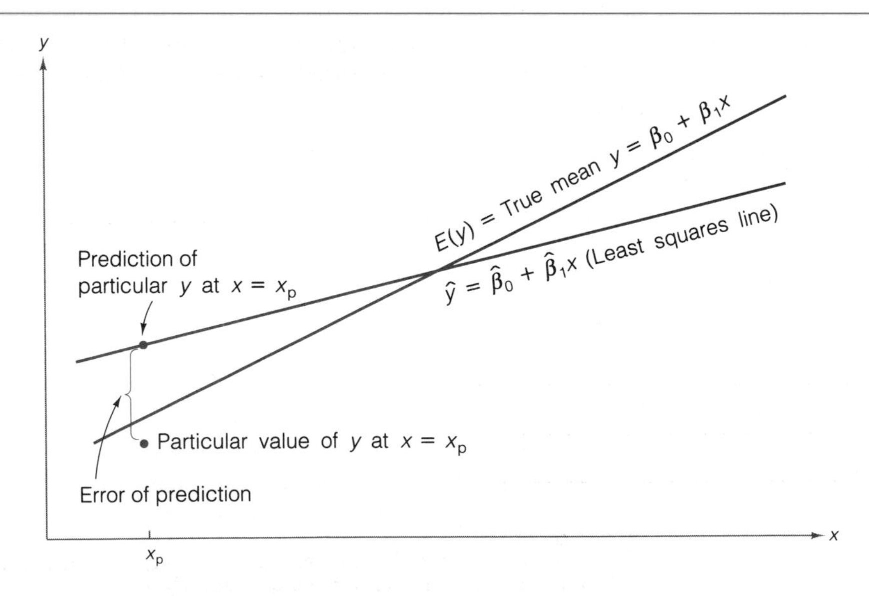

Both the confidence intervals for mean values and the prediction intervals for new values are depicted over the entire range of the regression line in Figure 10.18. You can see that the confidence interval is always narrower than the prediction interval, and that they are both narrowest at the mean $\bar{x}$, increasing steadily as the distance $|x - \bar{x}|$ increases.

FIGURE 10.18 ▶
Confidence intervals for mean values and prediction intervals for new values

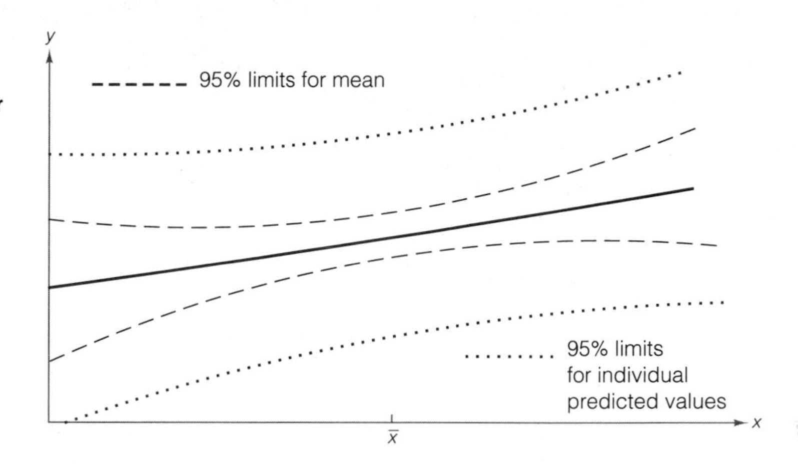

The confidence interval width grows smaller as n is increased; thus, in theory, you can obtain as precise an estimate of the mean value of y as desired (at any given x) by selecting a large enough sample. The prediction interval for a new value of y also grows smaller as n increases, but there is a lower limit on its width. If you examine the formula for the prediction interval (page 532), you will see that the interval can get no smaller than $\hat{y} \pm z_{\alpha/2}\sigma$.* Thus, the only way to obtain more accurate predictions for new values of y is to reduce the standard deviation of the regression model, σ. This can be accomplished only by improving the model, either by using a curvilinear (rather than linear) relationship with x or by adding new independent variables to the model, or both. Consult the references to learn more about these methods of improving the model.

Exercises 10.54–10.65

Learning the Mechanics

10.54 Consider the following pairs of measurements:

x	1	2	3	4	5	6	7
y	3	5	4	6	7	7	10

*The result follows from the facts that, for large n, $t_{\alpha/2} \approx z_{\alpha/2}$, $s \approx \sigma$, and the last two terms under the radical in the standard error of the predictor are approximately 0.

a. Construct a scattergram for these data.
b. Find the least squares line, and plot it on your scattergram.
c. Find s^2.
d. Find a 90% confidence interval for the mean value of y when $x = 4$. Plot the upper and lower bounds of the confidence interval on your scattergram.
e. Find a 90% prediction interval for a new value of y when $x = 4$. Plot the upper and lower bounds of the prediction interval on your scattergram.
f. Compare the widths of the intervals you constructed in parts **d** and **e**. Which is wider and why?

10.55 Consider the following pairs of measurements:

x	y	x	y
1	−1	−2	−6
−1	−5	3	4
2	1	−1	−4
0	−3	5	4
4	7	1	0

Given that $SS_{xx} = 47.6$, $SS_{yy} = 168.1$, $SS_{xy} = 85.6$, and $\hat{y} = -2.458 + 1.7983x$:
a. Construct a scattergram for the data.
b. Plot the least squares line on your scattergram.
c. Use a 95% confidence interval to estimate the mean value of y when $x = 5$. Plot the upper and lower bounds of the interval on your scattergram.
d. Repeat part **c** for $x = 1.2$ and $x = -2$.
e. Compare the widths of the three confidence intervals you constructed in parts **c** and **d** and explain why they differ.

10.56 Refer to Exercise 10.55.
a. Using no information about x, estimate and calculate a 95% confidence interval for the mean value of y. [*Hint:* Use the one-sample t methodology of Section 6.3.]
b. Plot the estimated mean value and the confidence interval as horizontal lines on your scattergram of Exercise 10.55.
c. Compare the confidence intervals you calculated in parts **c** and **d** of Exercise 10.55 with the one you calculated in part **a** of this exercise. Does x appear to contribute information about the mean value of y?
d. Check the answer you gave in part **c** with a statistical test of the null hypothesis H_0: $\beta_1 = 0$ against H_a: $\beta_1 \neq 0$. Use $\alpha = .05$.

10.57 Consider the following pairs of measurements:

x	4	6	0	5	2	3	2	6	2	1
y	3	5	−1	4	3	2	0	4	1	1

For these data, $SS_{xx} = 38.900$, $SS_{yy} = 33.600$, $SS_{xy} = 32.8$, and $\hat{y} = -.414 + .843x$.
a. Construct a scattergram for these data.
b. Plot the least squares line on your scattergram.
c. Use a 95% confidence interval to estimate the mean value of y when $x_p = 6$. Plot the upper and lower bounds of the interval on your scattergram.

d. Repeat part **c** for $x_p = 3.2$ and $x_p = 0$.
e. Compare the widths of the three confidence intervals you constructed in parts **c** and **d** and explain why they differ.

10.58 In fitting a least squares line to $n = 10$ data points, the following quantities were computed:

$$SS_{xx} = 32 \quad \bar{x} = 3 \quad SS_{yy} = 26 \quad \bar{y} = 4 \quad SS_{xy} = 28$$

a. Find the least squares line.
b. Graph the least squares line.
c. Calculate SSE.
d. Calculate s^2.
e. Find a 95% confidence interval for the mean value of y when $x_p = 2.5$.
f. Find a 95% prediction interval for y when $x_p = 4$.

Applying the Concepts

10.59 Certain dosages of a new drug developed to reduce a smoker's reliance on nicotine may reduce one's pulse rate to dangerously low levels. To investigate the drug's effect on pulse rate, different dosages of the drug were administered to six randomly selected patients, and 30 minutes later the decrease in each patient's pulse rate was recorded.

Patient	*Dosage* x *(cubic centimeters)*	*Decrease in Pulse Rate* y *(beats/minute)*
1	2.0	15
2	1.5	9
3	3.0	18
4	2.5	16
5	4.0	23
6	3.0	20

a. Is there evidence of a linear relationship between drug dosage and change in pulse rate? Test at $\alpha = .10$.
b. Find a 95% confidence interval for the slope β_1.
c. Find a 99% prediction interval for the decrease in pulse rate corresponding to a dosage of 3.5 cubic centimeters.

10.60 Refer to Exercises 10.19 and 10.37. The woolly-bear caterpillar data are shown in the table.

Day	*Temperature (°C)*		*Day*	*Temperature (°C)*	
	Air	*Cocoon*[a]		*Air*	*Cocoon*[a]
1	10.4	15.1	7	1.7	3.6
2	9.2	14.6	8	2.0	5.3
3	2.2	6.8	9	3.0	7.0
4	2.6	6.8	10	3.5	7.1
5	4.1	8.0	11	4.5	9.6
6	3.7	8.7	12	4.4	9.5

[a]Each cocoon temperature is the average of the temperatures of two cocoons.

a. Find a 95% confidence interval for the mean cocoon temperature when the air temperature is 7°C. Interpret the interval.
b. Suppose you were to place a single woolly-bear caterpillar cocoon in a controlled environment of 7°C. Find a 95% prediction interval for the cocoon temperature. Interpret the interval.

10.61 The reasons given by workers for quitting their jobs generally fall into one of two categories: (1) worker quits to seek or take a different job; or (2) worker quits to withdraw from the labor force. Economic theory suggests that wages and quit rates are related. The table lists quit rates (quits per 100 employees) and the average hourly wage in a sample of 15 manufacturing industries.

Industry	*Quit Rate*	*Average Wage*	*Industry*	*Quit Rate*	*Average Wage*
1	1.4	$ 8.20	9	2.0	$ 7.99
2	.7	10.35	10	3.8	5.54
3	2.6	6.18	11	2.3	7.50
4	3.4	5.37	12	1.9	6.43
5	1.7	9.94	13	1.4	8.83
6	1.7	9.11	14	1.8	10.93
7	1.0	10.59	15	2.0	8.80
8	.5	13.29			

a. Construct a scattergram for these data.
b. Use the method of least squares to model the quit rate as a function of average hourly wage.
c. Do the data present sufficient evidence to conclude that average hourly wage rate contributes useful information for the prediction of quit rates? What does your model suggest about the relationship between quit rates and wages?
d. Find a 95% prediction interval for the quit rate in an industry with an average hourly wage of $9.00.
e. Estimate the mean quit rate for industries with an average hourly wage of $6.50. Use a 95% confidence interval.

10.62 One of the many variables that influence the sales of existing single-family homes is the interest rate charged for mortgage loans. Shown in the table are the total number of existing single-family homes sold annually, y, and average annual conventional mortgage interest rate, x, for 1982–1991.

	Existing Single-Family Homes Sold y *(thousands)*	*Average Conventional Mortgage Interest Rate* x *(%)*
1982	1,990	14.8
1983	2,719	12.3
1984	2,868	12.0
1985	3,214	11.2
1986	3,565	9.8
1987	3,526	8.9
1988	3,594	9.0
1989	3,346	9.8
1990	3,211	9.8
1991	3,220	9.2

Source: *Statistical Abstract of the United States: 1992*, pp. 502, 712.

a. Plot the data points on graph paper.
b. Find the least squares line relating y to x. As a check on your calculations, plot the line on your graph from part **a** to see if the line appears to model the relationship between y and x.
c. Do the data provide sufficient evidence to indicate that mortgage interest rates contribute information for the prediction of the number of existing single-family homes sold annually? Use $\alpha = .05$.
d. Calculate r and r^2, and interpret their values.
e. Find a 90% confidence interval for the mean annual number of existing single-family homes sold if the average annual mortgage interest rate is 10.0%.
f. Find a 90% prediction interval for the annual number of existing single-family homes sold if the average annual mortgage interest rate is 10.0%.
g. Explain why the widths of the intervals found in parts **e** and **f** differ.

10.63 In Exercise 10.20, we found the least squares line relating the age x of a sand lance to its length y. The data are reproduced in the table.

Mean Length (millimeters)	176	194	212	226	236	244	254
Age (years)	2	3	4	5	6	7	8

a. Find a 95% confidence interval for the mean length of sand lances that are 4 years old. Interpret the interval.
b. Find a 95% prediction interval for the length of a sand lance that is 4 years old. Interpret the interval. Explain the difference between this interval and the interval obtained in part **a**.

10.64 Refer to Exercise 10.39. Find a 95% confidence interval for the mean weight gain of rainbow trout (percentage of body weight) over a 20-day period when the ration level is 2.5% of body weight and the water temperature is 12°C. Interpret the confidence interval.

10.65 Refer to Exercise 10.28. The data are shown in the accompanying table.

Cutting Speed (meters per minute)	*Useful Life (hours) Brand A*	*Brand B*	*Cutting Speed (meters per minute)*	*Useful Life (hours) Brand A*	*Brand B*
30	4.5	6.0	50	1.0	3.7
30	3.5	6.5	60	4.0	3.8
30	5.2	5.0	60	2.0	3.0
40	5.2	6.0	60	1.1	2.4
40	4.0	4.5	70	1.1	1.5
40	2.5	5.0	70	.5	2.0
50	4.4	4.5	70	3.0	1.0
50	2.8	4.0			

a. Use a 90% confidence interval to estimate the mean useful life of a brand A cutting tool when the cutting speed is 45 meters per minute. Repeat for brand B. Compare the widths of the two intervals, and comment on the reasons for any difference.
b. Use a 90% prediction interval to predict the useful life of a brand A cutting tool when the cutting speed is 45 meters per minute. Repeat for brand B. Compare the widths of the two intervals to each other, and to the two intervals you calculated in part **a**. Comment on the reasons for any differences.

c. Note that the estimation and prediction you performed in parts **a** and **b** were for a value of x that was not included in the original sample. That is, the value $x = 45$ was not part of the sample. However, the value is within the range of x values in the sample, so that the regression model spans the x value for which the estimation and prediction were made. In such situations, estimation and prediction represent **interpolations**.

Suppose you were asked to predict the useful life of a brand A cutting tool for a cutting speed of $x =$ 100 meters per minute. Since the given value of x is outside the range of the sample x values, the prediction is an example of **extrapolation**. Predict the useful life of a brand A cutting tool that is operated at 100 meters per minute, and construct a 95% confidence interval for the actual useful life of the tool. What additional assumption do you have to make in order to assure the validity of an extrapolation?

10.9 Simple Linear Regression: An Example

In the preceding sections we have presented the basic elements necessary to fit and use a straight-line regression model. In this section we will assemble these elements by applying them in an example.

Suppose a fire insurance company wants to relate the amount of fire damage in major residential fires to the distance between the residence and the nearest fire station. The study is to be conducted in a large suburb of a major city; a sample of 15 recent fires in this suburb is selected. The amount of damage, y, and the distance, x, between the fire and the nearest fire station are recorded for each fire. The results are given in Table 10.8.

TABLE 10.8 Fire Damage Data

Distance from Fire Station	Fire Damage
x (miles)	y (thousands of dollars)
3.4	26.2
1.8	17.8
4.6	31.3
2.3	23.1
3.1	27.5
5.5	36.0
.7	14.1
3.0	22.3
2.6	19.6
4.3	31.3
2.1	24.0
1.1	17.3
6.1	43.2
4.8	36.4
3.8	26.1

Step 1 First, we hypothesize a model to relate fire damage, y, to the distance from the nearest fire station, x. We hypothesize a straight-line probabilistic model:

$$y = \beta_0 + \beta_1 x + \epsilon$$

Step 2 Next, we use the data to estimate the unknown parameters in the deterministic component of the hypothesized model. We make some preliminary calculations:

$$SS_{xx} = \sum x^2 - \frac{\left(\sum x\right)^2}{n} = 196.16 - \frac{(49.2)^2}{15}$$
$$= 196.160 - 161.376 = 34.784$$
$$SS_{yy} = \sum y^2 - \frac{\left(\sum y\right)^2}{n} = 11{,}376.48 - \frac{(396.2)^2}{15}$$
$$= 11{,}376.480 - 10{,}464.96267 = 911.517334$$
$$SS_{xy} = \sum xy - \frac{\left(\sum x\right)\left(\sum y\right)}{n} = 1{,}470.65 - \frac{(49.2)(396.2)}{15}$$
$$= 1{,}470.650 - 1{,}299.536 = 171.114$$

Then the least squares estimate of the slope β_1 and intercept β_0 are

$$\hat{\beta}_1 = \frac{SS_{xy}}{SS_{xx}} = \frac{171.114}{34.784} = 4.919331$$
$$\hat{\beta}_0 = \bar{y} - \hat{\beta}_1\bar{x} = \frac{396.2}{15} - 4.919331\left(\frac{49.2}{15}\right) = 26.413333 - (4.919331)(3.28)$$
$$= 26.413333 - 16.135406 = 10.277927$$

and the least squares equation is

$$\hat{y} = 10.278 + 4.919x$$

This prediction equation is graphed in Figure 10.19 along with a plot of the data points.

FIGURE 10.19 ▶ **Least squares model for the fire damage data**

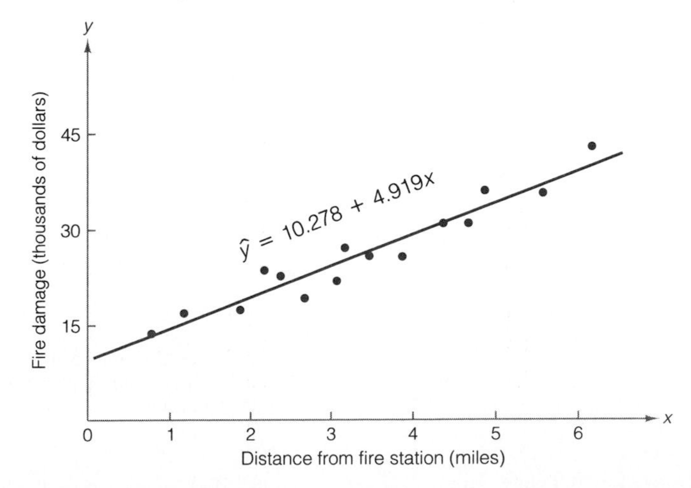

The least squares estimate of the slope β_1 is 4.919, which implies that the estimated mean damage increases by $4,919 for each additional mile from the fire station. This interpretation is valid over the range of x, or from .7 to 6.1 miles from the station. The estimated y-intercept, $\hat{\beta}_0 = 10.278$, has the interpretation that a fire 0 miles from the fire station has an estimated mean damage of $10,278. Although this would seem to apply to the fire station itself, remember that the y-intercept is meaningfully interpretable only if $x = 0$ is within the sampled range of the independent variable.

Step 3 Now we specify the probability distribution of the random error component ϵ. The assumptions about the distribution are identical to those listed in Section 10.3. Although we know that these assumptions are not completely satisfied (they rarely are for practical problems), we are willing to assume they are approximately satisfied for this example. We have to estimate the variance σ^2 of ϵ, so we calculate

$$\text{SSE} = \sum (y - \hat{y})^2 = \text{SS}_{yy} - \hat{\beta}_1 \text{SS}_{xy}$$

where the last expression represents a shortcut formula for SSE. Thus,

$$\text{SSE} = 911.517334 - (4.919331)(171.114) = 911.517334 - 841.766405$$
$$= 69.750929^*$$

To estimate σ^2, we divide SSE by the degrees of freedom available for error, $n - 2$. Thus,

$$s^2 = \frac{\text{SSE}}{n-2} = \frac{69.750929}{15-2} = 5.3655 \qquad s = \sqrt{5.3655} = 2.32$$

Step 4 We can now check the usefulness of the hypothesized model—that is, whether x really contributes information for the prediction of y using the straight-line model. First test the null hypothesis that the slope β_1 is 0, i.e., that there is no linear relationship between fire damage and the distance from the nearest fire station, against the alternative hypothesis that fire damage increases as the distance increases. We test

H_0: $\beta_1 = 0$

H_a: $\beta_1 > 0$

Test statistic: $t = \dfrac{\hat{\beta}_1 - 0}{s_{\hat{\beta}_1}} = \dfrac{\hat{\beta}_1}{s/\sqrt{\text{SS}_{xx}}}$

Assumptions: The same assumptions made about ϵ in Section 10.3

Rejection region: For $\alpha = .05$, we will reject H_0 if $t > t_\alpha$, where, for $n - 2 = 13$ df, we find $t_{.05} = 1.771$.

*The values for SS_{yy}, $\hat{\beta}_1$, and SS_{xy} used to calculate SSE are exact for this example. For other problems where rounding is necessary, at least six significant figures should be carried for these quantities. Otherwise, the calculated value of SSE may be substantially in error.

We then calculate the t statistic:

$$t = \frac{\hat{\beta}_1}{s_{\hat{\beta}_1}} = \frac{\hat{\beta}_1}{s/\sqrt{SS_{xx}}} = \frac{4.919}{2.32/\sqrt{34.784}} = \frac{4.919}{.393} = 12.5$$

This large t value leaves little doubt that mean fire damage and distance between the fire and the station are at least linearly related, with mean fire damage increasing as the distance increases.

We gain additional information about the relationship by forming a confidence interval for the slope β_1. A 95% confidence interval is

$$\hat{\beta}_1 \pm t_{.025}s_{\hat{\beta}_1} = 4.919 \pm (2.160)(.393) = 4.919 \pm .849 = (4.070,\ 5.768)$$

We estimate that the interval from $4,070 to $5,768 encloses the mean increase (β_1) in fire damage per additional mile distance from the fire station.

Another measure of the usefulness of the model is the coefficient of correlation r. We have

$$r = \frac{SS_{xy}}{\sqrt{SS_{xx}\,SS_{yy}}} = \frac{171.114}{\sqrt{(34.784)(911.517)}} = \frac{171.114}{178.062} = .96$$

The high correlation confirms our conclusion that β_1 is greater than 0; it appears that fire damage and distance from the fire station are positively correlated.

The coefficient of determination is

$$r^2 = (.96)^2 = .92$$

which implies that 92% of the sum of squares of deviations in the sample of y values about $\bar{y}$ is explained by the distance x between the fire and the fire station. All signs point to a strong relationship between x and y.

Step 5 We are now prepared to use the least squares model. Suppose the insurance company wants to predict the fire damage if a major residential fire were to occur 3.5 miles from the nearest fire station. The predicted value is

$$\hat{y} = \hat{\beta}_0 + \hat{\beta}_1 x = 10.278 + (4.919)(3.5) = 10.278 + 17.216 = 27.5$$

(We round to the nearest tenth to be consistent with the units of the original data in Table 10.8.) If we want a 95% prediction interval, we calculate

$$\hat{y} \pm t_{.025}s\sqrt{1 + \frac{1}{n} + \frac{(x-\bar{x})^2}{SS_{xx}}} = 27.5 \pm (2.16)(2.32)\sqrt{1 + \frac{1}{15} + \frac{(3.5-3.28)^2}{34.784}}$$
$$= 27.5 \pm (2.16)(2.32)\sqrt{1.0681}$$
$$= 27.5 \pm 5.2 = (22.3,\ 32.7)$$

The model yields a 95% prediction interval of $22,300 to $32,700 for fire damage in a major residential fire 3.5 miles from the nearest station.

One caution before closing: We would not use this prediction model to make predictions for homes less than .7 mile or more than 6.1 miles from the nearest fire station. A look at the data in Table 10.8 reveals that all the x values fall between .7 and 6.1. It is dangerous to use the model to make predictions outside the region in which the sample data fall. A straight line might not provide a good model for the relationship between the mean value of y and the value of x when stretched over a wider range of x values.

10.10 Using the Computer for Simple Linear Regression

All of the examples of simple linear regression that we have presented thus far in Chapter 10 have required rather tedious calculations involving SS_{yy}, SS_{xx}, SS_{xy}, and so forth. Even with the use of a pocket calculator, the process is laborious and susceptible to error. Fortunately, the use of computers can significantly reduce the labor involved in regression calculations. In this final section, we introduce the regression output from one statistical software package, SAS. Though this is just one of the many statistical packages available that provide simple linear regression output, most produce essentially the same quantities, differing only in format and labeling.

The SAS output for the fire damage example of Section 10.9 is presented in Figure 10.20. We have shaded the parts of the printout corresponding to most of the key simple linear regression quantities introduced in this chapter.

First, the estimates of the y-intercept and the slope are found about halfway down the printout on the left-hand side, under the column labeled Parameter Estimate and in the rows labeled INTERCEP and X, respectively. The values are $\hat{\beta}_0 = 10.277929$ and $\hat{\beta}_1 = 4.919331$. When rounded to three decimal places, these quantities agree with our calculations in Section 10.9.

Next, we find the measures of variability: SSE, s^2, and s. They are shaded in the upper portion of the printout. SSE is found under the column heading Sum of Squares and in the row labeled Error: SSE = 69.75098. The estimate of the error variance σ^2 is under the column heading Mean Square and in the row labeled Error: $s^2 = 5.36546$. The estimate of the standard deviation σ is next to the heading Root MSE: $s = 2.31635$. Again, all values (after rounding) agree with the corresponding quantities we calculated in Section 10.9.

The coefficient of determination is shown (shaded) in the upper portion of the printout next to the heading R-Square: $r^2 = .9235$. Again, to two decimal places, this agrees with our calculation in Section 10.9. The coefficient of correlation r is not given on the printout.

The t statistic for testing H_0: $\beta_1 = 0$ against H_a: $\beta_1 \neq 0$ is given (shaded) in the center of the page under the column heading T for H0: Parameter = 0 in the row corresponding to X. The value $t = 12.525$ agrees with our computed value when we used the formula

$$t = \frac{\hat{\beta}_1}{s_{\hat{\beta}_1}} = \frac{\hat{\beta}_1}{s/\sqrt{SS_{xx}}}$$

FIGURE 10.20 ►
SAS printout for fire damage regression analysis

Dep Variable: Y

Analysis of Variance

Source	DF	Sum of Squares	Mean Square	F Value	Prob>F
Model	1	841.76636	841.76636	156.886	0.0001
Error	13	69.75098	5.36546		
C Total	14	911.51733			

Root MSE	2.31635	R-Square	0.9235
Dep Mean	26.41333	Adj R-Sq	0.9176
C.V.	8.76961		

Parameter Estimates

Variable	DF	Parameter Estimate	Standard Error	T for H0: Parameter=0	Prob > ¦T¦
INTERCEP	1	10.277929	1.42027781	7.237	0.0001
X	1	4.919331	0.39274775	12.525	0.0001

Obs	X	Y	Predict Value	Residual	Lower95% Predict	Upper95% Predict
1	3.4	26.2000	27.0037	-0.8037	21.8344	32.1729
2	1.8	17.8000	19.1327	-1.3327	13.8141	24.4514
3	4.6	31.3000	32.9068	-1.6068	27.6186	38.1951
4	2.3	23.1000	21.5924	1.5076	16.3577	26.8271
5	3.1	27.5000	25.5279	1.9721	20.3573	30.6984
6	5.5	36.0000	37.3342	-1.3342	31.8334	42.8351
7	0.7	14.1000	13.7215	0.3785	8.1087	19.3342
8	3	22.3000	25.0359	-2.7359	19.8622	30.2097
9	2.6	19.6000	23.0682	-3.4682	17.8678	28.2686
10	4.3	31.3000	31.4311	-0.1311	26.1908	36.6713
11	2.1	24.0000	20.6085	3.3915	15.3442	25.8729
12	1.1	17.3000	15.6892	1.6108	10.1999	21.1785
13	6.1	43.2000	40.2858	2.9142	34.5906	45.9811
14	4.8	36.4000	33.8907	2.5093	28.5640	39.2175
15	3.8	26.1000	28.9714	-2.8714	23.7843	34.1585
16	3.5	.	27.4956	.	22.3239	32.6672

Sum of Residuals	-3.73035E-14
Sum of Squared Residuals	69.7510
Predicted Resid SS (Press)	93.2117

To determine which hypothesis this test statistic supports, we can establish a rejection region using the t table (Table V of Appendix A), just as we did in Section 10.9. However, the printout makes this unnecessary because the *observed significance level*, or *p-value*, is shown (shaded) immediately to the right of the t statistic, under the column heading Prob > |T|. Remember that if the observed significance level is less than the α-value we select, then the test statistic supports the alternative hypothesis at that level. For example, if we select $\alpha = .05$ in this example, the observed significance level of .0001 given on the printout indicates that we should reject H_0. We can conclude that there is sufficient evidence at $\alpha = .05$ to infer that a linear relationship between fire damage and distance from the station is useful for predicting damage. If we wish to conduct a one-tailed test, the observed significance level is half that given on the printout (assuming the sign of the test statistic agrees with the alternative hypothesis). Thus, if we were testing H_0: $\beta_1 = 0$ against H_a: $\beta_1 > 0$ in this example, the t statistic is positive, agreeing with H_a, so that the observed significance level would be $.0001/2 = .00005$.

Predicted y values and the corresponding prediction intervals are given in the lower portion of the SAS printout. To find the 95% prediction interval for the fire damage y when the distance from the fire station is $x = 3.5$ miles, first locate the value 3.5 in the column labeled X (the last value in the column). The prediction is given in the center column labeled Predict Value in the row corresponding to 3.5, $\hat{y} = 27.4956$. The lower and upper confidence bounds are given in the columns headed Lower and Upper 95% Predict, respectively:

$$\text{Lower} = 22.3239 \quad \text{and} \quad \text{Upper} = 32.6672$$

Again, all values agree after rounding with our calculations in Section 10.9.

Although much more information is given on the SAS printout, we have discussed only those aspects that have been presented in the chapter. The point here is that the computer can alleviate much of the burden of calculation involved in a regression analysis, and enable us to spend more time on the interpretation of the model. The time spent in learning how to read computer regression output is a good investment.

Exercises 10.66–10.67

Learning the Mechanics

10.66 Refer to Exercise 10.15, in which the number of wins for 14 American League baseball teams was related to their batting averages for the 1991 season using a simple linear regression model. The SAS printout for this regression model is given at the top of page 547.

a. Find the least squares estimates of the intercept β_0 and the slope β_1. Interpret them.

b. Identify the values of SSE, s^2, and s. Based on the standard deviation, how accurately do you expect to be able to predict the number of games won based on the team batting average?

c. Find and interpret the coefficient of determination, r^2.

d. Find and interpret the p-value for the test of the null hypothesis H_0: $\beta_1 = 0$. Is this the p-value for a one-tailed or two-tailed test? What is the p-value for the one-tailed alternative, H_a: $\beta_1 > 0$? Interpret it.

e. Suppose this model is used to estimate the mean number of games teams with .262 batting averages win. What is the point estimate of the mean? Use a 95% confidence interval to estimate this mean (on the printout).

```
Dependent Variable: WINS

                          Analysis of Variance

                          Sum of          Mean
Source            DF     Squares        Square       F Value        Prob>F

Model              1   244.93569     244.93569         2.907        0.1139
Error             12  1011.06431      84.25536
C Total           13  1256.00000

    Root MSE       9.17907     R-square       0.1950
    Dep Mean      81.00000     Adj R-sq       0.1279
    C.V.          11.33219

                          Parameter Estimates

                     Parameter       Standard     T for H0:
Variable  DF         Estimate          Error    Parameter=0     Prob > |T|

INTERCEP   1       -37.435445    69.50644929        -0.539         0.6000
AVERAGE    1       455.270793   267.01922126         1.705         0.1139

                            The SAS System

                  Dep Var    Predict    Std Err   Lower95%   Upper95%
Obs  AVERAGE        WINS      Value    Predict       Mean       Mean   Residual

  1    0.254     57.0000    78.2033      2.951    71.7736    84.6331   -21.2033
  2    0.256     71.0000    79.1139      2.691    73.2505    84.9773    -8.1139
  3    0.269     84.0000    85.0324      3.408    77.6079    92.4569    -1.0324
  4    0.257     91.0000    79.5691      2.593    73.9200    85.2183    11.4309
  5     0.27     85.0000    85.4877      3.598    77.6482    93.3271    -0.4877
  6    0.247     84.0000    75.0164      4.282    65.6871    84.3458     8.9836
  7     0.28     95.0000    90.0404      5.842    77.3112      102.8     4.9596
  8    0.254     67.0000    78.2033      2.951    71.7736    84.6331   -11.2033
  9    0.255     81.0000    78.6586      2.811    72.5331    84.7841     2.3414
 10    0.271     83.0000    85.9429      3.798    77.6684    94.2175    -2.9429
 11    0.255     83.0000    78.6586      2.811    72.5331    84.7841     4.3414
 12    0.264     82.0000    82.7560      2.661    76.9590    88.5531    -0.7560
 13    0.248     84.0000    75.4717      4.066    66.6130    84.3305     8.5283
 14    0.262     87.0000    81.8455      2.503    76.3923    87.2987     5.1545

Sum of Residuals                      0
Sum of Squared Residuals      1011.0643
Predicted Resid SS (Press)    1338.0375
```

f. If you were to predict the number of wins for a team with a .262 batting average, how would your prediction differ from the point estimate of the mean in part **d**? If you were to form a 95% prediction interval (not shown on printout) for the number of wins for this team, how would the interval compare to the confidence interval from part **d**: narrower, same width, or wider?

10.67 Refer to Exercise 10.19, in which the cocoon temperature of a woolly-bear caterpillar was related to the outside air temperature. The SAS simple linear regression printout for these data is given at the top of page 548.

Dep Variable: COCOON

Analysis of Variance

Source	DF	Sum of Squares	Mean Square	F Value	Prob>F
Model	1	120.18486	120.18486	164.090	0.0001
Error	10	7.32431	0.73243		
C Total	11	127.50917			

Root MSE	0.85582	R-Square	0.9426
Dep Mean	8.50833	Adj R-Sq	0.9368
C.V.	10.05863		

Parameter Estimates

Variable	DF	Parameter Estimate	Standard Error	T for H0: Parameter=0	Prob > |T|
INTERCEP	1	3.374666	0.47079265	7.168	0.0001
AIR	1	1.200858	0.09374540	12.810	0.0001

Obs	AIR	COCOON	Predict Value	Residual	Lower95% Predict	Upper95% Predict
1	10.4	15.1000	15.8636	-0.7636	13.5022	18.2250
2	9.2	14.6000	14.4226	0.1774	12.1870	16.6581
3	2.2	6.8000	6.0166	0.7834	3.9850	8.0481
4	2.6	6.8000	6.4969	0.3031	4.4815	8.5123
5	4.1	8.0000	8.2982	-0.2982	6.3131	10.2833
6	3.7	8.7000	7.8178	0.8822	5.8294	9.8062
7	1.7	3.6000	5.4161	-1.8161	3.3598	7.4725
8	2	5.3000	5.7764	-0.4764	3.7355	7.8172
9	3	7.0000	6.9772	0.0228	4.9747	8.9798
10	3.5	7.1000	7.5777	-0.4777	5.5863	9.5690
11	4.5	9.6000	8.7785	0.8215	6.7932	10.7638
12	4.4	9.5000	8.6584	0.8416	6.6735	10.6434
13	7	.	11.7807	.	9.7159	13.8454

Sum of Residuals	9.325873E-15
Sum of Squared Residuals	7.3243
Predicted Resid SS (Press)	11.6790

a. Find the least squares equation, and compare it to the value you calculated in Exercise 10.19.
b. Find the standard deviation s, and interpret the value in terms of this exercise.
c. Find and interpret r^2.
d. Is there sufficient evidence to indicate that the model is useful for predicting y? What is the observed significance level of the test?
e. Locate and interpret the predicted cocoon temperature and corresponding 95% prediction interval when the air temperature is 7°C.

Summary

We have introduced an extremely useful tool in this chapter—the **method of least squares** for fitting a prediction equation to a set of data. This procedure, along with the associated statistical tests and estimations, is called a **regression analysis**. In five steps we showed how to use sample data to build a model relating a dependent variable y to a single independent variable x:

1. The first step is to hypothesize a **probabilistic model**. In this chapter, we confined our attention to the straight-line model, $y = \beta_0 + \beta_1 x + \epsilon$.
2. The second step is to use the method of least squares to estimate the unknown parameters in the **deterministic component**, $\beta_0 + \beta_1 x$. The least squares estimates yield a model $\hat{y} = \hat{\beta}_0 + \hat{\beta}_1 x$ with a sum of squared errors (SSE) that is smaller than that produced by any other straight-line model.
3. The third step is to specify the probability distribution of the **random error component**, ϵ.
4. The fourth step is to assess the usefulness of the hypothesized model by making inferences about the slope, β_1, calculating the **coefficient of correlation**, r, and calculating the **coefficient of determination**, r^2.
5. Finally, if we are satisfied with the model, we are prepared to use it. We used the model to estimate the mean y value, $E(y)$, for a given x value and to predict an individual y value for a new value of x.

Supplementary Exercises 10.68–10.79

Learning the Mechanics

10.68 In fitting a least squares line to $n = 15$ data points, the following quantities were computed: $SS_{xx} = 55$, $SS_{yy} = 198$, $SS_{xy} = -88$, $\bar{x} = 1.3$, and $\bar{y} = 35$.

a. Find the least squares line.
b. Graph the least squares line.
c. Calculate SSE.
d. Calculate s^2.
e. Find a 90% confidence interval for β_1. Interpret this estimate.
f. Find a 90% confidence interval for the mean value of y when $x = 15$.
g. Find a 90% prediction interval for y when $x = 15$.

10.69 Consider the following sample data:

y	5	1	3
x	5	1	3

a. Construct a scattergram for the data.
b. It is possible to find many lines for which $\Sigma(y - \hat{y}) = 0$. For this reason, the criterion $\Sigma(y - \hat{y}) = 0$ is not used for identifying the "best-fitting" straight line. Find two lines that have $\Sigma(y - \hat{y}) = 0$.
c. Find the least squares line.
d. Compare the value of SSE for the least squares line to that of the two lines you found in part **b**. What principle of least squares is demonstrated by this comparison?

10.70 Consider the following 10 data points:

x	3	5	6	4	3	7	6	5	4	7
y	4	3	2	1	2	3	3	5	4	2

a. Plot the data on a scattergram.
b. Calculate the values of r and r^2.
c. Is there sufficient evidence to indicate that x and y are linearly correlated? Test at the $\alpha = .10$ level of significance.

Applying the Concepts

10.71 A study was conducted to determine whether the final grade in an introductory sociology course was related to a student's performance on the verbal ability test administered before college entrance. The verbal test scores and final grades for a random sample of 10 students are shown in the table.

Student	*Verbal Ability Test Score, x*	*Final Sociology Grade, y*
1	39	65
2	43	78
3	21	52
4	64	82
5	57	92
6	47	89
7	28	73
8	75	98
9	34	56
10	52	75

Note that $SS_{yy} = 2{,}056$, $SS_{xy} = 1{,}894$, $SS_{xx} = 2{,}474$, $\bar{y} = 76$, and $\bar{x} = 46$.
a. Assuming that a linear relationship exists between verbal scores and final grades, find the least squares line relating y to x.
b. Plot the data points and graph the least squares line.
c. Do the data provide sufficient evidence to indicate that a positive correlation exists between verbal score and final grade? (Use $\alpha = .01$.)
d. Find a 95% confidence interval for the slope β_1.
e. Predict a student's final grade in the introductory course when his or her verbal test score is 50. (Use a 90% prediction interval.)
f. Find a 95% confidence interval for the mean final grade for students scoring 35 on the college entrance verbal exam.

10.72 In placing a weekly order, a concessionaire who provides services at a baseball stadium must know what size crowd is expected during the coming week in order to know how much food, etc., to order. Since advance ticket sales give an indication of expected attendance, food needs might be predicted on the basis of the advance sales. Data from 7 previous weeks of games are given in the table.

Hot Dogs Purchased During Week y *(thousands)*	*Advance Ticket Sales for Week* x *(thousands)*
39.1	54.0
35.9	48.1
20.8	28.8
42.4	62.4
46.0	64.4
40.7	59.5
29.9	42.3

a. Use the data in the table to develop a simple linear model for hot dogs purchased as a function of advance ticket sales.
b. Plot the data and graph the line as a check on your calculations.
c. Do the data provide sufficient information to indicate that advance ticket sales provide information for the prediction of hot dog demand?
d. Calculate r^2 and interpret its value.
e. Find a 90% confidence interval for the mean number of hot dogs purchased when the advance ticket sales equal 50,000.
f. If the advance ticket sales this week equal 55,000, find a 90% prediction interval for the number of hot dogs that will be purchased this week at the game.

10.73 At temperatures approaching absolute zero (273° below 0°C), helium exhibits traits that defy many laws of conventional physics. An experiment has been conducted with helium in solid form at various temperatures near absolute zero. The solid helium is placed in a dilution refrigerator along with a solid impure substance, and the fraction (in weight) of the impurity passing through the solid helium is recorded. (This phenomenon of solids passing directly through solids is known as *quantum tunneling*.) The data are given in the table. The least squares printout for the simple linear regression model $E(y) = \beta_0 + \beta_1 x$ is also shown at the top of page 552.

Temperature x(°C)	*Proportion of Impurity Passing Through Helium* y	*Temperature* x(°C)	*Proportion of Impurity Passing Through Helium* y
−262.0	.315	−272.0	.935
−265.0	.202	−272.4	.957
−256.0	.204	−272.7	.906
−267.0	.620	−272.8	.985
−270.0	.715	−272.9	.987

a. Find the least squares estimates of the intercept and slope. Interpret them.
b. Use a 95% confidence interval to estimate the slope β_1. Interpret the interval in terms of this application. Does the interval support the hypothesis that temperature contributes information about the proportion of impurity passing through helium?
c. Interpret the coefficient of determination for this model.

```
Dep Variable: IMPURITY

                          Analysis of Variance

                         Sum of          Mean
Source          DF      Squares        Square       F Value       Prob>F

Model            1      0.83089       0.83089        46.728       0.0001
Error            8      0.14225       0.01778
C Total          9      0.97315

    Root MSE            0.13335     R-Square         0.8538
    Dep Mean            0.68260     Adj R-Sq         0.8356
    C.V.               19.53519

                          Parameter Estimates

                    Parameter        Standard     T for H0:
Variable   DF        Estimate           Error    Parameter=0    Prob > |T|

INTERCEP    1      -13.490347      2.07377160         -6.505        0.0002
TEMP        1       -0.052829      0.00772828         -6.836        0.0001

                                  Predict                   Lower95%    Upper95%
  Obs     TEMP       IMPURITY       Value     Residual      Predict     Predict

    1     -262         0.3150      0.3508      -0.0358      0.00946      0.6922
    2     -265         0.2020      0.5093      -0.3073       0.1816      0.8371
    3     -256         0.2040      0.0339       0.1701      -0.3559      0.4236
    4     -267         0.6200      0.6150      0.00502       0.2917      0.9383
    5     -270         0.7150      0.7735      -0.0585       0.4495      1.0974
    6     -272         0.9350      0.8791       0.0559       0.5499      1.2084
    7   -272.4         0.9570      0.9003       0.0567       0.5695      1.2310
    8   -272.7         0.9060      0.9161      -0.0101       0.5841      1.2481
    9   -272.8         0.9850      0.9214       0.0636       0.5890      1.2538
   10   -272.9         0.9870      0.9267       0.0603       0.5938      1.2595
   11     -273              .      0.9320            .       0.5987      1.2653

Sum of Residuals                    6.578071E-15
Sum of Squared Residuals                  0.1423
Predicted Resid SS (Press)                0.3402
```

d. Find the 95% prediction interval for the percentage of impurity passing through solid helium at −273°C. (Note that this value of x is outside the experimental region, where use of the model for prediction may be unreliable.)

10.74 Refer to Exercise 10.40, in which managerial success, y, was modeled as a function of the number of contacts a manager makes with people outside his or her work unit, x, during a specific period of time. The data are repeated in the table.

Manager	Manager Success Index	Number of Interactions with Outsiders	Manager	Manager Success Index	Number of Interactions with Outsiders
1	40	12	11	70	20
2	73	71	12	47	81
3	95	70	13	80	40
4	60	81	14	51	33
5	81	43	15	32	45
6	27	50	16	50	10
7	53	42	17	52	65
8	66	18	18	30	20
9	25	35	19	42	21
10	63	82			

Recall that $SS_{yy} = 7{,}006.6316$, $SS_{xx} = 10{,}824.5263$, $SS_{xy} = 2{,}561.2632$, $\bar{y} = 54.5789$, and $\bar{x} = 44.1579$.

a. A particular manager was observed for 2 weeks as in the Luthans, Rosenkrantz, and Hennessey (1985) study. She made 55 contacts with people outside her work unit. Predict the value of the manager's success index. Use a 90% prediction interval.

b. A second manager was observed for 2 weeks. This manager made 110 contacts with people outside his work unit. Give two reasons why caution should be exercised in using the least squares model developed from the given data set to construct a prediction interval for this manager's success index.

c. In the context of this problem, determine the value of x for which the associated prediction interval for y is the narrowest.

10.75 Firms planning to build new plants or make additions to existing facilities have become very conscious of the energy efficiency of proposed new structures, and are interested in the relation between yearly energy consumption and the number of square feet of building shell. The table lists the energy consumption in British thermal units (a BTU is the amount of heat required to raise 1 pound of water 1°F) for 1990 for 22 buildings that were all subjected to the same climatic conditions. The SAS printout that fits the straight-line model relating BTU consumption, y, to building shell area, x, is also given.

BTU/Year (thousands)	Shell Area (square feet)	BTU/Year (thousands)	Shell Area (square feet)
3,870,000	30,001	2,680,000	23,680
1,371,000	13,530	337,500	5,650
2,422,000	26,060	567,500	8,001
672,200	6,355	555,300	6,147
233,100	4,576	239,400	2,660
218,900	24,680	2,629,000	19,240
354,000	2,621	1,102,000	10,700
3,135,000	23,350	423,500	9,125
1,470,000	18,770	423,500	6,510
1,408,000	12,220	1,691,000	13,530
2,201,000	25,490	1,870,000	18,860

Printout for Exercise 10.75

Dep Variable: BTU

Analysis of Variance

Source	DF	Sum of Squares	Mean Square	F Value	Prob>F
Model	1	1.658498E+13	1.658498E+13	42.028	0.0001
Error	20	7.89232E+12	394616010047		
C Total	21	2.44773E+13			

Root MSE	628184.69422	R-Square	0.6776
Dep Mean	1357904.54545	Adj R-Sq	0.6614
C.V.	46.26133		

Parameter Estimates

Variable	DF	Parameter Estimate	Standard Error	T for H0: Parameter=0	Prob > |T|
INTERCEP	1	-99045	261617.65980	-0.379	0.7090
AREA	1	102.814048	15.85924082	6.483	0.0001

Obs	AREA	BTU	Predict Value	Residual	Lower95% Predict	Upper95% Predict
1	30001	3870000	2985479	884521	1546958	4424000
2	13530	1371000	1292029	78971.2	-47949.3	2632007
3	26060	2422000	2580289	-158289	1183940	3976637
4	6355	672200	554338	117862	-810192	1918868
5	4576	233100	371432	-138332	-1005463	1748327
6	24680	218900	2438405	-2219505	1054223	3822588
7	2621	354000	170430	183570	-1222796	1563657
8	23350	3135000	2301663	833337	927871	3675455
9	18770	1470000	1830774	-360774	482352	3179196
10	12220	1408000	1157342	250658	-184021	2498706
11	25490	2201000	2521685	-320685	1130530	3912840
12	23680	2680000	2335591	344409	959345	3711838
13	5650	337500	481854	-144354	-887287	1850995
14	8001	567500	723570	-156070	-631698	2078838
15	6147	555300	532953	22347.3	-832898	1898804
16	2660	239400	174440	64959.9	-1218433	1567313
17	19240	2629000	1879097	749903	528832	3229362
18	10700	1102000	1001065	100935	-343656	2345786
19	9125	423500	839133	-415633	-511035	2189301
20	6510	423500	570274	-146774	-793294	1933842
21	13530	1691000	1292029	398971	-47949.3	2632007
22	18860	1870000	1840028	29972.3	491266	3188789
23	8000	.	723467	.	-631806	2078740

Sum of Residuals	1.6298145E-9
Sum of Squared Residuals	7.89232E+12
Predicted Resid SS (Press)	1.012747E+13

a. Find the least squares estimates of the intercept β_0 and the slope β_1.
b. Investigate the usefulness of the model you developed in part **a**. Is yearly energy consumption positively linearly related to the shell area of the building? Test using $\alpha = .10$.
c. Calculate the observed significance level of the test of part **b** using the printout. Interpret its value.
d. Find the coefficient of determination r^2 and interpret its value.
e. A company wishes to build a new warehouse that will contain 8,000 square feet of shell area. Find the predicted value of energy consumption and a 95% prediction interval on the printout. Comment on the usefulness of this interval.
f. The application of the model you developed in part **a** to the warehouse problem of part **e** is appropriate only if certain assumptions can be made about the new warehouse. What are these assumptions?

10.76 A study was conducted to determine whether there is a linear relationship between the breaking strength, y, of wooden beams and the specific gravity, x, of the wood. Ten randomly selected beams of the same cross-sectional dimensions were stressed until they broke. The breaking strengths and the density of the wood are shown in the table for each of the 10 beams.

Beam	*Specific Gravity, x*	*Strength, y*	*Beam*	*Specific Gravity, x*	*Strength, y*
1	.499	11.14	6	.528	12.60
2	.558	12.74	7	.418	11.13
3	.604	13.13	8	.480	11.70
4	.441	11.51	9	.406	11.02
5	.550	12.38	10	.467	11.41

Given that $\bar{x} = .4951$, $\bar{y} = 11.8760$, $SS_{xx} = .03776$, $SS_{xy} = .4089$, and $SS_{yy} = 5.3102$:
a. Fit the model $y = \beta_0 + \beta_1 x + \epsilon$, and interpret the estimates.
b. Calculate and interpret the standard error of regression for the model in part **a**.
c. Test H_0: $\beta_1 = 0$ against the alternative hypothesis H_a: $\beta_1 \neq 0$.
d. Calculate and interpret the coefficient of determination.
e. Estimate the mean strength for beams with specific gravity .590. Use a 90% confidence interval.

10.77 The data in the accompanying table were collected to calibrate a new instrument for measuring interocular pressure. The interocular pressure for each of 10 glaucoma patients was measured by the new instrument and by a standard, reliable, but more time-consuming method.

Patient	*Reliable Method, x*	*New Instrument, y*	*Patient*	*Reliable method, x*	*New Instrument, y*
1	20.2	20.0	6	21.8	22.1
2	16.7	17.1	7	19.1	18.9
3	17.1	17.2	8	22.9	22.2
4	26.3	25.1	9	23.5	24.0
5	22.2	22.0	10	17.0	18.1

a. Fit a least squares line to the data.
b. Calculate r and r^2. Interpret each of these quantities.
c. Predict the pressure measured by the new instrument when the reliable method gives a reading of 20.0. Use a 90% prediction interval.

10.78 In Exercise 10.21, we discussed an experiment conducted to determine the effect of ration level on the growth rate of rainbow trout. Twenty fish were fed at each of four ration levels at 6°C water temperature. The experiment was repeated for water temperatures of 12°C and 18°C. The table gives the means and standard deviations (in parentheses) for each sample of 20 fish. Refer only to data collected on fish fed in water maintained at 18°C.

Ration (% body weight per day)	*Mean Wet Weight Gain (%)* 6°C	12°C	18°C
.0	−8.14 (3.27)	−10.33 (2.35)	−13.21 (1.96)
.8	12.31 (3.37)		
1.2		14.29 (5.32)	
1.5	28.19 (5.39)		15.51 (4.48)
2.5	29.12 (5.67)	38.05 (6.99)	
3.5			51.13 (7.14)
4.0		60.13 (12.71)	
4.5			65.49 (13.04)
Maintenance ration	.32	.48	.69

a. Find SSE and s^2 for the $n = 4$ data points.

b. Find a 95% confidence interval for the mean gain (percentage of body weight) for a 1% increase in ration level.

c. State the assumptions required for your inference in part **b** to be valid.

d. Find a 95% confidence interval for the mean weight gain of rainbow trout (percentage of body weight) over a 20-day period when the ration level is 4% of body weight.

10.79 *Comparable worth* is a compensation plan designed to eliminate pay inequities among jobs of similar worth. A number of state and municipal governments have adopted comparable-worth plans, and some unions have attempted to negotiate comparable-worth clauses into their contracts. To develop such a plan, a sample of benchmark jobs are evaluated and assigned points, x, based on factors such as responsibility, skill, effort, and working conditions. A market survey is conducted to determine the market rates (or salaries), y, of the benchmark jobs. A regression analysis is then used to characterize the relationship between salary and job evaluation points (Scholl and Cooper, 1991). The table gives the job evaluation points and salaries for a set of 21 benchmark jobs.

Job Evaluation Points, x	*Salary, y*	
970	$15,704	Electrician
500	13,984	Semiskilled laborer
370	14,196	Motor equipment operator
220	13,380	Janitor
250	13,153	Laborer
1,350	18,472	Senior engineering technician
470	14,193	Senior janitor
2,040	20,642	Revenue agent
370	13,614	Engineering aide
1,200	16,869	Electrician supervisor
820	15,184	Senior maintenance technician
1,865	$17,341	Registered nurse
1,065	15,194	Licensed practical nurse
880	13,614	Principal clerk typist
340	12,594	Clerk typist
540	13,126	Senior clerk stenographer
490	12,958	Senior clerk typist
940	13,894	Principal clerk stenographer
600	13,380	Institutional attendant
805	15,559	Eligibility technician
220	13,844	Cook's helper

a. Construct a scattergram for these data. What does it suggest about the relationship between salary and job evaluation points?

b. Using SAS, a straight-line model was fit to these data and the printout is given here. Identify and interpret the least squares equation.

Dependent Variable: Y

Analysis of Variance

Source	DF	Sum of Squares	Mean Square	F Value	Prob>F
Model	1	66801750.334	66801750.334	74.670	0.0001
Error	19	16997968.904	894629.94232		
C Total	20	83799719.238			

Root MSE	945.84879	R-square	0.7972
Dep Mean	14804.52381	Adj R-sq	0.7865
C.V.	6.38892		

Parameter Estimates

Variable	DF	Parameter Estimate	Standard Error	T for H0: Parameter=0	Prob > ¦T¦
INTERCEP	1	12024	382.31829064	31.449	0.0001
X	1	3.581616	0.41448305	8.641	0.0001

Obs	X	Dep Var Y	Predict Value	Std Err Predict	Lower95% Predict	Upper95% Predict	Residual
1	970	15704.0	15497.8	221.447	13464.6	17531.0	206.2
2	500	13984.0	13814.5	236.070	11774.1	15854.9	169.5
3	370	14196.0	13348.9	266.420	11292.1	15405.6	847.1
4	220	13380.0	12811.6	309.502	10728.6	14894.6	568.4
5	250	13153.0	12919.1	300.351	10842.0	14996.2	233.9
6	1350	18472.0	16858.8	314.833	14772.4	18945.3	1613.2
7	470	14193.0	13707.0	242.349	11663.4	15750.6	486.0
8	2040	20642.0	19330.2	562.933	17026.4	21633.9	1311.8
9	370	13614.0	13348.9	266.420	11292.1	15405.6	265.1
10	1200	16869.0	16321.6	270.968	14262.3	18380.9	547.4
11	820	15184.0	14960.6	207.190	12934.0	16987.2	223.4
12	1865	17341.0	18703.4	496.163	16467.8	20938.9	-1362.4
13	1065	15194.0	15838.1	238.553	13796.4	17879.8	-644.1
14	880	13614.0	15175.5	210.818	13147.2	17203.7	-1561.5
15	340	12594.0	13241.4	274.451	11180.1	15302.7	-647.4
16	540	13126.0	13957.7	228.483	11921.1	15994.4	-831.7
17	490	12958.0	13778.6	238.109	11737.2	15820.1	-820.6
18	940	13894.0	15390.4	217.251	13359.1	17421.6	-1496.4
19	600	13380.0	14172.6	218.972	12140.6	16204.7	-792.6
20	805	15559.0	14906.9	206.741	12880.4	16933.3	652.1
21	220	13844.0	12811.6	309.502	10728.6	14894.6	1032.4
22	800	.	14888.9	206.632	12862.6	16915.3	.

c. Interpret the value of r^2 for this least squares equation.

d. Is there sufficient evidence to conclude that a straight-line model provides useful information about the relationship in question? Interpret the p-value for this test.

e. A job outside the set of benchmark jobs is evaluated and receives a score of 800 points. Under the comparable-worth plan, what is a reasonable range within which a fair salary for this job should be found?

On Your Own

There are many dependent variables in all areas of research that are the subject of regression modeling efforts. We list five such variables here:

1. Crime rate in various communities
2. Daily maximum temperature in your town
3. Grade-point average of students who have completed one academic year at your college
4. Gross Domestic Product of the United States
5. Points scored by your favorite football team in a single game

Choose one of these dependent variables that is of particular interest to you or choose some other dependent variable for which you want to construct a prediction model. There may be a large number of independent variables that should be included in a prediction equation for the dependent variable you choose. List three potentially important independent variables, x_1, x_2, and x_3, that you think might be (individually) strongly related to your dependent variable. Next, obtain 10 data values, each of which consists of a measure of your dependent variable y and the corresponding values of x_1, x_2, and x_3.

a. Use the least squares formulas given in this chapter to fit three straight-line models—one for each independent variable—for predicting y.
b. Interpret the sign of the estimated slope coefficient $\hat{\beta}_1$ in each case, and test the utility of each model by testing H_0: $\beta_1 = 0$ against H_a: $\beta_1 \neq 0$. What assumptions must be satisfied to assure the validity of these tests?
c. Calculate the coefficient of determination r^2 for each model. Which of the independent variables predicts y best for the 10 sampled sets of data? Is this variable necessarily best in general (i.e., for the entire population)? Explain.

Using the Computer

Suppose we wish to estimate the relationship between the median household income y and the percentage of college graduates x in a zip code using a straight-line model.

a. Draw a random sample of 100 zip codes from the 1,000 described in Appendix B, and extract y and x for each zip code sampled. Use a software package that includes a regression program to obtain the least squares fit of the straight-line model of interest.
   1. Graph the fitted model.
   2. Interpret the estimated intercept and slope.
   3. Interpret the estimated standard deviation of the error term.
   4. Evaluate the usefulness of the model.
   5. Estimate the mean median income for all zip codes having 15% college graduates using a 95% confidence interval.
   6. Predict the mean median income for a particular zip code having 15% college graduates using a 95% prediction interval.
b. Repeat part **a** using the entire set of 1,000 zip codes. Compare the results to those you obtained in part **a**.

References

Bergin, D. A. "Leisure activity, motivation, and academic achievement in high school students." *Journal of Leisure Research*, 1992, Vol. 24, No. 3, pp. 225–239.

Deming, W. E. *Out of the Crisis*. Cambridge, Mass.: MIT Center for Advanced Engineering Study, 1986.

Graybill, F. *Theory and Application of the Linear Model*. North Scituate, Mass.: Duxbury, 1976, Chapter 5.

Heaukulani, D. "The normal distribution of crime." *Journal of Police Science and Administration*, 1975, Vol. 3, No. 3, pp. 312–318.

Kevan, P. G., Jensen, T. S., and Shorthouse, J. D. "Body temperatures and behavioral thermoregulation of High Arctic woolly-bear caterpillars and pupae (*Gynaephora rossii*, Lymantridae: Lepidoptera) and the importance of sunshine." *Arctic and Alpine Research*, 1982, 54.

Kovacs, T. G., and Leduc, G. "Sublethal toxicity to rainbow trout (*Salmo gairdnerii*) at different temperatures." *Canadian Journal of Fisheries and Aquatic Sciences*, 1982, 39.

Luthans, F., Rosenkrantz, S. A., and Hennessey, H. W. "What do successful managers really do? An observational study of managerial activities." *Journal of Applied Behavioral Science*, Aug. 1985, Vol. 21, No. 3, pp. 255–270.

Mendenhall, W. *Introduction to Linear Models and the Design and Analysis of Experiments*. Belmont, Ca.: Wadsworth, 1968.

Mendenhall, W., and Sincich, T. A. *Second Course in Business Statistics: Regression Analysis*, 4th ed. San Francisco: Dellen, 1993.

Mintzberg, H. *The Nature of Managerial Work*. New York: Harper & Row, 1973.

Neter, J., and Wasserman, W. *Applied Linear Statistical Models*. Homewood, Ill.: Richard Irwin, 1974, Chapters 2–6.

Scholl, R. W., and Cooper, E. "The use of job evaluation to eliminate gender based pay differentials." *Public Personnel Management*, Spring 1991, Vol. 20, No. 1, p. 1.

Waldman, S., and Springen, K. "Too old, too fast." *Newsweek*, Nov. 16, 1992, p. 80.

Winters, G. H. "Analysis of the biological and demographic parameters of northern sand lance, *Ammodytes dubins*, from the Newfoundland Grand Bank." *Canadian Journal of Fisheries and Aquatic Sciences*, 1983, 40.

Younger, M. S. A *First Course in Linear Regression*, 2d ed. Boston, Mass.: Duxbury, 1985.

APPENDIX A

Tables

Contents

TABLE I Random Numbers

Row \ Column	1	2	3	4	5	6	7	8	9	10	11	12	13	14
1	10480	15011	01536	02011	81647	91646	69179	14194	62590	36207	20969	99570	91291	90700
2	22368	46573	25595	85393	30995	89198	27982	53402	93965	34095	52666	19174	39615	99505
3	24130	48360	22527	97265	76393	64809	15179	24830	49340	32081	30680	19655	63348	58629
4	42167	93093	06243	61680	07856	16376	39440	53537	71341	57004	00849	74917	97758	16379
5	37570	39975	81837	16656	06121	91782	60468	81305	49684	60672	14110	06927	01263	54613
6	77921	06907	11008	42751	27756	53498	18602	70659	90655	15053	21916	81825	44394	42880
7	99562	72905	56420	69994	98872	31016	71194	18738	44013	48840	63213	21069	10634	12952
8	96301	91977	05463	07972	18876	20922	94595	56869	69014	60045	18425	84903	42508	32307
9	89579	14342	63661	10281	17453	18103	57740	84378	25331	12566	58678	44947	05585	56941
10	85475	36857	53342	53988	53060	59533	38867	62300	08158	17983	16439	11458	18593	64952
11	28918	69578	88231	33276	70997	79936	56865	05859	90106	31595	01547	85590	91610	78188
12	63553	40961	48235	03427	49626	69445	18663	72695	52180	20847	12234	90511	33703	90322
13	09429	93969	52636	92737	88974	33488	36320	17617	30015	08272	84115	27156	30613	74952
14	10365	61129	87529	85689	48237	52267	67689	93394	01511	26358	85104	20285	29975	89868
15	07119	97336	71048	08178	77233	13916	47564	81056	97735	85977	29372	74461	28551	90707
16	51085	12765	51821	51259	77452	16308	60756	92144	49442	53900	70960	63990	75601	40719
17	02368	21382	52404	60268	89368	19885	55322	44819	01188	65255	64835	44919	05944	55157
18	01011	54092	33362	94904	31273	04146	18594	29852	71585	85030	51132	01915	92747	64951
19	52162	53916	46369	58586	23216	14513	83149	98736	23495	64350	94738	17752	35156	35749
20	07056	97628	33787	09998	42698	06691	76988	13602	51851	46104	88916	19509	25625	58104
21	48663	91245	85828	14346	09172	30168	90229	04734	59193	22178	30421	61666	99904	32812
22	54164	58492	22421	74103	47070	25306	76468	26384	58151	06646	21524	15227	96909	44592
23	32639	32363	05597	24200	13363	38005	94342	28728	35806	06912	17012	64161	18296	22851
24	29334	27001	87637	87308	58731	00256	45834	15398	46557	41135	10367	07684	36188	18510
25	02488	33062	28834	07351	19731	92420	60952	61280	50001	67658	32586	86679	50720	94953

(continued)

TABLE I Continued

Row \ Column	1	2	3	4	5	6	7	8	9	10	11	12	13	14
26	81525	72295	04839	96423	24878	82651	66566	14778	76797	14780	13300	87074	79666	95725
27	29676	20591	68086	26432	46901	20849	89768	81536	86645	12659	92259	57102	80428	25280
28	00742	57392	39064	66432	84673	40027	32832	61362	98947	96067	64760	64584	96096	98253
29	05366	04213	25669	26422	44407	44048	37937	63904	45766	66134	75470	66520	34693	90449
30	91921	26418	64117	94305	26766	25940	39972	22209	71500	64568	91402	42416	07844	69618
31	00582	04711	87917	77341	42206	35126	74087	99547	81817	42607	43808	76655	62028	76630
32	00725	69884	62797	56170	86324	88072	76222	36086	84637	93161	76038	65855	77919	88006
33	69011	65795	95876	55293	18988	27354	26575	08625	40801	59920	29841	80150	12777	48501
34	25976	57948	29888	88604	67917	48708	18912	82271	65424	69774	33611	54262	85963	03547
35	09763	83473	73577	12908	30883	18317	28290	35797	05998	41688	34952	37888	38917	88050
36	91576	42595	27958	30134	04024	86385	29880	99730	55536	84855	29080	09250	79656	73211
37	17955	56349	90999	49127	20044	59931	06115	20542	18059	02008	73708	83517	36103	42791
38	46503	18584	18845	49618	02304	51038	20655	58727	28168	15475	56942	53389	20562	87338
39	92157	89634	94824	78171	84610	82834	09922	25417	44137	48413	25555	21246	35509	20468
40	14577	62765	35605	81263	39667	47358	56873	56307	61607	49518	89656	20103	77490	18062
41	98427	07523	33362	64270	01638	92477	66969	98420	04880	45585	46565	04102	46880	45709
42	34914	63976	88720	82765	34476	17032	87589	40836	32427	70002	70663	88863	77775	69348
43	70060	28277	39475	46473	23219	53416	94970	25832	69975	94884	19661	72828	00102	66794
44	53976	54914	06990	67245	68350	82948	11398	42878	80287	88267	47363	46634	06541	97809
45	76072	29515	40980	07391	58745	25774	22987	80059	39911	96189	41151	14222	60697	59583
46	90725	52210	83974	29992	65831	38857	50490	83765	55657	14361	31720	57375	56228	41546
47	64364	67412	33339	31926	14883	24413	59744	92351	97473	89286	35931	04110	23726	51900
48	08962	00358	31662	25388	61642	34072	81249	35648	56891	69352	48373	45578	78547	81788
49	95012	68379	93526	70765	10592	04542	76463	54328	02349	17247	28865	14777	62730	92277
50	15664	10493	20492	38391	91132	21999	59516	81652	27195	48223	46751	22923	32261	85653
51	16408	81899	04153	53381	79401	21438	83035	92350	36693	31238	59649	91754	72772	02338
52	18629	81953	05520	91962	04739	13092	97662	24822	94730	06496	35090	04822	86774	98289
53	73115	35101	47498	87637	99016	71060	88824	71013	18735	20286	23153	72924	35165	43040
54	57491	16703	23167	49323	45021	33132	12544	41035	80780	45393	44812	12515	98931	91202
55	30405	83946	23792	14422	15059	45799	22716	19792	09983	74353	68668	30429	70735	25499
56	16631	35006	85900	98275	32388	52390	16815	69298	82732	38480	73817	32523	41961	44437
57	96773	20206	42559	78985	05300	22164	24369	54224	35083	19687	11052	91491	60383	19746
58	38935	64202	14349	82674	66523	44133	00697	35552	35970	19124	63318	29686	03387	59846
59	31624	76384	17403	53363	44167	64486	64758	75366	76554	31601	12614	33072	60332	92325
60	78919	19474	23632	27889	47914	02584	37680	20801	72152	39339	34806	08930	85001	87820
61	03931	33309	57047	74211	63445	17361	62825	39908	05607	91284	68833	25570	38818	46920
62	74426	33278	43972	10119	89917	15665	52872	73823	73144	88662	88970	74492	51805	99378

(*continued*)

Row \ Column	1	2	3	4	5	6	7	8	9	10	11	12	13	14
63	09066	00903	20795	95452	92648	45454	09552	88815	16553	51125	79375	97596	16296	66092
64	42238	12426	87025	14267	20979	04508	64535	31355	86064	29472	47689	05974	52468	16834
65	16153	08002	26504	41744	81959	65642	74240	56302	00033	67107	77510	70625	28725	34191
66	21457	40742	29820	96783	29400	21840	15035	34537	33310	06116	95240	15957	16572	06004
67	21581	57802	02050	89728	17937	37621	47075	42080	97403	48626	68995	43805	33386	21597
68	55612	78095	83197	33732	05810	24813	86902	60397	16489	03264	88525	42786	05269	92532
69	44657	66999	99324	51281	84463	60563	79312	93454	68876	25471	93911	25650	12682	73572
70	91340	84979	46949	81973	37949	61023	43997	15263	80644	43942	89203	71795	99533	50501
71	91227	21199	31935	27022	84067	05462	35216	14486	29891	68607	41867	14951	91696	85065
72	50001	38140	66321	19924	72163	09538	12151	06878	91903	18749	34405	56087	82790	70925
73	65390	05224	72958	28609	81406	39147	25549	48542	42627	45233	57202	94617	23772	07896
74	27504	96131	83944	41575	10573	08619	64482	73923	36152	05184	94142	25299	84387	34925
75	37169	94851	39117	89632	00959	16487	65536	49071	39782	17095	02330	74301	00275	48280
76	11508	70225	51111	38351	19444	66499	71945	05422	13442	78675	84081	66938	93654	59894
77	37449	30362	06694	54690	04052	53115	62757	95348	78662	11163	81651	50245	34971	52924
78	46515	70331	85922	38329	57015	15765	97161	17869	45349	61796	66345	81073	49106	79860
79	30986	81223	42416	58353	21532	30502	32305	86482	05174	07901	54339	58861	74818	46942
80	63798	64995	46583	09785	44160	78128	83991	42865	92520	83531	80377	35909	81250	54238
81	82486	84846	99254	67632	43218	50076	21361	64816	51202	88124	41870	52689	51275	83556
82	21885	32906	92431	09060	64297	51674	64126	62570	26123	05155	59194	52799	28225	85762
83	60336	98782	07408	53458	13564	59089	26445	29789	85205	41001	12535	12133	14645	23541
84	43937	46891	24010	25560	86355	33941	25786	54990	71899	15475	95434	98227	21824	19585
85	97656	63175	89303	16275	07100	92063	21942	18611	47348	20203	18534	03862	78095	50136
86	03299	01221	05418	38982	55758	92237	26759	86367	21216	98442	08303	56613	91511	75928
87	79626	06486	03574	17668	07785	76020	79924	25651	83325	88428	85076	72811	22717	50585
88	85636	68335	47539	03129	65651	11977	02510	26113	99447	68645	34327	15152	55230	93448
89	18039	14367	61337	06177	12143	46609	32989	74014	64708	00533	35398	58408	13261	47908
90	08362	15656	60627	36478	65648	16764	53412	09013	07832	41574	17639	82163	60859	75567
91	79556	29068	04142	16268	15387	12856	66227	38358	22478	73373	88732	09443	82558	05250
92	92608	82674	27072	32534	17075	27698	98204	63863	11951	34648	88022	56148	34925	57031
93	23982	25835	40055	67006	12293	02753	14827	23235	35071	99704	37543	11601	35503	85171
94	09915	96306	05908	97901	28395	14186	00821	80703	70426	75647	76310	88717	37890	40129
95	59037	33300	26695	62247	69927	76123	50842	43834	86654	70959	79725	93872	28117	19233
96	42488	78077	69882	61657	34136	79180	97526	43092	04098	73571	80799	76536	71255	64239
97	46764	86273	63003	93017	31204	36692	40202	35275	57306	55543	53203	18098	47625	88684
98	03237	45430	55417	63282	90816	17349	88298	90183	36600	78406	06216	95787	42579	90730
99	86591	81482	52667	61582	14972	90053	89534	76036	49199	43716	97548	04379	46370	28672
100	38534	01715	94964	87288	65680	43772	39560	12918	86537	62738	19636	51132	25739	56947

Source: Abridged from W. H. Beyer (ed.). *CRC Standard Mathematical Tables*, 24th edition. (Cleveland: The Chemical Rubber Company), 1976. Reproduced by permission of the publisher.

TABLE II Binomial Probabilities

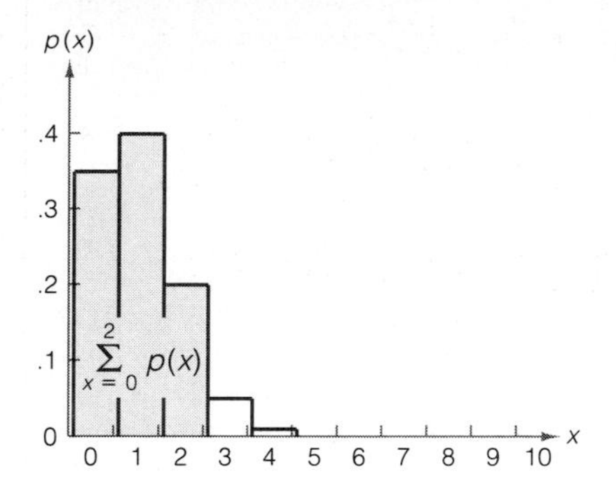

Tabulated values are $\sum_{x=0}^{k} p(x)$. *(Computations are rounded at the third decimal place.)*

a. $n = 5$

k \ p	.01	.05	.10	.20	.30	.40	.50	.60	.70	.80	.90	.95	.99
0	.951	.774	.590	.328	.168	.078	.031	.010	.002	.000	.000	.000	.000
1	.999	.977	.919	.737	.528	.337	.188	.087	.031	.007	.000	.000	.000
2	1.000	.999	.991	.942	.837	.683	.500	.317	.163	.058	.009	.001	.000
3	1.000	1.000	1.000	.993	.969	.913	.812	.663	.472	.263	.081	.023	.001
4	1.000	1.000	1.000	1.000	.998	.990	.969	.922	.832	.672	.410	.226	.049

b. $n = 6$

k \ p	.01	.05	.10	.20	.30	.40	.50	.60	.70	.80	.90	.95	.99
0	.941	.735	.531	.262	.118	.047	.016	.004	.001	.000	.000	.000	.000
1	.999	.967	.886	.655	.420	.233	.109	.041	.011	.002	.000	.000	.000
2	1.000	.998	.984	.901	.744	.544	.344	.179	.070	.017	.001	.000	.000
3	1.000	1.000	.999	.983	.930	.821	.656	.456	.256	.099	.016	.002	.000
4	1.000	1.000	1.000	.998	.989	.959	.891	.767	.580	.345	.114	.033	.001
5	1.000	1.000	1.000	1.000	.999	.996	.984	.953	.882	.738	.469	.265	.059

c. $n = 7$

k \ p	.01	.05	.10	.20	.30	.40	.50	.60	.70	.80	.90	.95	.99
0	.932	.698	.478	.210	.082	.028	.008	.002	.000	.000	.000	.000	.000
1	.998	.956	.850	.577	.329	.159	.063	.019	.004	.000	.000	.000	.000
2	1.000	.996	.974	.852	.647	.420	.227	.096	.029	.005	.000	.000	.000
3	1.000	1.000	.997	.967	.874	.710	.500	.290	.126	.033	.003	.000	.000
4	1.000	1.000	1.000	.995	.971	.904	.773	.580	.353	.148	.026	.004	.000
5	1.000	1.000	1.000	1.000	.996	.981	.937	.841	.671	.423	.150	.044	.002
6	1.000	1.000	1.000	1.000	1.000	.998	.992	.972	.918	.790	.522	.302	.068

TABLE II Continued

d. $n = 8$

k \ p	.01	.05	.10	.20	.30	.40	.50	.60	.70	.80	.90	.95	.99
0	.923	.663	.430	.168	.058	.017	.004	.001	.000	.000	.000	.000	.000
1	.997	.943	.813	.503	.255	.106	.035	.009	.001	.000	.000	.000	.000
2	1.000	.994	.962	.797	.552	.315	.145	.050	.011	.001	.000	.000	.000
3	1.000	1.000	.995	.944	.806	.594	.363	.174	.058	.010	.000	.000	.000
4	1.000	1.000	1.000	.990	.942	.826	.637	.406	.194	.056	.005	.000	.000
5	1.000	1.000	1.000	.999	.989	.950	.855	.685	.448	.203	.038	.006	.000
6	1.000	1.000	1.000	1.000	.999	.991	.965	.894	.745	.497	.187	.057	.003
7	1.000	1.000	1.000	1.000	1.000	.999	.996	.983	.942	.832	.570	.337	.077

e. $n = 9$

k \ p	.01	.05	.10	.20	.30	.40	.50	.60	.70	.80	.90	.95	.99
0	.914	.630	.387	.134	.040	.010	.002	.000	.000	.000	.000	.000	.000
1	.997	.929	.775	.436	.196	.071	.020	.004	.000	.000	.000	.000	.000
2	1.000	.992	.947	.738	.463	.232	.090	.025	.004	.000	.000	.000	.000
3	1.000	.999	.992	.914	.730	.483	.254	.099	.025	.003	.000	.000	.000
4	1.000	1.000	.999	.980	.901	.733	.500	.267	.099	.020	.001	.000	.000
5	1.000	1.000	1.000	.997	.975	.901	.746	.517	.270	.086	.008	.001	.000
6	1.000	1.000	1.000	1.000	.996	.975	.910	.768	.537	.262	.053	.008	.000
7	1.000	1.000	1.000	1.000	1.000	.996	.980	.929	.804	.564	.225	.071	.003
8	1.000	1.000	1.000	1.000	1.000	1.000	.998	.990	.960	.866	.613	.370	.086

f. $n = 10$

k \ p	.01	.05	.10	.20	.30	.40	.50	.60	.70	.80	.90	.95	.99
0	.904	.599	.349	.107	.028	.006	.001	.000	.000	.000	.000	.000	.000
1	.996	.914	.736	.376	.149	.046	.011	.002	.000	.000	.000	.000	.000
2	1.000	.988	.930	.678	.383	.167	.055	.012	.002	.000	.000	.000	.000
3	1.000	.999	.987	.879	.650	.382	.172	.055	.011	.001	.000	.000	.000
4	1.000	1.000	.998	.967	.850	.633	.377	.166	.047	.006	.000	.000	.000
5	1.000	1.000	1.000	.994	.953	.834	.623	.367	.150	.033	.002	.000	.000
6	1.000	1.000	1.000	.999	.989	.945	.828	.618	.350	.121	.013	.001	.000
7	1.000	1.000	1.000	1.000	.998	.988	.945	.833	.617	.322	.070	.012	.000
8	1.000	1.000	1.000	1.000	1.000	.998	.989	.954	.851	.624	.264	.086	.004
9	1.000	1.000	1.000	1.000	1.000	1.000	.999	.994	.972	.893	.651	.401	.096

TABLE II Continued

g. $n = 15$

k \ p	.01	.05	.10	.20	.30	.40	.50	.60	.70	.80	.90	.95	.99
0	.860	.463	.206	.035	.005	.000	.000	.000	.000	.000	.000	.000	.000
1	.990	.829	.549	.167	.035	.005	.000	.000	.000	.000	.000	.000	.000
2	1.000	.964	.816	.398	.127	.027	.004	.000	.000	.000	.000	.000	.000
3	1.000	.995	.944	.648	.297	.091	.018	.002	.000	.000	.000	.000	.000
4	1.000	.999	.987	.838	.515	.217	.059	.009	.001	.000	.000	.000	.000
5	1.000	1.000	.998	.939	.722	.403	.151	.034	.004	.000	.000	.000	.000
6	1.000	1.000	1.000	.982	.869	.610	.304	.095	.015	.001	.000	.000	.000
7	1.000	1.000	1.000	.996	.950	.787	.500	.213	.050	.004	.000	.000	.000
8	1.000	1.000	1.000	.999	.985	.905	.696	.390	.131	.018	.000	.000	.000
9	1.000	1.000	1.000	1.000	.996	.966	.849	.597	.278	.061	.002	.000	.000
10	1.000	1.000	1.000	1.000	.999	.991	.941	.783	.485	.164	.013	.001	.000
11	1.000	1.000	1.000	1.000	1.000	.998	.982	.909	.703	.352	.056	.005	.000
12	1.000	1.000	1.000	1.000	1.000	1.000	.996	.973	.873	.602	.184	.036	.000
13	1.000	1.000	1.000	1.000	1.000	1.000	1.000	.995	.965	.833	.451	.171	.010
14	1.000	1.000	1.000	1.000	1.000	1.000	1.000	1.000	.995	.965	.794	.537	.140

h. $n = 20$

k \ p	.01	.05	.10	.20	.30	.40	.50	.60	.70	.80	.90	.95	.99
0	.818	.358	.122	.012	.001	.000	.000	.000	.000	.000	.000	.000	.000
1	.983	.736	.392	.069	.008	.001	.000	.000	.000	.000	.000	.000	.000
2	.999	.925	.677	.206	.035	.004	.000	.000	.000	.000	.000	.000	.000
3	1.000	.984	.867	.411	.107	.016	.001	.000	.000	.000	.000	.000	.000
4	1.000	.997	.957	.630	.238	.051	.006	.000	.000	.000	.000	.000	.000
5	1.000	1.000	.989	.804	.416	.126	.021	.002	.000	.000	.000	.000	.000
6	1.000	1.000	.998	.913	.608	.250	.058	.006	.000	.000	.000	.000	.000
7	1.000	1.000	1.000	.968	.772	.416	.132	.021	.001	.000	.000	.000	.000
8	1.000	1.000	1.000	.990	.887	.596	.252	.057	.005	.000	.000	.000	.000
9	1.000	1.000	1.000	.997	.952	.755	.412	.128	.017	.001	.000	.000	.000
10	1.000	1.000	1.000	.999	.983	.872	.588	.245	.048	.003	.000	.000	.000
11	1.000	1.000	1.000	1.000	.995	.943	.748	.404	.113	.010	.000	.000	.000
12	1.000	1.000	1.000	1.000	.999	.979	.868	.584	.228	.032	.000	.000	.000
13	1.000	1.000	1.000	1.000	1.000	.994	.942	.750	.392	.087	.002	.000	.000
14	1.000	1.000	1.000	1.000	1.000	.998	.979	.874	.584	.196	.011	.000	.000
15	1.000	1.000	1.000	1.000	1.000	1.000	.994	.949	.762	.370	.043	.003	.000
16	1.000	1.000	1.000	1.000	1.000	1.000	.999	.984	.893	.589	.133	.016	.000
17	1.000	1.000	1.000	1.000	1.000	1.000	1.000	.996	.965	.794	.323	.075	.001
18	1.000	1.000	1.000	1.000	1.000	1.000	1.000	.999	.992	.931	.608	.264	.017
19	1.000	1.000	1.000	1.000	1.000	1.000	1.000	1.000	.999	.988	.878	.642	.182

(*continued*)

TABLE II Continued

i. $n = 25$

k \ p	.01	.05	.10	.20	.30	.40	.50	.60	.70	.80	.90	.95	.99
0	.778	.277	.072	.004	.000	.000	.000	.000	.000	.000	.000	.000	.000
1	.974	.642	.271	.027	.002	.000	.000	.000	.000	.000	.000	.000	.000
2	.998	.873	.537	.098	.009	.000	.000	.000	.000	.000	.000	.000	.000
3	1.000	.966	.764	.234	.033	.002	.000	.000	.000	.000	.000	.000	.000
4	1.000	.993	.902	.421	.090	.009	.000	.000	.000	.000	.000	.000	.000
5	1.000	.999	.967	.617	.193	.029	.002	.000	.000	.000	.000	.000	.000
6	1.000	1.000	.991	.780	.341	.074	.007	.000	.000	.000	.000	.000	.000
7	1.000	1.000	.998	.891	.512	.154	.022	.001	.000	.000	.000	.000	.000
8	1.000	1.000	1.000	.953	.677	.274	.054	.004	.000	.000	.000	.000	.000
9	1.000	1.000	1.000	.983	.811	.425	.115	.013	.000	.000	.000	.000	.000
10	1.000	1.000	1.000	.994	.902	.586	.212	.034	.002	.000	.000	.000	.000
11	1.000	1.000	1.000	.998	.956	.732	.345	.078	.006	.000	.000	.000	.000
12	1.000	1.000	1.000	1.000	.983	.846	.500	.154	.017	.000	.000	.000	.000
13	1.000	1.000	1.000	1.000	.994	.922	.655	.268	.044	.002	.000	.000	.000
14	1.000	1.000	1.000	1.000	.998	.966	.788	.414	.098	.006	.000	.000	.000
15	1.000	1.000	1.000	1.000	1.000	.987	.885	.575	.189	.017	.000	.000	.000
16	1.000	1.000	1.000	1.000	1.000	.996	.946	.726	.323	.047	.000	.000	.000
17	1.000	1.000	1.000	1.000	1.000	.999	.978	.846	.488	.109	.002	.000	.000
18	1.000	1.000	1.000	1.000	1.000	1.000	.993	.926	.659	.220	.009	.000	.000
19	1.000	1.000	1.000	1.000	1.000	1.000	.998	.971	.807	.383	.033	.001	.000
20	1.000	1.000	1.000	1.000	1.000	1.000	1.000	.991	.910	.579	.098	.007	.000
21	1.000	1.000	1.000	1.000	1.000	1.000	1.000	.998	.967	.766	.236	.034	.000
22	1.000	1.000	1.000	1.000	1.000	1.000	1.000	1.000	.991	.902	.463	.127	.002
23	1.000	1.000	1.000	1.000	1.000	1.000	1.000	1.000	.998	.973	.729	.358	.026
24	1.000	1.000	1.000	1.000	1.000	1.000	1.000	1.000	1.000	.996	.928	.723	.222

TABLE III Poisson Probabilities

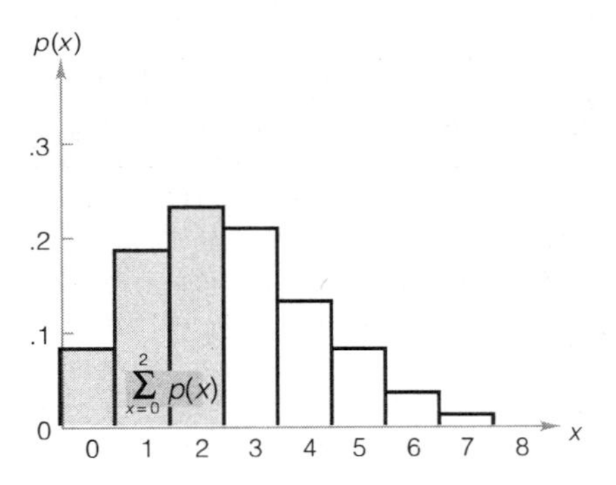

Tabulated values are $\sum_{x=0}^{k} p(x)$. *(Computations are rounded at the third decimal place.)*

λ \ x	0	1	2	3	4	5	6	7	8	9
.02	.980	1.000								
.04	.961	.999	1.000							
.06	.942	.998	1.000							
.08	.923	.997	1.000							
.10	.905	.995	1.000							
.15	.861	.990	.999	1.000						
.20	.819	.982	.999	1.000						
.25	.779	.974	.998	1.000						
.30	.741	.963	.996	1.000						
.35	.705	.951	.994	1.000						
.40	.670	.938	.992	.999	1.000					
.45	.638	.925	.989	.999	1.000					
.50	.607	.910	.986	.998	1.000					
.55	.577	.894	.982	.998	1.000					
.60	.549	.878	.977	.997	1.000					
.65	.522	.861	.972	.996	.999	1.000				
.70	.497	.844	.966	.994	.999	1.000				
.75	.472	.827	.959	.993	.999	1.000				
.80	.449	.809	.953	.991	.999	1.000				
.85	.427	.791	.945	.989	.998	1.000				
.90	.407	.772	.937	.987	.998	1.000				
.95	.387	.754	.929	.981	.997	1.000				
1.00	.368	.736	.920	.981	.996	.999	1.000			
1.1	.333	.699	.900	.974	.995	.999	1.000			
1.2	.301	.663	.879	.966	.992	.998	1.000			
1.3	.273	.627	.857	.957	.989	.998	1.000			
1.4	.247	.592	.833	.946	.986	.997	.999	1.000		
1.5	.223	.558	.809	.934	.981	.996	.999	1.000		

(continued)

TABLE III Continued

λ \ x	0	1	2	3	4	5	6	7	8	9
1.6	.202	.525	.783	.921	.976	.994	.999	1.000		
1.7	.183	.493	.757	.907	.970	.992	.998	1.000		
1.8	.165	.463	.731	.891	.964	.990	.997	.999	1.000	
1.9	.150	.434	.704	.875	.956	.987	.997	.999	1.000	
2.0	.135	.406	.677	.857	.947	.983	.995	.999	1.000	
2.2	.111	.355	.623	.819	.928	.975	.993	.998	1.000	
2.4	.091	.308	.570	.779	.904	.964	.988	.997	.999	1.000
2.6	.074	.267	.518	.736	.877	.951	.983	.995	.999	1.000
2.8	.061	.231	.469	.692	.848	.935	.976	.992	.998	.999
3.0	.050	.199	.423	.647	.815	.916	.966	.988	.996	.999
3.2	.041	.171	.380	.603	.781	.895	.955	.983	.994	.998
3.4	.033	.147	.340	.558	.744	.871	.942	.977	.992	.997
3.6	.027	.126	.303	.515	.706	.844	.927	.969	.988	.996
3.8	.022	.107	.269	.473	.668	.816	.909	.960	.984	.994
4.0	.018	.092	.238	.433	.629	.785	.889	.949	.979	.992
4.2	.015	.078	.210	.395	.590	.753	.867	.936	.972	.989
4.4	.012	.066	.185	.359	.551	.720	.844	.921	.964	.985
4.6	.010	.056	.163	.326	.513	.686	.818	.905	.955	.980
4.8	.008	.048	.143	.294	.476	.651	.791	.887	.944	.975
5.0	.007	.040	.125	.265	.440	.616	.762	.867	.932	.968
5.2	.006	.034	.109	.238	.406	.581	.732	.845	.918	.960
5.4	.005	.029	.095	.213	.373	.546	.702	.822	.903	.951
5.6	.004	.024	.082	.191	.342	.512	.670	.797	.886	.941
5.8	.003	.021	.072	.170	.313	.478	.638	.771	.867	.929
6.0	.002	.017	.062	.151	.285	.446	.606	.744	.847	.916
	10	**11**	**12**	**13**	**14**	**15**	**16**			
2.8	1.000									
3.0	1.000									
3.2	1.000									
3.4	.999	1.000								
3.6	.999	1.000								
3.8	.998	.999	1.000							
4.0	.997	.999	1.000							
4.2	.996	.999	1.000							
4.4	.994	.998	.999	1.000						
4.6	.992	.997	.999	1.000						
4.8	.990	.996	.999	1.000						
5.0	.986	.995	.998	.999	1.000					
5.2	.982	.993	.997	.999	1.000					
5.4	.977	.990	.996	.999	1.000					
5.6	.972	.988	.995	.998	.999	1.000				
5.8	.965	.984	.993	.997	.999	1.000				
6.0	.957	.980	.991	.996	.999	.999	1.000			

TABLE III Continued

λ \ x	0	1	2	3	4	5	6	7	8	9
6.2	.002	.015	.054	.134	.259	.414	.574	.716	.826	.902
6.4	.002	.012	.046	.119	.235	.384	.542	.687	.803	.886
6.6	.001	.010	.040	.105	.213	.355	.511	.658	.780	.869
6.8	.001	.009	.034	.093	.192	.327	.480	.628	.755	.850
7.0	.001	.007	.030	.082	.173	.301	.450	.599	.729	.830
7.2	.001	.006	.025	.072	.156	.276	.420	.569	.703	.810
7.4	.001	.005	.022	.063	.140	.253	.392	.539	.676	.788
7.6	.001	.004	.019	.055	.125	.231	.365	.510	.648	.765
7.8	.000	.004	.016	.048	.112	.210	.338	.481	.620	.741
8.0	.000	.003	.014	.042	.100	.191	.313	.453	.593	.717
8.5	.000	.002	.009	.030	.074	.150	.256	.386	.523	.653
9.0	.000	.001	.006	.021	.055	.116	.207	.324	.456	.587
9.5	.000	.001	.004	.015	.040	.089	.165	.269	.392	.522
10.0	.000	.000	.003	.010	.029	.067	.130	.220	.333	.458

λ \ x	10	11	12	13	14	15	16	17	18	19
6.2	.949	.975	.989	.995	.998	.999	1.000			
6.4	.939	.969	.986	.994	.997	.999	1.000			
6.6	.927	.963	.982	.992	.997	.999	.999	1.000		
6.8	.915	.955	.978	.990	.996	.998	.999	1.000		
7.0	.901	.947	.973	.987	.994	.998	.999	1.000		
7.2	.887	.937	.967	.984	.993	.997	.999	.999	1.000	
7.4	.871	.926	.961	.980	.991	.996	.998	.999	1.000	
7.6	.854	.915	.954	.976	.989	.995	.998	.999	1.000	
7.8	.835	.902	.945	.971	.986	.993	.997	.999	1.000	
8.0	.816	.888	.936	.966	.983	.992	.996	.998	.999	1.000
8.5	.763	.849	.909	.949	.973	.986	.993	.997	.999	.999
9.0	.706	.803	.876	.926	.959	.978	.989	.995	.998	.999
9.5	.645	.752	.836	.898	.940	.967	.982	.991	.996	.998
10.0	.583	.697	.792	.864	.917	.951	.973	.986	.993	.997

λ \ x	20	21	22
8.5	1.000		
9.0	1.000		
9.5	.999	1.000	
10.0	.998	.999	1.000

(continued)

TABLE III Continued

λ \ x	0	1	2	3	4	5	6	7	8	9
10.5	.000	.000	.002	.007	.021	.050	.102	.179	.279	.397
11.0	.000	.000	.001	.005	.015	.038	.079	.143	.232	.341
11.5	.000	.000	.001	.003	.011	.028	.060	.114	.191	.289
12.0	.000	.000	.001	.002	.008	.020	.046	.090	.155	.242
12.5	.000	.000	.000	.002	.005	.015	.035	.070	.125	.201
13.0	.000	.000	.000	.001	.004	.011	.026	.054	.100	.166
13.5	.000	.000	.000	.001	.003	.008	.019	.041	.079	.135
14.0	.000	.000	.000	.000	.002	.006	.014	.032	.062	.109
14.5	.000	.000	.000	.000	.001	.004	.010	.024	.048	.088
15.0	.000	.000	.000	.000	.001	.003	.008	.018	.037	.070
	10	**11**	**12**	**13**	**14**	**15**	**16**	**17**	**18**	**19**
10.5	.521	.639	.742	.825	.888	.932	.960	.978	.988	.994
11.0	.460	.579	.689	.781	.854	.907	.944	.968	.982	.991
11.5	.402	.520	.633	.733	.815	.878	.924	.954	.974	.986
12.0	.347	.462	.576	.682	.772	.844	.899	.937	.963	.979
12.5	.297	.406	.519	.628	.725	.806	.869	.916	.948	.969
13.0	.252	.353	.463	.573	.675	.764	.835	.890	.930	.957
13.5	.211	.304	.409	.518	.623	.718	.798	.861	.908	.942
14.0	.176	.260	.358	.464	.570	.669	.756	.827	.883	.923
14.5	.145	.220	.311	.413	.518	.619	.711	.790	.853	.901
15.0	.118	.185	.268	.363	.466	.568	.664	.749	.819	.875
	20	**21**	**22**	**23**	**24**	**25**	**26**	**27**	**28**	**29**
10.5	.997	.999	.999	1.000						
11.0	.995	.998	.999	1.000						
11.5	.992	.996	.998	.999	1.000					
12.0	.988	.994	.997	.999	.999	1.000				
12.5	.983	.991	.995	.998	.999	.999	1.000			
13.0	.975	.986	.992	.996	.998	.999	1.000			
13.5	.965	.980	.989	.994	.997	.998	.999	1.000		
14.0	.952	.971	.983	.991	.995	.997	.999	.999	1.000	
14.5	.936	.960	.976	.986	.992	.996	.998	.999	.999	1.000
15.0	.917	.947	.967	.981	.989	.994	.997	.998	.999	1.000

TABLE III Continued

λ \ x	4	5	6	7	8	9	10	11	12	13
16	.000	.001	.004	.010	.022	.043	.077	.127	.193	.275
17	.000	.001	.002	.005	.013	.026	.049	.085	.135	.201
18	.000	.000	.001	.003	.007	.015	.030	.055	.092	.143
19	.000	.000	.001	.002	.004	.009	.018	.035	.061	.098
20	.000	.000	.000	.001	.002	.005	.011	.021	.039	.066
21	.000	.000	.000	.000	.001	.003	.006	.013	.025	.043
22	.000	.000	.000	.000	.001	.002	.004	.008	.015	.028
23	.000	.000	.000	.000	.000	.001	.002	.004	.009	.017
24	.000	.000	.000	.000	.000	.000	.001	.003	.005	.011
25	.000	.000	.000	.000	.000	.000	.001	.001	.003	.006

λ \ x	14	15	16	17	18	19	20	21	22	23
16	.368	.467	.566	.659	.742	.812	.868	.911	.942	.963
17	.281	.371	.468	.564	.655	.736	.805	.861	.905	.937
18	.208	.287	.375	.469	.562	.651	.731	.799	.855	.899
19	.150	.215	.292	.378	.469	.561	.647	.725	.793	.849
20	.105	.157	.221	.297	.381	.470	.559	.644	.721	.787
21	.072	.111	.163	.227	.302	.384	.471	.558	.640	.716
22	.048	.077	.117	.169	.232	.306	.387	.472	.556	.637
23	.031	.052	.082	.123	.175	.238	.310	.389	.472	.555
24	.020	.034	.056	.087	.128	.180	.243	.314	.392	.473
25	.012	.022	.038	.060	.092	.134	.185	.247	.318	.394

λ \ x	24	25	26	27	28	29	30	31	32	33
16	.978	.987	.993	.996	.998	.999	.999	1.000		
17	.959	.975	.985	.991	.995	.997	.999	.999	1.000	
18	.932	.955	.972	.983	.990	.994	.997	.998	.999	1.000
19	.893	.927	.951	.969	.980	.988	.993	.996	.998	.999
20	.843	.888	.922	.948	.966	.978	.987	.992	.995	.997
21	.782	.838	.883	.917	.944	.963	.975	.985	.991	.994
22	.712	.777	.832	.877	.913	.940	.959	.973	.983	.989
23	.635	.708	.772	.827	.873	.908	.936	.956	.971	.981
24	.554	.632	.704	.768	.823	.868	.904	.932	.953	.969
25	.473	.553	.629	.700	.763	.818	.863	.900	.929	.950

λ \ x	34	35	36	37	38	39	40	41	42	43
19	.999	1.000								
20	.999	.999	1.000							
21	.997	.998	.999	.999	1.000					
22	.994	.996	.998	.999	.999	1.000				
23	.988	.993	.996	.997	.999	.999	1.000			
24	.979	.987	.992	.995	.997	.998	.999	.999	1.000	
25	.966	.978	.985	.991	.991	.997	.998	.999	.999	1.000

TABLE IV Normal Curve Areas

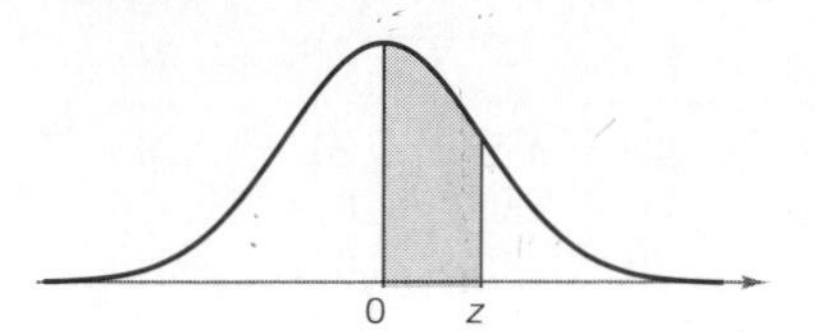

z	.00	.01	.02	.03	.04	.05	.06	.07	.08	.09
.0	.0000	.0040	.0080	.0120	.0160	.0199	.0239	.0279	.0319	.0359
.1	.0398	.0438	.0478	.0517	.0557	.0596	.0636	.0675	.0714	.0753
.2	.0793	.0832	.0871	.0910	.0948	.0987	.1026	.1064	.1103	.1141
.3	.1179	.1217	.1255	.1293	.1331	.1368	.1406	.1443	.1480	.1517
.4	.1554	.1591	.1628	.1664	.1700	.1736	.1772	.1808	.1844	.1879
.5	.1915	.1950	.1985	.2019	.2054	.2088	.2123	.2157	.2190	.2224
.6	.2257	.2291	.2324	.2357	.2389	.2422	.2454	.2486	.2517	.2549
.7	.2580	.2611	.2642	.2673	.2704	.2734	.2764	.2794	.2823	.2852
.8	.2881	.2910	.2939	.2967	.2995	.3023	.3051	.3078	.3106	.3133
.9	.3159	.3186	.3212	.3238	.3264	.3289	.3315	.3340	.3365	.3389
1.0	.3413	.3438	.3461	.3485	.3508	.3531	.3554	.3577	.3599	.3621
1.1	.3643	.3665	.3686	.3708	.3729	.3749	.3770	.3790	.3810	.3830
1.2	.3849	.3869	.3888	.3907	.3925	.3944	.3962	.3980	.3997	.4015
1.3	.4032	.4049	.4066	.4082	.4099	.4115	.4131	.4147	.4162	.4177
1.4	.4192	.4207	.4222	.4236	.4251	.4265	.4279	.4292	.4306	.4319
1.5	.4332	.4345	.4357	.4370	.4382	.4394	.4406	.4418	.4429	.4441
1.6	.4452	.4463	.4474	.4484	.4495	.4505	.4515	.4525	.4535	.4545
1.7	.4554	.4564	.4573	.4582	.4591	.4599	.4608	.4616	.4625	.4633
1.8	.4641	.4649	.4656	.4664	.4671	.4678	.4686	.4693	.4699	.4706
1.9	.4713	.4719	.4726	.4732	.4738	.4744	.4750	.4756	.4761	.4767
2.0	.4772	.4778	.4783	.4788	.4793	.4798	.4803	.4808	.4812	.4817
2.1	.4821	.4826	.4830	.4834	.4838	.4842	.4846	.4850	.4854	.4857
2.2	.4861	.4864	.4868	.4871	.4875	.4878	.4881	.4884	.4887	.4890
2.3	.4893	.4896	.4898	.4901	.4904	.4906	.4909	.4911	.4913	.4916
2.4	.4918	.4920	.4922	.4925	.4927	.4929	.4931	.4932	.4934	.4936
2.5	.4938	.4940	.4941	.4943	.4945	.4946	.4948	.4949	.4951	.4952
2.6	.4953	.4955	.4956	.4957	.4959	.4960	.4961	.4962	.4963	.4964
2.7	.4965	.4966	.4967	.4968	.4969	.4970	.4971	.4972	.4973	.4974
2.8	.4974	.4975	.4976	.4977	.4977	.4978	.4979	.4979	.4980	.4981
2.9	.4981	.4982	.4982	.4983	.4984	.4984	.4985	.4985	.4986	.4986
3.0	.4987	.4987	.4987	.4988	.4988	.4989	.4989	.4989	.4990	.4990

Source: Abridged from Table I of A. Hald, *Statistical Tables and Formulas* (New York: Wiley), 1952. Reproduced by permission of A. Hald and the publisher, John Wiley & Sons, Inc.

TABLE V Critical Values of t

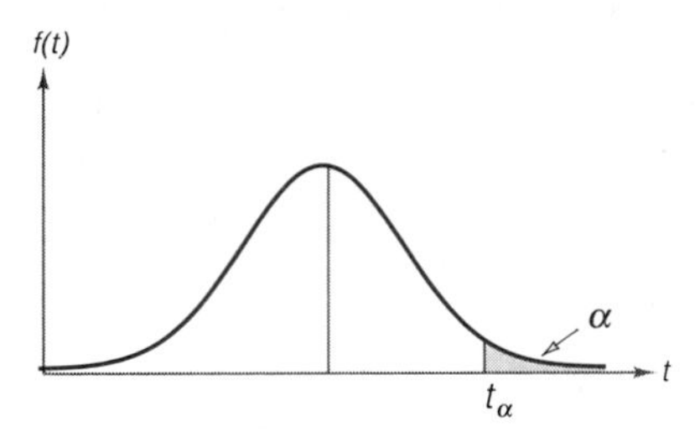

ν	$t_{.100}$	$t_{.050}$	$t_{.025}$	$t_{.010}$	$t_{.005}$	$t_{.001}$	$t_{.0005}$
1	3.078	6.314	12.706	31.821	63.657	318.31	636.62
2	1.886	2.920	4.303	6.965	9.925	22.326	31.598
3	1.638	2.353	3.182	4.541	5.841	10.213	12.924
4	1.533	2.132	2.776	3.747	4.604	7.173	8.610
5	1.476	2.015	2.571	3.365	4.032	5.893	6.869
6	1.440	1.943	2.447	3.143	3.707	5.208	5.959
7	1.415	1.895	2.365	2.998	3.499	4.785	5.408
8	1.397	1.860	2.306	2.896	3.355	4.501	5.041
9	1.383	1.833	2.262	2.821	3.250	4.297	4.781
10	1.372	1.812	2.228	2.764	3.169	4.144	4.587
11	1.363	1.796	2.201	2.718	3.106	4.025	4.437
12	1.356	1.782	2.179	2.681	3.055	3.930	4.318
13	1.350	1.771	2.160	2.650	3.012	3.852	4.221
14	1.345	1.761	2.145	2.624	2.977	3.787	4.140
15	1.341	1.753	2.131	2.602	2.947	3.733	4.073
16	1.337	1.746	2.120	2.583	2.921	3.686	4.015
17	1.333	1.740	2.110	2.567	2.898	3.646	3.965
18	1.330	1.734	2.101	2.552	2.878	3.610	3.922
19	1.328	1.729	2.093	2.539	2.861	3.579	3.883
20	1.325	1.725	2.086	2.528	2.845	3.552	3.850
21	1.323	1.721	2.080	2.518	2.831	3.527	3.819
22	1.321	1.717	2.074	2.508	2.819	3.505	3.792
23	1.319	1.714	2.069	2.500	2.807	3.485	3.767
24	1.318	1.711	2.064	2.492	2.797	3.467	3.745
25	1.316	1.708	2.060	2.485	2.787	3.450	3.725
26	1.315	1.706	2.056	2.479	2.779	3.435	3.707
27	1.314	1.703	2.052	2.473	2.771	3.421	3.690
28	1.313	1.701	2.048	2.467	2.763	3.408	3.674
29	1.311	1.699	2.045	2.462	2.756	3.396	3.659
30	1.310	1.697	2.042	2.457	2.750	3.385	3.646
40	1.303	1.684	2.021	2.423	2.704	3.307	3.551
60	1.296	1.671	2.000	2.390	2.660	3.232	3.460
120	1.289	1.658	1.980	2.358	2.617	3.160	3.373
∞	1.282	1.645	1.960	2.326	2.576	3.090	3.291

Source: This table is reproduced with the kind permission of the Trustees of Biometrika from E. S. Pearson and H. O. Hartley (eds.), *The Biometrika Tables for Statisticians*, Vol. 1, 3d ed., *Biometrika*, 1966.

TABLE VI Critical Values of T_L and T_U for the Wilcoxon Rank Sum Test: Independent Samples

Test statistic is the rank sum associated with the smaller sample (if equal sample sizes, either rank sum can be used).

a. $\alpha = .025$ one-tailed; $\alpha = .05$ two-tailed

n_2 \ n_1	3		4		5		6		7		8		9		10	
	T_L	T_U	T_L	T_U	T_L	T_U	T_L	T_U	T_L	T_U	T_L	T_U	T_L	T_U	T_L	T_U
3	5	16	6	18	6	21	7	23	7	26	8	28	8	31	9	33
4	6	18	11	25	12	28	12	32	13	35	14	38	15	41	16	44
5	6	21	12	28	18	37	19	41	20	45	21	49	22	53	24	56
6	7	23	12	32	19	41	26	52	28	56	29	61	31	65	32	70
7	7	26	13	35	20	45	28	56	37	68	39	73	41	78	43	83
8	8	28	14	38	21	49	29	61	39	73	49	87	51	93	54	98
9	8	31	15	41	22	53	31	65	41	78	51	93	63	108	66	114
10	9	33	16	44	24	56	32	70	43	83	54	98	66	114	79	131

b. $\alpha = .05$ one-tailed; $\alpha = .10$ two-tailed

n_2 \ n_1	3		4		5		6		7		8		9		10	
	T_L	T_U	T_L	T_U	T_L	T_U	T_L	T_U	T_L	T_U	T_L	T_U	T_L	T_U	T_L	T_U
3	6	15	7	17	7	20	8	22	9	24	9	27	10	29	11	31
4	7	17	12	24	13	27	14	30	15	33	16	36	17	39	18	42
5	7	20	13	27	19	36	20	40	22	43	24	46	25	50	26	54
6	8	22	14	30	20	40	28	50	30	54	32	58	33	63	35	67
7	9	24	15	33	22	43	30	54	39	66	41	71	43	76	46	80
8	9	27	16	36	24	46	32	58	41	71	52	84	54	90	57	95
9	10	29	17	39	25	50	33	63	43	76	54	90	66	105	69	111
10	11	31	18	42	26	54	35	67	46	80	57	95	69	111	83	127

Source: From F. Wilcoxon and R. A. Wilcox, "Some Rapid Approximate Statistical Procedures," 1964, 20–23. Reproduced with the permission of American Cyanamid Company.

TABLE VII Critical Values of T_0 in the Wilcoxon Paired Difference Signed Rank Test

One-Tailed	Two-Tailed						
		$n = 5$	$n = 6$	$n = 7$	$n = 8$	$n = 9$	$n = 10$
$\alpha = .05$	$\alpha = .10$	1	2	4	6	8	11
$\alpha = .025$	$\alpha = .05$		1	2	4	6	8
$\alpha = .01$	$\alpha = .02$			0	2	3	5
$\alpha = .005$	$\alpha = .01$				0	2	3
		$n = 11$	$n = 12$	$n = 13$	$n = 14$	$n = 15$	$n = 16$
$\alpha = .05$	$\alpha = .10$	14	17	21	26	30	36
$\alpha = .025$	$\alpha = .05$	11	14	17	21	25	30
$\alpha = .01$	$\alpha = .02$	7	10	13	16	20	24
$\alpha = .005$	$\alpha = .01$	5	7	10	13	16	19
		$n = 17$	$n = 18$	$n = 19$	$n = 20$	$n = 21$	$n = 22$
$\alpha = .05$	$\alpha = .10$	41	47	54	60	68	75
$\alpha = .025$	$\alpha = .05$	35	40	46	52	59	66
$\alpha = .01$	$\alpha = .02$	28	33	38	43	49	56
$\alpha = .005$	$\alpha = .01$	23	28	32	37	43	49
		$n = 23$	$n = 24$	$n = 25$	$n = 26$	$n = 27$	$n = 28$
$\alpha = .05$	$\alpha = .10$	83	92	101	110	120	130
$\alpha = .025$	$\alpha = .05$	73	81	90	98	107	117
$\alpha = .01$	$\alpha = .02$	62	69	77	85	93	102
$\alpha = .005$	$\alpha = .01$	55	61	68	76	84	92
		$n = 29$	$n = 30$	$n = 31$	$n = 32$	$n = 33$	$n = 34$
$\alpha = .05$	$\alpha = .10$	141	152	163	175	188	201
$\alpha = .025$	$\alpha = .05$	127	137	148	159	171	183
$\alpha = .01$	$\alpha = .02$	111	120	130	141	151	162
$\alpha = .005$	$\alpha = .01$	100	109	118	128	138	149
		$n = 35$	$n = 36$	$n = 37$	$n = 38$	$n = 39$	
$\alpha = .05$	$\alpha = .10$	214	228	242	256	271	
$\alpha = .025$	$\alpha = .05$	195	208	222	235	250	
$\alpha = .01$	$\alpha = .02$	174	186	198	211	224	
$\alpha = .005$	$\alpha = .01$	160	171	183	195	208	
		$n = 40$	$n = 41$	$n = 42$	$n = 43$	$n = 44$	$n = 45$
$\alpha = .05$	$\alpha = .10$	287	303	319	336	353	371
$\alpha = .025$	$\alpha = .05$	264	279	295	311	327	344
$\alpha = .01$	$\alpha = .02$	238	252	267	281	297	313
$\alpha = .005$	$\alpha = .01$	221	234	248	262	277	292
		$n = 46$	$n = 47$	$n = 48$	$n = 49$	$n = 50$	
$\alpha = .05$	$\alpha = .10$	389	408	427	446	466	
$\alpha = .025$	$\alpha = .05$	361	379	397	415	434	
$\alpha = .01$	$\alpha = .02$	329	345	362	380	398	
$\alpha = .005$	$\alpha = .01$	307	323	339	356	373	

Source: From F. Wilcoxon and R. A. Wilcox, "Some Rapid Approximate Statistical Procedures," 1964, p. 28. Reproduced with the permission of American Cyanamid Company.

TABLE VIII Critical Values of χ^2

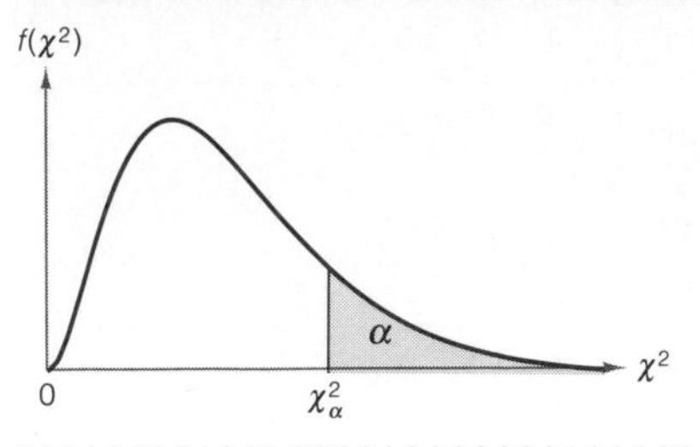

Degrees of Freedom	$\chi^2_{.995}$	$\chi^2_{.990}$	$\chi^2_{.975}$	$\chi^2_{.950}$	$\chi^2_{.900}$
1	.0000393	.0001571	.0009821	.0039321	.0157908
2	.0100251	.0201007	.0506356	.102587	.210720
3	.0717212	.114832	.215795	.351846	.584375
4	.206990	.297110	.484419	.710721	1.063623
5	.411740	.554300	.831211	1.145476	1.61031
6	.675727	.872085	1.237347	1.63539	2.20413
7	.989265	1.239043	1.68987	2.16735	2.83311
8	1.344419	1.646482	2.17973	2.73264	3.48954
9	1.734926	2.087912	2.70039	3.32511	4.16816
10	2.15585	2.55821	3.24697	3.94030	4.86518
11	2.60321	3.05347	3.81575	4.57481	5.57779
12	3.07382	3.57056	4.40379	5.22603	6.30380
13	3.56503	4.10691	5.00874	5.89186	7.04150
14	4.07468	4.66043	5.62872	6.57063	7.78953
15	4.60094	5.22935	6.26214	7.26094	8.54675
16	5.14224	5.81221	6.90766	7.96164	9.31223
17	5.69724	6.40776	7.56418	8.67176	10.0852
18	6.26481	7.01491	8.23075	9.39046	10.8649
19	6.84398	7.63273	8.90655	10.1170	11.6509
20	7.43386	8.26040	9.59083	10.8508	12.4426
21	8.03366	8.89720	10.28293	11.5913	13.2396
22	8.64272	9.54249	10.9823	12.3380	14.0415
23	9.26042	10.19567	11.6885	13.0905	14.8479
24	9.88623	10.8564	12.4011	13.8484	15.6587
25	10.5197	11.5240	13.1197	14.6114	16.4734
26	11.1603	12.1981	13.8439	15.3791	17.2919
27	11.8076	12.8786	14.5733	16.1513	18.1138
28	12.4613	13.5648	15.3079	16.9279	18.9392
29	13.1211	14.2565	16.0471	17.7083	19.7677
30	13.7867	14.9535	16.7908	18.4926	20.5992
40	20.7065	22.1643	24.4331	26.5093	29.0505
50	27.9907	29.7067	32.3574	34.7642	37.6886
60	35.5346	37.4848	40.4817	43.1879	46.4589
70	43.2752	45.4418	48.7576	51.7393	55.3290
80	51.1720	53.5400	57.1532	60.3915	64.2778
90	59.1963	61.7541	65.6466	69.1260	73.2912
100	67.3276	70.0648	74.2219	77.9295	82.3581

Source: From C. M. Thompson, "Tables of the Percentage Points of the χ^2-Distribution," *Biometrika*, 1941, 32, 188–189. Reproduced by permission of the *Biometrika* Trustees.

Degrees of Freedom	$\chi^2_{.100}$	$\chi^2_{.050}$	$\chi^2_{.025}$	$\chi^2_{.010}$	$\chi^2_{.005}$
1	2.70554	3.84146	5.02389	6.63490	7.87944
2	4.60517	5.99147	7.37776	9.21034	10.5966
3	6.25139	7.81473	9.34840	11.3449	12.8381
4	7.77944	9.48773	11.1433	13.2767	14.8602
5	9.23635	11.0705	12.8325	15.0863	16.7496
6	10.6446	12.5916	14.4494	16.8119	18.5476
7	12.0170	14.0671	16.0128	18.4753	20.2777
8	13.3616	15.5073	17.5346	20.0902	21.9550
9	14.6837	16.9190	19.0228	21.6660	23.5893
10	15.9871	18.3070	20.4831	23.2093	25.1882
11	17.2750	19.6751	21.9200	24.7250	26.7569
12	18.5494	21.0261	23.3367	26.2170	28.2995
13	19.8119	22.3621	24.7356	27.6883	29.8194
14	21.0642	23.6848	26.1190	29.1413	31.3193
15	22.3072	24.9958	27.4884	30.5779	32.8013
16	23.5418	26.2962	28.8454	31.9999	34.2672
17	24.7690	27.5871	30.1910	33.4087	35.7185
18	25.9894	28.8693	31.5264	34.8053	37.1564
19	27.2036	30.1435	32.8523	36.1908	38.5822
20	28.4120	31.4104	34.1696	37.5662	39.9968
21	29.6151	32.6705	35.4789	38.9321	41.4010
22	30.8133	33.9244	36.7807	40.2894	42.7956
23	32.0069	35.1725	38.0757	41.6384	44.1813
24	33.1963	36.4151	39.3641	42.9798	45.5585
25	34.3816	37.6525	40.6465	44.3141	46.9278
26	35.5631	38.8852	41.9232	45.6417	48.2899
27	36.7412	40.1133	43.1944	46.9630	49.6449
28	37.9159	41.3372	44.4607	48.2782	50.9933
29	39.0875	42.5569	45.7222	49.5879	52.3356
30	40.2560	43.7729	46.9792	50.8922	53.6720
40	51.8050	55.7585	59.3417	63.6907	66.7659
50	63.1671	67.5048	71.4202	76.1539	79.4900
60	74.3970	79.0819	83.2976	88.3794	91.9517
70	85.5271	90.5312	95.0231	100.425	104.215
80	96.5782	101.879	106.629	112.329	116.321
90	107.565	113.145	118.136	124.116	128.299
100	118.498	124.342	129.561	135.807	140.169

APPENDIX B

Demographic Data Set

A demographic data set was assembled based on a systematic random sample of 1,000 United States zip codes. To obtain the sample, the more than 30,000 zip codes were sorted, and approximately every 30th was selected. The map in Figure B.1 shows the number of zip codes selected in each state. Note that each state is classified according to its census region: North Central, Northeast, South, and West.

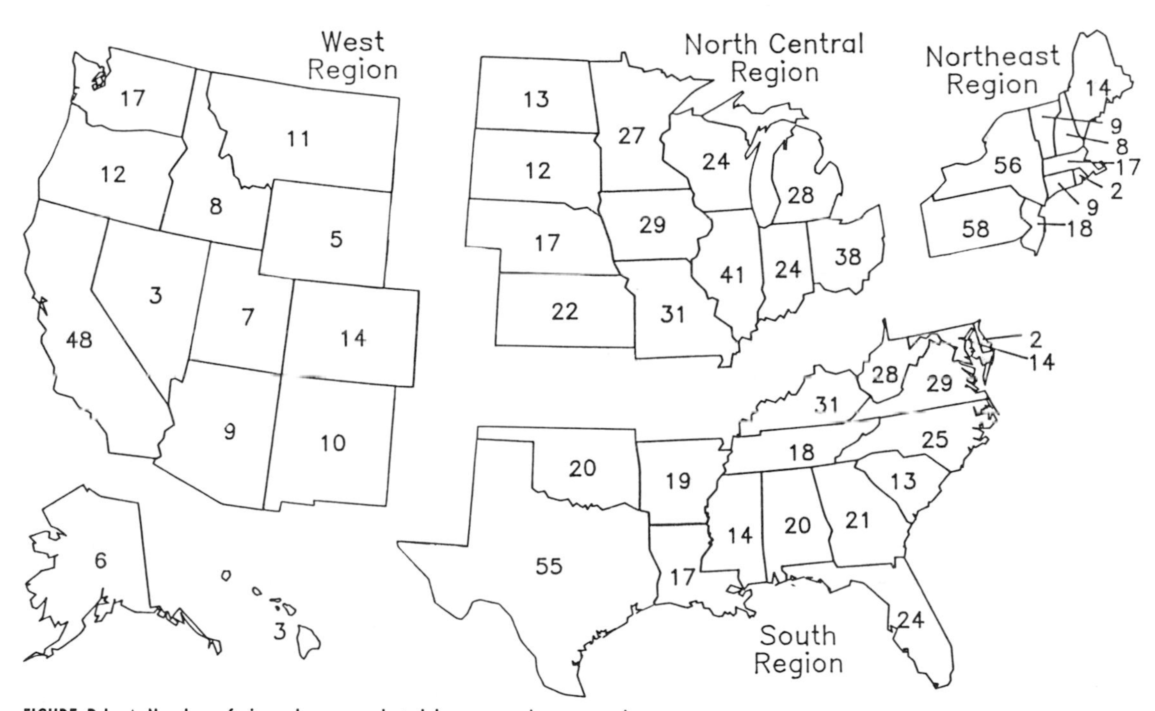

FIGURE B.1 ▲ Number of zip code areas selected by state and census region

Demographic data for each zip code area selected were supplied by CACI, an international demographic and market information firm, and are reproduced with its permission. CACI produces interim estimates of many of the demographic variables measured decennially by the Bureau of the Census. CACI also produces market information based on the Bureau of the Census Consumer Expenditure Survey (such as the four purchasing indexes given below for each zip code), which measures the relative propensity of a given market (zip code) to purchase the goods, a higher index value indicating a higher propensity to buy such goods as compared to other similar zip codes in the same census region. In general, the 1980 measurements are official U.S. Bureau of the Census estimates, while the 1986 measurements are CACI estimates.

Fifteen demographic measurements are presented for each zip code area. Portions of the data are referenced at the end of each chapter in **Using the Computer**. The objectives are to enable the student to analyze real data in a relatively large sample using the computer, and to gain experience using statistical techniques and concepts on real data. Of course, neither the student nor the instructor need be bound by the suggestions in **Using the Computer**; the data are rich enough to support many more analyses than could be listed (or imagined) by the authors.

The following 15 measurements are reported for each zip code in the sample:

1. Population (1986): Total population for the zip code, 1986.
2. Number of Households (1986): Number of households for the zip code in 1986.
3. Age (Median, 1986): Median age for the zip code in 1986.
4. Household Income (Median, 1986): Median household income for the zip code in 1986.
5. Home Value (Median, 1980): Median residential owner-occupied home value for the zip code in 1980.
6. Monthly Cost of Housing (Median, 1980): Median monthly homeowner cost, including mortgage payments, real estate taxes, property insurance, utilities, and fuels, for the zip code in 1980.
7. Household Size (Average, 1980): Average number of persons per household in the zip code, 1980.
8. Years of Education (Median, 1980): Median years of education for adults (persons 25 years of age and over) in the zip code, 1980.
9. College Education (Percentage, 1980): Percentage of adults in the zip code having a college education.
10. Women in Labor Force (Percentage, 1980): Percentage of women age 16 and over in the zip code who are in the labor force (either having jobs or actively seeking one).
11. Unemployment (Percentage, 1980): Percentage of the labor force in the zip code that is unemployed and actively seeking work.
12. Average Purchasing Potential (1986): Purchasing potential index for all consumer items over the zip code area, based on the Bureau of Labor Statistics Consumer Expenditure Survey. The index compares a zip code to similar urban or rural zip codes in the same census region, with 100 set as the average value.
13. Sporting Goods Purchasing Potential (1986): Purchasing potential index for sporting goods over the zip code area.

14. Groceries Purchasing Potential (1986): Purchasing potential index for groceries over the zip code area.
15. Home Improvement Purchasing Potential (1986): Purchasing potential index for home improvements over the zip code area.

Numerical codes have been assigned to each state and census region to facilitate computer utilization of these data. The codes were assigned alphabetically as shown in Tables B.1 and B.2.

TABLE B.1 Census Region Codes

Code	Census Region
1	North Central region
2	Northeast region
3	South region
4	West region

TABLE B.2 State Codes

Code	State		Code	State	
01	AK:	Alaska	26	MT:	Montana
02	AL:	Alabama	27	NC:	North Carolina
03	AR:	Arkansas	28	ND:	North Dakota
04	AZ:	Arizona	29	NE:	Nebraska
05	CA:	California	30	NH:	New Hampshire
06	CO:	Colorado	31	NJ:	New Jersey
07	CT:	Connecticut	32	NM:	New Mexico
08	DE:	Delaware	33	NV:	Nevada
09	FL:	Florida	34	NY:	New York
10	GA:	Georgia	35	OH:	Ohio
11	HI:	Hawaii	36	OK:	Oklahoma
12	IA:	Iowa	37	OR:	Oregon
13	ID:	Idaho	38	PA:	Pennsylvania
14	IL:	Illinois	39	RI:	Rhode Island
15	IN:	Indiana	40	SC:	South Carolina
16	KS:	Kansas	41	SD:	South Dakota
17	KY:	Kentucky	42	TN:	Tennessee
18	LA:	Louisiana	43	TX:	Texas
19	MA:	Massachusetts	44	UT:	Utah
20	MD:	Maryland	45	VA:	Virginia
21	ME:	Maine	46	VT:	Vermont
22	MI:	Michigan	47	WA:	Washington
23	MN:	Minnesota	48	WI:	Wisconsin
24	MO:	Missouri	49	WV:	West Virginia
25	MS:	Mississippi	50	WY:	Wyoming

The demographic database is contained in two ASCII files, which are both sorted by census region and state within census region. The file names and corresponding layouts are given in Tables B.3 and B.4. The data files are available on magnetic tape or diskette from the publisher.

TABLE B.3 ZIPCOD01.DAT

Columns	Description
1–4	Observation number
7	Census region number
9–10	State number
12–16	Zip code
18–22	Population in '86
24–28	Number of households in '86
30–33	Median age in '86
35–39	Median household income in '86
41–46	Median home value in '80
48–50	Median monthly homeowner cost in '80
52–54	Median household size in '80
56–59	Median years of education in '80

TABLE B.4 ZIPCOD02.DAT

Columns	Description
1–4	Observation number
7	Census region number
9–10	State number
12–16	Zip code
18–21	Percent college education in '80
23–26	Percent women in work force in '80
28–31	Percent unemployed in '80
33–37	Purchasing potential index '86: Overall average
39–43	Purchasing potential index '86: Sporting goods
45–49	Purchasing potential index '86: Grocery
51–55	Purchasing potential index '86: Home improvement

APPENDIX C

ASP Tutorial

This appendix provides an overview of the ASP program. It gives the minimal hardware requirements and start-up procedures necessary to begin an ASP session on a personal computer (PC). This tutorial is not intended to replace any of the ASP documentation manuals available from the publisher or DMC Software, Inc.

Hardware Requirements

ASP must be run on an IBM-compatible PC with at least 512K of memory, two disk drives (either one hard drive and one floppy drive, or two floppy drives), and DOS 2.0 or higher. A blank formatted floppy disk is also required for data storage, unless your PC has a hard drive (i.e., fixed disk) available for storing data.

Getting Started

To use the ASP program, you must first load it into the memory of the computer. To accomplish this when starting ASP from a floppy disk:

1. Insert your copy of ASP into either of your two disk drives, drive A or drive B. (Assume drive A.)
2. Type **A:** and press **ENTER** to make drive A the current drive:

   A: ⟨ENTER⟩

3. Type **ASP** and press **ENTER** to load the ASP program into memory.

   ASP ⟨ENTER⟩

The ASP disk must remain in drive A for as long as you are using the program.

To start ASP from a fixed disk or hard drive (e.g., drive C), it is first necessary to install ASP on the fixed disk. This is accomplished by placing your copy of the ASP disk into floppy drive A and entering the following commands at the DOS prompt:

```
C:           ⟨ENTER⟩
MD \ASP      ⟨ENTER⟩
CD \ASP      ⟨ENTER⟩
COPY A:*.*   ⟨ENTER⟩
```

(This sequence of DOS commands assumes the drive letter of the fixed disk is **C** and that the subdirectory in which the ASP program resides is **\ASP**.) Once ASP has been installed on the fixed disk, it need not be installed again. The ASP program can then be started at any point in the future by entering the following commands at the DOS prompt:

C: ⟨ENTER⟩
CD \ASP ⟨ENTER⟩
ASP ⟨ENTER⟩

The Main Menu

The initial screen to appear as the ASP program is loaded into memory displays copyright and licensing information. After reading this information, press any key to obtain the MAIN MENU shown in Figure C.1.

The MAIN MENU is a typical ASP "bounce bar" menu. The highlighted bar can be moved from option to option by pressing the SPACE BAR, the cursor control keys (→ ← ↑ ↓), or the TAB key. Once your selection is made, press **ENTER** to display submenus associated with the option. (You can also make a selection by pressing the letter of the desired option.)

Table C.1 gives a brief description of each of the MAIN MENU options and the corresponding chapters in the text. Several of these options contain statistical procedures that are beyond the scope of the text. Only the statistical routines covered in the text are described in the table.

FIGURE C.1 ►
The ASP main menu

```
************************   MAIN MENU   **********************
A Statistical Package for Business, Economics, and The Social Sciences
Copyright 1992 by DMC Software, Inc. (Version 2.xx)

A. Analysis of Variance   B. Regression Analysis     C. Correlation Matrix
D. Summary Statistics     E. Probability Dists.      F. File Management Menu
G. Time Series Analysis   H. Hypothesis Tests        I. INSTRUCTIONS
J. Factor Analysis        K. Miscellaneous Plots     L. Crosstab/Contingency
M. Auxiliary Programs     N. Enter a DOS Command     O. Scr./Data Dir. Dflts

F1=ALT COMMANDS MENU  F2=CALCULATOR  F3=TOGGLE PRINT (OFF)  X=EXIT
```

TABLE C.1 Options on the Main Menu

Option	Description	Chapter(s)
A. Analysis of Variance	(Beyond the scope of this text.)	—
B. Regression Analysis	Simple linear regression	10
C. Correlation Matrix	Bivariate correlations	10
D. Summary Statistics	Mean, median, standard deviation, etc.	2
E. Probability Dists.	Binomial, Poisson, and normal distributions	4
F. File Management Menu	Creating, saving, editing data	—
G. Time Series Analysis	(Beyond the scope of this text)	—
H. Hypothesis Tests	Confidence intervals and hypothesis tests for means and proportions; one-way table χ^2 test; nonparametric tests	6–9
I. INSTRUCTIONS	A short tutorial on the use of ASP	—
J. Factor Analysis Menu	(Beyond the scope of this text)	—
K. Miscellaneous Plots	Stem-and-leaf display, box plot, normal probability plot, scatter plot	2, 4, 10
L. Crosstab/Contingency	Two-way (contingency) table χ^2 test	9
M. Auxiliary Programs	(Beyond the scope of this text)	—
N. Enter a DOS Command	Enter and execute DOS commands within an ASP session	—
O. Scr./Data Dir. Dflts.	Set the color scheme on the monitor; set the default directory and printer port	—

Alternate Commands Menu

All of the statistical routines in ASP are accessible through the MAIN MENU. However, additional commands can be executed through the ALT COMMANDS MENU. The ALT COMMANDS MENU is called by pressing the **F1** function key from the main menu or from any menu one level below the main menu. The ALT COMMANDS MENU appears as shown in Figure C.2 (page 588).

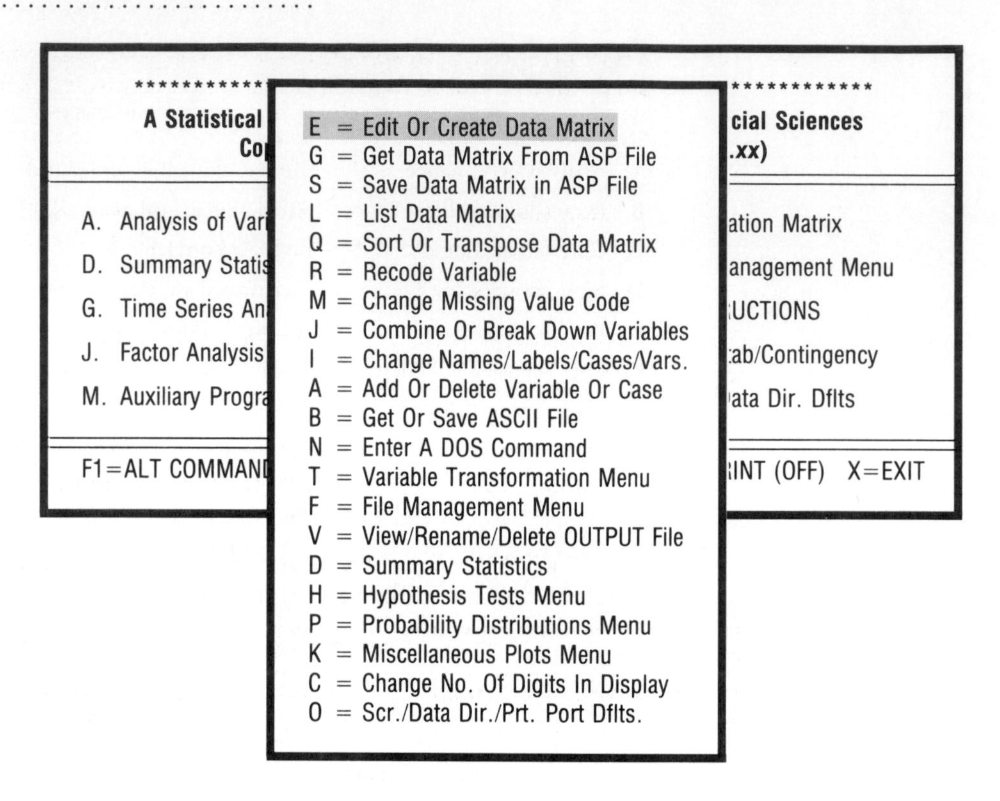

FIGURE C.2 ► The Alt commands menu

You can execute the commands on this menu either by moving the cursor to the desired option and pressing **ENTER** or by pressing the letter associated with the option. You will find this menu most useful for:

creating or editing data sets (option **E**)
listing data (option **L**)
getting data from an already created ASP data set (option **G**)
creating new variables for a data set (option **T**)
adding or deleting variables and/or cases (option **A**)
changing the names of variables (option **I**)
getting or saving data in an external ASCII file (option **B**)
saving an ASP data set (option **S**)

Creating a Data Matrix

Typically, you will use ASP to analyze a data set. To do this, you must first create an ASP "data matrix." Select **E = Edit or Create Data Matrix** on the ALT COMMANDS MENU, and ASP responds with a series of questions and prompts. The first question is:

EDIT or CREATE? E

Note that the ASP default answer is **E** for EDIT. This is used when you want to edit an existing ASP data matrix. To create a data matrix, press the letter **C** (for CREATE). You are now prompted with the question:

Number of Variables? 1

Change the default to the correct number and press **ENTER.** ASP creates names for the variables using the convention Var1, Var2, Var3, etc., then asks:

Are Names OK? Y

To change the name, Press **N** (for No). ASP will then ask you to enter the new name of each variable. Once this is completed, ASP will prompt you, one case (i.e., one observation) and one variable at a time, to enter the data into the data matrix.

Important: The ASP data editor will not accept letters or special characters (e.g., dollar sign, comma) as data. Only whole numbers or numbers with decimals should be entered into the data matrix.

When data entry is complete, press **X** to exit the numerical data editor. Several questions will be asked, the most important being:

Do You Wish to Save the Data Matrix? Y

Answer "yes" by pressing **ENTER**. ASP will then ask for the drive letter (e.g., drive A) and directory of the disk where you want to save the data:

DATA DIRECTORY: A:\

If the default is correct, press **ENTER.** Otherwise, enter the correct drive/path. You will be asked to name the ASP data file, provide a file label (optional), and whether you want to save all variables and all cases.

Suppose you enter the following file name:

File Name: MYDATA

ASP will save your data matrix in the ASP file named MYDATA.ASP in the directory specified earlier. In future ASP sessions, you can access this data set by first selecting the option **G = Get Data Matrix from ASP File** from ALT COMMANDS MENU, and then selecting MYDATA from the resulting menu-list of ASP data files.

Analyzing a Data Matrix

To analyze an ASP data matrix that you have just created or accessed, return to the MAIN MENU by pressing **X** or **ESC.** From the MAIN MENU, select the desired statistical routine. Each choice will result in a series of submenus, prompts, and/or questions similar to those shown previously. After making your selections, ASP will perform the analysis and display the results immediately on the monitor screen. ASP menu selections at the bottom of the screen permit you to send the output directly to a printer or to save the output in a file for future use.

Available Documentation

ASP User's Manual (by DMC Software, Inc.)—available free to adopters of the text from the publisher of the text.

ASP Tutorial and Student Guide (by George Blackford)—can be purchased directly from DMC Software, Inc., or from your campus bookstore.

Answers to Exercises

Chapter 1

1.6a. Population: all the students in the class; variable: GPA of a student **b.** Census **c.** Sample **d.** 100% reliability **e.** No **1.7** Every member of the class must have an equal chance of being selected. **1.8a.** All citizens of the United States **b.** Opinion of each citizen as to whether the president is doing a good or bad job **c.** The 2,000 individuals selected **d.** Proportion of all citizens who believe the president is doing a good job **1.9a.** All citizens of the United States **b.** Rating of the presidential job performance from 0 to 100 by each citizen **c.** The 2,000 individuals selected for the poll **d.** Average job performance rating of the president **1.10a.** All grade-school children in the United States **b.** Number of cavities each child has **c.** The 1,000 grade-school children selected from across the country **1.11a.** All possible 2-ml portions of a basic solution **b.** Amount of hydrochloric acid necessary to neutralize the 2-ml of basic solution **c.** The five 2-ml portions of the solution prepared **1.12a.** All vacuum cleaners produced by an assembly line in one day **b.** Condition (defective or not) of the vacuum cleaners **c.** The 30 cleaners randomly selected from the day's production **d.** Predict the percentage of defective vacuum cleaners **1.13a.** All medical doctors **b.** Whether or not the doctor has been involved in one or more malpractice suits **c.** The 500 medical doctors selected **d.** Estimate the proportion of medical doctors who have been involved in one or more malpractice suits **1.14a.** The 505 teenagers selected **b.** All teenagers across the country **c.** Topic that most teenagers most want to discuss with their parents **d.** As estimates of the percentage of teenagers who selected each of three topics **e.** Give a measure of reliability **1.15a.** All department store executives **b.** Job satisfaction and Machiavellian rating of each executive **c.** The 218 department store executives selected **d.** Executives with higher job satisfaction scores are likely to have a lower "Mach" rating. **1.16a.** All individuals **b.** The individuals surveyed **c.** Religion and environmental concern of the individuals **1.17a.** All elderly people who have been treated by the new method of treating blindness; all elderly people who have not been treated **b.** The 224 patients treated with the new method; the selected group of untreated elderly people **c.** Condition of the patients (blind or not) **1.18a.** Medicare surgical cases performed in hospital outpatient departments and ambulatory surgical centers **b.** Quality of care provided for Medicare recipients **c.** 5% of all Medicare surgical cases performed **d.** Estimate the average quality of care provided for all Medicare recipients **e.** Measure of reliability

Chapter 2

2.2a. Interval **b.** Nominal **c.** Ordinal **d.** Ratio **2.3** Qualitative; qualitative **2.4a.** Interval **b.** Ordinal **c.** Interval **d.** Nominal **e.** Ratio **f.** Ratio **2.5a.** Nominal **b.** Ratio **c.** Ratio **d.** Ratio **2.6a.** Quantitative **b.** Qualitative **c.** Qualitative **d.** Quantitative **2.7a.** Nominal **b.** Ratio **c.** Nominal **d.** Ordinal **2.8a.** Ratio **b.** Nominal **c.** Ratio **d.** Ordinal **e.** Nominal **f.** Ratio **2.10** Frequencies: 50, 75, 125, 100, 25, 50, 50, 25 **2.11a.** 23 **b.** Stem: 0; leaves: 0, 1, 2; numbers: 0, 1, and 2 **2.12a.** 7.5–9.5 **b.** .15 **c.** .20 **d.** 20 **2.13a.** Largest: 11,968.23 **b.** Data cluster in the 0 to 3,000 range **2.14a.** Distribution is skewed to the right; 80%; 0 and 3,000

2.15a.

Stem	*Leaf*
6	0123456889
7	0334566
8	112679
9	02

b. Skewed to the right **2.16** Scores are lower in 1990 than in 1975 **2.17a.** Scores appear to be lower in 1990 than in 1975. **b.** 100; Alabama **2.18** Scores in 1990 tend to be lower than those in 1975. **2.21** 15, 14.545, 15 **2.22** 2.717, 2.65 **2.25a.** 8.5 **b.** 25 **c.** .78 **d.** 13.44 **2.26a.** 2.5, 3, 3 **b.** 3.08, 3, 3 **c.** 49.6, 49, 50 **2.27a.** Mean is less than the median. **b.** Mean is greater than the median. **c.** Mean and median are equal. **2.29a.** 39.19; 27.95 **b.** Skewed to the right **c.** 40%; 50% **2.30c.** 81.15, 83, 83 **2.32a.** 78.15, 84, 88 **b.** Median **c.** 83.06, 86, 88; mean **2.33a.** No **b.** Yes; 110

2.34a. Median **b.** Mean **2.35** $.35 **2.36a.** $1,268.60, $942.5 **b.** Eastern states: $2,292.78; Western states: $430.64 **c.** No **2.37b.** Both data sets are skewed to the right. **2.38b.** Both data sets are skewed to the right. **c.** Some trend **2.41** No; yes **2.42a.** 4, 2.30, 1.52 **b.** 6, 3.62, 1.90 **c.** 10, 7.11, 2.67 **d.** 5, 1.62, 1.27 **2.43a.** 4.8889, 2.211 **b.** 3.3333; 1.826 **c.** .1868; .432 **2.44a.** 5, 3.70, 1.92 **b.** 99, 1,949.25, 44.15 **c.** 98, 1,307.84, 36.16 **2.45a.** 2; 37.5; 6.124 **b.** 1.25; 1.8354; 1.355 **2.46a.** 5.6, 17.3, 4.1593 **b.** 13.75 feet, 152.25 square feet, 12.339 feet **c.** −2.5, 4.3, 2.0736 **d.** .33 ounce, .0587 square ounce, .2422 ounce **2.49a.** 3; 1.3; 1.1402 **b.** 3; 1.3; 1.1402 **c.** 3; 1.3; 1.1402 **d.** Variance and standard deviation are unaffected. **2.51** First professor **2.52b.** 10; 10 **c.** 4.5814; 2.3944 **d.** Standard deviation **2.54** Any data set; mound-shaped **2.55a.** Dollars; ratio **b.** At least 150; at least 178; nothing; nothing **2.56a.** Nothing **b.** At least 3/4 **c.** At least 8/9 **2.57a.** Approximately 68% **b.** Approximately 95% **c.** Essentially all **2.58a.** 8.24, 3.357, 1.83 **b.** $\bar{x} \pm s$: 18, 72%; $\bar{x} \pm 2s$: 24, 96%; $\bar{x} \pm 3s$: 25, 100% **d.** 7; 1.75 **2.59** Between 104.17 and 156.25; no **2.60** 1975: 70%, 98%, 100%; 1990: 68%, 98%, 100% **2.61a.** Adults: (3.56, 9.34) (.67, 12.23); adolescents: (8.41, 13.37), (5.93, 15.85) **b.** Adults: 31.3, 43.7; adolescents: 12.9, 18.1 **2.62** Approximately 68% within (131.16, 235.70); approximately 95% within (78.89, 287.97); almost all within (26.62, 340.24) **2.63a.** At most 25% **b.** Approximately 2.5% **2.64** Do not buy **2.65** .025; yes **2.66a.** At least 8/9 **b.** Approximately .84 **c.** Yes **2.67a.** 1; .9524; .98; 9.1% **b.** 1; .2439; .4939; 76.2% **c.** 1; .1091; .33; 89.3% **2.68** At least 5 times are within (21.30, 102.50); at least 6 are within (1.00, 122.80) **2.69a.** 2 **b.** .5 **c.** 0 **d.** −2.5 **e.** Sample; population; population; sample **f.** 2 standard deviations above the mean; .5 standard deviation above the mean; 0 standard deviations above the mean; 2.5 standard deviations below the mean **2.70a.** 25%, 75% **b.** 50%, 50% **c.** 80%, 20% **d.** 16%, 84% **2.71** Median **2.72a.** 2 **b.** −3 **c.** −2 **d.** 1.67 **2.73a.** 3.7; 2.2; 2.95; 1.45 **b.** 1.9 **c.** Summa cum laude: $z > 2$; cum laude: $z > 1$; distribution is mound shaped **2.74** Yes; $\sigma = 10$, $\mu = 60$ **2.75a.** Distribution is skewed to the right. **b.** 90% of the scores are below 660. **c.** 94% of the scores are below your score. **2.76a.** −3 **b.** Essentially none **c.** Above 90 **2.77a.** −3 **b.** Yes **c.** Yes, claim is even more unlikely. **d.** Yes, no reason to doubt the claim. **2.78** Yes; $z = 35$ **2.80** IQR = 25, lower inner fence = 22.5, upper inner fence = 122.5, lower outer fence = −15, upper outer fence = 160 **2.81a.** 39 **b.** $Q_L = 31.5$; $Q_U = 45$ **c.** 13.5 **d.** Skewed to the left **e.** 50%; 75% **2.83c.** A: 84 and 100 are outliers; B: 140.4 and 206.4 may be outliers. **2.84a.** $M = 1{,}540$; $Q_U = 3{,}078$; $Q_L = 770$ **b.** Skewed to the right **c.** 2 outliers are 9,234 and 12,312; suspect outlier is 6,800 **2.85a.** Median is greater for 1982 than for 1984. **b.** Approximately the same **c.** Yes **d.** No **2.86b.** Suspected outliers: 238, 268, 269, and 264 **c.** 238: 1.92; 268: 2.06; 269: 2.13; 264: 3.14 **2.87a.** Median for 1975 exceeds median for 1990. **b.** IQR for 1990 is greater than IQR for 1975. **c.** No **2.88a.** Honda **b.** Honda **c.** Ford **d.** Honda **2.89a.** Ratio **b.** Nominal **c.** Interval **d.** Ordinal **3.** (a) Quantitative; (b) Qualitative; (c) Quantitative; (d) Qualitative **2.92a.** Both the height and width of the bars (peanuts) change. **2.93a.** −1; 1; 2 **b.** −2; 2; 4 **c.** 1; 3; 4 **d.** .1; .3; .4 **2.94** 5 **2.95a.** 3.1234 **b.** 9.0233 **c.** 9.7857 **2.96a.** 6, 27, 5.20 **b.** 6.25, 28.25, 5.32 **c.** 7, 37.67, 6.14 **d.** 3, 0, 0 **e.** (a) (−4.4, 16.4), 100%; (b) (−4.39, 16.89), 100%; (c) (−5.28, 19.28), 100%; (d) (3, 3), 100% **2.97a.** 5.67; 1.0667; 1.03 **b.** −$1.5; 11.5 dollars squared, $3.39 **c.** .4125%; .0883% squared; .30% **d.** 3; $10; .7375% **2.99a.** Skewed to the right **b.**

Interval	Chebyshev's	Empirical
(7, 13)	At least 0	≈68%
(4, 16)	At least 75%	≈95%
(1, 19)	At last 88.9%	≈100%

; Chebyshev's rule **c.** 38 **d.** 3.333; no

2.100a.

Interval	Percent
(2.25, 3.67)	At least 0
(1.54, 4.38)	At least 75%
(.83, 5.09)	At least 88.9%

; skewed to the left **b.**

Interval	Percent
(7.31, 17.45)	At least 0
(2.24, 22.52)	At least 75%
(−2.83, 27.59)	At least 88.9%

; skewed to the right

c.

Interval	Percent
(.80, 4.74)	At least 0
(−1.17, 6.71)	At least 75%
(−3.14, 8.68)	At least 88.9%

; skewed to the right **d.** Number of academic leisure activities

2.101 No; 1.90 **2.102a.** All gas stations in the United States **b.** The 200 stations surveyed **c.** Price of unleaded gasoline **f.** $z = -2.5$ **2.103a.** No outliers **b.** 58.24; 58; 65.44; 8.0895 **c.** Mound-shaped **d.** 68%; 95%; all **e.** 64%; 96%; 100% **2.104a.** 68% **b.** 47.5% **c.** 16% **2.105b.** $\bar{x} = 16.48$, median = 15, mode = 14 **c.** Range = 22, $s \approx 5.5$ **d.** $s^2 = 33.9267$, $s = 5.82$ **e.** 1.00 **2.107** 52 minutes **2.108a.** Essentially none **b.** No **c.** Yes **2.109a.** At most 25% **b.** 2.5%

Chapter 3

3.1a. .5 **b.** .3 **c.** .6 **3.2a.** Venn diagram **b.** $P(A) = .3$, $P(B) = .2$ **c.** $P(A) = .25$, $P(B) = .3$ **3.3** $P(A) = .55$, $P(B) = .50$, $P(C) = .70$ **3.4c.** $P(A) = 1/3$, $P(B) = 1/2$ **d.** Divide the number of 3s by the total number of tosses **3.5a.** 4/9 **b.** 1/3 **c.** 0

3.6a. (1, 1) (1, 2) (1, 3) (1, 4) (1, 5) (1, 6)
(2, 1) (2, 2) (2, 3) (2, 4) (2, 5) (2, 6)
(3, 1) (3, 2) (3, 3) (3, 4) (3, 5) (3, 6)
(4, 1) (4, 2) (4, 3) (4, 4) (4, 5) (4, 6)
(5, 1) (5, 2) (5, 3) (5, 4) (5, 5) (5, 6)
(6, 1) (6, 2) (6, 3) (6, 4) (6, 5) (6, 6)
b. $P(A) = 1/36$, $P(B) = 18/36$, $P(C) = 6/36$, $P(D) = 11/36$, $P(E) = 6/36$

3.7 $P(A) = 1/12$, $P(B) = 1/4$, $P(C) = 1/12$, $P(D) = 1/2$ **3.8** $P(A) = 1/10$, $P(B) = 3/5$, $P(C) = 3/10$ **3.10** $P(A) = .81$, $P(B) = .19$, $P(C) = .18$ **3.11a.** 1/3 **b.** 1/3 **c.** 1/6 **d.** 1/6 **3.12a.** 1/4 **b.** 1/2 **c.** 0 **3.13** No, probabilities are the same. **3.14a.** {RW, RB, DW, DB, IW, IB} **b.** Sample space **c.** .30 **d.** .35 **e.** .55 **f.** .12 **3.15a.** 1/10 **b.** 3/10 **c.** 7/10

3.16a. (S, H, H) (S, H, M) (S, H, L)
(S, M, H) (S, M, M) (S, M, L)
(S, L, H) (S, L, M) (S, L, L)
(U, H, H) (U, H, M) (U, H, L)
(U, M, H) (U, M, M) (U, M, L)
(U, L, H) (U, L, M) (U, L, L)
b. 1/18 **c.** 6/18

3.17a. 1 to 2 **b.** .5 **c.** .4 **3.18a.** A = {HHH, HHT, HTH, THH, TTH, THT, HTT}, B = {HHH, TTH, THT, HTT}, $A \cup B$ = {HHH, HHT, HTH, THH, TTH, THT, HTT}, A' = {TTT}, $A \cap B$ = {HHH, TTH, THT, HTT} **b.** $P(A) = 7/8$, $P(B) = 1/2$, $P(A \cup B) = 7/8$, $P(A') = 1/8$, $P(A \cap B) = 1/2$ **c.** 7/8 **d.** No **3.20a.** A: {(1, 6), (2, 5), (3, 4), (4, 3), (5, 2), (6, 1)}, B: {(1, 4), (2, 4), (3, 4), (4, 4), (5, 4), (6, 4), (4, 1), (4, 2), (4, 3), (4, 5), (4, 6)}, $A \cap B$: {(3, 4), (4, 3)}, $A \cup B$: {(1, 4), (2, 4), (3, 4), (4, 4), (5, 4), (6, 4), (4, 1), (4, 2), (4, 3), (4, 5), (4, 6) (1, 6), (2, 5), (5, 2), (6, 1)}, A': {(1, 1), (1, 2), (1, 3), (1, 4), (1, 5), (2, 1), (2, 2), (2, 3), (2, 4), (2, 6), (3, 1), (3, 2), (3, 3), (3, 5), (3, 6), (4, 1), (4, 2), (4, 4), (4, 5), (4, 6), (5, 1), (5, 3), (5, 4), (5, 5), (5, 6), (6, 2), (6, 3), (6, 4), (6, 5), (6, 6)} **b.** $P(A) = 1/6$, $P(B) = 11/36$, $P(A \cap B) = 1/18$, $P(A \cup B) = 5/12$, $P(A') = 5/6$ **c.** 5/12 **d.** No **3.21a.** 3/4 **b.** 13/20 **c.** 1 **d.** 2/5 **e.** 1/4 **f.** 7/20 **g.** 1 **h.** 1/4 **3.22a.** .5 **b.** .19 **c.** .5 **d.** 1.00 **e.** .31 **f.** .69 **g.** No **3.23a.** .5 **b.** .31875 **c.** .125 **d.** .75 **e.** .75 **f.** .3125 **g.** A and B; A and C; B and C; D and E; D and F; E and F **3.24a.** $A \cap F$ **b.** $B \cup C$; A' **3.25a.** E_1 = {3 H's}, E_2 = {2 H's}, E_3 = {1 H}, E_4 = {0 H's}; $A = E_1 \cup E_2 \cup E_3$; 7/8 **b.** $A = E_4'$; 7/8 **3.26a.** {G_1P_1, G_2P_1} **b.** {G_1G_2, G_1P_1, G_1P_2, G_2P_1, G_2P_2, P_1P_2} **c.** {P_1P_2} **d.** $P(A) = 5/6$, $P(B) = 1/2$, $P(A \cap B) = 1/3$, $P(A \cup B) = 1$, $P(A') = 1/6$ **e.** 1; no **3.27a.** $B \cap C$ **b.** A' **c.** $C \cup B$ **d.** $A \cup C'$ **3.28a.** {11, 13, 15, 17, 29, 31, 33, 35} **b.** {2, 4, 6, 8, 10, 11, 13, 15, 17, 20, 22, 24, 26, 28, 29, 31, 33, 35, 1, 3, 5, 7, 9, 19, 21, 23, 25, 27} **c.** $P(A) = 9/19$, $P(B) = 9/19$, $P(A \cap B) = 4/19$, $P(A \cup B) = 14/19$, $P(C) = 9/19$ **d.** {11, 13, 15, 17} **e.** 14/19; no **f.** 2/19 **g.** {1, 2, 3, … , 29, 31, 33, 35} **h.** 16/19 **3.29a.** (1, R), (1, S), (1, E), (2, R), (2, S), (2, E), (3, R), (3, S), (3, E) **b.** Sample space **c.** .46 **d.** .05 **e.** .30 **f.** .61 **g.** .34 **3.30a.** .03; no **b.** .62; no **c.** .28; no **d.** .32; no **e.** .72; no **3.31a.** Yes **b.** $P(A) = .26$; $P(B) = .35$; $P(C) = .72$; $P(D) = .28$; $P(E) = .05$ **c.** $P(A \cup B) = .56$; $P(A \cap B) = .05$; $P(A \cup C) = .77$ **d.** .74 **e.** No pairs are mutually exclusive. **3.32a.** $P(A) = .5$, $P(B) = .5$, $P(A \cap B) = .4$ **b.** $P(E_1 \mid A) = .2$, $P(E_2 \mid A) = .2$, $P(E_3 \mid A) = .6$ **c.** .8 **3.33** $P(A \mid B) = .25$; $P(B \mid A) = .5$ **3.34a.** $P(A) = .50$, $P(B) = .35$, $P(C) = .40$ **b.** $P(A \cap B) = .25$, $P(A \cap C) = 0$, $P(B \cap C) = 0$ **c.** $P(1 \mid A) = .4$, $P(2 \mid A) = .1$, $P(3 \mid A) = .5$ **d.** .5 **e.** $P(C \mid A) = 0$, $P(C \mid A') = .80$ **3.35b.** $P(A) = 3/4$; $P(B) = 1/2$; $P(A \cap B) = 1/2$ **c.** $P(A \mid B) = 1$; $P(B \mid A) = 2/3$ **3.36** $P(B \mid A) = 1/3$, $P(B \mid A') = 0$, $P(B \mid C) = 1/14$, $P(A \mid C) = 1/7$, $P(C \mid A') = 1$ **3.37a.** .557 **b.** .075 **c.** .028 **d.** .504 **3.38a.** $P(A) = .09$, $P(B) = .51$, $P(A \cap B) = .09$ **b.** $P(A \mid B) = .18$, $P(B \mid A) = 1$ **3.39a.** 1/15 **b.** 2/5 **c.** 14/15 **d.** 3/7 **e.** 5/14 **3.40** .60 **3.41a.** .5128 **b.** .24 **c.** .60

d. 1 **3.42a.** .105 **b.** .638 **c.** .517 **3.43a.** .37 **b.** .68 **c.** .15 **d.** .221 **e.** 0 **f.** 0 **g.** No **3.44a.** $P(A) = 7/8$, $P(B) = 3/8$, $P(C) = 3/8$, $P(D) = 1/2$, $P(A \cap B) = 3/8$, $P(A \cap D) = 3/8$, $P(B \cap C) = 0$, $P(B \cap D) = 0$ **b.** $P(B \mid A) = 3/7$, $P(A \mid D) = 3/4$, $P(C \mid B) = 0$ **c.** No pairs are independent. **3.45a.** .625 **b.** .5 **c.** .8 **d.** .625 **e.** Not independent **f.** No **3.46** Yes **3.47a.** A and C; B and C **b.** None are independent. **c.** $P(A \cup B) = .65$; $P(A \cup C) = .90$ **3.48a.** False **b.** True **c.** False **3.49a.** $P(A \cap B) = .08$; $P(A \mid B) = .4$; $P(A \cup B) = .52$ **b.** $P(A \cap B) = .12$; $P(B \mid A) = .30$ **3.50a.** .48 **b.** .32 **c.** .16 **d.** .64 **e.** .5 **f.** .47 **3.51a.** $A \cap B$; $A \cap B'$; A' **b.** $P(A \cap B) = .891$; $P(A) = .9$; $P(A \cap B') = .009$; **c.** .891 **3.52** .40 **3.53a.** $(1/2{,}780)^4$ **b.** .9986 **c.** .0014 **d.** $(1/2{,}920)^4$; .9986; .0014 **3.54a.** .008 **b.** .001 **c.** .125 **3.55a.** 2/9 **b.** 1/36 **c.** 11/36 **3.56** .04; .38 **3.57a.** 27/64 **b.** 1/64 **c.** 27/64 **3.58a.** .25 **b.** .016 **c.** .422 **3.59** .000000001; .997 **3.60a.** .4096 **b.** .0016 **c.** .9984 **3.61a.** *ABC, ACE, BCD, BEF, ABD, ACF, BCE, CDE, ABE, ADE, BCF, CDF, ABF, ADF, BDE, CEF, ACD, AEF, BDF, DEF*; 20 **b.** 1/20 **3.62a.** 35,820,200 **b.** 1/35,820,200

3.66a.

(1, *H*)	(2, 1)	(3, *H*)	(4, 1)	(5, *H*)	(6, 1)
(1, *T*)	(2, 2)	(3, *T*)	(4, 2)	(5, *T*)	(6, 2)
	(2, 3)		(4, 3)		(6, 3)
	(2, 4)		(4, 4)		(6, 4)
	(2, 5)		(4, 5)		(6, 5)
	(2, 6)		(4, 6)		(6, 6)

b. Each simple event whose first element is 1, 3, or 5 has probability 1/12; each simple event whose first element is 2, 4, or 6 has probability 1/36. **c.** $P(A) = 1/4$, $P(B) = 1/2$ **d.** A': all except (1, *H*), (3, *H*), and (5, *H*); B': {(2, 1), (2, 2), (2, 3), (2, 4), (2, 5), (2, 6), (4, 1), (4, 2), (4, 3), (4, 4), (4, 5), (4, 6), (6, 1), (6, 2), (6, 3), (6, 4), (6, 5), (6, 6)}; $A \cap B$: {(1, *H*), (3, *H*), (5, *H*)}; $A \cup B$: {(1, *H*), (1, *T*), (3, *H*), (3, *T*), (5, *H*), (5, *T*)} **e.** $P(A') = 3/4$, $P(B') = 1/2$, $P(A \cap B) = 1/4$, $P(A \cup B) = 1/2$, $P(A \mid B) = 1/2$, $P(B \mid A) = 1$ **f.** No; no **3.67a.** $P(A \cap B) = 0$, $P(B \cap C) = .2$, $P(A \cup C) = .9$, $P(A \cup B \cup C) = 1$, $P(B') = .7$, $P(A' \cap B) = .3$, $P(B \mid C) = .4$, $P(B \mid A) = 0$ **b.** No; yes **c.** No; no **3.68a.** No **b.** Yes **c.** No **3.69a.** No **b.** $P(A \mid B) = .3$, $P(B \mid A) = .1$ **c.** .37 **3.70a.** 34/60 **b.** 4/34 **3.71.** .64; .32; .04 **b.** .72; .22; .06 **c.** Dependent **3.72a.** .40 **b.** .30 **c.** .375 **d.** No **3.73a.** Any package **b.** Any brand B package or a package from brand A with over .200 sodium **c.** A brand A package that contains between .100 and .149 sodium **d.** A brand B package that contains between .150 and .199 sodium **e.** A brand A package that contains less than .149 sodium **3.74a.** Independent **b.** Independent **c.** Dependent **d.** Dependent **e.** Independent **f.** Independent **3.75b.** .95 **c.** .25 **d.** .5 **3.76a.** 1/32 **b.** 5/32 **3.77a.** .27 **b.** .52 **c.** .2917 **d.** No **e.** No **3.78a.** 124/153; 111/153 **b.** 101/153; 19/153 **c.** 33/153; 120/153 **3.79a.** 6/30 **b.** 6/29 **c.** .634 **3.80** .79 **3.81** .993 **3.82a** .036 **b.** .035 **3.83a.** .7127 **b.** .2873 **3.84a.** .9639 **b.** .3078 **c.** .0361 **d.** 3 **3.85a.** .3886 **b.** .3049 **c.** .1720 **d.** No **3.86** .0458; .7467 **3.87a.** .24 **b.** .2083 **c.** .57 **3.88a.** (H_1, H_2, H_3), (H_1, H_2, L_1), (H_1, H_2, L_2), (H_1, H_2, L_3), (H_1, H_3, L_1), (H_1, H_3, L_2), (H_1, H_3, L_3), (H_1, L_1, L_2), (H_1, L_1, L_3), (H_1, L_2, L_3), (H_2, H_3, L_1), (H_2, H_3, L_2), (H_2, H_3, L_3), (H_2, L_1, L_2), (H_2, L_1, L_3), (H_2, L_2, L_3), (H_3, L_1, L_2), (H_3, L_1, L_3), (H_3, L_2, L_3), (L_1, L_2, L_3) **b.** 1/20 **c.** 1/20 **d.** 1/2

Chapter 4

4.3a. Discrete **b.** Continuous **c.** Continuous **d.** Discrete **e.** Continuous **f.** Continuous **4.4a.** Continuous **b.** Discrete **c.** Continuous **d.** Continuous **e.** Discrete **f.** Discrete **4.5a.** Continuous **b.** Discrete **c.** Discrete **d.** Discrete **e.** Discrete **f.** Continuous **4.6a.** −4, 0, 1, or 3 **b.** 1 **c.** .7 **d.** 0 **4.7a.** Valid **b.** Not valid; $\Sigma p(x) \neq 1$ **c.** Not valid; one of the probabilities is negative. **d.** Not valid; $\Sigma p(x) > 1$ **4.8a.** .7 **b.** .3 **c.** 1 **d.** .2 **e.** .8

4.9a.

Simple Events:	*HHH*	*HHT*	*HTH*	*THH*	*HTT*	*THT*	*TTH*	*TTT*
x = # heads:	3	2	2	2	1	1	1	0

b. $p(3) = 1/8$, $p(2) = 3/8$, $p(1) = 3/8$, $p(0) = 1/8$ **d.** 1/2 **e.** $\mu = 1.5$, $\sigma^2 = .75$ **4.10a.** $\mu = 3.05$, $\sigma^2 = 1.2475$, $\sigma = 1.117$ **c.** (1.933, 4.167); .80 **d.** (−.301, 6.401); 1 **4.11a.** 3.8 **b.** 10.56 **c.** 3.2496 **e.** No **f.** Yes **4.12a.** .66 **b.** $\mu = 0$, $\sigma^2 = 2.94$, $\sigma = 1.715$ **d.** .96

4.14a.

x	$p(x)$
\$700	.40
\$1,110	.10
\$3,320	.45
\$16,450	.05

b. 2,707.5 **c.**

y	$p(y)$
\$700	.80
\$1,110	.20

d. 782 **4.15** 7/8 **4.16a.** 2.9; 3 **b.** 3; 4 **c.** $\mu = 2.15$; median = 2; 3; 3 **4.17a.** .118 **b.** .302 **c.** .580 **4.18a.** 4.199 **b.** 1.128 **c.** .989 **4.19a.** .0016, .0256, .1536, .4096, .4096 **c.** .0272 **4.20a.** 1, 2, 3, 4, ... **b.** .99 **c.** .01

d.

x	$p(x)$
1	.9
2	.09
3	.009
4	.0009
5	.00009
6	.000009
7	.0000009
8	.00000009
9	.000000009
10	.0000000009

; yes **4.21** No **4.22** \$.25 **4.23a.** \$7,200 **b.** \$11,000 **c.** Too little **4.24** \$882 **4.25a.** No **b.** −.05 **4.26a.** 15 **b.** 10 **c.** 1 **d.** 1 **e.** 4 **4.27a.** Discrete **b.** Binomial distribution **d.** $\mu = 3.5$, $\sigma = 1.0247$ **e.** (1.4506, 5.5494) **4.28a.** .4096 **b.** .3456 **c.** .027 **d.** .0081 **e.** .3456 **f.** .027 **4.29a.** $p(0) = .343$, $p(1) = .441$, $p(2) = .189$, $p(3) = .027$

b.

x	$p(x)$
0	.343
1	.441
2	.189
3	.027

4.30a. 12.5, 6.25, 2.5 **b.** 20, 12.8, 3.578 **c.** 60, 24, 4.899 **d.** 63, 6.3, 2.510 **e.** 48, 9.6, 3.098 **f.** 40, 38.4, 6.197 **4.31a.** .121 **b.** .034 **c.** .081 **d.** 0 **e.** .998 **f.** .137 **4.32a.** *FFFFF, FFFFS, FFFSF, FFSFF, FSFFF, SFFFF, FFFSS, FFSFS, FSFFS, SFFFS, FFSSF, FSFSF, SFFSF, FSSFF, SFSFF, SSFFF, FFSSS, FSFSS, SFFSS, FSSFS, SFSFS, SSFFS, FSSSF, SFSSF, SSFSF, SSSFF, FSSSS, SFSSS, SSFSS, SSSFS, SSSSF, SSSSS*;

x	0	1	2	3	4	5
$p(x)$	1/32	5/32	10/32	10/32	5/32	1/32

4.33a. .5 **b.** $p < .5$ **c.** $p > .5$ **e.** $p = .5$; $p < .5$; $p > .5$ **4.34a.** Yes **b.** .9873 **c.** 9.749×10^{-10} **4.35a.** No **b.** Yes **c.** No **d.** b: $\mu = 4.0$, $\sigma^2 = 3.2$, $\sigma = 1.789$ **4.36a.** .001 **b.** $\mu = 3{,}000$; $\sigma = 45.826$; approximate probability is .5 **4.37a.** .013 **b.** .783 **4.38a.** 0 **b.** Yes; drug is effective. **4.39** .0005 **4.40** No; $z = 4.95$ **4.41a.** .20 **b.** 4 **c.** .196 **d.** No **4.42** 15 **4.43a.** 0 **b.** Drug is effective. **4.44a.** $\mu = 520$, $\sigma = 13.4907$ **b.** No; $z = -3.56$ **c.** No **4.45a.** Discrete **b.** Poisson probability distribtuion **d.** $\mu = 3$, $\sigma = 1.7321$ **e.** $\mu = 3$, $\sigma = 1.7321$ **4.46a.** .920 **b.** .677 **c.** .423 **d.** Decreases **4.47a.** .934 **b.** .191 **c.** .125 **d.** .223 **e.** .777 **f.** .001 **4.48b.** $\mu = 1$, $\sigma = 1$; $(-1, 3)$ **c.** .981 **4.49b.** $\mu = 3$, $\sigma = 1.7321$; $(-.4642, 6.4642)$ **c.** .966 **4.50** $p(0) = .287$, $p(1) = .358$, $p(2) = .224$ **4.51** .1088 **4.52a.** 2 **b.** No; $P(x > 10) = .003$ **4.53a.** .251 **b.** .558 **c.** .191 **d.** .011 **4.54** .040 **4.55a.** .193 **b.** .660 **c.** Numbers of monthly accidents are independent. **4.56a.** .0000075 **b.** $E(x) = 11.8$, $\sigma = 3.435$ **c.** $z = 2.39$ **4.57a.** .080 **b.** No **4.58a.** .010 **b.** Yes **4.59a.** .4772 **b.** .3413 **c.** .4987 **d.** .2190 **e.** .4772 **f.** .3413 **g.** .4545 **h.** .2190 **4.60a.** .6826 **b.** .9544 **c.** .6934 **d.** .6378 **e.** .9901 **f.** .9901 **4.61a.** .0721 **b.** .0594 **c.** .2434 **d.** .3457 **e.** .5 **f.** .9233 **4.62a.** 1.645 **b.** 1.96 **c.** −1.96 **d.** 1.28 **e.** 1.28 **4.63a.** 0 **b.** 2.53 **c.** 1.90 **d.** −1.76 **4.64a.** 1 **b.** −1 **c.** 0 **d.** −2.5 **e.** 3 **4.65a.** .8413 **b.** .3300 **c.** .6450 **d.** .2047 **e.** .9292 **f.** .0708 **4.66a.** 30 **b.** 14.32 **c.** 40.24 **d.** 16.84 **e.** 19.76 **f.** 36.72 **g.** 48.64 **4.67a.** .9544 **b.** .0228 **c.** .1587 **d.** .8185 **e.** .1498 **f.** .9974 **4.68** 182 **4.69a.** 25.14% **b.** 90.375 **4.70** .0351; .0080 **4.71a.** 15.87% **4.72a.** 1.5% **b.** 6.3 ± 1.2 **4.73** .0019 **4.74a.** .0307 **b.** .0893 **4.75** 5.068 **4.76** 398.8 seconds **4.77** 472.44 **4.78a.** 7,667 **b.** 45.62% **4.79a.** $z_L = -.67$; $z_U = .67$ **b.** −2.68; 2.68 **c.** −4.69; 4.69 **d.** .0074; 0 **4.82a.** Not appropriate **b.** Appropriate **c.** Not appropriate **d.** Appropriate **e.** Appropriate **f.** Appropriate **4.83a.** Yes **b.** $\mu = 10$, $\sigma^2 = 6$ **c.** .726 **d.** .7291 **4.84a.** .3446 **b.** .1151 **c.** .9224 **4.85a.** .1788 **b.** .5236 **c.** .6950 **4.86a.** .4880 **b.** .2334 **c.** 0 **4.87** 0 **4.88** .7823 **4.89a.** No **b.** .6026 **c.** No; yes; yes **d.** .7190 **4.90a.** .0011 **b.** Yes **c.** No **4.91a.** $E(x) = 6{,}000$, $\sigma^2 = 2{,}400$ **b.** .0202 **c.** No **4.92** .0559 **4.93a.** .0037 **b.** Yes **4.94a.** .9808 **b.** 0 **4.95** .2676; no **4.96a.** .4681 **b.** .0436 **c.** .9822 **4.97a.** .0559 **b.** Yes **4.98a.** Continuous **b.** Discrete **c.** Continuous **d.** Discrete **4.99a.** .243 **b.** .131 **c.** .36 **d.** .157 **e.** .128 **f.** .121 **4.100a.** No **b.** No **c.** No **d.** Yes

4.101a. $\mu = 15.4$, $\sigma^2 = 18.44$, $\sigma = 4.294$ **b.** .5 **c.** (6.812, 23.988) **d.** 1 **4.102a.** .4750 **b.** .95 **c.** .8664 **d.** .0934 **e.** .2521 **f.** .8980 **4.103a.** .9821 **b.** .0179 **c.** .9505 **d.** .3243 **e.** .9107 **f.** .0764 **4.104a.** 1.13 **b.** 1.62 **c.** 0 **d.** 1.33 **e.** −.85 **f.** 2.64 **4.105a.** .6915 **b.** .0228 **c.** .5328 **d.** .3085 **e.** 0 **f.** .9938 **4.106a.** 40 **b.** 54.22 **c.** 23.38 **d.** 52 **e.** 32.32 **f.** 34.96 **4.107a.** .192 **b.** .228 **c.** .772 **d.** .987 **e.** .960 **f.** $\mu = 14$, $\sigma^2 = 4.2$, $\sigma = 2.049$ **g.** .975 **4.108a.** .180 **b.** .015 **c.** .076 **4.109a.** .3821 **b.** .5398 **c.** 0 **d.** .1395 **e.** .0045 **f.** .4602 **4.110a.** .1112 **b.** .000017 **c.** $p < .104$ **4.111** 10.56% **4.112** 22,250 **4.113a.** .027 **b.** .376 **c.** No **4.114a.** .8790 **b.** .7967 **c.** 163 **4.115** 83.15% **4.116a.** $\mu = 1.25$, $\sigma = 1.09$, no **b.** .007 **c.** Not applicable **4.117** .982 **4.118a.** Company A: 4.60; company B: 3.70 **b.** Company A: $46,000; company B: $55,500 **c.** Company A: $\sigma^2 = 1.34$, $\sigma = 1.16$; company B: $\sigma^2 = 1.21$, $\sigma = 1.10$ **d.** Company A: (2.28, 6.92), .95; company B: (1.5, 5.9), .95 **4.119** .0721 **4.120a.** .8861 **b.** .0301 **4.121a.** $\mu = 5$, $\sigma^2 = 4$ **b.** .617 **c.** .006 **4.122a.** .006 **b.** Not as effective as claimed **4.123** .5369; 1 **4.124a.** .0901 **b.** .0384 **c.** .9573 **4.125** .059; program is effective **4.126a.** .0548 **b.** .6006 **c.** .3446 **d.** $6,503.80 **4.127a.** .265 **b.** .176 **4.128** .1056 **4.129a.** 1.0 **b.** .8830 **4.130** $P(x \geq 400) \approx 0$ **4.131a.** .230 **b.** .143 **c.** .100 **4.132a.** .1922 **b.** .4681 **c.** .1026

Chapter 5

5.1a.–b.

Possible Samples	$\bar{x}$	*Possible Samples*	$\bar{x}$
0, 0	0	4, 0	2
0, 2	1	4. 2	3
0, 4	2	4, 4	4
0, 6	3	4, 6	5
2, 0	1	6, 0	3
2, 2	2	6, 2	4
2, 4	3	6, 4	5
2, 6	4	6, 6	6

c. 1/16 **d.**

$\bar{x}$	$p(\bar{x})$
0	1/16
1	2/16
2	3/16
3	4/16
4	3/16
5	2/16
6	1/16

5.3a.

$\bar{x}$	$p(\bar{x})$
1	.04
1.5	.12
2	.17
2.5	.20
3	.20
3.5	.14
4	.08
4.5	.04
5	.01

c. .05 **d.** No **5.4** 2.7

5.5a.

$\bar{m}$	$p(\bar{m})$
1	.04
1.5	.12
2	.17
2.5	.20
3	.20
3.5	.14
4	.08
4.5	.04
5	.01

5.8a. $\mu = 1.667$, $\sigma^2 = 2.889$ **b.**

$\bar{x}$	*Probability*
0	1/9
.5	2/9
1	1/9
2	2/9
2.5	2/9
4	1/9

c. $E(\bar{x}) = 5/3$ **d.**

s^2	*Probability*
0	3/9
.5	2/9
4.5	2/9
8	2/9

e. $E(s^2) = 2.889$

5.9a. 5 **b.**

$\bar{x}$	$p(\bar{x})$
2	1/27
8/3	3/27
10/3	3/27
4	1/27
13/3	3/27
5	6/27
17/3	3/27
20/3	3/27
22/3	3/27
9	1/27

; $E(\bar{x}) = 5$ **c.**

m	$p(m)$
2	7/27
4	13/27
9	7/27

; $E(m) = 4.778$ **d.** $\bar{x}$

5.10a. 1 **b.**

$\bar{x}$	Probability
0	1/27
1/3	3/27
2/3	6/27
1	7/27
4/3	6/27
5/3	3/27
2	1/27

c.

m	Probability
0	7/27
1	13/27
2	7/27

d. $E(\bar{x}) = 1$, $E(m) = 1$ **e.** $\sigma^2_{\bar{x}} = 2/9$, $\sigma^2_m = 14/27$ **f.** $\bar{x}$

5.12a. $E(\bar{x}) = 2.7$ **b.** .805 **c.** .95 **5.13a.**

s^2	$p(s^2)$
0	.22
.5	.36
2	.24
4.5	.14
8	.04

b. 1.61 **c.** $E(s^2) = 1.61$ **d.**

s	$p(s)$
0	.22
.7071	.36
1.414	.24
2.121	.14
2.828	.04

e. $E(s) = 1.003976$

5.14 Yes **5.15a.** $\mu_x = 100$, $\sigma_x = 5$ **b.** $\mu_x = 100$, $\sigma_x = 2$ **c.** $\mu_x = 100$, $\sigma_x = 1$ **d.** $\mu_x = 100$, $\sigma_x = 1.414$ **e.** $\mu_{\bar{x}} = 100$, $\sigma_{\bar{x}} = .447$ **f.** $\mu_{\bar{x}} = 100$, $\sigma_{\bar{x}} = .316$ **5.16a.** $\mu_{\bar{x}} = 10$, $\sigma_{\bar{x}} = .6$ **b.** $\mu_{\bar{x}} = 100$, $\sigma_{\bar{x}} = 5$ **c.** $\mu_{\bar{x}} = 20$, $\sigma_{\bar{x}} = 8$ **d.** $\mu_{\bar{x}} = 10$, $\sigma_{\bar{x}} = 20$ **5.17a.** $\mu = 2.9$, $\sigma^2 = 3.29$, $\sigma = 1.814$ **b.**

$\bar{x}$	$p(\bar{x})$
1	.01
1.5	.08
2	.24
2.5	.32
3	.16
4.5	.02
5	.08
5.5	.08
8	.01

c. $\mu_{\bar{x}} = 2.9$, $\sigma_{\bar{x}} = 1.283$

5.18 No; only if the sample size is large **5.19a.** $\mu_{\bar{x}} = 20$, $\sigma_{\bar{x}} = 2$ **b.** Approximately normal **c.** −2.25 **d.** 1.50 **5.20a.** .0228 **b.** .0668 **c.** .0062 **d.** .8185 **e.** .0013 **5.21a.** .8944 **b.** .0228 **c.** .1292 **d.** .9699 **5.22a.** (99, 101) **b.** 1 **c.** No **5.24a.** $\mu_{\bar{x}} = 3.5$, $\sigma_{\bar{x}} = .05$ **b.** .9544 **c.** .0082 **d.** $\mu_{\bar{x}} = 3.5$, $\sigma_{\bar{x}} = .03536$; .9954; 0 **5.25a.** Approximately normal; no **b.** .0336 **c.** No **5.26a.** Not necessarily normal; yes **b.** .0025 **c.** No **5.27a.** $\mu_{\bar{x}} = 1.3$, $\sigma_{\bar{x}} = .240$ **b.** Yes **c.** .0156 **d.** .0062 **5.28a.** .0013 **b.** Program decreases mean number of sick days.

5.29a. .0031; $\mu < 157$ **b.** More likely; less likely **c.** Less likely; more likely **5.30a.** Probability distribution of the variable A **b.** $E(A) = \alpha$ **c.** Choose statistic whose standard deviation is smaller. **d.** No **5.31a.** Decrease **b.** Not very good **c.** $\bar{x}$ **d.** $\sigma_{\bar{x}} = 1.25$, $\sigma_A = 2.5$ **5.32a.** $\mu_{\bar{x}} = 2.5$, $\sigma_{\bar{x}} = .1904$ **b.** Approximately normal; yes **5.33a.** .5 **b.** .0606 **c.** .0985 **d.** .8436 **5.34a.** (2 ones), (2 zeros), (1 one, 1 zero) **b.** $\bar{x}_{2\ \text{ones}} = 1$; $\bar{x}_{2\ \text{zeros}} = 0$; $\bar{x}_{1\ \text{one},\ 1\ \text{zero}} = ½$ **c.**

$\bar{x}$	$p(\bar{x})$
0	¼
½	½
1	¼

5.38a. 36.5148 **b.** 4 times (120) **c.** 1.778 times (54) **5.39a.** Discrete **b.** Approximately normal; $\mu_{\bar{x}} = 1.50$, $\sigma_{\bar{x}} = .01764$ **c.** .0023 **d.** No **5.40a.** Approximately normal, with $\mu_{\bar{x}} = 400$ and $\sigma_{\bar{x}} = 11.859$ **b.** .0174 **c.** .5 **d.** Sample size is sufficiently large. **5.41a.** .2327 **b.** .2215 **5.42a.** .5 **b.** .0023 **5.43b.** $\mu_y = .5$, $\sigma_y = .29/\sqrt{n}$ **c.** Normal **5.44a.** .3830 **5.45** .9772 **5.46a.** .0082 **b.** Sample size is sufficiently large. **5.47a.** Approximately normal, with $\mu_{\bar{x}} = 38$ and $\sigma_{\bar{x}} = .5$ **b.** .0139 **c.** No **5.48a.** $\mu_{\bar{x}} = 42$, $\sigma_{\bar{x}} = .65$, approximately normal **b.** .0069 **c.** No **5.49** .9932

Chapter 6

6.1a. 1.645 **b.** 2.58 **c.** 1.96 **d.** 1.28 **6.2a.** 95% **b.** 90% **c.** 99% **d.** 80% **e.** 67.78% **6.3a.** 25.9 ± .56 **b.** 25.9 ± .47 **c.** 25.9 ± .73 **6.4a.** 83.2 ± 1.25 **b.** 95% of all confidence intervals constructed will include μ. **c.** 83.2 ± 1.65 **d.** Increases **e.** Yes **6.5a.** 28 ± .784 **b.** 102 ± .65 **c.** 15 ± .0588 **d.** 4.05 ± .163 **e.** No **6.8** Yes **6.9a.** 33.9 ± .647 **b.** 33.9 ± .323 **c.** Halve the width **6.10a.** 3.39 ± .0466 **b.** Students exhibit an awareness of risk involved in bicycling. **6.11** 22 ± 3.88 **6.12** 480 ± 3.02 **6.13** 72 ± .05 **6.14a.** 79.73 ± 2.132 **b.** 95% confident that the mean participation rate is between 77.598% and 81.862% **c.** Sample size is sufficiently large. **d.** Yes **e.** Center will be larger; width will decrease. **6.15** 19.8 ± 1.225 **6.16** 2.26 ± .04 **6.17a.** Younger: 4.17 ± .095; middle-age: 4.04 ± .057; older: 4.31 ± .062 **b.** More likely **6.18** 519 **6.19** 2075 **6.20a.** .98, .784, .56, .392, .196 **6.21a.** 482 **b.** 214 **6.22a.** No; $n = 139$ **b.** Yes; $n = 98$ **6.23** 1,351 **6.24** 97 **6.25** 55 **6.26** 139 **6.27** 271 **6.28a.** 1,083 **b.** Wider **c.** 38.3% **6.29a.** Normal **b.** Approximately normal if n is sufficiently large; unknown if n is small **6.30a.** $z_{.10} = 1.28$, $t_{.10} = 1.440$ **b.** $z_{.05} = 1.645$, $t_{.05} = 1.943$ **c.** $z_{.025} = 1.96$, $t_{.025} = 2.447$ **d.** $z_{.01} = 2.33$, $t_{.015} = 3.143$ **e.** $z_{.005} = 2.575$, $t_{.005} = 3.707$ **6.31a.** 2.228 **b.** 2.567 **c.** −3.707 **d.** −1.771 **6.32a.** 5 ± 1.876 **b.** 5 ± 2.394 **c.** 5 ± 3.754 **d.** (a) 5 ± .780; (b) 5 ± .941; (c) 5 ± 1.276 **6.33a.** 97.94 ± 4.240 **b.** 97.94 ± 6.737 **6.34a.** 52.6 ± 1.47 **b.** 95% confident that the mean pulse rate is between 51.13 and 54.07 beats per minute **c.** Population of pulse rates is normal. **6.35a.** 3.8 ± .464 **b.** 90% confident that the mean LOS will be between 3.336 and 4.264 days **c.** 90% of all intervals constructed will contain the population mean. **6.36a.** 23.8 ± 7.225 **b.** 99% confident that the mean percentage is between 16.575% and 31.025% **c.** Distributon of percentages is normal. **6.37** 49.70 ± .1498 **6.38a.** Costs of hiring each secretary hired by the corporation in the last 2 years **b.** 1,901.88 ± 213.74 **c.** 427.48; wider **6.39a.** Outstanding principal balances of all home mortgages foreclosed by the bank due to default; normal **b.** 67,648.583 ± 9,939.533 **c.** 90% confident that the mean outstanding principal balance is between $57,709.03 and $77,588.136 **6.40a.** 135 ± 13.839 **b.** Population of all health insurance costs is normally distributed. **6.41** Approximately normal with $\mu_{\hat{p}} = p$ and $\sigma_{\hat{p}} = \sqrt{\frac{pq}{n}}$ **6.42** $E(\hat{p}) = p$ **6.43a.** Yes **b.** .64 ± .067 **c.** 95% confident that the true value of p will fall beween .573 and .707 **6.44a.** Yes **b.** No **c.** Yes **d.** No **6.45a.** .3 ± .083 **b.** 80% confident that the proportion of all consumers who like the new snack food is between .217 and .383 **6.46a.** Poor: yes; average: yes; good: yes **b.** Poor: .155 ± .046; average .133 ± .046; good: .108 ± .062 **6.47a.** .623 ± .086 **b.** 95% confident that the proportion of all Illinois law firms that used microcomputers is between .537 and .709 **d.** Probably not **6.48a.** .775 ± .097 **c.** Wider **6.49** .633 ± .051 **6.50a.** Normal approximation is adequate for all three stations. **b.** KOCO: .773 ± .075; WCBS: .6 ± .162; WXIA: .446 ± .073 **c.** .58 ± .045 **6.51a.** .92 ± .029 **b.** 1,840,000 **c.** 383.67 **d.** 1,841,151 **6.52** 46%: ±.043; 37%: ±.042; 30%: ±.040 **6.53** .5325 ± .0457 **6.54** $p \neq .5$ **6.55a.** 1,083 **b.** 1,692 **6.56a.** 225 **b.** 267 **6.57** 34 **6.58** 425 **6.59** No; need $n = 984$ **6.60** 2,401 **6.61** 32,270 **6.62** 271 **6.63** No; $n = 457$ **6.64** 3,701 **6.65a.** t **b.** z **c.** z **d.** z **e.** Neither **6.66a.** 32.5 ± 5.16 **b.** 23,964 **6.67a.** .744 ± .024 **b.** 5,149 **6.68a.** Men: 7.4 ± .979; women: 4.5 ± .755 **b.** Men: 9.3 ± 1.185; women: 6.6 ± 1.138 **6.69** .268 ± .033; .145 ± .034 **6.70a.** 12.2 ± 1.645 **b.** 167 **6.71a.** 54 ± 6.135 **b.** Maximum **6.72** .125 ± .095 **6.73** .56 ± .02 **6.74** 26.4 ± 1.54 **6.75** 2 ± .47 **6.76** .613 ± .051 **6.77a.** $\bar{x}$ = 42,250 miles **b.** 42,250 ± 1,683.713 **c.** Interval estimation **6.78a.** $\bar{x}$ = 42,250 miles **b.** 42,250 ± 506.570

c. Interval estimation **6.79a.** $\hat{p}$ = .094 **b.** Yes **c.** .094 ± .037 **6.80** 818 **6.81a.** 34.76 ± 6.344 **b.** Population of all claim amounts is normally distributed **6.82a.** 856 **b.** Random sample

Chapter 7

7.1 Null **7.2** Test statistic **7.3** α **7.4** Type I error: reject H_0 when it is true; Type II error: accept H_0 when it is false; α = P(Type I error); β = P(Type II error) **7.5** Rejecting the null hypothesis when it is true; accepting the null hypothesis when it is true; rejecting the null hypothesis when it is false; accepting the null hypothesis when it is false
7.7 No **7.8a.** Type I error: conclude an individual is a liar when he or she is a truth-teller; Type II error: conclude an individual is a truth-teller when he or she is a liar **b.** .370; .240 **7.9a.** Unsafe; safe **b.** Type I error: conclude the drug is safe when it is not safe; Type II error: conclude the drug is not safe when it is **c.** α **7.10g.** .0250, .05, .0049, .1003, .10, .0049 **7.11a.** Reject H_0 if $z > 1.88$ **b.** .0301 **7.12a.** $z = 1.67$; reject H_0 **b.** $z = 1.67$; do not reject H_0 **7.13a.** $z = -1.60$; reject H_0 **b.** $z = -1.60$; do not reject H_0 **7.14a.** H_0: $\mu = 10$; H_a: $\mu > 10$ **b.** $z > 1.645$ **c.** $z = 1.29$; do not reject H_0 **7.15a.** H_0: $\mu = 35$; H_a: $\mu > 35$ **b.** Yes; $z = 1.8$ **7.16b.** $z = 1.8$; reject H_0 **7.17** Yes; $z = -4.34$ **7.18** Yes; $z = 2.00$ **7.19a.** H_0: $\mu = 10$; H_a: $\mu < 10$ **b.** Type I error: conclude the mean number of solder joints inspected per second is less than 10 when it is 10 or more; Type II error: conclude the mean number of solder joints inspected per second is at least 10 when it is less than 10 **c.** $z = -2.33$; reject H_0 **7.20a.** Skewed to the right **b.** Yes; $z = 1.75$ **c.** Yes; no **7.21a.** Do not reject H_0 **b.** Do not reject H_0 **c.** Reject H_0 **d.** Reject H_0 **e.** Do not reject H_0 **f.** Reject H_0 **7.22a.** Do not reject H_0 **b.** Reject H_0 **c.** Reject H_0 **d.** Do not reject H_0 **e.** Do not reject H_0 **7.23** .06 **7.24** .0150 **7.25** .0300 **7.26** .1075 **7.27** .9279 **7.28** .1096 **7.29** No; $z = -1.16$, p-value = .1230 **7.30a.** No; $z = -1.29$ **b.** .0985 **c.** Small **7.31a.** H_0: μ = \$8,446; H_a: μ > \$8,446 **b.** 0; reject H_0 **7.32b.** Do not reject H_0 for any $\alpha \leq .05$. **7.33a.** H_0: $\mu = 2.5$; H_a: $\mu > 2.5$ **b.** 0 **c.** Reject H_0 **7.34a.** p-value = .2892; do not reject H_0 **b.** Yes **7.35a.** H_0: $\mu = 250$; H_a: $\mu > 250$ **b.** .0455; reject H_0 for $\alpha = .05$ **c.** p-value = .0910; do not reject H_0 for $\alpha = .05$ **7.37** n small; population normal; variance unknown **7.38a.** $|t| > 2.160$ **b.** $t > 2.500$ **c.** $t > 1.397$ **d.** $t < -2.718$ **e.** $|t| > 1.729$ **f.** $t < -2.353$ **7.40a.** .10 **b.** .05 **c.** .05 **7.41a.** $t = -2.058$; do not reject H_0 **b.** $t = -2.058$; do not reject H_0 **c.** (a): .05 < p-value < .10; (b): .10 < p-value < .20 **7.42a.** $t = -1.40$; do not reject H_0 **b.** $t = -1.40$; do not reject H_0 **c.** (a): p-value > .10; (b): p-value = .20 **7.43a.** Normal population **b.** Reject H_0 **c.** .0764; reject H_0 for $\alpha > .0764$ **7.44a.** Yes; $t = -2.01$ **b.** Type II error **c.** Type I error **d.** 19.86 ± .13 **7.45a.** H_0: $\mu = 15$; H_a: $\mu < 15$ **c.** Reject H_0 for $\alpha > .0556$ **7.46a.** No; $t = .87$ **b.** p-value > .10 **c.** Lengths of great white sharks are normal. **d.** No **7.47a.** Yes; $t = 2.33$ **b.** Population of yields is normal. **7.48a.** Yes; $t = -3.63$ **b.** .001 < p-value < .002 **7.49a.** H_0: $\mu = 10$; H_a: $\mu > 10$ **c.** Reject H_0 for $\alpha > .0676$ **7.50** Population is not normal. **7.51a.** .5 **b.** .5 **c.** Cannot be determined **d.** .5 **7.52a.** .063 **b.** .500 **c.** .004 **d.** .151; .1515 **e.** .212; .2119 **7.53a.** $S = .7$, p-value = .172; do not reject H_0 **b.** $S = 7$, p-value = .344; do not reject H_0 **c.** $S = 9$, p-value = .011; reject H_0 **d.** $S = 9$, p-value = .022; reject H_0 **e.** p-value = .1911, p-value = .3422, p-value = .0136, p-value = .0272 **f.** Sample is selected randomly from a continuous probability distribution. **7.54** $S = 16$, p-value = .115; do not reject H_0 **7.55** $S = 6$, p-value = .063; reject H_0 for $\alpha > .063$ **7.56a.** H_0: $M = 60$; H_a: $M < 60$ **c.** Sample was randomly selected from a continuous probability distribution. **d.** $S = 14$, p-value = .058; do not reject H_0 **e.** .058 **7.57a.** $S = 17$, p-value = .0329; reject H_0 **b.** .0329 **c.** Sample of 24 tree heights was randomly selected from a continuous probability distribution. **7.58a.** Yes **b.** No **c.** Yes **d.** No **e.** No **7.59b.** $z = -1.39$; do not reject H_0 **c.** .0823 **7.60a.** $z = -2.00$ **c.** Reject H_0 **d.** .0228 **7.61a.** No **b.** p-value = .6600 **7.62a.** H_0: $p = .4$; H_a: $p < .4$ **b.** No; $z = -.19$ **c.** .4247 **7.63a.** No; $z = -1.51$ **b.** .0655 **7.64a.** Yes **b.** Yes; $z = 1.72$ **c.** .1209 ± .0065 **7.65** .0427 **7.66a.** No; $z = 1.25$ **b.** .1056 **7.67a.** H_0: $p = .5$; H_a: $p > .5$ **b.** Do not reject H_0 for $\alpha \leq .10$; **7.68** Power = $1 - \beta$ **7.69** Increase α; increase n; increase distance between μ_0 and μ_a **7.70b.** 1,032.9 **d.** .7422 **e.** .2578 **7.71a.** .3594 **b.** .6406 **7.72a.** Approximately normal with $\mu_{\bar{x}} = 50$ and $\sigma_{\bar{x}} = 2.5$ **b.** Approximately normal with $\mu_{\bar{x}} = 45$ and $\sigma_{\bar{x}} = 2.5$ **c.** .2358 **d.** .7642 **7.73** .3156 **7.74a.** Approximately normal with $\mu_{\bar{x}} = 10$ and $\sigma_{\bar{x}} = .1$ **b.** Approximately normal with $\mu_{\bar{x}} = 9.9$ and $\sigma_{\bar{x}} = .1$ **c.** .8300 **d.** .8300 **7.75a.** .8106, .5319, .2358, .0643, .0102 **c.** .67 **d.** .1894, .4681, .7642, .9357, .9898 **7.76** .1075 **7.77a.** .1251, .2578, .4404, .6406, .8051 **c.** .54; .5438 **d.** 1; 0 **7.78** Powers: .2090, .5080, .8051, .9545, .9941 **7.79a.** .1949; Type II error **b.** .05; Type I error **c.** .8051 **7.80** H_0, H_a, α **7.81** Alternative **7.83a.** H_0:

Individual does not have the disease; H_a: individual does have the disease **b.** Type I error: false positive; Type II error: false negative **c.** Minimize β **7.84** Null **7.85a.** $t = -7.51$; reject H_0 **b.** $t = -7.51$; reject H_0 **7.86a.** $z = -1.78$; reject H_0 **b.** $z = -1.78$; do not reject H_0 **7.87a.** $z = -1.67$; do not reject H_0 **b.** $z = -3.35$; reject H_0 **7.88a.** Do not reject H_0 **b.** Population is normal. **c.** .2576 **7.89a.** Yes; $z = -2.99$ **b.** p-value = .0028 **7.90b.** Reject H_0 for $\alpha >$ .0362 **c.** Population of lifetimes is normal. **d.** Type I error **7.91a.** Yes; $t = 1.56$ **b.** $.05 < p\text{-value} < .10$ **c.** Test scores are normal. **7.92a.** Yes; $z = 6.11$ **b.** 0 **c.** .4247 **7.93a.** H_0: $M = .75$; H_a: $M \neq .75$ **c.** Type I error: conclude the median level is not .75 when it is. Type II error: conclude the median level is .75 when it is not. **d.** $S = 18$, p-value = .044; reject H_0 **7.94a.** Yes; $t = 3.82$ **b.** No **7.95a.** No; $z = 1.41$ **b.** Small **7.96a.** Conclude the mean amount of PCB is less than or equal to 3 ppm when it is more than 3 parts ppm **b.** .8212 **c.** .1788 **d.** $\beta = .3121$, power = .6879 **7.97a.** No **b.** $\beta = .5910$, power = .4090 **c.** Increases **7.98** $z = 1.41$; do not reject H_0 **7.99** .0793 **7.100a.** Conclude the newsletter does not significantly increase the odds of winning when in fact it does. **b.** .8264 **c.** Decrease, $\beta = .7389$ **7.101** No; $z = 1.07$ **7.102** .1423 **7.103a.** H_0: $\mu = 1$; H_a: $\mu < 1$ **b.** $S = 7$; do not reject H_0 **c.** Random sample from continuous distribution **7.104** Yes; $z = -1.67$ **7.105a.** H_0: $\mu = 1$; H_a: $\mu > 1$ **b.** Rejection region: $t > 1.345$ **c.** Population of beta coefficients is normal. **d.** $t = 2.41$; reject H_0 **7.106a.** Reject H_0 for $\alpha > .0304$ **b.** .0152 **7.107** No; $z = -1.45$ **7.108a.** H_0: $p = .5$; H_a: $p > 5$ **b.** Reject H_0 for $\alpha = .05$

Chapter 8

8.1a. 150 ± 6 **b.** 150 ± 8 **c.** $\mu_{\bar{x}_1-\bar{x}_2} = 0$, $\sigma_{\bar{x}_1-\bar{x}_2} = 5$ **d.** 0 ± 10 **8.2a.** 14, .4 **b.** 10, .3 **c.** 4, .5 **d.** Yes **8.3a.** −2 **b.** .9772 **c.** .0228 **d.** .1815 **8.4a.** 1.2806 **c.** 20 **d.** $|z| > 1.96$ **e.** Reject H_0 **8.5** 20 ± 2.51; confidence interval **8.6a.** Approximately normal, $\mu_{\bar{x}_1-\bar{x}_2} = -70$, $\sigma_{\bar{x}_1-\bar{x}_2} = 3.1783$ **b.** $z = -3.46$, reject H_0 **c.** 0 **8.7a.** Do not reject H_0 **b.** .0575; reject H_0 for $\alpha = .05$ **8.8a.** No; $z = -.15$ **b.** $-.6 \pm 6.67$; yes **d.** .8808 **e.** Yes **8.9b.** Yes; $z = -3.76$ **c.** -2.9 ± 1.99 **d.** Narrower **8.10a.** Yes; $z = 2.83$ **b.** 30 ± 20.79 **8.11** 3.4 ± 8.76 **8.12** Yes; $z = -2.09$ **8.13a.** H_0: $\mu_1 - \mu_2 = 0$; H_a: $\mu_1 - \mu_2 \neq 0$ **b.** Reject H_0 **c.** No **d.** $-1.1 \pm .43$ **8.14a.** 2.9 ± 1.63 **b.** 2.7 ± 2.16 **c.** Yes **d.** Independent random samples **8.15a.** H_0: $\mu_1 - \mu_2 = 5$; H_a: $\mu_1 - \mu_2 > 5$ **b.** Yes; $z = 1.70$ **d.** .0446 **8.16** $z = 1.963$; do not reject H_0 at $\alpha = .05$ **8.17a.** H_0: $\mu_1 - \mu_2 = 0$; H_a: $\mu_1 - \mu_2 < 0$ **b.** Do not reject H_0 **c.** Independent random samples **8.19a.** No **b.** No **c.** No **d.** Yes **e.** No **8.21a.** 190 **b.** 29.8214 **c.** .2611 **d.** 2,138.7097 **e.** Nearer the variance with the larger sample size **8.22a.** .5989 **b.** Yes; $t = -2.39$ **c.** $-1.24 \pm .98$ **8.23a.** Do not reject H_0 for $\alpha \leq .10$ **b.** -2.50 ± 3.12 **8.24a.** No; $t = .013$ **b.** 10.025 ± 4.817 **8.25a.** Beginning: 6.09 ± 41.835; first follow-up: -52.24 ± 40.842; second follow-up: -48.41 ± 40.643; third follow-up: -37.68 ± 39.784; fourth follow-up: -38.54 ± 42.958 **8.26** Yes; $t = -4.22$ **8.27a.** No; $t = .445$ **b.** p-value > .20 **8.28a.** $t = 2.76$; reject H_0 **b.** $.005 < p\text{-value} < .01$ **8.29** No; $t = -1.57$ **8.30a.** Each count is an average of 5 measurements. **b.** H_0: $\mu_1 - \mu_2 = 0$; H_a: $\mu_1 - \mu_2 > 0$ **c.** Do not reject H_0 **8.31a.** -6.58 ± 16.88 **b.** No **c.** $t = -.85$; do not reject H_0; yes **8.32a.** H_0: $\mu_1 - \mu_2 = 0$; H_a: $\mu_1 - \mu_2 \neq 0$ **b.** Reject H_0 **c.** 95% confident that mean LOC score for obese females is between −5.93 and −2.95 **d.** Confidence interval **8.34a.** $t > 1.833$ **b.** $t > 1.328$ **c.** $t > 2.776$ **d.** $t > 2.896$ **8.35a.** $\bar{x}_D = 2$; $s_D^2 = 2$ **b.** $\mu_D = \mu_1 - \mu_2$ **c.** 2 ± 1.484 **d.** $t = 3.46$; reject H_0 **8.36a.** H_0: $\mu_D = 0$; H_a: $\mu_D < 0$ **b.** Reject H_0 **c.** 95% confident the difference in the two population means is between −5.258 and −2.116 **8.37a.** $t < -1.341$ **b.** $t = -3.5$; reject H_0 **d.** -7 ± 3.506 **e.** Confidence interval **8.38a.** $t = .81$; do not reject H_0 **b.** p-value > .20 **8.39a.** No; $t = -3.08$ **b.** -31.13 ± 32.17 **8.40** 5 ± 1.96 **8.41a.** H_0: $\mu_D = 0$; H_a: $\mu_D > 0$ **b.** Paired difference **c.** Random sample **d.** Leadership: do not reject H_0 for $\alpha = .05$; popularity: reject H_0 for $\alpha = .05$; intellectual self-confidence: reject H_0 for $\alpha = .05$ **8.42** $4.23 \pm .63$ **8.43a.** H_0: $\mu_D = 0$; H_a: $\mu_D > 0$ **b.** Reject H_0 **c.** 95% confident that decrease in the mean blood pressure is between 1.389 and 19.011 **8.44b.** No; $t = .80$ **c.** 84.17 ± 226.47 **8.45a.** Yes, $t = 7.68$ **c.** $.4167 \pm .1395$ **8.46a.** No; $t = .40$ **b.** $.417 \pm 2.2963$ **d.** No **8.47b.** No, $t = 1.18$ **8.48** 24 **8.49a.** 193 **b.** 47 **c.** 144 **8.50a.** 52 **b.** 118 **c.** 22 **8.51** 1,729 **8.52** 136 **8.53** 136 **8.54** 28 **8.55** 35 **8.56a.** T_A; $T_A \leq 41$ or $T_A \geq 71$ **b.** T_A; $T_A \geq 50$ **c.** T_A; $T_A \leq 43$ **d.** z; $|z| > 1.96$ **8.57a.** H_0: Populations have identical probability distributions; H_a: Probability distribution for population A is shifted to the left of that for B **b.** $T_B = 42.5$; reject H_0 **8.59a.** $z = -3.76$; reject H_0 **b.** 0 **8.60** Yes; $z = -2.47$ **8.61a.** $T_B = 53$; reject H_0 **8.62** Yes; $T_A = 39$ **8.63** No; $T_B = 29$ **8.64a.** No; $T_A = 82$ **b.** No; paired experiment **8.65** Yes; $T_A = 69.5$ **8.66** $T_A = 150.5$; reject H_0 **8.67** $z = 3.44$; reject H_0 **8.68a.** Smaller of T_+ and T_-; $T \leq 60$ **b.** T_-; $T_- \leq 271$ **c.** T_+; never reject H_0 **8.69a.** H_0: Populations have identical probability distributions; H_a: Population A is shifted to the right of population B **b.** $T_- = 3.5$; reject H_0 **8.72a.** $T_- = 2.5$; reject H_0 **b.** $T_- = 2.5$; reject H_0 **8.73a.** H_0: Populations have identical probability distributions; H_a: Population A is shifted to the right of population B **b.** $z = 2.499$; reject H_0 **c.** .0062 **8.74a.** Paired difference experiment **b.** H_0: Populations have identical probability distributions; H_a: Population A (face-to-face) is shifted to the left of population B (video conferencing) **c.** $T_+ = 3.5$; reject H_0 **d.** $.01 < p\text{-value} < .025$ **8.75a.** $T_+ = 7$; do not reject H_0 **b.** Distribution of the differences is normal. **8.76** Yes; $T_- = 4.5$ **8.77** No; $T_- = 13$ **8.78** $T_- = 3$; do not reject H_0 **8.79** No; $T_- = 24.5$ **8.80** $T_- = 36$, do not reject H_0 **8.81a.** $t = .78$, do not reject H_0 **b.** 2.50 ± 8.99 **c.** 225 **8.82a.** $3.90 \pm .31$ **b.** $z = 20.60$, reject H_0 **c.** 346 **8.83a.** $t = 5.73$, reject H_0 **b.** 3.8 ± 1.84 **c.** Population of differences has a normal distribution. **8.85** No; $T_B = 55.5$ **8.86a.** H_0: $\mu_1 - \mu_2 = 0$; H_a: $\mu_1 - \mu_2 \neq 0$ **b.** $z = -7.69$; reject H_0 **8.87a.** Paired difference **b.** $.24 \pm .073$ **c.** $t = 5.81$, reject H_0 **8.88** No; $z = .90$ **8.89a.** No, $t = .13$ **b.** Type I: decide that average R&D expenditures increased from 1988 to 1989 when no increase occurred; Type II: decide that there was no increase in the average R&D expenditures from 1988 to 1989 when an increase did occur **8.90** $.3 \pm 5.381$ **8.91** No; $T_A = 21$ **8.92a.** Yes, $T_- = 22$ **b.** .025 **8.93** No; $t = -.70$ **8.94a.** H_0: $\mu_1 - \mu_2 = 0$; H_a: $\mu_1 - \mu_2 \neq 0$ **b.** Reject H_0 **8.95a.** H_0: $\mu_1 - \mu_2 = 0$; H_a: $\mu_1 - \mu_2 \neq 0$ **b.** Hours of work: do not reject H_0; free time: reject H_0; breaks: rejects H_0 **8.96** Yes; $t = 6.14$ **8.97a.** H_0: $\mu_1 - \mu_2 = 0$; H_a: $\mu_1 - \mu_2 < 0$ **b.** $t = .92$, do not reject H_0 **8.98a.** H_0: $\mu_D = 0$; H_a: $\mu_D > 0$ **b.** Paired difference experiment **c.** Aad: do not reject H_0; Ab: reject H_0; Intention: do not reject H_0 **8.99** Yes; $t = -2.0$ **8.100a.** Reject H_0 **b.** Do not reject H_0 **c.** Reject H_0 **d.** Reject H_0 **e.** Do not reject H_0 **f.** Do not reject H_0 **8.101** 193

Chapter 9

9.3a. Yes **b.** No **c.** No **d.** No **e.** No **9.4a.** $z < -2.33$ **b.** $z < -1.96$ **c.** $z < -1.645$ **d.** $z < -1.28$ **9.5a.** $z = -4.02$, reject H_0 **b.** $z = -4.02$, reject H_0 **c.** $z = -4.02$, reject H_0 **d.** $-.10 \pm .04$ **9.6a.** $.07 \pm .067$ **b.** $.06 \pm .086$ **c.** $.15 \pm .131$ **9.7** Approximately normal $\mu_{(\hat{p}_1-\hat{p}_2)} = -.4$ and $\sigma_{(\hat{p}_1-\hat{p}_2)} = .046$ **9.8** $z = 1.16$; do not reject H_0 **9.9a.** $.41 \pm .070$ **b.** 95% confident that difference in proportion of married men and women managers is between .34 and .48 **9.10** No; $z = -1.67$ **9.11a.** Yes, $z = 3.00$ **b.** $.15 \pm .081$ **9.12a.** H_0: $p_1 - p_2 = 0$; H_a: $p_1 - p_2 \neq 0$ **b.** z **c.** Right tail **d.** Reject H_0 **9.13a.** No; $z = 1.27$ **b.** $z = .82$, do not reject H_0 **9.14a.** H_0: $p_1 - p_2 = 0$; H_a: $p_1 - p_2 < 0$ **b.** Reject H_0 **c.** No **9.16a.** $z = .79$, do not reject H_0 **9.17** $-.195 \pm .041$ **9.18** No; $z = 1.93$ **9.19a.** H_0: $p_1 - p_2 = 0$; H_a: $p_1 - p_2 > 0$ **b.** $z = .664$, do not reject H_0 **c.** H_0: $p_1 - p_3 = 0$; H_a: $p_1 - p_3 > 0$ **d.** No evidence to reject H_0 **9.20a.** 29,954 **b.** 2,165 **c.** 1,113 **9.21** 3,383 **9.22a.** $n_1 = n_2 = 911$ **b.** No **9.23** 4,802 **9.24** $n_1 = 520$, $n_2 = 260$ **9.25a.** 546 **b.** .0107; reject H_0 **9.26a.** 18.3070 **b.** 29.7067 **c.** 23.5418 **d.** 79.4900 **9.27a.** $\chi^2 > 5.99147$ **b.** $\chi^2 > 7.77944$ **c.** $\chi^2 > 11.3449$ **9.29** Expected cell counts of 5 or more **9.30** $X^2 = 8.075$, reject H_0 **9.31a.** No, $X^2 = 3.293$ **b.** Type I: concluding the multinomial probabilities differ when they do not, Type II: concluding the multinomial probabilities are equal, when they are not **9.32** $.288 \pm .062$ **9.33a.** Yes, $X^2 = 26.66$ **b.** No; $X^2 = 1$ **d.** $.55 \pm .082$ **9.34a.** $X^2 = 12.374$; do not reject H_0 **b.** $.05 < p\text{-value} < .10$ **9.35** Yes; $X^2 = 17.36$ **9.36** No; $X^2 = 4.4$ **9.37** Yes, $X^2 = 9.48$ **9.38** No, $X^2 = 1.5$ **9.40a.** $X^2 > 26.2962$ **b.** $X^2 > 15.9871$ **c.** $X^2 > 9.21034$ **9.41a.** H_0: The row and column classifications are independent; H_a: The row and column classifications are dependent **b.** Rejection region: $X^2 > 9.21034$ **c.**

		Column		
		1	2	3
Row	1	14.37	36.79	44.84
	2	10.63	27.21	33.16

d. $X^2 = 8.71$, do not reject H_0

9.42a.

		Column			
		1	2	3	Row Total
Row	1	36%	53.1%	67.9%	57.5%
	2	64%	46.9%	32.1%	42.5%

c. No **9.43** $X^2 = 12.36$, reject H_0

9.44

		Column			
		B_1	B_2	B_3	Totals
	A_1	29.9%	44.2%	29.6%	35.1%
Row	A_2	47.0%	32.5%	49.3%	42.4%
	A_3	23.1%	23.3%	21.1%	22.6%

9.45a. $X^2 = 39.22$; reject H_0 **b.** $X^2 = 2.84$, do not reject H_0 **9.46** Yes, $X^2 = 24.524$ **9.47** $X^2 = 19.10$, reject H_0 **9.48a.** Yes, $X^2 = 5.91$ **b.** $.010 < p\text{-value} < .025$ **9.49a.** Yes, $X^2 = 5.05$ **b.** $.01 < p\text{-value} < .025$ **9.50a.** Yes, $X^2 = 19.72$ **9.51** No, $X^2 = 1.35$ **9.52** Yes, $X^2 = 9.35$ **9.53** $X^2 = .599$, $.10 < p\text{-value} < .90$ **9.54a.** Union: $X^2 = 13.36$, reject H_0; nonunion: $X^2 = 9.16$, do not reject H_0 **b.** Union: $.025 \le p\text{-value} \le .05$; $.1 < p\text{-value} < .9$ **9.55a.** $z = -2.04$, reject H_0 **b.** $.10 \pm .096$ **c.** $n_1 = n_2 = 72{,}991$ **9.56a.** 27,292 **b.** 542 **c.** 325 **9.57a.** No, $X^2 = 2.133$ **b.** $.233 \pm .057$ **9.58a.** Yes, $X^2 = 54.14$ **b.** No **c.** Yes **d.**

		Column			
		1	2	3	Totals
	1	40%	22.2%	9.1%	20%
Row	2	20%	22.2%	63.6%	40%
	3	40%	55.6%	27.3%	40%

9.59a. $-.091 \pm .052$ **b.** $.308 \pm .040$ **9.60** $n_1 = n_2 = 780$ **9.61** Yes, $z = -2.61$ **9.62a.** Contra Costa: $.17 \pm .063$; Santa Clara: $-.004 \pm .066$; San Bernardino: $.102 \pm .070$ **9.63a.** H_0: Insurance status and county are independent; H_a: Insurance status and county are dependent **b.** Yes, $X^2 = 47.11$ **9.64a.** $p_1 - p_2$ **b.** $.07 \pm .08$ **9.65** $n_1 = n_2 = 1{,}905$ **9.66** Yes, $X^2 = 103.08$ **9.67** No, $X^2 = 7.384$ **9.68a.**

		Committee		
		Acceptable	Rejected	Totals
Inspector	Acceptable	101	23	124
	Rejected	10	19	29
	Totals	111	42	153

b. Yes **c.** $X^2 = 26.034$, reject H_0 **9.69** Yes, $X^2 = 180.874$ **9.70a.** $X^2 = 9.16$, reject H_0 **b.** $.90 \pm .06$ **9.71a.** No, $X^2 = 8.664$ **9.72a.** Yes, $X^2 = 87.38$

Chapter 10

10.2a. Slope = 1, y-intercept = 0 **b.** Slope -1, y-intercept = 3 **c.** Slope = .2, y-intercept = 1.2 **d.** Slope = 1.125, y-intercept = 3.75 **10.3** $\beta_1 = 1/3$, $\beta_0 = 14/3$; $y = 14/3 + 1/3x$ **10.4a.** $y = x$ **b.** $y = 3 - x$ **c.** $y = 6/5 + 1/5x$ **d.** $y = 30/8 + 9/8x$ **10.6a.** Slope = 1; intercept = 4 **b.** Slope = -2; intercept = 5 **c.** Slope = 3; intercept = -4 **d.** Slope = -2; intercept = 0 **e.** Slope = 1; intercept = 0 **f.** Slope = 1.5; intercept = .5 **10.9** No **10.10a.** $\Sigma x_i = 24$, $\Sigma y_i = 31$, $\Sigma x_i^2 = 116$, $\Sigma x_i y_i = 80$ **b.** -26.2857 **c.** 33.7143 **d.** $-.7797$ **e.** $\bar{x} = 3.4286$, $\bar{y} = 4.4286$ **f.** 7.102 **g.** $\hat{y} = 7.102 - .7797x$ **10.11a.** $\Sigma(y - \hat{y}) = 0$, SSE = 1.2204 **c.** SSE = 108.00 **10.12b.** $y = 1 + x$ **d.** $SSE_1 = 1.5$, $SSE_2 = 3.5$ **e.** $\hat{y} = 1 + x$ **10.13b.** Increasing **c.** $\hat{\beta}_1 = .918$, $\hat{\beta}_0 = .020$ **d.** Yes **10.14a.** $y = \beta_0 + \beta_1 x + \epsilon$ **b.** y-intercept = $-30{,}000$, slope = 70 **c.** Not meaningful **d.** For each additional square foot of living space, the mean selling price of a house is estimated to increase by \$70. **e.** \$180,000; yes **f.** \$320,000; no **10.15a.** Positive **c.** $\hat{y} = -37.435 + 455.271x$

e. Intercept has no meaning; for each 1-unit increase in team batting average, the estimated change in the mean number of games won is 455.271. **10.16a.** $\hat{y} = 36.351 + 2.8442x$ **c.** 79.014¢ **10.17a.** $\hat{y} = 57.912 - .811x$ **c.** 9.252 beats per minute **10.18b.** $\hat{\beta}_1 = 2.38$, $\hat{\beta}_0 = 24.45$ **c.** 33.375 **10.19a.** $\hat{y} = 3.375 + 1.201x$ **c.** Intercept has no meaning; for each 1° increase in outside air temperature, the mean cocoon temperature will increase by 1.201°. **10.20a.** $\hat{y} = 156.357 + 12.786x$ **c.** Intercept has no meaning; for each additional year of age, the mean length of the sand lance will increase by 12.786 millimeters. **10.21a.** 6°C: $\hat{y} = -2.792 + 15.135x$; 12°C: $\hat{y} = -8.368 + 17.612x$; 18°C: $\hat{y} = -12.068 + 17.599x$ **10.22a.** .0429 **b.** .414 **10.23a.** 57.5, 2.19444 **b.** 257.5, 6.7763 **c.** 9.288, 1.1610 **10.24** SSE = 1.2203, $s^2 = .2441$, $s = .4960$; SSE = 5.1349, $s^2 = 1.0270$, $s = 1.0134$ **10.25** Graph c **10.26a.** $\hat{y} = 3.068 - .015x$ **c.** 2.918; 2.828 **d.** SSE = .0193, $s^2 = .0064$, $s = .0801$ **e.** .1602 **10.27a.** $\hat{y} = 18.89 + .01087x$ **b.** 23.238 years **c.** SSE = 19.4286, $s^2 = 3.8857$, $s = 1.9712$ **d.** Expect data points to fall within 3.9424 years of the estimated regression line. **10.28b.** Brand A: $\hat{y} = 6.62 - .0727x$; Brand B: $\hat{y} = 9.31 - .0177x$ **c.** Brand A: SSE = 19.056, $s^2 = 1.466$, $s = 1.211$; Brand B: SSE = 4.833, $s^2 = .372$, $s = .610$ **d.** Brand A: (−.891, 3.593); Brand B: (.551, 2.991) **e.** Brand B **10.29a.** $\hat{y} = -.246 + 1.3152x$ **c.** 7.6452; 11.5908 **d.** SSE = 12.4352, $s^2 = 2.487$, $s = 1.577$ **e.** 3.154 **10.30a.** 31 ± 1.13; 31 ± .92 **b.** 64 ± 4.28; 64 ± 3.53 **c.** −8.4 ± .67; −8.4 ± .55 **10.31b.** $\hat{y} = .544 + .617x$ **d.** H_0: $\beta_1 = 0$; H_a: $\beta_1 \neq 0$ **e.** $t = 5.50$; 5 **f.** Reject H_0 **10.32** .617 ± .166; .617 ± .378 **10.33** No; $t = .627$ **10.34a.** .0010; reject H_0 **b.** For each additional square foot of living space, the price of the house is estimated to increase from \$49.1 to \$90.9; decrease $1 - \alpha$. **10.35a.** No; $t = .81$ **b.** .011 ± .027 **10.36a.** $y = \beta_0 + \beta_1 x + \epsilon$; $\hat{y} = 70.8 - .28x$ **b.** No; $t = -.48$ **10.37a.** SSE = 7.3243, $s^2 = .7324$, $s = .8558$ **b.** Expect most of the cocoon temperatures to be within 1.7116 of the least squares line. **c.** Yes, $t = 12.81$ **d.** p-value = .001 **10.38a.** Yes, $t = 13.69$ **b.** p-value < .001 **10.39a.** SSE = 15.6758, $s^2 = 7.8379$, $s = 2.7996$ **b.** 17.612 ± 4.045 **10.40b.** $\hat{y} = 44.1304 + .2366x$ **c.** SSE = 6,400.59, $s^2 = 376.505$, $s = 19.404$ **e.** $t = 1.27$, do not reject H_0 **f.** .2366 ± .3935 **10.41a.** Perfectly, positively related **b.** Perfectly, negatively related **c.** Not linearly related **d.** Strongly, positively related **e.** Weakly, positively related **f.** Strongly, negatively related **10.42a.** Positive **b.** Negative **c.** 0 slope **d.** Either positive or negative **10.43a.** .9853, .9709 **b.** −.9934, .9868 **c.** 0, 0 **d.** 0, 0 **10.44a.** .9438 **b.** .8769 **10.45** $r = .9046$, $r^2 = .8182$ **10.46a.** $r = .442$, $r^2 = .195$ **b.** No, $t = 1.71$ **10.47** $r = .9709$, $r^2 = .9426$ **10.48a.** $r = -.366$, $r^2 = .134$ **b.** No, $t = -.79$ **10.49a.** .6903 **b.** .4765 **c.** Yes, $t = 3.44$ **10.50c.** $r_1 = .9563$, $r_2 = .996$ **d.** Yes **10.51a.** $\hat{y} = -768.6 + 6.2033x$ **b.** Yes, $t = 3.35$ **c.** .584 **10.52a.** $\hat{y} = -235.1 + 1.273x_2$ **b.** $t = 18.29$; reject H_0 **c.** .9767 **d.** $\hat{y} = -235.1 + 1.273x_2$ **e.** Number of law enforcement employees **10.53a.** H_0: $\rho = 0$; H_a: $\rho \neq 0$ **b.** Reject H_0 for $\alpha = .05$ **10.54b.** $\hat{y} = 2 + x$ **c.** .8 **d.** 6 ± .681 **e.** 6 ± 1.927 **f.** Prediction interval is wider. **10.55c.** 6.5335 ± 1.9487 **d.** −.3 ± .9703; −6.0546 ± 1.7225 **10.56a.** −.3 ± 3.091 **c.** Yes **d.** $t = 9.32$, reject H_0 **10.57c.** 4.644 ± 1.118 **d.** 2.284 ± .629; −.414 ± 1.717 **10.58a.** $\hat{y} = 1.375 + .875x$ **c.** 1.5 **d.** .1875 **e.** 3.5625 ± .3279 **f.** 3.875 ± 1.062 **10.59a.** Yes, $t = 6.97$ **b.** 5.261 ± 2.096 **c.** 21.22 ± 7.90 **10.60a.** 11.782 ± .7918 **b.** 11.782 ± 2.0646 **10.61b.** $\hat{y} = 4.861 - .3466x$ **c.** Yes, $t = -5.91$ **d.** 1.7416 ± 1.0858 **e.** 2.6081 ± .3802 **10.62b.** $\hat{y} = 5{,}761.227 - 246.8097x$ **c.** Yes, $t = -8.63$ **d.** $r = -.9502$, $r^2 = .9029$ **e.** 3,293.13 ± 102.00 **f.** 3,293.13 ± 318.358 **10.63a.** 207.861 ± 5.370 **b.** 207.861 ± 13.795 **10.64** 35.662 ± 6.457 **10.65a.** Brand A: 3.349 ± .587; Brand B: 4.464 ± .296 **b.** Brand A: 3.349 ± 2.224; Brand B: 4.464 ± 1.120 **c.** −.65 ± 3.606 **10.66a.** $\hat{\beta}_0 = -37.435445$, $\hat{\beta}_1 = 455.270793$ **b.** SSE = 1,011.06431, $s^2 = 84.25536$, $s = 9.17907$ **c.** .1950 **d.** .1139; two-tailed; .05695 **e.** 81.8455; (76.3923, 87.2987) **f.** Wider **10.67a.** $\hat{y} = 3.374666 + 1.200858x$ **b.** .85582 **c.** .9426 **d.** Yes, $t = 12.810$ **e.** (9.7159, 13.8454) **10.68a.** $\hat{y} = 37.08 - 1.6x$ **c.** 57.2 **d.** 4.4 **e.** −1.6 ± .501 **f.** 13.08 ± 6.929 **g.** 13.08 ± 7.862 **10.69b.** $\hat{y} = x$; $\hat{y} = 3$ **c.** $\hat{y} = x$ **10.70b.** $r = -.1245$, $r^2 = .0155$ **c.** No, $t = -.35$ **10.71a.** $\hat{y} = 40.7842 + .7656x$ **c.** Yes, $t = 4.38$ **d.** .7656 ± .4035 **e.** 79.06±17.03 **f.** 67.58 ± 7.74 **10.72a.** $\hat{y} = 2.0841 + .6682x$ **c.** Yes, $t = 16.28$ **d.** .9815 **e.** 35.4941 ± .9786 **f.** 38.8351 ± 2.7762 **10.73a.** $\hat{\beta}_0 = 13.490347$, $\hat{\beta}_1 = -.052829$ **b.** −.0528 ± .0178 **c.** .8538 **d.** (.5987, 1.2653) **10.74a.** 57.1434 ± 34.8183 **c.** 44 **10.75a.** $\hat{\beta}_0 = -99{,}045$, $\hat{\beta}_1 = 102.814048$ **b.** $t = 6.483$, reject H_0 **c.** .00005 **d.** .6776 **e.** (−631,806, 2,078,740) **10.76a.** $\hat{y} = 6.51 + 10.83x$ **b.** $s = .3321$ **c.** $t = 6.34$, reject H_0 **d.** .834 **e.** 12.90 ± .36 **10.77a.** $\hat{y} = 2.758 + .866x$ **b.** $r = .985$, $r^2 = .971$ **c.** 20.078 ± 1.003 **10.78a.** SSE = 7.9416, $s^2 = 3.9708$ **b.** 17.599 ± 2.4561 **d.** 58.3279 ± 5.8575 **10.79b.** $\hat{y} = 12{,}024 + 3.581616x$ **c.** .7972 **d.** Yes; $t = 8.641$, p-value = .0001 **e.** (\$12,862.60, \$16,915.30)

Index

Symbol	Description
A′	Complement of event A 118
A, *B*	Events 103
A ∪ *B*	Union of events A and *B* 111
A ∩ *B*	Intersection of events A and *B* 111
α (alpha)	Probability of rejecting H_0 when H_0 is true (Type I error) 316
β (beta)	Probability of accepting H_0 when H_0 is false (Type II error) 316
β_0	*y*-intercept in regression models 490
$\hat{\beta}_0$	Least squares estimator of β_0 496
β_1	Slope of straight-line regression model 490
$\hat{\beta}_1$	Least squares estimator of β_1 496
χ^2 (chi-square)	Probability distribution of various test statistics 457
df	Degrees of freedom for the *t*- and χ^2-distributions 287
D_0	Hypothesized difference between population means or binomial proportions 373
$E(x)$	Expected value of the random variable *x* 170
ϵ (epsilon)	Random error component in regression models 490
$f(x)$	Probability density or frequency function for the continuous random variable *x* 199
H_a	Alternative (or research) hypothesis 317
H_0	Null hypothesis 317
IQR	Interquartile range 72
λ (lambda)	Mean number of events over given unit of time, area, volume, etc. (Poisson) 193
M	Middle quartile; median 72; 238
μ (mu)	Population mean 41
μ_D	Population mean of differences in a paired difference experiment 395
μ_0	Hypothesized value of a population mean 324
$\mu_{\bar{x}}$	Mean of the sampling distribution of $\bar{x}$ 250
n	(1) Sample size, total number of measurements in a sample 39 (2) Number of trials in a binomial experiment 180
n_D	Number of differences (or pairs) in a paired difference experiment 396
$n!$	*n* factorial; product of first *n* integers ($0! = 1$) 146
$\binom{n}{x}$	Shorthand for $n!/[x!(n-x)!]$ 146

(continued inside back cover)

Symbol	Description
p	(1) Observed significance level for a hypothesis test 328 (2) Probability of Success for the binomial distribution 180
$\hat{p}$	Sample estimator of binomial proportion p; proportion of successes in the sample 296
$p(x)$	Probability distribution for a discrete random variable x 169
p_0	Hypothesized value for binomial probability 346
$p_1 - p_2$	Difference between true proportions of success in two independent binomial experiments 445
$\hat{p}_1 - \hat{p}_2$	Difference between sample proportions of success in two independent binomial samples; sample estimator of $(p_1 - p_2)$ 445
P(A)	Probability of event A 102
P(A|B)	Probability of event A given that event B occurs 125
q	Probability of Failure for the binomial distribution 180
Q_L	Lower quartile of a data set 72
Q_U	Upper quartile of a data set 72
r	Pearson product moment coefficient of correlation between two populations 520
r^2	Coefficient of determination in simple linear regression 523
ρ (rho)	Pearson product moment coefficient of correlation between two populations 520
s	Sample standard deviation 50
s^2	Sample variance 50
$s_{\hat{\beta}_1}$	Sample estimator of $\sigma_{\hat{\beta}_1}$ 511
s_D	Sample standard deviation of differences in a paired difference experiment 396
s_p^2	Pooled sample variance in two-sample t test 383
S	Sample space 99
SS_{xx}	Sum of squares of the distances between x measurements and their mean, i.e., $\sum (x - \bar{x})^2$ 496
SS_{xy}	Sum of products of distances of x and y measurements from their means, i.e., $\sum (x - \bar{x})(y - \bar{y})$ 496
SS_{yy}	Sum of squares of the distances between y measurements and their mean, i.e., $\sum (y - \bar{y})^2$ 506
SSE	Sum of squared distances between observed and predicted values in a regression model, i.e., $\sum (y - \hat{y})^2$ 495